Society of Exploration Geophysicists

DECONVOLUTION

Edited by
Gerald M. Webster

Series Editor
Franklyn K. Levin

Geophysics reprint series
No. 1

ISBN 0-931830-01-X
0-931830-00-1

Library of Congress Catalog Card Number: 78-55218

Society of Exploration Geophysicists
P.O. Box 3098
Tulsa, Oklahoma 74101

Published 1978
Reprinted 1981

Printed in the United States of America.

CONTENTS

INTRODUCTION

To the geophysicist, deconvolution means removing, through data processing, undesirable effects which have occurred in the earth: attenuation, reverberation, and ghosting are prime examples of these effects. The undesirable effects in the earth can be described in terms of linear filters. To deconvolve data, we must first estimate the parameters of the earth filter (usually from the recorded data), then design and apply the required inverse filter.

The recovery of undistorted data from a given set of measurements has a long history in physics and engineering: one of the first research papers published in GEOPHYSICS is entitled "True Ground Motion from Mechanical Seismograph Records" (Bryan, 1936). Even in the days of galvanometer cameras and paper records, "wavelet contractor" electronic input filters were designed to enhance resolution (Ricker, 1953). The results obtained were not entirely satisfactory, probably due to an inadequate compromise between increased resolution and noise amplification. (Choosing post-deconvolution filters to achieve a good compromise is still the subject of research.) A few years later, the shift of emphasis from land to marine exploration almost immediately created a need for filters to cope with the reverberation in shallow water. In 1956 one company was doing seismic surveys in Lake Maracaibo with "notch" prerecord filters. Since the frequency of the notch had to be set (before each shot) according to the estimated water depth, this was a tricky operation! The difficulties with prerecord filtering did, however, provide great incentive for the development of digital recorders with a dynamic range large enough to allow effective postrecord filtering. The development of digital recorders was followed promptly by a period of great progress in digital processing, and as a result it now seems likely that even if analog equipment provided more faithful recording than digital equipment, data processing would still be done digitally.

Geophysical exploration has led to data acquisition and processing on a scale whose sheer volume is startling. According to a recent survey (Espey, 1977) some 800,000 miles of seismic line were recorded in 1976. If one assumes an average of 25 records/mile, an average of 50 traces/record, and an average of 6 seconds of data/trace, it follows that 6×10^9 seconds of seismic data were recorded in 1976. Probably the majority of these records were subjected to deconvolution by the industry. Data processing on this scale was envisioned only weakly (if at all) by the geophysicist of 1949, the year of unrestricted publication of Norbert Wiener's work on filtering.

In a classified report to the National Defense Research Council, first published in February 1942, Wiener developed a mathematical theory of smoothing and prediction that later became the foundation for seismic deconvolution. The purpose of his book (Wiener, 1949) is stated in the introduction:

> This book represents an attempt to unite the theory and practice of two fields of work which are of vital importance in the present emergency, and which have a complete natural methodological unity, but which have up to the present drawn their inspiration from two entirely distinct traditions, and which are widely different in their vocabulary and the training of their personnel. These two fields are those of time series in statistics and of communication engineering.

Besides long sentences, the book unfortunately involves some rather formidable mathematics. According to Bode and Shannon (1950), Wiener's yellow-bound report soon came to be known among bewildered engineers as "The Yellow Peril." Further, Wiener emphasized the use of an ensemble of data to design an optimum filter which could then be approximated by a hardware device to be manufactured for subsequent use. He says on page 102: "Much less important, though of real interest, is the problem of the numerical filter for statistical work, as contrasted with the filter as a physically active piece of engineering apparatus."

On the other hand, Levinson (1947), in a very readable discussion, presents the least-squares filter design for a specific discrete-time signal; this is the procedure followed in seismic deconvolution. Levinson is primarily recognized for the fast algorithm which he developed for the solution of the "normal equations" of least-squares filter design. Rather surprisingly, Levinson himself rather deprecates his work: "A few months after Wiener's work appeared, the author, in order to facilitate computational procedure, worked out an approximate, and one might say, mathematically trivial procedure." Despite these words, his work has clearly had great impact on geophysics, and also has been widely used by statisticians. A very interesting account of the contributions of Wiener and Levinson (and a great deal more) will be found in an excellent survey paper by Kailath (1974); this paper also includes a bibliography of 390 entries from the literature of filter theory.

In the geophysical literature, deconvolution has developed in a series of papers beginning with the classic work done at MIT. The purpose of this volume is to select from the complete set a subset that illustrates the development of and the many aspects of deconvolution. The papers that follow were selected from those published in GEOPHYSICS through June 1977. Regrettably, because of the bulk of this volume, it was necessary to omit several very fine papers.[1] I wish to thank the Exxon Production Research Company for permission to undertake this project, and to thank Dr. F. K. Levin for his continual encouragement.

[1]Another source of information that it was necessary to omit is that of patents. In particular, the fundamental patent of John P. Burg, U. S. No. 3,512,127, "Deconvolution Seismic Filtering," may be the first publication which describes seismic deconvolution in sufficient detail that it could be carried out by a novice.

REFERENCES

Bode, H. W., and Shannon, C. E., 1950, A simplified derivation of linear least-squares smoothing and prediction theory: Proceedings of the I.R.E., v. 38, p. 417-425.

Bryan, A. B., 1936, True ground motion from mechanical seismograph records: Geophysics, v. 1, p. 340-346.

Espey, H. R., 1977, Geophysical activity in 1976: Geophysics, v. 42, p. 1070-1084.

Kailath, Thomas, 1974, A view of three decades of linear filtering theory: IEEE Trans. on Information Theory, v. IT-20, p. 146-181.

Levinson, Norman, 1947, The Wiener rms (root mean square) error criterion in filter design and prediction: Journal of Mathematics and Physics, v. 25, p. 261-278. Reprinted as Appendix B of Wiener's book.

Ricker, Norman, 1953, Wavelet contraction, wavelet expansion, and the control of seismic resolution: Geophysics, v. 18, p. 769-792.

Wiener, Norbert, 1949, Extrapolation, interpolation, and smoothing of stationary time series, with engineering applications: New York, MIT Technology Press and John Wiley & Sons, Inc.

I. PREDICTION ERROR FILTERING

The paper by G. P. Wadsworth et al was the starting gun of the digital revolution that was to develop in the 1960s; this paper introduces predictive deconvolution to seismic data processing. The model chosen for the seismic record is one of zones of predictable noise that can be used for operator design plus regions during which unpredictable reflections occur. For a single channel, an ideal situation would be occasional broadband reflections in the midst of narrowband noise. A note of nostalgia: all the computations were at first done by girls using desk calculators (although later the MIT Whirlwind Computer was used). As it turned out, predictive deconvolution has not been widely used for the detection of reflections (but has found use in the detection of teleseismic events); the application of predictive deconvolution was eventually found to be in the removal of reverberations, where it is invaluable.

The cause of energy conservation may perhaps be aided by the comment that it was conversations in a carpool (between Professors Wadsworth and Hurley) that led to the formation of the Geophysical Analysis Group, and eventually to the award-winning paper by E. A. Robinson. First published as MIT GAG Report no. 7, 12 July 1954, this was Robinson's Ph.D. thesis. As noted in a review by E. A. Flinn[1], it could serve as the framework for a logical development of almost the entire subject of digital processing.

The reader should note that "predictive deconvolution" has now replaced the older term "predictive decomposition"; also that the *Z*-transform is often defined with the opposite sign in the exponent (which inverts the location of poles and zeros with respect to the unit circle). The model chosen for the seismic trace is that of the convolution of a random spike series with a minimum-delay waveform; this is still regarded as a useful model of seismic data. In current thinking, however, the model for marine data might be expanded to be a random spike series first convolved with a realizable (but possibly nonminimum-phase) short-period waveform representing the source, and then convolved again with a long-period minimum-delay reverberation wavelet.

It is worth emphasizing that for a time series which is the convolution of a random spike series with a wavelet, the autocorrelation of the time series is the same function as the autocorrelation of the wavelet [equation (5.232), p. 111]; this equality makes possible the design of deconvolution operators from the observed time series. There are, of course, difficulties in practice; we have available only a finite length of data, and the observations invariably contain noise.

Many readers will want to turn directly to Chapter VI for a discussion of deconvolution which is less formidable than the detailed treatment of the earlier chapters. The theoretical procedure for noiseless infinite time series is reviewed on p. 111-112, and relations given for the wavelet, inverse wavelet, wavelet spectrum, and inverse wavelet spectrum. Incidentally, the application of predictive deconvolution to spectral estimation was to become a major area of research. Robinson's paper (in abbreviated form) and the paper by Wadsworth et al have rightly been recognized as classics of GEOPHYSICS.

J. F. Claerbout's paper describes the application of predictive deconvolution to the detection of weak seismic waves from distant sources in the presence of ambient seismic noise. In this application, it is important to preserve the first motion; fortunately, with predictive deconvolution, signal distortion can become a problem only after the prediction span. Prediction error filtering is shown to be substantially better than band-pass filtering, which indicates that the prediction error filters had notches in their frequency response at those frequencies where the noise power was strong. With signals from an array of detectors, and noise with strong directionality, matrix prediction error filtering would similarly produce notches corresponding to the direction and velocity properties of the noise.

[1]GEOPHYSICS, v. 32, p. 411-413. Drs. Flinn, Robinson, and Treitel also assembled the expanded bibliography of Appendix. 2

Reprinted from Geophysics v. 18, no. 3, p. 539-586

DETECTION OF REFLECTIONS ON SEISMIC RECORDS BY LINEAR OPERATORS*

G. P. WADSWORTH,† E. A. ROBINSON,‡ J. G. BRYAN,§ AND P. M. HURLEY||

ABSTRACT

Linear operators are used as a statistical tool in the analysis of seismic records. It was found that when an operator is chosen for a non-reflection interval and applied to regions during which reflections occur, the effectiveness of the operator is disturbed and large errors of prediction result. This effect provides a method for the discrimination of reflections. The technique used has its basis in the theory of time series, relevant concepts of which are reviewed. The mechanics of application are given, and illustrative examples on ten seismograms are presented.

INTRODUCTION

Any variable which is generated sequentially in time constitutes a time series. This paper deals with modern methods of time-series analysis as applied to seismology. Although the underlying principles and basic techniques have been treated elsewhere in the literature, their potential contribution to seismology has not hitherto been exploited.

Credit for the first systematic study of time series belongs to Schuster, whose work on periodicities in geophysical data appeared at the turn of the century. Following Schuster, Yule (Ref. 2, 3) investigated economic time series from the viewpoint of difference equations, the use of which freed the subject from Schuster's assumption of strict periodicities.

In the late 1920's, Norbert Wiener studied this problem profoundly and produced his important paper on Generalized Harmonic Analysis in 1930. Some 10 years later, the concepts embodied in this paper were further developed by Wiener in the direction of prediction and filtering (Ref. 1).

Almost immediately upon the restricted publication of Wiener's work in 1942, a large-scale application of his methods to weather forecasting was undertaken in a project headed by Wadsworth at MIT. From 1942 until the present, Wadsworth and Bryan have been concerned with the application of suitable prediction techniques to non-stationary time series of relatively short duration. These results appear in a sequence of reports from 1942 to the present time. (Ref. 10, 11, 13, 14, 15, 16). In 1947 Wadsworth, collaborating with H. R. Seiwell, applied these statistical techniques to ocean wave research (Ref. 12). The success of the application of these techniques to this type of research led to the symposium on Applications of Autocorrelation Analysis[1] at Woods Hole, Mass. in June 1949.

* Manuscript received by the Editor April 28, 1953.

† Associate Professor of Mathematics, Massachusetts Institute of Technology, Cambridge, Mass.

‡ Research Associate in Department of Geology and Geophysics, MIT, Cambridge, Mass.

§ Division of Industrial Cooperation, MIT, Cambridge, Mass.

|| Professor of Geology, MIT, Cambridge, Mass.

[1] An important contribution to this symposium was a paper by J. W. Tukey on sampling theory of power spectrum estimation.

In 1949 Hurley and Wadsworth discussed the problem of the statistical analysis of seismic information and preliminary examination was made of data then available. These results were sufficiently interesting so that more computations were made in the spring and summer of 1950 on seismic data supplied by the Magnolia Petroleum Company. During 1950 and 1951 Robinson had become interested in this subject and it was at that time that concentrated effort commenced in the study of linear operators and their usefulness in seismic record analysis.

The technique that is used in this paper is not exactly that of Wiener, since we are dealing with non-stationary processes. In some ways our approach is more general since some of the assumptions can be relaxed.

THE PROBABILISTIC APPROACH IN SEISMOLOGY

The Deterministic Approach and the Probabilistic Approach

There are two basic approaches to treating data observed in nature, and in particular the data represented on a seismogram. One is the deterministic approach and the other is the probabilistic approach. Many people think of these two approaches as conflicting, but actually this is not the case. Recent investigations indicate that each approach is fundamentally equivalent to the other.

Up to the present time the approach in seismology has been almost exclusively deterministic. In this approach deterministic methods are used to investigate laws connecting seismological phenomena. These laws are considered to be precise in action even though the observations on the quantities involved may be inaccurate and are certain to be incomplete.

On the other hand, the probabilistic approach utilizes quantities in the form in which they are observed. Distributions and statistical functions of these quantities are examined in such combinations as one chooses. Of course, one has considerable freedom in the selection of the quantities which are to form the subject of a statistical investigation.

Actually, in an ideally complete survey, one should investigate all possible statistical parameters and combinations of parameters and not merely a selection from among them. Unfortunately, such an undertaking would be impossible because of its sheer magnitude.

Therefore in a statistical investigation one should look for groups of parameters which are connected by rigid dynamic laws with each other and with the nature of the desired information. For such a group, some of the parameters would be determined by a knowledge of the remaining ones and the dynamics would be expressed as a statistical fact. If the dynamics are not so expressed, one can conclude either that a sufficient number of the significant statistical parameters have not been considered to give a true picture of the situation, or that these parameters have been observed so inaccurately that they can not give the true picture.

It is impossible for a significant dynamic relationship not to be brought out by a proper statistical examination of the relevant quantities. In fact, if certain simplifying assumptions have to be made in the derivation of dynamic laws by a deterministic approach, there are frequent cases where the probabilistic approach actually yields more information. Such a case might be as likely in seismology as it is in quantum theory and in statistical mechanics.

The Basic Problems

Exploration seismology can be broken down into logical steps which are presented graphically in Figure 1. We see that there are two mathematical approaches for the treatment of data, the deterministic approach and the probabilistic approach. The deterministic approach consists of utilizing physical theories of wave propagation involving the solutions of integral and differential

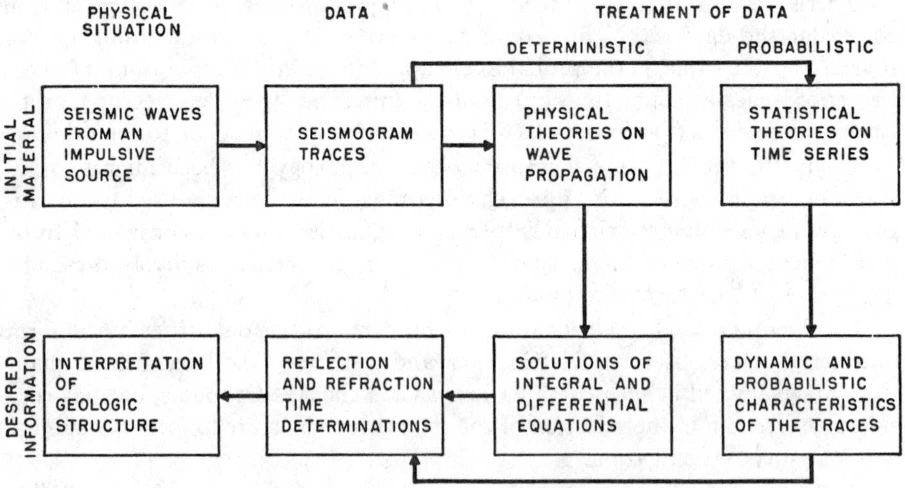

FIG. 1. The deterministic and probabilistic approaches to seismology.

equations satisfying boundary and initial conditions. The probabilistic approach consists of utilizing statistical theories of time series leading to the expression of the dynamics as a statistical fact.

The basic problems of our research program have not been solved, but nevertheless steps have been taken in that direction. Briefly, the basic problems deal with the determination of (1) statistical methods adequate to separate desired information from the total information present on the seismogram, (2) the relationship of desired information in statistical form with significant seismologic variables and the geologic structure, and (3) the interrelation of the deterministic and probabilistic approaches.

The Introduction of Statistical Methods to Seismology

This section deals with the first basic problem, namely, the determination of statistical methods adequate to separate desired information from the total information present on the seismogram. We consider statistical methods because a seismogram trace as recorded is a statistical time series, and all the traces on a seismic record constitute a set of multiple time series. This set of multiple time series is tied down to a specific origin in time, namely the time of the explosion. Time series with such a time origin dependence are called non-stationary, as opposed to stationary time series which are not linked to a specific origin in time.

Our research program in seismology was concerned with the evaluation of valid statistical methods in order to obtain the goal set forth in the first basic problem. Of all the statistical methods evaluated it was found that the application of linear operators was the most satisfactory method with regard to seismogram analysis. The method of utilizing linear operators is mathematically equivalent to the solution of a system of differential equations in space and time, but is more powerful in that it can cope effectively with disturbing influences of a random or quasi-random nature. By the use of linear operators the dynamic elements of the physical situation can be brought into sharp focus and disturbing elements suppressed.

As an approximate method of treating the non-stationary phenomenon represented by a seismogram, we divide the record into time intervals which may be considered approximately stationary. The traces in a given time interval, which we shall call the operator time interval, are used to determine an optimum linear operator for this interval. This linear operator contains, inherently, the dynamic characteristics of the traces in the operator time interval.

One type of linear operator which may be used is a pure prediction operator. That is, the operator predicts the future value of one trace from its past values and from past values of adjacent traces. The past values used as a basis of prediction are always the actual values taken from the seismic record, and are not previously predicted values corresponding to these actual values. Hence the input into the operator is always representative of the seismogram at any given point.

The ability of an operator to detect disturbances in the dynamics of the seismogram depends directly upon its accuracy in reconstructing the trace in the operator time interval. It was found that a multiple prediction operator which utilizes the traces from several geophone positions is superior to an operator which utilizes only one trace in its prediction mechanism. Having established the dynamic characteristics of the operator time interval by means of a linear operator, we wish to compare the dynamics of this interval with the rest of the seismogram. In order to accomplish this end, the operator is used to predict over the length of the seismic record in which we are interested.

The prediction error is defined as the difference between the predicted value and the actual value of the trace. This error gives a measure of the dynamic uniformity of the seismogram in the sense that low error, or good predictability, indicates small dynamical change, and large error indicates considerable dynamical change. Each conceivable operator generates a sequence of errors which forms a time series, which we may call the error time series. An investigation of these error time series, as we shall see, exposes the dynamics of the seismogram as a statistical fact. Such information about the dynamic characteristics constitutes desired information.

The Relationship of Statistical Information to Geologic Information

Once the dynamic characteristics of a seismic record are reduced by statistical methods into a statistical form, the next problem is to translate this statistical information into meaningful geologic information. This makes up the second basic problem of our research.

By computing linear operators over different sections of the record, it was found that the dynamic characteristics remained fairly consistent in the sense that one operator would apply to several sections except during intervals corresponding to a reflection. In other words, it was found that when an operator chosen for a non-reflection interval is applied to regions during which reflections occur, the effectiveness of the operator is disturbed and large errors of prediction result. Apparently, however, the disturbance is quickly assimilated, for the mechanism rapidly reverts to approximately the former level of prediction accuracy in the region following the reflection. Thus the error of prediction gives a measure of the extent to which a reflection changes the dynamic properties of a seismogram over a short interval. Therefore, by examining error time series or statistics computed from these error time series, one is able to pick off the arrival of reflected energy at places of high error. Such a procedure of analysis has been found to have two advantages: (1) the qualitative characteristics of reflections are better defined and (2) quantitatively, more reflections may be distinguished than by visual inspection of the seismogram. Illustrations of the measurement of prediction errors from ten records are presented in the section on Applications.

The problems to be explored in this direction include a more complete analysis of the reason why the dynamics of the seismograms change as they do. Various types of operators should be explored to determine which types are more sensitive to the location and discrimination of reflections. The variables which come under consideration may be subdivided into geologic variables, instrumental variables and mathematical variables. The geologic variables include such factors as underground structure, physical constants of the earth, and shot-hole effects. Instrumental variables include geophone layouts and response characteristics of the instrument used. Mathematical variables include the parameters of the linear operator and the statistics used to characterize the time series under consideration. It is necessary to investigate the effects of all these variables and to optimize those under our control.

The Interrelation of the Deterministic and Probabilistic Approaches

The third basic problem, namely, the interrelation of the probabilistic and deterministic approaches, is fundamental and also the most difficult of the problems. We have not considered this problem in detail but have it before us for future work.

One method of approach is the following. Given a physical situation which can be treated by exact physical theory, we shall introduce some random effects. From such a model a theoretical time series analysis could be carried out. In particular, the form of such statistical functions as the autocorrelation and cross-correlation could be derived from the physical situation.

Another method of approach is to determine what makes up the predictable component of the seismogram traces. The degree of predictability is intimately tied up with the stability of the operator, which in turn is tied up with the stability of the autocorrelation and cross-correlation functions. The stability of these functions depends upon the type of non-stationary time series generated by the geologic situation. The non-stationary character of these time series means that any differential equation which is set up to explain this phenomenon has variable coefficients which depend either upon time or upon the phenomenon itself. For optimum results the solution of the proper differential equation might be incorporated into the prediction mechanism.

CONCEPTS FROM THE THEORY OF STATIONARY TIME SERIES

Time Series Analysis

We shall now present some concepts from the theory of stationary time series. We are concerned chiefly with stating definitions and important theorems, and we shall by no means present a complete summary of the theory of time-series analysis. In particular we shall not deal with the important work of the so-called "English school" of statisticians, which bases its analysis on stochastic difference equations leading to the autoregressive time series (Ref. 2, 3, 4, and 5). Those many readers who are familiar with the concepts we present in this section may proceed directly to the section on Applications without losing the continuity of this paper.

The concepts given in this section depend upon the time series being stationary. A time series is said to be stationary if the probabilities involved are not tied down to a specific origin in time and if the series is conceived to run from minus infinity to infinity in time.

The Linear Operator

The one particular feature which makes the application of traditional statistical methods to time series analysis difficult is the absence of independence between successive observations. This lack of independence between successive observations is the fundamental characteristic of a time series. In the time series observed in nature such interdependence is caused by neither completely random nor completely deterministic factors, but instead the motivating factors lie somewhere between these two extremes.

As a result the most direct procedure in the analysis of time series is one which exploits the lack of independence between successive observations. The use of an operator which depends upon the interdependence of observations in the time series is one such procedure. Operators may be linear or non-linear. Since the concept of linearity which is used is of the most general type and includes a wide range of time series, we deal with linear operators exclusively in this paper, although conceivably the use of non-linear operators may prove of greater applicability to seismogram analysis.

As an introduction to the concept of a linear operator we shall consider the pure sine series

$$u_t = A \sin (\omega_0 t + \theta). \tag{1}$$

Such a series is completely deterministic for it contains no random element. As a result there exists an identity connecting three consecutive observations which is given by

$$u_{t+2h} = au_{t+h} + bu_t. \tag{2}$$

The constant a is equal to $2 \cos \omega_0 h$ and the constant b is equal to -1. The constants a and b constitute a linear operator whereby, from two consecutive values of the time series, all future values may be obtained.

It should be noted that the constants a and b of the linear operator are independent of the phase θ of the sine series. Hence the linear operator reveals the stationary character of the sine series in that the operator is not tied down to any particular origin in time. Also the operator is independent of the amplitude A which means that it is independent of the measurement used in the observations. Therefore we see that the linear operator represents an intrinsic property of the sine series, and it is not linked to the time origin or scale of the individual observations. These properties of a linear operator carry over for a large group of functions other than sines and cosines.

The Relationship of the Linear Operator and the Autocorrelation

Such an operator, which is linear, invariant with respect to the origin in time, and dependent only on the past history of the time series, is the type which Prof. N. Wiener considers in his theory of prediction for stationary time series (Ref. 1, Chapter II). He approximates the future value $f(t+\alpha)$ of the time series by applying a linear operator $K(\tau)$ to the past values $f(t-\tau)$ by means of the expression

$$\int_0^\infty f(t - \tau) dK(\tau). \tag{3}$$

The linear operator $K(\tau)$ is determined by minimizing the mean square error, which is given by

$$\lim_{T\to\infty} \frac{1}{2T} \int_{-T}^{T} \left| f(t+\alpha) - \int_{0}^{\infty} f(t-\tau)dK(\tau) \right|^2 dt. \tag{4}$$

Utilizing the calculus of variations for this minimization process, he obtains the integral equation

$$\phi(\alpha+\tau) = \int_{0}^{\infty} \phi(\tau-\sigma)dK(\sigma) \quad \text{for} \quad \tau > 0. \tag{5}$$

Here $\phi(\tau)$ is defined by

$$\phi(\tau) = \lim_{T\to\infty} \frac{1}{2T} \int_{-T}^{T} f(t)f(t+\tau)dt \tag{6}$$

and is called the autocorrelation function. Hence equation (5) tells us that the linear operator $K(\tau)$ is determined from the autocorrelation function $\phi(\tau)$. The autocorrelation function represents intrinsic dynamic properties of the time series.

The Spectrum

The fundamental theorem of generalized harmonic analysis, due to Prof. N. Wiener (Ref. 1, Chapter I), relates the autocorrelation function $\phi(\tau)$ with a monotone increasing function $\Lambda(\omega)$. More precisely, if the autocorrelation function $\phi(\tau)$ exists, then there exists a monotone increasing function $\Lambda(\omega)$ which is given by the Fourier transform

$$\phi(\tau) = \int_{-\infty}^{\infty} e^{i\omega\tau} d\Lambda(\omega). \tag{7}$$

The function $\Lambda(\omega)$ is called the integrated spectrum or integrated periodogram of $f(t)$, and represents the total power in the spectrum of $f(t)$ between the frequency $\omega = -\infty$ and the frequency ω. The integrated spectrum $\Lambda(\omega)$ may have a series of jumps if there are exact frequencies or spectral lines in the function $f(t)$. Otherwise $\Lambda(\omega)$ will be continuous, and this is the usual case met in applications.

If $\Lambda(\omega)$ is differentiable, its derivative $\Lambda'(\omega) = \Phi(\omega)$ is called the spectrum of the function $f(t)$. We may then write the Fourier transform

$$\phi(\tau) = \int_{-\infty}^{\infty} e^{i\omega\tau}\Phi(\omega)d\omega \tag{8}$$

and, in the case of simple functions, the inverse transform

$$\Phi(\omega) = \frac{1}{2\pi}\int_{-\infty}^{\infty} e^{-i\omega\tau}\phi(\tau)d\tau. \tag{9}$$

Since in this case both $\phi(\tau)$ and $\Phi(\omega)$ are even functions we may rewrite equations (8) and (9) as

$$\phi(\tau) = 2\int_{0}^{\infty} \cos \omega\tau\Phi(\omega)d\omega \tag{10}$$

and

$$\Phi(\omega) = \frac{1}{\pi}\int_{0}^{\infty} \cos \omega\tau\phi(\tau)d\tau. \tag{11}$$

Thus the autocorrelation function gives information about $f(t)$ which is analogous to the spectrum. More precisely, information as to the frequencies of $f(t)$ is preserved, and information as to the phases of the individual frequencies is lost, both in the autocorrelation function and in the spectrum.

By setting $\tau = 0$, equations (6) and (8) reduce to

$$\phi(0) = \lim_{T\to\infty} \frac{1}{2T}\int_{-T}^{T} f^2(t)dt = \int_{-\infty}^{\infty} \Phi(\omega)d\omega. \tag{12}$$

Hence we see that the total power in the spectrum is given by $\phi(0)$. The customary statistical practice is to normalize the autocorrelation function and spectrum by normalizing $f(t)$ so that it has zero mean and unit variance in the time-average sense. Then $\phi(0) = 1$ and $|\phi(\tau)| \leqq 1$, and the total power in the spectrum is equal to one.

The Relationship of the Autocorrelation and the Spectrum

We shall now give a few examples illustrating the relationship as given in equations (10) and (11) between the autocorrelation function and the spectrum.

The first example is that of the pure sine series given by equation (1). This series is completely deterministic for it contains no random element. The normalized autocorrelation of this series, computed from equation (6), is given by

$$\phi(\tau) = \cos \omega_0\tau, \tag{13}$$

which is an undamped cosine wave. The Fourier transform does not exist, but from classical Fourier series methods it can be shown that the spectrum is a line spectrum in which all the power is concentrated at the frequency ω_0.

This example allows us to give an heuristic interpretation to the relationships of the autocorrelation function and spectrum given in equation (10). Consider the spectrum $\Phi(\omega)$ of an arbitrary time series $f(t)$. Each small band of frequencies between ω and $\omega + d\omega$ acts with the differential power $\Phi(\omega)d\omega$. In view of equation (13), the differential transform of the small band of frequencies is given by

$\cos \omega\tau \Phi(\omega) d\omega$. This differential is the contribution of the small band of frequencies between ω and $\omega + d\omega$ to the autocorrelation function. Summing these differential transforms from $\omega = -\infty$ to $\omega = \infty$ we obtain the integral for the autocorrelation $\phi(\tau)$ given in equation (10).

The second example is that of a random series. A random series is conceived to have a "white light" spectrum; that is, the spectrum is given by a rectangular distribution over a long range R. Let the spectrum be

$$\Phi(\omega) = \frac{1}{2R} \tag{14}$$

for the range given by $-(a+R) \leqq \omega \leqq -a \leqq 0$ and $0 \leqq a \leqq \omega \leqq a+R$, and let $\Phi(\omega) = 0$ for values of ω outside of this range. Then the normalized autocorrelation is given by

$$\phi(\tau) = \frac{2}{R\tau} \sin \frac{R\tau}{2} \cos \left[a\tau + \frac{R\tau}{2} \right]. \tag{15}$$

We note that as R tends toward zero, $\phi(\tau)$ tends toward $\cos a\tau$, which is expected in view of equation (13).

The third example is the case of pure persistence in a time series. In this case the normalized autocorrelation function is given by the exponential

$$\phi(\tau) = e^{-a|\tau|}, \qquad a > 0 \tag{16}$$

and the spectrum by the Cauchy curve

$$\Phi(\omega) = \frac{1}{\pi} \frac{a}{a^2 + \omega^2} \cdot \tag{17}$$

Hence we see that all frequencies exist from $\omega = -\infty$ to $\omega = +\infty$.

The last example is the case in which the spectrum is the mean of two normal (Gaussian) distributions, that is,

$$\Phi(\omega) = \frac{1}{2\sigma\sqrt{2\pi}} e^{-(\omega+a)^2/2\sigma^2} + \frac{1}{2\sigma\sqrt{2\pi}} e^{-(\omega-a)^2/2\sigma^2}. \tag{18}$$

Then the normalized autocorrelation is given by

$$\phi(\tau) = e^{-\sigma^2\tau^2/2} \cos a\tau, \tag{19}$$

which is a damped cosine wave.

The Relationship of the Linear Operator and the Cross-correlation

The discussion to this point has concerned itself with the statistical properties of a single stationary time series $f(t)$. We now wish to extend these concepts to the case where we have multiple stationary time series.

A linear operator for this case is defined in a way analogous to the case of single time series. It predicts the future of one time series from its past values and the past values of the other time series. The minimization of the mean square error for the general case is carried out by N. Wiener (Ref. 1, Chapter IV). It is shown that the linear operator depends only on the autocorrelation and cross-correlations of the time series considered. The cross-correlation function is a property of two time series $f_1(t)$ and $f_2(t)$, and is defined in a way similar to the autocorrelation function by

$$\phi_{12}(\tau) = \lim_{T\to\infty} \frac{1}{2T} \int_{-T}^{T} f_1(t) f_2(t+\tau)\, dt. \tag{20}$$

The cross-correlation between $f_2(t)$ and $f_1(t)$ is defined as

$$\phi_{21}(\tau) = \lim_{T\to\infty} \frac{1}{2T} \int_{-T}^{T} f_2(t) f_1(t+\tau)\, dt. \tag{21}$$

From the definitions (20) and (21) it follows that

$$\phi_{12}(\tau) = \phi_{21}(-\tau). \tag{22}$$

In statistical practice the cross-correlation function is usually normalized by letting both $f_1(t)$ and $f_2(t)$ have zero mean and unit variance in the time-average sense. Then we have $|\phi_{11}(\tau)| \leqq 1$ and $|\phi_{22}(\tau)| \leqq 1$. Using the Schwarz inequality, we have the desired normalization of the cross-correlation which is $|\phi_{12}(\tau)| \leqq 1$.

The Cross-spectrum

The cross-correlation function $\phi_{12}(\tau)$ of $f_1(t)$ and $f_2(t)$ may be expressed as the Fourier transform

$$\phi_{12}(\tau) = \int_{-\infty}^{\infty} e^{i\omega\tau} \Phi_{12}(\omega)\, d\omega. \tag{23}$$

Here $\Phi_{12}(\omega)$ is defined to be the cross-spectrum of $f_1(t)$ and $f_2(t)$. In the case of simple functions, the inverse transform may be written as

$$\Phi_{12}(\omega) = \frac{1}{2\pi} \int_{-\infty}^{\infty} e^{-i\omega\tau} \phi_{12}(\tau)\, d\tau. \tag{24}$$

In general, the cross-correlation function $\phi_{12}(\tau)$ is not an even function of τ, and hence equation (24) tells us that the cross-spectrum $\Phi_{12}(\omega)$ has real and imaginary parts. Equations analogous to (23) and (24) hold for the cross-correlation $\phi_{21}(\tau)$ and the cross-spectrum $\Phi_{21}(\omega)$ between the time series $f_2(t)$ and $f_1(t)$. From these relations we find that

$$\Phi_{12}(\omega) = \overline{\Phi_{21}(\omega)}, \tag{25}$$

where the bar indicates the complex conjugate.

Since the cross-spectrum $\Phi_{12}(\omega)$ is a complex-valued function of the real variable ω we may write

$$\Phi_{12}(\omega) = \text{Re}\,[\Phi_{12}(\omega)] + i\,\text{Im}\,[\Phi_{12}(\omega)] \tag{26}$$

where $Re[\Phi_{12}(\omega)]$ designates the real part, and $Im[\Phi_{12}(\omega)]$ designates the imaginary part, of the cross-spectrum. We may also express the cross-spectrum by

$$\Phi_{12}(\omega) = |\Phi_{12}(\omega)|\,e^{i\theta(\omega)}. \tag{27}$$

Here $|\Phi_{12}(\omega)|$ designates the absolute value of the cross-spectrum, and is given by

$$|\Phi_{12}(\omega)| = \sqrt{[\text{Re}\,(\Phi_{12})]^2 + [\text{Im}\,(\Phi_{12})]^2}. \tag{28}$$

The argument $\theta(\omega)$ of the cross-spectrum is a function of the frequency ω, and is given by

$$\theta(\omega) = \text{arc tan}\,\frac{\text{Im}\,[\Phi_{12}(\omega)]}{\text{Re}\,[\Phi_{12}(\omega)]}\,. \tag{29}$$

Let the spectrum of $f_1(t)$ be $\Phi_{11}(\omega)$ and the spectrum of $f_2(t)$ be $\Phi_{22}(\omega)$. It can be shown (see Appendix) that the absolute value of the cross-spectrum is given by the geometric mean of the individual spectra, that is

$$|\Phi_{12}(\omega)| = \sqrt{\Phi_{11}(\omega)}\,\sqrt{\Phi_{22}(\omega)}. \tag{30}$$

Hence we see that the cross-spectrum preserves only the common frequencies of $f_1(t)$ and $f_2(t)$.

Let $\theta_1(\omega)$ represent the phase angle of the frequency ω in the time series $f_1t)$, and let $\theta_2(\omega)$ represent the phase angle of the frequency ω in the time series $f_2(t)$. The phase angles $\theta_1(\omega)$ and $\theta_2(\omega)$ depend upon the time origin of the individual time series $f_1(t)$ and $f_2(t)$ respectively. It can be shown (see Appendix) that the argument $\theta(\omega)$ of the cross-spectrum $\Phi_{12}(\omega)$ is given by the difference of the phase angles of the individual time series, that is

$$\theta(\omega) = \theta_1(\omega) - \theta_2(\omega). \tag{31}$$

The difference $\theta_1(\omega)-\theta_2(\omega)$ is independent of the time origin of the individual time series $f_1(t)$ and $f_2(t)$, and hence the stationary character of the cross-correlation and the cross-spectrum is indicated. Thus the phase differences of common frequencies of the two individual time series are preserved in the cross-spectrum.

Equations (23) and (24) show that the cross-spectrum $\Phi_{12}(\omega)$ gives information about $f_1(t)$ and $f_2(t)$ which is analogous to the cross-correlation $\Phi_{12}(\tau)$. Hence, information as to frequencies common to $f_1(t)$ and $f_2(t)$ is preserved, and information as to the phase differences of these common frequencies is preserved, both in the cross-spectrum and cross-correlation.

The Relationship of the Cross-correlation and the Spectra

Substituting equations (30) and (31) into equation (27), we see that the cross-spectrum is given by

$$\Phi_{12}(\omega) = \sqrt{\Phi_{11}(\omega)}\sqrt{\Phi_{22}(\omega)}\, e^{i[\theta_1(\omega)-\theta_2(\omega)]}. \tag{32}$$

Substituting this equation into equation (23) we find that the desired relationship between the cross-correlation and the spectra is given by

$$\phi_{12}(\tau) = \int_{-\infty}^{\infty} \sqrt{\Phi_{11}(\omega)}\sqrt{\Phi_{22}(\omega)} \cos\,[\omega\tau + \theta_1(\omega) - \theta_2(\omega)]d\omega. \tag{33}$$

This relationship may be interpreted heuristically as follows. Let us consider the two time series given by the sine functions A sin $(\omega_1 t+\theta_1)$ and B sin $(\omega_2 t+\theta_2)$. From the definition (20) of the cross-correlation, we see that the cross-correlation of these two sine functions is zero unless $\omega_1=\omega_2$. If $\omega_1=\omega_2$ then their cross-correlation, when normalized, becomes

$$\cos\,[\omega_1\tau + \theta_1 - \theta_2]. \tag{34}$$

Hence we see that their cross-correlation depends on the phase difference of their common frequency ω_1.

Let us now consider two arbitrary time series $f_1(t)$ and $f_2(t)$. Each small band of frequencies between ω and $\omega+d\omega$ acts with the differential power

$$|\,\Phi_{12}(\omega)\,|\,d\omega = \sqrt{\Phi_{11}(\omega)}\sqrt{\Phi_{22}(\omega)}\,d\omega. \tag{35}$$

Let the phase difference of the small band of frequencies in $f_1(t)$ and in $f_2(t)$ be given by $\theta_1(\omega)-\theta_2(\omega)$. In view of equation (34), we see that the differential transform of the small band of frequencies is given by

$$\sqrt{\Phi_{11}(\omega)}\sqrt{\Phi_{22}(\omega)} \cos\,[\omega\tau + \theta_1(\omega) - \theta_2(\omega)]d\omega. \tag{36}$$

This differential is the contribution of the small band of frequencies between ω and $\omega+d\omega$ to the cross-correlation function. Summing these differential transforms from $\omega=-\infty$ to $\omega=\infty$ we obtain the integral for the cross-correlation $\phi_{12}(\tau)$ given in equation (33).

In closing this section, which deals with concepts from the theory of stationary time series, we mention the following interesting example. Consider the purely random series $u_1, u_2, u_3, \cdots$ and the purely random series $v_1, v_2, v_3, \cdots$, in which the v_1 series is defined by the relationship $v_i=u_{i-j}$. Then it is seen that the cross-correlation of the two series is zero everywhere except at the jth lag, where the cross-correlation is equal to one. Such an example illustrates the value of the cross-correlation function in determining phase relationships.

APPLICATIONS TO SEISMOGRAM ANALYSIS

The Linear Operator Employed

We now present the mechanics of the statistical method described in the section on the Probabilistic Approach. As pointed out there, seismogram traces are non-stationary time series. Our statistical method for the analysis of these time series is the method of multiple correlation. Such a technique yields nearly the maximum amount of linear predictability in a finite time interval. This method differs from the Fourier methods developed by Professor Wiener for stationary time series in that he takes the entire past of the function as the basis of prediction.

In order to simplify notation we shall consider only two traces on the seismogram, but our presentation should allow the reader to extend the analysis to more than two traces. Let two traces be represented by $f_1(t)$ and $f_2(t)$ where t is time in seconds from the explosion. For computational purposes a discrete series is usually preferred, and by picking the time interval h between observations small enough a continuous trace may be transformed into a discrete series without losing essential information. Let $x_i = f_1(ih)$ and $y_i = f_2(ih)$ where h is the time unit of a discrete time series.

By the method of multiple correlation we wish to fit a linear operator to a time interval of the seismogram. This time interval, which we shall call the operator time interval, has a time duration of nh seconds and consists of the values x_i and y_i from $i = N$ to $i = N + n - 1$.

The linear operator is one which predicts the future of trace x from its past values and from corresponding past values of trace y. Algebraically the operator is expressed as

$$\hat{x}_{i+k} = c + \sum_{s=0}^{M} (a_s x_{i-s} + b_s y_{i-s}). \qquad (37)$$

In this expression $\hat{x}_{i+k}$ is the approximated value of the future value x_{i+k}, where k is the number of time units ahead which the operator predicts. That is, the prediction distance is given by kh. The $M+1$ past values of trace x are given by x_{i-s} where $s = 0, 1, \cdots, M$, and the corresponding $M+1$ past values of trace y are given by y_{i-s} where $s = 0, 1, \cdots, M$. The $2M+3$ constants of the operator given by c, a_s, and b_s $(s = 0, 1, \cdots, M)$ are constants determined in an optimum sense.

As is customary in the method of multiple correlation, these constants are determined by the Gauss method of least squares. Hence we wish to find the values of c, a_s, b_s $(s = 0, 1, \cdots, M)$ which minimize the sum of square errors between the actual value x_{i+k} and the predicted value $\hat{x}_{i+k}$. The summation is taken over the operator time interval; that is, the summation index $i+k$ should run from $i+k = N$ to $i+k = N+n-1$. Here we wish to minimize

$$I = \sum_{i=N-k}^{N+n-1-k} (x_{i+k} - \hat{x}_{i+k})^2 \tag{38}$$

with respect to c, a_s, b_s ($s=0, 1, \cdots, M$). Substituting equation (37) into equation (38) we have

$$I = \sum_i \left[x_{i+k} - c - \sum_s (a_s x_{i-s} + b_s y_{i-s}) \right]^2. \tag{39}$$

In equation (39), and from now on all summations on the index i are for $i=N-k$ to $i=N+n-1-k$, and all summations on the index s are for $s=0$ to $s=M$.

In order to carry out the minimization we set the partial derivatives of I with respect to c, a_s, b_s ($s=0, 1, \cdots, M$) equal to zero. Since there are $(M+1)a$'s, $(M+1)b$'s, and one c, we obtain $2M+3$ linear algebraic equations in the $2M+3$ unknowns given by a_s, b_s, c (where $s=0, 1, \cdots, M$). This set of simultaneous equations is

$$cn + \sum_s \left(a_s \sum_i x_{i-s} + b_s \sum_i y_{i-s} \right) = \sum_i x_{i+k}$$

$$c \sum_i x_{i-r} + \sum_s \left(a_s \sum_i x_{i-r} x_{i-s} + b_s \sum_i x_{i-r} y_{i-s} \right) = \sum_i x_{i-r} x_{i+k}$$

$$\text{for } r = 0, 1, \cdots, M \tag{40}$$

$$c \sum_i y_{i-r} + \sum_s \left(a_s \sum_i y_{i-r} x_{i-s} + b_s \sum_i y_{i-r} y_{i-s} \right) = \sum_i y_{i-r} x_{i+k}$$

$$\text{for } r = 0, 1, \cdots, M.$$

Expanding equation (39) and using equations (40) we see that the minimum value I_m of I is given by

$$I_m = \sum_i x_{i+k}^2 - c \sum_i x_{i+k} - \sum_s \left[a_s \sum_i x_{i-s} x_{i+k} + b_s \sum_i y_{i-s} x_{i+k} \right]. \tag{41}$$

The minimum value I_m is one measure of how well the operator reproduces the x series in the operator time interval.

We define the sample variance I_0 to be

$$I_0 = \sum_i (x_{i+k} - \bar{x})^2 \tag{42}$$

where $\bar{x} = 1/n \sum x_{i+k}$ is the sample mean of the x series in the operator time interval. Then the percent reduction of the sample variance about the sample mean is defined to be

$$\text{percent } R = 100 \left[1 - \frac{I_m}{I_0} \right]. \tag{43}$$

The percent reduction is another measure of how well the operator reproduces the x trace in the operator time interval, where one hundred percent reduction is perfect reproduction.

The Measure of Error of Prediction

Once numerical values of the constants a_s, b_s, c ($s=0, 1, \cdots, M$) of the operator are determined, they are used to predict the whole of trace x by means of equation (37). In using equation (37) it must be remembered that the past values of x_{i-s} and y_{i-s} (where $s=0, 1, \cdots, M$) are the actual values taken from the seismogram traces.

In order to determine how well trace x is being predicted by the operator the error between the actual value x_{i+k} and the predicted value $\hat{x}_{i+k}$ should be measured. These errors, or residuals, may be studied individually or statistically in order to measure the effectiveness of the operator in reconstructing the x trace.

One such statistical quantity which may be used is the running average E_i of square errors given by

$$E_i = \frac{1}{2p} \sum_{j=i-p}^{i+p-1} (x_j - \hat{x}_j)^2 \tag{44}$$

where $2p$ is the number of elements in each average. If we let

$$V_i^{(x)} = \frac{1}{2p} \sum_{j=i-p}^{i+p-1} (x_j - \bar{x})^2 \tag{45}$$

denote the running variance of trace x, then another measure of the goodness of the prediction of the x trace is the running reduction R_i given by

$$R_i = 1 - \frac{E_i}{V_i^{(x)}} \cdot \tag{46}$$

A third measure of the goodness of prediction of the x trace is given by

$$T_i = \frac{E_i}{\alpha V_i^{(x)} + \beta V_i^{(y)}} \tag{47}$$

where $V_i^{(x)}$ is the running variance of trace x and $V_i^{(y)}$ is the running variance of trace y. Here α and β are weighting factors which should be determined in an optimum sense with respect to the operator coefficients.

The three measures of the effectiveness of prediction given by equations (44), (46), and (47) were used in the computational studies in the summer and fall of 1951. Nevertheless, since the seismograms were A.V.C. (automatic volume control) records, the error curves E_i given by equation (44) were considered satisfactory as a first approximation to the measurement of prediction error.

The averages given in equations (44), (46), and (47) represent time averages of a single error time series. If we have many operators predicting several traces another type of averaging is possible on the "ensemble" of error time series generated by these operators. An ensemble average, or an average across these series rather than along an individual one, is another measure of prediction error.

Let us suppose that we have taken a series of operators on a record which consists of traces from equally spaced seismometers. Suppose there are T traces, and on the lth trace ($l=1, 2, \cdots T$) we have chosen N_l operators. For the kth operator on this trace ($k=1, \cdots N_l$) there is an associated error time series which we define as $e_i(kl)$. Then, for example, we may construct a single error time series $\epsilon_i{}^l$ to be associated with the lth trace by the expression

$$\epsilon_i{}^l = \sum_{k=1}^{Nl} \left[e_i{}^{(kl)}\right]^2. \tag{48}$$

We may then average these error time series over the various traces. Now since we are interested not only in finding reflection times but also the associated step-outs, we construct the error time series $\delta_i{}^{(\alpha)}$ with an arbitrary lag or lead α

$$\delta_i{}^{(\alpha)} = \sum_{l=1}^{T} \epsilon_{i-\alpha l}{}^{(l)} \qquad \alpha = 0, \pm 1, \pm 2, \cdots \tag{49}$$

with the expectation that a peak on this error time series, corresponding to a certain reflection, should be highest and narrowest for that value of α most closely corresponding to the true step-out of the given reflection. At the present time a study is being made on the use of ensemble averaging of this type.

COMPUTATIONAL RESULTS

The Tukey-Hamming Computational Procedure to Estimate the Spectrum

In 1950 and 1951 computational studies were undertaken to determine the statistical properties of the correlation functions and the spectra taken over various intervals of seismic records. We shall not reproduce the specific results from these studies in this paper. Nevertheless, we should like to present Tukey's method of computing the spectrum. It should be pointed out that this method was developed by Tukey (Ref. 6), and Tukey and Hamming (Ref. 7), so that the spectra of short time series could be estimated with a knowledge of the variability of the estimate.

The unnormalized autocorrelation of a group of data at equally spaced intervals of time, say $x_1, x_2 \cdots x_n$, are computed by the standard formulae. The formula which Tukey prefers is given by

$$R_p = \frac{1}{N-p} \sum_{i=1}^{N-p} x_i x_{i+p} \tag{50}$$

Here the R_p's are estimates of the unnormalized autocorrelation function at the discrete lag p, and are called the sample serial products.

The basic problem is to obtain an approximation to the spectrum from a given number m of these serial products computed from a finite time series. Frequencies ω and frequencies $2\pi \pm \omega$, $4\pi \pm \omega$, etc., are equivalent as far as the method of Tukey and Hamming is concerned. Thus the effect is to fold over the last part of the frequency scale where $2\pi/\omega < 2$ into that portion of the scale where $2\pi/\omega \geq 2$. This means of course that on the ω scale the distribution of frequencies will now run from $-\pi$ to π, and we shall confine our attention to this region. By choosing discrete values of angular frequency $\omega = s\pi/m$ ($s = 0, 1, 2, \cdots, m$), we may perform numerical integration of (11) by the trapezoidal rule and write

$$\Phi\left(\frac{s\pi}{m}\right) \approx L_s \tag{51}$$

where

$$L_s = \frac{1}{\pi}\left[\frac{1}{2} R_0 \cos 0 + \sum_{j=1}^{m-1} R_j \cos \frac{s\pi j}{m} + \frac{1}{2} R_m \cos s\pi\right]. \tag{52}$$

By letting $r_j = R_j/R_0$ ($j = 0, 1, \cdots, m$), we have

$$L_s = \frac{R_0}{2\pi}\left[1 + 2\sum_{j=1}^{m-1} r_j \cos \frac{s\pi j}{m} + r_m \cos s\pi\right]. \tag{53}$$

An approximate integration of $\Phi(\omega)$ from $-\pi$ to π by the trapezoidal rule yields

$$\int_{-\pi}^{\pi} \Phi(\omega) d\omega = 2\int_{0}^{\pi} \Phi(\omega) d\omega \approx \frac{\pi}{m}\left[L_0 + 2\sum_{1}^{m-1} L_s + L_m\right]. \tag{54}$$

Now by standard summation formulae, we find that

$$L_0 + 2\sum_{1}^{m-1} L_s + L_m = \frac{m}{\pi} R_0. \tag{55}$$

Thus, for all values of m, the area given by the trapezoidal rule is R_0, the serial product of lag zero. Since R_0 is the computed estimate of $\phi(0)$, the trapezoidal approximation of equation (11) yields the statistical analogue of equation (12). Moreover we see, once again, that in the estimated spectrum the total power has been compressed into the interval $(-\pi, \pi)$.

Because the values of L_s are subject to systematic statistical errors, they must be smoothed in order to obtain a satisfactory estimate of the spectral density. After research into feasible smoothing techniques, Tukey and Hamming settled upon the following simple scheme. The smoothed estimate U_s of the spectral density is given by

$$U_s = .23L_{s-1} + .54L_s + .23L_{s+1} \quad (s = 0, 1, 2, \cdots, m). \tag{56}$$

Since U_0 and U_m respectively involve L_{-1} and L_{m+1}, which have not been defined, set $L_{-1}=L_1$ and $L_{m+1}=L_{m-1}$. Or what amounts to the same thing, smooth the end points thus:

$$\begin{aligned} U_0 &= .54L_0 + .46L_1 \\ U_m &= .54L_m + .46L_{m-1}. \end{aligned} \tag{57}$$

Because of the identity

$$U_0 + 2\sum_{1}^{m-1} U_s + U_m = L_0 + 2\sum_{1}^{m-1} L_s + L_m \tag{58}$$

we see that the smoothing process is area-preserving, and that the statistical analogue of equation (12) also holds for the smoothed estimates U_s.

Examples of Correlation Functions and Spectra

In Figures 2 and 3 we present examples of correlation functions and spectra. The solid curve in the left hand side of Figure 2 shows an autocorrelation function for trace N750 of MIT Record No. 2 (Figure 4) from time equal to .2 seconds to 1.2 seconds; and the dashed curve shows corresponding autocorrelations for trace S750. Both of these autocorrelation functions have been normalized so that their zeroth lag is equal to one. Utilizing the computational procedure of Tukey and Hamming the spectra of each of these traces were computed from their autocorrelation functions. These spectra, which are shown in the right hand side of Figure 2, have not been normalized; hence their areas represent the relative energies present in the traces.

A second example of correlation functions and spectra is presented in Figure 3. We let $f_1(t)$ represent trace N750 on MIT Record No. 1 (Figure 4) from time equal to 1.05 seconds to 1.225 seconds, and $f_2(t)$ represent trace N250 over the same interval of time. These two traces were divided into a discrete series of observations, with a spacing of 2.5 milliseconds between observations. These traces were measured in arbitrary units of amplitude about their mean values.

In diagram A of Figure 3 the solid curve shows the autocorrelation function $\phi_{11}(\tau)$ of trace N750, and the dashed curve shows the autocorrelation function $\phi_{22}(\tau)$ of trace N250. These autocorrelation functions, which are not normalized, were estimated by the serial products of the traces. In diagram B of Figure 3 the solid curve shows the spectrum $\Phi_{11}(\omega)$ of trace N750; and the dashed curve shows the spectrum $\Phi_{22}(\omega)$ of trace N250. These spectra were estimated from the autocorrelation functions given in diagram A by use of the Tukey-Hamming computational procedure.

Diagram C of Figure 3 gives the cross-correlation function $\phi_{12}(\tau)$ of trace N750 and trace N250. The cross-correlation function, which is not normalized, was estimated by the cross products of the two traces. Diagram D of Figure 3 gives the cross spectrum $\Phi_{12}(\omega)$ of trace N750 and trace N250. The cross-spectrum

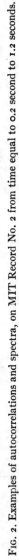

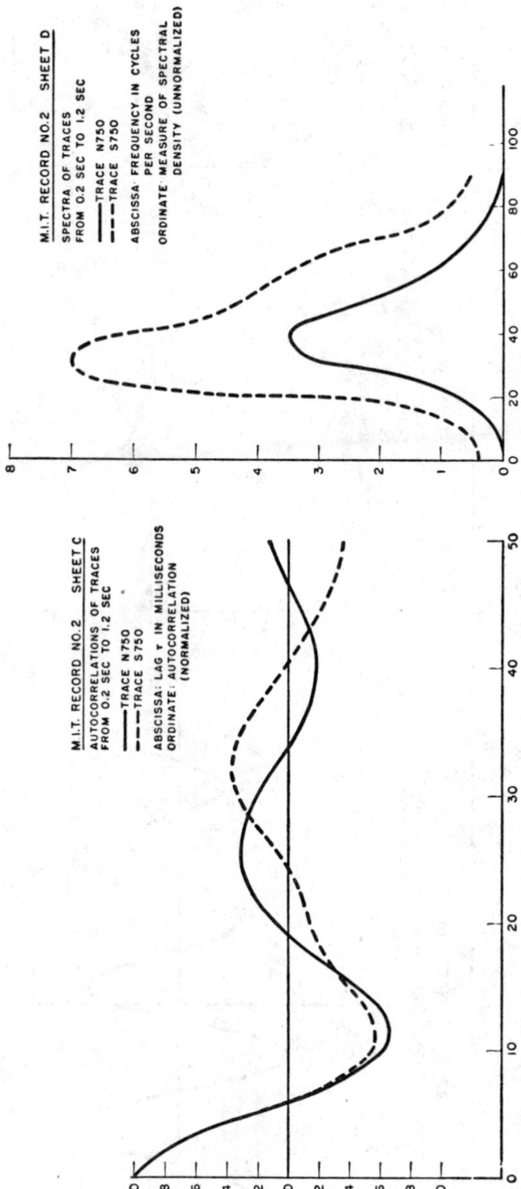

FIG. 2. Examples of autocorrelations and spectra, on MIT Record No. 2 from time equal to 0.2 second to 1.2 seconds.

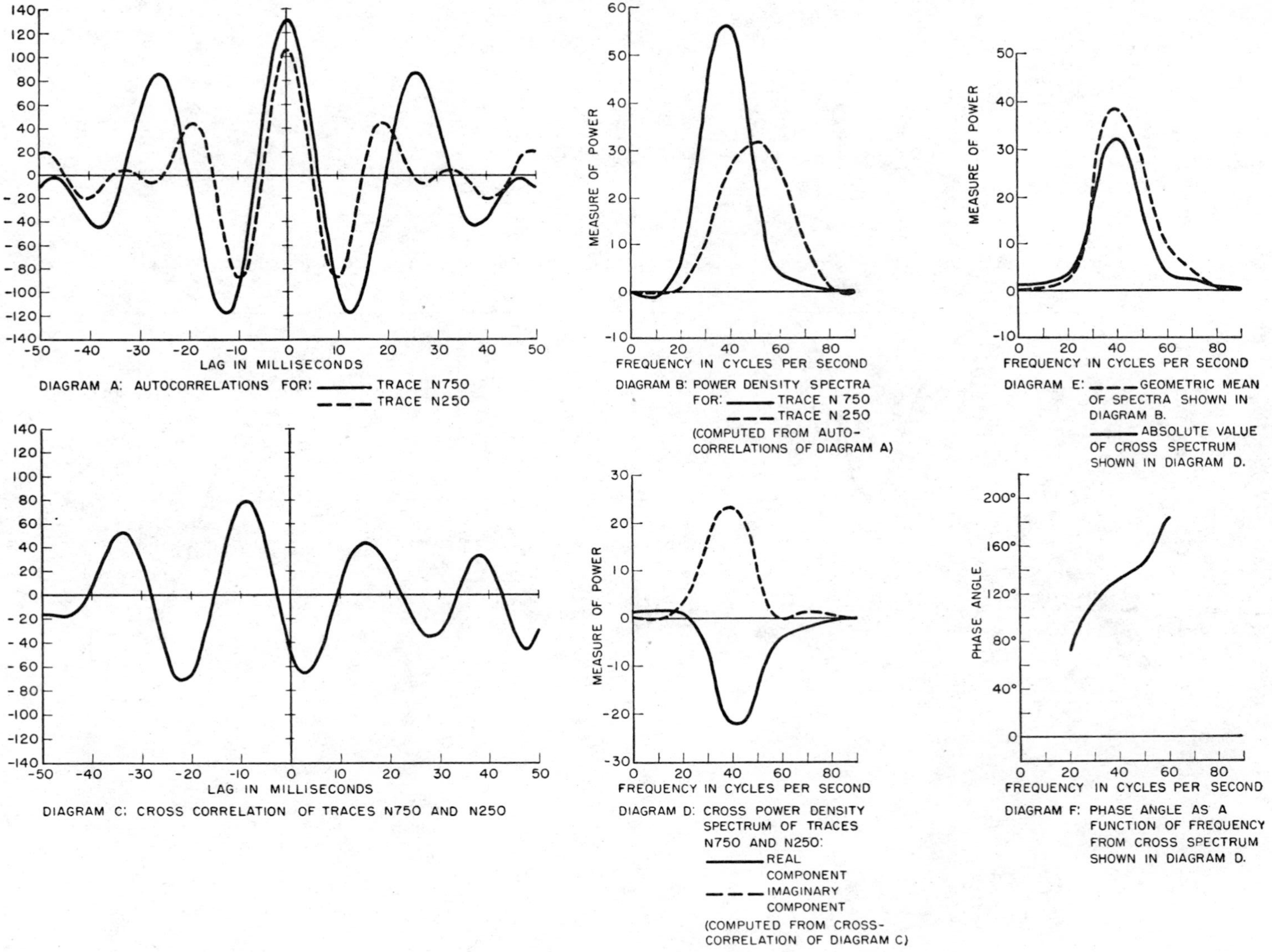

FIG. 3. Examples of correlation functions and spectra, on MIT Record No. 1 from time equal to 1.05 seconds to 1.225 seconds.

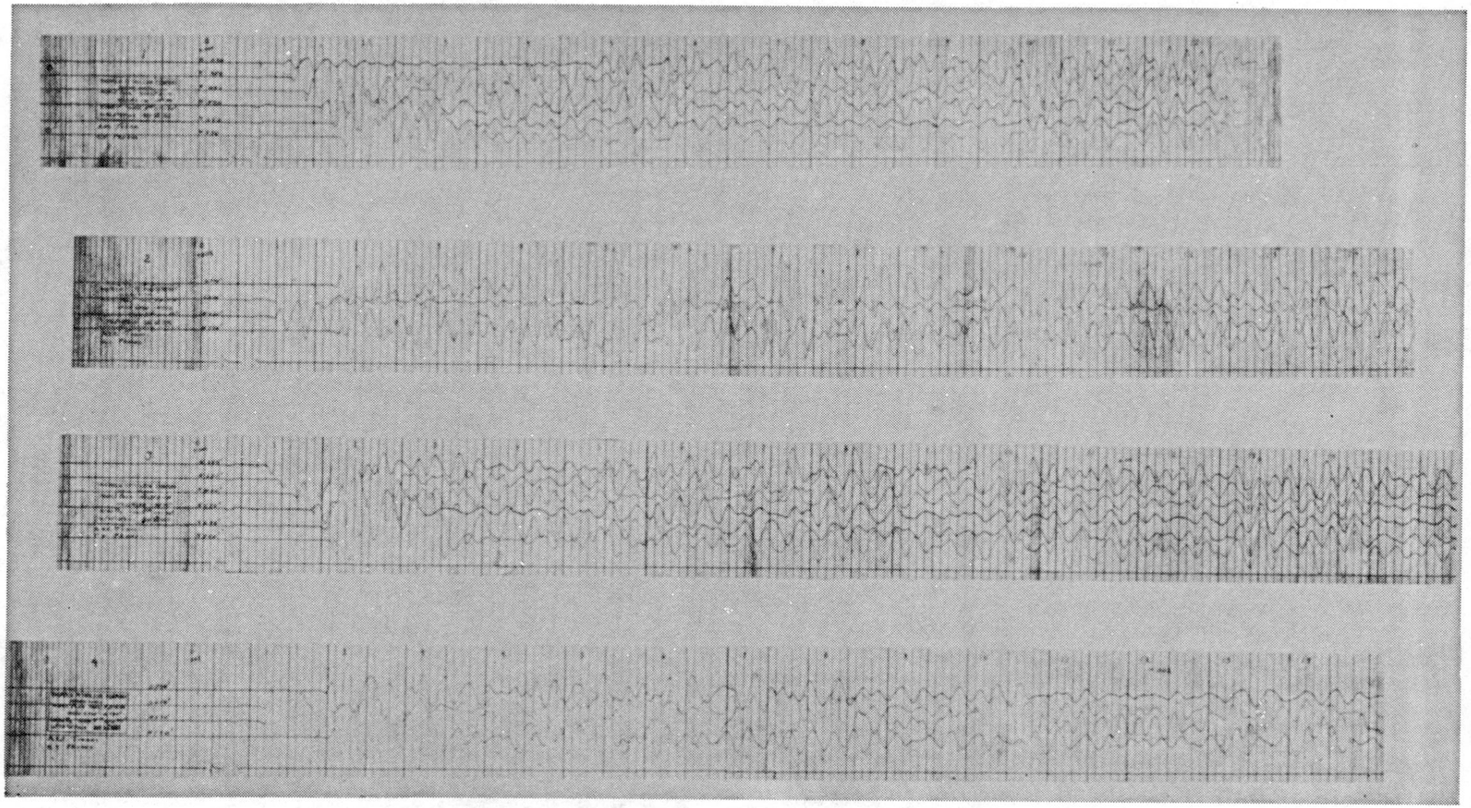

Fig. 4. MIT Records No. 1, 2, 3, and 4. (Furnished by Magnolia Petroleum Co.)

was estimated from the cross-correlation given in diagram C by an extension of the Tukey-Hamming computational procedure to equation (24). Since the cross-correlation function is not an even function of the lag τ, the cross-spectrum has real and imaginary parts. In diagram D of Figure 3 the real part of the cross-spectrum is shown by the solid curve, and the imaginary part is shown by the dashed curve.

The dashed curve of diagram E shows the geometric mean of the individual spectra shown in diagram B. The solid curve in diagram E shows the absolute value of the cross-spectrum which is shown in diagram D. The absolute value is computed from equation (28). According to equation (30) the two curves of diagram E should be identical. The discrepancy between the two curves is probably attributable to statistical errors in the computations and to errors due to a finite number of observations. Diagram F gives the argument $\theta(\omega)$ of the cross-spectrum which according to equation (31) represents the difference of the phase angles of the time series $f_2(t)$ and $f_1(t)$. The phases for high and low frequencies are not plotted because the low power at these frequencies (Diagram D) is considered to be unreliable for use in equation (31) due to statistical errors in computation.

Results of the Application of Linear Operators to Seismic Records

We shall now present figures which describe the results of applying linear operators to ten records of the Magnolia Petroleum Company. The actual mechanics of the method are presented in the previous section. For all the results presented here the form of the operator which was used is given by equation (37), in which the following values of the parameters were taken. The traces were transformed into discrete series, with the time interval h between observations equal to 2.5 milliseconds. The value of M, which represents number of past lags on each trace, was chosen equal to 3. The value of n, which is the number of observations from each trace in the operator time interval, was chosen equal to 50. The value of the prediction distance k, which was allowed to take values of 2, 4, and 6, corresponding to 5 milliseconds, 10 milliseconds, and 15 milliseconds respectively, is given in the figures. The x trace, which is the predicted trace, and the y trace, which is the other trace used in the operator, are marked on the figures.

The values of these parameters were chosen subject to severe limitations in computational facilities. A much more complete analysis of the behavior of the auto- and cross-correlations would be desirable in order to understand better the correct choice of these parameters. Several operators which utilized three traces in their prediction mechanism showed promising results. At the present time further studies are being undertaken on the optimization of these parameters with respect to the analysis of seismic records and the physical characteristics of the phenomenon.

In all the figures which are labeled "Error Curves," the measure of error was computed from equation (44) with p equal to 10, which means there are 20 squared errors in each average. In Figures 5 through 11 and Figures 13 through 16, the operator time interval is represented by an arrow with the word "Operator" written on it. In these same figures reflections are represented by cross-hatched areas. The cross-hatching runs northeast for reflections which were marked by the Magnolia Petroleum Company, and the cross-hatching runs northwest for reflections which were not marked by the Magnolia Petroleum Company but were determined from company-marked reflection times on corresponding seismic records. In Figures 18 through 21, on the other hand, operator time intervals are represented by cross-hatching, and the reflection times of the first trace are represented by heavy vertical lines.

The four records from Prospect Pace, which we designate as MIT Records Nos. 1, 2, 3, and 4, are shown in Figure 4. These records were taken by a Magnolia Petroleum Company seismic party on March 15, 1950, with A.V.C. traces, filter band $F28^2$-89, and CV phones. Information as to profile, spread, charge, and depth for these records is given in the figures. Table I gives the approximate time durations of reflections marked by the Magnolia Petroleum Company. Time is measured in seconds from the time of the explosion.

TABLE I

REFLECTION TIMES ON RECORDS NO. 1–4

Record	Reflection a (Seconds)	Reflection b (Seconds)	Reflection c (Seconds)	Reflection d (Seconds)
No. 1	.51 to .54	1.00 to 1.04	1.16 to 1.24	Record too short
No. 2	Not marked	Not marked	Not marked	Not marked
No. 3	Not marked	1.00 to 1.04	1.16 to 1.24	1.29 to 1.32
No. 4	Not marked	Not marked	Not marked	Not marked

Figure 5 shows actual and predicted values of trace N750 of MIT Record No. 1. The solid curves in Figure 5 show a section of trace N750 from time equal to .85 seconds to 1.25 seconds. The three dashed curves represent the predicted values of trace N750 for three operators with prediction distances of 5, 10, and 15 milliseconds. In the equation for these operators the x trace was N750 and the y trace was N250. Error Curves for these three operators are shown in Figure 6 (continued) and reduction curves for them are shown in Figure 7. The measure of reduction used in Figure 7 is $1-R_i$, where R_i is given by equation (46).

Figures 6, 8, 9, 10, and 11, give Error Curves for operators taken on MIT Records No. 1, 2, 3, and 4 respectively. For these operators the percent reductions were computed from equation (43) and are given in Table II.

The four records from Prospect T-218, which we designate as MIT Records No. 5, 6, 7, and 8, are shown in Figure 12. These records were taken by a Magnolia Petroleum Company seismic party on March 17, 1950 with A.V.C. traces and filter band $F28^2$-89. Information as to profile, spread, charge, and depth is

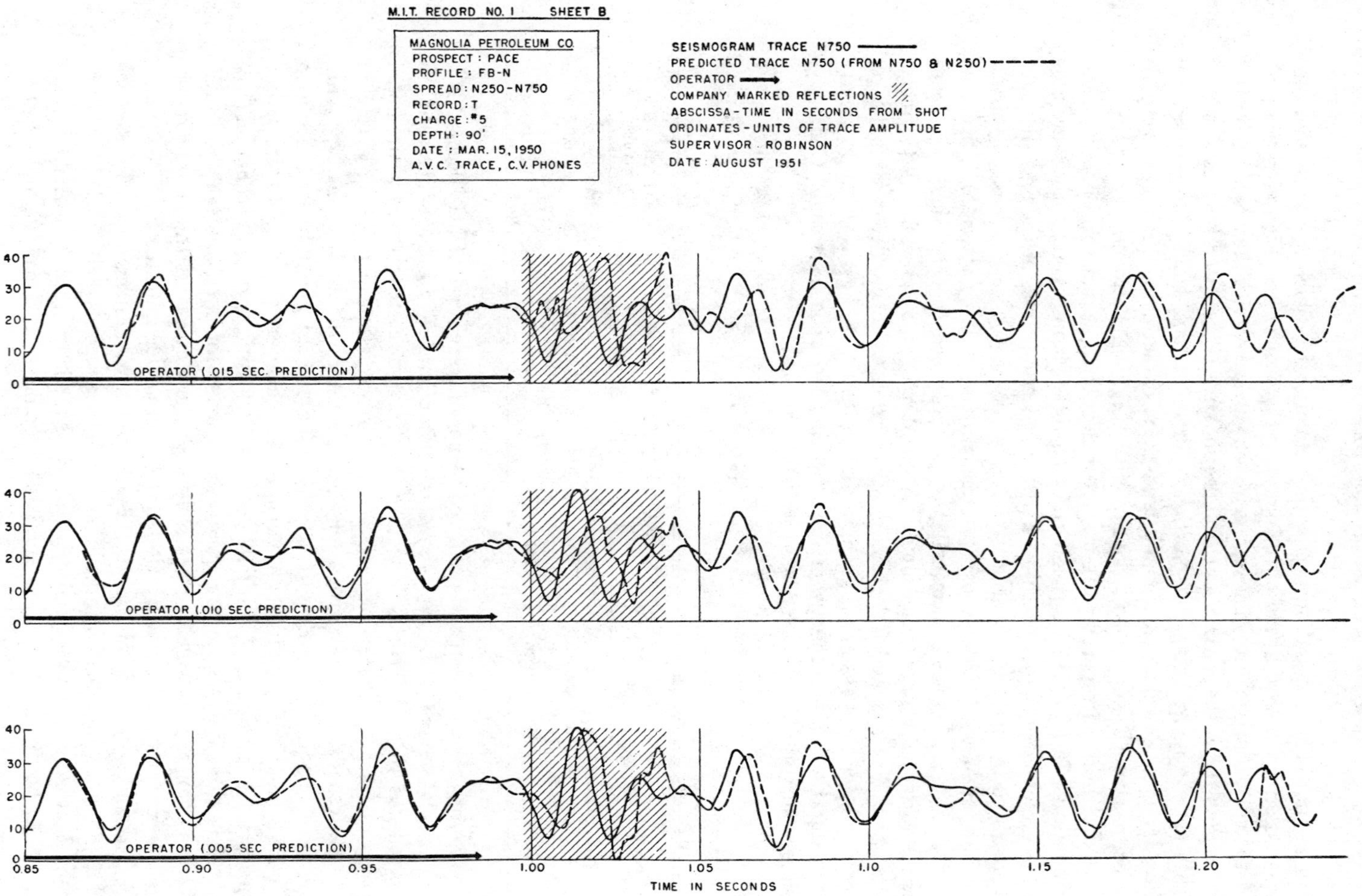

FIG. 5. Actual and predicted values of trace N750, MIT Record No. 1, for operators with various prediction distances.

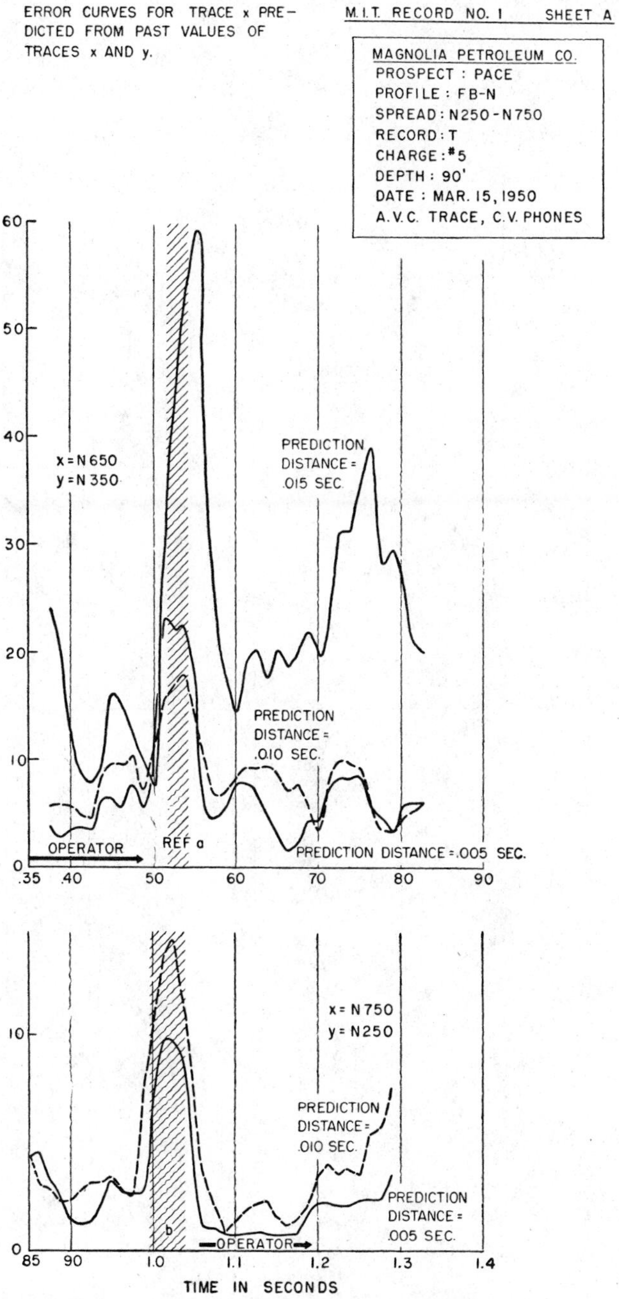

FIG. 6. Error curves for operators taken on MIT Record No. 1.

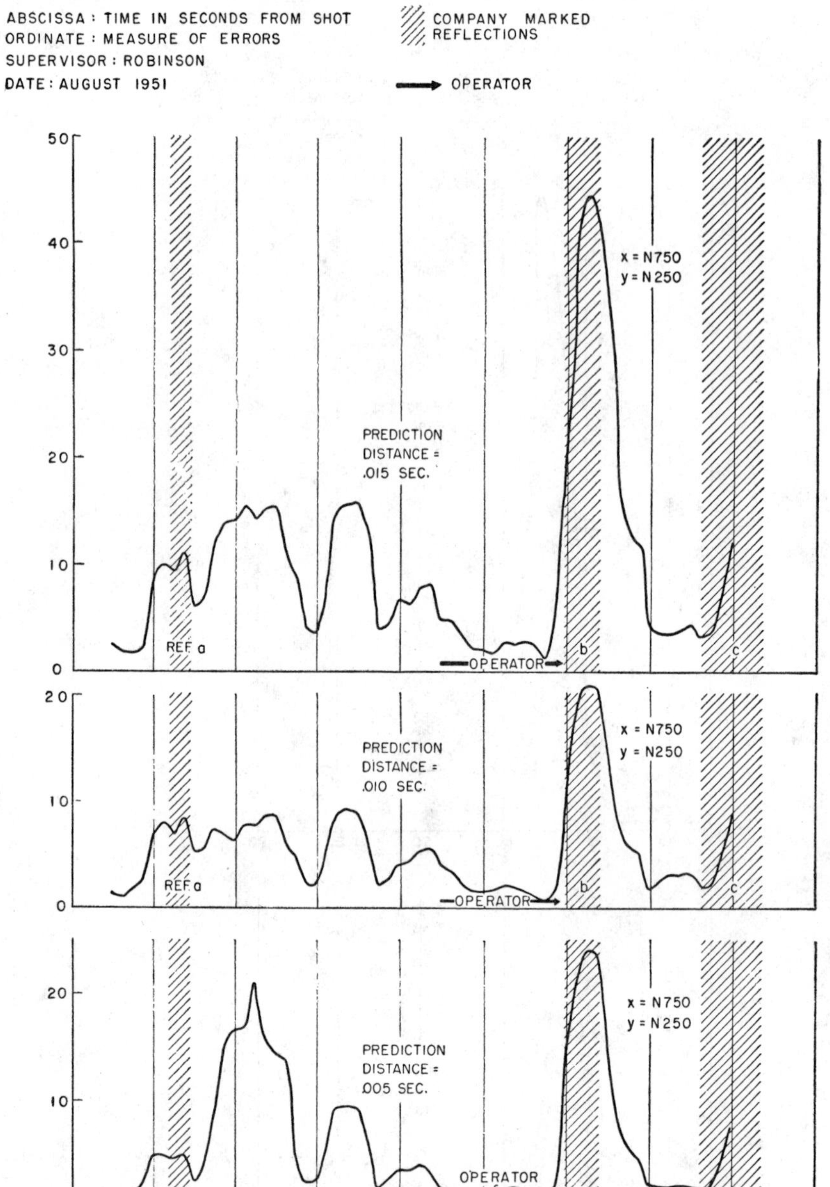

FIG. 6. (Continued). Error curves for operators taken on MIT Record No. 1.

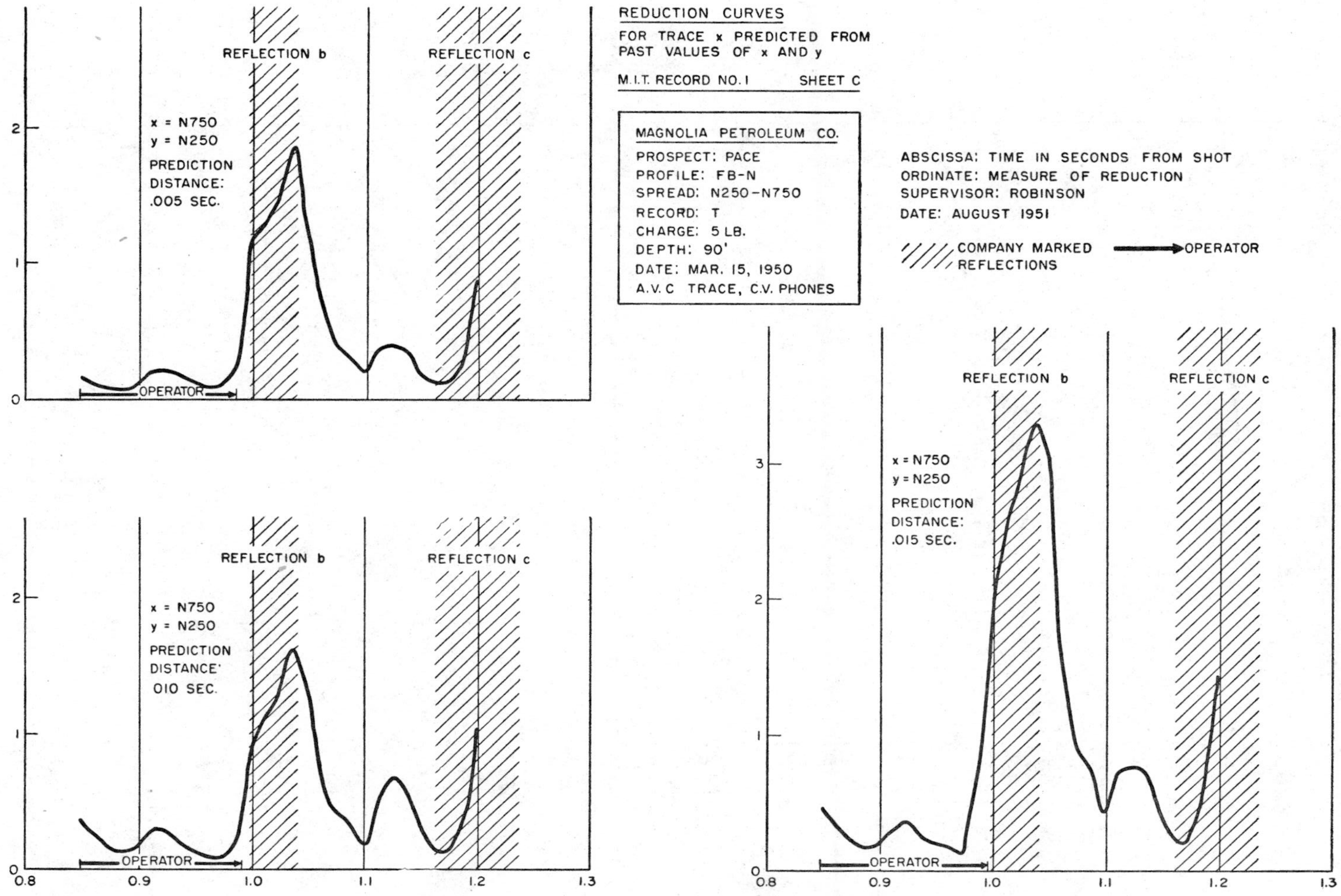

FIG. 7. Reduction curves for operators taken on MIT Record No. 1. (A reduction curve is an error curve from which effects due to variance of trace amplitude have been removed.)

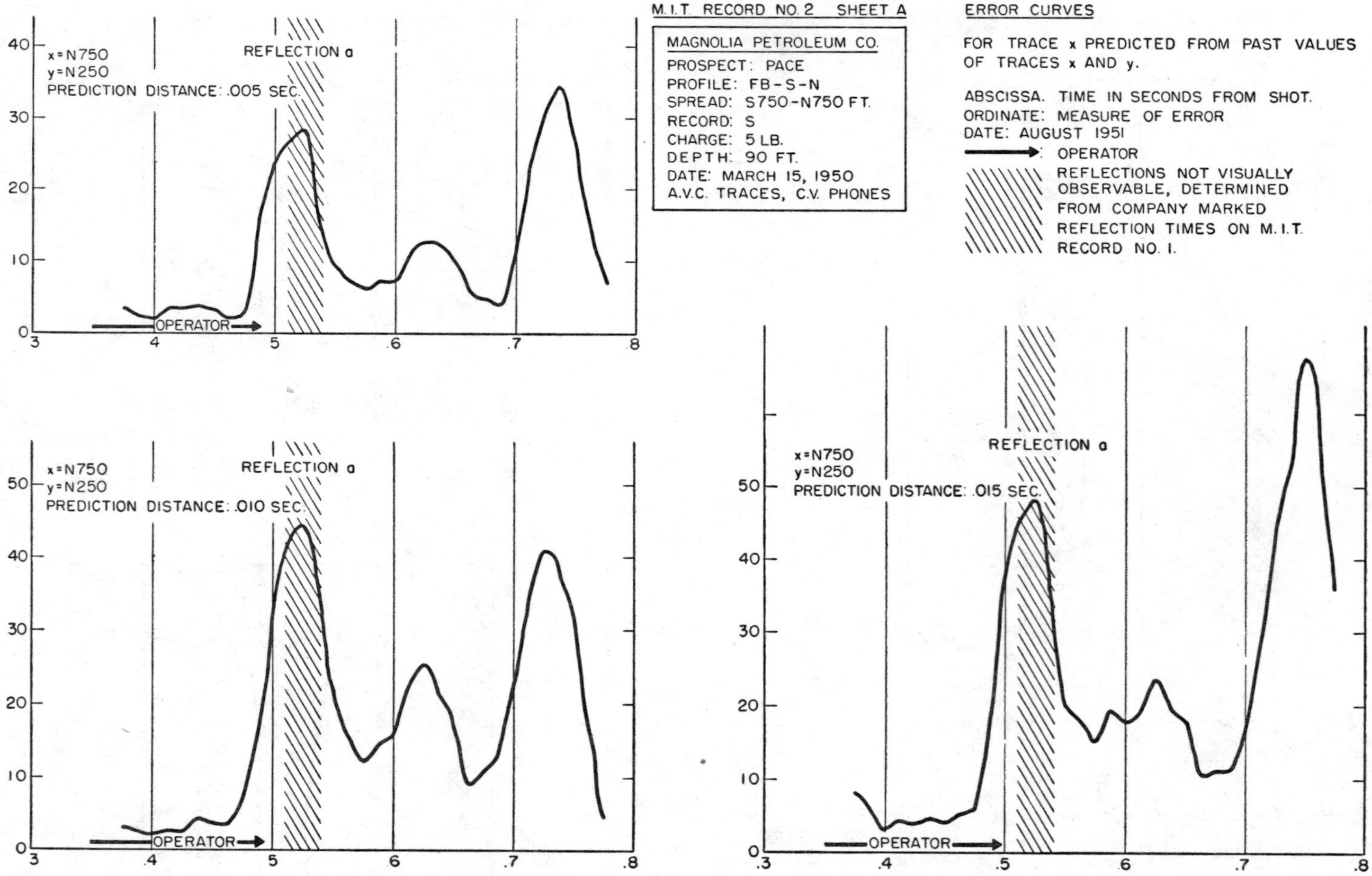
M.I.T. RECORD NO. 2 SHEET A
MAGNOLIA PETROLEUM CO.
PROSPECT: PACE
PROFILE: FB-S-N
SPREAD: S750-N750 FT.
RECORD: S
CHARGE: 5 LB.
DEPTH: 90 FT.
DATE: MARCH 15, 1950
A.V.C. TRACES, C.V. PHONES
ERROR CURVES
FOR TRACE x PREDICTED FROM PAST VALUES OF TRACES x AND y.
ABSCISSA. TIME IN SECONDS FROM SHOT.
ORDINATE: MEASURE OF ERROR
DATE: AUGUST 1951
: OPERATOR
REFLECTIONS NOT VISUALLY OBSERVABLE, DETERMINED FROM COMPANY MARKED REFLECTION TIMES ON M.I.T. RECORD NO. 1.
x=N750
y=N250
PREDICTION DISTANCE: .005 SEC.
REFLECTION a
OPERATOR
x=N750
y=N250
PREDICTION DISTANCE: .010 SEC.
REFLECTION a
OPERATOR
x=N750
y=N250
PREDICTION DISTANCE: .015 SEC.
REFLECTION a
OPERATOR

FIG. 8. Error curves for operators taken on MIT Record No. 2.

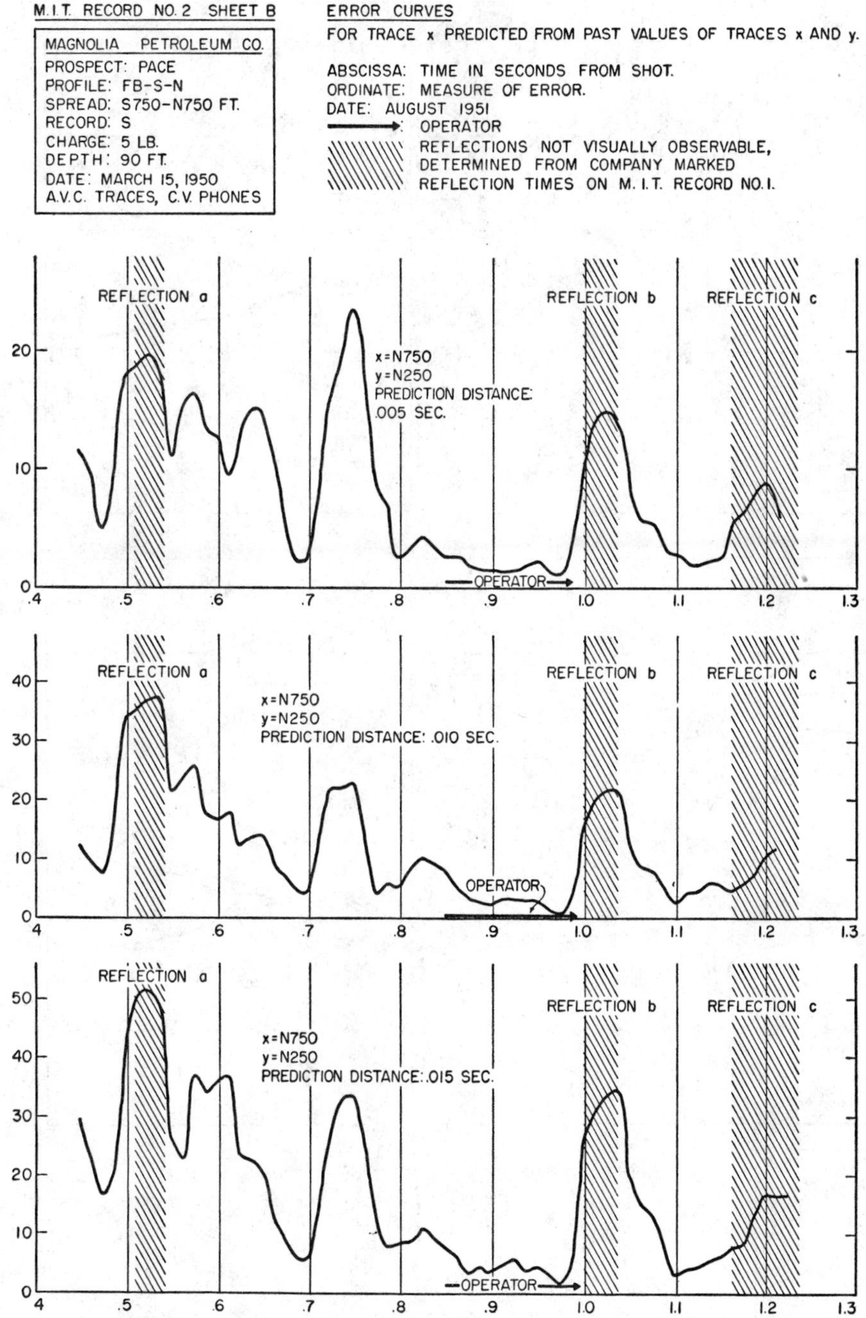

Fig. 9. Error curves for operators taken on MIT Record No. 2.

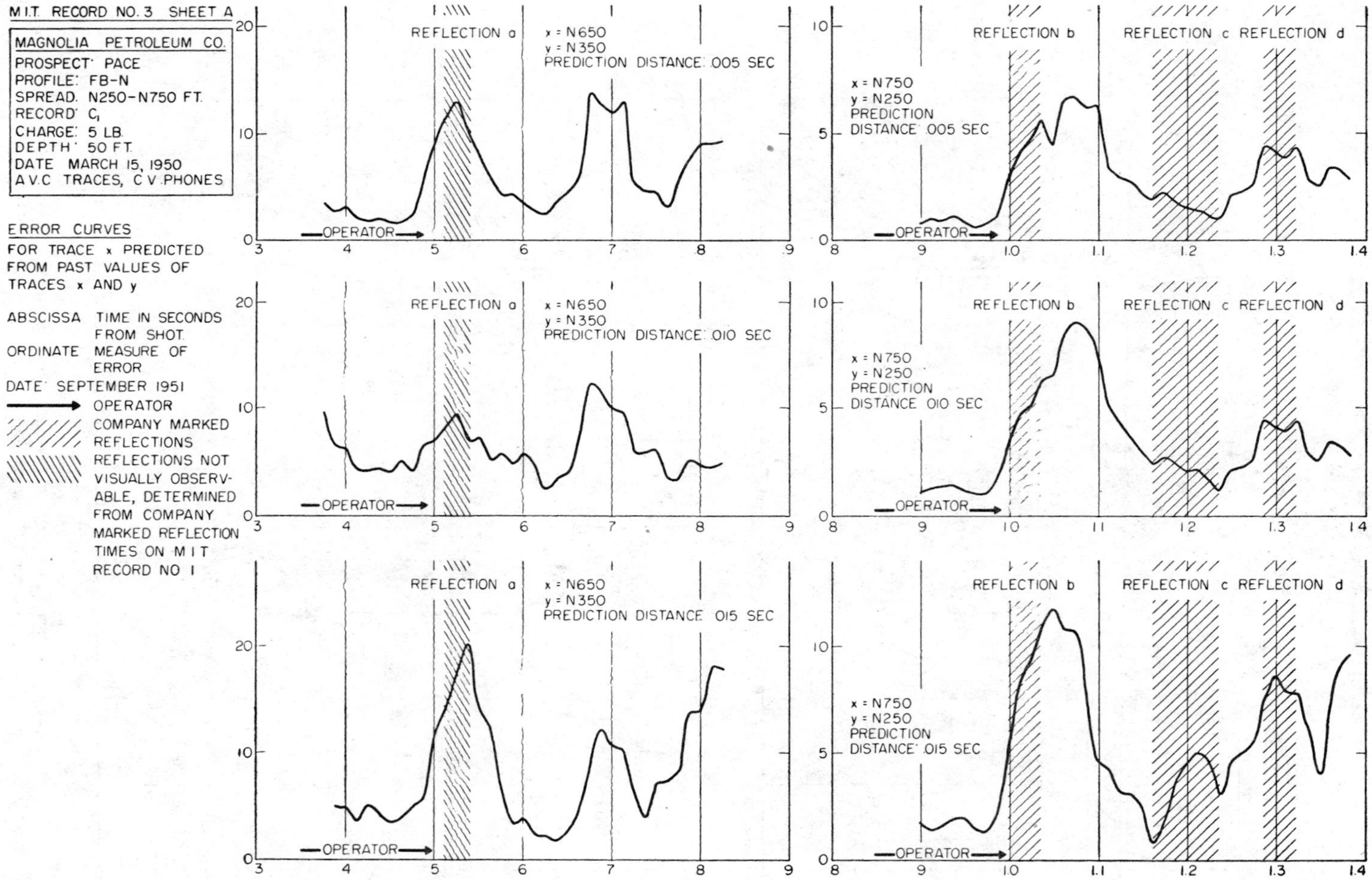
M.I.T. RECORD NO. 3 SHEET A
MAGNOLIA PETROLEUM CO.
PROSPECT: PACE
PROFILE: FB-N
SPREAD. N250-N750 FT.
RECORD: C_1
CHARGE: 5 LB.
DEPTH: 50 FT.
DATE MARCH 15, 1950
A.V.C. TRACES, C.V. PHONES
ERROR CURVES
FOR TRACE x PREDICTED FROM PAST VALUES OF TRACES x AND y
ABSCISSA TIME IN SECONDS FROM SHOT.
ORDINATE MEASURE OF ERROR
DATE: SEPTEMBER 1951
OPERATOR
COMPANY MARKED REFLECTIONS
REFLECTIONS NOT VISUALLY OBSERVABLE, DETERMINED FROM COMPANY MARKED REFLECTION TIMES ON M.I.T. RECORD NO. 1
REFLECTION a
x = N650
y = N350
PREDICTION DISTANCE: .005 SEC
PREDICTION DISTANCE: .010 SEC
PREDICTION DISTANCE .015 SEC
REFLECTION b
REFLECTION c
REFLECTION d
x = N750
y = N250
PREDICTION DISTANCE: .005 SEC
PREDICTION DISTANCE .010 SEC
PREDICTION DISTANCE: .015 SEC
OPERATOR
0
5
10
20
3
4
5
6
7
8
9
1.0
1.1
1.2
1.3
1.4

FIG. 10. Error curves for operators taken on MIT Record No. 3.

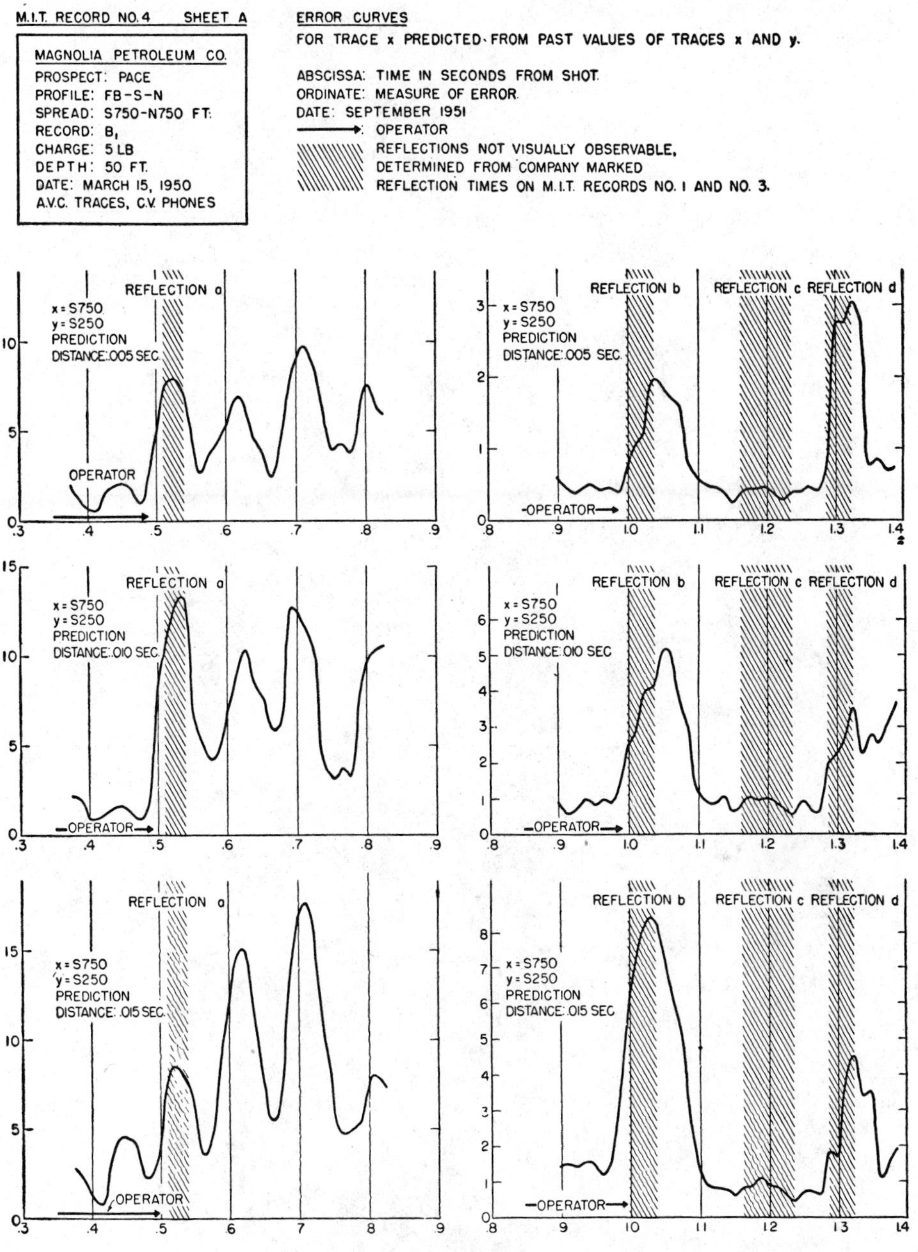

FIG. 11. Error curves for operators taken on MIT Record No. 4.

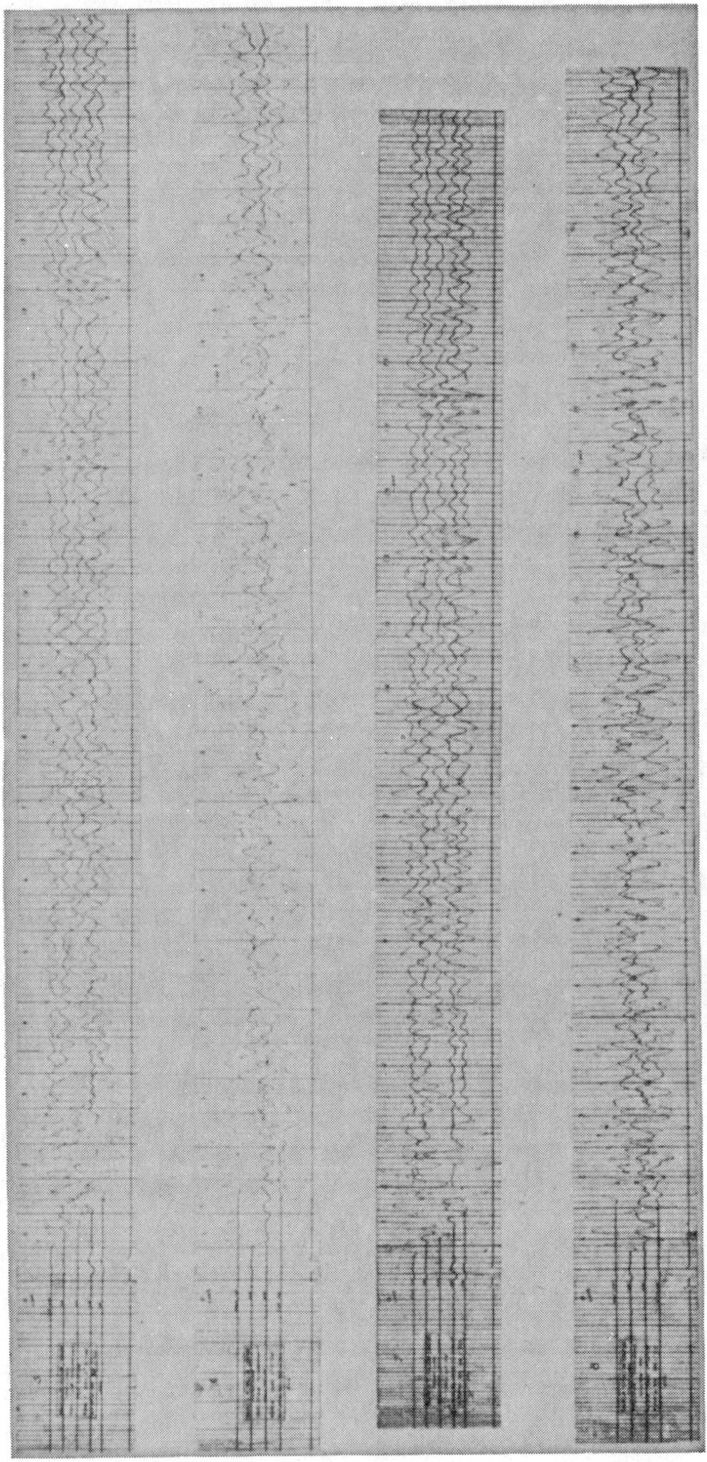

FIG. 12. MIT Records No. 5, 6, 7, and 8. (Furnished by Magnolia Petroleum Co.)

TABLE II

PERCENT REDUCTIONS FOR OPERATORS ON RECORDS NO. 1 TO 4

Figure Number and Description	Prediction Distance (Milliseconds)	Percent Reduction
6. (Upper graph)	5	66
	10	61
	15	59
(Lower graph)	5	93
	10	88
6. (Right-hand side)	5	87
	10	84
	15	78
8.	5	61
	10	49
	15	50
9.	5	90
	10	84
	15	80
10. (Left-hand graphs)	5	82
	10	62
	15	70
(Right-hand graphs)	5	88
	10	87
	15	77
11. (Left-hand graphs)	5	79
	10	83
	15	60
(Right-hand graphs)	5	84
	10	67
	15	80

given in the figures. Table III gives the approximate time durations of reflections marked by the Magnolia Petroleum Company. Time is measured in seconds from the time of explosion.

TABLE III

REFLECTION TIMES ON RECORDS NO. 5 TO 8

Record No.	Reflection e (Seconds)	Reflection f (Seconds)
5	1.17 to 1.23	1.40 to 1.45
6	1.17 to 1.23	1.40 to 1.45
7	1.17 to 1.23	1.40 to 1.45
8	not marked	1.40 to 1.45

Figures 13, 14, 15, and 16 show Error Curves for operators taken on MIT Records No. 5, 6, 7, and 8. Percent reductions, computed from equation (43), for these operators are given in Table IV.

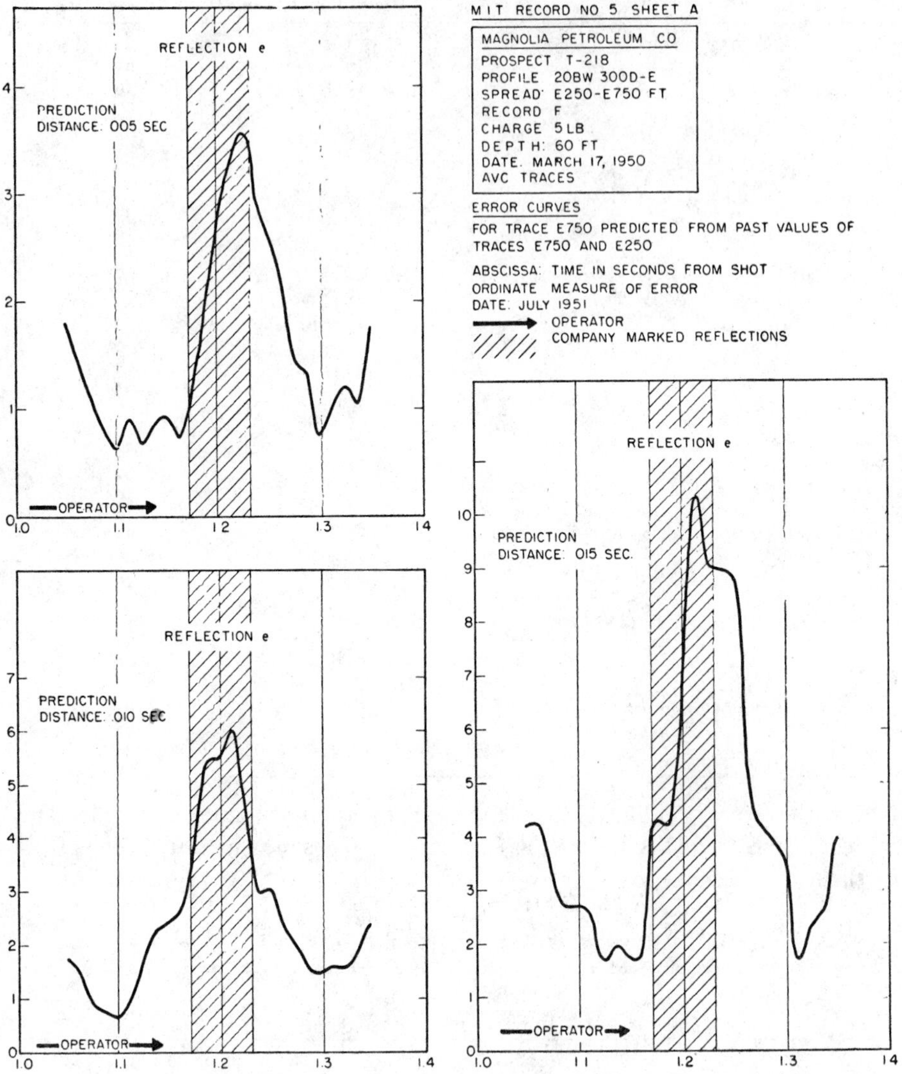

Fig. 13. Error curves for operators taken on MIT Record No. 5.

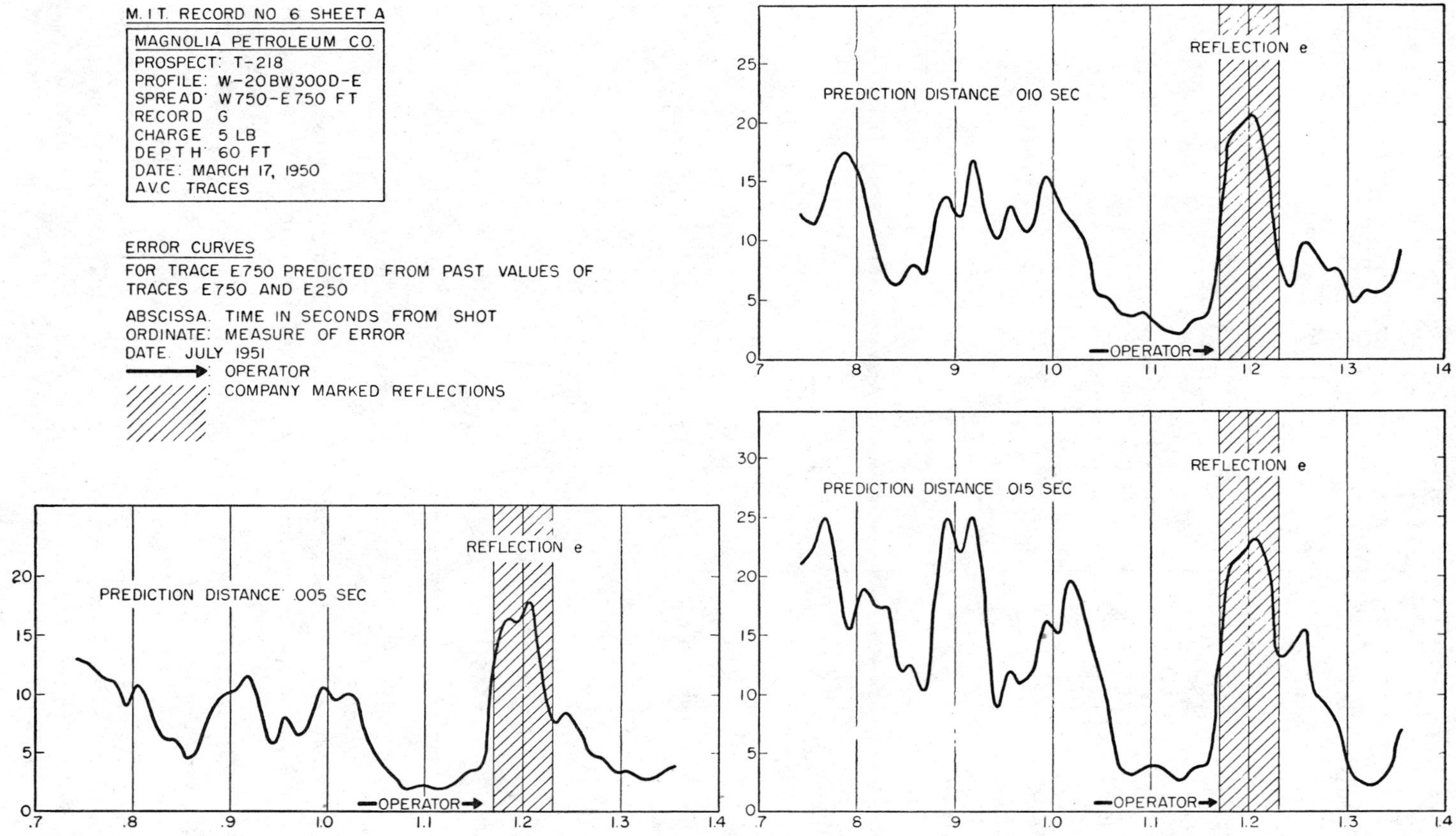

FIG. 14. Error curves for operators taken on MIT Record No. 6.

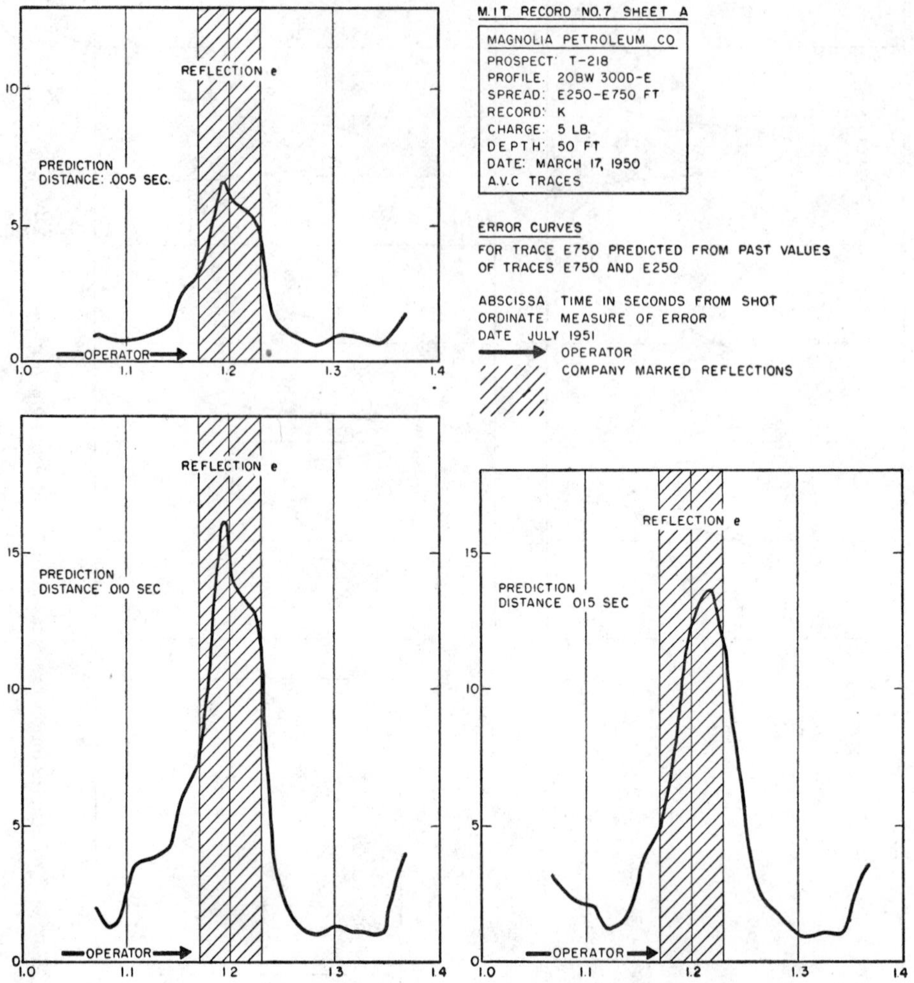

FIG. 15. Error curves for operators taken on MIT Record No. 7.

The percent reductions as shown in Tables II and IV, computed from equation (43), were high for the operators chosen on the records analyzed in comparison with typical percent reductions for operators used in meteorological and economic applications. In spite of the excellent reproduction of the predicted trace in the operator time interval the prediction errors in this interval were highly autocorrelated. Therefore, even better reproductions of the predicted trace can be expected by using more traces and more lags on these traces in the form of the operator. For the errors presented in the figures, our tentative judgment is that the magnitude of errors at reflections are in most cases significantly different from the magnitude of errors in non-reflection intervals. Any statistical test of

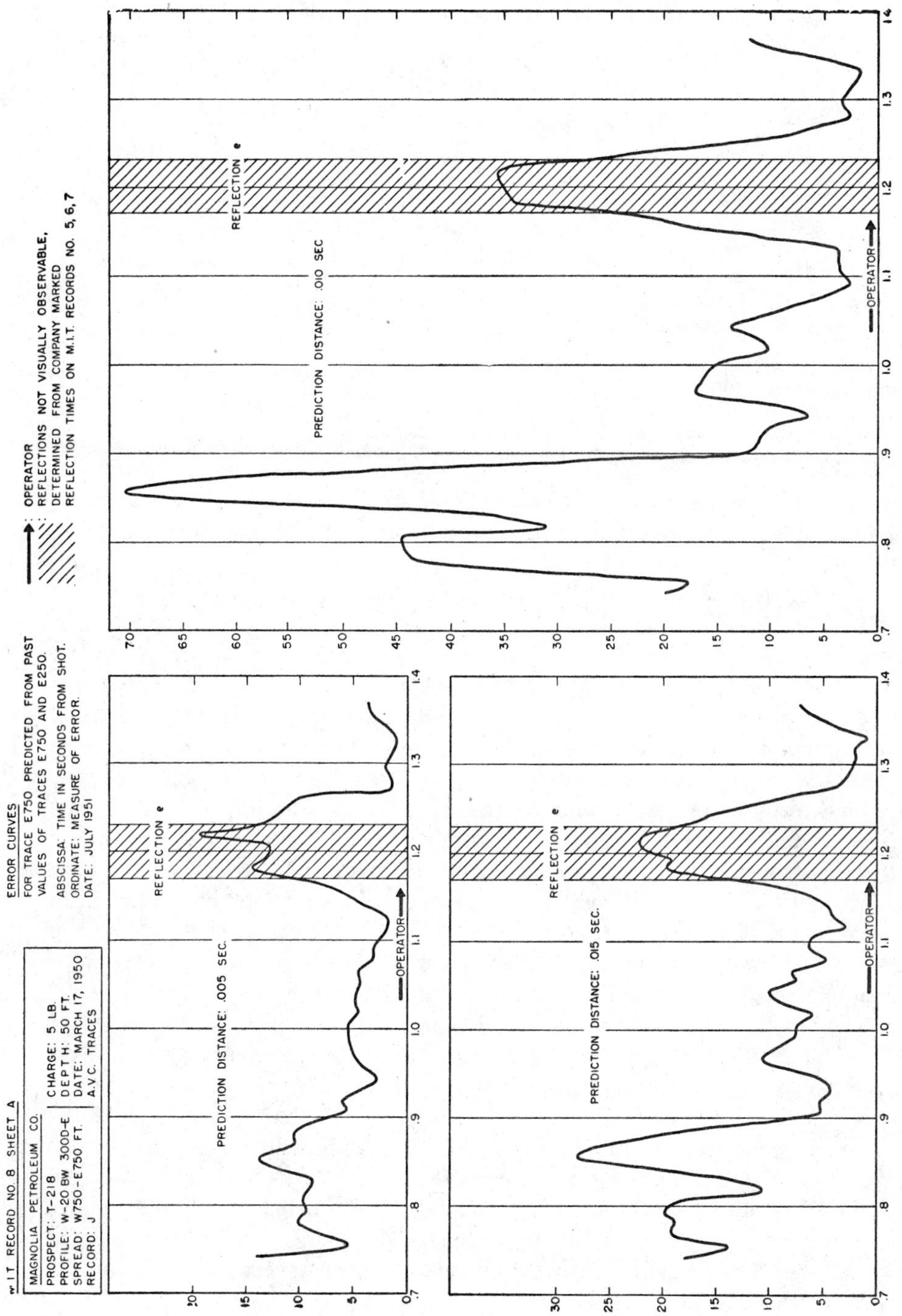
M.I.T. RECORD NO. 8 SHEET A
MAGNOLIA PETROLEUM CO.
PROSPECT: T-218
PROFILE: W-20 BW 3000-E
SPREAD: W750-E750 FT.
RECORD: J
CHARGE: 5 LB.
DEPTH: 50 FT.
DATE: MARCH 17, 1950
A.V.C. TRACES
ERROR CURVES
FOR TRACE E750 PREDICTED FROM PAST VALUES OF TRACES E750 AND E250.
ABSCISSA: TIME IN SECONDS FROM SHOT.
ORDINATE: MEASURE OF ERROR.
DATE: JULY 1951
: OPERATOR
: REFLECTIONS NOT VISUALLY OBSERVABLE, DETERMINED FROM COMPANY MARKED REFLECTION TIMES ON M.I.T. RECORDS NO. 5, 6, 7
PREDICTION DISTANCE: .005 SEC.
PREDICTION DISTANCE: .015 SEC.
PREDICTION DISTANCE: .010 SEC
REFLECTION e
OPERATOR

FIG. 16. Error curves for operators taken on MIT Record No. 8.

TABLE IV

PERCENT REDUCTIONS FOR OPERATORS ON RECORDS NO. 5–8

Figure No.	Prediction Distance (Milliseconds)	Percent Reduction
13	5	82
	10	82
	15	66
14	5	87
	10	83
	15	82
15	5	90
	10	81
	15	81
16	5	93
	10	85
	15	85

significance which might be used must take into account the autocorrelation of errors, and hence the usual chi-square test of significance is invalid. The whole problem of tests of significance, tests of hypotheses, and other questions regarding estimation is now under investigation.

The Magnolia Petroleum Company Record H from prospect 4-E was shot on the Forehand Farm in Henderson County, Texas, where about sixty feet of loose Carrizo sand is underlain by Wilcox sandy clay with lignite stringers. The record is shown in Figure 17. The Magnolia Petroleum Company considers this to be a problem area because of the loose sand. The spread is rather far from the shot, as indicated on the record. This causes early reflections to have appreciable "step-out" because of the non-vertical path. Reflection times on the top trace are marked by ϕ, and times on lower traces are later, depending on the amount of departure from vertical travel. A different shooting procedure showed these reflections much better and provided the basis for indicating reflection times. Figures 18 and 19 give Error Curves for operators taken on Record H from Prospect 4-E. The heavy vertical lines indicate the reflection times of the top trace, which are marked by ϕ. In the equation for the operator the x trace, which is the predicted trace, was trace 4, and the y trace was trace 1.

The Magnolia Petroleum Company Record K (profile 65 A-W-1) and Record K (profile 65 A-W-2) (Figure 17) were prepared from an original shot in West Texas where high-speed materials are near the surface. On Record K (profile 65 A-W-1) the top six traces represent a relatively wide band recording and the bottom six represent the same stations with proper measures taken to develop reflections. The first four traces of Record K (profile 65 A-W-1) are reproduced at higher amplitudes on Record K (profile 65 A-W-2). Figures 20 and 21 give Error Curves for operators taken on Record K (profile 65 A-W-2). In the equation for the operator the x trace, which is the predicted trace, was trace 4 and the y trace was trace 1.

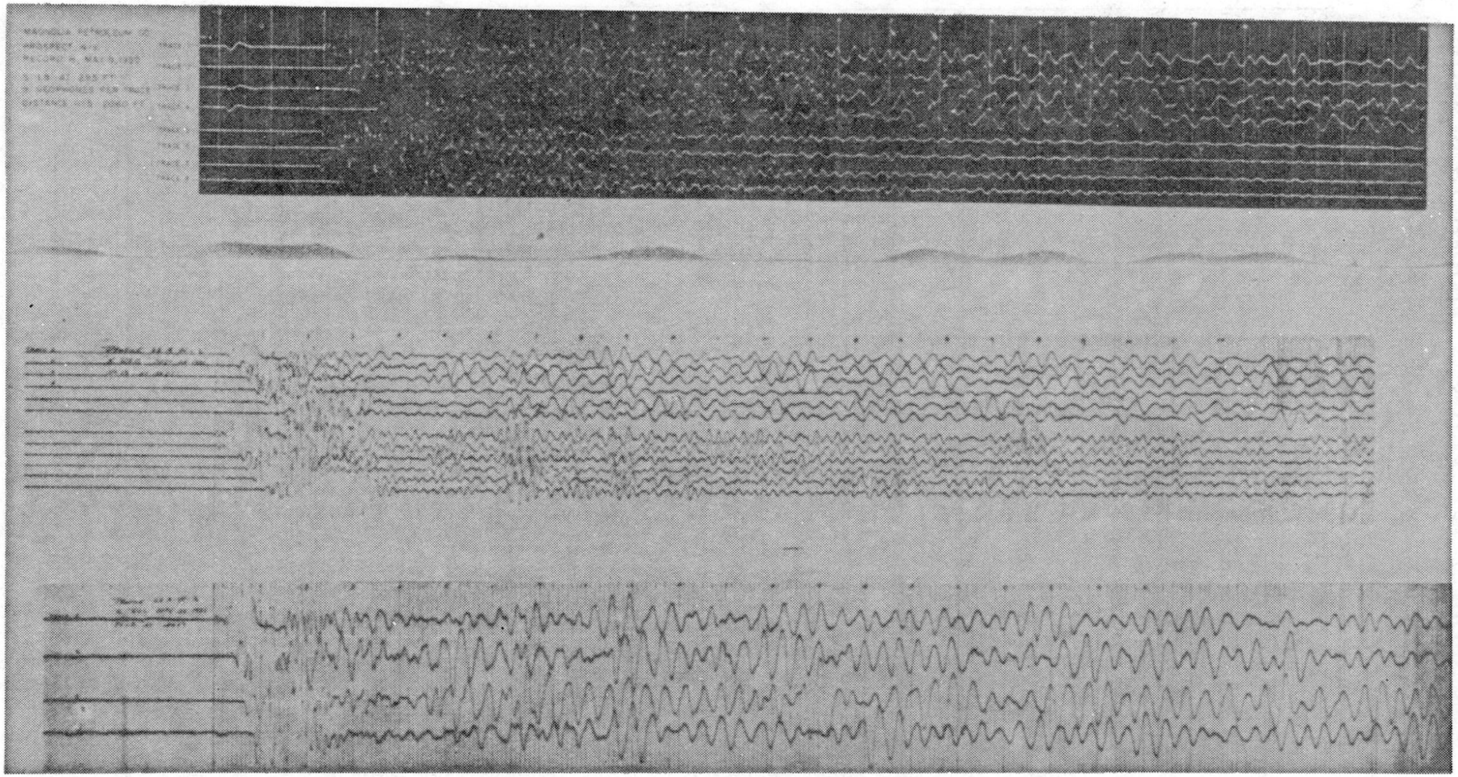

FIG. 17. Record H (prospect 4-E), Record K (profile 65 A-W-1), and Record K (profile 65 A-W-2). (Furnished by Magnolia Petroleum Co.)

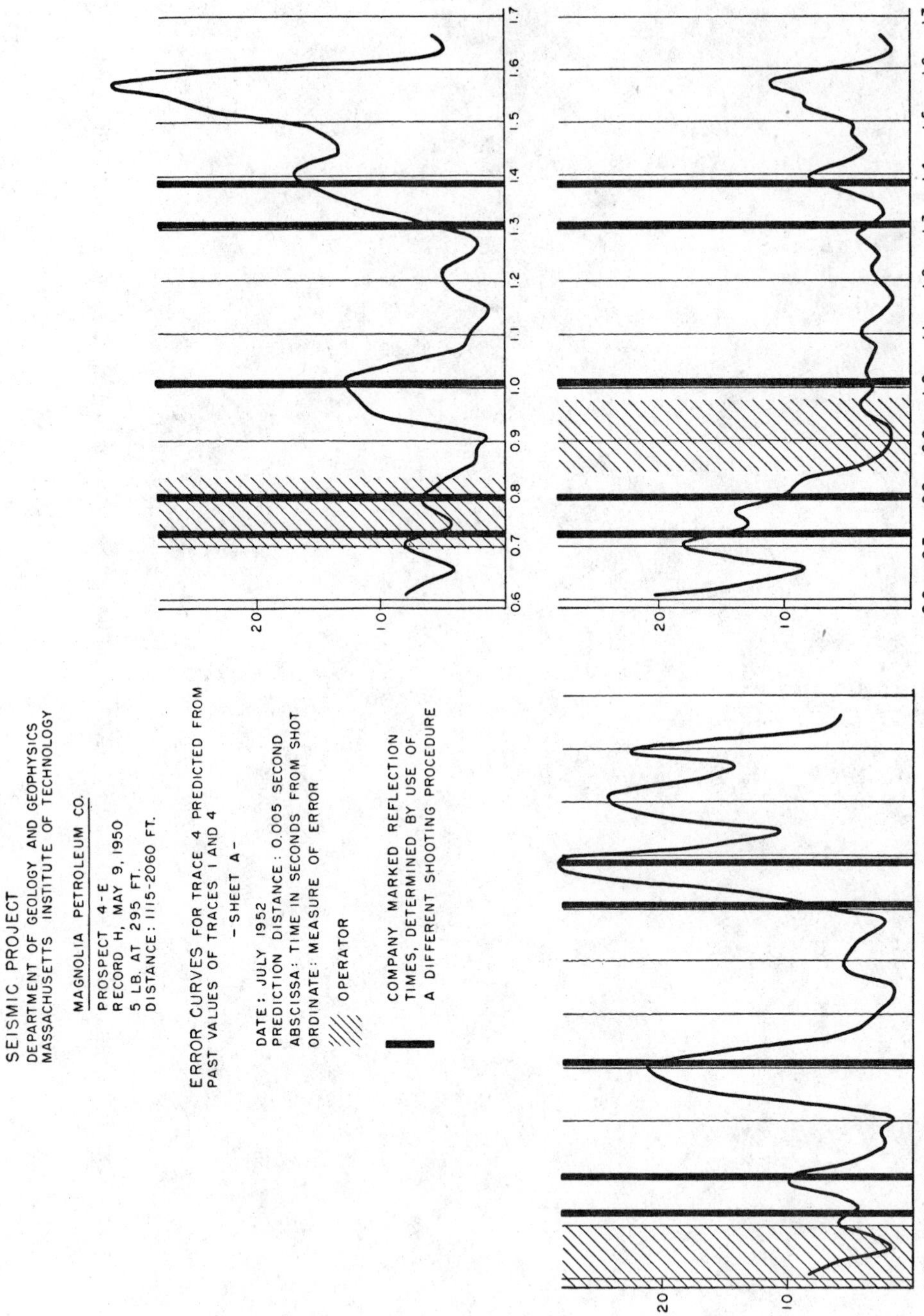

FIG. 18. Error curves for operators taken on Record H (prospect 4-E).

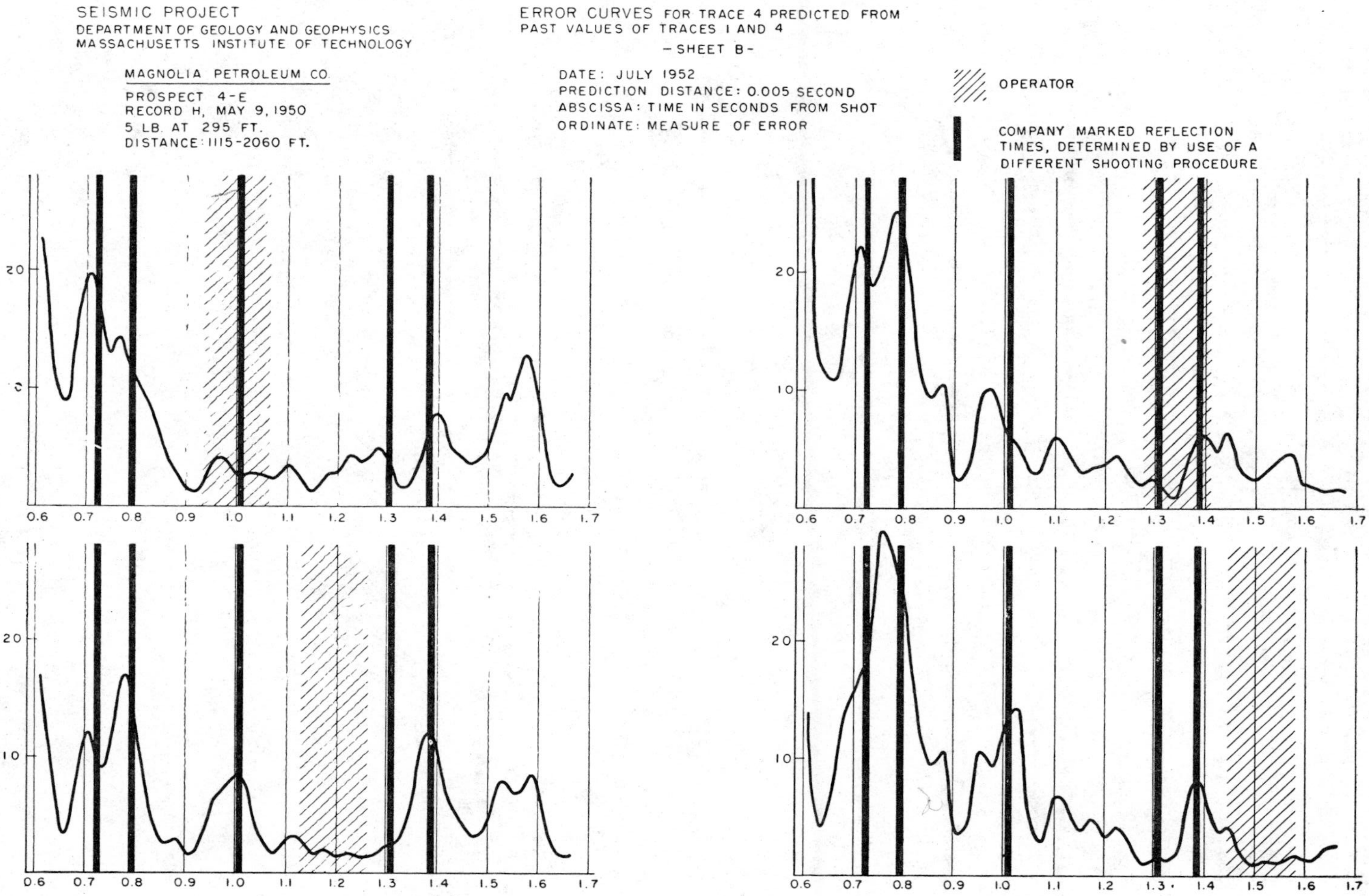

FIG. 19. Error curves for operators taken on Record H (prospect 4-E).

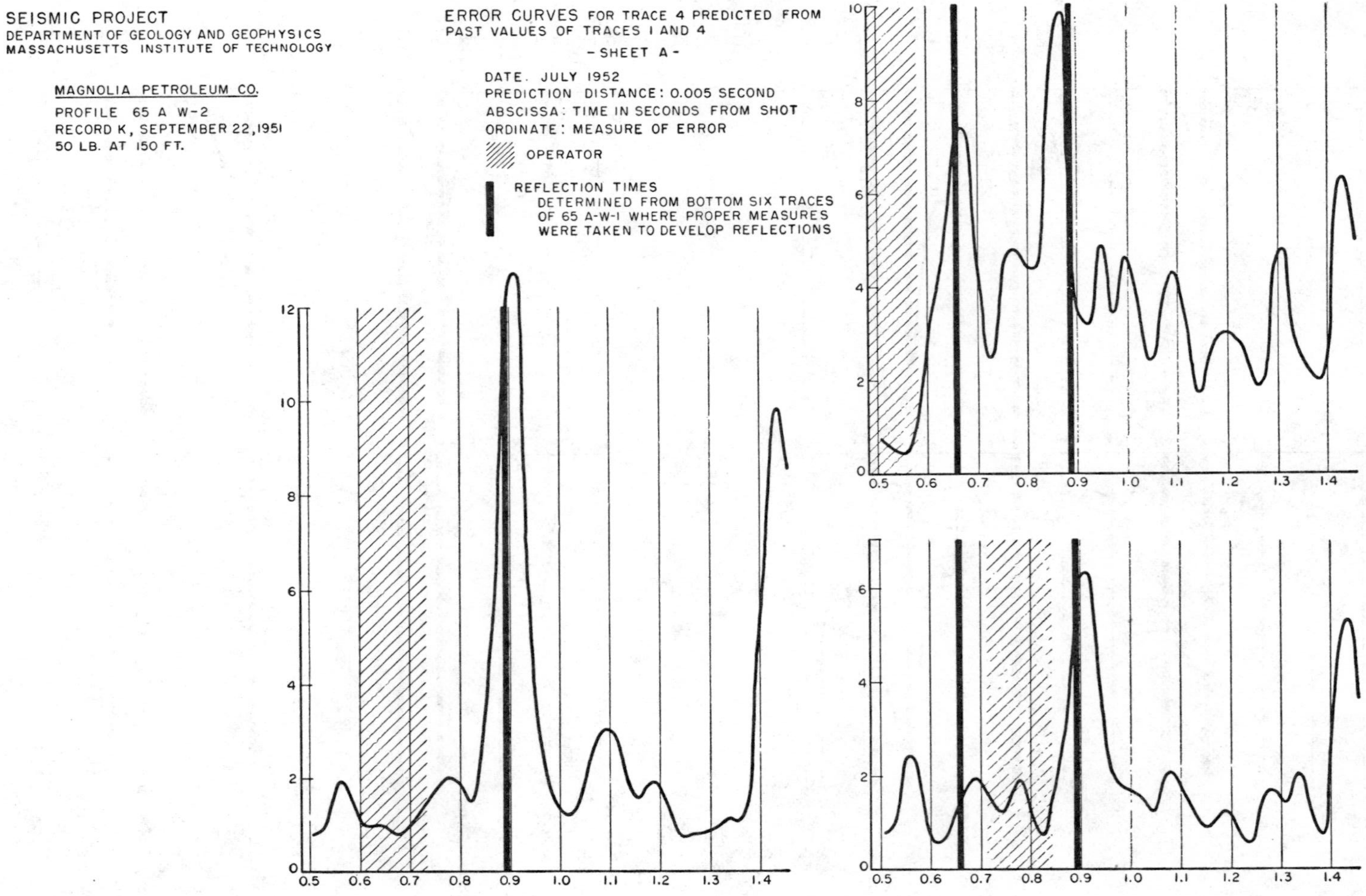

FIG. 20. Error curves for operators taken on Record K (profile 65 A-W-2).

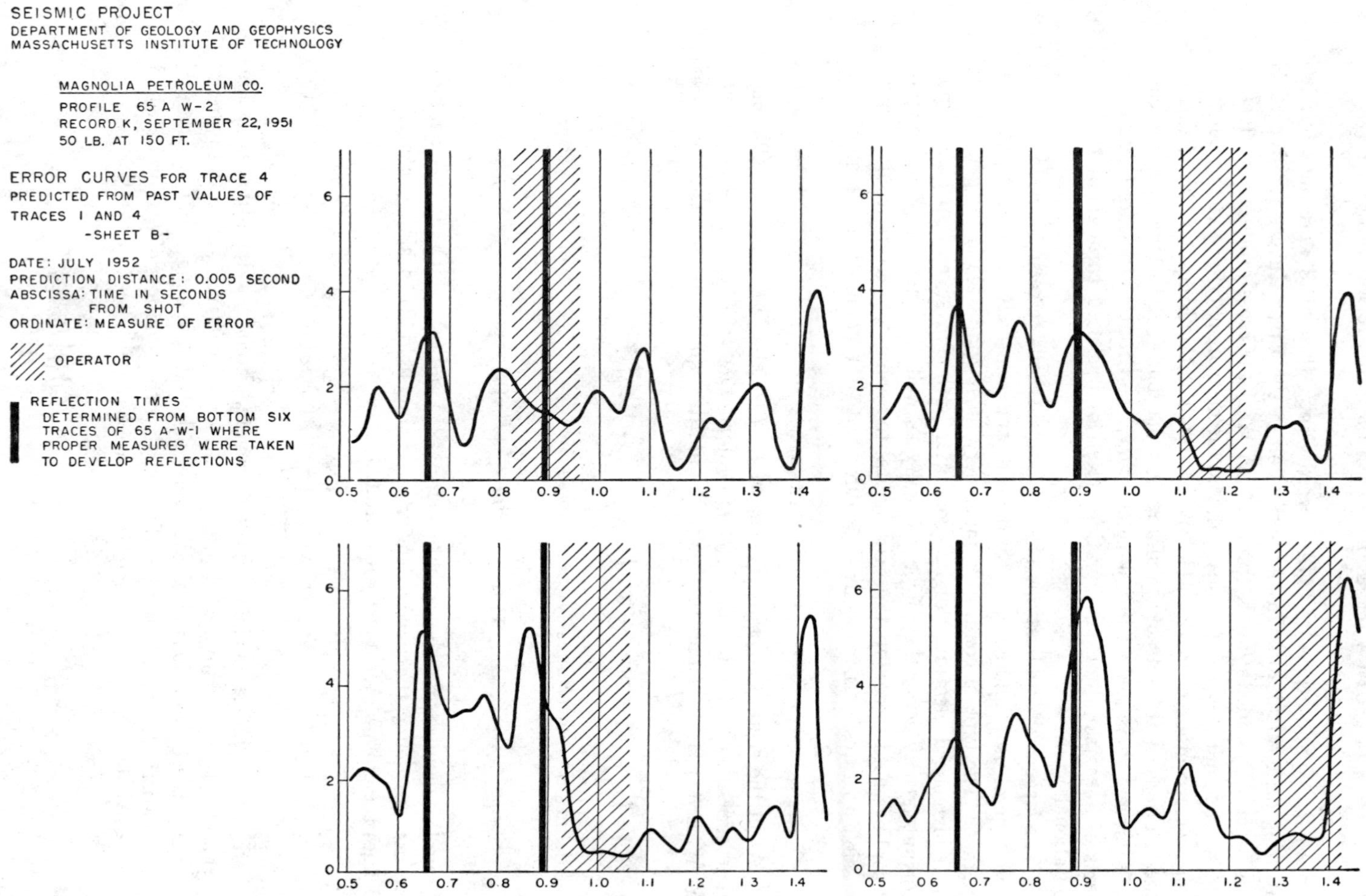

FIG. 21. Error curves for operators taken on Record K (profile 65 A-W-2).

All the computations on MIT Records No. 1 through 8 were done by girls utilizing desk calculators. In the derivation of the Error Curves for Record H (prospect 4-E) and Record K (profile 65 A-W-2) the cross products involved in the set of simultaneous equations (40), and the measure of error given by equation (44), were computed on the MIT Whirlwind Digital Computer. Nevertheless, the solutions of the sets of simultaneous equations (40) for the operator coefficients were computed with desk calculators. The Crout method for solving systems of linear simultaneous equations (Ref. 9) was used throughout the computational work.

The large scale applications of statistical methods to seismology in order to find general use require rapid, automatic, and relatively cheap computational methods. The type of computing machinery most applicable is the general purpose high-speed electronic digital computer, although several special purpose computers may serve the purpose equally as well. Magnetic tape seismogram recordings used in conjunction with digital-to-analog converters would make it possible to transfer the seismic data directly to the computing machines.

APPENDIX

As in the section on Time Series, we let $f_1(t)$ and $f_2(t)$ represent stationary time series. Then by letting $f_{1T}(t)=f_1(t)$ for $-T<t<T$ and $f_{1T}=0$ elsewhere Y. W. Lee (Ref. 8) defines the complex amplitude and phase spectrum as

$$F_{1T}(\omega) = \frac{1}{2\pi}\int_{-T}^{T} f_{1T}(t)e^{-i\omega t}dt. \tag{59}$$

He defines the power density spectrum as

$$\Phi_{11}(\omega) = \lim_{T\to\infty} \frac{\pi}{T}\,|F_{1T}(\omega)|^2. \tag{60}$$

From this definition he shows that equations (8) and (9), which are the equations we used to define the power density spectrum, hold. In a similar manner he defines the cross power density spectrum as

$$\Phi_{12}(\omega) = \lim_{T\to\infty}\left[\frac{\pi}{T}F_{1T}(\omega)\overline{F_{2T}(\omega)}\right] \tag{61}$$

where the bar indicates the complex conjugate. From this definition, equations (23) and (24), which were used to define the cross spectrum follow as theorems.

Using these definitions of Y. W. Lee, we shall derive equations (30) and (31) in an heuristic manner. Using the theorem that the products of two limits which exist is equal to the limit of the products, we have

$$\Phi_{11}(\omega)\Phi_{22}(\omega) = \lim_{T\to\infty}\frac{\pi^2}{T^2}|F_{1T}(\omega)|^2|F_{2T}(\omega)|^2. \tag{62}$$

By taking the positive square roots of this expression it becomes

$$\sqrt{\Phi_{11}(\omega)}\,\sqrt{\Phi_{22}(\omega)} = \lim_{T\to\infty} \frac{\pi}{T}\,|F_{1T}(\omega)|\,|F_{2T}(\omega)|. \tag{63}$$

By writing the complex amplitude and phase spectra as

$$F_{1T}(\omega) = |F_{1T}(\omega)|\,e^{i\theta_{1T}(\omega)} \tag{64}$$

and similarly for $F_{2T}(\omega)$, we see that the cross power spectrum, given by definition (61), is

$$\Phi_{12}(\omega) = \lim_{T\to\infty} \left[\frac{\pi}{T}\,|F_{1T}(\omega)|\,|F_{2T}(\omega)|\,e^{i[\theta_{1T}(\omega)-\theta_{2T}(\omega)]}\right]. \tag{65}$$

If we assume that the limits

$$\theta_1(\omega) = \lim_{T\to\infty} \theta_{1T}(\omega) \quad \text{and} \quad \theta_2(\omega) = \lim_{T\to\infty} \theta_{2T}(\omega) \tag{66}$$

exist, then by using equation (63), we have

$$\Phi_{12}(\omega) = \sqrt{\Phi_{11}(\omega)}\,\sqrt{\Phi_{22}(\omega)}\,e^{i[\theta_1(\omega)-\theta_2(\omega)]}. \tag{67}$$

From this equation we see that the absolute value of the cross-spectrum is given by equation (30) and the argument of the cross-spectrum is given by equation (31). The limits given in equation (66) do not necessarily exist for arbitrary stationary time series and hence, equations (67), (30), and (31) are not valid in general. Nevertheless, for time series $f_{1T}(t)$ and $f_{2T}(t)$ which are of finite duration $2T$, we see that equations (60) through (66) are valid if we do not take the limits as T tends to infinity. Hence equations (67), (30), and (31) are valid for time series of finite duration.

ACKNOWLEDGEMENTS

We should like to express indebtedness to the Magnolia Petroleum Company for supplying us with the seismograms and for the many helpful discussions with members of their staff through the years. Grateful acknowledgement is made to S. M. Simpson, and to our other colleagues H. W. Briscoe, M. K. Smith, and W. P. Walsh, who are associated with the MIT Geophysical Analysis Group. Also we should like to thank Miss Virginia Woodward who did a large part of the computing work presented in this paper and Miss Barbara Halpern for typing the manuscript.

REFERENCES

1. Wiener, Norbert, *Extrapolation, Interpolation, and Smoothing of Stationary Time Series*, The Technology Press of MIT and John Wiley and Sons, Inc. (1950).
2. Yule, G. U., "Why Do We Sometimes Get Nonsense Correlations Between Time Series?," *Journal Royal Statistical Society*, Vol. 89 (1926) p. 1.
3. Yule, G. U., "On a Method of Investigating Periodicities in Disturbed Series, with Special Application to Wolfort's Sun Spot Numbers," *Phil. Trans. Royal Soc. Ser. A*, Vol. 226 (1926) pp. 267 ff.
4. Wold, Herman, *A Study in the Analysis of Stationary Time Series*, Uppsala (1938), Almquist and Wiksells Boktryekeri A-B.
5. Kendall, Maurice G., *The Advanced Theory of Statistics Volume II*, Charles Griffin and Company Limited, London (1947) (especially Chapter 30, pp. 396–437).
6. Tukey, J. W., "The Sampling Theory of Power Spectrum Estimates," *Woods Hole Symp. Autocorr. Applic. to Phys. Prob.* (1949) pp. 46–67, D.N.R. NAVEXOS-P-735.
7. Tukey, J. W. and Hamming, R. W., "Measuring Noise Color," *Bell Tel. Lab. Memo*, mm-49-110-119 (1949).
8. Lee, Y. W., "Application of Statistical Method to Communication Problems," Technical Report No. 181, Massachusetts Institute of Technology Research Laboratory of Electronics (1950).
9. Crout, P. D., "A Short Method for Evaluating Determinants and Solving Systems of Linear Equations with Real or Complex Coefficients," *Trans. Am. Inst. of Elect. Eng.*, Vol. 60 (1941).
10. Wadsworth, G. P., Director, MIT Division of Industrial Cooperation Contract No. 6083, Reports to Army Air Forces (1942–1946):
    - A- 2 Preliminary Report on Prediction of Pacific High.
    - A- 3 Temperature Variations Study.
    - A- 4 Preliminary Study of Predictions of Daily Temperatures.
    - A- 5 Path of Pacific High Study.
    - A- 6 Methods of Combining Auto-Correlation Coefficients.
    - A- 7 Properties of Auto-Correlations.
    - A- 8 Long Range Temperature Auto-Correlation Functions.
    - A- 9 Temperature Correlations.
    - A-11 Prediction by Continuous Revision of Coefficients.
    - C- 3 Prediction of Daily Mean Temperatures for Columbus, Ohio.
    - C- 4 Prediction of Daily Mean Pressures for Columbus, Ohio.
    - C- 5 Preliminary Pressure Map Prediction.
    - C- 6 Linear Predictability of Polar Processes.
    - C- 8 Alaskan Trial Study.
    - C- 9 Dutch Harbor Periodicities.
    - C-11 Considerations of Limitations of Linear Predictions of Pressures.
    - C-12 Auto- and Cross-correlations of Surface Pressure over the United States.
    - D- 1 The Lag Correlator.
11. Wadsworth, G. P., "Short Range and Extended Forecasting by Statistical Methods," Air Weather Service, Tech. Report 105-37 Headquarters Air Weather Service, Washington, D. C., Feb. 1948.
12. Seiwell, H. R. and Wadsworth, G. P., "A New Development in Ocean Wave Research, *Science* 109, 1949, pp. 271–274.
13. Wadsworth, G. P., Director; MIT Division of Industrial Cooperation Contract No. W28-099-ac-398, Geophysical Research Directorate:
    - Reps. No. 1, 2, and 3, 1948: Analogue Techniques in the Forecasting of Rainfall
    - Rep. No. 4, 1949: Dynamics of the Semi-permanent Pressure Cells
    - Rep. No. 5, 1949: Further Analysis of Dynamics of Major Pressure Cells
    - Rep. No. 6, 1949: Position of Major Pressure Cells in Relation to Rainfall
    - Rep. No. 7, 1949: A Search for Pertinent Parameters to Classify Pressure Distributions

Rep. No. 8, 1950: Statistical Prediction of Migratory North American Anticyclones: An Introductory Attack Upon Non-Stationary Time Series in Weather

Rep. No. 9, 1950: Preliminary Studies on Variations of Sea Surface Temperatures

14. Wadsworth, G. P., "Application of Statistical Methods to Weather Forecasting," *Compendium of Meteorology*, Boston, edited by Thomas F. Malone, 1951.
15. Wadsworth, G. P., "Power Spectra of Surface Pressure at 35 U. S. Weather Stations," MIT, D.I.C. Report No. 3, Nov. 1951.
16. Wadsworth, G. P., "An Investigation of Non-Linearity in the Prediction of Surface Pressure 12 Hours in Advance," MIT, D.I.C. Report No. 8, Feb. 1953.
17. Wadsworth, G. P., and Robinson, E. A., Directors, MIT Geophysical Analysis Group, "Results of an Auto-Correlation and Cross-Correlation Analysis of Seismic Records," presented at MIT Industrial Liaison Conference of August 6, 1952, Oct. 1, 1952.
18. Wadsworth, G. P., and Robinson, E. A., Directors, MIT Geophysical Analysis Group, "A Prospectus on the Applications of Linear Operators to Seismology," Nov. 5, 1952.

Reprinted from Geophysics v. 32, no. 3, p. 418-484

PREDICTIVE DECOMPOSITION OF TIME SERIES WITH APPLICATION TO SEISMIC EXPLORATION†

ENDERS A. ROBINSON*

TABLE OF CONTENTS

† MIT GAG Report No. 7, 12 July 1954.

* Digital Consultants, Inc., Houston, Texas.

CHAPTER I

INTRODUCTION AND SUMMARY

1.1 Introduction

In exploration seismology, a charge of dynamite is exploded under controlled conditions, and the resulting vibrations at various points on the surface of the ground are detected by geophones and are recorded as seismic traces on the seismogram. The analysis of such seismic records yields valuable information about the structure of the sedimentary rock layers in potential oil producing areas, and such information is of considerable economic value in increasing the probability of locating new oil fields.

Present day techniques still ultimately require the visual examination and mental interpretation of seismograms, with considerable importance placed on the detection of reflected energy or "reflections" which indicate reflecting horizons of subsurface sedimentary layers. From this information the geologic structure of an area may be estimated.

Although reflection seismology was only begun after the First World War, it has played an important role in the discovery of many of the world's oil fields. The credit for much of this success belongs to the oil companies and geophysical exploration companies which have the practical task of locating new petroleum reserves. It was the working geophysicist of these companies who developed a large part of the seismic method of today. To help him in his job, the engineer, who in many instances is the geophysicist himself, has developed the instrumentation needed for the ever increasing demands of seismic exploration. In addition, the research scientist has taken an active role in the development of the basic scientific theory of exploration seismology.

For a further discussion of the seismic method, together with references to the literature, the reader is referred to books by Dix (1952), Dobrin (1952), Heiland (1940), Jakosky (1950) and Nettleton (1940); to past issues of GEOPHYSICS, and to *Geophysical Prospecting*.

A large part of basic seismic research is directed toward a better understanding of the physical processes involved in the seismic method. Such an approach is fundamentally sound. From this point of view, the seismic trace is the response of the system consisting of the earth and recording apparatus to the impulsive source, the explosion. This system, although usually very complicated, is susceptible to a deterministic (nonrandom) approach toward its analysis. To this end, controlled experiments may be carried out, and mathematical and physical models may be set up from the resulting data. Careful replication of the experiment and high precision of measurement can render such data very accurate.

On the other hand, large numbers of seismic records, which have twenty or more traces per record, are needed to carry out an exploration program over a geographic area. This quantity of data necessarily requires the consideration of each record as a member of a larger group or ensemble of records. Thus the reliability of a single record is considerably less than the reliability of the ensemble of records in connection with the description of the geologic conditions existing in that area. Also from an economic standpoint, the amount of control in such an exploration program must be kept at the bare minimum consistent with worthwhile results. Thus, as a rule, the controlled experiment aspect of exploration seismology, although possible, falls short of the needs of a research scientist who wishes to set up a mathematical or physical model. As a result, in these cases the working geophysicist must proceed to fit his empirical information into the larger overall framework without the aid of elaborate mathematical or physical models. In particular, he is faced with the general problems of description, analysis, and prediction (Cramér, 1946).

That is, first, the working seismologist is faced with problems of the description of the overall exploration seismic picture. In particular, he wishes to replace the mass of original data, which is of a complicated nature, by a small number of descriptive characteristics; that is, he is faced with the problem of the reduction of data.

Next, he is concerned with the problems of analysis in which he wishes to argue from the sample, the evidence from a limited number of seismograms, to the population—the geologic structure of the area. In other words, from the sample data he wishes to find estimates of the true values which describe the geologic structure.

Finally, the working geophysicist is concerned with the problem of prediction, that is, from knowledge of past experience what course of action should he take in the future. In particular one of the goals of an exploration program is to determine favorable drilling sites.

Since the geologic structure is physically fixed and constant in nature, and has no intrinsic random characteristics, any statistical approach to these problems immediately encounters difficulties which are commonly associated with Bayes' Theorem in the statistical literature (Cramér, 1946; Jeffreys, 1939). Nevertheless modern statistical theory admits the bypassing of these difficulties, although with reservation, and hence the working geophysicist may be considered to be faced with a situation which is essentially statistical. For example, a reflection which may be followed from trace to trace, record to record, usually has more value to the seismic interpreter, and hence is statistically more significant, than a reflection which appears only on a few traces. Such a procedure in picking reflections does not imply that the reflection which appears only on a few traces is necessarily spurious information, but only that economic limitations preclude further examination and experimentation which may render it in a more useful form.

This thesis deals with the analysis of seismic records from the statistical point of view. In those years in which the exploration seismic method was first being developed, the English statistician, G. Udny Yule, was developing methods of time series analysis which proved to open a new epoch in the analysis of time functions. The concept which Yule introduced is that a large class of disturbed motions are characterized by a random shift of phase and change of amplitude as time progresses. Yule applied this hypothesis with success to empirical data, and thus the analysis of time series was freed, for the first time, from either the hypothesis of a strictly periodic variation or aperiodic variation, or the counter-hypothesis of a purely random variation. Yule's concept was formulated on a firm axiomatic basis in the founding of the theory of stochastic processes by the Russian mathematicians A. Kolmogorov (1933) and A. Khintchine (1933), and in the definition and basic work in theory of stationary processes by A. Khintchine (1934).

In the field of generalized harmonic analysis, the work of Norbert Wiener stands in the foremost place. His classic paper "Generalized Harmonic Analysis" (1930) represents a true discontinuity in the normal development of ideas. Not only did that paper discover the field of generalized harmonic analysis, but it also built the framework upon which subsequent work could be developed.

The Swedish statistician Harald Cramér taught a course on time series analysis in 1933 which laid the foundation for a thesis prepared by his student, the Swedish statistician and economist, Herman Wold (1938). Wold describes his work as a trial to subject the fertile methods of empirical analysis proposed by Yule to an examination and a development by the use of the mathematically strict tools supplied by the modern theory of probability. In his thesis Wold deals, among other things, with the decomposition of stationary time series.

These developments in time series analysis culminated in the parallel and independent work of André Kolmogorov (1941) and Norbert Wiener (1942) on the prediction and filtering of stationary time series.

In the final analysis, the potential usefulness of the statistical approach depends upon the coordination of statistical methods with knowledge of practical and theoretical seismology.

1.2 Summary of chapters

In this section we present a summary of the chapters which follow. Since detailed references to the literature are given in these chapters, we shall not state any references in this summary.

In Chapter II we discuss the properties of finite discrete linear operators. We distinguish between extrapolation or one-sided operators on the one hand, and interpolation or two-sided operators on the other hand. A one-sided operator has an inherent one-sidedness in that it operates on the past values of a time series, but not on the future values. Consequently a one-sided operator is computationally realizable, and may represent the impulsive response of a realizable linear system.

A two-sided operator, however, operates on both past and future values of a time series, and thereby is not computationally realizable. Nevertheless, finite two-sided operators can be made realizable by introducing a time delay in the computations so that all the necessary data are available at the time the computations are to be carried out. The fact that a seismic disturbance is recorded on paper or tape in the form of a seismogram means that we have waited until all the pertinent information is available. Consequently, the necessary time delay has been introduced to utilize finite time-delay two-sided operators which are computationally realizable. That is, a finite time-delay two-sided operator has the same mathematical form as a one-sided operator. In the remaining parts of this thesis we deal mainly with operators of the one-sided type.

The transfer function is defined as the Fourier transform of the linear operator, and corresponds to the transfer function, the system function, or filter characteristics of a linear system. By analytic continuation we may extend the transfer function into the complex plane, where the real axis represents real angular frequency.

A discrete time function is said to be stable if the sum of squares of the coefficients is finite. A linear difference equation is called a stable difference equation if its general solution for time $t \geq 0$ is a stable time function.

We give the condition that the general solution (for $t \geq 0$) of the linear difference equation formed by the coefficients of a finite one-sided operator be stable; that is, the condition that the difference equation be a stable difference equation: viz., all the zeros of the z-transform of the operator have modulus greater than unity. A finite one-sided operator that satisfies this condition is called *minimum-delay*. We show that this minimum-delay condition is precisely the condition that the transfer function have all its zeros above the axis of real frequency. Thus an operator which is both realizable and minimum-delay has a transfer function with no singularities or zeros below the axis of real frequency: this is the requirement that its phase characteristic be of the minimum phase-shift type familiar to electrical engineers. Moreover, we extend this concept of minimum-delay to one-sided operators with an infinite number of coefficients.

We show that each minimum-delay operator has a unique inverse one-sided operator which is also minimum-delay, and that their respective transfer functions are reciprocals of each other. We see that the inverse operator may be readily computed in the time domain from a given linear operator.

Finally in Chapter II, we show that in order to design a minimum-delay operator one needs only the absolute gain characteristic of the desired filtering properties, and not the phase characteristic. That is, the phase characteristic of the resulting minimum-delay operator is the minimum phase characteristic, which is determined uniquely from the absolute gain characteristic. We give a direct computational procedure which may be readily programmed for a digital computer.

In time series analysis, there are two lines of approach, the nonstatistical and the statistical. In the nonstatistical approach the given time series is interpreted as a mathematical function, and in the statistical approach as a random specimen out of a population of mathematical functions.

In Chapter III, we treat the methodology used in the nonstatistical or deterministic approach to the study of time series. Such an approach leads to a perfect functional representation of the observational or theoretical data. In particular, the methods of this chapter are applicable to the determination of linear operations on transient time functions of finite energy, such as seismic wavelets. Further we observe that even under a deterministic hypothesis it may be necessary to utilize averaging methods for certain purposes such as approximations. Although methods of averaging may be developed without recourse to the theory of probability, in many applications it is not until probability theory is introduced that certain averaging operations become meaningful in a physical sense. With this situation in mind, in the following chapters we consider the statistical approach to the study of time series in which methods of averaging play a central role.

In Chapter IV, we present concepts from the theory of discrete stationary time series which represents a statistical approach to the study of time series. We consider stationary stochastic or random processes which generate stationary time series, and give properties of the autocorrelation and spectrum. In particular we consider time series which are "white noise." We see that any time series with an absolutely continuous spectral distribution is a process of moving summation.

In Chapter V, we give an heuristic exposition of the method of the factorization of the power spectrum. We show how this factorization leads to the predictive decomposition of a stationary time series. The Predictive Decomposition Theorem shows that any stationary time series (with an absolutely continuous spectral distribution) can be considered to be additively composed of many overlapping wavelets. All these wavelets have the same minimum-delay shape; and the arrival times and strengths of these wavelets are random and uncorrelated with each other. If the wavelet shape represents the impulsive response of a minimum-phase system, and if the uncorrelated arrival times and strengths represent a white noise input to the linear system, then the stationary time series represents the output of the linear system.

We then show that the solutions of the prediction and filtering problems for single time series follows directly from the Predictive Decomposition Theorem. We examine those stochastic processes having power spectra which are rational functions, and see that the autoregressive process and the process of finite moving averages are special cases of such processes. We deal with the theory of multiple time series, in which we see that the concept of coherency plays an important role, and we treat the general technique of discrete prediction for multiple time series.

In Chapter VI we deal with applications to seismic exploration. In particular we consider the situation in which a given section of seismic trace (recorded with automatic volume control) is additively composed of seismic wavelets, where each wavelet has the same minimum-delay shape, and where the strengths and arrival times of the wavelets are random and uncorrelated with each other. Under these assumptions, the Predictive Decomposition Theorem tells us that the section of trace may be considered to be a section of a stationary time series.

To illustrate the probabilistic approach, we consider the problem in which we wish to separate the dynamic component (the wavelet shape) from the random components (the arrival times and strengths of the wavelets).

The *theoretical* solution of this problem consists of the following steps:

(1) Average out the random components of the trace so as to yield the wavelet shape.

(2) From the wavelet shape thus found, compute the inverse wavelet shape, which is the prediction error operator for unit prediction distance. Apply this prediction error operator to the trace in order to yield the random components, which are the prediction errors. That is, the prediction error operator contracts the wavelets to impulses, which are the prediction errors. If one wishes to filter the seismic trace, one further step is added, namely:

(3) Refilter the prediction errors by means of a stable linear operator so as to approximate the desired output.

The *practical* solution of this problem consists of the following steps:

(1) Compute the prediction error operator[1] (the inverse wavelet shape) directly by the Gauss method of least squares, which yields operator coefficients which satisfy certain statistical optimum properties under general conditions.

(2) Use the prediction error operator on the seismic trace to determine the prediction errors,[2] which are the random unpredictable components.

Alternatively, from other considerations, one may have available the shape of the seismic wavelet. Then the prediction error operator or inverse wavelet shape may be computed from this given wavelet shape. In summary, then, the prediction error operator for unit prediction distance is the inverse seismic wavelet, and the inverse of the prediction error operator for unit prediction distance is the seismic wavelet.

Finally, we note that multitrace operators take into account the coherency between the traces, which is important in that seismic traces become more coherent at major reflections.

CHAPTER II

THEORY OF FINITE DISCRETE LINEAR OPERATORS

2.1 *The finite discrete linear operator*

In this chapter we wish to consider properties of the finite discrete linear operator for a single time series, that is, a linear operator containing a finite number of discrete operator coefficients which perform linear operations on single discrete time series.

In this thesis we deal almost exclusively with discrete time series. A discrete time series is a sequence of equidistant observations x_t which are associated with the discrete time parameter t. Without loss of generality we may take the spacing between each successive observation to be one unit of time, and thus we may represent the time series as

$$\cdots, x_{t-2}, x_{t-1}, x_t, x_{t+1}, x_{t+2}, \cdots \quad (2.11)$$

[1] In current (1967) terminology, the prediction error operator is called the *deconvolution operator* (the editors).
[2] In current (1967) terminology, the prediction error trace is called the *deconvolved seismic trace* (the editors).

where t takes on integer values. As a result the minimum period which may be observed is equal to two units, and consequently the maximum frequency which may be observed is equal to one half, which is an angular frequency of π. Thus, we may require all frequencies f of the discrete time series to lie between $-\frac{1}{2}$ and $\frac{1}{2}$, and all angular frequencies $\omega = 2\pi f$ to lie between $-\pi$ and π.

The time series x_t may be finite or infinite in extent, and may be treated from a deterministic or statistical point of view. In this chapter we shall develop those properties of the finite discrete linear operator which are independent of the nature of the x_t time series, whereas in the following chapters we shall be mainly concerned with the nature of the x_t time series.

A discrete operator is one which transforms a discrete input into a discrete output.

A *linear time-invariant discrete operator* may be characterized by a sequence of numerical coefficients $(\cdots, a_{-1}, a_0, a_1, a_2, \cdots)$, which is called the *impulse response function* of a system which transforms the input x_t into the output y_t by means of the *moving summation*, or (discrete) *convolution*, formula

$$y_t = \cdots + a_{-1}x_{t+1} + a_0 x_t + a_1 x_{t-1} + a_2 x_{t-2} + \cdots \quad (2.12)$$

which holds for all discrete time points (integers) t. Briefly we will denote this convolution formula as $y_t = a_t * x_t$. This formula may be divided into two terms, namely

$$y_t = \sum_{s=-\infty}^{-1} a_s x_{t-s} + \sum_{s=0}^{\infty} a_s x_{t-s}. \quad (2.121)$$

If we regard the time instant t as the present time, then the first term involves the future values $\cdots, x_{t+2}, x_{t+1}$ of the input time series, whereas the second term involves the present and past values $x_t, x_{t-1}, x_{t-2}, \cdots$ of the input time series. As a result the coefficients involved in the first term, namely $(\cdots, a_{-2}, a_{-1})$, represent the *anticipation component* of the impulse response function, and the coefficients involved in the second term, namely $(a_0, a_1, a_2, \cdots)$, represent the *memory component* of the impulse response function. If time t represents historical or so-called nominal time, in which case all time points are in actuality past events, then an operator with both an anticipation component and a memory component may be used in order to transform an input time series to an output time series. However, if time t represents real time, then of course the future values $\cdots, x_{t+2}, x_{t+1}$ of the input time series are not available at the present time instant t, and hence an operator with an anticipation component cannot be used. Of course, if we are willing to tolerate a time delay between input and output, then those future values which occur during this time delay become available, and hence that portion of the anticipation component corresponding to this delay can be used in order to transform the input into the output. In radar systems such a time delay is usually introduced so that although the system is operating in real time, some portion of the anticipation component may be used, thereby improving the information handling capacity of the system. An operator with only a memory component is said to be *realizable* (in real time) whereas an operator which involves an anticipation component is *nonrealizable*. This is the same definition of realizability as that given in electrical engineering.

Prediction operators by their very nature are designed to operate in real time and hence must be realizable. On the other hand, various types of smoothing operators are designed to operate in historical time, or else in real time with a sufficiently long time-delay, and hence need not be realizable in real time.

2.2 Prediction operators

The extrapolation or prediction operator (Kolmogorov, 1939, 1941) is defined by

$$\hat{x}_{t+\alpha} = k_0 x_t + k_1 x_{t-1} + \cdots + k_M x_{t-M} = \sum_{s=0}^{M} k_s x_{t-s}, \quad \alpha \geq 0 \quad (2.21)$$

where α is the prediction distance. The operator coefficients, $k_0, k_1, \cdots, k_M$ are chosen so that the actual output $\hat{x}_{t+\alpha}$ approximates the desired output $x_{t+\alpha}$ according to some criterion. The operator is discrete since its coefficients are discrete values, and the operator is finite since there are only a finite number $M+1$ of such coefficients.

In Section 5.3 we discuss the solution of the prediction problem in which the operator coefficients are determined by the least-squares criterion. In general, we shall see that an infinite number of operator coefficients are required, although for autoregressive time series (see Section

5.5-B) only a finite number of operator coefficients are required.

The prediction operator (2.21) is one from which a specific time function x_t may be generated if a sufficient number of initial values of the time function are specified. That is, if we are given the set of operator coefficients $k_0, k_1, \cdots, k_M$, and the initial values

$$x_0, x_1, x_2, \cdots, x_{M+\alpha-1} \qquad (2.22)$$

and if we define

$$\hat{x}_t = x_t \text{ for } t = M+\alpha, M+\alpha+1, M+\alpha+2, \cdots \quad (2.221)$$

then we may generate the time function x_t by means of equation (2.21). For example, let us consider the case given by $\alpha=1$, $M=0$, $k_0=\frac{1}{2}$, and take $x_0=1$. We have

$$\hat{x}_{t+1} = .5x_t. \qquad (2.23)$$

Then we may generate the time function

$$x_t = \hat{x}_t = .5x_{t-1} \quad \text{for } t = 1, 2, 3, \cdots \qquad (2.24)$$

which is the sequence

$$(.5), (.5)^2, (.5)^3, (.5)^4, (.5)^5, \cdots (.5)^t, (.5)^{t+1}, \cdots . \qquad (2.241)$$

In Section 2.6 we shall see that such a sequence must form a damped motion in order for the operator to be minimum-delay.

Further, the prediction operator (2.21) has the property that only the values $x_t, x_{t-1}, x_{t-2}, \cdots$ of the time series at and prior to time t, and none of the values $x_{t+1}, x_{t+2}, \cdots$, subsequent to time t, are required in order to compute the actual output $\hat{x}_{t+\alpha}$. Thus a prediction operator has an inherent one-sidedness in that it operates on present and past values, but no future values, of the time series. As a result, if time t represents the present calendar time, as is the case for a meteorologist who makes daily weather predictions, then only observations of the time series at the present time and at past times, and no observations at future times are required to carry out the necessary computation.

The prediction operator coefficients represent the impulsive response of an equivalent electric network. Thus those impulses which have arrived at time t or prior to time t will make the network respond, whereas those impulses which have not yet arrived at time t, i.e., those impulses which arrive subsequent to time t, cannot make the network respond. In summary, then, we may say that a finite prediction operator is computationally realizable to the statistician, and physically realizable to the engineer.

Instead of considering the predicted values $\hat{x}_{t+\alpha}$ as the output of the prediction operator, one may consider the prediction error $\xi_{t+\alpha} = x_{t+\alpha} - \hat{x}_{t+\alpha}$ (Wadsworth, et al., 1953). We have

$$\begin{aligned} \xi_{t+\alpha} &= x_{t+\alpha} - \hat{x}_{t+\alpha} \\ &= x_{t+\alpha} - \sum_{s=0}^{M} k_s x_{t-s} \qquad (2.25) \end{aligned}$$

which may be rewritten

$$\begin{aligned} \xi_t &= x_t - \sum_{s=0}^{M} k_s x_{t-s-\alpha} \\ &= x_t - k_0 x_{t-\alpha} - k_1 x_{t-\alpha-1} - \cdots \\ &\quad - k_M x_{t-\alpha-M}. \qquad (2.26) \end{aligned}$$

Let us define $m = M+\alpha$ and

$$\begin{aligned} a_0 = 1,\ a_1 &= 0,\ a_2 = 0, \cdots, \\ a_{\alpha-1} &= 0,\ a_\alpha = -k_0, \\ a_{\alpha+1} = -k_1, \cdots, a_{\alpha+M} &= a_m = -k_M. \qquad (2.27) \end{aligned}$$

Then the prediction error ξ_t may be written as the (discrete) convolution of the operator coefficients with the time series, that is

$$\begin{aligned} \xi_t &= a_0 x_t + a_1 x_{t-1} + \cdots + a_m x_{t-m} \\ &= \sum_{s=0}^{m} a_s x_{t-s}, \quad a_0 = 1. \qquad (2.271) \end{aligned}$$

In the sequel we shall be primarily concerned with the prediction and prediction-error operators for prediction distance α equal to one. In this case, equation (2.271) for the prediction error becomes

$$\begin{aligned} \xi_t &= \sum_{s=0}^{m} a_s x_{t-s} \\ &= x_t + a_1 x_{t-1} + \cdots \\ &\quad + a_m x_{t-m}, \quad a_0 = 1 \qquad (2.28) \end{aligned}$$

and equation (2.27) becomes

$$\begin{aligned} a_0 = 1,\ a_1 &= -k_0, \\ a_2 = -k_1, \cdots, \quad a_m &= -k_M. \qquad (2.281) \end{aligned}$$

We shall regard equation (2.28) as the basic form of the prediction error operator with the coefficients $a_0, a_1, a_2, \cdots, a_m$. The prediction error ξ_t is the actual output at time t of the operator (2.28).

2.3 Smoothing operators

The interpolation or smoothing operator (Kolmogorov, 1939, 1941) is given by

$$\begin{aligned}\hat{x}_t &= d_{-M}x_{t+M} + d_{-M+1}x_{t+M-1} + \cdots \\ &\quad + d_{-1}x_{t+1} + d_1x_{t-1} \\ &\quad + d_2x_{t-2} + \cdots + d_Mx_{t-M} \\ &= \sum_{\substack{s=-M\\ s\neq 0}}^{M} d_sx_{t-s}. \end{aligned} \tag{2.31}$$

We may consider the smoothing error given by

$$\gamma_t = x_t - \hat{x}_t = -\sum_{s=-M}^{-1} d_sx_{t-s} + x_t - \sum_{s=1}^{M} d_sx_{t-s} \tag{2.32}$$

and by letting $m=M$, $c_0=1$, and $c_s=-d_s$ $(s=\pm 1, \pm 2, \cdots, \pm M)$ we have

$$\gamma_t = \sum_{s=-m}^{m} c_sx_{t-s},\ c_0 = 1. \tag{2.33}$$

In the sequel we shall regard this equation as the basic form of the smoothing error operator with operator coefficients $c_{-m}, c_{-m+1}, \cdots, c_0 = 1, c_1, \cdots, c_m$. The smoothing error γ_t is the actual output of the operator.

The smoothing error operator does not have the same properties in regard to computational or physical realizability as the prediction error operator. The smoothing error operator (2.33) has the property that values of the time series $x_{t+m}, \cdots, x_{t+2}, x_{t+1}$ at times subsequent to time t, as well as values $x_t, x_{t-1}, x_{t-2}, \cdots, x_{t-m}$ at and prior to time t, are required to compute γ_t. Consequently, if time t represents the present calendar time, as in the example of the meteorologist, then γ_t given by equation (2.33) can not be computed, since it involves observations $x_{t+1}, x_{t+2}, \cdots, x_{t+m}$ at future times which therefore are not observable at the present time t. Similarly, the network which would be equivalent to a smoothing operator would be one which would respond to impulses which have not yet arrived at the present time t. Consequently, smoothing operators are neither computationally realizable to the statistician, nor physically realizable to the engineer.

Nevertheless, we make finite smoothing error operators both computationally and physically realizable very simply: to compute

$$\begin{aligned}\gamma_t &= \sum_{s=-m}^{m} c_sx_{t-s} = c_{-m}x_{t+m} \\ &\quad + c_{-m+1}x_{t+m-1} + \cdots \\ &\quad + c_{-1}x_{t+1} + c_0x_t + c_1x_{t-1} + \cdots \\ &\quad + c_mx_{t-m} \end{aligned} \tag{2.33}$$

we must delay computation until at least time $t+m$ or later (a time delay of m or greater), at which time all the values needed in the computation will have occurred. That is, we delay computation at least until time $\tau=t+m$, and then compute

$$\begin{aligned}\gamma_t &= \sum_{s=-m}^{m} c_sx_{t-s} = \sum_{s=-m}^{m} c_sx_{\tau-m-s} \\ &= c_{-m}x_\tau + c_{-m+1}x_{\tau-1} + \cdots + c_{-1}x_{\tau-m-1} \\ &\quad + c_0x_{\tau-m} + c_1x_{\tau-m-1} + \cdots \\ &\quad + c_mx_{\tau-2m} \end{aligned} \tag{2.34}$$

which we shall call the time-delay form of the smoothing error operator with coefficients c_s $(s=0, \pm 1, \cdots, \pm m)$. Such an operator is in the form of a realizable operator with the present time now being τ. That is, τ is the time at which computations are to be carried out.

The fact that a seismic disturbance is recorded on paper or magnetic tape means that we have indeed waited until all the pertinent information is available. Consequently, we have introduced the time delay necessary to utilize finite time-delay smoothing error operators, which are computationally realizable. On the other hand, if computations were to be carried out at the same time that the seismic disturbance is occurring, then we would not be able to compute such smoothing operations.

2.4 The transfer function or filter characteristics

As Wiener (1942) points out, the linear operator is the approach from the standpoint of time to a filter which is essentially an instrument for the separation of different frequency ranges. The filtering action of the linear operator is brought out by its transfer function, which is the analog of the transfer or system function of the linear system of which the linear operator is the unit impulse response.

Smith (1954) gives the following interpretation to the transfer function. Let the input information x_t be samples of a sine wave of angular frequency ω. Since the system is linear the output will be a sine wave of the same frequency, which will generally differ in phase and amplitude. Using the complex notation for a sine wave of angular frequency ω, $x_t = e^{i\omega t}$, the transfer ratio at angular frequency ω is the output of the linear operator (which is a complex sine wave of angular frequency ω) divided by the input $x_t = e^{i\omega t}$. The transfer function is the totality of these transfer ratios for $-\pi < \omega \leq \pi$, and represents the filter characteristics of the linear operator. As we shall now see, the transfer function is the Fourier transform of the linear operator. Since the operator is discrete, the transfer function is in the form of a Fourier series rather than a Fourier integral.

For the prediction operator, equation (2.21),

$$\hat{x}_{t+\alpha} = \sum_{s=0}^{M} k_s x_{t-s}, \qquad \alpha \geq 0, \qquad (2.21)$$

by letting $x_t = e^{i\omega t}$ we obtain the transfer ratio

$$\frac{\hat{x}_{t+\alpha}}{x_t} = \frac{\sum_{s=0}^{M} k_s x_{t-s}}{x_t} = \frac{\sum_{s=0}^{M} k_s e^{i\omega(t-s)}}{e^{i\omega t}}$$

$$= \sum_{s=0}^{M} k_s e^{-i\omega s}. \qquad (2.41)$$

The totality of these transfer ratios yields the transfer function

$$K(\omega) = \sum_{s=0}^{M} k_s e^{-i\omega s} \qquad (2.441)$$

which is the Fourier transform of the operator coefficients k_s.

By letting $x_t = e^{i\omega t}$ be the input for the operator for the prediction error, equation (2.28),

$$\xi_t = \sum_{s=0}^{m} a_s x_{t-s}, \qquad a_0 = 1, \qquad (2.28)$$

we obtain the transfer ratio

$$\frac{\xi_t}{x_t} = e^{-i\omega t} \sum_{s=0}^{m} a_s e^{i\omega(t-s)} = \sum_{s=0}^{m} a_s e^{-i\omega s}. \qquad (2.42)$$

The transfer function is then

$$A(\omega) = \sum_{s=0}^{m} a_s e^{-i\omega s} \qquad (2.421)$$

which is the Fourier transform of the operator coefficients.

For the smoothing operator, equation (2.31),

$$\hat{x}_t = \sum_{\substack{s=-M \\ s\neq 0}}^{M} d_s x_{t-s}, \qquad (2.31)$$

by letting the input be $x_t = e^{i\omega t}$, the transfer ratio is

$$\frac{\hat{x}_t}{x_t} = e^{-i\omega t} \sum_{\substack{s=-M \\ s\neq 0}}^{M} d_s e^{i\omega(t-s)}$$

$$= \sum_{\substack{s=-M \\ s\neq 0}}^{M} d_s e^{-i\omega s} \qquad (2.43)$$

and the transfer function is

$$D(\omega) = \sum_{\substack{s=-M \\ s\neq 0}}^{M} d_s e^{-i\omega s}. \qquad (2.431)$$

Similarly letting $x_t = e^{i\omega t}$ be the input for the operator for the smoothing error

$$\gamma_t = \sum_{s=-m}^{m} c_s x_{t-s} \qquad (2.33)$$

we obtain the transfer ratio

$$\frac{\gamma_t}{x_t} = e^{-i\omega t} \sum_{s=-m}^{m} c_s e^{i\omega(t-s)}$$

$$= \sum_{s=-m}^{m} c_s e^{-i\omega s} \qquad (2.44)$$

so that the transfer function is

$$C(\omega) = \sum_{s=-m}^{m} c_s e^{-i\omega s}. \qquad (2.441)$$

Using the same operator coefficients c_s ($s=0, \pm 1, \cdots, \pm m$) but introducing a time delay m, the time-delay smoothing operator, equation (2.34)

$$\gamma_{\tau-m} = \gamma_t \sum_{s=-m}^{m} c_s x_{\tau-m-s} \qquad (2.34)$$

is realizable at the time instant $\tau = t+m$. Since τ now represents the time instant at which computations are to be carried out the input is $x_\tau = e^{i\omega\tau}$. The transfer ratio is

$$\frac{\sum_{s=-m}^{m} c_s e^{i\omega(\tau-m-s)}}{e^{i\omega\tau}} = \sum_{s=-m}^{m} c_s e^{-i\omega(s+m)} \qquad (2.45)$$

and its transfer function is

$$\sum_{s=-m}^{m} c_s e^{-i\omega(s+m)} = e^{-i\omega m} \sum_{s=-m}^{m} c_s e^{-i\omega s} = e^{-i\omega m} C(\omega). \qquad (2.451)$$

2.5 *The z-transform*

The *z-transform* of a discrete sequence ($\cdots, a_{-1}, a_0, a_1, a_2, \cdots$) is defined as the expression

$$A(z) = \sum_{t=-\infty}^{\infty} a_t z^t$$

whose coefficient for z^t is the value a_t of the sequence at the tth time index. In case the sequence is one-sided (that is, $a_t=0$ for $t<0$) then the z-transform becomes the power series

$$A(z) = \sum_{t=0}^{\infty} a_t z^t.$$

In the case of a realizable operator ($a_0, a_1, a_2, \cdots, a_m$) with a finite number ($m+1$) of coefficients, the z-transform is the polynomial

$$A(z) = a_0 + a_1 z + a_2 z^2 + \cdots + a_m z^m.$$

Two important properties of the z-transform which are used frequently are:

(1) Convolution in the time-domain corresponds to multiplication in the z-domain; that is, if $X(z)$, $A(z)$, and $Y(z)$ are the z-transforms of the sequences x_t, a_t, and y_t respectively, then the convolution $y_t = a_t * x_t$ in the time-domain corresponds to the multiplication $Y(z) = A(z)X(z)$ in the z-domain.

(2) The z-transform evaluated on the unit circle $z = e^{-i\omega}$ corresponds to the Fourier transform, that is, the (discrete) Fourier transform of the sequence a_t is

$$A(e^{-i\omega}) = \sum_{t=-\infty}^{\infty} a_t e^{-i\omega t},$$

where ω is angular frequency in radians per time unit. For simplicity we will usually write $A(\omega)$ for $A(e^{-i\omega})$.

As we have seen, this expression is called the *transfer function* or *filter characteristics* of the operator. If we write the transfer function $A(\omega)$ in polar form as

$$A(\omega) = |A(\omega)| e^{i\theta(\omega)}$$

then $|A(\omega)|$ is called the gain, or amplitude spectrum, of the operator, and $\theta(\omega)$ is called the phase characteristic, or phase spectrum.

In order to study the characteristic properties of a finite discrete realizable operator, we may factor the operator into simpler components and then use the properties of these components to classify the given operator. The z-transform of the finite discrete operator ($a_0, a_1, a_2, \cdots, a_m$) can be factored into the form

$$A(z) = a_0 + a_1 z + a_2 z^2 + \cdots + a_m z^m = a_m(z - z_1)\cdot(z - z_2) \cdots (z - z_m)$$

where z_i (for $i = 1, 2, \cdots, m$) are the zeros (or roots) of the polynomial $A(z)$. Some or all of these roots may be complex; in the case when the coefficients ($a_0, a_1, \cdots, a_m$) are real, then all complex roots must occur in complex conjugate pairs. In the time-domain, this factorization exhibits the operator as the result of cascaded convolutions of *two-coefficient operators*, or *couplets*, as given by

$$(a_0, a_1, a_2, \cdots, a_m) = a_m(-z_1, 1) * (-z_2, 1) * \cdots * (-z_m, 1).$$

A couplet (c_0, c_1) is said to be *minimum-delay* provided the magnitude of its first coefficient equals or exceeds the magnitude of its second coefficient, that is, provided $|c_0| \geq |c_1|$. On the other hand, a couplet (c_0, c_1) is said to be *maximum-delay* provided the magnitude of its second coefficient equals or exceeds the magnitude of its first coefficient, that is, provided $|c_1| \geq |c_0|$. Thus an *equi-delay* couplet (that is, one whose two coefficients have the same magnitude) may be regarded as both minimum-delay and maximum-delay. For example, the couplet (2, 1) is minimum-delay, the couplet (3, 4) is maximum-delay, and the couplet (1, −1) is equi-delay. The concepts of minimum-delay and maximum-delay can be extended to longer operators as follows. If each couplet appearing in the factorization of a realizable operator is minimum-delay, then the operator is called *minimum-delay*. If each couplet appearing in the factorization of a realizable operator is maximum-delay, then the operator is called *maximum-delay*. However, if the couplets appearing in the factorization of a realizable operator are a mixture of minimum-delay and maximum-delay ones, then the operator is called *mixed-delay*. As we have defined them, the concepts of minimum-delay, maximum-delay, and mixed-delay apply only in the case of realizable operators.

In summary let us give the following definitions. A sequence a_t of numerical coefficients defined for integer time indices t represents

(1) a *realizable operator* provided that it is one-sided in the sense that its coefficients for negative time indices vanish (that is, provided $a_t=0$ for $t<0$).

(2) a *stable operator* provided that it has finite energy in the sense that the sum of squares of its coefficients is finite (that is, provided $\cdots + a_{-2}^2 + a_{-1}^2 + a_0^2 + a_1^2 + a_2^2 + \cdots$ is finite).

(3) a *wavelet* provided that it is realizable and stable.

(4) a *minimum-delay operator*[3] provided that it is realizable and stable and provided that all the zeros of its z-transform lie on or outside the unit circle in the z-plane (that is, provided that $A(z) \neq 0$ for $|z| < 1$). A minimum-delay operator is said to be *strictly minimum-delay* provided that $A(z) \neq 0$ for $|z| \leq 1$.

For example, the operator $(a_0, a_1, a_2) = (6, 5, 1)$ is minimum-delay because $A(z) = 6+5z+z^2$ has zeros $z_1 = -2$ and $z_2 = -3$, both of which have magnitude greater than unity. On the other hand, the operator (1, 5, 6) is maximum-delay because $A(z) = 1+5z+6z^2$ has zeros $z_1 = -\frac{1}{3}$ and $z_2 = -\frac{1}{2}$, both of which have magnitude less than unity. The operator (3, 7, 2) is mixed-delay because $A(z) = 3+7z+2z^2$ has zeros $z_1 = -\frac{1}{2}$ and $z_2 = -3$, one of which has magnitude less than unity and the other greater than unity.

[3] In this definition we rule out the presence of the so-called Type 3 all-pass system. For the general definition of minimum-delay, see Robinson (1962).

2.6 The minimum-delay operator

In the remaining sections of this chapter we shall consider only the operator form of the realizable type, say

$$\xi_t = \sum_{s=0}^{m} a_s x_{t-s}. \tag{2.28}$$

Let us now consider the linear difference equation

$$\sum_{s=0}^{m} a_s x_{t-s} = 0 \tag{2.61}$$

obtained from equation (2.28) by requiring $\xi_t = 0$. We see that the constant coefficients a_s of the difference equation are the operator coefficients. There is no loss of generality in assuming that a_0 is equal to one, in conformity with our usual convention.

The theory of difference equations is presented by various authors, and the reader is referred especially to Wold (1938) and Samuelson (1947). Let us now examine the condition that the difference equation (2.61) have a stable solution. The characteristic equation of the difference equation (2.61) is defined to be

$$P(\zeta) = a_0\zeta^m + a_1\zeta^{m-1} + \cdots + a_{m-1}\zeta + a_m = \sum_{s=0}^{m} a_s \zeta^{m-s}. \tag{2.62}$$

Since $P(\zeta)$ is a polynomial of degree m, it follows that $P(\zeta)$ has m roots or zeros ζ_j such that

$$P(\zeta_j) = 0, \quad \text{for } j = 1, 2, \cdots, m. \tag{2.621}$$

As a result $P(\zeta)$ may be written in the form

$$P(\zeta) = a_0(\zeta - \zeta_1) \cdot (\zeta - \zeta_2) \cdots (\zeta - \zeta_m) \tag{2.622}$$

Since the operator coefficients a_s ($s=0, 1, \cdots, m$) are real, the roots or zeros $\zeta_1, \zeta_2, \cdots, \zeta_m$ must be real or occur in complex conjugate pairs. Let the distinct real roots of $P(\zeta)$ be represented by α_j, $j=1, 2, 3, \cdots, h$ where each distinct root α_j is repeated γ_j times ($j=1, 2, 3, \cdots, h$); that is, the zero α_j is a zero of order γ_j. Let the distinct complex roots and their conjugate roots be represented by $\beta_j e^{i\theta_j}$ and $\beta_j e^{-i\theta_j}$ ($j=1, 2, \cdots, k$), where each distinct complex root is repeated ρ_j times ($j=1, 2, \cdots, k$); that is, the zero $\beta_j e^{i\theta_j}$ is a zero of order ρ_j, and the zero $\beta_j e^{-i\theta_j}$ is also a zero of order ρ_j. Here β_j represents the modulus, and θ_j or $-\theta_j$ represents the argument, of the complex root. Consequently, equation (2.62) becomes

$$\begin{aligned} P(\zeta) &= a_0\zeta^m + a_1\zeta^{m-1} + \cdots + a_{m-1}\zeta + a_m \\ &= a_0(\zeta - \zeta_1)(\zeta - \zeta_2) \cdots (\zeta - \zeta_m) \\ &= a_0 \prod_{j=1}^{m} (\zeta - \zeta_j) \\ &= a_0 \prod_{j=1}^{h} (\zeta - \alpha_j)^{\gamma_j} \\ &\quad \cdot \prod_{j=1}^{k} (\zeta - \beta_j e^{i\theta_j})^{\rho_j}(\zeta - \beta_j e^{-i\theta_j})^{\rho_j} \\ &= a_0 \prod_{j=1}^{h} (\zeta - \alpha_j)^{\gamma_j} \\ &\quad \cdot \prod_{j=1}^{k} (\zeta^2 - 2\beta_j \cos\theta_j \zeta + \beta_j^2)^{\rho_j}. \quad (2.623) \end{aligned}$$

For an arbitrary set of initial values

$$x_0, x_1, x_2, \cdots, x_{t-1}, \quad (2.63)$$

the series $x_t, x_{t+1}, x_{t+2}, \cdots$, may be generated by recursive deductions from the difference equation (2.61), and this series will form the general solution of the difference equation. Explicitly, the general solution is given by

$$x_t = \sum_{j=1}^{h} P^{(1)}_{\gamma_j-1}(t)\alpha_j^t + \sum_{j=1}^{k} [P^{(2)}_{\rho_j-1}(t) \cdot \cos\theta_j t + P^{(3)}_{\rho_j-1}(t) \sin\theta_j t]\beta_j^t \quad (2.631)$$

where the $P_r^{(s)}(t)$ denotes a polynomial of order r with arbitrary coefficients, and the h, k, α_j, β_j, γ_j, and ρ_j are given by the characteristic equation (2.623).

The asymptotic behavior of the general solution x_t in equation (2.631) is dependent on the exponential factors α_j^t and β_j^t. A necessary and sufficient condition that

$$\sum_{t=0}^{\infty} x_t^2 \quad \text{and} \quad \sum_{t=0}^{\infty} |x_t| \quad (2.64)$$

converge for any values taken on by the $P_r^{(s)}(t)$ is that the magnitude of the roots $|\zeta_j|$, $j=1, 2, \cdots, m$, of the characteristic equation (2.623) be less than one, that is, $|\zeta_j| < 1, j=1, 2, \cdots, m$, which is

$$\begin{aligned} |\alpha_j| &< 1, \quad j = 1, 2, \cdots, h \\ |\beta_j| &< 1, \quad j = 1, 2, \cdots, k. \quad (2.641) \end{aligned}$$

In this case the solution x_t for $t \geq 0$ of the difference equation (2.61) describes a damped oscillation, or in other words the solution x_t for $t \geq 0$ is a stable one-sided time sequence. That is, the solution x_t for $t \geq 0$ is *stable* if and only if the zeros ζ_k of its associated characteristic equation $P(\zeta)$ lie within the periphery $|\zeta| = 1$ of the unit circle, that is $|\zeta_k| < 1$. Without loss of generality, we assume $a_m \neq 0$, so that $|\zeta_k| \neq 0$ for $k=1, 2, \cdots, m$. Thus the characteristic equation (2.623) may be written (where we let $a_0=1$)

$$\begin{aligned} P(\zeta) &= (\zeta-\zeta_1)(\zeta-\zeta_2) \cdots (\zeta-\zeta_m) \\ &= \zeta^m \zeta_1 \zeta_2 \cdots \zeta_m(\zeta_1^{-1} - \zeta^{-1}) \\ &\quad \cdot (\zeta_2^{-1} - \zeta^{-1}) \cdots (\zeta_m^{-1} - \zeta^{-1}) \\ &= \zeta^m \zeta_1 \zeta_2 \cdots \zeta_m (-1)^m (\zeta^{-1} - \zeta_1^{-1}) \\ &\quad \cdot (\zeta^{-1} - \zeta_2^{-1}) \cdots (\zeta^{-1} - \zeta_m^{-1}) \\ &= a_m \zeta^m (\zeta^{-1} - \zeta_1^{-1})(\zeta^{-1} - \zeta_2^{-1}) \cdots \\ &\quad \cdot (\zeta^{-1} - \zeta_m^{-1}) \quad (2.65) \end{aligned}$$

since

$$a_m = (-1)^m \zeta_1 \zeta_2 \cdots \zeta_m. \quad (2.651)$$

Under the transformation $z = \zeta^{-1}$, the function $P(\zeta)$, given by equation (2.65), becomes

$$\begin{aligned} P(\zeta = z^{-1}) &= \sum_{s=0}^{m} a_s z^{-m+s} \\ &= z^{-m} a_m (z - z_1)(z - z_2) \cdots (z - z_m) \quad (2.652) \end{aligned}$$

where we define

$$z_1 = \zeta_1^{-1}, \; z_2 = \zeta_2^{-1}, \cdots, z_m = \zeta_m^{-1}. \quad (2.653)$$

Thus the function

$$\begin{aligned} z^m P(\zeta = z^{-1}) &= \sum_{s=0}^{m} a_s z^s \\ &= a_m(z - z_1)(z - z_2) \cdots (z - z_m) \quad (2.654) \end{aligned}$$

is the z-transform

$$A(z) = \sum_{s=0}^{m} a_s z^s$$

$$= a_m(z - z_1)(z - z_2) \cdots (z - z_m) \quad (2.655)$$

where the roots $z_k (k=1, 2, \cdots, m)$ are given by equation (2.653). The condition that the roots ζ_k have modulus $|\zeta_k|$ less than one is equivalent to the condition that the roots z_k of $A(z)$ have modulus greater than one.

$$|z_k| = |\zeta_k^{-1}| > 1. \quad (2.66)$$

We have said that the realizable operator $a_0, a_1, \cdots, a_m$ is minimum-delay provided that the condition holds that the roots z_k of the polynomial $A(z)$ all have modulus greater than one. We have seen that the homogeneous difference equation (2.61) formed from the operator coefficients $a_0, a_1, \cdots, a_m$ has a stable solution provided the same condition holds. Hence it follows that the operator $a_0, a_1, \cdots, a_m$ is minimum-delay if and only if the homogeneous difference equation (2.61) formed from the operator coefficients has a stable solution. This result links up the theory of difference equations with the concept of minimum-delay.

For example, let us consider the so-called cosine operator (Simpson, 1953)

$$\xi_t = x_t + a_1 x_{t-1} + a_2 x_{t-2}$$

$$= x_t - (2 \cos \omega_0) x_{t-1} + x_{t-2}. \quad (2.67)$$

The coefficients of this operator were chosen by the requirement that the prediction error $\xi_t = 0$ for $x_t = \cos \omega_0 t$. The z-transform is then

$$A(z) = 1 + a_1 z + a_2 z^2$$

$$= 1 - (2 \cos \omega_0) z + z^2$$

$$= (e^{i\omega_0} - z)(e^{-i\omega_0} - z) \quad (2.671)$$

so that the roots of $A(z)$ are

$$z_1 = e^{i\omega_0}, \quad z_2 = e^{-i\omega_0}. \quad (2.672)$$

Since $|z_1| = |z_2| = 1$, the cosine operator is just on the borderline of minimum-delay; that is, the cosine operator is equi-delay.

Let us consider the minimum-delay condition (2.66) in terms of the transfer function $A(\lambda)$ in the complex λ-plane. The minimum-delay condition (2.66) becomes, under the transformation $e^{-i\lambda} = z$, that the transfer function

$$A(\lambda) = \sum_{s=0}^{m} a_s e^{-i\lambda s} \quad (\lambda = \omega + i\sigma) \quad (2.651)$$

has zeros only in the upper half λ-plane (see Figure 1). Then $A(\lambda)$ has no singularities or zeros below the axis of real frequency ω and its logarithm in that half plane is as small as possible at infinity.

A linear system has a minimum phase-shift characteristic if its transfer function has no singularities or zeros in the lower half λ-plane (Bode and Shannon, 1950; Goldman, 1953). Thus if we consider $A(\lambda)$ to be a transfer function of a linear system and if $A(\lambda)$ satisfies the conditions of a minimum phase-shift characteristic, then the linear operator

$$a_t = \frac{1}{2\pi} \int_{-\pi}^{\pi} A(\omega) e^{i\omega t} d\omega \quad (2.67)$$

is minimum-delay.

In summary, then, we see that the minimum-delay condition is that the difference equation (2.61) formed by the operator coefficients a_s $(s=0, 1, \cdots, m)$ is to have a solution which describes a damped oscillation. This minimum-delay condition is also that the transfer function $A(\lambda)$ of the linear operator is to have no singularities or zeros below the axis of real frequency ω, and that its logarithm in that half plane is to be as small as possible at infinity. Briefly, a minimum-delay operator represents a minimum-phase network, and conversely.

2.7 *The inverse linear operator*

The *inverse* of a finite discrete realizable operator $(a_0, a_1, a_2, \cdots, a_m)$ is defined to be the *stable* operator $(\cdots, b_{-2}, b_{-1}, b_0, b_1, b_2, \cdots)$ such that

$$a_t * b_t = \delta_t$$

where δ_t is the Kronecker delta function, which is defined as $\delta_t = 1$ when $t=0$ and $\delta_t = 0$ when $t \neq 0$; hence δ_t represents a sequence $(\cdots, 0, 0, 1, 0, 0, \cdots)$ where the unit spike 1 occurs at time index 0. In terms of z-transforms, the inverse relationship is

$$A(z)B(z) = 1$$

since the z-transform of δ_t is equal to unity. Solving this equation we have the expression for the z-transform of the inverse operator given by

Table 1.

Finite realizable operator $(a_0, a_1, \cdots, a_m)$	Inverse operator b_t
Minimum-delay (Roots $z_1, z_2, \cdots, z_m$ have magnitude greater than unity.)	Stable, realizable (one-sided memory), minimum-delay $b_s = \begin{cases} -\sum_{j=1}^{m} u_j z_j^{-s-1} & \text{for } s \geq 0 \\ 0 & \text{for } s < 0 \end{cases}$
Mixed-delay (Roots $z_1, \cdots, z_h$ have magnitude greater than unity and roots $z_{h+1}, \cdots, z_m$ have magnitude less than unity.)	Stable, nonrealizable (two-sided) $b_s = \begin{cases} -\sum_{j=1}^{h} u_j z_j^{-s-1} & \text{for } s \geq 0 \\ \sum_{j=h+1}^{m} u_j z_j^{-s-1} & \text{for } s < 0 \end{cases}$
Maximum-delay (Roots $z_1, z_2, \cdots, z_m$ have magnitude less than unity.)	Stable, purely nonrealizable (one-sided anticipation) $b_s = \begin{cases} 0 & \text{for } s \geq 0 \\ \sum_{j=1}^{m} u_j z_j^{-s-1} & \text{for } s < 0 \end{cases}$

$$B(z) = \frac{1}{A(z)} \cdot$$

Using the factorization of $A(z)$ we can write this expression as

$$B(z) = \frac{1}{a_m(z - z_1)(z - z_2) \cdots (z - z_m)} \cdot$$

For convenience, let us assume that all the zeros $z_1, z_2, \cdots, z_m$ are distinct. Then $B(z)$ has the partial fraction expansion

$$B(z) = \frac{u_1}{z - z_1} + \frac{u_2}{z - z_2} + \cdots + \frac{u_m}{z - z_m}$$

where the constant u_1 is defined as

$$u_1 = (z - z_1)B(z)\big|_{z=z_1} = \frac{1}{a_m(z_1 - z_2)(z_1 - z_3) \cdots (z_1 - z_m)}$$

and where the constants $u_2, u_3, \cdots, u_m$ are similarly defined. Relabeling the zeros if necessary, we may suppose that $z_1, z_2, \cdots, z_h$ have magnitudes greater than unity, and $z_{h+1}, z_{h+2}, \cdots, z_m$ have magnitudes less than unity (for simplicity, we assume that no roots have magnitude unity). In order to obtain a stable sequence we must make use of one or the other of two possible expansions for each term $u_j/(z-z_j)$ of $B(z)$. If the root z_j has magnitude greater than unity, then we should expand in positive powers of z, that is, we should use the expansion

$$\frac{u_j}{z-z_j} = -u_j(z_j^{-1} + z_j^{-2}z + z_j^{-3}z^2 + \cdots).$$

On the other hand, if the root z_j has magnitude less than unity, then we should expand in negative powers of z, that is, we should use the expansion

$$\frac{u_j}{z-z_j} = u_j(z^{-1} + z_j z^{-2} + z_j^2 z^{-3} + \cdots).$$

In other words, roots greater than one in magnitude contribute to the memory component of the inverse, whereas roots less than one in magnitude contribute to the anticipation component of the inverse. From the partial fraction expansion of $B(z)$ we obtain Table 1.

Let us now examine in more detail the case when the operator $a_0, a_1, \cdots, a_m$ is strictly minimum-delay; that is,

$$A(z) \neq 0 \qquad \text{for } |z| \leq 1. \quad (2.71)$$

Consequently the function

$$A^{-1}(z) = \frac{1}{A(z)} = \frac{1}{\sum_{s=0}^{m} a_s z^s} = B(z) \quad (2.711)$$

is a function which is analytic for $|z| \leq 1$ and has no zeros for $|z| \leq 1$. As a result we may expand this function in the power series given by the polynomial division

$$\frac{1}{A(z)} = \frac{1}{\sum_{s=0}^{m} a_s z^s} = \sum_{t=0}^{\infty} b_t z^t = B(z) \qquad (2.712)$$

which converges for $|z| < 1$ and has no zeros for $|z| \leq 1$. The values of b_t for $t=0, 1, 2, \cdots$, may be found by direct division of the polynomial $a_0 + a_1 z + a_2 z^2 + \cdots + a_m z^m$ into unity. Hence the inverse operator is made up of the coefficients b_t defined by equation (2.712) and

$$b_t = 0 \qquad \text{for } t < 0. \qquad (2.713)$$

From the equation (2.712) we have

$$A(z)B(z) = 1 = \sum_{s=0}^{m} a_s z^s \sum_{t=0}^{\infty} b_t z^t$$

$$= \sum_{s=0}^{m} a_s \sum_{t=0}^{\infty} b_t z^{s+t}$$

$$= \sum_{s=0}^{m} a_s \sum_{n=s}^{\infty} b_{n-s} z^n \qquad (2.72)$$

where we have let $n = s + t$. Recalling that b_t is equal to zero for $t < 0$ we have

$$A(z)B(z) = 1 = \sum_{s=0}^{m} a_s \sum_{n=0}^{\infty} b_{n-s} z^n$$

$$= \sum_{n=0}^{\infty} \left(\sum_{s=0}^{m} a_s b_{n-s} \right) z^n. \qquad (2.721)$$

In order for this equation to hold, we see that

$$a_0 b_0 = 1 \qquad (2.722)$$

$$\sum_{s=0}^{m} a_s b_{n-s} = 0, \quad \text{for } n = 1, 2, 3, \cdots. \qquad (2.723)$$

Since we let $a_0 = 1$, we have $b_0 = 1$. Thus, given the operator coefficients $a_0, a_1, \cdots, a_m$, the $b_1, b_2, b_3, \cdots$, may be uniquely determined by recursive deductions from the difference equation

$$\sum_{s=0}^{m} a_s b_{t-s} = 0, \quad t = 1, 2, 3, \cdots \qquad (2.723)$$

subject to the initial values $b_t = 0$ for $t < 0$ and $b_0 = (a_0)^{-1} = 1$.

The minimum-delay operator $a_0, a_1, \cdots, a_m$ acts on the input x_t to yield the output ξ_t according to the discrete convolution formula

$$\xi_t = \sum_{s=0}^{m} a_s x_{t-s}.$$

The inverse operator $b_0, b_1, b_2, \cdots$, which is also minimum-delay, acts on the ξ_t according to the discrete convolution formula

$$\sum_{r=0}^{\infty} b_r \xi_{t-r}.$$

Inserting into this equation the output $\cdots, \xi_{t-2}, \xi_{t-1}, \xi_t$ produced by the operator $a_0, a_1, \cdots, a_m$, we have

$$\sum_{r=0}^{\infty} b_r \xi_{t-r} = \sum_{r=0}^{\infty} b_r \sum_{s=0}^{m} a_s x_{t-r-s}$$

$$= \sum_{s=0}^{m} a_s \sum_{r=0}^{\infty} b_r x_{t-r-s}. \qquad (2.76)$$

Letting $n = r + s$, we have

$$\sum_{r=0}^{\infty} b_r \xi_{t-r} = \sum_{s=0}^{m} a_s \sum_{n=s}^{\infty} b_{n-s} x_{t-n} \qquad (2.761)$$

and recalling $b_r = 0$ for $r < 0$, we have

$$\sum_{r=0}^{\infty} b_r \xi_{t-r} = \sum_{s=0}^{m} a_s \sum_{n=0}^{\infty} b_{n-s} x_{t-n}\,,$$

$$= \sum_{n=0}^{\infty} \left(\sum_{s=0}^{m} a_s b_{n-s} \right) x_{t-n}. \qquad (2.762)$$

Therefore, because of equations (2.722) and (2.723) we have

$$\sum_{r=0}^{\infty} b_r \xi_{t-r} = x_t. \qquad (2.77)$$

which shows that the inverse operator $b_0, b_1, b_2, \cdots$, acts on the ξ_t time series in order to yield the x_t time series. In summary, the mimimum-delay operator $a_0, a_1, \cdots, a_m$ operates on the present and past values $\cdots, x_{t-2}, x_{t-1}, x_t$ to yield the present and past values $\cdots, \xi_{t-2}, \xi_{t-1}, \xi_t$, and the minimum-delay inverse operator $b_0, b_1, b_2, \cdots$, operates on the present and past values $\cdots, \xi_{t-2}, \xi_{t-1}, \xi_t$ to yield the present and past values $\cdots, x_{t-2}, x_{t-1}, x_t$. Hence the transformation from the initial input x_t through the intermediate output and input ξ_t to the final output x_t takes place *with no overall time delay.*

More generally, the infinite linear operators

$$\xi_t = \sum_{s=0}^{\infty} a_s x_{t-s} \quad (2.78)$$

$$x_t = \sum_{s=0}^{\infty} b_s \xi_{t-s} \quad (2.781)$$

are minimum-delay and inverse to each other if the coefficients a_t satisfy

$$a_t = 0 \qquad \text{for } t < 0$$

(the realizability condition)

$$\sum_{t=0}^{\infty} a_t^2 < \infty$$

(the stability condition)

$$A(z) = \sum_{t=0}^{\infty} a_t z^t \neq 0 \quad \text{for } |z| \leq 1,$$

(the minimum-delay condition)[4] (2.782)

and if the coefficients b_t satisfy

$$b_t = 0 \quad \text{for } t < 0$$

(the realizability condition)

$$\sum_{t=0}^{\infty} b_t^2 < \infty$$

(the stability condition)

$$B(z) = \sum_{t=0}^{\infty} b_t z^t \neq 0 \quad \text{for } |z| \leq 1$$

(the minimum-delay condition) (2.783)

and if

$$A(z)B(z) = 1.$$

(the inverse condition). (2.784)

Hence the a_t and b_t are related by

$$a_0 b_0 = 1$$

$$\sum_{s=0}^{t} a_s b_{t-s} = 0, \quad \text{for } t = 1, 2, 3, \cdots . \quad (2.785)$$

[4] See footnote 3 on page 428.

Thus, given the set a_s, the set b_s may be uniquely determined, and vice versa. For example equation (2.785) for $t=1, 2, 3$, yields

$$a_1 = -b_1$$
$$a_2 = -b_2 + b_1^2$$
$$a_3 = -b_3 + 2b_1b_2 - b_1^3 \quad (2.786)$$

on the one hand, and

$$b_1 = -a_1$$
$$b_2 = -a_2 + a_1^2$$
$$b_3 = -a_3 + 2a_1a_2 - a_1^3 \quad (2.787)$$

on the other hand. Both the a_t series and the b_t series ($t=0, 1, 2, \cdots$) form damped oscillations.

The transfer functions of the mutually inverse operators

$$A(\omega) = \sum_{s=0}^{\infty} a_s e^{-i\omega s} \quad (2.791)$$

and

$$B(\omega) = \sum_{s=0}^{\infty} b_s e^{-i\omega s} \quad (2.792)$$

are free from singularities and zeros in the lower half λ-plane, $\lambda=\omega+i\sigma$, and have minimum phase shift characteristics. They are related by

$$A(\omega)B(\omega) = 1 \quad (2.793)$$

$$\frac{1}{A(\omega)} = A^{-1}(\omega) = B(\omega) \quad (2.794)$$

$$\frac{1}{B(\omega)} = B^{-1}(\omega) = A(\omega) \quad (2.795)$$

$$A^{-1}(\omega)A(\omega) = 1 \quad (2.796)$$

and

$$B^{-1}(\omega)B(\omega) = 1. \quad (2.797)$$

2.8 The power transfer function and its minimum-delay operator

The power transfer function $\Psi(\omega)$ is defined to be the square of the absolute value $|A(\omega)|$ of the transfer function $A(\omega)$; that is, the power transfer function is given by

$$\Psi(\omega) = |A(\omega)|^2 = A(\omega)\overline{A(\omega)} = \{\mathrm{Re}[A(\omega)]\}^2 + \{\mathrm{Im}[A(\omega)]\}^2 \geq 0 \qquad (2.81)$$

where the bar indicates the complex conjugate. Here $|A(\omega)|$ is called the gain of the linear operator, and it is given by the square root of the power transfer function; that is

$$|A(\omega)| = [\Psi(\omega)]^{1/2} \qquad (2.811)$$

so that knowledge of the gain of the linear operator and knowledge of the power transfer function are equivalent.

The power transfer function $\Psi(\omega)$ of a finite linear operator $a_0, a_1, \cdots a_m$ may be expressed by the finite trigonometric series

$$\Psi(\omega) = \sum_{s=0}^{m} a_s e^{-i\omega s} \cdot \sum_{t=0}^{m} a_t e^{i\omega t} = |A(\omega)|^2 \geq 0 \qquad (2.812)$$

which is nonnegative for $-\pi \leq \omega \leq \pi$. We may rewrite this expression in the following way

$$\Psi(\omega) = \sum_{s=0}^{m} \sum_{t=0}^{m} a_s a_t e^{-i\omega(s-t)} = \sum_{\tau=-m}^{m} e^{-i\omega\tau} \sum_{t=0}^{m} a_t a_{t+\tau} \qquad (2.813)$$

where $\tau = s - t$. Let us define r_τ to be

$$r_\tau = \sum_{t=0}^{m} a_t a_{t+\tau}; \quad r_\tau = r_{-\tau} \qquad (2.814)$$

that is,

$$r_0 = a_0^2 + a_1^2 + a_2^2 + \cdots + a_{m-2}^2 + a_{m-1}^2 + a_m^2$$

$$r_1 = r_{-1} = a_0 a_1 + a_1 a_2 + a_2 a_3 + \cdots + a_{m-2} a_{m-1} + a_{m-1} a_m$$

$$r_2 = r_{-2} = a_0 a_2 + a_1 a_3 + a_2 a_4 + \cdots + a_{m-2} a_m$$

$$\cdots\cdots\cdots\cdots$$

$$r_{m-1} = r_{1-m} = a_0 a_{m-1} + a_1 a_m$$

$$r_m = r_{-m} = a_0 a_m. \qquad (2.814)$$

Therefore if we are given the linear operator with coefficients $a_0, a_1, \cdots, a_m$, we may find the r_τ by means of equation (2.814) and thereby determine the power transfer function

$$\Psi(\omega) = \left| \sum_{s=0}^{m} a_s e^{-i\omega s} \right|^2 = \sum_{\tau=-m}^{m} r_\tau e^{-i\omega\tau} = r_0 + 2\sum_{\tau=0}^{m} r_\tau \cos \omega\tau \geq 0. \qquad (2.815)$$

In this section we consider the inverse problem; that is, given the power transfer function $\Psi(\omega)$, find the coefficients $a_0, a_1, \cdots, a_m$ of the linear operator which yields this power transfer function. This inverse problem, as it stands, is not unique in that many different linear operators do yield the same power transfer function $\Psi(\omega)$. As we shall see, however, all except one of these linear operators are non-minimum-delay. In other words, given the power transfer function $\Psi(\omega)$, we wish to find the one and only realizable minimum-delay operator which yields the power transfer function $\Psi(\omega)$. This minimum-delay operator has the transfer function with gain equal to $[\Psi(\omega)]^{1/2}$ and minimum-phase characteristic. The import of this section resides in the fact that if one wishes to design a minimum-delay operator, he needs only to have information about the desired absolute gain characteristics $|A(\omega)| = [\Psi(\omega)]^{1/2}$, and needs no information concerning the phase characteristics. In this section, we give a direct procedure for the determination of such minimum-delay operators. This procedure may be readily programmed for automatic computation on a digital computer.

Since we wish to consider finite linear operators with the coefficients $a_0, a_1, \cdots, a_m$, it is necessary to express the power transfer function in terms of a finite trigonometric series

$$\Psi(\omega) = \sum_{\tau=-m}^{m} r_\tau e^{-i\omega\tau} = r_0 + 2\sum_{\tau=1}^{m} r_\tau \cos \omega\tau,$$

$$r_0 > 0, \; r_\tau = r_{-\tau} \qquad (2.82)$$

which is nonnegative for $-\pi < \omega \leq \pi$, and the r_τ are real.

If the power transfer function is given by the infinite Fourier series

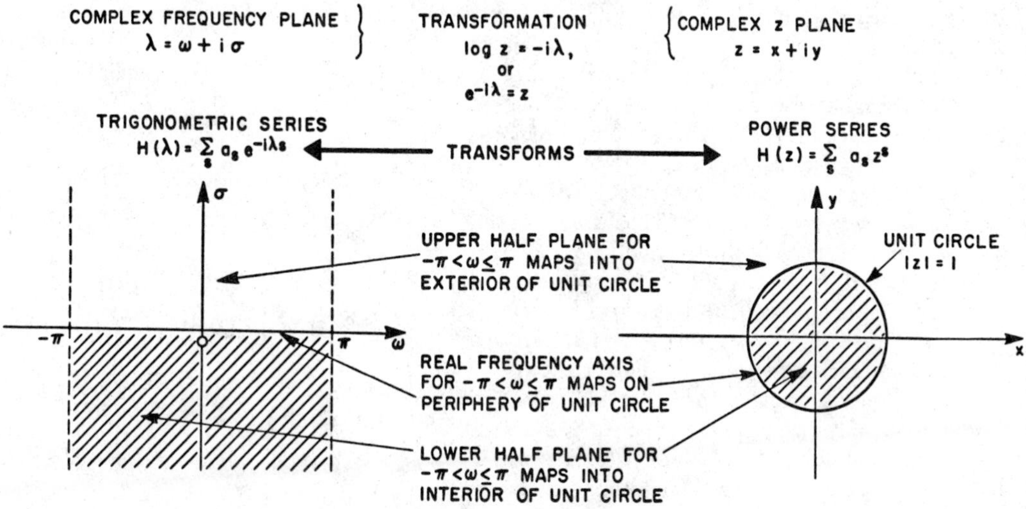

FIG. 1. Transformation of trigonometric series into power series.

$$\Psi(\omega) = \rho_0 + 2\sum_{\tau=1}^{\infty} \rho_\tau \cos \omega\tau \geq 0, \qquad -\pi < \omega \leq \pi, \quad (2.821)$$

which is nonnegative for $-\pi<\omega\leq\pi$, then the Césaro partial sum

$$\rho_0 + 2\sum_{\tau=1}^{N}\left(1-\frac{\tau}{N}\right)\rho_\tau \cos \omega\tau, \quad (2.822)$$

which is also nonnegative for $\pi\leq\omega<\pi$, may be used as the finite series approximation to $\Psi(\omega)$. We then have

$$\Psi(\omega) \approx r_0 + 2\sum_{\tau=1}^{m} r_\tau \cos \omega\tau,$$

$$r_\tau = \left(1-\frac{\tau}{N}\right)\rho_\tau \quad (2.823)$$

which may be used for equation (2.82).

Let $Q(u)$ be the polynomial of order m obtained from

$$r_0 + 2\sum_{\tau=1}^{m} r_\tau(z^{-\tau}+z^\tau) \quad (2.824)$$

by the substitution $z^{-1}+z=u$. Wold (1938), in connection with his study of the process of moving averages, shows that a necessary and sufficient condition that the finite series

$$r_0 + 2\sum_{\tau=1}^{m} r_\tau(e^{i\omega\tau}+e^{-i\omega\tau}) \quad (2.825)$$

be nonnegative for $-\pi\leq\omega\leq\pi$ is that the equation $Q(u)=0$ should have no real root of odd multiplicity in the interval $-2<u<2$.

The method which we give in this section was used by Fejér (1915) and Wold (1938), working in different, but mathematically equivalent, settings, Fejér in terms of trigonometric series, Wold in terms of stationary stochastic processes.

Returning now to the power transfer function let us suppose that the $r_\tau=r_{-\tau}$ are such that $\Psi(\omega)$ given by equation (2.82) is nonnegative for $-\pi<\omega\leq\pi$. Under the transformation $z=e^{-i\lambda}$, where $\lambda=\omega+i\sigma$ (see Figure 1), the power transfer function $\Psi(\omega)$ becomes

$$\Psi(z) = \sum_{\tau=-m}^{m} r_\tau z^\tau \quad (2.83)$$

where $\Psi(z)$ is a rational function in z.

We see that

$$z^m\Psi(z) = \sum_{\tau=-m}^{m} r_\tau z^{\tau+m} \quad (2.831)$$

is a polynomial of order $2m$. Because $r_\tau=r_{-\tau}$ we may express this polynomial as

$$z^m\Psi(z) = r_m(z-z_1)(z-z_2)\cdots(z-z_{2m})$$

$$= z^m \sum_{\tau=-m}^{m} r_{-\tau} z^\tau$$

$$= z^m \sum_{s=-m}^{m} r_s (z^{-1})^s = z^m\Psi(z^{-1}).$$

Hence we see that $z_i^m\Psi(z_i)=0$ if and only if $z_i^m\Psi(z_i^{-1})=0$, so z_i is a root of this polynomial if and only if z_i^{-1} is also a root. Moveover, since the polynomial has real coefficients, it follows that for every complex root z_i there must occur a root given by the complex conjugate $\bar{z}_i$.

In summary, then, if z_k is a root of $z^m\Psi(z)$ then $\bar{z}_k$, z_k^{-1}, and $\bar{z}_k^{-1}$ are also roots. Thus if α_k is a complex root of $z^m\Psi(z)$ with modulus $|\alpha_k|\neq 1$, then α_k, α_k^{-1}, $\bar{\alpha}_k^{-1}$ are distinct from each other, and are all roots of $z^m\Psi(z)$. If β_k is a real root of $z^m\Psi(z)$ with modulus $|\beta_k|\neq 1$, then β_k and β_k^{-1} are distinct from each other, and are both roots of $z^m\Psi(z)$. If γ_k is a complex root of $z^m\Psi(z)$ with modulus $|\gamma_k|=1$, then γ_k and $\bar{\gamma}_k$ are distinct from each other, and are both roots of $z^m\Psi(z)$. Let ρ_k represent the real roots of $z^m\Psi(z)$ with modulus $|\rho_k|=1$.

Accordingly, the polynomial $z^m\Psi(z)$ may be expressed as

$$z^m\Psi(z) = r_m \prod_{k=1}^{h} (z-\alpha_k)(z-\bar{\alpha}_k)(z-\alpha_k^{-1}) \cdot (z-\bar{\alpha}_k^{-1}) \prod_{k=1}^{j} (z-\beta_k)(z-\beta_k^{-1}) \cdot \prod_{k=1}^{l} (z-\gamma_k)(z-\bar{\gamma}_k) \prod_{k=1}^{n} (z-\rho_k) \quad (2.85)$$

where any root of order p is repeated p times.

Let us now turn our attention to

$$\Psi(\omega) = \sum_{s=0}^{m} a_s e^{-i\omega s} \sum_{t=0}^{m} a_t e^{i\omega t} \quad (2.812)$$

which expresses the relationship of the power transfer function $\Psi(\omega)$ with the coefficients a_0, $a_1, \cdots, a_m$ of the operator which yields $\Psi(\omega)$. Under the transformation $z=e^{-i\lambda}$, $\lambda=\omega+i\sigma$, we have

$$\Psi(z) = \sum_{s=0}^{m} a_s z^s \sum_{t=0}^{m} a_t z^{-t}. \quad (2.86)$$

We thus have

$$z^m\Psi(z) = \sum_{\tau=-m}^{m} r_\tau z^{\tau+m} = \sum_{s=0}^{m} a_s z^s \sum_{t=0}^{m} a_t z^{m-t} \quad (2.861)$$

The z-transform $A(z)$ of the operator $(a_0, a_1, \cdots, a_m)$ is

$$A(z) = \sum_{s=0}^{m} a_s z^s \quad (2.652)$$

so equation (2.86) becomes

$$\Psi(z) = A(z)A(z^{-1}) \quad (2.862)$$

and equation (2.861) becomes

$$z^m\Psi(z) = [A(z)][z^m A(z^{-1})] \quad (2.863)$$

which is

$$r_m z^{2m} + r_{m-1}z^{2m-1} + \cdots + r_1 z^{m+1} + r_0 z^m + r_1 z^{m-1} + \cdots + r_{m-1}z + r_m = (a_0 + a_1 z + \cdots + a_{m-1}z^{m-1} + a_m z^m)(a_0 z^m + a_1 z^{m-1} + \cdots + a_{m-1}z + a_m). \quad (2.864)$$

In order to factor $z^m\Psi(z)$ into the two real polynomials $A(z)$ and $z^mA(z^{-1})$, we see that one of the real polynomials $(z-\alpha_k)$ $(z-\bar{\alpha}_k)$ or $(z-\alpha_k^{-1})$ $(z-\bar{\alpha}_k^{-1})$ must be a factor in the polynomial $A(z)$. Since β_k is real, then either $(z-\beta_k)$ or $(z-\beta_k^{-1})$ is a factor in $A(z)$. On the other hand, since the factors $(z-\gamma_k)$ and $(z-\bar{\gamma}_k)$ are complex, both of them must be contained in $A(z)$. Likewise $(z-\rho_k)$ must appear in $A(z)$. Thus it is necessary that the roots γ_k and ρ_k, which have modulus 1, appear an even number of times, that is $l=2l'$ and $n=2n'$. This condition that roots of modulus one appear an even number of times is satisfied since $\Psi(\omega)\geq 0$. Thus we have

$$A(z) = a_m \prod_{k=1}^{h} (z-\alpha_k)(z-\bar{\alpha}_k) \prod_{k=1}^{j} (z-\beta_k) \cdot \prod_{k=1}^{l'} (z-\gamma_k)(z-\bar{\gamma}_k) \prod_{k=1}^{n'} (z-\rho_k) \quad (2.87)$$

and

$$z^m A(z^{-1}) = \prod_{k=1}^{h} (z-\alpha_k^{-1})(z-\bar{\alpha}_k^{-1}) \cdot \prod_{k=1}^{j} (z-\beta_k^{-1}) \prod_{k=1}^{l'} (z-\gamma_k^{-1})(z-\bar{\gamma}_k^{-1}) \cdot \prod_{k=1}^{n'} (z-\rho_k^{-1}) \quad (2.871)$$

since $\bar{\gamma}_k = \gamma_k^{-1}$, $\gamma_k = \bar{\gamma}_k^{-1}$, and $\rho_k = \rho_k^{-1}$.

Thus if the zeros of $A(z)$ are z_k, then the zeros of $z^m A(z^{-1})$ are z_k^{-1}. In order for the linear operator with coefficients $a_0, a_1, \cdots, a_m$ to be strictly minimum-delay, then all the roots of

$$A(z) = \sum_{s=0}^{m} a_s z^s \qquad (2.652)$$

must have modulus $|z_k| > 1$. Thus if the transfer function $\Psi(\omega)$ yields roots γ_k and ρ_k which do have modulus $|\gamma_k| = |\rho_k|$ equal to one, then there is no strictly minimum-delay operator which yields this transfer function, although there is a linear operator on the borderline of minimum-delay, such as the cosine operator (2.67).

Let us suppose that $\Psi(\omega)$ yields no roots γ_k and ρ_k with modulus equal to one. Then we have

$$A(z) = a_m \prod_{k=1}^{h} (z - \alpha_k)(z - \bar{\alpha}_k) \prod_{k=1}^{j} (z - \beta_k)$$

$$= \sum_{s=0}^{m} a_s z^s \qquad (2.88)$$

and

$$z^m A(z^{-1}) = \prod_{k=1}^{h} (z - \alpha_k^{-1})(z - \bar{\alpha}_k^{-1})$$

$$\cdot \prod_{k=1}^{j} (z - \beta_k^{-1}) = \sum_{s=0}^{m} a_s z^{m-s} \qquad (2.881)$$

where $|\alpha_k| \neq 1$ for $k = 1, 2, \cdots, h$ and $|\beta_k| \neq 1$ for $k = 1, 2, \cdots, j$. If the zeros of $A(z)$ are z_k for $k = 1, 2, \cdots, m$, then the zeros of $z^m A(z^{-1})$ are z_k^{-1} for $k = 1, 2, \cdots, m$, and since $|z_k| \neq 1$, it follows that half of the $2m$ roots of

$$z^m \Psi(z) = A(z) z^m A(z^{-1}) \qquad (2.863)$$

have modulus greater than one, and the other half of the $2m$ roots have modulus less than one.

Thus if we choose those m roots of $z^m \Psi(z)$ which have modulus greater than one, and call these roots the $\alpha_k (k = 1, 2, \cdots, h)$, the $\bar{\alpha}_k (k = 1, 2, \cdots, h)$, and the $\beta_k (k = 1, 2, \cdots, j)$ which appear in equation (2.88), then $A(z)$ will represent the z-transform of a minimum-delay operator.

On the other hand, if we did not choose the roots in the above fashion, there being at most 2^m different ways of choosing the roots, then $A(z)$ would have roots, some of which have modulus greater than one, and some of which have modulus less than one. Consequently the linear operator would not be minimum-delay.

Let us summarize the computational procedure required to determine the minimum-delay operator coefficients $a_0, a_1, \cdots, a_m$ from the power transfer function

$$\Psi(\omega) = \sum_{\tau=-m}^{m} r_\tau e^{-i\omega\tau} \geq 0, \quad r_0 > 0. \qquad (2.82)$$

Form the polynomial

$$z^m \Psi(z) = \sum_{\tau=-m}^{m} r_\tau z^{\tau+m} \qquad (2.831)$$

and solve for its roots $z_1, z_2, \cdots, z_{2m}$. Let $z_1', z_2', \cdots, z_m'$ be those $z_k (k = 1, 2, \cdots, 2m)$ of modulus greater than one and also those z_k of modulus one counted half as many times. (In order for there to be a strictly minimum-delay operator, there may be no z_k of modulus one.) Then we form the polynomial

$$A(z) = (z - z_1')(z - z_2') \cdots (z - z_m')$$

$$= \sum_{s=0}^{m} a_s z^s, \qquad (2.89)$$

and the operator coefficients are given by the a_s. They represent a minimum-delay operator, the transfer function of which has minimum phase characteristic. Thus we have shown that the power transfer function $\Psi(\omega)$, equation (2.82), may be factored into

$$\Psi(\omega) = A(\omega)\overline{A(\omega)}$$

where the transfer function $A(\omega)$ is minimum-delay.

CHAPTER III

THE NONSTATISTICAL ANALYSIS OF TIME SERIES

3.1 *The functional approach*

In the last chapter those properties of the finite discrete linear operator were developed which are independent of the properties of the time series x_t under consideration. In this chapter we wish to consider the methodology of the nonstatistical or deterministic approach to the study of discrete time series x_t. Since this approach leads to perfect functional representations of the empirical data, the various schemes of analysis are called functional schemes. As we shall see, the harmonic analysis of the time series, that is the analysis in terms of $e^{i\omega t}$, plays an important role in these functional schemes.

As an introduction to the concept of the functional approach, let us for the moment consider the continuous time series $x(t)$. If the integral which represents the total energy of $x(t)$, given by

$$\int_{-\infty}^{\infty} x^2(t)dt \tag{3.11}$$

is finite, then the Fourier integral representation of $x(t)$,

$$x(t) = \frac{1}{2\pi}\int_{-\infty}^{\infty} X(\omega)e^{i\omega t}d\omega, \tag{3.12}$$

is a perfect functional representation of $x(t)$. Here the function $X(\omega)$, given by the inverse Fourier transform

$$X(\omega) = \int_{-\infty}^{\infty} x(t)e^{-i\omega t}dt, \tag{3.13}$$

is a function which contains the same information as $x(t)$, but is in the frequency domain rather than in the time domain as $x(t)$.

In this chapter we consider only discrete time series. We first indicate the methodology of the periodic functional scheme. We then consider the aperiodic functional scheme for transient time series, especially in connection with the so-called linear filtering problem, which is the transformation of one transient time series into another transient by linear operations. Finally we see that the functional approach in certain cases requires methods of averaging in order to approximate certain functions for computational purposes. Since, in many cases, methods of averaging become more physically meaningful from the standpoint of probability theory, in the last three chapters we shall consider the analysis of time series from the statistical point of view.

In particular, in Chapter V, we shall see that a stationary time series (with an absolutely continuous spectral distribution) may be considered to be additively composed of wavelets, all with the same shape, and with random strengths and arrival times. As we shall see, methods of averaging play an important role in the determination of the wavelet shape. Since the shape of such a wavelet represents a deterministic transient time function, it may be treated by the methodology of this chapter.

3.2 The periodic functional scheme

The various periodic functional schemes assume that the time series under consideration are composed of strictly periodic components. Since such schemes are treated in detail in the literature and since they have limited application in seismology, we shall only briefly indicate the methodology used. For more detailed discussions, the reader is referred to Schuster (1898, 1900) and Whittaker and Robinson (1926).

Let an observational time series be given by

$$x_1, x_2, \cdots, x_T \tag{3.21}$$

for the time range $1 \leq t \leq T$, where the mean value of x_t is zero. We shall consider the case in which it is assumed that the observational data are strictly periodic, say with period T. Then the infinite time series is given by the sequence

$$\cdots x_T, x_1, x_2, \cdots, x_T, x_1, x_2, \cdots,$$
$$x_T, x_1, x_2, \cdots, x_T, x_1, \cdots. \tag{3.22}$$

Representing this time series by $x_t(-\infty < t < \infty)$, the difference equation

$$x_t - x_{t-T} = 0 \tag{3.23}$$

holds for any t. The solution of this difference equation (where for simplicity we let T be even) is

$$x_t = \sum_{n=1}^{T/2}\left[A_n \cos\left(\frac{2\pi}{T}nt\right) - B_n \sin\left(\frac{2\pi}{T}nt\right)\right]$$
$$= \sum_{n=1}^{T/2} C_n \cos\left(\frac{2\pi}{T}nt + \theta_n\right) \tag{3.24}$$

where

$$C_n^2 = A_n^2 + B_n^2, \qquad \tan\theta_n = \frac{B_n}{A_n}. \tag{3.25}$$

The Fourier analysis of the time series x_t will yield the A_n and B_n, so that equation (3.24) will be a perfect functional representation of the original data.

3.3 The aperiodic functional scheme

The aperiodic functional scheme deals with that class of discrete time series x_t whose total energy

$$\sum_{t=-\infty}^{\infty} x_t^2 \qquad (3.31)$$

is finite. Examples of such time series are transient time functions, such as the seismic wavelets of Ricker (1940, 1941, 1943, 1944, 1945, 1949, 1953a, 1953b).

We see that all observational time series of finite time duration, say from $t=0$ to $t=T$, fall into this class, if we define the time series x_t to be zero outside the basic time interval, that is,

$$x_t = 0 \quad \text{for } t < 0 \text{ and } t > T. \qquad (3.311)$$

Hence the time series x_t defined for all time has finite total energy given by

$$\sum_{t=-\infty}^{\infty} x_t^2 = \sum_{t=0}^{T} x_t^2. \qquad (3.312)$$

For the remaining parts of this chapter we shall deal with finite time series which are defined to be zero outside of their basic time interval. Although the nonstatistical methods of this chapter can be applied to any time series of finite duration we do not wish to imply that they should be applied to the analyses of all time series of finite time duration. Instead, the methodology to be used, statistical or nonstatistical, should depend upon the type of problem to be solved, and should be chosen with consideration to all prior knowledge and experience about the problem.

In particular, the methodology of this chapter is applicable to wavelets which damp toward zero sufficiently rapidly so that they may be approximated by zero outside a finite time interval.

As we shall see in the following chapters, a stationary time series (with an absolutely continuous spectral distribution function) may be considered to be additively composed of wavelets, all of the same shape, but with random strengths and arrival times. Let us outline heuristically how one, in effect, determines a certain linear operation for such a time series. First the random elements of the time series are destroyed by averaging, (i.e. the computation of the auto-correlation function), and a unique minimum-delay wavelet shape (i.e. the $\psi_1(t)$ of Levinson (1947b) which is our b_t in the following chapters) is preserved. Then the particular linear operation may be determined by nonstatistical methods on this deterministic wavelet shape. Since the time series is additively composed of such wavelets, this linear operation applies equally as well to the time series itself.

The aperiodic functional scheme is a functional scheme because it leads to a perfect functional representation of the observational or theoretical data. This functional representation is given by the Fourier integral representation of the function

$$x_t = \frac{1}{2\pi}\int_{-\pi}^{\pi} X(\omega)e^{i\omega t}d\omega \qquad (3.32)$$

where $X(\omega)$, the (complex) phase and amplitude spectrum of x_t, is given by

$$X(\omega) = \sum_{t=-\infty}^{\infty} x_t e^{-i\omega t} = \sum_{t=0}^{T} x_t e^{-i\omega t} \qquad (3.321)$$

since $x_t=0$ for $t<0$ and $t>T$. Equation (3.32) gives a perfect functional representation of x_t, and we see that the function $X(\omega)$ contains the same information as the function x_t. The energy spectrum $\Phi_{xx}(\omega)$ of x_t is defined to be

$$\Phi_{xx}(\omega) = \overline{X(\omega)}X(\omega) = |X(\omega)|^2 \geq 0, \quad -\pi \leq \omega \leq \pi, \qquad (3.33)$$

which is equal to

$$\begin{aligned}\Phi_{xx}(\omega) &= \sum_{t=0}^{T} x_t e^{i\omega t} \sum_{s=0}^{T} x_s e^{-i\omega s} \\ &= \sum_{t=-\infty}^{\infty}\sum_{s=-\infty}^{\infty} x_t x_s e^{-i\omega(s-t)} \\ &= \sum_{\tau=-\infty}^{\infty} e^{-i\omega\tau} \sum_{t=-\infty}^{\infty} x_t x_{t+\tau} \\ &= \sum_{\tau=-T}^{T} e^{-i\omega\tau} \sum_{t=0}^{T-\tau} x_t x_{t+\tau} \qquad (3.331)\end{aligned}$$

where $\tau=s-t$. Let us define the autocorrelation function of x_t to be

$$\phi_{xx}(\tau) = \sum_{t=0}^{T-\tau} x_t x_{t+\tau}, \qquad (3.332)$$

where we see

$$\phi_{xx}(-\tau) = \phi_{xx}(\tau)$$
$$\phi_{xx}(\tau) = 0 \quad \text{for } |\tau| > T. \quad (3.333)$$

Then equation (3.331) becomes

$$\Phi_{xx}(\omega) = \sum_{\tau=-T}^{T} \phi_{xx}(\tau)e^{-i\omega\tau}$$
$$= \phi(0) + 2\sum_{\tau=1}^{T} \phi(\tau) \cos \omega\tau \qquad (3.334)$$

which expresses the energy spectrum as the Fourier transform of the autocorrelation. On the other hand, the transform of the energy spectrum is the autocorrelation, as seen by

$$\frac{1}{2\pi}\int_{-\pi}^{\pi} \Phi_{xx}(\omega)e^{i\omega t}d\omega$$
$$= \frac{1}{2\pi}\int_{-\pi}^{\pi} \sum_{\tau=-T}^{T} \phi_{xx}(\tau)e^{-i\omega\tau}e^{i\omega t}d\omega$$
$$= \phi_{xx}(t) \qquad (3.34)$$

because

$$\int_{-\pi}^{\pi} e^{-i\omega\tau}e^{i\omega t}d\omega = \begin{cases} 0 & \tau \neq t \\ 2\pi & \tau = t \end{cases} \qquad (3.341)$$

for integer values of t and τ. We see that the total energy of x_t is given by

$$\phi_{xx}(0) = \sum_{t=0}^{T} x_t^2 = \frac{1}{2\pi}\int_{-\pi}^{\pi} \Phi_{xx}(\omega)d\omega. \quad (3.342)$$

Equation (3.33), which may be written

$$\Phi_{xx}(\lambda) = X(\lambda)\overline{X(\lambda)}, \quad \lambda = \omega + i\sigma \quad (3.35)$$

generally does not represent a factorization of the spectrum which satisfies the Kolmogorov conditions (see Section 5.1) In other words, the factor $X(\lambda)$ generally will not be free from singularities and zeros in the lower half λ-plane. This condition will hold only if the finite time series x_t satisfies the same minimum-delay conditions which we gave for a finite linear operator a_t in Section 2.6. Since $\Phi_{xx}(\omega)$ in equation (3.334) is expressed in a finite nonnegative trigonometric series, the method of Section 2.8 may be used to factor $\Phi_{xx}(\omega)$ into the product

$$\Phi_{xx}(\omega) = B(\omega)\overline{B(\omega)}$$

where the transfer function $B(\omega)$ is minimum-delay. The function $B(\omega)$ may then be used in preference to $X(\omega)$ in many applications, although we shall not explicitly make use of $B(\omega)$ in the remainder of this chapter.

Suppose that we have another time series y_t of finite time duration, say from $t=0$ to $t=T$. As in the case of x_t, we let

$$y_t = 0 \quad \text{for } t < 0, t > T. \qquad (3.36)$$

Similar relations hold for y_t as for the function x_t. As a matter of notation, the complex phase-amplitude spectrum is denoted by $Y(\omega)$, the energy spectrum by $\Phi_{yy}(\omega)$, and the auto-correlation by $\phi_{yy}(\tau)$.

The cross energy spectrum of x_t and y_t is

$$\Phi_{xy}(\omega) = \overline{X(\omega)}\,Y(\omega) \qquad (3.37)$$

which is equal to

$$\Phi_{xy}(\omega) = \sum_{t=0}^{T} x_t e^{i\omega t} \sum_{s=0}^{T} y_s e^{-i\omega s}$$
$$= \sum_{\tau=-T}^{T} e^{-i\omega\tau} \sum_{t=0}^{T} x_t y_{t+\tau} \quad (3.371)$$

where $\tau = s-t$. The crosscorrelation function of x_t and y_t is defined to be

$$\phi_{xy}(\tau) = \sum_{t=0}^{T-\tau} x_t y_{t+\tau} \qquad (3.38)$$

where

$$\phi_{xy}(-\tau) = \phi_{yx}(\tau)$$
$$\phi_{xy}(\tau) = 0 \quad \text{for } |\tau| > T. \quad (3.381)$$

Equation (3.371) becomes

$$\Phi_{xy}(\omega) = \sum_{\tau=-T}^{T} \phi_{xy}(\tau)e^{-i\omega\tau}. \quad (3.382)$$

We have

$$\frac{1}{2\pi}\int_{-\pi}^{\pi} \Phi_{xy}(\omega)e^{i\omega t}d\omega = \phi_{xy}(t). \quad (3.39)$$

3.4 The filter problem for transients: finite linear operators

The filter problem is concerned with the determination of the linear operator which transforms the transient time series $x_t(t=0, 1, 2, \cdots, T)$ into the transient time series $v_t(t=0, 1, 2, \cdots, T)$. The time series x_t may be called the input, the signal plus noise, or the perturbed signal. The time series v_t may be called the output, the signal, or the desired information. Under the aperiodic functional scheme, the observational values of x_t $(t=0, 1, 2, \cdots T)$ and $v_t(t=0, 1, 2, \cdots T)$ are assumed to be known, and not all equal to zero.

In this section we shall consider the case in which the linear operator is required to have a finite number of coefficients, and more particularly, the same number of coefficients as the number of terms in the time series x_t. Thus the desired linear operation is represented by

$$v_t = \sum_{s=0}^{T} a_s x_{t-s}, \quad t = 0, 1, 2, \cdots, T. \quad (3.41)$$

This equation represents the system of simultaneous linear equations, given by

$$\begin{aligned} v_0 &= a_0 x_0 \\ v_1 &= a_0 x_1 + a_1 x_0 \\ v_2 &= a_0 x_2 + a_1 x_1 + a_2 x_0 \\ &\cdots\cdots\cdots\cdots \\ v_T &= a_0 x_T + a_1 x_{T-1} \\ &\quad + a_2 x_{T-2} + \cdots + a_T x_0, \quad (3.42) \end{aligned}$$

in which the operator coefficients $a_0, a_1, a_2, \cdots a_T$ are the unknowns. Without loss of generality, we may assume $x_0 \neq 0$, and consequently the determinant of this system, which is equal to x_0^T, does not vanish. As a result, the system of equations has a unique solution which yields the values $a_0, a_1, a_2, \cdots, a_T$ for the operator coefficients.

Under the aperiodic functional scheme we set $x_t=0$ for t less than zero and for t greater than T. As a result if we assume the equation

$$v_t = \sum_{s=0}^{T} a_s x_{t-s} \quad (3.43)$$

holds for all $t(-\infty < t < \infty)$, then we see that this equation specifies the values of v_t outside of the range $0 \leq t \leq T$, in which range the values of v_t were given in the original statement of the problem. In particular, this specification is

$$\begin{aligned} v_t &= 0, \quad t < 0, \\ v_t &= \sum_{s=0}^{T} a_s x_{t-s}, \text{ for} \\ t &= T+1, T+2, \cdots, 2T \\ v_t &= 0, \quad t > 2T. \quad (3.44) \end{aligned}$$

Consequently, for these values of v_t the operator equation (3.43) is valid for all integer values of t. Thus by multiplying each side of equation (3.43) by $e^{-i\omega t}$ and summing over t, we have

$$\begin{aligned} \sum_{t=-\infty}^{\infty} v_t e^{-i\omega t} &= \sum_{t=-\infty}^{\infty} e^{-i\omega t} \sum_{s=0}^{T} a_s x_{t-s} \\ &= \sum_{s=0}^{T} a_s e^{-i\omega s} \\ &\quad \cdot \sum_{t=-\infty}^{\infty} x_{t-s} e^{-i\omega(t-s)} \quad (3.45) \end{aligned}$$

which is

$$\sum_{t=0}^{2T} v_t e^{-i\omega t} = \sum_{s=0}^{T} a_s e^{-i\omega s} \sum_{n=0}^{T} x_n e^{-i\omega n} \quad (3.451)$$

Letting

$$V(\omega) = \sum_{t=0}^{2T} v_t e^{-i\omega t} \quad (3.452)$$

$$A(\omega) = \sum_{s=0}^{T} a_s e^{-i\omega s} \quad (3.453)$$

and

$$X(\omega) = \sum_{n=0}^{T} x_n e^{-i\omega n} \quad (3.454)$$

equation (3.45) becomes

$$V(\omega) = A(\omega) X(\omega) \quad (3.46)$$

which is

$$A(\omega) = \frac{V(\omega)}{X(\omega)} = \sum_{s=0}^{m} a_s e^{-i\omega s}. \quad (3.461)$$

The linear operator a_s $(s=0, 1, \cdots, m)$, determined from $X(\omega)$ and $V(\omega)$ by this equation, is not necessarily minimum delay.

In Figure 2, a symmetric smoothing type linear operator is given which contracts a symmetric Ricker wavelet (Ricker, 1953b) into one of lesser breadth. The respective Fourier transforms are shown to the right of these time functions.

3.5 The filter problem for transients: infinite linear operators

In this section we wish to consider the filtering problem, which is the transformation of the time

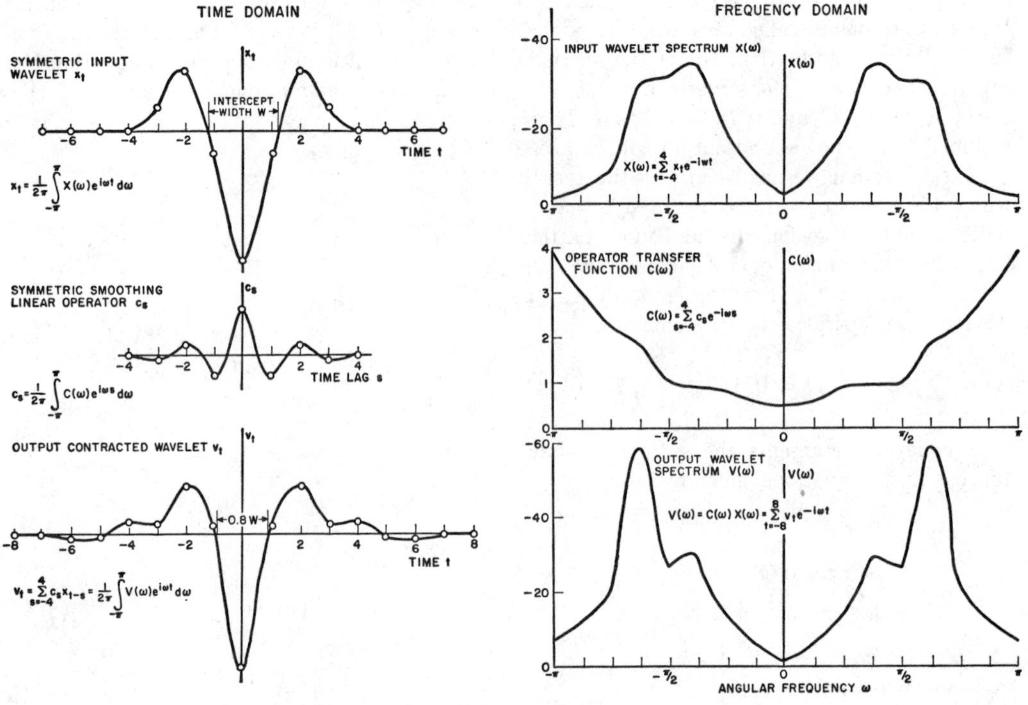

FIG. 2. Ricker wavelet contraction by linear operators.

series $x_t(t=0, 1, 2, \cdots T)$ into the time series $v_t(t=0, 1, 2, \cdots T)$ under the aperiodic hypothesis that

$$v_t = x_t = 0, \quad \text{for } t < 0, t > T. \quad (3.51)$$

That is, the values of v_t and x_t are specified for all values of t. Thus we wish to find the linear operator, with coefficients h_s, which we shall not restrict in number, such that

$$v_t = \sum_{s=-\infty}^{\infty} h_s x_{t-s}, \quad -\infty < t < \infty. \quad (3.52)$$

Since this equation represents an infinite system of simultaneous equations, that is, one equation for each value of t, this system in general will require an infinite number of unknowns h_s for a solution. That is, the linear operator is allowed to be a two-sided operator of infinite extent.

The formal solution for this linear operator may be found in terms of its transfer function

$$H(\omega) = \sum_{s=-\infty}^{\infty} h_s e^{-i\omega s}. \quad (3.53)$$

Equation (3.52) becomes

$$\sum_{t=-\infty}^{\infty} v_t e^{-i\omega t} = \sum_{t=-\infty}^{\infty} e^{-i\omega t} \sum_{s=-\infty}^{\infty} h_s x_{t-s}$$

$$= \sum_{s=-\infty}^{\infty} h_s e^{-i\omega s} \sum_{t=-\infty}^{\infty} x_{t-s} e^{-i\omega(t-s)} \quad (3.54)$$

which is

$$\sum_{t=0}^{T} v_t e^{-i\omega t} = \sum_{s=-\infty}^{\infty} h_s e^{-i\omega s} \sum_{n=0}^{T} x_n e^{-i\omega n}. \quad (3.541)$$

Letting

$$X(\omega) = \sum_{n=-\infty}^{\infty} x_n e^{-i\omega n} = \sum_{n=0}^{T} x_n e^{-i\omega n} \quad (3.542)$$

and

$$V(\omega) = \sum_{t=-\infty}^{\infty} v_t e^{-i\omega t} = \sum_{t=0}^{T} v_t e^{-i\omega t} \quad (3.543)$$

equation (3.541) becomes

$$V(\omega) = H(\omega) X(\omega). \quad (3.55)$$

Thus the formal solution, given by the transfer function of the desired linear operator, is

$$\sum_{s=-\infty}^{\infty} h_s e^{-i\omega s} = H(\omega) = \frac{V(\omega)}{X(\omega)}$$

$$= \frac{V\overline{X}}{X\overline{X}} \cdot \qquad (3.551)$$

The infinitely long two-sided linear operator with coefficients h_s will be stable provided that $X^{-1}(z)$ is expanded as the stable inverse of $X(z)$ as described in Section 2.7. The real part of $H(\omega)$ is

$$\mathrm{Re}[H(\omega)] = \frac{1}{2}[H + \overline{H}]$$

$$= \frac{1}{2}\left[\frac{V}{X} + \frac{\overline{V}}{\overline{X}}\right] = \frac{\overline{V}X + V\overline{X}}{2X\overline{X}} \qquad (3.552)$$

and the imaginary part

$$\mathrm{Im}[H(\omega)] = -\frac{i}{2}[H - \overline{H}]$$

$$= -\frac{i}{2}\left[\frac{V}{X} - \frac{\overline{V}}{\overline{X}}\right]$$

$$= i\,\frac{\overline{V}X - V\overline{X}}{2X\overline{X}} \cdot \qquad (3.553)$$

This result corresponds to that given by Smith (1954), obtained by requiring that the sum of squared errors

$$I = \sum_{t=0}^{T}\left(v_t - \sum_{s=-\infty}^{\infty} h_s x_{t-s}\right)^2$$

$$= \frac{1}{2\pi}\int_{-\pi}^{\pi} | V(\omega)$$

$$- H(\omega)X(\omega)|^2 d\omega \geq 0 \qquad (3.56)$$

be a minimum. We see that the linear operator thus found, with transfer function $H = V/X$ yields the minimum value

$$I_{\min} = \frac{1}{2\pi}\int_{-\pi}^{\pi}\left| V(\omega) - \frac{V(\omega)}{X(\omega)} X(\omega)\right|^2 d\omega$$

$$= \frac{1}{2\pi}\int_{-\pi}^{\pi} | V(\omega) - V(\omega)|^2 d\omega = 0 \qquad (3.561)$$

which is zero.

In other words, the linear operator h_s precisely transforms the input x_t into the signal v_t. Consequently, the addition of more information, say in the form of a second input time series y_t, cannot improve this transformation in the sense of reducing the error, which is already zero for all values of t. That is, let us consider the transformation

$$v_t = \sum_{s=-\infty}^{\infty} a_s x_{t-s} + \sum_{s=-\infty}^{\infty} b_s y_{t-s} \qquad (3.57)$$

where v_t, x_t, y_t are specified to be equal to zero for $t<0$ and $t>T$. The formal solution of this equation is given by the $A(\omega)$ and $B(\omega)$ for which the equation

$$V(\omega) = A(\omega)X(\omega) + B(\omega)Y(\omega) \qquad (3.571)$$

holds. This equation holds for $A(\omega)$ and $B(\omega)$ given by

$$A(\omega) = \gamma\frac{V(\omega)}{X(\omega)} \qquad (3.572)$$

$$B(\omega) = (1-\gamma)\frac{V(\omega)}{Y(\omega)}$$

where γ is any number. The solution given by equation (3.551) is the case for which $\gamma = 1$. Since γ is arbitrary there is no unique solution for the a_s and b_s in equation (3.57).

This condition is reflected in the fact that the coherency matrix, given by

$$\begin{bmatrix} \Phi_{xx}(\omega) & \Phi_{xy}(\omega) \\ \Phi_{yx}(\omega) & \Phi_{yy}(\omega) \end{bmatrix} = \begin{bmatrix} \overline{X}X & \overline{X}Y \\ \overline{Y}X & \overline{Y}Y \end{bmatrix} \qquad (3.58)$$

is singular; that is, its determinant is equal to zero. That is, since we have complete knowledge of v_t, and indeed may represent v_t in functional form in terms of x_t alone, as given by equation (3.52), we can introduce no new information about v_t into its representation in the form of an additional time series y_t.

3.6 Averaging and the probabilistic point of view

For practical purposes one may only utilize linear operators with a finite number of coefficients. As a result any linear operator with an infinite number of coefficients must be approximated by one with a finite number of coefficients. Thus, for example, we must approximate the infinite linear operator h_s $(-\infty < s < \infty)$ of the preceding section by a finite linear operator, say

h_s' $(-m \leq s \leq m)$, or, in other words, approximate the transfer function

$$H(\omega) = \sum_{s=-\infty}^{\infty} h_s e^{-i\omega s} = \frac{V(\omega)}{X(\omega)} \quad (3.551)$$

by the transfer function

$$H'(\omega) = \sum_{s=-m}^{m} h_s' e^{-i\omega s}. \quad (3.61)$$

Such an approximation procedure requires that a certain amount of information contained in the infinite linear operator h_s be lost in order to obtain the approximate finite linear operator h_s', and consequently we shall need to use some type of averaging process to carry out this approximation. One such averaging process is to require that the sum of squared errors

$$I = \sum_{t=-\infty}^{\infty} \left(v_t - \sum_{s=-m}^{m} h_s' x_{t-s} \right)^2$$

$$= \frac{1}{2\pi} \int_{-\pi}^{\pi} | V(\omega) - H'(\omega) X(\omega) |^2 d\omega \quad (3.62)$$

be a minimum. Because of equation (3.51), equation (3.62) becomes

$$I = \sum_{t=-m}^{T+m} \left(v_t - \sum_{s=-m}^{m} h_s' x_{t-s} \right)^2.$$

Setting the partial derivative of I with respect to h_τ' equal to zero, we obtain

$$-2 \sum_{t=-m}^{T+m} \left(v_t - \sum_{s=-m}^{m} h_s' x_{t-s} \right) x_{t-\tau} = 0$$

$$\text{for } \tau = -m, -m+1, \cdots, m \quad (3.63)$$

If we define $r_{\tau-s}$ and g_τ as

$$r_{\tau-s} = \sum_{t=-m}^{T+m} x_{t-s} x_{t-\tau} \quad \text{and}$$

$$g_\tau = \sum_{t=-m}^{T+m} v_t x_{t-\tau} \quad (3.64)$$

then the set of equations (3.63) may be written as

$$\sum_{s=-m}^{m} h_s' r_{\tau-s} = g_\tau$$

$$\text{for } \tau = -m, -m+1, \cdots, m. \quad (3.65)$$

The solution of these equations, called the *normal equations*, yields the required coefficients h_s' (for $s = -m, -m+1, \cdots, m$) of the approximate finite linear operator.

Thus, in many cases, even if one uses a purely functional approach it is necessary to develop methods of averaging in order to obtain certain desired goals. These methods of averaging may be carried out with respect to certain desirable criteria, where such criteria may be established and justified from a purely mathematical or functional point of view.

Although one may work with various averaging procedures with no recourse to the theory of probability, in many applications it is not until probability theory is introduced that certain procedures become meaningful in a physical sense. The theory of stationary time series, as conceived by Yule (1921, 1926, 1927), and established in full generality by Khintchine (1934), makes use of averaging procedures which were arrived at from the probability point of view. This theory has found many applications in pure and applied science. In fact, Wiener (1942) emphasizes the probabilistic or statistical point of view for engineering problems and applies the theory of stationary time series toward their solution.

Although seismic records are not stationary in the sense of Khintchine (1934), nevertheless one may treat sections of these records as being approximately stationary and consequently apply statistical methods to their analyses (Wadsworth, et al., 1953). Further discussion of this problem is given in Chapter VI. For numerical computational purposes discrete time series must be utilized, and so in the next two chapters we shall discuss the theory of discrete stationary time series.

CHAPTER IV

THEORY OF DISCRETE STATIONARY TIME SERIES

4.1 *Random processes*

Wadsworth et al. (1953) presented concepts from the theory of stationary time series. In this chapter and the next we extend this presentation to discrete stationary time series, and in particular we develop the concept of the predictive decomposition of stationary time series. For other more comprehensive presentations of the theory of stationary time series, the reader is referred to Wiener (1930), Khintchine (1934), Wold (1938), Kolmogorov (1941), Wiener (1942), and Doob (1953).

A discrete time series is a sequence of equidistant observations x_t which are associated with the discrete time parameter t. Without loss of generality we may take the spacing between each successive observation to be one unit of time, and thus we may represent the time series as

$$\cdots, x_{t-2}, x_{t-1}, x_t, x_{t+1}, x_{t+2}, \cdots \qquad (4.11)$$

where t takes on all integer values from minus infinity to infinity ($-\infty < t < \infty$). Thus all angular frequencies ω may be required to lie between $-\pi$ and π.

Any observational times series x_t ($-\infty < t < \infty$) may be considered as a realization of a so-called random process, or stochastic process, which is a mathematical abstraction defined with respect to a probability field. In many phenomena, the number of observational time series supplied by a random process is limited. It is often the case, especially in economic applications, that only one time series is generated by a random process. Such a case, nevertheless, is in full accord with the frequency interpretation of probability. (The scientific word "stochastic" is used in English as a synonym of "random"; hence the terms "random process" and "stochastic process" may be used interchangeably.)

4.2 *Stationary time series*

A time series is said to be stationary if the probabilities involved in the stochastic process are not tied down to a specific origin in time; that is, the probability of any event associated with the time t is equal to the probability of the corresponding event associated with the time $t+\tau$, where t and τ are any integer values.

For any stochastic process, one may form averages with respect to the statistical population or "ensemble" of realizations x_t for a fixed time t. Such averages are called ensemble averages or population averages, and we shall denote such an averaging process by the expectation symbol E. In particular the mean value $m = \mathrm{E}[x_t]$ and the variance $\sigma^2 = \mathrm{E}[(x_t - m)^2]$ of a stationary stochastic process are independent of time t. Likewise, the (unnormalized) autocorrelation coefficients

$$\phi(\tau) = \mathrm{E}[x_t x_{t+\tau}] \qquad (4.21)$$

are independent of t, and constitute an even function of the time lag τ, that is

$$\phi(\tau) = \phi(-\tau). \qquad (4.22)$$

Also we have

$$|\phi(\tau)| \leq \phi(0). \qquad (4.221)$$

The normalized autocorrelation function is defined to be

$$\phi(\tau) = \frac{\mathrm{E}[(x_t - m)(x_{t+\tau} - m)]}{\mathrm{E}[(x_t - m)]^2} \qquad (4.222)$$

so that

$$\phi(0) = 1, \quad |\phi(\tau)| \leq 1. \qquad (4.223)$$

In what follows we shall assume that the mean value m is equal to zero which we may do without loss of generality.

There is another type of average known as a time average or phase average in which the averaging process is carried out with respect to all values of time t for a fixed realization $x_t(-\infty < t < \infty)$ of the stochastic process. A stationary process is called an ergodic process if the ensemble averages and time averages are equal with probability one. As a result the autocorrelation of an ergodic process may be expressed as the time average

$$\phi(\tau) = \lim_{T\to\infty} \frac{1}{2T+1} \sum_{t=-T}^{T} x_{t+\tau} x_t. \qquad (4.23)$$

4.3 *The autocorrelation*

The autocorrelation function $\phi(\tau)$ is a nonnegative definite function, that is

$$\phi(\tau) = \phi(-\tau)$$

$$\sum_{j=1}^{N} \sum_{k=1}^{N} \phi(j-k) a_j a_k \geq 0$$

$$N = 1, 2, \cdots \qquad (4.31)$$

for every real set of $a_1, a_2, \cdots, a_N$. Thus, the autocorrelation matrix ($N = 1, 2, \cdots$)

$$\begin{bmatrix} \phi(0) & \phi(1) & \cdots & \phi(N) \\ \phi(-1) & \phi(0) & \cdots & \phi(N-1) \\ \cdot & \cdot & \cdots & \cdot \\ \phi(-N) & \phi(-N+1) & \cdots & \phi(0) \end{bmatrix} \qquad (4.32)$$

is symmetric, has its elements equal along its main diagonal and along all other diagonals, has nonnegative eigenvalues $\lambda_j \geq 0$ $(j=1, 2, \cdots, N)$, has a nonnegative determinant, and has a nonnegative definite quadratic form given by equation (4.31). The nonnegative definiteness of the autocorrelation follows from the inequality

$$\sum_{j=1}^{N}\sum_{k=1}^{N} \phi(j-k)a_j a_k = \sum_{j=1}^{N}\sum_{k=1}^{N} \phi(|j-k|)a_j a_k = \sum_{j=1}^{N}\sum_{k=1}^{N} \mathrm{E}[a_j a_k x_j x_k] = \mathrm{E}\left[\left(\sum_{j=1}^{N} a_j x_j\right)^2\right] \geq 0. \quad (4.33)$$

4.4 The spectrum

The property that the autocorrelation function $\phi(\tau)$ is nonnegative definite is equivalent to its representation by the Fourier transform

$$\phi(\tau) = \frac{1}{\pi}\int_0^{\pi} \cos \omega\tau \, d\Lambda(\omega) \quad (4.41)$$

where $\Lambda(\omega)$, called the integrated spectrum or the spectral distribution function, is a real monotone nondecreasing function of ω $(0 \leq \omega \leq \pi)$ with $\Lambda(0)=0$ and $\Lambda(\pi)=\pi$. This theorem (usually known as the Wiener-Khintchine theorem, used by Wiener, 1930) was first used in this setting by Khintchine (1934) in his development of the theory of continuous stationary time series.

The inversion formula expresses the integrated spectrum in terms of the autocorrelation, that is

$$\Lambda(\omega) = \omega + 2\sum_{\tau=1}^{\infty} \frac{\phi(\tau)}{\tau} \sin \omega\tau \qquad 0 \leq \omega \leq \pi. \quad (4.42)$$

Moreover, if

$$\sum_{\tau=0}^{\infty} |\phi(\tau)|$$

is convergent, then $\Lambda(\omega)$ will be absolutely continuous, with the continuous derivative

$$\Lambda'(\omega) = \frac{d\Lambda(\omega)}{d\omega} = \Phi(\omega) = 1 + 2\sum_{\tau=1}^{\infty} \phi(\tau) \cos \omega\tau = \sum_{\tau=-\infty}^{\infty} \phi(\tau) \cos \omega\tau. \quad (4.43)$$

In the remaining part of this thesis, we shall confine ourselves to stochastic processes for which the spectral distribution function $\Lambda(\omega)$ is absolutely continuous and for which $\sum|\phi(\tau)| < \infty$, unless it is otherwise stated. Wiener (1942) restricts himself to those processes where the spectral distribution function $\Lambda(\omega)$ is absolutely continuous.

The derivative $\Phi(\omega)=\Lambda'(\omega)$ is called the spectral density function, the power spectrum, or simply the spectrum. Since $\Phi(\omega)$ is the slope of a real monotone nondecreasing function $\Lambda(\omega)$, we have

$$\Phi(\omega) \geq 0 \quad (4.44)$$

and $\Phi(\omega)$ may be considered to represent the power density in the time series x_t $(-\infty < t < \infty)$. In order to have equal power at ω and $-\omega$, let us define $\Phi(\omega)$ to be an even function of ω, that is,

$$\Phi(-\omega) = \Phi(\omega), \quad -\pi < \omega \leq \pi. \quad (4.45)$$

Equation (4.41) thus becomes

$$\phi(\tau) = \frac{1}{\pi}\int_0^{\pi} \cos \omega\tau \, \Phi(\omega) d\omega = \frac{1}{2\pi}\int_{-\pi}^{\pi} e^{i\omega\tau}\Phi(\omega) d\omega \quad (4.46)$$

and, in particular, we have

$$\phi(0) = \frac{1}{2\pi}\int_{-\pi}^{\pi} \Phi(\omega) d\omega. \quad (4.461)$$

To show that the autocorrelation $\phi(\tau)$ as given by equation (4.46) is a nonnegative definite function, we let

$$H_N(\omega) = \sum_{k=1}^{N} a_k e^{-i\omega k} \quad (4.47)$$

where a_k $(k=1, 2, \cdots, N)$ is any arbitrary set of real numbers. Then the quadratic form of the autocorrelation matrix is

$$\sum_{j=1}^{N}\sum_{k=1}^{N}\phi(j-k)a_ja_k$$

$$= \int_{-\pi}^{\pi}\sum_{j=1}^{N}\sum_{k=1}^{N}a_ja_ke^{i\omega(j-k)}\Phi(\omega)d\omega \quad (4.48)$$

which is equal to

$$\int_{-\pi}^{\pi}H_N(\omega)\overline{H_N(\omega)}\Phi(\omega)d\omega$$

$$= \int_{-\pi}^{\pi}|H_N(\omega)|^2\Phi(\omega)d\omega \geq 0,$$

$$N = 1, 2, \cdots. \quad (4.49)$$

Thus equation (4.31) is verified for $\phi(\tau)$ given by equation (4.46).

The important role played by the harmonic analysis of a stationary time series is brought out by the Spectral Representation Theorem due to Harald Cramér (1942) and others. The theorem allows the time series x_t to be represented by a stochastic integral which involves the harmonic components of the time series. For a statement and a discussion of this theorem, the reader is referred to Doob (1953, p. 481).

4.5 Processes with white spectra

A process is said to have a white spectrum if its power spectrum has a constant value, that is

$$\Phi(\omega) = \text{constant}, \qquad -\pi \leq \omega \leq \pi. \quad (4.51)$$

In this section we consider two types of processes which have white spectra, namely, the purely random process and the mutually uncorrelated process. For a further discussion of these processes, see Doob (1953) and Wold (1953). As a matter of terminology, one should distinguish between the "purely random process" and a "random process." The "purely random" process is defined in this section. On the other hand, as we have seen, a "random" process designates any process which generates one or more observational time series, and such processes range from purely random processes to nonrandom or deterministic processes.

A realization from a purely random process is the time series $\xi_t(-\infty<t<\infty)$ where each ξ_t is an independent random variate from a single probability distribution function. Therefore the joint distribution functions of the ξ_t are simply products of this probability distribution function. Consequently, we have

$$E[\xi_t\xi_s] = E[\xi_t]E[\xi_s] \qquad s \neq t. \quad (4.52)$$

Such a process is stationary and ergodic. In order to normalize the process so that it will have zero mean and unit variance, we let

$$E[\xi_t] = 0, \qquad E[\xi_t^2] = 1. \quad (4.53)$$

The autocorrelation function is then given by

$$\phi(0) = E[\xi_t^2] = 1$$

$$\phi(\tau) = E[\xi_t\xi_{t+\tau}] = E[\xi_t]E[\xi_{t+\tau}] = 0$$

$$\tau = \pm 1, \pm 2, \pm 3, \cdots. \quad (4.531)$$

An alternative assumption as to the nature of the ξ_t leads to the definition of the so-called mutually uncorrelated process. Instead of assuming that the ξ_t and $\xi_{t+\tau}$ are independent, we assume for a mutually uncorrelated process that they are uncorrelated in pairs, that is

$$E[\xi_t\xi_s] = E[\xi_t]E[\xi_s] \qquad s \neq t \quad (4.54)$$

which is the same as equation (4.52). That is, independent random variables are uncorrelated, but the converse is not necessarily true. Again we shall normalize the ξ_t as in equation (4.53), in which case the uncorrelated random variables ξ_t are called orthogonal random variables. The orthogonality is illustrated by

$$E[\xi_t^2] = 1,$$

$$E[\xi_t\xi_s] = 0, \qquad t \neq s. \quad (4.55)$$

In what follows, we shall assume all uncorrelated random variables are normalized in this manner (equation 4.53), so alternatively we may call them orthogonal random variables. Since equation (4.531) holds for a (normalized) uncorrelated process we see that this process has the same autocorrelation coefficients as the purely random process. Also, an uncorrelated process is stationary and ergodic.

The spectral distribution function, equation (4.42), of a purely random process or of an uncorrelated process is given by

$$\Lambda(\omega) = \omega, \qquad 0 \leq \omega \leq \pi \quad (4.56)$$

and the power spectrum, by equation (4.43), is a constant given by

$$\Phi(\omega) = 1, \qquad -\pi < \omega \leq \pi. \quad (4.57)$$

These processes therefore have white spectra, and the ξ_t may be called white noise (Wiener, 1930).

In summary, then, the purely random process and the uncorrelated process both have white spectra and have autocorrelation coefficients which vanish except for lag zero.

4.6 Processes of moving summation

An important theorem states that a time series with an absolutely continuous spectral distribution function is generated by a process of moving summation, and conversely. In this section we shall define what is meant by a process of moving summation, and then indicate in an heuristic way why this theorem holds. For a rigorous proof, the reader is referred to Doob (1953).

For the fixed realization

$$\cdots, \xi_{t-1}, \xi_t, \xi_{t+1}, \xi_{t+2}, \cdots \quad (4.61)$$

of a purely random process or of a mutually uncorrelated process, the corresponding fixed realization of a process of moving summation is

$$\cdots, x_{t-1}, x_t, x_{t+1}, x_{t+2}, \cdots \quad (4.62)$$

where

$$x_n = \sum_{k=-\infty}^{\infty} c_k \xi_{n-k}$$

$$n = t, t \pm 1, t \pm 2, \cdots. \quad (4.621)$$

Since two random variables are uncorrelated if they are independent, whereas the converse is not always true, in what follows we shall impose only the weaker restriction on the ξ_t and thus only assume they are mutually uncorrelated (i.e. orthogonal) in the definition of the process of moving summation (see Section 4.5). In particular, we assume

$$E[\xi_t] = 0, \qquad E[\xi_t^2] = 1,$$

$$E[\xi_t \xi_s] = E[\xi_t]E[\xi_s] = 0 \quad \text{for } t \neq s. \quad (4.63)$$

The mean of the x_t process is

$$E[x_t] = \sum_{k=-\infty}^{\infty} c_k E[\xi_{t-k}] = 0 \quad (4.64)$$

and the variance is

$$E[x_t^2] = \sum_{k=-\infty}^{\infty} c_k^2 E[\xi_{t-k}^2]$$

$$= \sum_{k=-\infty}^{\infty} c_k^2. \quad (4.641)$$

The autocorrelation coefficients are

$$\phi(\tau) = E[x_t x_{t+\tau}] = \sum_{t=-\infty}^{\infty} c_{t-\tau} c_t$$

$$= \sum_{t=-\infty}^{\infty} c_t c_{t+\tau}. \quad (4.642)$$

We now wish to indicate why a process of moving summation has an absolutely continuous spectral distribution function. Let x_t be a process of moving summation given by equation (4.621). Define $F(\omega)$ by

$$F(\omega) = \sum_{k=-\infty}^{\infty} c_k e^{-i\omega k} \quad (4.65)$$

which is the transfer function of the infinite smoothing operator c_k. Then, by equation (4.642) we have

$$\phi(\tau) = \sum_{j=-\infty}^{\infty} c_j c_{j-\tau}$$

$$= \frac{1}{2\pi} \int_{-\pi}^{\pi} d\omega \sum_{j=-\infty}^{\infty} c_{j-\tau}$$

$$\cdot e^{i\omega j} \sum_{k=-\infty}^{\infty} c_k e^{-i\omega k} \quad (4.651)$$

since

$$\int_{-\pi}^{\pi} e^{i\omega j} e^{-i\omega k} d\omega = \begin{cases} 2\pi & j = k \\ 0 & j \neq k \end{cases} \quad (4.652)$$

for integer j and k, and by letting $j - \tau = l$, we have

$$\phi(\tau) = \frac{1}{2\pi} \int_{-\pi}^{\pi} e^{i\omega\tau} \sum_{l=-\infty}^{\infty} c_l e^{i\omega l}$$

$$\cdot \sum_{k=-\infty}^{\infty} c_k e^{-i\omega k} d\omega. \quad (4.653)$$

Then, using equation (4.65) we have

$$\phi(\tau) = \frac{1}{2\pi}\int_{-\pi}^{\pi} F(\omega)\overline{F(\omega)}e^{i\omega\tau}d\omega$$
$$= \frac{1}{2\pi}\int_{-\pi}^{\pi} | F(\omega)|^2 e^{i\omega\tau}d\omega \quad (4.654)$$

which is in the form of equation (4.46) with the power spectrum

$$\Phi(\omega) = | F(\omega)|^2. \quad (4.655)$$

Thus the spectral distribution function

$$\Lambda(\omega) = \int_0^{\omega} \Phi(u)du \quad (4.656)$$

is absolutely continuous.

Conversely, any process with absolutely continuous spectral distribution is a process of moving summation, and in this paragraph we wish to indicate some reasons for this theorem. Because of equation (4.44), which states that the spectrum $\Phi(\omega)$ is nonnegative for every value of ω, we may set

$$\Phi(\omega) = | F(\omega)|^2 = F(\omega)\overline{F(\omega)} \quad (4.66)$$

where $F(\omega)$ is the Fourier series of any square root of $\Phi(\omega)$. Let us represent this Fourier series by

$$F(\omega) = \sum_{k=-\infty}^{\infty} c_k e^{-i\omega k}. \quad (4.661)$$

Using the coefficients c_k we may define the process of moving summation

$$x_t = \sum_{k=-\infty}^{\infty} c_k \xi_{t-k}. \quad (4.662)$$

The autocorrelation of the process (4.662) will be given by equation (4.642). The Fourier transform of this autocorrelation function gives the power spectrum of the process (4.662) which is the same as the original power spectrum in equation (4.66). Thus the process x_t given by equation (4.662) is a process of moving summation which has the given power spectrum (4.66).

Hence, for the process of moving summation represented by

$$x_t = \sum_{k=-\infty}^{\infty} c_k \xi_{t-k} \quad (4.621)$$

the c_k represents a linear operator, and the ξ_{t-k} represents white noise. The transfer function is given by $F(\omega)$ in equation (4.65), and the power transfer function is then the power spectrum of the process, that is

$$\Phi(\omega) = | F(\omega)|^2. \quad (4.655)$$

Thus the time series x_t is the output of a linear system, with power transfer function $\Phi(\omega)$, into which white noise is passed. Since the c_k may be an infinite two-sided operator, this system need not necessarily be realizable or minimum-delay. In Section 5.1 we shall see that a unique realizable and minimum-delay operator may be found, which leads to the predictive decomposition of stationary time series given in Section 5.2.

Finally, let us note that processes of moving summation are ergodic; for a proof the reader is referred to Doob (1953).

CHAPTER V

THE PREDICTIVE DECOMPOSITION OF STATIONARY TIME SERIES

5.1 The factorization of the spectrum

In the preceding chapter we saw that the power spectrum of a time series x_t may be regarded as a power transfer function of a linear system into which white noise ξ_t is passed in order to obtain the time series x_t as output. The gain characteristic of this linear system is $[\Phi(\omega)]^{\frac{1}{2}}$. The problem of the factorization of the spectrum is the problem of determining the phase characteristic so that the system is physically realizable and stable, with minimum-phase characteristic for the gain $[\Phi(\omega)]^{\frac{1}{2}}$. Thus the transfer function of the desired physically realizable minimum-phase network may be given by

$$B(\omega) = | B(\omega) | e^{i\theta(\omega)} = [\Phi(\omega)]^{1/2} e^{i\theta(\omega)} \quad (5.11)$$

where $[\Phi(\omega)]^{\frac{1}{2}}$ is the gain, and $\theta(\omega)$ represents the desired minimum-phase characteristic.

Kolmogorov (1939) gave the general solution of this factorization problem. A rigorous exposition of his results may be found in Doob (1953), and in this section we give an heuristic exposition.

Let us first turn our attention to the properties of a realizable, stable linear system with minimum phase-shift characteristic. As we have seen in Section 2.7, the conditions that the transfer function be physically realizable and minimum phase is that it may be expressed as

$$B(\omega) = \sum_{s=0}^{\infty} b_s e^{-i\omega s} \tag{5.12}$$

where

$$b_s = 0 \qquad \text{for } s < 0 \tag{5.121}$$

$$\sum_{s=0}^{\infty} b_s^2 < \infty \tag{5.122}$$

and

$$B(\lambda) = \sum_{s=0}^{\infty} b_s e^{-i\lambda s}, \qquad \lambda = \omega + i\sigma \tag{5.123}$$

has no singularities or zeros in the lower half λ-plane ($\sigma < 0$). Under the transformation $z = e^{-i\lambda}$ (see Figure 1), this last condition becomes that

$$B(z) = \sum_{s=0}^{\infty} b_s z^s \tag{5.124}$$

have no singularities or zeros for $|z| \leq 1$. Under these conditions, $\log B(z)$ will be analytic for $|z| \leq 1$, and consequently has the power series representation

$$\log B(z) = \sum_{t=0}^{\infty} \beta_t z^t \quad \text{for } |z| \leq 1 \tag{5.125}$$

and, as $|z|$ approaches 1, this series converges to

$$\begin{aligned} \log B(\omega) &= \sum_{t=0}^{\infty} \beta_t e^{-i\omega t} \\ &= \beta_0 + \sum_{t=1}^{\infty} \beta_t \cos \omega t \\ &\quad - i \sum_{t=1}^{\infty} \beta_t \sin \omega t. \end{aligned} \tag{5.126}$$

Let us now turn our attention to the power spectrum $\Phi(\omega)$. The spectrum $\Phi(\omega)$ is a real function of ω such that

$$\begin{aligned} \Phi(\omega) &= \Phi(-\omega) \\ \Phi(\omega) &\geq 0, \quad -\pi < \omega \leq \pi. \end{aligned} \tag{5.13}$$

Moreover, the following conditions on $\Phi(\omega)$ must be satisfied:

$$\Phi(\omega) = 0 \tag{5.131}$$

at most on a set of Lebesgue measure zero;

$$\int_{-\pi}^{\pi} \Phi(\omega) d\omega < \infty \tag{5.132}$$

and

$$\int_{-\pi}^{\pi} \log \Phi(\omega) d\omega > -\infty. \tag{5.133}$$

Under these conditions, $\log [\Phi(\omega)]^{\frac{1}{2}}$, which is an even real function of ω, may be expressed in the real Fourier cosine series

$$\log [\Phi(\omega)]^{\frac{1}{2}} = \sum_{t=-\infty}^{\infty} \alpha_t \cos \omega t \tag{5.134}$$

where the Fourier coefficients α_t are given by

$$\begin{aligned} \alpha_t &= \frac{1}{2\pi} \int_{-\pi}^{\pi} \cos \omega t \log [\Phi(\omega)]^{1/2} d\omega \\ &= \frac{1}{2\pi} \int_{0}^{\pi} \cos \omega t \log \Phi(\omega) d\omega. \end{aligned} \tag{5.135}$$

From this equation we see that α_t is an even real function of t, that is

$$\alpha_t = \alpha_{-t}. \tag{5.136}$$

Consequently the Fourier expansion (5.134) of $\log [\Phi(\omega)]^{\frac{1}{2}}$ becomes

$$\log [\Phi(\omega)]^{1/2} = \alpha_0 + 2 \sum_{t=1}^{\infty} \alpha_t \cos \omega t. \tag{5.137}$$

Equation (5.11), which gives the transfer function $B(\omega)$ for the desired minimum-phase network, is

$$\begin{aligned} B(\omega) &= |B(\omega)| e^{i\theta(\omega)} \\ &= [\Phi(\omega)]^{1/2} e^{i\theta(\omega)} \end{aligned} \tag{5.11}$$

where $[\Phi(\omega)]^{\frac{1}{2}}$ is the gain, and $\theta(\omega)$ represents the minimum-phase characteristic. By taking the logarithm of each side of this equation, we have

$$\log B(\omega) = \log [\Phi(\omega)]^{1/2} + i\theta(\omega) \tag{5.14}$$

which, by equation (5.137) is

$$\begin{aligned} \text{og } B(\omega) &= \alpha_0 + 2 \sum_{t=1}^{\infty} \alpha_t \cos \omega t \\ &\quad + i\theta(\omega). \end{aligned} \tag{5.141}$$

Now equation (5.126) gives an expression for $\log B(\omega)$ which was derived from the knowledge that $\log B(z)$ be analytic for $|z| \le 1$, whereas equation (5.141) gives an expression for $\log B(\omega)$ derived from the knowledge that the gain $|B(\omega)|$ be equal to $[\Phi(\omega)]^{\frac{1}{2}}$. Setting these two equations equal to each other, we have

$$\log B(\omega) = \beta_0 + \sum_{t=1}^{\infty} \beta_t \cos \omega t - i \sum_{t=1}^{\infty} \beta_t \sin \omega t \qquad (5.126)$$

$$= \alpha_0 + \sum_{t=1}^{\infty} 2\alpha_t \cos \omega t + i\theta(\omega). \qquad (5.141)$$

We therefore have

$$\text{Re}\,[\log B(\omega)] = \log\,[\Phi(\omega)]^{1/2} = \beta_0 + \sum_{t=1}^{\infty} \beta_t \cos \omega t = \alpha_0 + \sum_{t=1}^{\infty} 2\alpha_t \cos \omega t \qquad (5.15)$$

so that

$$\beta_0 = \alpha_0, \qquad \beta_t = 2\alpha_t \qquad \text{for } t = 1, 2, 3, \cdots \qquad (5.151)$$

where α_t $(t=0, 1, 2, \cdots)$ is given by equation (5.135). We also have

$$\text{Im}\,[\log B(\omega)] = \theta(\omega) = -\sum_{t=1}^{\infty} \beta_t \sin \omega t \qquad (5.16)$$

which, by equation (5.151), is

$$\theta(\omega) = -2 \sum_{t=1}^{\infty} \alpha_t \sin \omega t. \qquad (5.161)$$

This equation expresses the minimum-phase characteristic $\theta(\omega)$ in terms of the α_t $(t=1, 2, \cdots)$, which are computed from knowledge of the power spectrum $\Phi(\omega)$ by means of equation (5.135).

As a result, the operator coefficients b_s may be determined in the following manner. Equations (5.12) and (5.11) give

$$B(\omega) = \sum_{s=0}^{\infty} b_s e^{-i\omega s} = [\Phi(\omega)]^{1/2} e^{i\theta(\omega)} \qquad (5.17)$$

which, because of equations (5.15) and (5.161), yields

$$\log B(\omega) = \text{Re}\,[\log B(\omega)] + i\,\text{Im}\,[\log B(\omega)] = \log\,[\Phi(\omega)]^{1/2} + i\theta(\omega) = \alpha_0 + \sum_{t=1}^{\infty} 2\alpha_t \cos \omega t - i \sum_{t=1}^{\infty} 2\alpha_t \sin \omega t = \alpha_0 + 2 \sum_{t=1}^{\infty} \alpha_t e^{-i\omega t}. \qquad (5.171)$$

Since

$$B(\omega) = e^{\log B(\omega)} \qquad (5.172)$$

and using equation (5.12) we have

$$B(\omega) = \sum_{s=0}^{\infty} b_s e^{-i\omega s} = \exp\left[\alpha_0 + 2 \sum_{t=1}^{\infty} \alpha_t e^{-i\omega t}\right] \qquad (5.173)$$

Letting $\lambda = \omega + i\sigma$ and making the substitution $z = e^{-i\lambda}$, we have

$$B(z) = \sum_{s=0}^{\infty} b_s z^s = \exp\left[\alpha_0 + 2 \sum_{t=1}^{\infty} \alpha_t z^t\right] \qquad |z| \le 1. \qquad (5.174)$$

By means of this equation, we may solve for the linear operator b_s in terms of the α_t. In particular, we have from equations (5.174) and (5.135) that

$$e^{\alpha_0} = b_0 = \exp\left[\frac{1}{2\pi} \int_0^{\pi} \log \Phi(\omega) d\omega\right] > 0. \qquad (5.175)$$

Therefore Kolmogorov (1939) shows that the power spectrum $\Phi(\omega)$ may be factored in the following manner:

$$\Phi(\omega) = |B(\omega)|^2 = \left| \sum_{t=0}^{\infty} b_t e^{-i\omega t} \right|^2 \qquad (5.181)$$

where the linear operator b_t may be determined from

$$b_t = 0, \qquad \text{for } t < 0 \qquad (5.121)$$

$$b_0 = \exp\left[\frac{1}{2\pi}\int_0^{\pi} \log \Phi(\omega)d\omega\right] > 0 \qquad (5.175)$$

and

$$\sum_{s=1}^{\infty} b_s z^s = b_0 \exp\left[2\sum_{t=1}^{\infty} \alpha_t z^t\right]. \qquad (5.182)$$

The transfer function of the linear operator b_t given by

$$B(\omega) = \sum_{t=0}^{\infty} b_t e^{-i\omega t} \qquad (5.12)$$

has gain

$$|B(\omega)| = [\Phi(\omega)]^{1/2} \qquad (5.183)$$

and minimum phase $\theta(\omega)$ given by equation (5.161). The transfer function $B(\omega)$ is the factor of the spectrum. Since

$$b_t = 0 \qquad \text{for } t < 0 \qquad (5.121)$$

$$b_0 > 0 \qquad (5.175)$$

$$\sum_{t=0}^{\infty} b_t^2 < \infty \qquad (5.122)$$

$$\sum_{t=0}^{\infty} |b_t| < \infty \qquad (5.184)$$

and

$$\sum_{t=0}^{\infty} b_t z^t \neq 0, \quad \text{for } |z| \leq 1, \quad (5.124)$$

the linear operator $b_0, b_1, b_2, \cdots$, is physically realizable and minimum-delay.

Kolmogorov (1939) normalizes the b_t so that $b_0 = 1$. That is, he gives

$$B(\omega) = \exp\left[\frac{1}{2\pi}\int_0^{\pi} \log \Phi(\omega)d\omega\right] \cdot \sum_{s=0}^{\infty} b_s e^{-i\omega s} \quad (b_0 = 1)$$

$$= \exp\left[\frac{1}{2\pi}\int_0^{\pi} \log \Phi(\omega)d\omega\right] \cdot (1 + b_1 e^{-i\omega} + b_2 e^{-2i\omega} + \cdots) \qquad (5.19)$$

where

$$\log \Phi(\omega) = 2\alpha_0 + 4\alpha_1 \cos \omega + 4\alpha_2 \cos 2\omega + \cdots \qquad (5.191)$$

and

$$\exp[2(\alpha_1 z + \alpha_2 z^2 + \cdots)] = 1 + b_1 z + b_2 z^2 + \cdots. \qquad (5.192)$$

We shall refer to $B(\omega)$ given in equation (5.181) as the minimum-delay factor of the power spectrum. The Kolmogorov method given in this section represents the most general method of factoring the spectrum in order to obtain the minimum-delay factor. As we shall see in Section 5.5, the Féjer factorization method of Section 2.8 can be used in the case of processes with rational spectra to obtain the minimum-delay factor. In practical work, the solution of the least-squares normal equations (6.261) can be used to find the inverse wavelet a_t which by (6.262) yields b_t, as described in Section 6.2.

5.2 *The predictive decomposition theorem*

In the preceding section we have seen that the power spectrum $\Phi(\omega)$ may be factored in the following manner:

$$\Phi(\omega) = B(\omega)\overline{B(\omega)} = |B(\omega)|^2 = \left|\sum_{s=0}^{\infty} b_s e^{-i\omega s}\right|^2 \qquad (5.181)$$

where

$$b_s = 0, \quad s < 0 \qquad (5.121)$$

$$b_0 > 0 \qquad (5.175)$$

$$\sum_{s=0}^{\infty} b_s^2 < \infty \qquad (5.122)$$

$$\sum_{s=0}^{\infty} |b_s| < \infty \qquad (5.184)$$

and

$$B(\lambda) = \sum_{s=0}^{\infty} b_s e^{-i\lambda s}, \quad \lambda = \omega + i\sigma \qquad (5.123)$$

has no singularities or zeros in the lower half λ plane ($\sigma < 0$). In other words, the linear operator represented by $b_0, b_1, b_2, \cdots$, is realizable and minimum-delay, and its transfer function $B(\omega)$ has minimum-phase characteristic.

In Section 4.6 we saw that a time series with an absolutely continuous spectral distribution function may be represented by a process of moving summation. We see that equation (5.181) may be used in place of equations (4.66) of Section 4.6. In other words, we may replace the linear operator c_t of Section 4.6 by the realizable and minimum-delay operator b_t of Section 5.1. Thus, the process of moving summation is given by

$$x_t = \sum_{s=0}^{\infty} b_s \xi_{t-s} = b_0\xi_t + b_1\xi_{t-1} + b_2\xi_{t-2} + \cdots \tag{5.21}$$

which replaces equation (4.662) of Section 4.6. In this equation ξ_t $(-\infty<t<\infty)$ represents a realization from a mutually uncorrelated process, the x_t is the time series with power spectrum $\Phi(\omega)$, and $b_0, b_1, b_2, \cdots$, is the realizable and minimum-delay operator determined as in Section 5.1. More particularly, the variables $\xi_t(-\infty<t<\infty)$ have zero mean $E(\xi_t)=0$, unit variance $E(\xi_t^2)=1$, and are mutually uncorrelated, that is $E[\xi_t\xi_s]=0$ for $t\neq s$, and consequently they have a white spectrum (see Section 4.5).

That all nondeterministic stationary processes with absolutely continuous spectral distribution functions may be represented in the form (5.21) is a special case of the more general Decomposition Theorem of Herman Wold (1938).[5] That is, the Wold Decomposition Theorem is not restricted to processes with absolutely continuous spectral distribution functions, nor is the realizable operator b_t required to the minimum-delay. The statement of the Wold Decomposition Theorem is: Given a nondeterministic stationary process y_t $(-\infty<t<\infty)$ with discrete time parameter t, then y_t $(-\infty<t<\infty)$ allows the decomposition

$$y_t = x_t + z_t$$

where the components x_t and z_t are two mutually uncorrelated stationary processes with the following properties:

(1) z_t is deterministic;

(2) x_t is nondeterministic with an absolutely continuous spectral distribution function (that is, x_t is nondeterministic with no deterministic component);

(3) $x_t=b_0\xi_t+b_1\xi_{t-1}+b_1\xi_{t-2}+\cdots$ where the components $(-\infty<t<\infty)$ have the following properties:

A. The variables ξ_t have zero mean $E[\xi_t]=0$; unit variance $E[\xi_t^2]=1$; and are mutually uncorrelated $E[\xi_s\xi_t]=0$ for $s\neq t$.
B. $b_0>0$ and $b_0^2+b_1^2+b_2^2+\cdots<\infty$
C. $E[\xi_s z_t]=0$ for all s, t.

Let us now state the Predictive Decomposition Theorem, which may be described as the minimum-delay specialization of the Wold Decomposition Theorem. Given a nondeterministic stationary process $x_t(-\infty<t<\infty)$ with discrete time parameter t, suppose x_t $(-\infty<t<\infty)$ has an absolutely continuous spectral distribution. Then $x_t(-\infty<t<\infty)$ allows the decomposition

$$x_t = b_0\xi_t + b_1\xi_{t-1} + b_2\xi_{t-2} + \cdots \tag{5.21}$$

where the components $(-\infty<t<\infty)$ have the following properties:

A. The variables ξ_t have zero mean $E[\xi_t]=0$; unit variance $E[\xi_t^2]=1$; and are mutually uncorrelated $E[\xi_s\xi_t]=0$ for $s\neq t$.
B. $b_0>0$ and $b_0^2+b_1^2+b_2^2+\cdots<\infty$.
C. The operator $b_0, b_1, b_2, \cdots$, is minimum-delay.

The Predictive Decomposition Theorem, as expressed by equation (5.21), renders the time series x_t in terms of a realizable and minimum-delay operator $b_0, b_1, b_2, \cdots$, operating on the present and past values $\xi_t, \xi_{t-1}, \xi_{t-2}, \cdots$, of a realization of a mutually uncorrelated process. Because the operator $b_0, b_1, b_2, \cdots$, is minimum-delay, there exists an inverse operator[6] $a_0, a_1, a_2, \cdots$, which is also realizable and minimum-delay (see Section 2.7). As a result ξ_t can be given in terms of the inverse operator $a_0, a_1, a_2, \cdots$, operating on the present and past values $x_t, x_{t-1}, x_{t-2}, \cdots$, that is

$$\xi_t = a_0x_t + a_1x_{t-1} + a_2x_{t-2} + \cdots.$$

Let us now consider the Predictive Decomposition Theorem in the language of the engineer

[5] Recent discussions of the Wold Decomposition Theorem are given in Wold (1959, p. 356), Hannan (1960, p. 21), Whittle (1963, p. 23), and Wold (1965, p. 61) (the editors).

[6] The inverse operator must in some cases be interpreted as a generalized function (see Robinson, 1963) (the editors).

(Bode and Shannon, 1950). The nonrandom or deterministic elements of a stochastic process may be represented by a physically realizable and stable electric or mechanical network or filter. This filter has minimum-phase characteristic. The time function $b_t(t=0, 1, 2, 3, \cdots)$ is equal to the output obtained from the filter in response to a unit impulse impressed upon the filter at time $t=0$. That is, the linear operator b_t is the impulsive response of the filter, and we shall call it the response function of the stochastic process.

The random or nondeterministic elements of the stochastic process are represented by the $\xi_t(-\infty<t<\infty)$, which may be considered to be the mutually uncorrelated impulses of wideband resistance noise or "white" noise. The time series $x_t(-\infty<t<\infty)$ is the response of the filter to the white noise input $\xi_s(-\infty<s\leq t)$. That is $\xi_s(-\infty<s\leq t)$ may be regarded as an impulse of strength ξ_s, which will produce a response $\xi_s b_{t-s}$ at the subsequent time t. By adding the contributions of all the impulses $\xi_s(-\infty<s\leq t)$, we obtain the total response, which is the time series x_t:

$$x_t = \sum_{s=t}^{-\infty} \xi_s b_{t-s}. \qquad (5.22)$$

Letting $t-s=n$, we have

$$x_t = \sum_{n=0}^{\infty} b_n \xi_{t-n} = b_0\xi_t + b_1\xi_{t-1} + b_2\xi_{t-2} + \cdots \qquad (5.21)$$

which is the predictive decomposition.

Since the impulsive response of the filter is given by $b_t(t\geq 0)$ with $b_t=0$ for $t<0$, its transfer function is the Fourier transform

$$B(\omega) = \sum_{s=0}^{\infty} b_s e^{-i\omega s} \qquad (5.23)$$

which has gain

$$|B(\omega)| = [\Phi(\omega)]^{1/2} \qquad (5.231)$$

and has minimum-phase characteristic, given by equation (5.161). The power spectrum of the ξ_t, which is equal to 1 $(-\pi<\omega\leq\pi)$, multiplied by the power transfer function of the filter, which is $|B(\omega)|^2$, yields the power spectrum $\Phi(\omega)$ of the time series x_t. The transfer function $B(\omega)$ is the minimum-delay factor of the power spectrum $\Phi(\omega)$.

We see that the Predictive Decomposition Theorem states that any stationary time series (with an absolutely continuous spectral distribution) can be considered to be composed of many overlapping pulses, or wavelets, or responses, all with the same minimum-delay shape b_n, where $b_n=0$ for $n<0$. The arrival times of these wavelets, and their relative weighting, are given by the impulses ξ_{t-n}. The response function b_n, which is the shape of these wavelets, reflects the dynamics of the process, whereas the mutually uncorrelated impulses ξ_{s-t} reflects the statistical character of the process.

The autocorrelation function of the stochastic process is

$$\begin{aligned}\phi(\tau) &= \mathrm{E}[x_t x_{t+\tau}] \\ &= \mathrm{E}\left[\sum_{s=0}^{\infty} b_s \xi_{t-s} \sum_{r=0}^{\infty} b_r \xi_{t+\tau-r}\right] \\ &= \sum_{s=0}^{\infty} b_s b_{s+\tau}. \end{aligned} \qquad (5.232)$$

Thus the autocorrelation function of the wavelet b_s (where $b_s=0$ for $s<0$) is the same as the autocorrelation function of the time series x_t. As a result the energy spectrum of the wavelet (see Section 3.3) is the same function as the power spectrum of the time series x_t, a fact of which we made use in Section 5.1, where we determined the shape of the wavelet b_n from the power spectrum $\Phi(\omega)$ of the time series x_t.

Since the filter is realizable and minimum-delay, there exists the inverse filter which is also realizable and minimum-delay. Let the response function of this inverse filter be $a_t(t\geq 0)$ with $a_t=0$ for $t<0$, so that its transfer function is

$$A(\omega) = B^{-1}(\omega) = \sum_{s=0}^{\infty} a_s e^{-i\omega s} = \frac{1}{B(\omega)} = \frac{1}{\sum_{s=0}^{\infty} b_s e^{-i\omega s}} \qquad (5.24)$$

so that

$$B^{-1}(\omega)B(\omega) = \sum_{s=0}^{\infty} a_s e^{-i\omega s} \cdot \sum_{s=0}^{\infty} b_s e^{-i\omega s} = 1. \qquad (5.241)$$

The relationship of the inverse filter, $A(\omega)$ with the response a_t, to the filter $B(\omega)$ with the response b_t, both of which are realizable and minimum-delay, are given in equations (2.78) through (2.797) of Section 2.7. In Section 2.7 the response functions a_t and b_t are referred to as linear operator coefficients.

Accordingly, the white noise $\xi_t(-\infty < t < \infty)$ is the total response at t of the inverse filter to the input $x_s(-\infty < s \leq t)$. That is, the $x_s(-\infty < s \leq t)$ may be regarded as an impulse of strength x_s, which will produce a response $x_s a_{t-s}$ at the subsequent time t. Adding all these contributions we have the total response

$$\xi_t = \sum_{s=t}^{-\infty} x_s a_{t-s} \tag{5.25}$$

which, by letting $n=t-s$, is

$$\xi_t = \sum_{n=0}^{\infty} a_n x_{t-n} = a_0 x_t + a_1 x_{t-1} + a_2 x_{t-2} + \cdots. \tag{5.26}$$

Since the present value x_t and the past values $x_{t-1}, x_{t-2}, \cdots$, yield the value of ξ_t, we see that knowledge of x_t up to the time t is equivalent to knowledge of ξ_t up to time t. The representation (5.26), which is the predictive decomposition of the impulse ξ_t, will be called the inverse predictive decomposition of the time series x_t.

5.3 *Prediction of Stationary Time Series*

The value $x_{t+\alpha}$ of the time series at time $t+\alpha$, because of the predictive decomposition (5.21), is given by

$$x_{t+\alpha} = \sum_{s=0}^{\infty} b_s \xi_{t+\alpha-s}$$
$$= (b_0\xi_{t+\alpha} + b_1\xi_{t+\alpha-1} + \cdots + b_{\alpha-1}\xi_{t+1}) + (b_\alpha \xi_t + b_{\alpha+1}\xi_{t-1} + \cdots). \tag{5.31}$$

Let us now consider time t to be the present time with respect to the filter; that is, time t is the time at which the computations are to be carried out. As a result, all values of the time series $x_s(-\infty < s \leq t)$ at and prior to time t are known at time t, and consequently all values of the white noise $\xi_s(-\infty < s \leq t)$ at and prior to time t may be found by means of the inverse filter, represented by

$$\xi_s = \sum_{n=0}^{\infty} a_n x_{s-n} \text{ for } -\infty < s \leq t. \tag{5.26}$$

Thus the component

$$(b_\alpha \xi_t + b_{\alpha+1}\xi_{t-1} + b_{\alpha+2}\xi_{t-2} + \cdots) \tag{5.311}$$

of the value of $x_{t+\alpha}$ given by equation (5.31) may be computed at time t, since the values $\xi_t, \xi_{t-1}, \xi_{t-2}, \cdots$, are available at time t. On the other hand, the component

$$(b_0\xi_{t+\alpha} + b_1\xi_{t+\alpha-1} + \cdots + b_{\alpha-1}\xi_{t+1}) \tag{5.312}$$

of $x_{t+\alpha}$ given by the predictive decomposition (5.31) cannot be computed at time t, since the values $\xi_{t+1}, \xi_{t+2}, \cdots, \xi_{t+\alpha}$ are not available at time t. In other words, the component (5.311) is the predictable component of $x_{t+\alpha}$ at time t, and the component (5.312) is the unpredictable component of $x_{t+\alpha}$ at time t. That is, the predictable part of $x_{t+\alpha}$ is made up of the response due to the impulses $\xi_t, \xi_{t-1}, \xi_{t-2}, \cdots$, which have occurred at and prior to time t, and the unpredictable part is made up of the impulses $\xi_{t+1}, \xi_{t+2}, \cdots, \xi_{t+\alpha}$ which occur between the present time t and the time $t+\alpha$.

Since the impulses ξ_t are uncorrelated, it follows from the Gram-Schmidt process for orthogonalizing vectors that the best linear forecast in the least-squares sense is

$$\hat{x}_{t+\alpha} = b_\alpha\xi_t + b_{\alpha+1}\xi_{t-1} + b_{\alpha+2}\xi_{t-2} + \cdots \qquad \alpha = 1, 2, \cdots \tag{5.32}$$

The forecast (5.32) is the predictable component (5.311).

For the forecast (5.32) the error is given by the unpredictable component (5.312), and the mean square error, given by the expectation

$$I_{\min} = \mathrm{E}[x_{t+\alpha} - \hat{x}_{t+\alpha}]^2 = \mathrm{E}[b_0\xi_{t+\alpha} + b_1\xi_{t+\alpha-1} + \cdots + b_{\alpha-1}\xi_{t+1}]^2 \tag{5.321}$$

is a minimum. Since $\mathrm{E}[\xi_t\xi_s]=0$ for $t \neq s$, this minimum value is

$$I_{\min} = \mathrm{E}[x_{t+\alpha} - \hat{x}_{t+\alpha}]^2 = (b_0^2 + b_1^2 + \cdots + b_{\alpha-1}^2)\mathrm{E}[\xi_t^2]. \tag{5.322}$$

By letting $E[\xi_t^2]=1$, we have

$$I_{\min} = \sum_{n=0}^{\alpha-1} b_n^2, \qquad (5.323)$$

thereby showing that the efficiency of the forecast decreases as the prediction distance α increases. Since

$$E[x_t^2] = E\left[\left(\sum_{s=0}^{\infty} b_s \xi_{t-s}\right)^2\right]$$

$$= E[\xi_t^2] \sum_{t=0}^{\infty} b_t^2 \qquad (5.324)$$

the prediction error $I_{\min}$ tends toward $E[x_t^2]$ as the prediction distance α tends toward infinity and hence for large values of the prediction distance α the trivial forecast of $\hat{x}_{t+\alpha}=0$ has about the same efficiency as the forecast (5.32).

Kolmogorov (1939) generalizes the result (5.323) by showing that the minimum mean-square prediction error for a process with a non-absolutely continuous spectral distribution is given by

$$I_{\min} = \sum_{n=0}^{\alpha-1} b_n^2. \qquad (5.325)$$

Let us use the Kolmogorov normalization, which is that $b_0=1$. Then $B(\omega)$ is given by equation (5.19) and $I_{\min}$ becomes

$$I_{\min} = \exp\left[\frac{1}{\pi}\int_0^{\pi} \log \Phi(\omega)\,d\omega\right] \cdot (1 + b_1^2 + \cdots + b_{\alpha-1}^2). \qquad (5.326)$$

Kolmogorov states that

$$\exp\left[\frac{1}{\pi}\int_0^{\pi} \log \Phi(\omega)\,d\omega\right] = 0 \qquad (5.327)$$

and consequently $I_{\min}=0$ if $\Phi(\omega)=0$ on a set of positive measure and also if the integral

$$\int_0^{\pi} \log \Phi(\omega)\,d\omega \qquad (5.328)$$

diverges, referring to this situation as the singular case.

Thus we see the reasons for the restrictions (5.131) and (5.133) in Section 5.1.

Doob (1953, p. 584) points out that in the transformation of discrete time series with the spectrum $\Phi(\omega)$, to continuous time series with the spectrum $W(\omega)$, the integrals

$$\int_{-\pi}^{\pi} \log \Phi(\omega)\,d\omega$$

$$\int_{-\infty}^{\infty} \frac{\log W(\omega)}{1+\omega^2}\,d\omega \qquad (5.329)$$

are finite and infinite together. The condition that the second integral be finite may be referred to as the Paley-Wiener criterion (Paley and Wiener, 1934), and Wiener (1942) uses this condition in the same connection as Kolmogorov (1939).

Let us now summarize the solution of the prediction problem for stationary time series with absolutely continuous spectral distribution functions.

In order to obtain the predicted values $\hat{x}_{t+\alpha}$ from the values $x_t, x_{t-1}, x_{t-2}, \cdots$, we first apply the linear operator

$$\xi_s = \sum_{n=0}^{\infty} a_n x_{s-n} \quad \text{for } -\infty < s \leq t \quad (5.26)$$

to yield the values $\cdots, \xi_{t-2}, \xi_{t-1}, \xi_t$, and then apply the operator

$$\hat{x}_{t+\alpha} = \sum_{s=0}^{\infty} b_{\alpha+s}\xi_{t-s} \qquad (5.32)$$

to yield the predicted value $\hat{x}_{t+\alpha}$. The operations on the past values $x_t, x_{t-1}, x_{t-2}, \cdots$, represented by equations (5.26) and (5.32) may be combined into

$$\hat{x}_{t+\alpha} = \sum_{s=0}^{\infty} b_{\alpha+s}\xi_{t-s}$$

$$= \sum_{s=0}^{\infty} b_{\alpha+s} \sum_{n=0}^{\infty} a_n x_{t-s-n} \qquad (5.33)$$

which becomes, by letting $r=s+n$, and recalling that $a_t=0$ for $t<0$,

$$\hat{x}_{t+\alpha} = \sum_{s=0}^{\infty} b_{\alpha+s} \sum_{r=0}^{\infty} a_{r-s} x_{t-r}$$

$$= \sum_{r=0}^{\infty}\left(\sum_{s=0}^{\infty} b_{\alpha+s} a_{r-s}\right) x_{t-r}. \qquad (5.331)$$

Let us define

$$k_\tau(\alpha) = \sum_{s=0}^{\infty} b_{\alpha+s}a_{\tau-s} \qquad (5.332)$$

so that equation (5.331) becomes

$$\hat{x}_{t+\alpha} = \sum_{\tau=0}^{\infty} k_\tau(\alpha)x_{t-\tau}. \qquad (5.333)$$

This equation thereby expresses the predicted value $\hat{x}_{t+\alpha}$ in terms of the present value x_t and past values $x_{t-1}, x_{t-2}, x_{t-3}, \cdots$, and has the same form as the pure prediction operator (2.21, Chapter II) except that we now allow the number of operator coefficients k_s to become infinite. Also the dependence of the operator coefficients k_s on the value α of the prediction distance, is indicated. We note that $k_s(\alpha)=0$ for s less than zero, so that the operator is realizable.

Further,

$$\begin{aligned}\hat{x}_{t+\alpha} = &- a_1\hat{x}_{t+\alpha-1} - a_2\hat{x}_{t+\alpha-2} - \cdots \\ &- a_{\alpha-1}\hat{x}_{t+1} - a_\alpha x_t - a_{\alpha+1}x_{t-1} \\ &- a_{\alpha+2}x_{t-2} - \cdots \end{aligned} \qquad (5.34)$$

and the operator coefficients $k_\tau(\alpha)$ satisfy

$$\begin{aligned} &k_0(\alpha) + a_1k_0(\alpha-1) + a_2k_0(\alpha-2) + \cdots \\ &\qquad + a_{\alpha-1}k_0(1) + a_\alpha = 0 \\ &k_1(\alpha) + a_1k_1(\alpha-1) + a_2k_1(\alpha-2) + \cdots \\ &\qquad + a_{\alpha-1}k_1(1) + a_{\alpha+1} = 0. \end{aligned} \qquad (5.35)$$

. . . .

Figure 3 illustrates the least-squares linear operator for unit prediction distance. That is, the operator coefficients are determined by the condition that the mean square prediction error be a minimum. The operator is in the form of equation (2.28). From the expression for $\Phi(\omega)$ in Figure 3, we see that time series is an autoregressive time series (see Section 5.5-B) and thus the solution of the simultaneous equations given in Figure 3 may be found by the method of the factorization of the autoregressive spectrum given in Section 5.5-B.

5.4 *The filtering problem*

The solution of the filtering problem for stationary time series, continuous and discrete, with absolutely continuous spectral distribution functions is given by Wiener (1942, 1948). As in the case of the prediction problem the solution of the filtering problem is a direct consequence of the Predictive Decomposition Theorem. We shall translate the solution of Wiener (1948) for continuous time series to discrete time series, with his equation numbers on the left. The predictive decomposition of the time series x_t consisting of message plus noise is

(3.89) $$x_t = m_t + n_t = \sum_{\tau=0}^{\infty} b_\tau\xi_{t-\tau}. \qquad (5.41)$$

Since the message has an absolutely continuous spectral distribution function it may be represented by the process of moving summation (see Section 4.6) given by

(3.90) $$m_t = \sum_{\tau=-\infty}^{\infty} q_\tau\xi_{t-\tau} + \sum_{\tau=-\infty}^{\infty} r_\tau\gamma_{t-\tau} \qquad (5.411)$$

where the random variables ξ_t and γ_t are mutually uncorrelated, that is

$$\begin{aligned} &\mathrm{E}[\xi_t^2] = 1, \quad \mathrm{E}[\xi_t\xi_{t+k}] = 0 \quad \text{for } k \neq 0, \\ &\mathrm{E}[\xi_t\gamma_s] = 0, \qquad \mathrm{E}[\gamma_t^2] = 1, \\ &\mathrm{E}[\gamma_t\gamma_{t+k}] = 0 \quad \text{for } k \neq 0. \end{aligned} \qquad (5.412)$$

The predictable part of the message $m_{t+\alpha}$, where α is the lead, is

(3.901)

$$\hat{m}_{t+\alpha} = \sum_{\tau=0}^{\infty} q_{\tau+\alpha}\xi_{t-\tau} = \sum_{\tau=\alpha}^{\infty} q_\tau\xi_{t-\tau-\alpha} \qquad (5.413)$$

and the (minimum) mean-square error of prediction is

(3.902) $$I_{\min} = \sum_{\tau=-\infty}^{\alpha} q_\tau^2 + \sum_{\tau=-\infty}^{\infty} r_\tau^2. \qquad (5.414)$$

From equations (5.41) and (5.411) we see that noise n_t is given by

$$\begin{aligned} n_t = &\sum_{\tau=0}^{\infty} b_\tau\xi_{t-\tau} - \sum_{\tau=-\infty}^{\infty} q_\tau\xi_{t-\tau} \\ &- \sum_{\tau=-\infty}^{\infty} r_\tau\gamma_{t-\tau}. \end{aligned} \qquad (5.415)$$

GIVEN THE POWER SPECTRUM $\Phi(\omega)$ OF THE TIME SERIES x_t $(-\infty < t < \infty)$ WITH $E[x_t] = 0$, WE WISH TO FIND THE LINEAR OPERATOR WITH OPERATOR COEFFICIENTS a_0, a_1, a_2, a_3, a_4 SUCH THAT THE PREDICTION ERROR VARIANCE $E\left[\xi_t^2\right]$ IS A MINIMUM WHERE THE PREDICTION ERROR IS GIVEN BY $\xi_t = a_0 x_t + a_1 x_{t-1} + a_2 x_{t-2} + a_3 x_{t-3} + a_4 x_{t-4}$

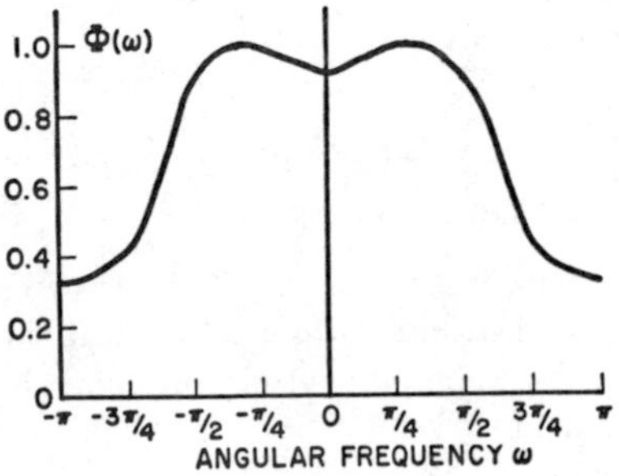

POWER SPECTRUM $\Phi(\omega)$

$$\Phi(\omega) = \frac{1}{r_0 + 2\sum_{\tau=1}^{4} r_\tau \cos \omega\tau}$$

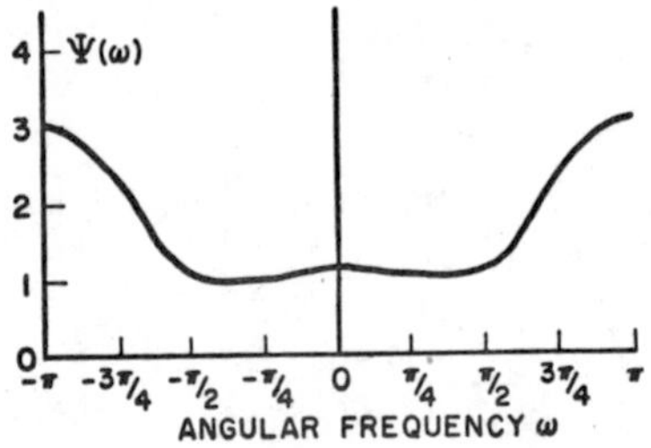

POWER TRANSFER FUNCTION $\Psi(\omega) = \dfrac{1}{\Phi(\omega)}$

$$\Psi(\omega) = |A(\omega)|^2 = r_0 + 2\sum_{\tau=1}^{4} r_\tau \cos \omega\tau = \sum_{\tau=-4}^{4} r_\tau e^{-i\omega\tau}$$

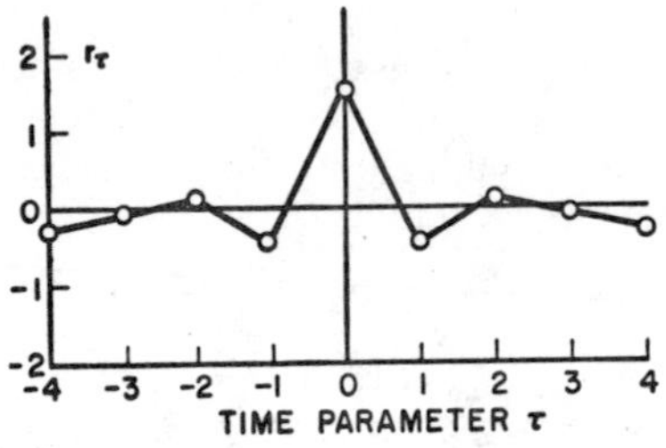

FOURIER TRANSFORM OF POWER TRANSFER FUNCTION

$$r_\tau = \frac{1}{2\pi}\int_{-\pi}^{\pi} \Psi(\omega) e^{i\omega\tau}\, d\omega$$

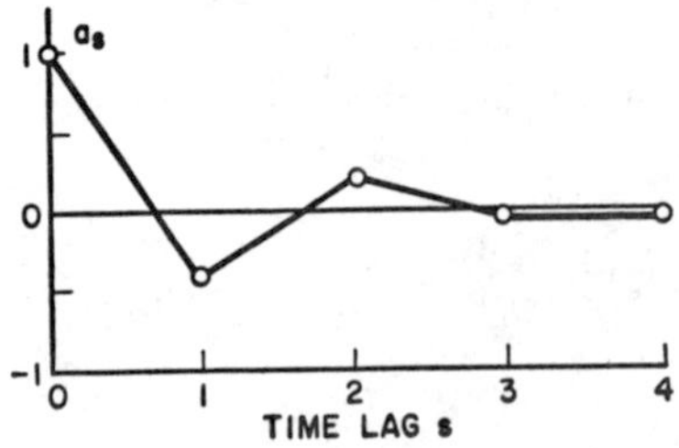

STABLE OPERATOR COEFFICIENTS a_s

DETERMINED FROM SOLUTION OF THE SIMULTANEOUS EQUATIONS

$$r_0 = a_0^2 + a_1^2 + a_2^2 + a_3^2 + a_4^2$$
$$r_1 = a_0 a_1 + a_1 a_2 + a_2 a_3 + a_3 a_4$$
$$r_2 = a_0 a_2 + a_1 a_3 + a_2 a_4$$
$$r_3 = a_0 a_3 + a_1 a_4$$
$$r_4 = a_0 a_4$$

SUBJECT TO THE CONDITION THAT THE TRANSFER FUNCTION $A(\omega) = \sum_{t=0}^{4} a_t e^{-i\omega t}$ HAVE NO SINGULARITIES OR ZEROS IN THE LOWER HALF λ PLANE, WHERE $\lambda = \omega + i\sigma$

FIG. 3. The least squares linear operator.

Thus the autocorrelation $\phi_{22}(t)$ of the noise is given by

(3.903)

$$\phi_{22}(t) = \mathrm{E}[n_{t+\tau}n_\tau]$$
$$= \sum_{\tau=0}^{\infty} (b_{|t|+\tau} - q_{|t|+\tau})(b_\tau - q_\tau)$$
$$+ \sum_{\tau=-|t|}^{-1} (b_{|t|+\tau} - q_{|t|+\tau})(-q_\tau)$$
$$+ \sum_{-\infty}^{-|t|-1} q_{|t|+\tau}q_\tau + \sum_{-\infty}^{\infty} r_{|t|+\tau}r_\tau$$
$$= \sum_{\tau=0}^{\infty} b_{|t|+\tau}b_\tau - \sum_{\tau=0}^{\infty} q_{|t|+\tau}b_\tau$$
$$- \sum_{\tau=-|t|}^{\infty} q_\tau b_{|t|+\tau}$$
$$+ \sum_{\tau=-\infty}^{\infty} q_{|t|+\tau}q_\tau$$
$$+ \sum_{\tau=-\infty}^{\infty} r_{|t|+\tau}r_\tau. \quad (5.42)$$

The autocorrelation $\phi_{11}(t)$ of the message is

$$(3.904) \quad \phi_{11}(t) = \mathrm{E}[m_{|t|+\tau}m_\tau]$$
$$= \sum_{\tau=-\infty}^{\infty} q_{|t|+\tau}q_\tau$$
$$+ \sum_{\tau=-\infty}^{\infty} r_{|t|+\tau}r_\tau. \quad (5.421)$$

The crosscorrelation $\phi_{12}(t)$ of the message and noise is

(3.905)

$$\phi_{12}(t)$$
$$= \mathrm{E}[m_{t+\tau}n_\tau]$$
$$= \mathrm{E}[m_{t+\tau}(m_\tau + n_\tau) - m_{t+\tau}m_\tau]$$
$$= \mathrm{E}[m_{t+\tau}(m_\tau + n_\tau)] - \phi_{11}(\tau)$$
$$= \mathrm{E}\left[\sum_{\sigma=-1}^{\infty} b_{\sigma+t}\xi_{\tau-\sigma} \sum_{\sigma=-t}^{\infty} q_\sigma \xi_{\tau-\sigma}\right] - \phi_{11}(\tau)$$
$$= \sum_{\tau=-t}^{\infty} b_{t+\tau}q_\tau - \phi_{11}(\tau). \quad (5.422)$$

Let us define

$$(3.907) \quad B(\omega) = \sum_{s=0}^{\infty} b_s e^{-i\omega s}$$
$$Q(\omega) = \sum_{s=-\infty}^{\infty} q_s e^{-i\omega s}$$
$$R(\omega) = \sum_{s=-\infty}^{\infty} r_s e^{-i\omega s}. \quad (5.423)$$

The Fourier transforms of the correlation functions $\phi_{22}(t)$, $\phi_{11}(t)$, and $\phi_{12}(t)$ are the power spectrum $\Phi_{22}(\omega)$ of the noise, the power spectrum $\Phi_{11}(\omega)$ of the message, and the cross spectrum $\Phi_{12}(\omega)$ of the message and noise respectively, and are given by

(3.906)

$$\Phi_{22}(\omega) = B\bar{B} + Q\bar{Q} - Q\bar{B} - B\bar{Q} + R\bar{R}$$
$$\Phi_{11}(\omega) = Q\bar{Q} + R\bar{R}$$
$$\Phi_{12}(\omega) = B\bar{Q} - Q\bar{Q} - R\bar{R}$$
$$= B\bar{Q} - \Phi_{11}(\omega). \quad (5.43)$$

Therefore, we have

(3.908)

$$\Phi_{11}(\omega) + \Phi_{12}(\omega) + \overline{\Phi_{12}(\omega)} + \Phi_{22}(\omega)$$
$$= B\bar{B} = |B(\omega)|^2. \quad (5.431)$$

In order to compute $B(\omega)$, we must have the sum of spectra given by the left hand side of this equation. Let us call this sum $\Phi(\omega)$, that is,

$$\Phi(\omega) = \Phi_{11}(\omega) + 2\mathrm{Re}[\Phi_{12}(\omega)]$$
$$+ \Phi_{22}(\omega). \quad (5.432)$$

We see that $\Phi(\omega)$ is the power spectrum of the time series x_t, equation (5.41), and thus $\Phi(\omega)$ may be computed directly from this time series. We factor $\Phi(\omega)$ into $B(\omega)\overline{B(\omega)}$ according to the method of Section 5.1. In addition, from equations (5.43) we have

$$B\bar{Q} = \Phi_{12}(\omega) + \Phi_{11}(\omega) \quad (5.433)$$

so that

$$(3.909) \quad Q\bar{B} = \overline{\Phi_{12}(\omega)} + \Phi_{11}(\omega)$$
$$= \Phi_{11}(\omega) + \Phi_{21}(\omega) \quad (5.434)$$

because

$$\Phi_{21}(\omega) = \overline{\Phi_{12}(\omega)}. \quad (5.435)$$

Thus we have

$$(3.910)\quad Q(\omega) = \frac{\Phi_{11}(\omega) + \Phi_{21}(\omega)}{\overline{B(\omega)}} \qquad (5.436)$$

and

$$q_t = \frac{1}{2\pi}\int_{-\pi}^{\pi} Q(\omega)e^{i\omega t}d\omega$$

$$= \frac{1}{2\pi}\int_{-\pi}^{\pi} \frac{\Phi_{11}(\omega) + \Phi_{21}(\omega)e^{i\omega t}d\omega}{\overline{B(\omega)}} \cdot (5.437)$$

We let the inverse predictive decomposition of the time series $x_t = m_t + n_t$ be

$$\xi_t = \sum_{s=0}^{\infty} a_s x_{t-s}$$

$$= \sum_{s=0}^{\infty} a_s(m_{t-s} + n_{t-s}) \qquad (5.44)$$

which gives the prediction errors ξ_t (for unit prediction distance) of the time series x_t. Thus equation (5.413), which gives the predictable part of $m_{t+\alpha}$, consists of reaveraging these prediction errors by the realizable linear operation

$$\hat{m}_{t+\alpha} = \sum_{\tau=0}^{\infty} q_{\tau+\alpha}\xi_{t-\tau} = \sum_{\tau=0}^{\infty} q_{\tau+\alpha}$$

$$\cdot \sum_{s=0}^{\infty} a_s(m_{t-\tau-s} + n_{t-\tau-s}). \qquad (5.441)$$

The transfer function $H_\alpha(\omega)$ of this linear operator is the totality of transfer ratios obtained by letting $m_t + n_t = e^{i\omega t}$ in equation (5.441) and dividing by $e^{i\omega t}$, that is,
(3.913), (3.941)

$$H_\alpha(\omega) = \sum_{\tau=0}^{\infty} q_{\tau+\alpha} \sum_{s=0}^{\infty} a_s e^{-i\omega(\tau+s)}$$

$$= \sum_{\tau=0}^{\infty} q_{\tau+\alpha}e^{-i\omega\tau} \sum_{s=0}^{\infty} a_s e^{-i\omega s}$$

$$= \frac{1}{B(\omega)} \sum_{n=\alpha}^{\infty} q_n e^{-i\omega(n-\alpha)}$$

$$= \frac{1}{2\pi B(\omega)} \sum_{t=\alpha}^{\infty} e^{-i\omega(t-\alpha)}$$

$$\cdot \int_{-\pi}^{\pi} \frac{\Phi_{11}(\omega) + \Phi_{21}(\omega)}{\overline{B(\omega)}} e^{i\omega t}d\omega. \quad (5.45)$$

As Wiener (1948) points out, the equation for continuous time series (Wiener's equation 3.913), which corresponds to our equation (5.45) for discrete time series is the transfer function of a filter. The quantity α is the lead of the filter, and may be positive or negative. When it is negative, $-\alpha$ is known as the lag. Wiener also points out that apparatus corresponding to this equation may always be constructed with as much accuracy as we like, and he refers to papers by Y. W. Lee.

The mean-square filtering error (5.414) is

$$I_{\min} = \sum_{-\infty}^{\alpha} q_\tau^2 + \sum_{-\infty}^{\infty} r_\tau^2 \qquad (5.414)$$

where the first term on the right depends on the lag $-\alpha$, whereas the second term does not. For infinite lag, that is, $\alpha = -\infty$, the error becomes

$$(3.914)\quad I_{\min}(\alpha = -\infty) = \sum_{\tau=-\infty}^{\infty} r_\tau^2$$

$$= \phi_{11}(0) - \sum_{\tau=-\infty}^{\infty} q_\tau^2 \quad (5.46)$$

because of equation (5.421), with $t=0$. Equation (5.46) becomes

$$(3.914)\quad I_{\min}(\alpha = -\infty) = \frac{1}{2\pi}\int_{-\pi}^{\pi}$$

$$\cdot \frac{\begin{vmatrix} \Phi_{11}(\omega) & \Phi_{12}(\omega) \\ \Phi_{21} & \Phi_{22}(\omega) \end{vmatrix} d\omega}{\Phi_{11}(\omega) + \Phi_{12}(\omega) + \Phi_{21}(\omega) + \Phi_{22}(\omega)} \qquad (5.461)$$

which is

$$I_{\min}(\alpha = -\infty)$$

$$= \frac{1}{2\pi}\int_{-\pi}^{\pi} \frac{\{\text{Determinant of Coherency Matrix of Message and of Noise}\}}{\{\text{Power Spectrum of Message plus Noise}\}} d\omega. \quad (5.462)$$

Thus, if message and noise were completely coherent, then $I_{\min}(\alpha=-\infty)=0$. The part of $I_{\min}$ depending on lag is (3.915)

$$\sum_{\tau=-\infty}^{\alpha} q_\tau^2 = \sum_{\tau=-\infty}^{\alpha} \left| \frac{1}{2\pi} \int_{-\pi}^{\pi} \cdot \frac{\Phi_{11}(\omega) + \Phi_{21}(\omega)}{\overline{B(\omega)}} e^{i\omega t} d\omega \right|^2. \quad (5.47)$$

In conclusion, then, we see that the general solution of the filtering problem for the time series $x_t = m_t + n_t$ consists of refiltering the prediction errors ξ_t (for unit prediction distance) of the time series x_t by means of the operator

$$\hat{m}_{t+\alpha} = \sum_{\tau=0}^{\infty} q_{\tau+\alpha}\xi_{t-\tau}. \quad (5.441)$$

Here the operator coefficients q_t are determined from $\Phi_{11}(\omega)$ which is the power spectrum of the message m_t, from $\Phi_{21}(\omega)$ which is the cross spectrum of the noise n_t and the message m_t, and from $\overline{B(\omega)}$ which is the minimum-delay factor of the power spectrum $\Phi(\omega)$ of the time series $x_t = m_t + n_t$.

5.5 *Time series with rational power spectra*

Of particular interest to the working statistician are those time series with power spectra $\Phi(\omega)$ which are rational functions in $z = e^{-i\omega}$ (see Figure 1).

A. Process of finite moving averages.—Let the minimum-delay response function b_t be of finite extent, that is

$$b_t = 0 \quad \text{for } t < 0 \text{ and } t > M,$$
$$b_0 \neq 0, \quad b_M \neq 0. \quad (5.51)$$

Then the predictive decomposition of the time series x_t, with this response function, is

$$x_t = b_0\xi_t + b_1\xi_{t-1} + \cdots + b_M\xi_{t-M} \quad (5.511)$$

which represents a stationary process with an absolutely continuous spectral distribution.

Strictly speaking, the process x_t given by this decomposition is defined to be a process of finite moving averages if white noise $\xi_t(-\infty < t < \infty)$ is a purely random process, that is, if all of the components ξ_t are independent random variables with the same distribution function. Nevertheless, we shall also include in this definition those processes (5.511) for which the $\xi_t(-\infty < t < \infty)$ represents a mutually uncorrelated process (see Section 4.5). We shall let ξ_t have mean value zero and unit variance. Thus, the $\xi_t(-\infty < t < \infty)$ is a process such that

$$E(\xi_t) = 0$$
$$E(\xi_t^2) = 1$$
$$E(\xi_t\xi_s) = 0, \quad s \neq t. \quad (5.512)$$

The autocorrelation is given by

$$\phi(\tau) = E[x_t x_{t+\tau}] = \sum_{t=0}^{M-\tau} b_t b_{t+\tau} \quad (5.513)$$

so that

$$\phi(\tau) = 0 \quad \text{for } |\tau| > M. \quad (5.514)$$

The power spectrum of the process of finite moving averages is then

$$\Phi(\omega) = \phi(0) + 2\sum_{\tau=1}^{M} \phi(\tau)\cos\omega\tau$$
$$= \sum_{\tau=-M}^{M} \phi(\tau)e^{-i\omega\tau} \quad (5.515)$$

which is a rational function in $z = e^{-i\omega}$. Let $B(\omega)$ be the transfer function

$$B(\omega) = \sum_{s=0}^{M} b_s e^{-i\omega s} \quad (5.516)$$

so we see that

$$\Phi(\omega) = B(\omega)\overline{B(\omega)} = |B(\omega)|^2$$
$$= \sum_{\tau=-M}^{M} \phi(\tau)e^{-i\omega\tau} \geq 0 \quad (5.517)$$

is nonnegative.

In general, an arbitrary set of coefficients

$$\phi(0), \phi(1), \phi(2), \cdots, \phi(M) \quad (5.52)$$

will not be such that the rational function

$$\phi(0) + 2\sum_{\tau=1}^{M} \phi(\tau)\cos\omega\tau$$
$$= \sum_{\tau=-M}^{M} \phi(\tau)e^{-i\omega\tau} \quad (5.521)$$

is nonnegative, and hence it is not an acceptable function to represent the power spectrum of a process of moving averages.

Wold (1938) gives the following theorem:

Theorem. Let $Q(u)$ be the polynomial of order M obtained from

$$\phi(0) + \sum_{\tau=1}^{M} \phi(\tau)[z^{-\tau} + z^{\tau}]$$

by the substitution $z^{-1}+z=u$. A necessary and sufficient condition that $\phi(0)$, $\phi(1)$, $\phi(2)$, $\cdots$, $\phi(M)$ be the correlogram of a process of moving averages (5.511) is that the polynomial should have no real root of odd multiplicity in the interval $-2<u<2$.

Thus for a sequence (5.52) which does fulfill the conditions of Wold's Theorem, the function

$$\phi(0) + 2\sum_{\tau=1}^{M} \phi(\tau)\cos\omega\tau \geq 0 \quad (5.522)$$

is nonnegative and hence may represent the power spectrum $\Phi(\omega)$ of a process of moving averages. In order to determine the response function $b_t (t=0, 1, 2, \cdots, M)$ of the process with this power spectrum (5.522), it is necessary to factor this power spectrum into

$$\Phi(\omega) = \phi(0) + 2\sum_{\tau=1}^{M} \phi(\tau)\cos\omega\tau$$
$$= B(\omega)\overline{B(\omega)} \quad (5.523)$$

where

$$B(\omega) = \sum_{s=0}^{M} b_s e^{-i\omega s} \quad (5.524)$$

is free from singularities and zeros in the lower half λ plane. Then the minimum-delay operator b_t is given by

$$b_t = \frac{1}{2\pi}\int_{-\pi}^{\pi} B(\omega)e^{i\omega t}d\omega,$$
$$t = 0, 1, 2, \cdots, M. \quad (5.525)$$

In order to carry out the factorization (5.523), let $\Phi(\omega)$, given by equation (5.523) be the $\Psi(\omega)$ of Section 2.8, given by equation (2.82). Then we determine the minimum-delay operator b_t from $\Phi(\omega)$ in the same manner as given in Section 2.8. The inverse predictive decomposition is given by

$$\xi_t = a_0x_t + a_1x_{t-1} + a_2x_{t-2} + \cdots$$
$$= \sum_{s=0}^{\infty} a_s x_{t-s} \quad (5.526)$$

where the inverse linear operator a_t (see Section 2.7) is realizable, minimum-delay, and infinite in extent.

The process of finite moving averages was introduced by Yule (1921, 1926) and Slutsky (1927), and consequently was the first stochastic process studied which was neither a purely random or an uncorrelated process (Section 4.5) nor a deterministic strictly periodic process (Section 3.2).

The solution of the prediction problem for such processes for the special case in which

$$b_t = (M+1)^{-1/2} \quad (5.527)$$

is given by Kosulajeff (1941).

B. The autoregressive process.—The autoregressive process is a stochastic process for which the response function b_t $(t=0, 1, 2, \cdots)$ is of infinite extent, but the inverse response function a_t $(t=0, 1, 2, \cdots, m)$ is of finite extent and strictly minimum-delay. Thus the inverse predictive decomposition of an autoregressive time series x_t of the mth order is

$$\xi_t = a_0x_t + a_1x_{t-1} + \cdots + a_mx_{t-m}$$
$$= \sum_{s=0}^{m} a_s x_{t-s} \quad (5.53)$$

and the predictive decomposition is

$$x_t = b_0\xi_t + b_1\xi_{t-1} + b_2\xi_{t-2} + \cdots$$
$$= \sum_{s=0}^{\infty} b_s \xi_{t-s} \quad (5.531)$$

where the operators a_t and b_t are minimum-delay, and inverse to each other (see Section 2.7). Such a process is stationary with an absolutely continuous spectral distribution.

Strictly speaking, in the definition of the autoregressive process, the impulses ξ_t are independent with the same distribution function, but we shall also admit ξ_t which are mutually uncorrelated. We shall let ξ_t be a process such that

$$E(\xi_t) = 0$$
$$E(\xi_t^2) = 1$$
$$E(\xi_t\xi_s) = 0, \qquad s \neq t. \quad (5.532)$$

The autoregressive process, or the process of disturbed harmonics, was introduced by Yule (1927), and was a major step forward toward the establishment of the general theory of stochastic processes by Kolmogorov (1933) and Khintchine (1933). Because equation (5.53) for the prediction error ξ_t (for unit prediction distance) has only a finite number of operator coefficients, $a_0, a_1, a_2, \cdots, a_m$, the prediction operator for any prediction distance will require only a finite number of coefficients, as seen by equation (5.34).

The autocorrelation function is given by

$$\phi(\tau) = \mathrm{E}[x_t x_{t+\tau}] = \sum_{t=0}^{\infty} b_t b_{t+\tau} \quad (5.533)$$

and the power spectrum by

$$\Phi(\omega) = \sum_{\tau=-\infty}^{\infty} \phi(\tau) e^{-i\omega\tau} = |B(\omega)|^2$$

$$= \frac{1}{|A(\omega)|^2} \geq 0 \quad (5.534)$$

where

$$B(\omega) = \sum_{s=0}^{\infty} b_s e^{-i\omega s} \quad (5.535)$$

and

$$A(\omega) = B^{-1}(\omega) = \sum_{s=0}^{\infty} a_s e^{-i\omega s}. \quad (5.536)$$

Since

$$\Phi(\omega) = \frac{1}{|A(\omega)|^2} = \frac{1}{\left| \sum_{s=0}^{m} a_s e^{-i\omega s} \right|^2}$$

$$= \frac{1}{\sum_{\tau=-m}^{m} r_\tau e^{-i\omega\tau}} \quad (5.537)$$

where

$$r_\tau = \sum_{s=0}^{M} a_s a_{s+\tau} \quad (5.538)$$

we see that the power spectrum is a rational function in $e^{-i\omega}$.

In general the rational function in $e^{-i\omega}$

$$\left[\sum_{\tau=-m}^{m} r_\tau e^{-i\omega\tau} \right]^{-1} \quad (5.54)$$

will not be a nonnegative integrable function for $-\pi < \omega \leq \pi$. The function (5.54) will be nonnegative if the $r_0, r_1, r_2, \cdots, r_m$ satisfy the conditions of Wold's Theorem, which we stated in Part A of this Section in connection with the process of finite moving averages. The function (5.54) will be integrable for $-\pi \leq \omega \leq \pi$ if the polynomial

$$z^m \sum_{\tau=-m}^{m} r_\tau z^\tau \quad (5.541)$$

has no roots of modulus one. If these conditions are satisfied, then we may let the rational function (5.54) represent the power spectrum $\Phi(\omega)$ of an autoregressive process. Then the inverse power spectrum

$$\Psi(\omega) = \frac{1}{\Phi(\omega)} = \sum_{\tau=-m}^{m} r_\tau e^{-i\omega\tau}$$

$$= r_0 + 2 \sum_{\tau=1}^{m} r_\tau \cos \omega\tau \quad (5.542)$$

may be factored according to the method given in Section 2.8 so that

$$\Psi(\omega) = A(\omega)\overline{A(\omega)} \quad (5.543)$$

where

$$A(\omega) = \sum_{s=0}^{m} a_s e^{-i\omega s} \quad (5.545)$$

has no zeros or singularities in the lower half λ-plane. Then the spectrum is given by

$$\Phi(\omega) = \frac{1}{A(\omega)} \frac{1}{\overline{A(\omega)}} = B(\omega)\overline{B(\omega)} \quad (5.546)$$

where the factor

$$\frac{1}{A(\omega)} = B(\omega) \quad (5.547)$$

is also free of singularities and zeros in the lower half λ-plane. In Figures 4 and 5, the various time and frequency functions of a particular second order autoregressive process are given. Two fundamental sets of difference equations (Wold,

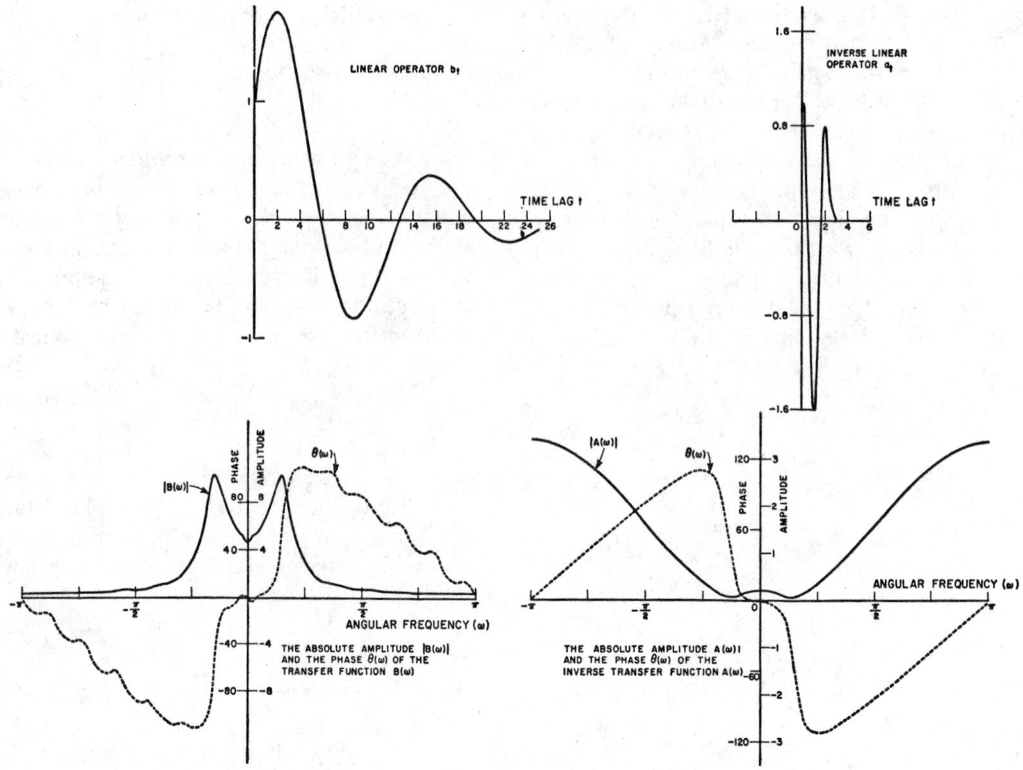

FIG. 4. The autoregressive process $a_0x_t+a_1x_{t-1}+a_2x_{t-2}=\xi_t$.

1938; Kendall, 1946) exist for the autoregressive process

$$x_t = \sum_{r=0}^{\infty} b_r\xi_{t-r}. \qquad (5.55)$$

Let us multiply this equation by the equation for the inverse decomposition at $t+n$:

$$\sum_{s=0}^{m} a_s x_{t+n-s} = \xi_{t+n}, \qquad n > 0 \qquad (5.551)$$

which yields

$$x_t \sum_{s=0}^{m} a_s x_{t+n-s} = \xi_{t+n} \sum_{r=0}^{\infty} b_r\xi_{t-r}, \qquad n > 0 \qquad (5.552)$$

and, by taking the expectation of each side, we have

$$\sum_{s=0}^{m} a_s \mathrm{E}[x_t x_{t+n-s}] = \sum_{r=0}^{\infty} b_r \mathrm{E}[\xi_{t+n}\xi_{t-r}], \qquad n > 0 \qquad (5.553)$$

which yields the set of difference equations

$$\sum_{s=0}^{m} a_s \phi(n-s) = 0, \qquad n > 0. \qquad (5.554)$$

Let us multiply the process (5.55) by the inverse decomposition at $t-n$:

$$\sum_{s=0}^{m} a_s x_{t-n-s} = \xi_{t-n}, \qquad n \geq 0 \qquad (5.555)$$

which yields

$$x_t \sum_{s=0}^{m} a_s x_{t-n-s} = \xi_{t-n} \sum_{r=0}^{\infty} b_r \xi_{t-r}, \qquad n \geq 0. \qquad (5.556)$$

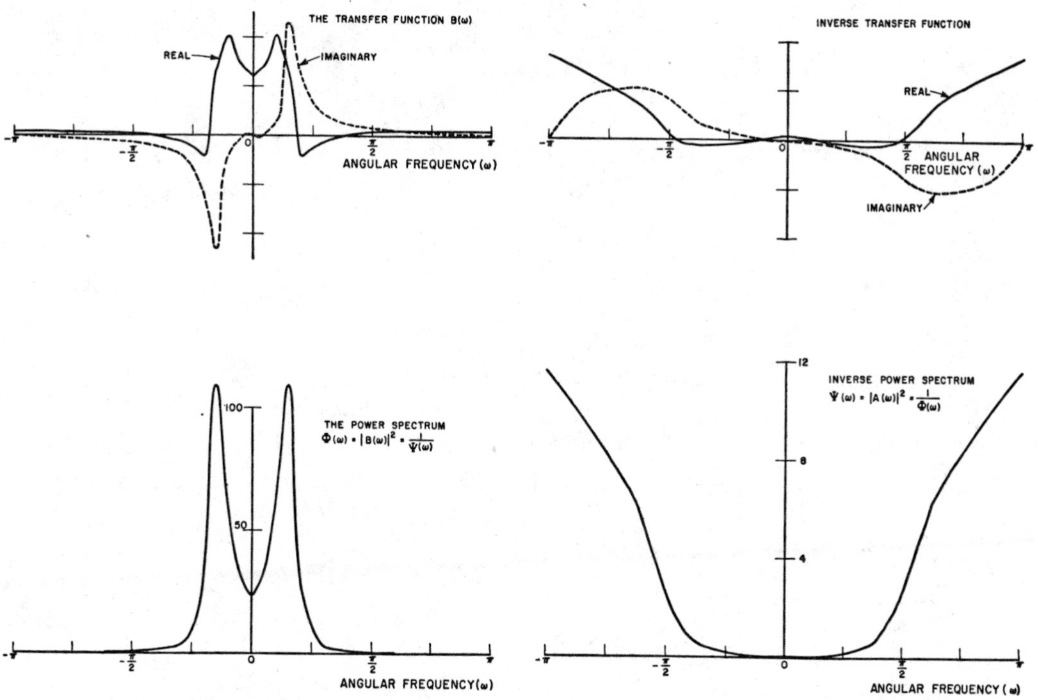

FIG. 5. The autoregressive process $a_0x_t+a_1x_{t-1}+a_2x_{t-2}=\xi_t$.

Taking the expectation of each side, we obtain the other set of difference equations

$$\sum_{s=0}^{m} a_s\phi(n+s) = b_n, \qquad n \geq 0. \qquad (5.557)$$

The first m linear equations of the set (5.554) are

$$\phi(1) + a_1\phi(0) + a_2\phi(1) + \cdots + a_m\phi(m-1) = 0$$

$$\phi(2) + a_1\phi(1) + a_2\phi(0) + \cdots + a_m\phi(m-2) = 0$$

$$\phi(3) + a_1\phi(2) + a_2\phi(1) + \cdots + a_m\phi(m-3) = 0$$

$$\cdots\cdots\cdots\cdots\cdots\cdots$$

$$\phi(m-1) + a_1\phi(m-2) + a_2\phi(m-3) + \cdots + a_m\phi(1) = 0$$

$$\phi(m) + a_1\phi(m-1) + a_2\phi(m-2) + \cdots + a_m\phi(0) = 0 \quad (5.56)$$

which correspond to the normal equations of the Gauss method of least squares (equations (40) of Wadsworth et al., 1953). Thus, if we know the values of $\phi(0), \phi(1), \cdots, \phi(m)$, that is, the first m lags of the autocorrelation function, we may compute $a_0=1, a_1, a_2, \cdots, a_m$. It is this property which makes the autoregressive process a fundamental model in statistical work. That is, the statistician need only estimate the first m lags of the autocorrelation to specify an mth order autoregressive process. Under the autoregressive hypothesis, the higher lags of the autocorrelation function may be found by successive recursions of equation (5.554).

C. The Markov process.—An autoregressive process of the first order as defined in Section 5.5-B is a Markov process. For a more general definition of the Markov process, see Feller (1950) and Doob (1953). Applications of the autoregressive process and Markov process to economic analysis are given by the author (1952).

In order to obtain a Markov process, we let $m=1$ in equation (5.53). Hence the impulse ξ_t has the representation

$$\xi_t = x_t + a_1 x_{t-1},$$
$$a_0 = 1, \ |a_1| < 1. \quad (5.57)$$

The response function b_t ($t=0, 1, 2, \cdots$) may be found by repeated iterations of the difference equation

$$0 = b_t + a_1 b_{t-1},$$
$$t = 1, 2, \cdots; b_0 = 1 \quad (5.571)$$

which yields

$$b_t = -a_1 b_{t-1} = (-a_1)^t. \quad (5.572)$$

Thus the predictive decomposition of the Markov process is

$$x_t = \sum_{s=0}^{\infty} b_s \xi_{t-s}$$
$$= \sum_{s=0}^{\infty} (-a_1)^s \xi_{t-s}. \quad (5.573)$$

The autocorrelation function is

$$\phi(\tau) = \mathrm{E}[x_t x_{t+\tau}] = \sum_{t=0}^{\infty} b_t b_{t+\tau}$$
$$= \sum_{t=0}^{\infty} (-a_1)^{2t+\tau}, \qquad \tau > 0 \quad (5.574)$$

which is

$$\phi(\tau) = (-a_1)^{|\tau|} \sum_{t=0}^{\infty} (-a_1)^{2t}$$
$$= \frac{(-a_1)^{|\tau|}}{(1-a_1^2)}. \quad (5.575)$$

Thus we see that the autocorrelation of such a Markov process is an exponential which is the case of pure persistence. Equation (5.575) is the discrete time series analogue of equation (16) in Wadsworth et al. (1953).

The spectrum $\Phi(\omega)$ is given by equation (5.537) with $m=1$; that is

$$\Phi(\omega) = \frac{1}{|1 + a_1 e^{-i\omega}|^2}$$
$$a_0 = 1, \ |a_1| < 1. \quad (5.576)$$

The prediction operator $k_t(\alpha)$, given by equation (5.332), becomes

$$k_0(\alpha) = b_\alpha = (-a_1)^\alpha$$
$$k_t(\alpha) = 0, \qquad \text{for } t = 1, 2, 3, \cdots \quad (5.58)$$

so that the optimum prediction is given by

$$x_{t+\alpha} = (-a_1)^\alpha x_t. \quad (5.581)$$

The filter characteristics of the prediction operator are

$$K_\alpha(\omega) = \sum_{t=0}^{\infty} k_t(\alpha) e^{-i\omega t} = (-a_1)^\alpha. \quad (5.582)$$

The mean square error of prediction given by equation (5.323) becomes

$$I_{\min} = 1 + b_1^2 + b_2^2 + \cdots + b_{\alpha-1}^2$$
$$= 1 + a_1^2 + a_1^4 + \cdots + a_1^{2\alpha-2}. \quad (5.583)$$

D. Hybrid processes.—Let x_t be a process with power spectrum, which is a rational function in $e^{-i\omega}$

$$\Phi(\omega) = \frac{|B(\omega)|^2}{|A(\omega)|^2} \quad (5.59)$$

where

$$A(\omega) = \sum_{s=0}^{m} a_s e^{-i\omega s} \quad (5.591)$$

and

$$B(\omega) = \sum_{s=0}^{M} b_s e^{-i\omega s}. \quad (5.592)$$

Here the polynomials $A(z)$ and $B(z)$, $z=e^{-i\omega}$, are required to have no common factors, the roots of $A(z)$ to have modulus greater than one, and the roots of $B(z)$ to have modulus greater than or equal to one (see Figure 1). Then Doob (1949, 1953) shows that the process x_t, with spectrum $\Phi(\omega)$, is an hybrid between an autoregressive process and a finite moving average process. That is, we have

$$\sum_{s=0}^{m} a_s x_{t-s} = \gamma_t \quad (5.593)$$

where γ_t is the moving average

$$\gamma_t = \sum_{s=0}^{M} b_s \xi_{t-s} \tag{5.594}$$

with

$$E(\xi_t^2) = 1$$
$$E(\xi_t \xi_s) = 0 \qquad \text{for } s \neq t. \tag{5.595}$$

5.6 Multiple time series

In this section and the next we consider multiple discrete stationary ergodic time series with absolutely continuous spectral distribution functions. We shall have to modify our previous notation to some extent in order to accommodate the bulk of notation required. Let us consider the set of stationary processes $x_j(t)$, $(j=1, 2, \cdots, n)$ which we take to be real functions. From now on, where two or more secondary symbols appear, subscripts will denote the particular time series under consideration, whereas the time parameter will appear in the parentheses following the symbol for the function.

We define the correlation functions (Cramér, 1940) to be

$$\phi_{jk}(\tau) = E[x_j(t+\tau)x_k(t)]$$
$$= \lim_{T\to\infty} \frac{1}{2T+1} \sum_{t=-T}^{T} x_j(t+\tau)x_k(t). \tag{5.61}$$

For $j=k$ this equation gives the autocorrelation function of $x_j(t)$, whereas for $j\neq k$ it gives the crosscorrelation function of $x_j(t)$ and $x_k(t)$. We have

$$\phi_{jj}(\tau)=\phi_{jj}(-\tau), \quad j=1, 2, \cdots, n \tag{5.611}$$

which states that the autocorrelation function is an even function of τ. For the crosscorrelation functions, we have

$$\phi_{jk}(\tau) = \phi_{kj}(-\tau), \qquad k \neq j,$$
$$j, k = 1, 2, \cdots, n. \tag{5.612}$$

Since the time series $x_j(t)$ are real functions of time t, the correlation functions are real functions of τ. From their definition (5.61), the Schwarz inequality gives

$$|\phi_{jk}(\tau)| \leq [\phi_{jj}(0)\phi_{kk}(0)]^{1/2} \tag{5.613}$$

which provides a basis for normalizing the correlation functions.

Cramér (1940) shows that these correlation functions may be expressed as Fourier integrals of the form:

$$\phi_{jk}(\tau) = \frac{1}{2\pi}\int_{-\pi}^{\pi} e^{i\omega\tau}\Phi_{jk}(\omega)d\omega,$$
$$\text{for } j, k = 1, 2, \cdots, n. \tag{5.62}$$

The inverse transforms may be written as

$$\Phi_{jk}(\omega) = \sum_{\tau=-\infty}^{\infty} e^{-i\omega\tau}\phi_{jk}(\tau). \tag{5.63}$$

Here the $\Phi_{jk}(\omega)$ are the spectra of the set of stationary processes. For $j=k$, we have the power spectrum of $x_j(t)$:

$$\Phi_{jj}(\omega)=\Phi_{jj}(-\omega), \quad j=1, 2, \cdots, n \tag{5.631}$$

which is a positive real function of ω. For $j\neq k$, we have the cross-power spectrum $\Phi_{jk}(\omega)$ of $x_j(t)$ and $x_k(t)$. The cross-power spectrum, which is a complex valued function of the real variable ω, satisfies

$$\Phi_{jk}(\omega) = \Phi_{kj}(-\omega) \tag{5.632}$$

and

$$\Phi_{jk}(\omega) = \overline{\Phi_{kj}(\omega)} \tag{5.633}$$

where the bar indicates the complex conjugate. Consequently, we have

$$\Phi_{jk}(\omega) = \overline{\Phi_{jk}(-\omega)}. \tag{5.634}$$

Thus we see that the real part of the cross spectrum

$$\text{Re}\,[\Phi_{jk}(\omega)] = \text{Re}\,[\Phi_{jk}(-\omega)] \tag{5.635}$$

is an even function of the real variable ω, whereas the imaginary part

$$\text{Im}\,[\Phi_{jk}(\omega)] = -\,\text{Im}\,[\Phi_{jk}(-\omega)] \tag{5.636}$$

is an odd function of the real variable ω.

By letting $\tau=0$ in equation (5.62), we see that

$$\phi_{jk}(0) = \frac{1}{2\pi}\int_{-\pi}^{\pi} \Phi_{jk}(\omega)d\omega$$
$$= \frac{1}{\pi}\int_{0}^{\pi} \text{Re}\,[\Phi_{jk}(\omega)]d\omega. \tag{5.64}$$

Let us consider the linear combination

$$x(t) = \sum_{1}^{n} a_j x_j(t) \qquad (5.65)$$

which defines the time series $x(t)$. Here the weighting coefficients a_j are real. If $x_j(t), j=1, 2, \cdots, n$, represent the traces on a seismogram, and if $a_j=1$, $j=1, 2, \cdots, n$, then $x(t)$ is the stacked trace. The autocorrelation of $x(t)$ is

$$\begin{aligned}\phi(\tau) &= \lim_{T\to\infty} \frac{1}{2T+1} \sum_{t=-T}^{T} x(t+\tau)x(t) \\ &= \sum_{j=1}^{n}\sum_{k=1}^{n} a_j a_k \lim_{T\to\infty} \frac{1}{2T+1} \\ &\quad \cdot \sum_{t=-T}^{T} x_j(t+\tau)x_k(t) \qquad (5.66)\end{aligned}$$

which is

$$\phi(\tau) = \sum_{j=1}^{n}\sum_{k=1}^{n} a_j a_k \phi_{jk}(\tau). \qquad (5.661)$$

Since $\phi(\tau)$ is the autocorrelation function of the time series $x(t)$ it is an even function of τ:

$$\phi(\tau) = \phi(-\tau). \qquad (5.662)$$

The spectrum of $x(t)$ is given by the hermitian form

$$\begin{aligned}\Phi(\omega) &= \sum_{\tau=-\infty}^{\infty} \phi(\tau)e^{-i\omega\tau}d\tau \\ &= \sum_{j=1}^{n}\sum_{k=1}^{n} a_j a_k \Phi_{jk}(\omega) \geq 0 \qquad (5.67)\end{aligned}$$

which is a nonnegative function of ω (Cramér, 1940). The matrix of the hermitian form (5.67) may be represented by the hermitian matrix

$$[\Phi_{jk}(\omega)], \text{ for } j, k = 1, 2, \cdots, n \qquad (5.671)$$

which is

$$\begin{bmatrix} \Phi_{11}(\omega) & \Phi_{12}(\omega) & \cdots & \Phi_{1n}(\omega) \\ \Phi_{21}(\omega) & \Phi_{22}(\omega) & \cdots & \Phi_{2n}(\omega) \\ \cdot & \cdot & \cdots & \cdot \\ \Phi_{n1}(\omega) & \Phi_{n2}(\omega) & \cdots & \Phi_{nn}(\omega) \end{bmatrix}. \qquad (5.672)$$

We shall call this matrix, which determines the spectra of all possible linear combinations of $x_1(t), \cdots, x_n(t)$, the coherency matrix. The elements of our coherency matrix are the derivatives of the elements of Wiener's coherency matrix (Wiener, 1930, 1942).

Further, for the time series $x_1(t)$ and $x_2(t)$ with the coherency matrix

$$\begin{bmatrix} \Phi_{11}(\omega) & \Phi_{12}(\omega) \\ \Phi_{21}(\omega) & \Phi_{22}(\omega) \end{bmatrix} \qquad (5.673)$$

the significant invariants of this hermitian matrix are

$$\mathrm{coh}_{12}(\omega) = \frac{\Phi_{12}(\omega)}{[\Phi_{11}(\omega)\Phi_{22}(\omega)]^{1/2}} \qquad (5.674)$$

which Wiener (1930) calls the coefficient of coherency of $x_1(t)$ and $x_2(t)$ for frequency ω, and

$$\sigma_1(\omega) = \frac{\Phi_{12}(\omega)[\Phi_{11}(\omega)]^{1/2}}{\Phi_{22}(\omega)} \qquad (5.675)$$

and

$$\sigma_2(\omega) = \frac{\Phi_{21}(\omega)[\Phi_{22}(\omega)]^{1/2}}{\Phi_{11}(\omega)} \qquad (5.676)$$

which Wiener (1930) calls the coefficients of regression respectively of x_1 on x_2 and of x_2 on x_1. Wiener (1930) points out that the modulus of the coefficient of coherency represents the amount of linear coherency between $x_1(t)$ and $x_2(t)$ and the argument, the phase-lag of this coherency. The coefficients of regression determine in addition the relative scale for equivalent changes of $x_1(t)$ and $x_2(t)$.

Cramér (1940) shows that the determinant

$$\Phi_{11}(\omega)\Phi_{22}(\omega) - \Phi_{12}(\omega)\Phi_{21}(\omega) \qquad (5.677)$$

of the coherency matrix (5.673) is nonnegative. Therefore we have

$$\begin{aligned}\Phi_{12}(\omega)\Phi_{21}(\omega) &= |\Phi_{12}(\omega)|^2 \\ &\leq \Phi_{11}(\omega)\Phi_{22}(\omega) \qquad (5.678)\end{aligned}$$

so that the magnitude of the coefficient of coherency

$$|\mathrm{coh}_{12}(\omega)| = \frac{|\Phi_{12}(\omega)|}{[\Phi_{11}(\omega)\Phi_{22}(\omega)]^{1/2}} \qquad (5.679)$$

lies between zero and one.

Inequality (5.678) may be written

$$\sum_{\tau=-\infty}^{\infty} \phi_{12}(\tau)e^{-i\omega\tau} \sum_{s=-\infty}^{\infty} \phi_{21}(s)e^{-i\omega s}$$

$$\leq \sum_{\tau=-\infty}^{\infty} \phi_{11}(\tau)e^{-i\omega\tau} \sum_{s=-\infty}^{\infty} \phi_{22}(s)e^{-i\omega s} \quad (5.68)$$

which is

$$\sum_{r=-\infty}^{\infty} \sum_{\tau=-\infty}^{\infty} \phi_{12}(\tau)\phi_{12}(r)e^{-i\omega(\tau-r)}$$

$$\leq \sum_{r=-\infty}^{\infty} \sum_{\tau=-\infty}^{\infty} \phi_{11}(\tau)\phi_{22}(r)e^{-i\omega(\tau-r)} \quad (5.681)$$

or

$$\sum_{n=-\infty}^{\infty} e^{-i\omega n} \sum_{r=-\infty}^{\infty} \phi_{12}(n+r)\phi_{12}(r)$$

$$\leq \sum_{n=-\infty}^{\infty} e^{-i\omega n} \sum_{r=-\infty}^{\infty} \phi_{11}(n+r)\phi_{22}(r). \quad (5.682)$$

This inequality states that for two stationary time series the Fourier transform of the autocorrelation of their crosscorrelation cannot exceed the Fourier transform of the crosscorrelation of their autocorrelations.

As Wold (1953, Chapter 12.7, p. 202) observes, in view of the abundance of possible variations and combinations available in the extention of the theory of one dimensional stochastic processes to multiple stochastic processes, the main difficulty does not lie in developing formulas of great generality, but rather in picking out those processes for further study that merit interest from the point of view of applications. This observation is particularly pertinent concerning seismic applications, where any statistical approach to the study of multiple seismic traces should originate from considerations of the physical phenomena which generate these traces in time and space.

As an example of multiple time series let us consider the special case in which two stationary time series have the predictive decompositions (see Section 5.2)

$$x_1(t) = \sum_{s=0}^{\infty} b_s \xi_{t-s}$$

$$x_2(t) = \sum_{s=0}^{\infty} d_s \gamma_{t-s} \quad (5.683)$$

where the prediction error ξ_t of $x_1(t)$ represents a mutually uncorrelated process, that is

$$E[\xi_t^2] = 1, \; E[\xi_t \xi_s] = 0, \quad t \neq s.$$

where the prediction error of $x_2(t)$, γ_t, represents a mutually uncorrelated process, that is

$$E[\gamma_t^2] = 1, \quad E[\gamma_t \gamma_s] = 0, \; t \neq s, \quad (5.684)$$

and where the crosscorrelation of the prediction errors, ξ_t and γ_t, is given by

$$E[\xi_t \gamma_{t+\tau}] = \begin{cases} \rho & \text{for } \tau = \alpha \\ 0 & \text{otherwise.} \end{cases} \quad (5.685)$$

The inequality (5.682) becomes

$$\rho^2 \leq 1. \quad (5.686)$$

The autocorrelation $\phi_{11}(\tau)$ of $x_1(t)$ is

$$\phi_{11}(\tau) = E[x_1(t)x_1(t+\tau)]$$

$$= \sum_{s=0}^{\infty} b_s b_{s+\tau} \quad (5.687)$$

and the spectrum is

$$\Phi_{11}(\omega) = B(\omega)\overline{B(\omega)} = |B(\omega)|^2 \quad (5.688)$$

where

$$B(\omega) = \sum_{s=0}^{\infty} b_s e^{-i\omega s} \quad (5.689)$$

Likewise we have

$$\phi_{22}(\tau) = \sum_{s=0}^{\infty} d_s d_{s+\tau} \quad (5.69)$$

and

$$\Phi_{22}(\tau) = D(\omega)\overline{D(\omega)} = |D(\omega)|^2 \quad (5.691)$$

where

$$D(\omega) = \sum_{s=0}^{\infty} d_s e^{-i\omega s}. \quad (5.692)$$

The crosscorrelation of $x_1(t)$ and $x_2(t)$ for this special case is

$$\phi_{12}(\tau) = E[x_1(t)x_2(t)]$$

$$= \rho \sum_{s=0}^{\infty} b_s d_{s+\tau-\alpha} \quad (5.693)$$

and the cross spectrum is

$$\Phi_{12}(\omega) = \rho \sum_{\tau=-\infty}^{\infty} \left(\sum_{s=0}^{\infty} b_s d_{s+\tau-\alpha} \right) e^{-i\omega\tau}$$

$$= \rho e^{-i\omega\alpha} \sum_{n=-\infty}^{\infty} \sum_{s=0}^{\infty} b_s d_{s+n} e^{-i\omega n}$$

$$= \rho e^{-i\omega\alpha} \overline{B(\omega)} D(\omega). \qquad (5.694)$$

Thus the coherency matrix, equation (5.673), for this special case is

$$\begin{bmatrix} B\overline{B} & \rho e^{-i\omega\alpha} \overline{B} D \\ \rho e^{i\omega\alpha} B \overline{D} & D\overline{D} \end{bmatrix} \qquad (5.695)$$

and the coefficient of coherency is

$$\mathrm{coh}_{12}(\omega) = \rho e^{-i\omega\alpha} \frac{\overline{B(\omega)}}{|B(\omega)|} \cdot \frac{D(\omega)}{|D(\omega)|} \qquad (5.696)$$

the magnitude of which is

$$|\rho| \leq 1. \qquad (5.697)$$

We see that the coherency of the two time series, $x_1(t)$ and $x_2(t)$, depends on the crosscorrelation of their respective prediction errors, ξ_t and γ_t.

In general, if the magnitude of the coefficient of coherency is equal to one, we say the two time series are completely coherent; if equal to zero, completely incoherent. For completely coherent time series, the coherency matrix is singular.

The concept of coherency is an important one in the study of seismic records. Computations carried out by the MIT Geophysical Analysis Group indicate that seismic traces are more coherent on the average in an interval containing a major reflection than in an adjacent nonreflection interval. This coherency property of reflections assists the visual detection of reflections on a seismogram, and hence may be exploited in the detection of weak reflections by statistical methods. In these computations, the coherency was estimated through the estimation of the various power spectra and cross spectra. The problem of how to estimate spectra from finite time series is a major problem. In Figure 6, we show examples of correlation functions and spectra computed according to the method of Tukey (1949) and Tukey and Hamming (1949) from MIT Record No. 1 (supplied by the Magnolia Petro-

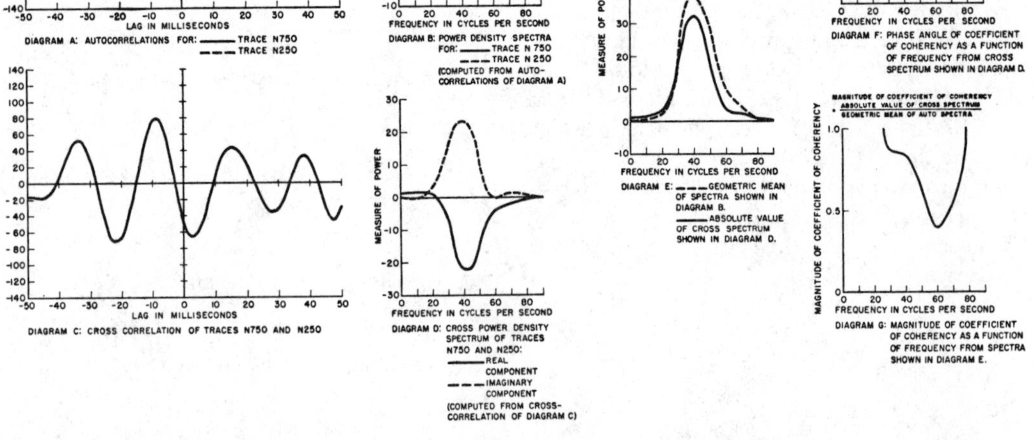

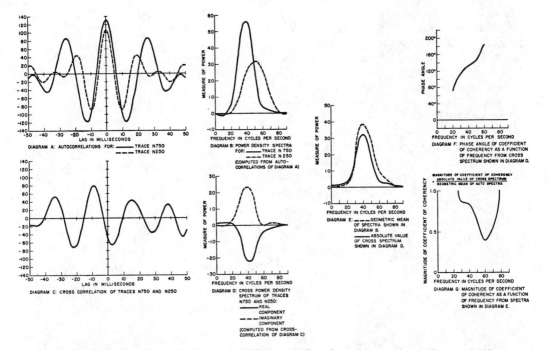

FIG. 6. Correlation functions and spectra, on MIT Record No. 1 from time equal to 1.05 sec to 1.225 sec.

leum Co.) over the time interval from 1.05 seconds to 1.225 seconds. Further discussion of this figure and of MIT Record No. 1 is given in Wadsworth et al. (1953), where also the method of Tukey and Hamming for the estimation of power spectra is presented. Also the reader is referred to MIT GAG Report No. 5 for an extension of the method of Tukey and Hamming to the estimation of cross spectra.

Under the hypothesis that a finite section of a seismic trace is a section of a stationary time series, which we discuss in the next chapter, a good estimate of the prediction operator for a single trace is one which yields prediction errors which satisfy statistical tests of being mutually uncorrelated. Then another approach to the study of coherency involves the examination of the cross-correlation existing between the sets of prediction errors for the various traces. A simple test to determine whether the empirical operator for a stationary time series is a good estimate of the theoretical operator consists of testing whether the empirical prediction errors are mutually uncorrelated.

5.7 General Technique of Discrete Prediction

Let us now examine the solution to the prediction problem for multiple discrete time series given by Wiener (1942, Section 4.6). We shall again write the equation numbers of Wiener (1942) to the left of our corresponding equation. Thus we start with the fundamental set of equations (discrete analogue of Wiener-Hopf integral equation)

$$\phi_{1k}(\alpha+t) = \sum_{j=1}^{n}\sum_{\tau=0}^{\infty}\phi_{jk}(t-\tau)q_j(\tau)$$

$$t \geq 0 (4.835), \quad k = 1, 2, \cdots, n \qquad (5.73)$$

the solution of which yields the linear operator $q_j(\tau)$ where $\tau=0, 1, 2, \cdots$ Since this set of equations need not hold for $t<0$, we shall define the residual for $t<0$ to be β_{-t}. (The β_{-t} for $t<0$ used here is Wiener's function b_t for $t>0$, in his Section 4.6. To avoid confusion in this section, we shall follow Wiener's notation quite closely except, as before, we let small letters refer to time functions and capital letters refer to frequency functions.) We have

$$\phi_{1k}(\alpha+t) - \sum_{j=1}^{n}\sum_{\tau=0}^{\infty}\phi_{jk}(t-\tau)q_j(\tau)$$

$$(4.635) \qquad = \begin{cases} 0, & t \geq 0 \\ \beta_{-t}, & t<0 \end{cases} \qquad (5.731)$$

where the residual β_{-t} for $t<0$ may be arbitrarily chosen so that the equation (5.73) is satisfied for $t \geq 0$. Let us put

$$(4.64) \qquad \sum_{n=-\infty}^{\infty}\phi_{jk}(n)e^{-i\omega n} = \Phi_{jk}(\omega) \qquad (5.732)$$

and

$$(4.64) \qquad \sum_{n=0}^{\infty} q_j(n)e^{-i\omega n} = Q_j(\omega). \qquad (5.733)$$

Then we have

$$\sum_{t=-\infty}^{\infty}\phi_{1k}(\alpha+t)e^{-i\omega t}$$

$$- \sum_{j=1}^{n}\sum_{\tau=0}^{\infty}\sum_{t=-\infty}^{\infty} e^{-i\omega t}\phi_{jk}(t-\tau)q_j(\tau)$$

$$= \sum_{t=-\infty}^{-1}\beta_{-t}e^{-i\omega t} \qquad (5.734)$$

which is

$$\sum_{t=-\infty}^{\infty}\phi_{1k}(t)e^{-i\omega(t-\alpha)} - \sum_{j=1}^{n}\sum_{\tau=0}^{\infty}e^{-i\omega\tau}$$

$$\cdot q_j(\tau)\sum_{-\infty}^{\infty}\phi_{jk}(t-\tau)e^{-i\omega(t-\tau)}$$

$$= \sum_{t=-\infty}^{-1}\beta_{-t}e^{-i\omega t} \qquad (5.735)$$

or

$$(4.645) \quad e^{i\omega\alpha}\Phi_{1k}(\omega) - \sum_{j=1}^{n}Q_j(\omega)\Phi_{jk}(\omega)$$

$$= \sum_{t=1}^{\infty}\beta_t e^{i\omega t} \quad \text{for } k = 1, 2, \cdots, n. \qquad (5.736)$$

For $n=1$, (the prediction problem for a single time series), we have

$$e^{i\omega\alpha}\Phi_{11}(\omega) - \Phi_{11}(\omega)Q_1(\omega) = \sum_{t=1}^{\infty}\beta_t e^{i\omega t}. \qquad (5.74)$$

Let us factor the power spectrum so that

$$\Phi_{11}(\omega) = |B(\omega)|^2 = B(\omega)\overline{B(\omega)} \quad (5.741)$$

where

$$B(\omega) = b_0 + \sum_1^n b_s e^{-i\omega s}. \quad (5.742)$$

(The $b_0, b_1, \cdots, b_n$ which we use here is our minimum-delay response function b_t of section 5.2, which in Wiener's notation is $d_0, d_1, \cdots, d_m$.) That is, Wiener assumes that $\Phi_{11}(\omega)$ is given by a terminating series in positive and negative powers of $e^{i\omega}$ so that it is the spectrum of a process of finite moving averages. The factorization may then be carried out as described in Section 5.5-A. Here $B(\omega)$, given by equation (5.742), is to have no zeros or singularities below the real axis; that is, $B(\omega)$ is to have minimum phase-shift characteristic.

Then we have

$$[e^{i\omega\alpha} - Q_1(\omega)]B(\omega) = \sum_{t=1}^{\infty} \beta_t e^{i\omega t} \frac{1}{B(\omega)}. \quad (5.743)$$

Let us define h_t $(t=1, 2, 3, \cdots)$ by

$$\frac{\sum_{t=1}^{\infty} \beta_t e^{i\omega t}}{\sum_{t=0}^{\infty} b_t e^{i\omega t}} = \sum_{t=1}^{\infty} h_t e^{i\omega t}. \quad (5.744)$$

Then equation (5.743) becomes

$$\sum_{s=\alpha}^{\infty} b_s e^{-i\omega(s-\alpha)} - Q_1(\omega)B(\omega) = \sum_{t=1}^{\infty} h_t e^{i\omega t} - \sum_{s=0}^{\alpha-1} b_s e^{-i\omega(s-\alpha)} \quad (5.745)$$

where the left hand side contains only nonpositive powers of $e^{i\omega}$, whereas the right hand side contains only positive powers of $e^{i\omega}$. Thus each side is respectively equal to zero, so we have

$$\sum_{s=\alpha}^{\infty} b_s e^{-i\omega(s-\alpha)} - Q_1(\omega)B(\omega) = 0 \quad (5.746)$$

and solving for the filter characteristic $Q_1(\omega)$ we have

$$Q_1(\omega) = \frac{1}{B(\omega)} \sum_{s=\alpha}^{\infty} b_s e^{-i\omega(s-\alpha)}. \quad (5.747)$$

Since

$$b_s = \frac{1}{2\pi}\int_{-\pi}^{\pi} B(u)e^{ius}du \quad (5.748)$$

we have

$$Q_1(\omega) = \frac{1}{2\pi B(\omega)} \sum_{s=\alpha}^{\infty} e^{-i\omega(s-\alpha)}$$

$$(2.630) \qquad \cdot \int_{-\pi}^{\pi} B(u)e^{ius}du \quad (5.749)$$

which is equation (5.373).

For $n=2$, (the prediction problem for double time series) we have (4.65)

$$[e^{i\alpha\omega} - Q_1(\omega)]\Phi_{11}(\omega) - \Phi_{21}(\omega)Q_2(\omega) = \sum_{s=1}^{\infty} \beta_s e^{i\omega s}$$

$$[e^{i\alpha\omega} - Q_1(\omega)]\Phi_{12}(\omega) - \Phi_{22}(\omega)Q_2(\omega) = \sum_{s=1}^{\infty} c_s e^{i\omega s} \quad (5.75)$$

where the residuals of the Wiener-Hopf equation (5.73) are β_s and c_s $(s=1, 2, 3, \cdots)$ for $k=1$ and $k=2$, respectively. The determinant of these equations is

$$\Phi(\omega) = \det\begin{bmatrix} \Phi_{11}(\omega) & \Phi_{21}(\omega) \\ \Phi_{12}(\omega) & \Phi_{22}(\omega) \end{bmatrix} = \Phi_{11}(\omega)\Phi_{22}(\omega) - \Phi_{21}(\omega)\Phi_{12}(\omega) \quad (5.751)$$

which is the determinant of the coherency matrix of the time series $x_1(t)$ and $x_2(t)$.

Thus for time series which are not completely coherent, the determinant of the coherency matrix does not vanish, and equations (5.75) will have a solution for the coefficients of the multiple linear operator as given by Wiener (1942) in the remainder of his section 4.6.

For finite observational time series, the Gauss method of least squares (Wadsworth et al. 1953) takes into account the empirical coherency existing between finite time series, and thus yields a unique solution for the empirical operator coefficients.

Finally, we note that Michel Loève (1946) has obtained a Predictive Decomposition Theorem for a nonstationary random process which generates time series $x(t)$ of finite time duration $0 \leq t \leq T$. Karhunen (1947) also treats this decomposition problem. Davis (1952) applies this predictive decomposition to the prediction problem for nonstationary time series. Since the time series are nonstationary, ensemble averages are used instead of time averages. Thus the autocorrelation function, which plays a central role in this Predictive Decomposition Theorem, is given by

$$\phi(t, s) = \mathrm{E}[x(t)x(s)] \qquad (5.76)$$

which now is a function of the two time instants t and s, and is no longer a function of only their difference $\tau = t - s$ as for a stationary process. Toward the determination of the applicability of this Predictive Decomposition Theorem to seismic data, an exploratory step would be to carry out computations to determine what degree of statistical regularity exists for estimates of the autocorrelation function (5.76).

CHAPTER VI

APPLICATIONS TO SEISMIC EXPLORATION

6.1 The response function

From a physical point of view, the seismic trace is the response of the system consisting of the earth and recording apparatus to the impulsive source, the explosion. This system, although usually very complicated, is susceptible to a deterministic approach toward its analysis. The explosion may be considered to yield an impulse ξ_t of relatively short time duration, so that the impulse ξ_t is equal to zero before the explosion and is equal to zero a short time after the explosion. In some instances the impulse function ξ_t may be considered to be a sharp impulsive disturbance of very short duration. In other instances, for example, the occurrence of bubble pulses in seismic prospecting over bodies of water (Worzel and Ewing, 1948), the shape of the impulse ξ_t may have a more complicated form.

The impulse ξ_t yields the energy of the seismic disturbance. This energy is dissipated in various ways as it spreads out from the source. Some of this energy is transmitted to the geophones and recorded in the form of seismic traces. Such recorded energy may be considered to be the response function to the impulse ξ_t. The study of this response function, that is, the response of the earth and recording system to the seismic explosion, has led to many important contributions to theoretical and practical seismology, and the reader is referred to past issues of GEOPHYSICS. Instead of dealing with the response function as such, which is a time function, one may deal with its Fourier transform which is the frequency and phase components of this function in the form of spectra.

Nevertheless, the complicated nature of seismograms taken in seismic exploration often precludes the study of the overall response of the earth and recording system taken as a whole. Also in the final analysis one is interested in the various components of this total response; for example, one wishes to separate components of reflected energy from those of nonreflected energy.

In a sequence of fundamental papers, Norman Ricker (1940, 1941, 1943, 1944, 1945, 1949, 1953a, 1953b) proposes the wavelet theory of seismogram structure. A seismogram, according to Ricker, is an elaborate wavelet complex, and the analysis of a seismogram consists in breaking the record down into its components.

Ricker (1940) points out that, according to the classical theory of the propagation of elastic waves in homogeneous, isotropic media, a wave form remains unchanged as it is transmitted. Thus a wave due to a sharp impulse, such as an explosion, should be propagated without change in form and received at a distance as the same waveform. Consequently in media strictly obeying the elastic equations a seismogram should consist of a succession of disturbances, due to waves which have traveled different paths by refractions and reflections. Ricker goes on to state that if such a sharp and clear-cut series of impulses did constitute a seismogram, many of the difficulties in seismic prospecting would disappear. As we know, however, no such simple seismogram is received in the propagation of seismic waves through the earth. Instead he points out that we obtain more complicated seismograms, which are familiar to every geophysicist.

In order to explain this complicated nature of a seismogram, Ricker proposes his wavelet theory of seismogram structure. The reader is referred to Ricker's work in which he demonstrates mathematically and experimentally that a sharp seismic disturbance, or impulse, gives rise to a traveling wavelet, the shape of which is determined by the nature of the absorption spectrum of the earth for elastic waves. The shape of this wavelet, which is a time function, is the response of the earth to the sharp seismic disturbance, or impulse. A seismogram, then, consists of many of these wavelets, with different strengths and arrival times, due to disturbances which have traveled different paths by refractions and reflections.

6.2 The statistical determination of seismic wavelets and their arrival times

Thus the seismogram may be visualized as the totality of responses to impulses, each impulse being associated with a disturbance which has traveled a certain path by refractions and reflections. These responses, or response functions, are the seismic wavelets. The analysis of a seismogram consists in breaking down this elaborate wavelet complex into its component wavelets. In particular we desire the arrival times of the theoretical sharp impulses which produce these wavelets or responses.

There are two basic approaches which one may use toward the solution of this problem, the deterministic approach and the probabilistic or statistical approach. In the deterministic approach one utilizes basic physical laws, for example, in order to determine the shape of the wavelet, or the absorption spectrum of the earth. At all stages in such an investigation, one may compare mathematical results with direct and indirect observation of the physical phenomenon.

In this thesis we are concerned with the statistical approach. Such an approach in no way conflicts with the deterministic approach, although each approach has certain advantages and disadvantages which do not necessarily coincide. The emphasis we place on the probabilistic approach is due to its being the subject of investigation of this thesis. In practice the two approaches may be utilized in such a manner so as to complement each other.

Let us apply the probabilistic approach to one specific problem. Let us set up a hypothetical situation. Let us assume that a given section of seismic trace is additively composed of seismic wavelets, where each wavelet has the same shape or form. We shall assume that the shape of the wavelet is minimum-delay, that is, the discrete representation of the wavelet shape is a solution of a stable difference equation. Further, we assume that from knowledge of the arrival time of one wavelet we cannot predict the arrival time of another wavelet; and, we assume that from knowledge of the strength of one wavelet we cannot predict the strength of another wavelet. Finally, let us assume that the seismic trace is an automatic volume control (AVC) recording so that the strengths of these wavelets have a constant standard deviation (or variance) with time.

The specific problem which we wish to consider is the following: given the seismic trace described in the above paragraph, determine the arrival times and strengths of the seismic wavelets, and determine the basic wavelet shape. We shall discuss a theoretical solution of this problem, and shall also discuss a practical solution which involves statistical estimation.

Let us translate our assumptions about the seismic trace into mathematical notation for discrete time t. First let the shape of the fundamental seismic wavelet be given by the discrete minimum-delay time function b_t where $b_t=0$ for t less than zero. That is, b_0 is the initial (nonzero) amplitude of the wavelet. Discrete minimum-delay time functions are discussed in Section 2.6.

Let the strength, or weighting factor, of the wavelet which arrives at time t be given by ξ_t. That is, ξ_t is a constant weighting factor which weights the entire wavelet whose arrival time is time t. The variable ξ_t is the theoretical impulse of which the particular wavelet (i.e. the one which arrives at time t) is the response. For example, if no wavelet arrives at a particular time t, then $\xi_t=0$.

In our discussion of the nature of the seismic trace, we shall call the impulses ξ_t "random variables." Our use of the term "random variable ξ_t" does not imply that the variable ξ_t is one whose value is uncertain and can be determined by a "chance" experiment. That is, the variable ξ_t is not random in the sense of the frequency interpretation of probability (Cramér, 1946), but is fixed by the geologic structure. Frechet (1937) describes this type of variable as "nombre certain" and "function certaine" and Neyman (1941) translates these terms by "sure number" and "sure function." Another example of a "sure number" is the ten millionth digit of the expansion $e=2.71828\ldots$, which, although unknown, is a definite fixed number. Since the knowledge of working geophysicist about the entire deterministic setting is far from complete, we shall treat this incomplete knowledge from a statistical point of view. We thus call ξ_t a "random variable," although we keep in mind that it is a "sure number." Further discussions about this general type of problem may be found in the statistical literature with discussions about the theorem of the English clergyman Thomas Bayes and with discussions about statistical estimation (Cramér, 1946; Jeffreys, 1939). The relationship of the use of Bayes' Theorem in statistical estimation to other methods of statistical estimation is discussed by the author (1950).

Without loss of generality, we may center the impulses ξ_t so that their mean $E[\xi_t]$ is equal to zero. Nevertheless, the following discussions may be carried out, by some minor modifications, without centering the ξ_t.

Our assumption about the unpredictability of the arrival times and strengths of wavelets means mathematically that the impulses ξ_t are mutually uncorrelated random variables, that is

$$E[\xi_t \xi_s] = 0, \qquad s \neq t. \qquad (6.21)$$

An explanation of the expectation symbol E is given in Section 4.2; of mutually uncorrelated variables, Section 4.5. Our assumption that the impulses ξ_t are mutually uncorrelated with each other is an orthogonality assumption, and is a weaker assumption than the assumption that the ξ_t are statistically independent, which we need not make.

Returning again, for the moment, to our discussion about the "sure" nature of the impulses ξ_t, we see that the assumption that they are mutually uncorrelated in time and in strength does not hold in a completely deterministic system. Nevertheless, such an assumption is a reasonable one again for the working geophysicist whose knowledge of the entire deterministic setting is far from complete, and who is faced with essentially a statistical problem.

In other words, we assume that knowledge of the arrival time and strength of one wavelet does not allow us to predict the arrival time and strength of any other wavelets. In particular, we assume that an arrival time and magnitude of a reflection from a certain reflecting horizon does not allow us to predict the arrival time and magnitude of a reflection from a deeper reflecting horizon.

The use of AVC recordings means mathematically that the strengths ξ_t have a constant variance, which without loss of generality we shall take to be unity,

$$E[\xi_t^2] = 1. \qquad (6.211)$$

Finally, since we assume that the seismogram trace x_t is additively composed of wavelets, all with shape b_t and strengths ξ_t, we may write this wavelet complex mathematically as

$$x_t = \sum_{s=0}^{\infty} b_s \xi_{t-s} \qquad \text{for } t_1 \leq t \leq t_2 \qquad (6.22)$$

where the time interval ($t_1 \leq t \leq t_2$) comprises our basic section of seismic trace. This equation includes tails of wavelets with shape b_t, these wavelets being due to impulses $\xi_{t1-1}, \xi_{t1-2}, \cdots$, which occur before time t_1. Equation (6.22) is illustrated in Figure 7, in which the top diagram shows the impulses ξ_t, the center diagram shows the wavelets b_t weighted by these impulses, and the bottom diagram shows the seismic trace x_t which is obtained by adding the wavelets of the center diagram.

For the purposes of our theoretical discussion, let us assume that our assumptions about the time series x_t, equation (6.22), now hold for all time. That is, we consider the mathematical abstraction in which equation (6.22) holds for all t, where ξ_t now represents a stationary mutually uncorrelated process (Section 4.5). Thus equation (6.22) becomes

$$x_t = \sum_{s=0}^{\infty} b_s \xi_{t-s} \quad \text{for } -\infty < t < \infty. \qquad (6.221)$$

Equation (6.221) is the mathematical representation of the Predictive Decomposition Theorem for a stationary time series with an absolutely continuous spectral distribution function. For further discussion of this theorem, see Section 5.2. Thus the infinite time series x_t given by equation (6.221) is a stationary time series with an absolutely continuous spectral distribution, and the finite time series x_t given by equation (6.22) represents a finite section of the infinite time series (6.221).

In other words, the Predictive Decomposition Theorem states that a stationary time series is the summation of the responses of minimum-delay system to impulses ξ_t which have uncorrelated

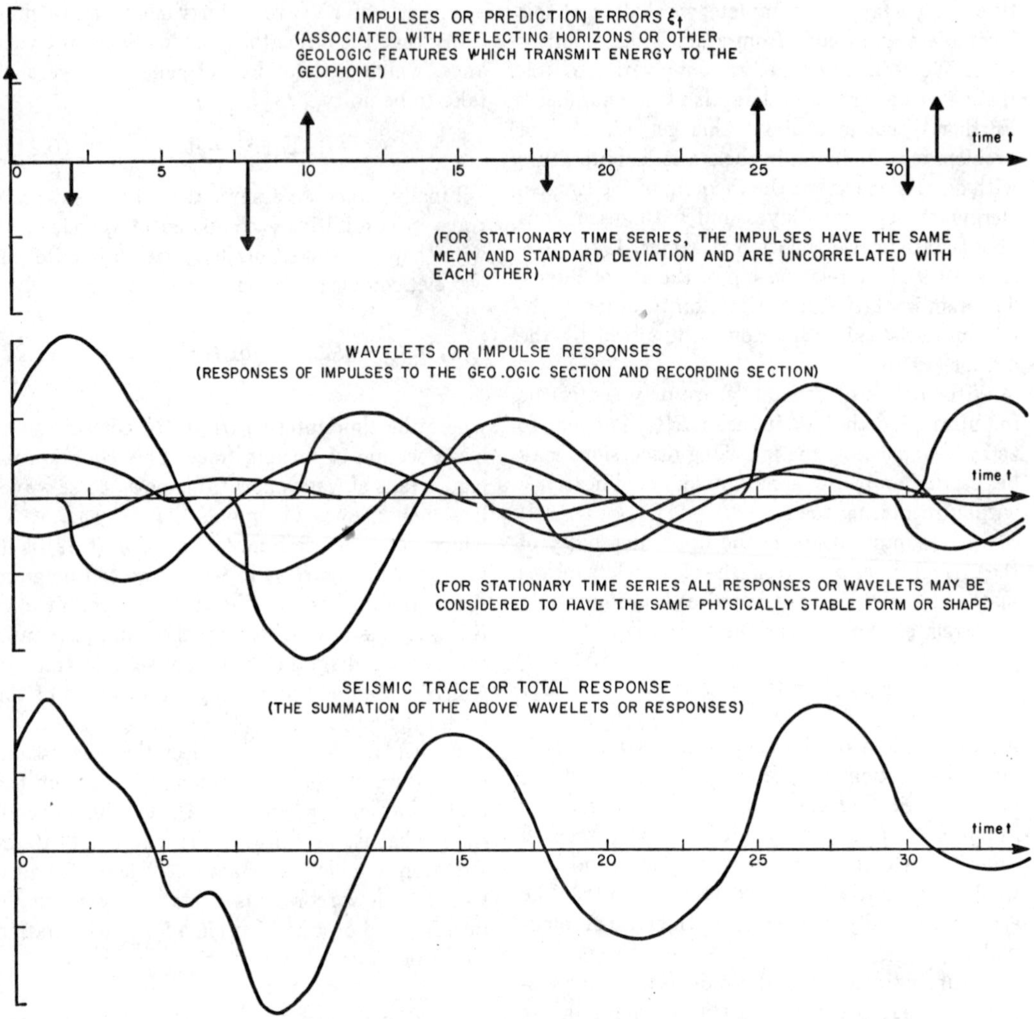
IMPULSES OR PREDICTION ERRORS ξ_t
(ASSOCIATED WITH REFLECTING HORIZONS OR OTHER GEOLOGIC FEATURES WHICH TRANSMIT ENERGY TO THE GEOPHONE)
time t
0
5
10
15
20
25
30
(FOR STATIONARY TIME SERIES, THE IMPULSES HAVE THE SAME MEAN AND STANDARD DEVIATION AND ARE UNCORRELATED WITH EACH OTHER)
WAVELETS OR IMPULSE RESPONSES
(RESPONSES OF IMPULSES TO THE GEOLOGIC SECTION AND RECORDING SECTION)
time t
(FOR STATIONARY TIME SERIES ALL RESPONSES OR WAVELETS MAY BE CONSIDERED TO HAVE THE SAME PHYSICALLY STABLE FORM OR SHAPE)
SEISMIC TRACE OR TOTAL RESPONSE
(THE SUMMATION OF THE ABOVE WAVELETS OR RESPONSES)
time t
0
5
10
15
20
25
30

FIG. 7. The predictive decomposition of a seismic trace.

strengths and arrival times. The response to each impulse has the minimum-delay shape b_t, and the variance $E[\xi_t^2]$ of the impulses is constant with time.

In equation (6.221), the wavelet b_t represents the dynamics of the time series, whereas the impulses ξ_t represents the "random" nature of the time series. The basic problem which we wish to consider consists of the separation of the dynamic from the random components of the time series, or seismic trace.

Will the computation of the Fourier transform of the trace effect this separation? The answer is no, because that merely transforms time information into equivalent frequency information. As an illustration, let us consider the following example.

To avoid difficulties with end effects, let us assume, for this example, that the wavelet b_t damps sufficiently rapidly so that we may let

$$b_t = 0 \qquad \text{for } t > M. \qquad (6.23)$$

Then the predictive decomposition becomes

$$x_t = \sum_{s=0}^{M} b_s \xi_{t-s}. \qquad (6.231)$$

Also, for this example, let us assume the trace for $t_1 \leq t \leq t_2$ consists of only those responses to impulses ξ_t which arrive for times $t_1 \leq t \leq t_2 - M$. The

Fourier transform of this section of the trace becomes

$$X(\omega) = \sum_{t_1}^{t_2} x_t e^{-i\omega t} = \sum_{t=t_1}^{t_2} \sum_{s=0}^{M} b_s \xi_{t-s} e^{i\omega t}$$

$$= \sum_{s=0}^{M} b_s e^{-i\omega s} \sum_{t=t_1}^{t_2} \xi_{t-s} e^{-i\omega(t-s)}$$

$$= \left(\sum_{s=0}^{M} b_s e^{-i\omega s} \right) \left(\sum_{t=t_1}^{t_2-M} \xi_t e^{-i\omega t} \right)$$

$$= B(\omega) I(\omega) \qquad (6.232)$$

where $B(\omega)$ is the Fourier transform, or spectrum, of the wavelet and $I(\omega)$ is the Fourier transform of a realization of the uncorrelated impulses. Although the Fourier transform $X(\omega)$ contains the dynamic and random elements of a seismic trace, it does not help us to separate the dynamic component $B(\omega)$ from the random component $I(\omega)$ since $X(\omega)$ is the product of the two.

In order to separate the random components ξ_t from the dynamic component b_t of the seismic trace one may use statistical method of averaging. The basic probabilistic approach from a theoretical point of view consists of the following operations on the mathematical abstraction of the seismic trace (i.e., the stationary time series x_t, given by equation 6.221):

(1) Average out the random components ξ_t so as to yield the wavelet shape b_t.

(2) Using the wavelet shape thus found, remove this wavelet shape from the trace, thereby leaving, as a residual, the random components ξ_t (which are the prediction errors for prediction distance $\alpha = 1$).

If one wishes to filter the seismic trace (see Section 5.4) one further step is added, namely:

(3) Re-filter the prediction errors ξ_t by means of a stable linear operator q_t so as to approximate the desired output or message $m_{t+\alpha}$. That is, compute

$$\hat{m}_{t+\alpha} = \sum_{\tau=0}^{\infty} q_{\tau+\alpha} \xi_{t-\tau} \qquad (5.441)$$

which is the optimum filtered time series in the least-squares sense. In Section 5.4 we describe how the linear operator q_t is determined from the spectra and cross spectra of message and noise.

The theoretical procedure for carrying out these operations has been treated in detail in our discussion of stationary time series (with absolutely continuous spectral distributions) in the preceding chapters. Let us review this theoretical procedure for infinite stationary time series:

(1) Compute the autocorrelation function of the time series

$$\phi(\tau) = \lim_{T\to\infty} \frac{1}{2T+1} \sum_{t=-T}^{T} x_t x_{t+\tau}$$

$$= \mathrm{E}[x_t x_{t+\tau}]$$

$$= \sum_{t=0}^{\infty} b_t b_{t+\tau}$$

$$= b_0 b_\tau + b_1 b_{\tau+1} + b_2 b_{\tau+2} \cdots. \qquad (5.232)$$

This computation averages out the random elements ξ_t and preserves the dynamic elements b_t in the form of the autocorrelation function

$$\sum_{t=0}^{\infty} b_t b_{t+\tau}$$

of the wavelet. That is, the autocorrelation of the time series x_t is the same function as the autocorrelation function of the wavelet b_t.

From this autocorrelation function, compute the shape b_t of the wavelet in the following manner. Take the Fourier transform of the autocorrelation function to get the power spectrum $\Phi(\omega)$ of the time series x_t, which is also the energy spectrum $|B(\omega)|^2$ of the wavelet b_t; that is

$$\sum_{\tau=-\infty}^{\infty} \phi(\tau) e^{-i\omega\tau} = \Phi(\omega) = |B(\omega)|^2 \qquad (6.24)$$

where

$$B(\omega) = \sum_{t=0}^{\infty} b_t e^{-i\omega t} \qquad (6.241)$$

is the Fourier transform of the wavelet b_t. Thus we have determined $|B(\omega)|^2$ but not $B(\omega)$. Although there are many wavelet shapes which yield the energy spectrum $|B(\omega)|^2$, only one of these wavelet shapes is minimum-delay. Therefore it is not unreasonable to assume that this unique minimum-delay wavelet is the wavelet generated by a physical phenomenon, and that the others are not. The Fourier transform $B(\omega)$ of this minimum-delay wavelet may then be found by the Féjer or Kolmogorov method of factoring the power spectrum (see Section 5.1) expressed by

$$\Phi(\omega) = |B(\omega)|^2 = B(\omega)\overline{B(\omega)} \quad (5.181)$$

where $B(\lambda)$ is required to have no singularities or zeros in the lower half λ plane, where $\lambda = \omega + i\sigma$. In the language of the engineer, $B(\omega)$ is a transfer function with minimum phase-shift characteristic. Having thus determined $B(\omega)$, the minimum-delay wavelet (or linear operator) b_t is given by

$$b_t = \frac{1}{2\pi}\int_{-\pi}^{\pi} B(\omega)e^{i\omega t}d\omega. \quad (6.242)$$

(2) From this wavelet shape b_t, we find the inverse wavelet shape a_t, where a_t is equal to zero for t less than zero. If we let the b_t represent the coefficients of a linear operator, then the a_t are the coefficients of the inverse linear operator (Section 2.7). Thus the values of a_t are found by

$$a_t = 0 \qquad \text{for } t < 0$$

$$a_0 b_0 = 1$$

$$\sum_{s=0}^{t} a_s b_{t-s} = 0 \quad \text{for } t = 1, 2, 3 \cdots. \quad (2.785)$$

Since the wavelet b_t is minimum-delay, the inverse wavelet a_t is also minimum-delay. Let $A(\omega)$ be the Fourier transform of a_t:

$$A(\omega) = \sum_{s=0}^{\infty} a_s e^{-i\omega s}. \quad (2.791)$$

Then $A(\omega)$ and $B(\omega)$ are related by

$$A(\omega) = \frac{1}{B(\omega)} \quad (2.795)$$

and $A(\omega)$ also has minimum phase-shift characteristic. The reciprocal of $|A(\omega)|$ then gives $|B(\omega)|$ which is the absolute value of the wavelet spectrum.

We use the inverse wavelet shape a_t to remove the wavelets, which are of shape b_t, from the time series x_t, by compressing the wavelets into the impulses ξ_t. That is, the linear operator a_t is the prediction error operator for unit prediction distance, and the prediction errors ξ_t are yielded by the computation

$$\sum_{s=0}^{\infty} a_s x_{t-s}. \quad (6.25)$$

To see that this computation does yield the ξ_t we use the predictive decomposition (6.221) for x_{t-s} and thus obtain

$$\sum_{s=0}^{\infty} a_s x_{t-s} = \sum_{s=0}^{\infty} a_s \sum_{\tau=0}^{\infty} b_\tau \xi_{t-s-\tau}$$

$$= \sum_{s=0}^{\infty} a_s \sum_{n=s}^{\infty} b_{n-s}\xi_{t-n}. \quad (6.251)$$

Recalling that $b_r = 0$ for $r < 0$, and using equation (2.785), we have

$$\sum_{s=0}^{\infty} a_s x_{t-s}$$

$$= \sum_{n=0}^{\infty}\left[\sum_{s=0}^{\infty} a_s b_{n-s}\right]\xi_{t-n} = \xi_t. \quad (6.252)$$

Thus we have

$$\sum_{s=0}^{\infty} a_s x_{t-s} = \xi_t \quad (6.253)$$

which is the prediction error.

Thus by these theoretical steps we may separate the dynamic component, represented by the response function or wavelet shape b_t, from the random component, represented by the impulses ξ_t which represent arrival times and strengths of the wavelets which comprise the time series. These theoretical steps are illustrated in Figure 8. In this figure, as in others, we plot the discrete time functions as points and then draw smooth curves through these points.

The practical solution of the problem of separating the dynamic and random components of a finite section of seismic trace involves statistical estimation. One method consists of estimating the prediction error operator, or inverse wavelet shape, directly from the finite section of seismic trace. For this purpose one may use the Gauss method of least squares as described in Wadsworth et al. (1953). Since the method described there is more general, let us write down the equations to be used for our specific problem in which we consider only one trace x_t for a prediction distance equal to one.

Let us note these differences in notation: the prediction distance k of Wadsworth et al. (1953) is our α; the operator coefficients $a_s (s = 0, 1, \cdots, M)$ of Wadsworth et al., are, respectively, our operator coefficients k_s $(s = 0, 1, \cdots, M)$ of equa-

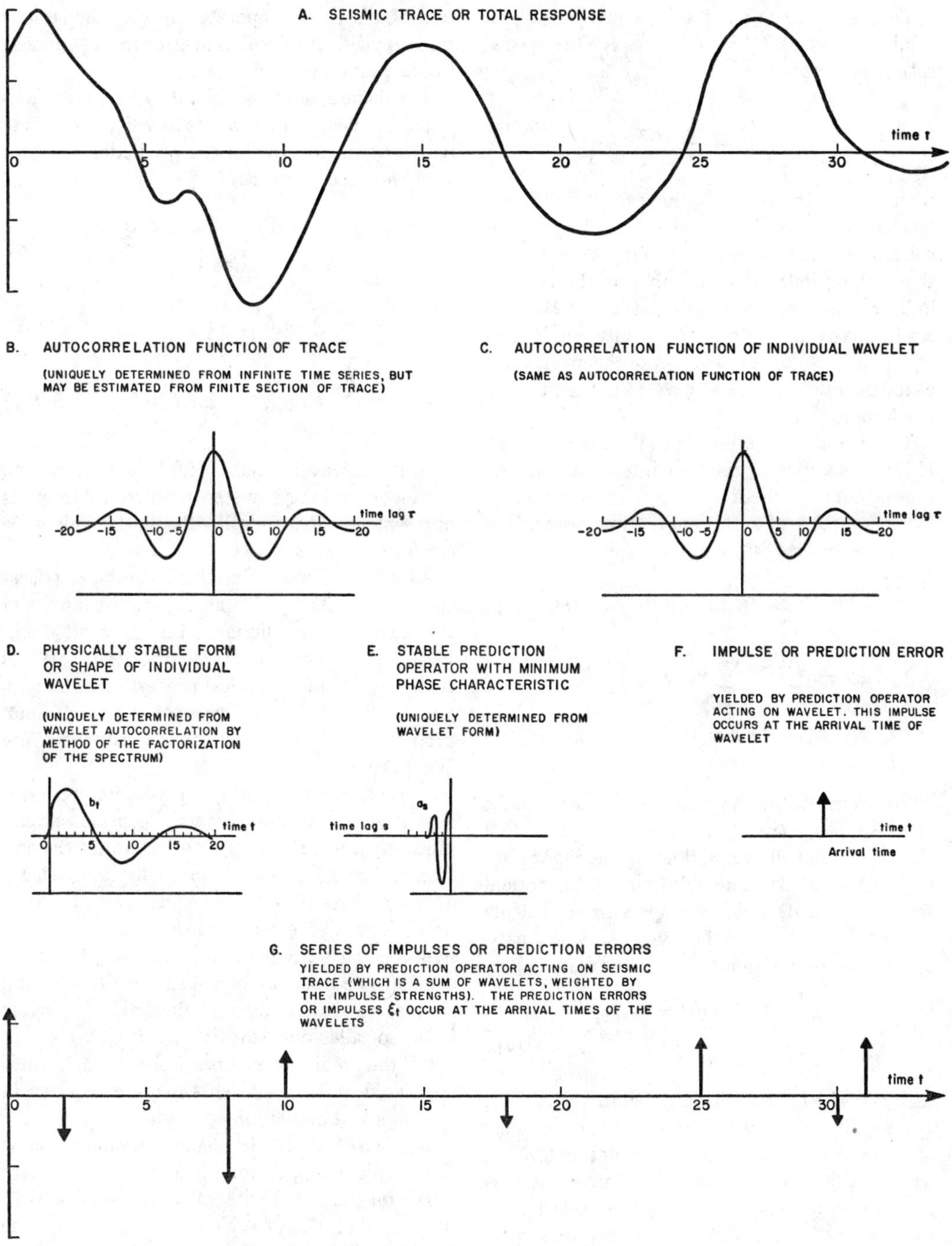

FIG. 8. Analysis of a seismic trace by a prediction operator.

tion (2.21); and the operator coefficients b_s for the y-trace of Wadsworth et al., are not our operator coefficients b_s in this thesis. In other words, our use of the symbols a_s and b_s is different from the use of the symbols a_s and b_s in Wadsworth et al. (1953).

Then utilizing our notation, equation (37) of Wadsworth et al. (1953) becomes for the special case of our problem:

$$(37) \qquad \hat{x}_{t+1} = c + \sum_{s=0}^{M} k_s x_{t-s} \qquad (6.26)$$

where we use the same notation except for the differences we have just noted. According to our convention, we have let the spacing $h=1$, so that the running index t is the same as the running index i. Equation (6.26) is equation (2.21) with $\alpha=1$, except that a constant c also appears in equation (6.26) to take account of the mean value of the time series, since now we do not require the mean to be zero.

The operator time interval (Wadsworth et al. 1953) is chosen to be the time interval of the section of our hypothetical trace which we assume to be a section of a stationary time series. The normal equations (40) become

$$cn + \sum_s k_s \sum_t x_{t-s} = \sum_t x_{t+1}$$

$$c \sum_t x_{t-r} + \sum_s \left(k_s \sum_t x_{t-r} x_{t-s} \right)$$
$$= \sum_t x_{t-r} x_{t+1}, \quad \text{for } r = 0, 1, \cdots, M \quad (6.261)$$

where, as in Wadsworth et al. (1953), summations on the index t are for $t=N-1$ to $t=N+n-2$ since $\alpha=1$, and all summations on the index s are for $s=0$ to $s=M$. The solution of the normal equations (6.261) yields the operator coefficients $c, k_0, k_1, \cdots, k_M$. Then the inverse wavelet shape a_t is given by equation (2.281) which is

$$a_0 = 1, \qquad a_1 = -k_0,$$
$$a_2 = -k_1, \cdots, a_m = -k_M \quad (2.281)$$

with $m=M+1$. As we have noted in Section 2.2, although both a_t and k_t represent the coefficients of the same operator, as seen by equation (2.281), we call a_t the standard form of the prediction error operator. Also, as we have noted, our $k_0, k_1, k_2, \cdots, k_M$ in equation (2.281) are respectively the $a_0, a_1, a_2, \cdots, a_M$ of Wadsworth et al. (1953). The constant c of equation (6.26), which adjusts for the mean value of the empirical trace, is not used in determining the shape a_t of the wavelet. Since $a_0=1$, the inverse wavelet a_t may be called a "unit" inverse wavelet. The convolution of the inverse operator a_t with the seismic trace x_t yields the prediction error trace ξ_t, as seen from equation (6.252).[7]

The shape b_t of the wavelet may be readily computed by means of equations (2.785), which may be rewritten in terms of the prediction operator k_s of equation (6.26) as

$$b_t = 0 \qquad \text{for } t < 0$$
$$b_0 = 1$$
$$b_{t+1} = -\sum_{s=1}^{m} a_s b_{t+1-s}$$
$$= \sum_{s=0}^{M} k_s b_{t-s}, \qquad \text{for } t > 0. \qquad (6.262)$$

That is, the wavelet shape b_t for $t>0$ is determined by successive step-by-step predictions from its past values, where we let the initial values be $b_t=0$ for $t<0$ and $b_0=1$.

As we have seen, the Gauss method of least squares described yields an empirical estimate of the theoretical prediction error operator, or inverse wavelet shape. This empirical estimate has certain optimum statistical properties under general conditions. For a treatment of the optimum properties of linear least-squares estimates, see Wold (1953).

A good estimate of the prediction operator should yield prediction errors which are not significantly autocorrelated. In other words, the prediction errors ξ_t should be mutually uncorrelated at some preassigned level of significance. Let it be noted that we are confining our attention to the hypothetical section of the trace which we assumed to be a section of a stationary time series; that is, we are dealing with the prediction errors in the so-called operator time interval. For example, if the prediction errors are significantly autocorrelated, more coefficients may be required in the empirical prediction operator.

In Figure 9, in the left hand diagram, we show the prediction operator a_s computed for trace N650 for the time interval 0.350 sec to 0.475 sec on MIT Record No. 1 (supplied by the Magnolia Petroleum Co.). This seismogram is illustrated and described in Wadsworth et al. (1953). In the computation of this inverse wavelet

[7] In current (1967) terminology, the inverse operator is called the *deconvolution operator*, and the prediction error trace is called the *deconvolved trace* (the editors).

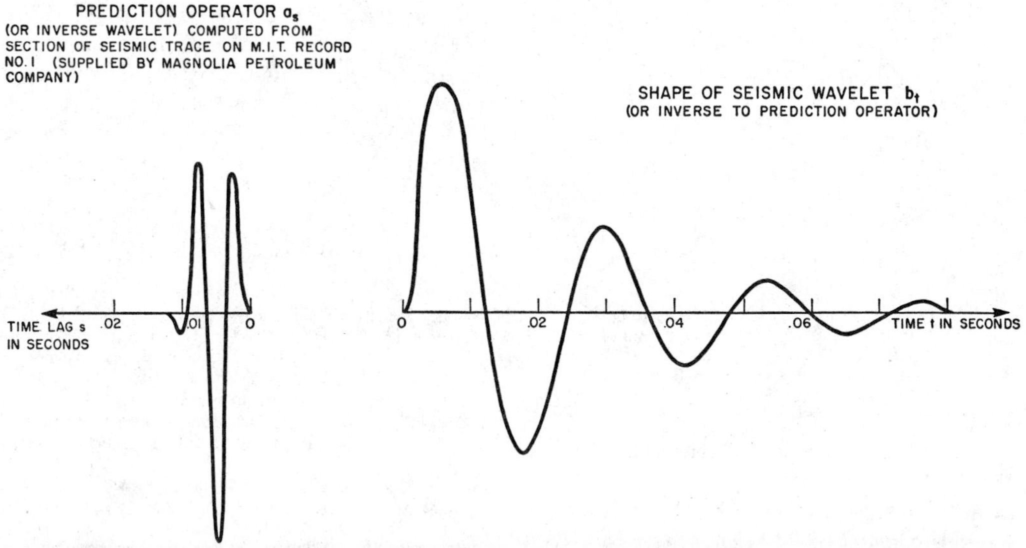

FIG. 9. Computation of wavelet from section of trace on MIT Record No. 1 (supplied by the Magnolia Petroleum Company).

shape a_s we used equations (6.261) to find the k_s, and then used equations (2.281) to find the a_s. In the right hand diagram of Figure 9, we show the inverse prediction operator, which is the shape of the wavelet b_t. The shape of the wavelet was "predicted" by means of equation (6.262). In these computations, we used discrete time series where the spacing $h=2.5$ ms. In plotting a_s and b_t we followed our usual procedure which is to plot discrete time functions, such as a_s and b_t, as discrete points, and then to draw a smooth curve through these discrete points. Also in Figure 9, the time axes are shifted by one discrete time unit (which is 2.5 ms), which is not the convention we have used in our other figures. Thus in Figure 9, $a_{-1}=0$ is plotted at time lag $s=0$, and $a_0=1$ is plotted at time lag $s=0.0025$ sec. Similarly $b_{-1}=0$ is plotted at time $t=0$ sec and $b_0=1$ is plotted at time $t=0.0025$ sec. As is our usual convention, the prediction operator a_s is plotted in the reverse manner, as described by Swartz and Sokoloff (1954); that is, the time lag s runs in the positive direction toward the left. Swartz and Sokoloff (1954) also describe the filtering action of discrete linear operators, and their relation to the continuous response functions of electric filters.

Here we have described a statistical method to determine the shape of a seismic wavelet. Alternatively, from other considerations, one may know the shape of the seismic wavelet. Then the prediction error operator, or inverse wavelet shape, may be computed by means of equations (2.785).

So far we have confined ourselves to a section of trace which we assume to be approximately stationary. The prediction error operator transforms this section of trace into the uncorrelated prediction errors ξ_t, the mean square value of which is a minimum. As we have seen the operator cannot predict from past values of the trace the initial arrival of a new wavelet, and thus a prediction error ξ_t is introduced at the arrival time of each wavelet. Nevertheless, for times subsequent to the arrival time of the wavelet, the prediction error operator which is the inverse to the wavelet can perfectly predict this wavelet, thereby yielding zero error of prediction.

Nevertheless, a seismic trace is not made up of wavelets which have exactly the same form and which differ only in amplitudes and arrival times. Thus if a prediction error operator, which is the unique inverse of a certain wavelet shape, encounters a different wavelet shape, the prediction error will no longer be an impulse, but instead will be a transient time function. Thus the prediction errors yielded by this prediction error operator acting on a time series additively composed of wavelets of different shapes will not have a minimum mean square value. Since reflected wavelets

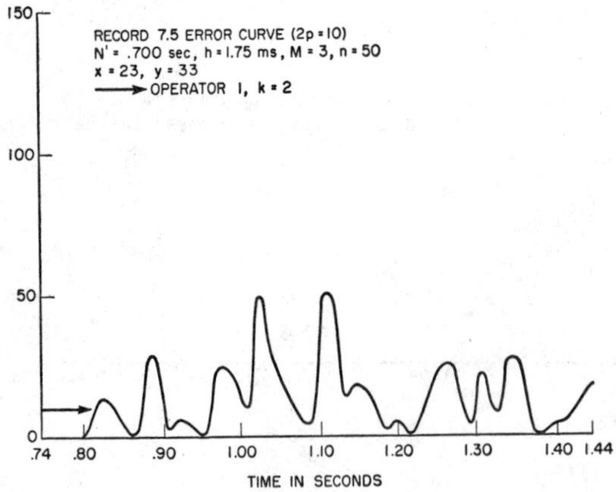

FIG. 10. Prediction error curve.

in many cases have different shapes than the wavelets comprising the seismic trace in a given nonreflection interval, a prediction error operator determined from this nonreflection interval will yield high errors of prediction at such reflections. Such a procedure provides a method for the detection of reflections (Wadsworth et al., 1953). In Figures 10 and 11, running averages of the squared prediction errors are plotted. The peaks on these prediction error curves indicate reflections on the seismogram. Since two-trace operators were used, the empirical coherency existing between the two traces was utilized in the determination of these prediction errors. The arrow indicates the operator time interval. Further description of these figures is given in MIT GAG Report No. 6. Since only the information existing in the operator time interval is used in the determination of linear operators by this method, one

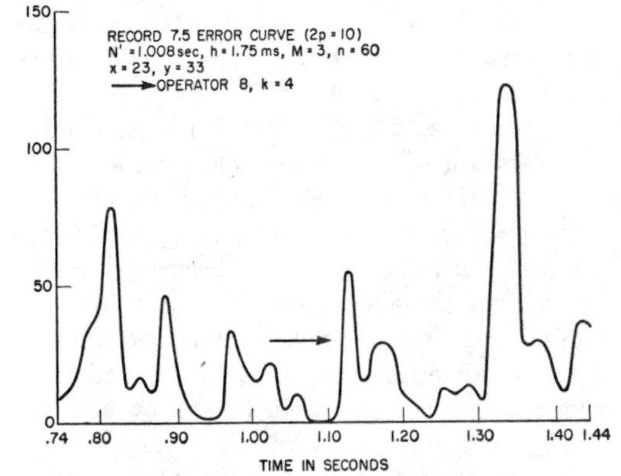

FIG. 11. Prediction error curve.

may expect greater resolving power if more information on the seismogram is utilized in the determination of various other types of operators. Further research on this general subject is now being carried out by the Geophysical Analysis Group.

REFERENCES

Aitken, A. C., 1935, On least squares and linear combination of observations: Proc. Roy. Soc. Edinburgh, v. 55, p. 42–48.

Bartlett, M. S., 1936, The information available in small samples; Proc. Camb. Phil. Soc., v. 32, p. 560.

Bartlett, M., 1946, On the theoretical specification and sampling properties of autocorrelated time-series: J. Roy. Statist. Soc. (B) v. 8, p. 27–41.

Bode, H. W., and Shannon, C. E., 1950, A simplified derivation of linear least-square smoothing and prediction theory: Proc. Inst. Radio. Eng., v. 38, p. 417–425.

Booten, R. C., 1952, Nonstationary statistical theory associated with time-varying linear systems: Sc.D. Thesis, M.I.T.

Bullen, K. E., 1953, An introduction to the theory of seismology (Second edition): Cambridge, Cambridge University Press.

Champernowne, D. G., 1948, Sampling theory applied to auto-regressive sequence: J. Roy. Statist. Soc. (B), v. 10, p. 204–231 (Discussion, p. 232–242).

Cochrane, D., and Orcutt, G. H., 1949, Application of least-squares regression to relationships containing autocorrelated error terms: J. Amer. Statist. Assoc., v. 44, p. 32–61.

Cohen, R., 1948, Some analytical and practical aspects of Wiener's theory of prediction: Technical Report No. 69, Res. Lab. Electronics, M.I.T.

Cramér, H., 1940, On the theory of stationary random processes: Ann. Math., v. 41, p. 215–230.

——— 1942, On harmonic analysis in certain functional spaces: Ark. Mat. Astr. Fys., v. 28B, no. 12.

——— 1945, Mathematical methods of statistics: Princeton, Princeton University Press.

——— 1947, Problems in probability theory: Ann. Math. Statist., v. 18, p. 165–193.

——— 1951, A contribution to the theory of stochastic processes: Proc. Second Berkeley Symp. Math. Statistics and Prob., Berkeley, California, p. 329–339.

Davis, R. C., 1952, On the theory of prediction of nonstationary stochastic processes: J. Appl. Phys., v. 23, p. 1047.

Dix, C. H., 1952, Seismic prospecting for oil: New York, Harper and Brothers.

Dobrin, M. B., 1952, Introduction to geophysical prospecting: New York, McGraw-Hill Book Company, Inc.

Doob, J. L., 1949, Time series and harmonic analysis: Proc. Second Berkeley Symp. Math. Statistics and Prob., Berkeley, California, p. 303–343.

——— 1953, Stochastic processes: New York, John Wiley and Sons, Inc.

Féjer, L., 1915, Über trigonometrische Polynome: J. reine ang. Math., v. 146, p. 53–82.

Feller, W., 1950, An introduction to probability theory and its applications, v. 1: New York, John Wiley and Sons.

Fisher, R. A., 1921, On the mathematical foundations of theoretical statistics, Phil. Trans. Roy. Soc., p. 222–309.

——— 1925, Statistical methods for research workers

(Eleventh edition, 1950): Edinburgh, Oliver and Boyd.

Frank, H. R., and Doty, W. E. M., 1953, Signal-to-noise ratio improvements by filtering and mixing: Geophysics, v. 23, p. 587.

Frechet, M., 1937, Recherches théoriques modernes sur la théorie des probabilities: Paris, Gauthier-Villars.

Goldman, S., 1953, Information theory: New York, Prentice-Hall, Inc.

Hamerle, J. F., 1951, Linear prediction of discrete stationary time series: S.M. Thesis, M.I.T.

Hardy, G. H., Littlewood, J. E., and Polya, G., 1934, Inequalities: Cambridge, Cambridge University Press.

Heiland, C. A., 1940, Geophysical exploration: New York, Prentice-Hall, Inc.

Jakosky, J. J., 1950, Exploration geophysics: Los Angeles, Trija Publishing Co.

Jeffreys, H., 1939, Theory of probability: Oxford, Oxford University Press.

——— 1952, The earth (Third edition): Cambridge, Cambridge University Press.

Kac, M., and Siegert, A. J. F., 1947a, An explicit representation of a stationary Gaussian process: Ann. Math. Statist., v. 18, p. 438–442.

——— 1947b, On the theory of noise in radio receivers with square-law detectors: J. Appl. Phys., v. 18, p. 383–397.

Karhunen, K., 1947, Über lineäre Methoden in der Wahrscheinlichkeitsrechnung: Ann. Acad. Sci. Fennicae, Ser. A, I. Math. Phys., v. 37.

Kendall, M., 1946, The advanced theory of statistics, v. 2: London, Chas. Griffin and Co., Ltd.

——— 1946, Contributions to the study of oscillatory time-series: Natl. Inst. of Econ. Soc. Res., Occasional Papers, 9, Cambridge, Cambridge University Press.

Khintchine, A., 1933, Asymptotische Gesetze der Wahrscheinlichkeitsrechnung. Ergeb. der Math. und ihrer Grenz. v. 2, no. 4.

——— 1934, Korrelationstheorie der stationären stochastischen Prozesse: Math. Ann., v. 109, p. 604–615.

Kolmogorov, A., 1933, Grundbegriffe der Wahrscheinlichkeitsrechnung: Ergeb. Math. und ihrer Grenzgeb. (Also separately at Berlin, Springer; American edition, 1950, New York, Chelsea Pub. Co.)

——— 1939, Sur l'interpolation et extrapolation des suites stationnaires: C. R. Acad. Sci. Paris, v. 208, p. 2043–2045.

——— 1941, Interpolation und Extrapolation von stationären zufalligen Folgen: Bull. Acad. Sci. U.S.S.R., Ser. Math., v. 5, p. 3–14.

Koopmans, T. C., Ed., 1950, Statistical inference in dynamic economic models: Cowles Commission Monograph No. 10, New York, John Wiley and Sons.

——— 1951, Activity analysis of production and allocation, Cowles Commission Monograph No. 13: New York, John Wiley and Sons.

Kosulajeff, P. A., 1941, Sur les problemes d'interpolation et d'extrapolation des suites stationnaires: Comptes rendus de l'academie des sciences de l'U.R.S.S., v. 30, p. 13–17.

Lawson, J. E., and Uhlenbeck, G. E., 1950, Threshold signals: New York, McGraw-Hill Book Co., Inc.

Lee, Y. W., 1949, Communication applications of correlation analysis: Woods Hole Symp. Appl. Autocorr. Anal. Phys. Prob., p. 4–23, Office of Naval Research, NAVEXOS-P-735.

——— 1950, Application of statistical methods to communication problems: Technical Report No. 181, M.I.T. Res. Lab. Electronics.

Lee, Y. W., and Stutts, C. A., 1949, Statistical prediction of noise: Technical Report No. 129, M.I.T. Res. Lab. Electronics.

Levinson, N., 1947a, The Wiener rms (root mean square) error criterion in filter design and prediction: J. Math. Phys., v. 25, p. 261–278.

——— 1947b, A heuristic exposition of Wiener's mathematical theory of prediction and filtering: J. Math. Phys., v. 26, p. 110–119.

Loève, Michael, 1946, Quelques propriétés des fonctions aléatoires de seconde ordre: C. R. Acad. Sci. Paris, v. 222, p. 469–470.

Middleton, D., 1953, Statistical criteria for the detection of pulsed carriers in noise; I, II: J. Appl. Phys., v. 24, p. 371–391.

MIT Geophysical Analysis Group Reports[8]

Wadsworth, G. P., Robinson, E. A., Bryan, J. G., and Hurley, P. M., Oct. 1, 1952, Results of an autocorrelation and cross-correlation analysis of seismic records: Geophysical Analysis Group, Report No. 1, M.I.T. (Reprinted in Wadsworth, et al., 1953).

——— Nov. 5, 1952, A prospectus on the application of linear operators to seismology: Geophysical Analysis Group, Report No. 2, M.I.T. (Reprinted in Wadsworth, et al., 1953).

Robinson, E. A., Simpson, S. M., and Smith, M. K., July 8, 1953, Case study of Henderson County seismic record: Geophysical Analysis Group, Report No. 3, M.I.T.

——— July 21, 1953, Linear operator study of a Texas Company seismic profile, Part I: Geophysical Analysis Group, Report No. 4, M.I.T.

——— August 4, 1953, On the theory and practice of linear operators in seismic analysis: Geophysical Analysis Group, Report No. 5, M.I.T.

——— March 10, 1954, Further research on linear operators in seismic analysis: Geophysical Analysis Group, Report No. 6, M.I.T.

Nettleton, L. L., 1940, Geophysical prospecting for oil: New York, McGraw-Hill Book Co., Inc.

Neyman, J., 1941, Fiducial argument and the theory of confidence: Biometrika, v. 32, p. 128.

Neyman, J., and Pearson, E. S., 1933, On the problem of the most efficient tests of statistical hypotheses: Phil. Trans. Roy. Soc. (A), v. 231, p. 289–338.

——— 1936, Sufficient statistics and uniformly most powerful tests of statistical hypothesis: Statist. Res. Mem., v. 1.

Paley, R. E. A. C., and Wiener, N., 1934, Fourier transforms in the complex domain: Amer. Math. Soc. Colloq., Pub. v. 19.

Piety, R. G., 1942, Interpretation of the transient behavior of the reflection seismograph: Geophysics, v. 7, p. 123–132.

Quenouille, M. H., 1947, A large-sample test for the goodness of fit of autoregressive schemes: J. Roy. Statist. Soc. (A), v. 110, p. 123–129.

Rice, S. O., 1944–1945, Mathematical analysis of random noise: Bell Sys. Tech. J., v. 23, p. 282–332; v. 24, p. 46–156.

Ricker, N., 1940, The form and nature of seismic waves and the structure of seismograms: Geophysics, v. 5, p. 348–366.

——— 1941, A note on the determination of the viscosity of shale from the measurement of wavelet breadth: Geophysics, v. 6, p. 254–258.

——— 1943, Further development in the wavelet

[8] Available from the MIT Library, Cambridge, Mass.

theory of seismogram structure: Bull. Seis. Soc. Amer., v. 33, p. 197–228.
——— 1945, The computation of output disturbances from amplifiers for true wavelet inputs: Geophysics, v. 10, p. 207–220.
——— 1949, Attenuation and amplitudes of seismic waves: Trans. Amer. Geophys. Union, v. 30, p. 184–186.
——— 1953a, The form and laws of propagation of seismic wavelets: Geophysics, v. 18, p. 10–40.
——— 1953b, Wavelet contraction, wavelet expansion, and the control of seismic resolution: Geophysics, v. 18, p. 769–792.
Robinson, E. A., 1950, The theories of fiducial limits and confidence limits: S.B. Thesis, M.I.T.
——— 1952, Eight new applications of the theory of stochastic processes to economic analysis: S.M. Thesis, M.I.T.
Samuelson, P. A., 1941a, The stability of equilibrium: Comparative statics and dynamics: Econometrica, v. 9, p. 97–120.
——— 1941b, Conditions that the roots of a polynomial be less than unity in absolute value: Ann. of Math. Statist., v. 12, p. 360–366.
——— 1941c, A method of determining explicitly the coefficients of the characteristic equation: Ann. of Math. Statist., v. 12, p. 424–429.
——— 1947, Foundations of economic analysis: Cambridge, Harvard University Press.
Schuster, A., 1898, On the investigation of hidden periodicities with application to a supposed 26-day period of meteorological phenomena: Terrest. Magnet., v. 3.
——— 1900, The periodogram of the magnetic declination as obtained from the records of the Greenwich Observatory during the years 1871–1895: Trans. Cambridge Phil. Soc., v. 18.
Seiwell, H. R., and Wadsworth, G. P., 1949, A new development in ocean wave research: Science, v. 109, p. 271–274.
Simpson, S. M., 1953, Statistical approaches to certain problems in geophysics: Ph.D. Thesis, M.I.T.
Slutsky, E., 1927, The summation of random causes at the source of cyclic processes (Russian with English summary): Problems of Economic Conditions (Inst. Econ. Conjucture, Moscow), v. 3 (Revised English edition, 1937: Econometrica, v. 5, p. 105–146).
——— 1928, Sur les fonctions eventuelles continues. integrables et derivables dans le sens stochastique:, C. R. Acad. Sci. Paris, v. 187, p. 878–880.
Smith, M. K., 1954, Filter theory of linear operators with seismic applications: Ph.D. Thesis, M.I.T. (Sections to which references are made here are reprinted in M.I.T. Geophysical Analysis Group Report No. 6).
Swartz, C. A., 1954, Some geometrical properties of residual maps: Geophysics, v. 19, p. 46–70.
Swartz, C. A., and Sokoloff, V. M., 1954, Filtering associated with selective sampling of geophysical data: Geophysics, v. 19, p. 402–419.
Titner, G., 1952, Econometrics: New York, John Wiley and Sons.
Tukey, J. W., 1949, The sampling theory of power spectrum estimates: Woods Hole Symp. Appl. Autocorr. Phys. Prob., p. 47–67, Office of Naval Research, NAVEXOS-P-735.
Tukey, J. W., and Hamming, R. W., 1949, Measuring noise color, I; Bell Telephone Laboratories Mem. MM-49-110-119.
von Neumann, J., 1941, Distribution of the mean square succession difference to the variance: Ann. Math. Statist., v. 12, p. 367–395.
von Neumann, J., and Morgenstern, O., 1944, Theory of games and economic behaviour (Second edition, 1947): Princeton, Princeton University Press.
Wadsworth, G. P., 1947, Short-range and extended forecasting by statistical methods: Air Weather Service Tech. Rep. 105-37.
Wadsworth, G. P., Robinson, E. A., Bryan, J. G., and Hurley, P. M., 1953, Detection of reflections on seismic records by linear operators: Geophysics, v. 18, p. 539–586.
Whittaker, E. T., and Robinson, G., 1924, The calculus of observations (Fourth edition, 1944): London, Blackie.
Wiener, N., 1923, Differential space: J. Math. Phys., M.I.T., v. 2, p. 131–174.
——— 1930, Generalized harmonic analysis: Acta Math., v. 55, p. 117–258.
——— 1933, The Fourier integral and certain of its applications: Cambridge, Cambridge University Press.
——— 1942, The extrapolation, interpolation, and smoothing of stationary time series with engineering applications: Cambridge, M.I.T. DIC Contract 6037, National Defense Research Council (Section D2) (Reprinted 1949: New York, John Wiley and Sons, Inc., to which edition references to page numbers refer).
Wiener, N., 1948, Cybernetics: New York, John Wiley and Sons, Inc.
Wold, H., 1938, A study in the analysis of stationary time series: Uppsala, Almqvist and Wisksells.
——— 1948, On prediction in stationary time series: Ann. Math. Statist., v. 19, p. 559–567.
Wold, H., 1953 (in association with L. Jureen), Demand Analysis: New York, John Wiley and Sons.
Worzel, J. L., and Ewing, M., 1948, Explosion sounds in shallow water: Geol. Soc. Amer. Mem. 27.
Yule, G. U., 1907, On the theory of correlation for any number of variables treated by a new system of notation: Proc. Roy. Soc., v. 79, p. 182–193.
——— 1921, On the time-correlation problem, with special reference to the variate-difference correlation method: J. Roy. Stat. Soc., v. 84, p. 497–526.
——— 1926, Why do we sometimes get nonsense-correlations between time-series? J. Roy. Stat. Soc., v. 89, p. 1–64.
——— 1927, On a method of investigating periodicities in disturbed series, with special reference to Wolfer's sunspot numbers: Phil. Trans. Roy. Soc., v. 226, p. 267–298.
Zadeh, L. A., 1950, The determination of the impulsive response of variable networks: J. Appl. Phys., v. 21, p. 642–655.
——— 1953, Optimum nonlinear filters: J. Appl. Phys. v. 24, p. 396–404.
Zadeh, L. A., and Ragazzini, J. R., 1950, An extension of Wiener's theory of prediction: J. Appl. Phys., v. 21, p. 645–655.
Zygmund, Antoni, 1935, Trigonometrical series: Warsaw-Lwow.

Reprinted from Geophysics v. 29, no. 2, p. 197-211

DETECTION OF P-WAVES FROM WEAK SOURCES AT GREAT DISTANCES*

JON F. CLAERBOUT†

Optimum (Wiener sense) filters for suppression of noise in multiple time series are computed by a new method due to E. A. Robinson. Filters for prediction error and interpolation error have been used to detect P-wave signals from three teleseismic events. These filters facilitate detection of signals in noise with low signal-to-noise ratios. The instrumentation consists of short-period Benioff seismometers, both three-component stations and surface arrays of verticals.

It was found that microseismic noise in the pass band of these instruments is more accurately termed "Brownian motion of a surface" than "random waveforms with characteristic direction(s) of propagation." Thus, single time-series filters work almost as well as multiple time-series matrix filters. Prediction-error filters gave results substantially more satisfactory than simple band-pass filters.

INTRODUCTION

The basic problem in the detection of weak seismic waves from distant sources is recognition of signals in the presence of ambient microseismic noise. In filtering to improve the signal-to-noise ratio, one characteristic of the signal we would like to preserve, if possible, is the first motion.

One approach to optimum filtering which has to preserve the first motion is prediction filtering. A prediction-error filter predicts the noise at a future time and subtracts the prediction from the actual noise present at that time. The filter does not predict the signal so that with the noise highly predictable a large output indicates a signal. A prediction-error filter will not begin to distort the signal until the time equal to the prediction interval after the onset of the signal. Thus, the first cycle of the signal, or the first fraction of the first cycle, will be undistorted and disguised only to the extent that the filter fails to predict the noise exactly.

The prediction of the noise on a particular seismic trace from an array of seismometers may be markedly improved if the prediction filter makes use of not only the noise on that trace, but also the crosscorrelation of the noise on that trace with the noise on other seismic traces. Thus, a prediction filter for N time series consists of an $N \times N$ matrix of filters. Since one equation is required for each filter coefficient, the use of methods of multiple time series was not practical heretofore because of the large amount of computing required to solve the basic equations.

Due to a new computational algorithm or method of E. A. Robinson (1963), it is now practical to apply much of the mathematical theory of multiple time series to the problem of detecting weak signals. This paper presents an attempt to exploit this new technique.

PREDICTION AND INTERPOLATION ERROR IN MULTIPLE TIME SERIES

A prediction-error filter for an array of seismometers will use the history of the ith series and the history of all the other time series in the array to predict and cancel the noise on the ith trace. A block diagram for the prediction-error filtering of two time series is given in Figure 1.

The filters for this study were derived as follows: A sample of noise just prior to the onset of a teleseism is selected, a sample long compared to the signal duration. The noise statistics are then determined from the auto- and crosscorrelations of the noise sample. From the correlation functions, a set of prediction filters for multiple inputs is derived which best predicts the noise traces in the sample interval. Then, under the assumption that the noise statistics remain stationary for a time outside the sample interval, these fixed mul-

* Manuscript received by the Editor November 11, 1963.
† Data Analysis and Technique Development Center, United ElectroDynamics, Inc., Alexandria, Virginia. Currently studying in the Institute of Seismology and the Institute of Statistics, Uppsala University, Uppsala, Sweden.

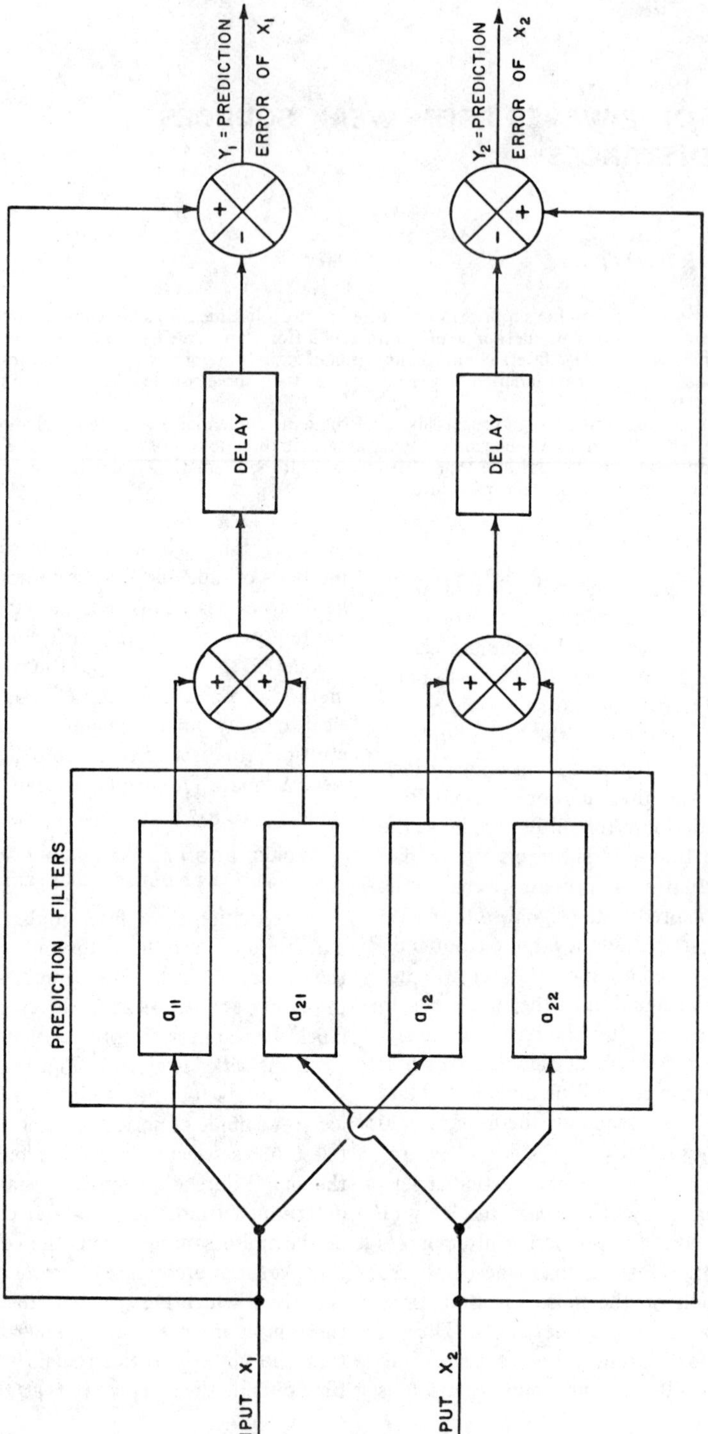

FIG. 1. Schematic of prediction-error filtering of a time series with bivariate structure.

tichannel filters are used to improve the signal-to-noise ratio for subsequent data.

The prediction span is a preset parameter. There is a different set of prediction filters for each prediction span. As the span increases, the filters predict more and more conservatively, until finally the prediction error tends to the unfiltered trace, i.e. the prediction tends to zero.

If the correlation matrix of the prediction error is computed, it will be found that both the auto- and crosscorrelations are zero for lags greater than the prediction span. This is because any correlation, plus or minus, at a lag greater than the prediction span can be used to enhance the least-squares prediction. If one computed the "prediction error of the prediction error," one would find that nothing was improved by the second stage of prediction since there was no correlation upon which to base the second prediction. (The inhomogeneous part of the prediction equations vanishes, leading to a zero-predicting filter.)

If the noise prediction were perfect, the desired signal starting from its first motion and lasting for a length of time equal to the prediction span would be noise-free and undistorted by filtering. Since the noise is not perfectly predictable, that segment is still undistorted by filtering but has a certain (reduced) noise content. After a time equal to the prediction span, the predictors will be operating on not just the noise, but the noise plus the signal. Thus, signal distortion can become a problem only after the prediction span. If we are primarily interested in the time or shape of the first motion, this need not concern us, at least not when the desired signal does not affect other traces before the filtered trace.

Another type of filtering closely related to prediction-error filtering is interpolation-error filtering. Interpolation-error filtering is like prediction-error filtering except that the diagonal-term filters are left out. This means that the history of the signal being predicted is not used in the prediction. Then the prediction need not even be a prediction into the future but might simply be a prediction (best guess) as to the present or past. In this case, one signal is said to be interpolated from the other signals. For example, if we are detecting teleseisms (distant events) whose angle of emergence is almost vertical we could assume that all the signal is on the vertical component and hence that the horizontal components are relatively pure samples of the noise. Then we can construct filters to interpolate what the vertical noise should be and subtract it from the actual vertical component. Like the prediction error, the result is expected to be small until a real signal arrives. In principle, the signal cannot be distorted by interpolation filtering. In practice, the amount of the noise which can be subtracted depends upon the interpolatability of the noise. This type of filtering involves no assumptions whatever of the waveform or the spectrum of the event being detected.

FORMULAS AND COMPUTATIONS

Formulas for prediction of multiple time series are in the literature (Robinson, 1962). Therefore, only a cursory derivation will be given here.

Let x_t be an input time series and $x_{t,i}$ be the ith such series. The time parameter, t, takes on integral values. Let $y_{t,i}$ be the corresponding output of the filter a. Generalizing Figure 1 to filtering an arbitrary number of time series, the definition of convolution in standard tensor algebra notation is:

$$y_{t,i} = a_{\tau,i,j} x_{t-\tau,j}.$$

Thus, the filter, $a_{\tau,i,j}$, is seen to be a three-dimensional matrix as is the correlation matrix which is defined as:

$$XY_{\tau,i,j} = x_{(t+\tau),i} y_{t,j}.$$

One minimizes the sum-squared difference between the actual filter output, $y_{t,i}$, and some desired output $d_{t,i}$. The minimization is done by differentiating with respect to each of the elements of the filter matrix.

To minimize $(y_{t,i} - d_{t,i})^2$ using the product rule for differentiation, we get

$$0 = 2(y_{t,i} - d_{t,i}) \frac{\partial}{\partial a_{pqr}} (y_{t,i} - d_{t,i})$$

$$0 = (a_{s,i,l}\, x_{(t-s),l} - d_{t,i}) \frac{\partial}{\partial a_{pqr}} (a_{\tau,i,k} x_{(t-\tau),k})$$

$$0 = (a_{s,i,l}\, x_{(t-s),l}\, x_{(t-\tau),k} - d_{t,i}\, x_{(t-\tau),k})$$
$$\delta_{pt}\delta_{qi}\delta_{kr},$$

or using the definition of correlation we get the tensor algebraic equation

$$0 = a_{s,i,l} XX_{(\tau-s),l,k} - DX_{\tau,i,k}$$

which must be solved for the three-dimensional array of filter coefficients $a_{s,i,l}$. These equations, often called normal equations, are well known (Robinson, 1962).

Solutions are computed by solving n sets of simultaneous equations each of which has the same $(n \times T) \times (T \times n)$ matrix of coefficients where n is the number of time series and T is the (discrete) time duration of the impulse response of the filters.

The biggest set of simultaneous equations which can now be solved by standard methods in a reasonable length of time on a full-scale digital computer is about 100×100. If one were filtering three-component seismograms, this would imply filters of about 33 coefficients. Since this encompasses only a few cycles of the data being filtered, the filter can hardly do anything spectacular which the human eye could not do just looking at the data.

Now a new method (Robinson, 1963) of solving the normal equations leads to more satisfactory results. The method is a generalization of a method given by Levinson (1947). Levinson's method solves the normal equations for single time series by a recurrence technique. Robinson wrote the normal equations for multiple time series in a form where they resemble those of single time series, the difference being that the elements of the simultaneous equations, both coefficients and variables, are scalars in the single time series case, whereas the elements are partitioned submatrices for multiple time series. Then he showed that one could use Levinson's recurrence method simply by replacing scalar multiplication and division with matrix multiplication and inversion and making a few other changes since matrices do not commute. The result is a practical method of solving the normal equations. The time increases only quadratically with the filter length instead of cubically as with the usual methods of simultaneous equations. Examples of typical filter lengths and their corresponding computation time are given in Figure 2.

Number of time series	1	3	10
Number of filters	1	9	100
Number of coefficients per filter	500	150	20
Total number of regression coefficients	500	1350	2,000

FIG. 2. Typical times required for exact computation of the solution filters of the normal equations in five minutes on the CDC 1604 computer.

A MODEL FOR THE OBSERVED NOISE

Although it was not the primary purpose of this work to study the microseismic noise, something was learned of its nature. There is evidence that noise sources are distributed fairly homogeneously in azimuth. This seems to be particularly true at the frequencies which are important for detecting distant events. Spatial correlation drops very low with separations greater than about a wavelength. This implies that simple velocity filtering of the noise will not work. If a man were able to hear microseisms, he would not hear distinct sounds from certain directions, but rather he would hear a fairly steady buzz coming from all directions. This type of noise might be called Brownian motion of a surface.

These statements must be qualified to some extent. At periods which are too long (greater than 1.5 sec) to be of great importance in the detection of weak events at great distances the noise definitely has some directional properties. At 4-sec periods for example, it is often possible to see a few (one or two) wave lengths move across an array of seismometers. In fact the phase velocity of microseisms has been measured at these periods in this way (Toksöz, 1964). An example of this is the microseismic noise three minutes before the arrival of the southern Algeria event of March 18, 1963, at the Tonto Forest Observatory. A few waveforms of about 4-sec period are clearly propagating from the southwest (Pacific Ocean). But two minutes later, just before the teleseism, there seems to be no directional character to the noise. This time-varying nature of the noise has disappointed expectations of improvement by crosscorrelation over long samples (more than a minute). This situation is worse at periods of one sec or shorter. Here it is impossible to find with any certainty even one wavelet traverse a spatial array.

The first evidence to support these conclusions is this: In the construction of the prediction filters, it is necessary to compute the crosscorrelation of the noise seismograms at each pair of stations. If the noise really has a predominant direction, the

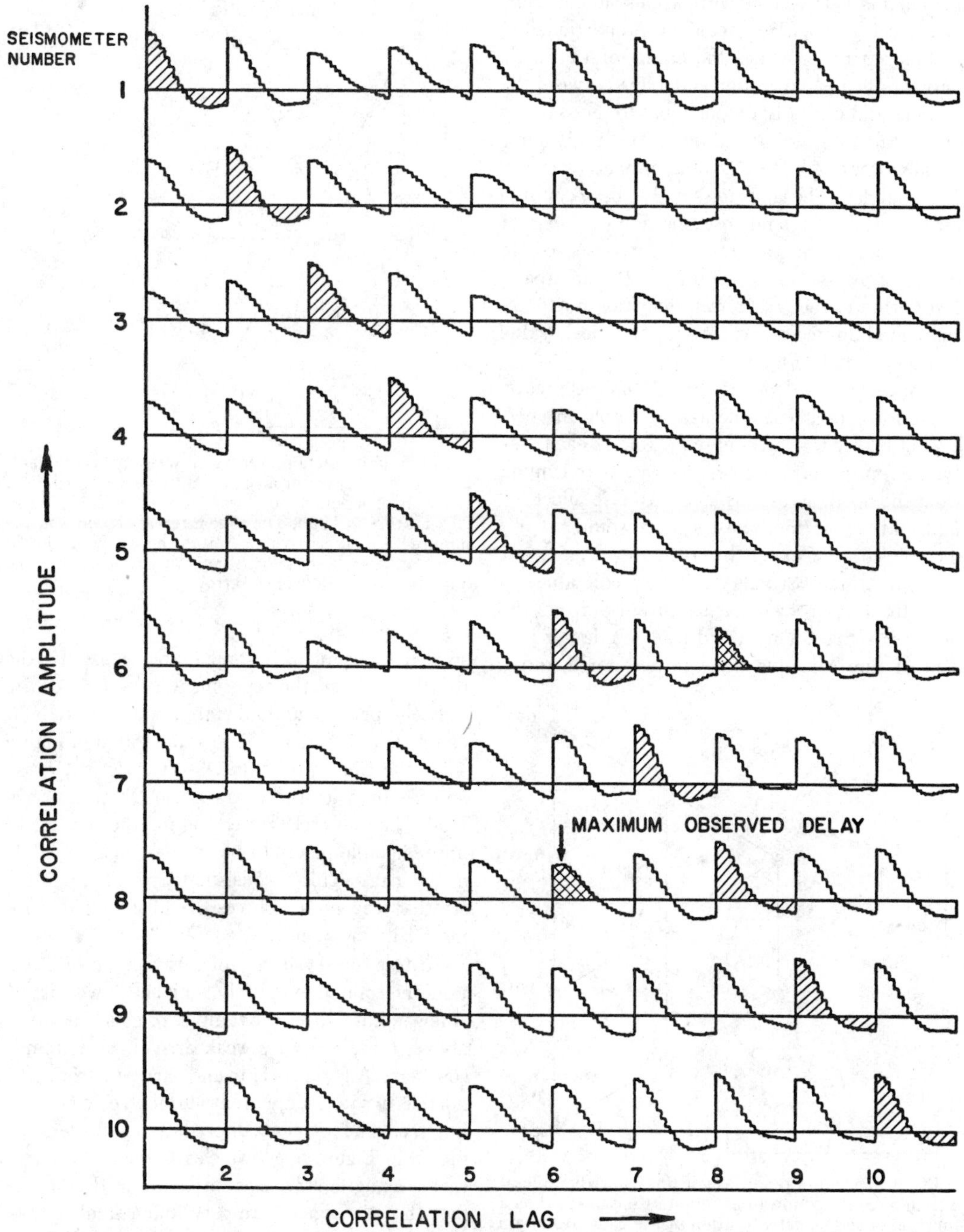

FIG. 3. Correlation matrix of a 60-sec noise sample on the ten vertical seismometers at Uinta Basin, Utah. Autocorrelations are on the main diagonal. One side of the crosscorrelation is above the main diagonal, the other is below. The digital interval is 0.1 sec. The crosscorrelation between seismometer no. 8 and seismometer no. 6 (shaded on the figure) shows a peak at .1-sec delay. These seismometers are about one mile apart. The predominant period of this noise is about 4 sec as evidenced by the correlations going to the first zero crossing in about one sec. The prediction filters are computed from this matrix.

maximums of these crosscorrelations should come at some lag. This lag should be proportional to station-separation times the cosine of the angle between the noise direction and a line between the stations and inversely proportional to the velocity of propagation. In actual fact, the maximums usually occurred at zero lag and only occasionally at a small lag. The lag is too small to be accounted for by surface-wave microseisms. For example, at Uinta Basin, the lag was rarely observed to exceed 2/10 sec on instruments $2\frac{1}{2}$ km apart. Furthermore, the crosscorrelation was much less than one hundred percent even at its peak value (see for example, Figure 3).

The second evidence is this: After the prediction-error experiment, one has available the percent of power which is unpredictable at each station. Suppose, for example, the noise were coming from the north across the array. One should then find that the northernmost signals would be predictable to some extent from their own past, but the southernmost signals would be predictable not only from their own pasts, but also from the northern signals. This would lead to a larger percent of unpredictable noise in the north and a smaller percent of unpredictable noise in the south. However, the experimental results indicate that the percent unpredictable power was about equal in all directions. This is illustrated in Figure 4 for the Blue Mountain Observatory in Oregon and Figure 5 for the Uinta Basin Observatory in Utah. The only clear trend is for the percent of unpredictable power to be smallest in the middle of the array. This is consistent with the hypothesis that the noise sources are distributed symmetrically in azimuth.

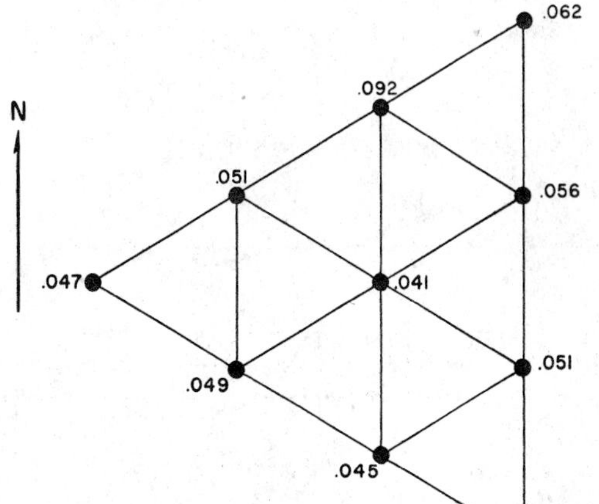

FIG. 4. Percent of power which was unpredictable at a span of 3/10 sec for a one-minute time interval as a function of seismometer location at the Blue Mountain Observatory in Oregon. The .092 is anomalously high due to some high-frequency local noise. The time interval chosen was one minute before the arrival of P from the Novaya Zemlya event on September 18, 1962. There appears to be no preferred direction, but the seismic noise from the middle of the array seems to be slightly more predictable than the noise from peripheral instruments.

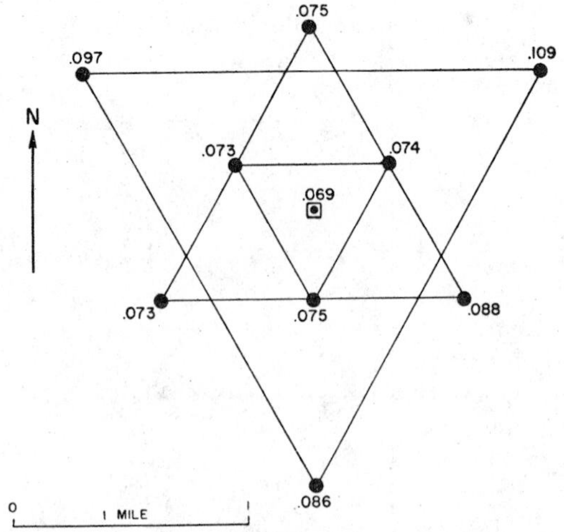

FIG. 5. Fraction of power which was unpredictable at a span of 3/10 sec for a one-minute time interval as a function of seismometer location at the Uinta Basin Observatory in Utah. The time interval chosen was one minute before the arrival of P from the southern Algeria event on March 18, 1963. Again the most predictable noise is in the middle of the array.

To test the sensitivity of this method of direction-determination the following check was made. A few minutes after the samples of noise discussed above, there is a very weak arrival. It is from a southern Algeria event and arrives from the northeast with an apparent surface velocity of 25 km/sec. The percent unpredictable power, depicted in Figure 6, shows clearly that the northeast seismograms are least predictable, despite the fact that the signals are very much smaller than the noise and the surface velocity very high. Thus, under even rather stringent conditions, this method of direction-determination is quite sensitive to signals in the important frequency band.

If the noise is supposed to consist of Rayleigh waves, a third kind of evidence can be given:

According to the theory of plane-wave Rayleigh waves a vertical component should be the same as a horizontal except for a scale factor and a 90-degree phase shift.

Interpolation of the vertical noise components from the horizontals was attempted. In the data sample where the filter coefficients were determined, about half of the power in the vertical component was interpolatable from the horizontals. When the filter was outside the fitting interval, even less of the power was interpolatable. This is illustrated in Figure 7. The result may have been suspected by the fact that the crosscorrelation between vertical and horizontals seems to be small at all lags. The presence of Love waves on the horizontals could be responsible for the poor interpolation in this case.

The results of interpolation-error filtering are more satisfactory near seacoast areas. Figure 8 is a cross correlation between a vertical and a horizontal seismogram for a noise sample of two minutes duration at Delhi, New York. The antisymmetry of the crosscorrelation about zero-lag indicates the 90-degree phase shift of Rayleigh waves.

RESULTS

Figure 9 shows seismograms from the Uinta Basin Observatory in Utah for the southern Al-

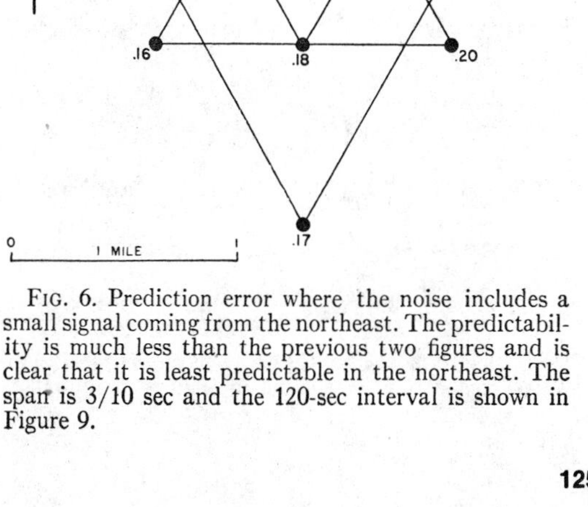

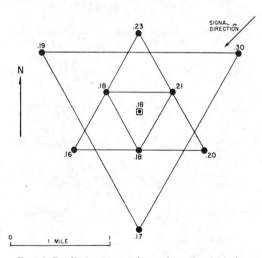

FIG. 6. Prediction error where the noise includes a small signal coming from the northeast. The predictability is much less than the previous two figures and is clear that it is least predictable in the northeast. The span is 3/10 sec and the 120-sec interval is shown in Figure 9.

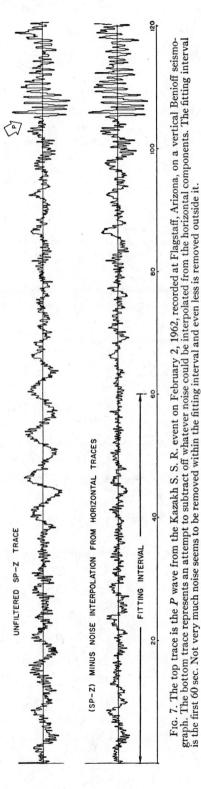

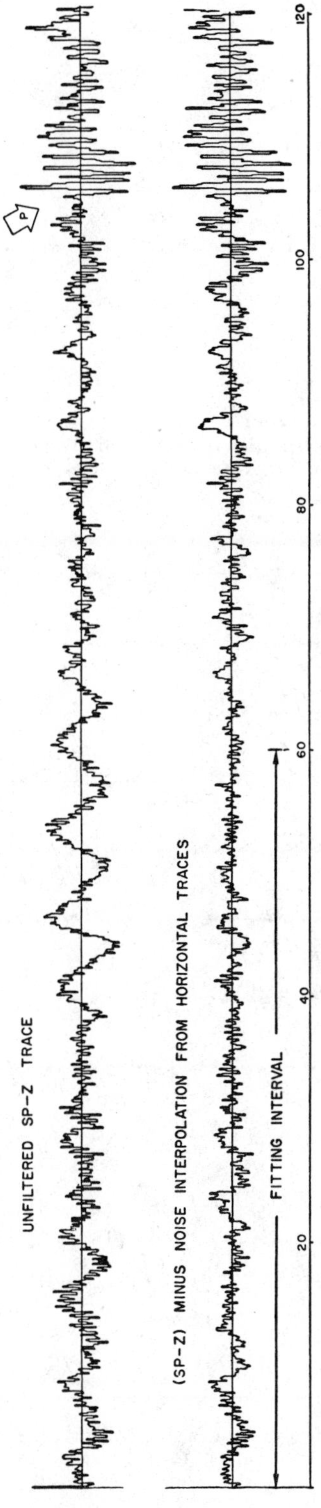

FIG. 7. The top trace is the P wave from the Kazakh S. S. R. event on February 2, 1962, recorded at Flagstaff, Arizona, on a vertical Benioff seismograph. The bottom trace represents an attempt to subtract off whatever noise could be interpolated from the horizontal components. The fitting interval is the first 60 sec. Not very much noise seems to be removed within the fitting interval and even less is removed outside it.

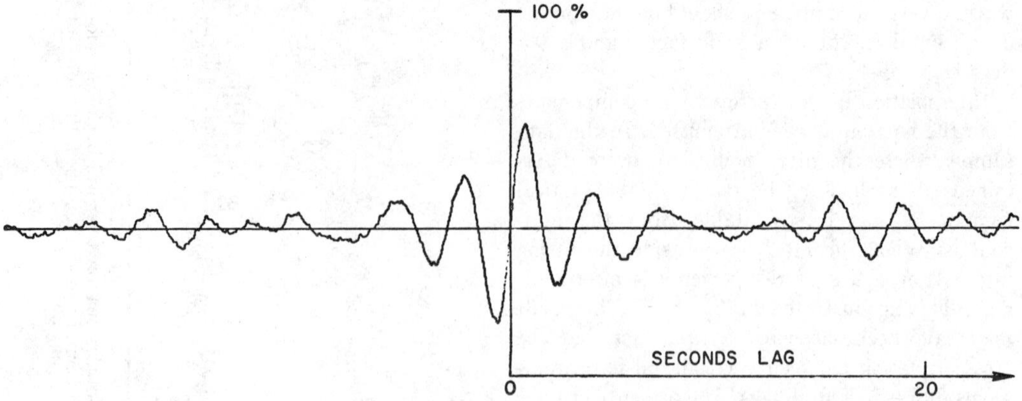

FIG. 8. Crosscorrelation of noise vertical and horizontal components at Delhi, New York. The anti-symmetry of this function is indicative of the 90 degree phase shift associated with Rayleigh waves.

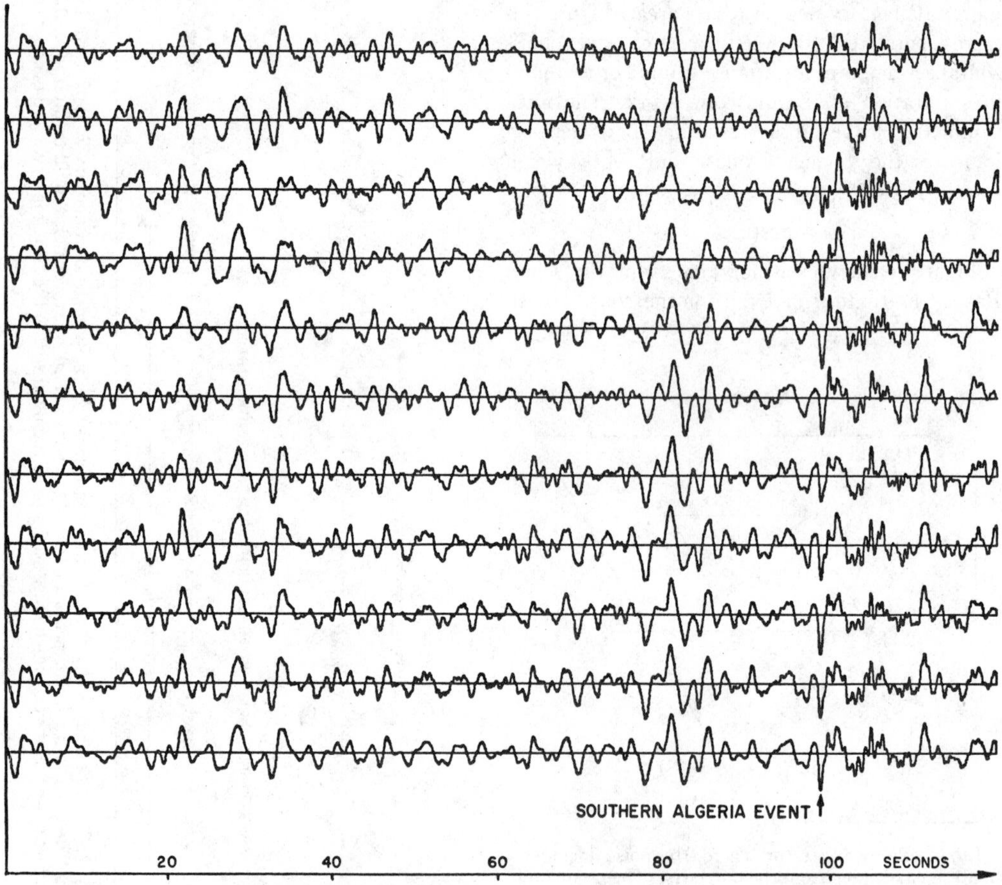

FIG. 9. Seismograms at the Uinta Basin Observatory in Utah from the southern Algeria event on March 18, 1963. The *P*-phase from the southern Algeria event arrives at 97 sec. The bottom trace is the unweighted sum of all the others.

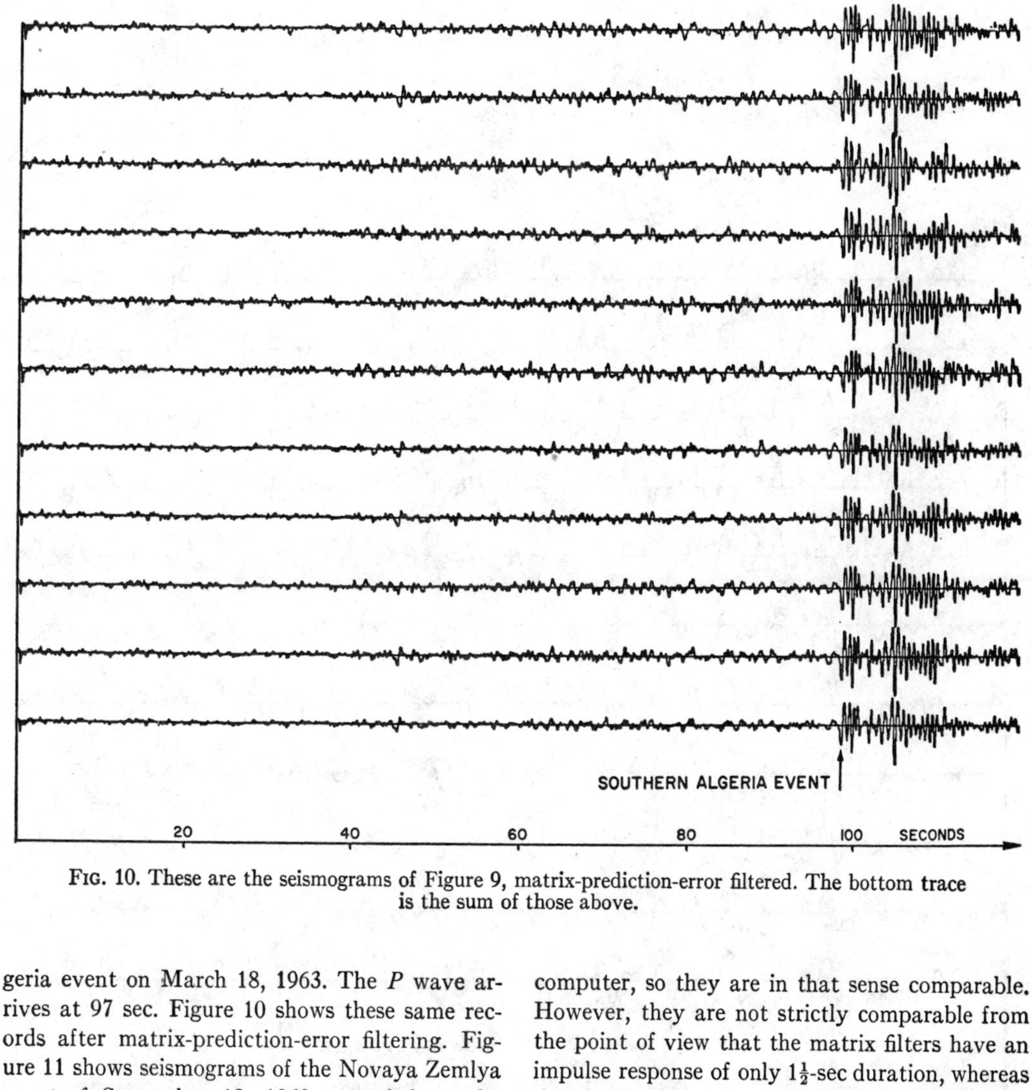

FIG. 10. These are the seismograms of Figure 9, matrix-prediction-error filtered. The bottom trace is the sum of those above.

geria event on March 18, 1963. The P wave arrives at 97 sec. Figure 10 shows these same records after matrix-prediction-error filtering. Figure 11 shows seismograms of the Novaya Zemlya event of September 18, 1962, recorded at the Blue Mountain Observatory in Oregon. The filtered seismograms are shown in Figure 12. It is readily apparent from these examples that filtering greatly improves the visual quality of the seismograms. First motion time is much easier to pick. At Uinta Basin the polarity of first motion is obviously down on each trace. Also, three separate phase arrivals can be picked in the filtered records of the southern Algeria event.

One can judge how much improvement is obtained by matrix filters over the usual single time series prediction-error filters by comparing Figures 10 and 13. Each of these figures took about ten minutes of computation on the CDC 1604 computer, so they are in that sense comparable. However, they are not strictly comparable from the point of view that the matrix filters have an impulse response of only $1\frac{1}{2}$-sec duration, whereas the single filters have duration of 15 sec. Ultimately there is a point of diminishing returns, but generally long filters work better than short ones. Therefore, the use of more computing time should improve the matrix filtering more than single-series filtering.

Figure 14 shows the Uinta Basin data bandpass filtered with various pass bands. The prediction-error-filtered signal is shown for comparison. The best of these pass bands, a surprisingly high 1.3 to 5.0 cps with a roll-off of 0.2 cps, was then applied to each of the arrays of seismograms. This is shown in Figure 15.

It can be compared with the same data, Figure

(*Text continued on page 210*)

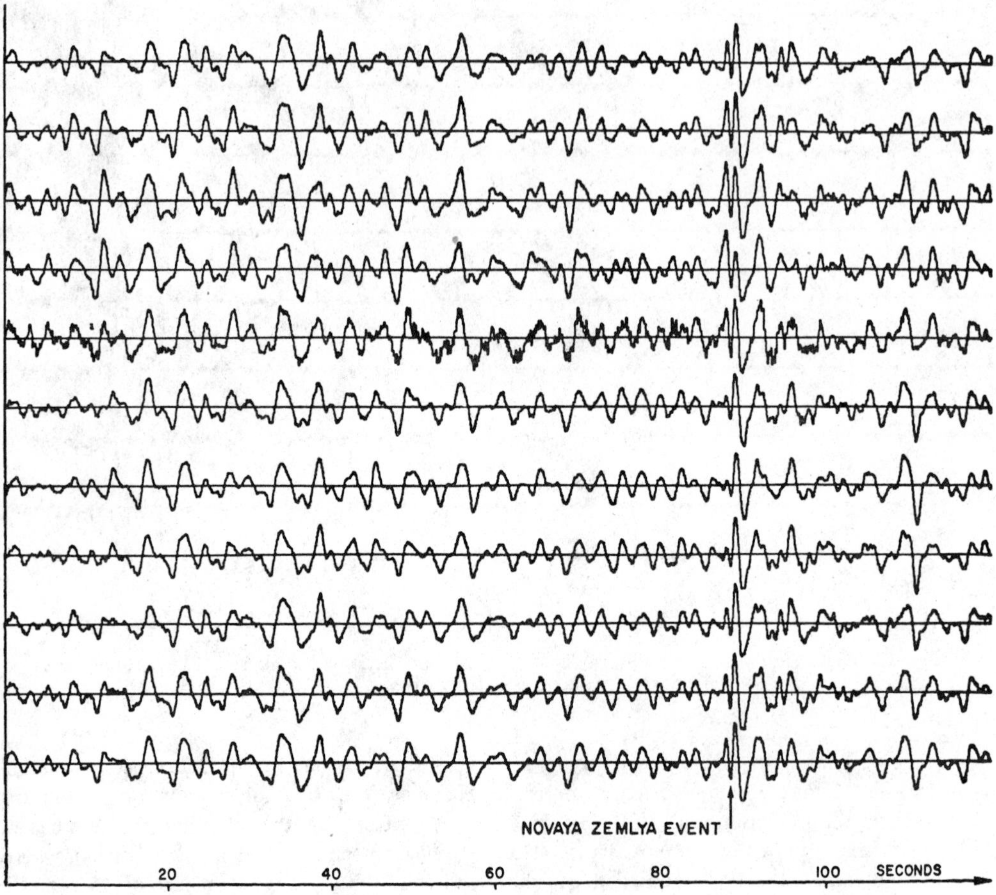

FIG. 11. Seismograms of the Novaya Zemlya event of September 18, 1962. Recorded at the Blue Mountain Observatory in Oregon.

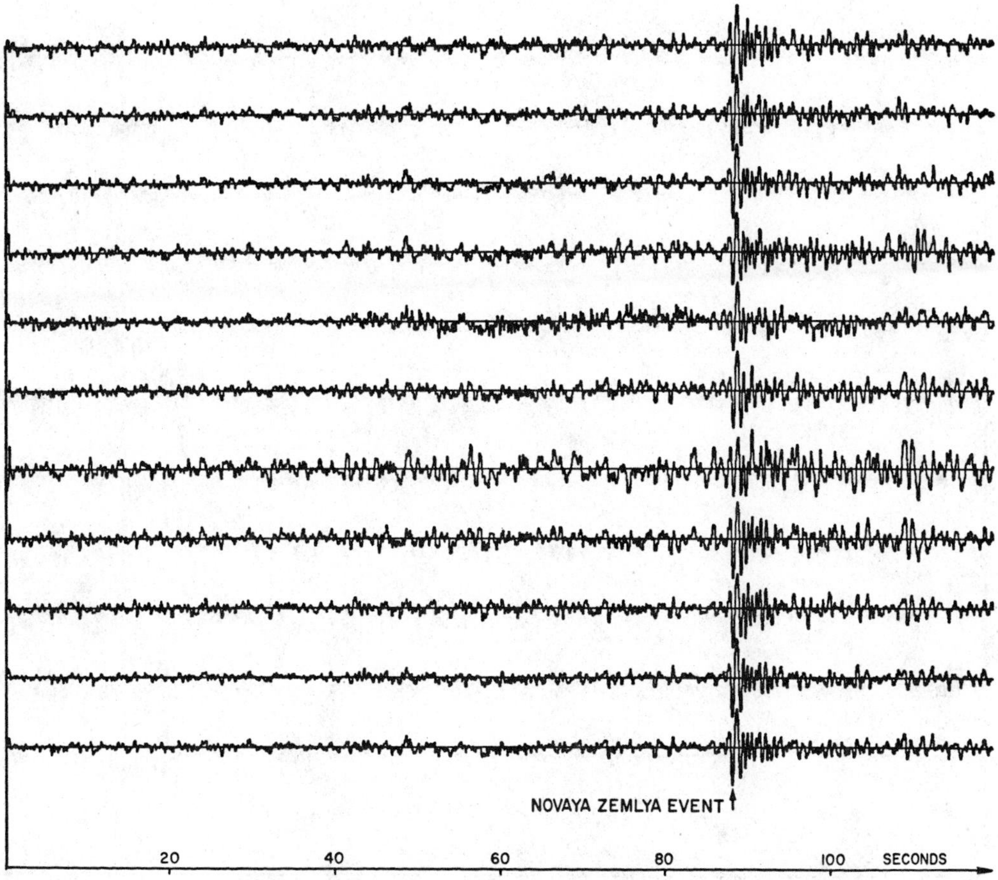

FIG. 12. Seismograms of Figure 11 filtered by the prediction-error method.

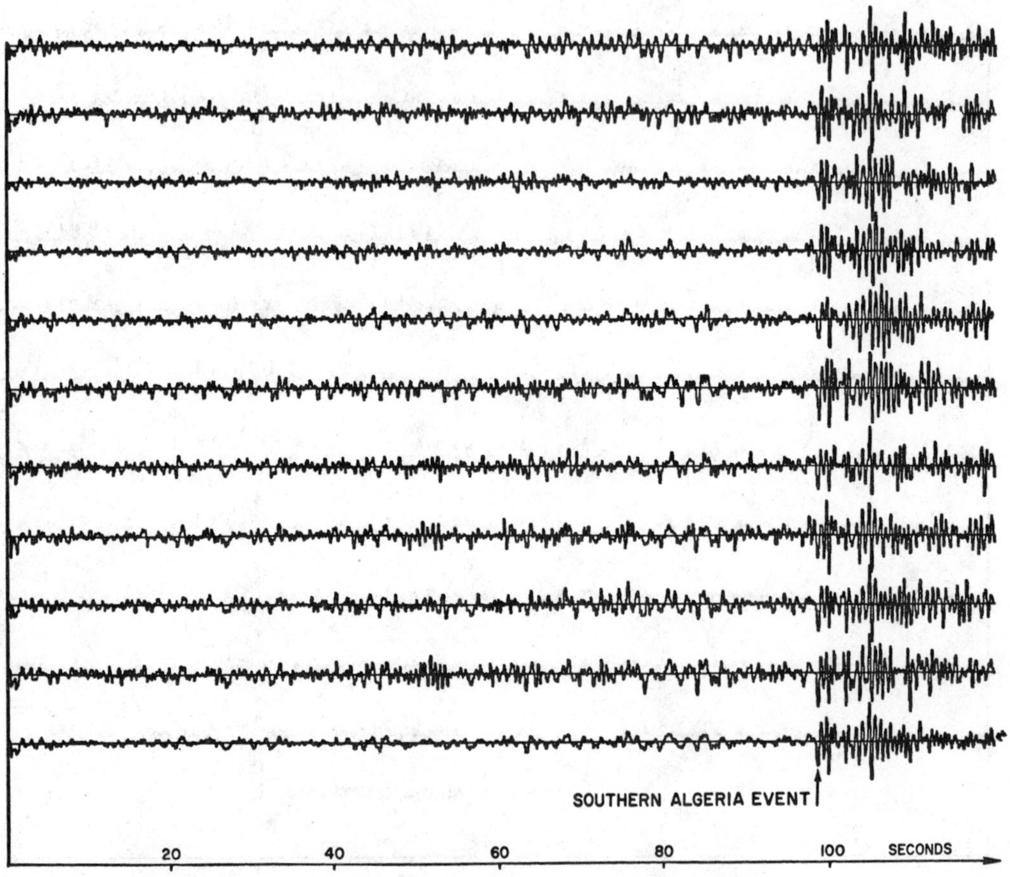

FIG. 13. Seismograms of Figure 9 filtered by the method of prediction error without cross-terms.

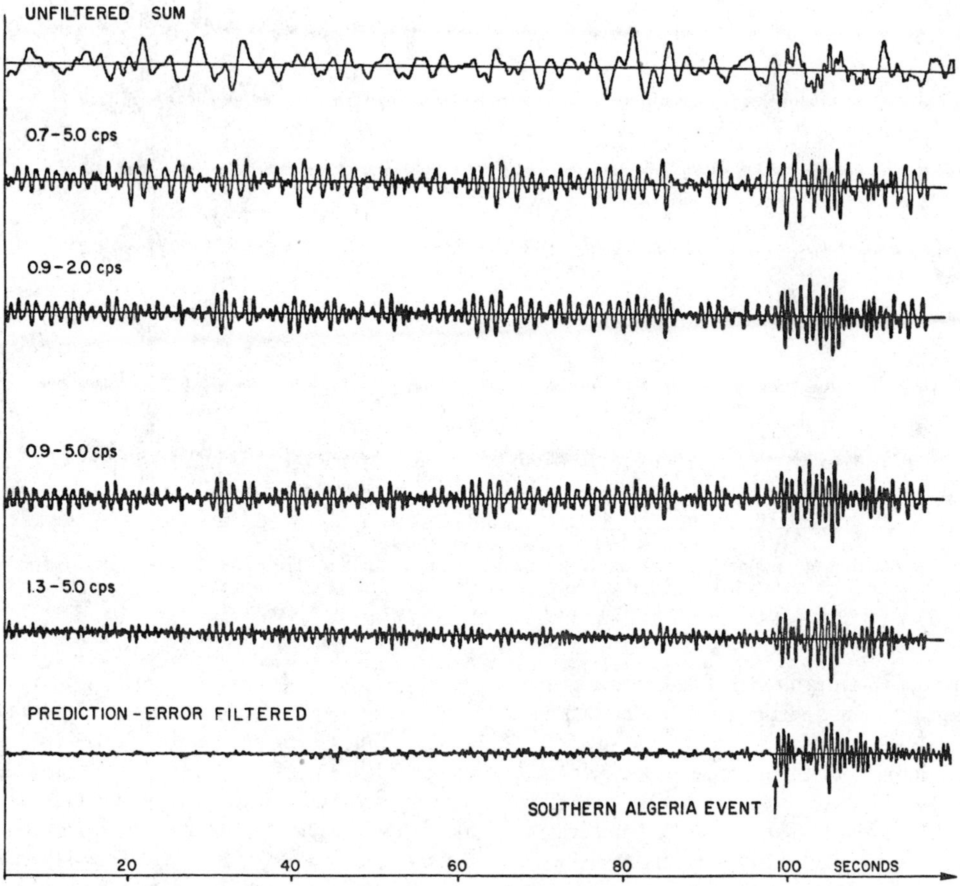

FIG. 14. The data of Figure 9 are shown in the top trace. Successive traces show the results of filtering with different pass bands. The bottom trace shows multiple prediction-error filtering.

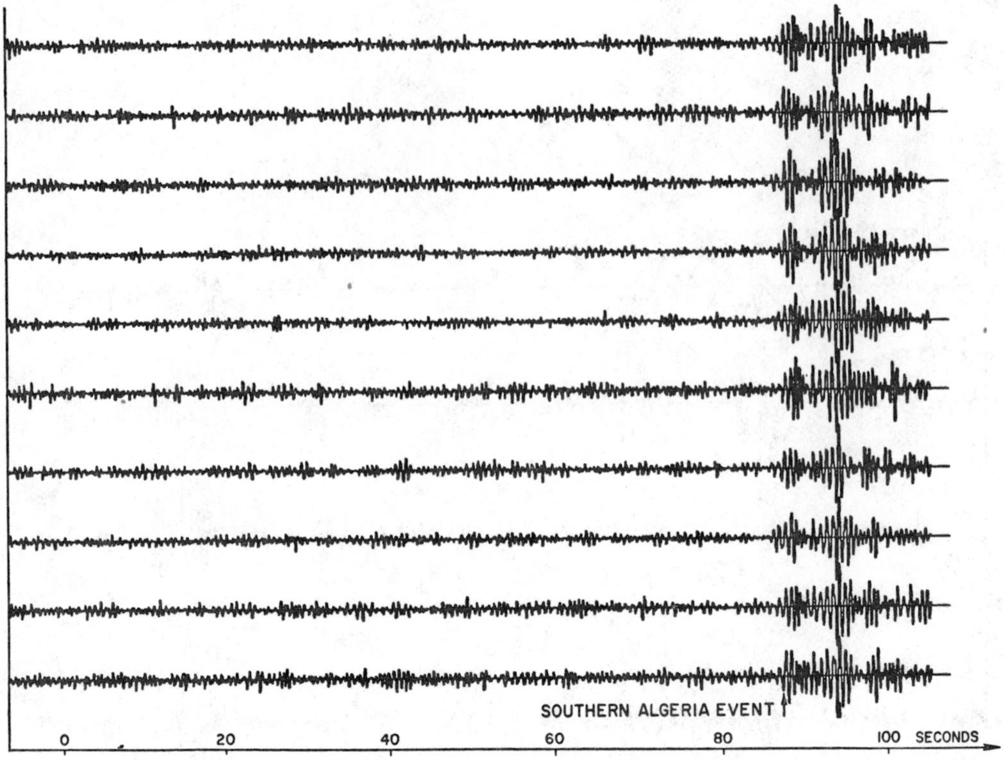

FIG. 15. The best of the pass bands shown in Figure 14 was used to filter all of the elements of the array. This pass band is 1.3 to 5.0 cps with a roll-off of 0.2 cps. Notice that this filtering has made the quarter cycle of the first motion on the southern Algeria event ambiguous, as can be expected in phaseless filtering.

10, prediction-error filtered. The improvement in the signal-to-noise energy ratio for the bandpass filtering looks almost as good as the prediction-error filtering, but there are several very notable differences in favor of the prediction-error filtering. These are: (1) The frequency band was determined by trying different bands to see which would work the best. Prediction-error filtering requires no assumptions on the frequency spectrum of the signal; (2) after bandpass filtering, both the event and the noise are limited to the same frequency range. This is not true of the prediction-error filtering. In picking records, a change in predominant frequency is an important criterion which tends to be lost in bandpass filtering but not completely so in prediction-error filtering; and (3) precursors inherent in phaseless bandpass filtering make it impossible in this case to pick the time of the half-cycle of the first motion, to say nothing of the sign of first motion. The first motion is clearer on the prediction-error record.

CONCLUSIONS

Prediction-error filtering enables one to detect weak events which may be overlooked without filtering. It is superior in several respects to bandpass filtering. For surface arrays prediction-error filtering without cross-terms is almost as good as matrix prediction-error filtering. This is because the microseisms which interfere with detection can often be thought of as the Brownian motion of a surface with low coherence further than about a wavelength. Since matrix filtering includes velocity filtering as a special case, it would seem that velocity filtering of the noise will not work (at quiet stations for the frequency range from .5 to 2.0 cps). Interpolating the noise on a vertical component from the horizontals does not work well enough to be justified.

One advantage of prediction-error filtering over other methods of "optimum" filtering is that it is unnecessary to assume a model of either the noise or the signal. Any properties of the noise such as directionality, velocity of propagation, or types of waves should be observable through the correlation matrix of the seismograms.

The main objective of prediction-error filtering is to improve the first arrival. In difficult cases of epicenter determination for example, the time of the first arrival will be determined more accurately if prediction-error filtering is used.

ACKNOWLEDGMENT

The work reported herein was sponsored by the Air Force Technical Applications Center under the VELA UNIFORM Project of the Advanced Research Projects Agency.

I wish to thank A. Booker, W. Dean, E. Flinn, E. Robinson, and R. Wiggins for helpful comments and discussions. I wish also to thank W. C. Dean for presenting this paper at the October, 1963, meeting of the SEG.

REFERENCES

Levinson, N., 1947, The Wiener RMS (root-mean-square) error criterion in filter design and prediction: Jour. Math. Phy., v. 25, p. 261–278. Also appears as an appendix to Wiener, N, 1950, Extrapolation, interpolation, and smoothing of stationary time series: New York, John Wiley & Sons, Inc., p. 129–148.

Robinson, E. A., 1963, Mathematical development of discrete filters for the detection of nuclear explosions: Jour. Geoph. Res., v. 68, p. 5559–5567.

———, 1962, Random wavelets and cybernetic systems: New York, Hafner Publishing Co.

Toksöz, M. N., 1964, Microseisms and an attempted application to exploration: Geophysics, v. 29, p. 154–177 (see also, Toksöz, M. N., 1962, Ph.D. thesis, California Institute of Technology).

II. ANALOG INVERSE FILTERING

Early in the development of deconvolution, the frequently cited paper by M. M. Backus clarified an important point: Water layer reverberations should be regarded as a linear filtering effect rather than as an additive noise. Accordingly, if no nonlinear processes (specifically automatic gain control, AGC) are involved in recording, the desired primary reflections can be recovered in their original form by passing the recorded data through an inverse filter. It is true that an undesirable component of the records is due to energy which never leaves the water layer, but analysis shows that subsurface-reflected energy will always eventually become dominant over water-confined energy.[1] Further, for reasonable values of the reflection coefficients, subsurface-reflected energy should become dominant early on most marine records.

This conclusion is supported by an analysis of experimental data from the Persian Gulf on the basis of both normal moveout and amplitude decay. It is shown that the water layer filter and its inverse are both series of impulses, although all discussion is in terms of continuous time. Presumably, for the examples presented, the inverse filtering was actually performed using a delay line. The water bottom time was estimated from the observed ringing frequencies, and (in one example) the water bottom reflection coefficient was somewhat arbitrarily set equal to one. The need for high-fidelity recording in order to make effective application of the inverse filtering technique is clearly stated, and indeed at the time of publication digital recording was not far in the future. Backus's pioneering paper has been recognized as a classic of GEOPHYSICS.

Turning now to the analysis of land data, J. P. Lindsey describes the ghost reflection in terms of a linear filter, and its inverse as a system with feedback. Had the inverse filter been described in digital terms, it would have been called a recursive (or Infinite Impulse Response) filter. A significant feature of this paper is the careful use of a sample autocorrelation function to estimate the needed parameters of the inverse filter (reflection coefficient and time lag). The user is cautioned not to apply the inverse ghosting filter to the first arrivals, and also to avoid the use of a very rapid AGC.

[1]A correction to equation (14) was found by Kalisvaart and Seriff; see GEOPHYSICS, 1961, v. 26, p. 242. The original equation is correct for energy which has been reflected from the deep interface less than three times.

Reprinted from Geophysics v. 24, no. 2, p. 233-261

WATER REVERBERATIONS—THEIR NATURE AND ELIMINATION*

MILO M. BACKUS†

ABSTRACT

In offshore shooting the validity of previously recorded seismic data has been severely limited by multiple reflections within the water layer. The magnitude of this problem is dependent on the thickness and the nature of the boundaries of the water layer.

The effect of the water layer is treated as a linear filtering mechanism, and it is suggested that most apparent water reverberation records probably contain some approximate subsurface structural information, even in their present form.

The use of inverse filtering techniques for the removal or attenuation of the water reverberation effect is discussed. Examples show the application of the technique to conventional magnetically recorded offshore data. It has been found that the effectiveness of the method is strongly dependent on the instrumental parameters used in the recording of the original data.

INTRODUCTION

In marine seismic operations, the water-air interface is a flat, strong reflector, with a reflection coefficient close to -1. In many areas the water-bottom interface is also a strong reflector. We then have an energy trap—a non-attenuating medium bounded by two strong reflecting interfaces. A pulse generated in the trap, or entering the trap from below, will be successively reflected between the two interfaces, with a time interval equal to the two-way travel time, and an amplitude decay dependent on the reflection coefficients. As a result, valid primary reflections from depth are obscured by previously established reverberations.

This water reverberation problem was first recognized on the basis of the apparent periodicity of a suite of seismograms from the Persian Gulf and was treated for the one-layer case by wave-guide theory (Burg et al., 1951). Since then, singing records have been recognized as a serious and widespread limitation to the acquisition of valid subsurface structural information, particularly in the Persian Gulf and in Lake Maracaibo. The problem has also been recognized in the Gulf of Mexico and off the coast of California, and it is present to some degree in any marine operation.

The water reverberation problem has been examined experimentally in model studies by Sarrafian (1956). The equivalent problem for a dipping bottom was studied and utilized in the interpretation by Poulter (1950) in his Antarctic studies.

In the first part of this paper, the water reverberation problem is examined approximately as a linear filtering problem. The predicted effect of the water layer on seismic data is discussed on the basis of this analysis. In the second

* Paper read at the 28th Annual International Meeting of the Society at San Antonio on October 16, 1958. Manuscript received by the Editor December 3, 1958.

† Geophysical Service Inc., Dallas, Texas.

section, experimental data on the nature of water reverberation records is presented. Finally, the use of inverse filtering to eliminate or reduce the water reverberation effect is discussed and some examples are presented.

MARINE SHOOTING IN TERMS OF LINEAR FILTERING

It is useful to regard the reflection seismic method as an attempt to obtain approximately the impulse response of the subsurface to vertically travelling energy. The amplitude peaks in the impulse response are then interpreted in terms of subsurface layering. Filtering of the impulse response in the instruments, necessary for a satisfactory signal-to-noise ratio, results in a degradation of the data

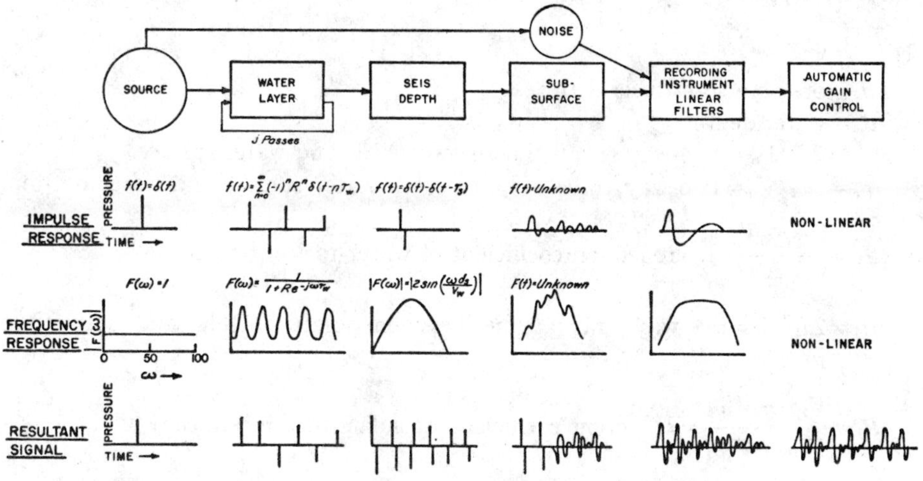

FIG. 1. Summary of marine-reflection technique in terms of linear filtering.

as does filtering in the ground due to frequency dependent attenuation. The water layer in marine work may be regarded as an additional undesirable and extremely sharp filter which is acting on the data. A picture of the factors acting on the data in marine work which may be considered approximately as filtering effects is provided in Figure 1. The actual order of occurrence is shifted for clarity in developing the final form of the seismic signal.

The marine seismic source, typically 15 to 25 lb of high velocity dynamite detonated 4 to 6 ft below the surface, produces a disturbance which may, on the seismic time scale, be regarded essentially as a positive pressure impulse (Cole, 1948). The reflection from the water-air interface is presumably small due to the loss of energy in cavitation. The pressure pulse reverberates within the water layer, producing a series of pulses. The signal is further shaped according to the position of the transducer with respect to the free surface, and this effect will be treated as a separate filtering effect. The ringing signal is also transmitted into the subsurface and simply reflected back into the water layer. This is the portion of the filtering process which provides the desired data. Upon their re-entry into the water layer, arrivals from the subsurface ring again. Primary reflections thus essentially pass through two sections of the water-layer filter. Similarly, section multiples may essentially pass through a number of sections of the water-layer filter. The data then passes through the instrumental filters and conventionally is nonlinearly treated by automatic gain control.

The water-layer filter not only constitutes a sharply ringing filter but one with characteristics which vary along the line. The records we obtain are then similar (except for a compounding effect on section multiples) to what we could obtain by using a very ringing filter on land records, varying the filter from profile to profile.

The Water-Layer Filter

d_s = pressure transducer depth.

d_w = water depth.

$\tau_s = 2d_s/V_w$ = two-way travel time between seis and water surface.

$\tau_w = 2d_w/V_w$ = two-way travel time between water surface and water bottom.

$R = \dfrac{V_1\rho_1 - V_w\rho_w}{V_1\rho_1 + V_w\rho_w}$ = reflection coefficient of water-rock interface.

$\tau_1 = 2d_1/V_1$ = two-way travel time between ocean bottom and subsurface reflection.

$R_1 = \dfrac{V_2\rho_2 - V_1\rho_1}{V_2\rho_2 + V_1\rho_1}$ = reflection coefficient at subsurface interface.

Consider an upward travelling impulsive plane wave, entering the water layer from below, arriving at $Z = d_s$ at $t = 0$ (Figure 2). The transducer output would represent the successive reflections between the two interfaces and would be of the form,

$$
\begin{aligned}
g(t) &= \delta(t) - \delta(t - \tau_s) - R\delta(t - \tau_w) + R\delta(t - \tau_w - \tau_s) \\
&\quad + R^2\delta(t - 2\tau_w) - \cdots \\
&= \sum_{n=0}^{\infty} (-1)^n R^n \delta(t - n\tau_w) - \sum_{n=0}^{\infty} (-1)^n R^n \delta(t - n\tau_w - \tau_s) \\
&= f(t) - f(t - \tau_s), \qquad (1)
\end{aligned}
$$

where

$$
f(t) = \sum_{n=0}^{\infty} (-1)^n R^n \delta(t - n\tau_w) \cdot \qquad (2)
$$

For an impulsive input to the water layer, $g(t)$ is measured as the output. Equation (1) thus represents the impulse response of the filter equivalent to the effect of the water layer and the measurement position. This filter may be broken

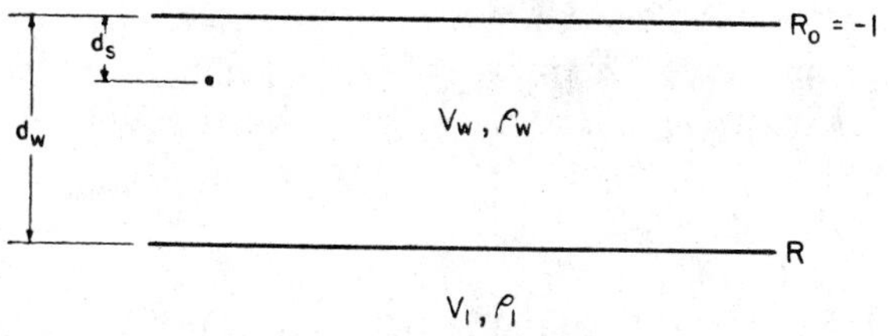

FIG. 2. Schematic diagram of a water layer (V_w, ρ_w) bounded above by a perfect reflector $R_0=-1$, and below by a hard bottom of reflection coefficient R.

up into two parts: (a) that which is independent of transducer depth [equation (2)], and (b) the filtering effect due to the relationship between seismometer and the free surface. We shall consider that part of the filter which is independent of the means of recording as the water-layer filter.

The equivalent frequency response may be obtained from (2).

$$F(\omega) = \int_{-\infty}^{\infty} \sum_{n=0}^{\infty} (-1)^n R^n \delta(t - n\tau_w) e^{-j\omega t} dt$$
$$= \sum_{n=0}^{\infty} (-1)^n R^n e^{-jn\omega\tau_w} \tag{3}$$

which is the binomial expansion for

$$F(\omega) = \frac{1}{1 + Re^{-j\omega\tau_w}}, \tag{4}$$

$$|F(\omega)| = (1 + R^2 + 2R \cos [\omega\tau_w])^{-1/2}, \tag{4a}$$

$$\phi(\omega) = \tan^{-1} \frac{R \sin (\omega\tau_w)}{1 + R \cos (\omega\tau_w)} \cdot \tag{4b}$$

In the limiting case of a perfectly rigid bottom, $R=1$ and equation (4a) becomes

$$|F(\omega)|_{R=1} = \left| 1/2 \sec \left(\frac{\omega\tau_w}{2} \right) \right|, \tag{5}$$

which blows up at the resonant frequencies,

$$f_n = \frac{(2n - 1)V_w}{4d_w} \tag{6}$$

corresponding to equation (2) in the article by Burg et al. (1951).

Equation (4a) is plotted in Figure 3 for a water depth, $d_w=100$ ft, for two cases: $R=1$, and $R=0.5$. The water layer is equivalent to a sharply peaked comb filter with resonance at a fundamental frequency of $12\frac{1}{2}$ cps and at the odd harmonics 37 cps, 62 cps, 87 cps, etc. If we recorded data which had been passed through this filter, using a pass-band from 20–55 cps, we would essentially obtain a 37-cps sine wave. Often the energy returning from the subsurface is enriched in certain frequencies, and the filtering due to seismometer depth is gently peaked. Thus we would expect often to obtain a nearly pure sine wave even on a wide-band recording.

For different water depths the frequency scale in Figure 3 is merely expanded or compressed. As R, the ocean-bottom reflection coefficient, changes, the degree of peaking at resonance changes. For an incident plane wave travelling at an angle i with the vertical, the effective filtering of the water layer is given by replacing τ_w in equation (4) by $\tau_w \cos i$.

If the ocean bottom is acoustically soft, that is, if $R<0$, inspection of equation (4) shows that the same shape filter is obtained except that the frequency scale must be shifted $f=V_w/4d_w$ cps to the right. For the limiting case, $R=-1$, equation (4a) becomes,

$$| F(\omega) |_{R=-1} = \left| \frac{1}{2} \csc \left(\frac{\omega \tau_w}{2} \right) \right| , \tag{7}$$

which blows up at the resonant frequencies,

$$f_n = \frac{nV_w}{2d_w} . \tag{8}$$

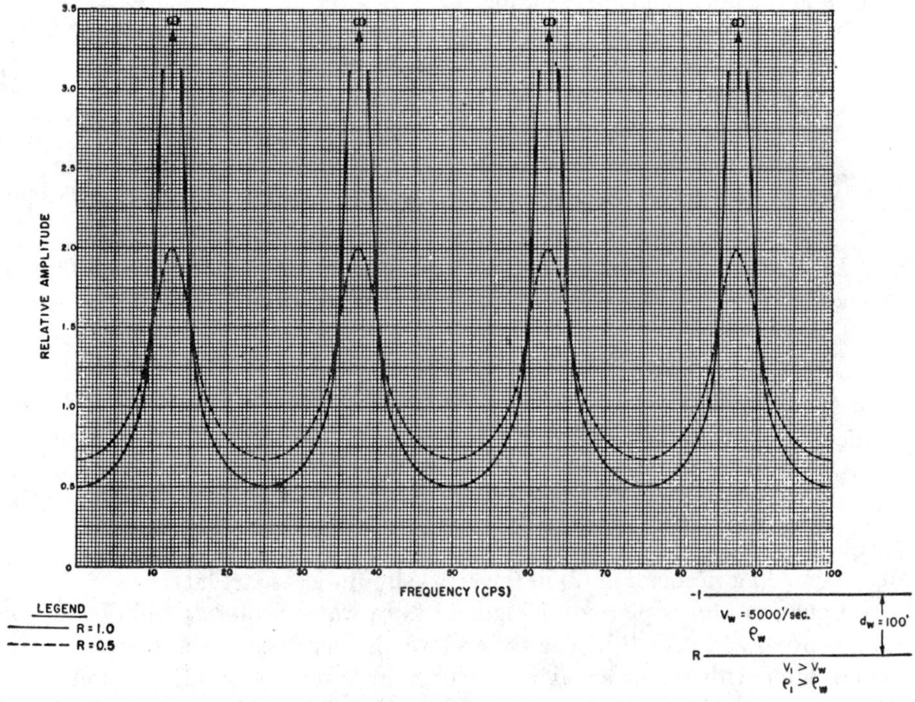

FIG. 3. Filtering effect of the water-layer trap.

In addition to a resonance at $f=0$, we have resonance at a fundamental and all harmonic frequencies. For the case illustrated in Figure 3, resonance occurs at 0, 25, 50, 75, 100 cps, etc. The nature of one example of a thin, acoustically soft layer on a lake bottom has been discussed by Jones et al. (1958).

The same results may be obtained in a straightforward manner by the use of the wave equation and the appropriate boundary conditions.

The Filtering Effect of Seismometer Depth

The complete alteration of the signal in equation (1) was broken up into two parts. We shall now examine the filtering which is associated with the transducer position relative to the free surface. As shown in equation (1), the effect of pressure-seismometer depth is equivalent to a filter with an impulse response,

$$f_p(t) = \delta(t) - \delta(t - \tau_s). \tag{9}$$

This merely expresses the fact that for an up-travelling wave, the pressure transducer in a homogeneous half-space sees the direct arrival and the inverted reflection from the free interface. The frequency response of this filter is,

$$\begin{aligned} F_p(\omega) &= \int_{-\infty}^{\infty} [\delta(t) - \delta(t - \tau_s)]e^{-j\omega t}dt \\ &= 1 - e^{-j\omega\tau_s} \end{aligned} \tag{10}$$

$$| F_p(\omega) | = \left| 2 \sin \left(\frac{\omega\tau_s}{2}\right) \right| \tag{10a}$$

$$\phi_p(\omega) = \frac{3}{2}\pi - \frac{\omega\tau_s}{2}. \tag{10b}$$

Similarly, for a velocity seismometer at depth d_s, the filtering effect would be of the form,

$$F_v(\omega) = 1 + e^{-j\omega\tau_s}. \tag{11}$$

The amplitude and phase response [equation (10)] is illustrated in Figure 4 for seismometer depths of 10 and 25 ft. Phase lag is positive. The effect of this filter on seismic reflection character is also illustrated in Figure 4, which shows vertical reflection energy recorded from a vertical pressure spread, using automatic gain control. The predicted increase in low frequency response with increasing seismometer depth is apparent. It may also be seen that small variations in seismometer depth are less significant when the peak of the seismometer response is somewhat to the right of the signal spectrum. For this reason, and because of the general shape of the filter, a pressure seismometer depth of 10–15 ft is most commonly used in marine work.

Effect of the Water-Layer Filter on Subsurface Reflections

We now shall consider the effect of the water layer on the seismic reflection

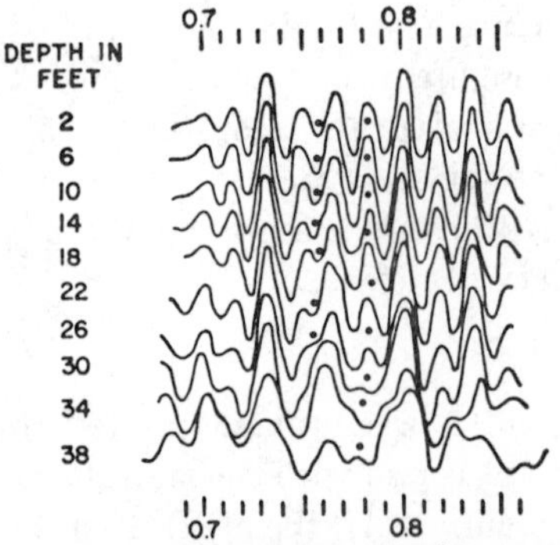

ILLUSTRATION OF EFFECT OF SEISMOMETER DEPTH ON REFLECTION CHARACTER

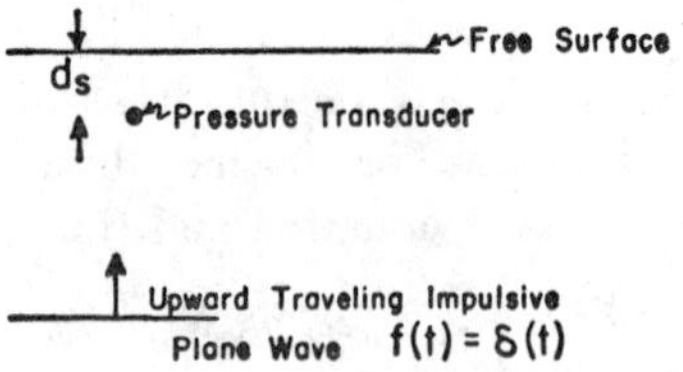

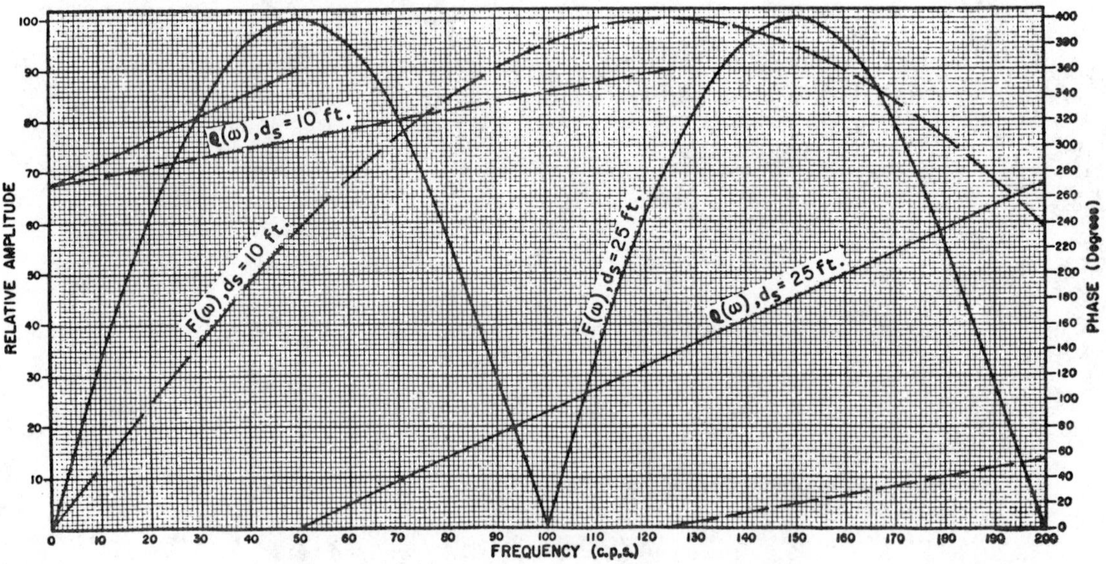

FIG. 4. Filtering effect of seismometer depth for a pressure seismometer.

process in the simple case in Figure 5 in which there is a single subsurface reflector. Consider the injection into the system of a downward travelling impulsive plane pressure wave, at $t=0$, $z=d_s$. In addition to the signal trapped in the water layer [of the form in equation (1)], we would measure energy reflected from the subsurface interface. The initial arrival will be of the form

$$f_1(t) = (1 - R^2)R_1\delta(t - \tau_1 - \tau_w + \tau_s), \tag{12}$$

where $(1-R^2)$ is the two-way transmission coefficient at the water-rock interface.

At $t=[\tau_1+(n+1)\tau_w+\tau_s]$, $(n+1)$ in phase arrivals, reflected up from the subsurface once and reflected upward from the water-bottom n times, will be de-

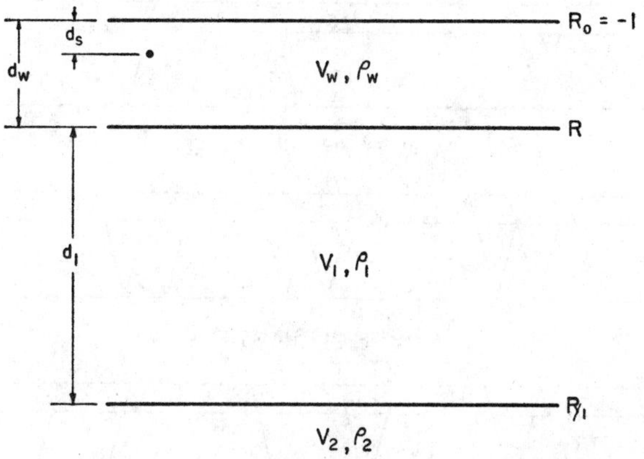

FIG. 5. Schematic diagram of the same conditions as shown in Figure 2 with the addition of a single subsurface reflector of reflection coefficient R_1.

tected. The pressure signal resulting from all events singly reflected from the subsurface interface will thus be of the form

$$g'(t) = f'(t) - f'(t - \tau_s)$$
$$f'(t) = (1 - R^2)R_1 \sum_{n=0}^{\infty} (-1)^n(n + 1)R^n\delta[t - \tau_1 - (n + 1)\tau_w + \tau_s]. \tag{13}$$

Similarly, by considering all possible ray-path segments with associated reflection and transmission coefficients, and summing in-phase arrivals, we can formulate the complete expression for the signal picked up by the transducer.

$$g(t) = \delta(t) + f(t) - f(t - \tau_s),$$
$$f(t) = \sum_{n=1}^{\infty} (-1)^{n+1}R^n\delta(t - n\tau_w + \tau_s) + \sum_{j=1}^{j=\infty} R_1{}^j \sum_{k=0}^{k=j-1} (-1)^k(1 - R^2)^{(j-k)}R^k \tag{14}$$
$$\cdot \sum_{n=0}^{\infty} \frac{(n + j - k)!}{(j - k)!n!} (-1)^{n+j-k+1}R^n\delta[t - j\tau_1 - (n + j - k)\tau_w + \tau_s].$$

In equation (14) the jth term represents all energy which has been reflected upward from the deep interface j times. The kth term for a particular j represents all energy which has been reflected upward j times from the deep reflector and has suffered k downward reflections from the water-rock interface. The nth term in the inner sum represents the energy for a given j and k which has been multiply reflected n times in the water-layer trap. The multiplier,

$$\frac{(n + j - k)!}{(j - k)!\, n!}$$

represents the number of different permutations of the different ray-path seg-

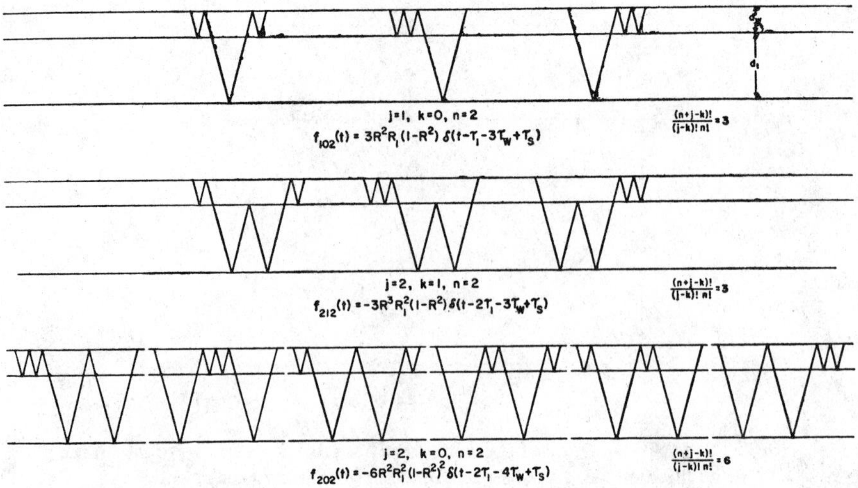

FIG. 6. Ray paths corresponding to particular terms in equation (14). This illustrates the basis for the initial increase in amplitude of the water reverberations due to an initial arrival from depth when the water-bottom coefficient is greater than 0.5. Looking at the upper example, the initial arrival from depth would have an amplitude of $R_1(1-R^2)$. The ratio of the second reverberation to the initial arrival would then be $3R^2$, since there are three independent permutations of the ray-path segments which constitute simultaneous in-phase arrivals. Thus if R^2 is greater than $\frac{1}{3}$, the second water multiple from a primary reflection will have higher amplitude than the initial arrival from the primary reflection.

ments for a given j, k, n. In Figure 6 the ray paths (distorted to a point source for clarity) corresponding to several particular terms in equation (14) are illustrated.

The only term in equation (14) which represents a desired signal is the ($j=1$, $k=0$, $n=0$) term of the second sum. The other terms all represent ringing due to the water-layer filter and section multiples.

This formulation becomes cumbersome in more complex multilayered cases, but it is quite convenient for examining the amplitude relationships for the different kinds of arrivals. Amplitude as a function of time for the arrivals corresponding to $j=0$, 1, 2, 3 is plotted for $R=0.3$, 0.5, and 0.7 respectively in Figures 7–9. The cases plotted are for $R_1=0.1$, $\tau_1=5\tau_w$. However, the results for different values of R_1 and τ_1 may be conveniently obtained merely by shifting the curves. For example, to obtain the results for $\tau_1=10.6\tau_w$, each curve is shifted $5.6j$ units to the right. To obtain results for $\tau_1=5\tau_w$, $R_1=0.2$, all curves are shifted $6j$ decibels upward.

The curve for $j=0$, which represents the energy which never leaves the water layer, shows simple exponential decay,

$$f_0(t)\alpha R^n = R^{t/\tau} = e^{-\lambda t},$$
$$\lambda = \frac{1}{\tau}\ln\left(\frac{1}{R}\right). \tag{15}$$

Actually, the curve for $j=1$, which represents energy reflected upward only once from the subsurface, initially increases with time for $R>0.5$. The rate of decay for $j=1$ is always less than that in equation (15), approaching (15) as $t\to\infty$. Thus we would expect that subsurface reflected energy will always eventually become dominant over water confined energy, no matter how high R is, except

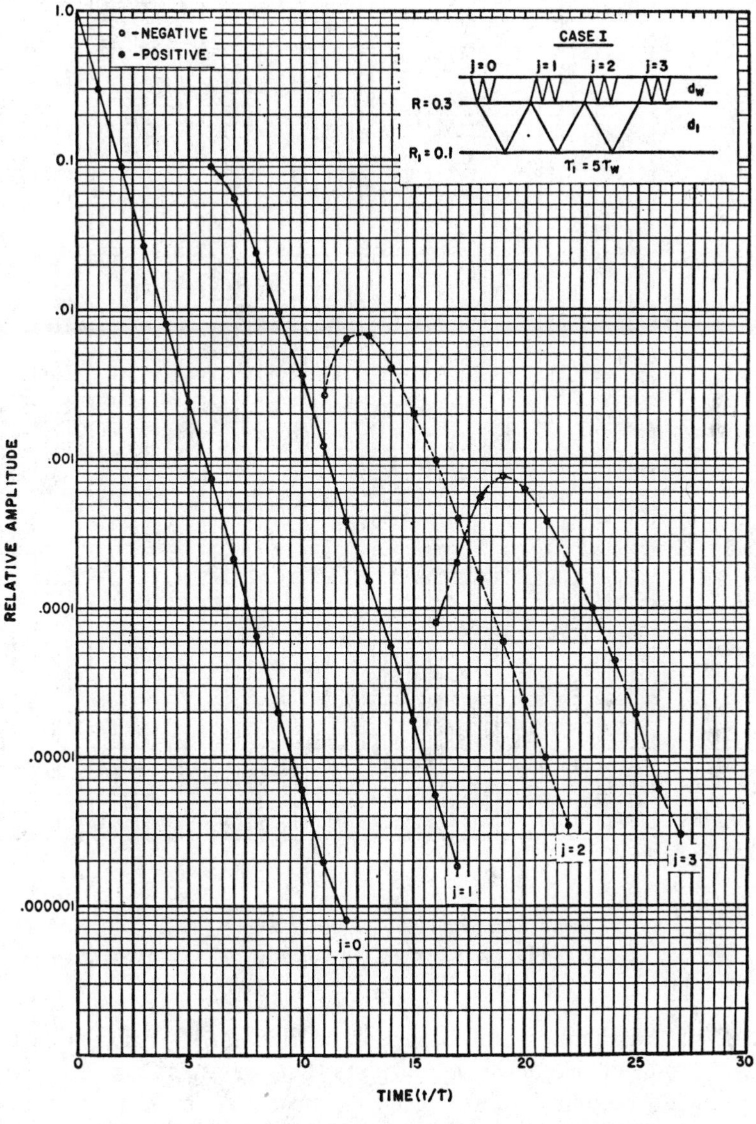

FIG. 7. Relative amplitudes and signs for terms $j=0$, 1, 2, 3, Case 1. Calculated for a plane wave, subsurface attenuation neglected.

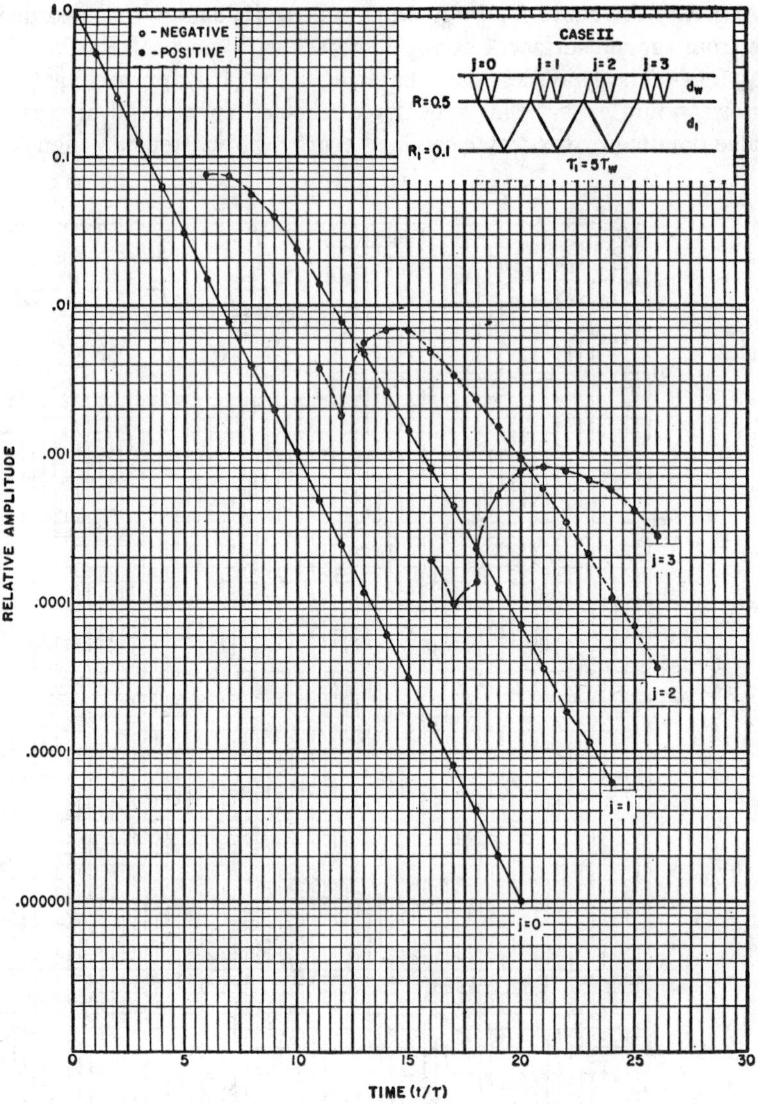

FIG. 8. Relative amplitudes and signs for terms $j=0, 1, 2, 3$, Case 2. Calculated for a plane wave, subsurface attenuation neglected.

when $R=1$. Furthermore, an inspection of Figures 7–9, considering reasonable values for reflection coefficients, suggests that, in general, subsurface reflected energy should become dominant early on most marine records.

Section multiple reflection energy, illustrated for $j=2$ and $j=3$, shows a more pronounced increase with time. The initial peculiar behavior is due to interference between the reverberations corresponding to different values of k. However, the energy for $k=0$ always rapidly becomes dominant, and it is the only significant energy involved at the point of maximum amplitude. Thus the dominant portion of the section multiple energy essentially passes through the water-layer filter j times. The mere presence of a perfect reflector at the air-water inter-

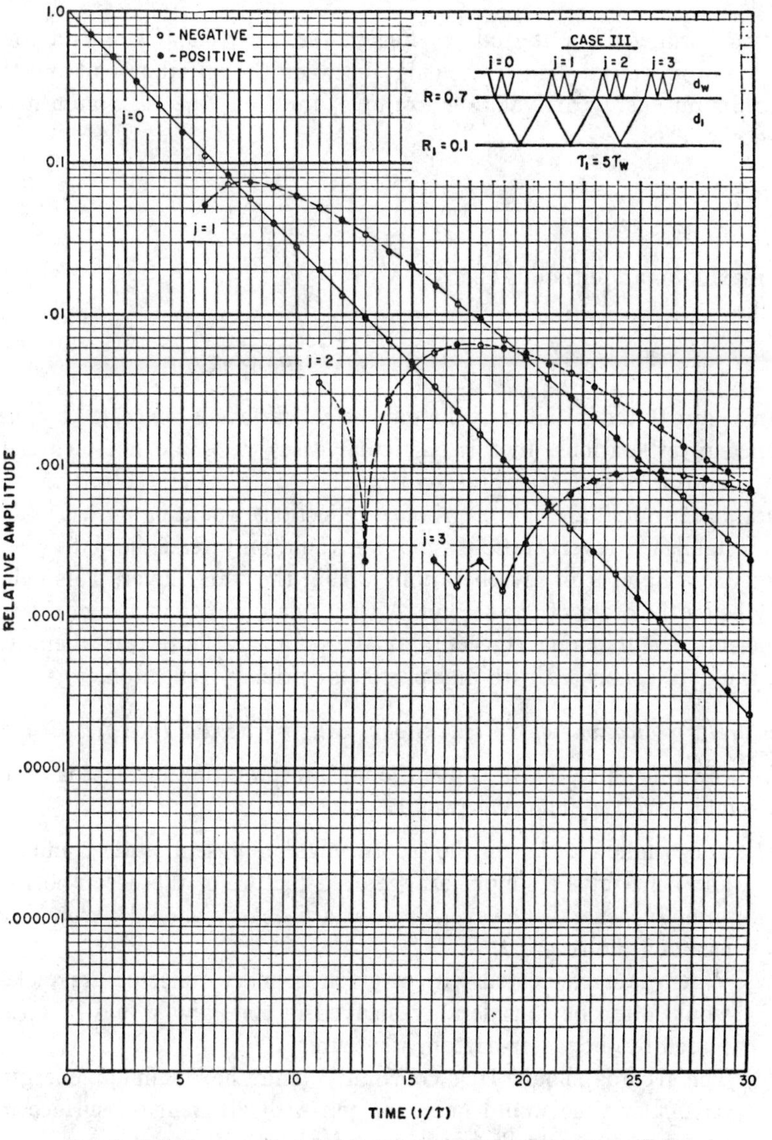

FIG. 9. Relative amplitudes and signs for terms $j=0, 1, 2, 3$, Case 3. Calculated for a plane wave, subsurface attenuation neglected.

face increases the expectation of section multiples over that encountered on land. However, the enhancement or compounding of section multiples by the water-layer filter is an even more significant factor. Section multiples involving more than one subsurface interface would be even more emphasized on marine records, due again to the existence of multiple in-phase ray paths. We should, therefore, expect section multiple energy to be very important on most marine singing records.

By regarding the amplitude coefficient for a particular j, k as a continuous function of n, we can predict the delay between the initial arrival and the point of maximum energy arrival. For example, for $j=1$, the maximum amplitude, y_{max} occurs at

$$n = n_{max} = \left[-1 - \frac{1}{\ln R}\right]. \tag{16}$$

The maximum amplitude is then,

$$y_{max} = R_1(1 - R^2)\left(-\frac{1}{Re \ln R}\right). \tag{17}$$

In Figure 9, the initial arrival from the subsurface is concealed by the water-confined reverberations. In this particular case, we would not pick the energy from the primary reflection until a time $2\tau_w$ to $3\tau_w$ after the initial arrival. In general, later initial arrivals of primary reflections would be buried under previously established reverberations, and we would only read the data after it had undergone a number of reverberations within the water layer. This delay could easily be on the order of one second. A very small change in any of the reflection coefficients would change considerably the point on the record where we would read the data connected with any particular primary reflection.

Structural Implications of a Conventional Singing Record Interpretation

We can thus infer that a conventional interpretation of records in a singing area should give the following results.

(1) Very shallow picks on the record might represent water-confined energy and hence give a simply exaggerated picture of the ocean-bottom structure. In general, however, subsurface energy should become dominant early on the record.

(2) When data connected with a given primary reflector are picked, they would often be considerably delayed in time and would be structurally biased by the water-bottom structure.

(3) The records should lack continuity, and the dominant energy at any particular time would be connected with different subsurface reflectors as one moves laterally across a section.

(4) Deeper structural indications would, in general, represent water rever-

berations excited by section multiples; hence they would show exaggerated shallow structure biased by exaggerated water-bottom structure.

(5) True deep unconformable structure would be masked by reverberations connected with shallow reflectors.

It is important to note, however, that in the case of a reasonably conformable subsurface, a flat water bottom, and large simple structures, sufficient structural information is included in the ringing records as they stand, to locate these structures approximately. The effect of the water layer on primary reflection energy is equivalent to the use of very tight ringing filters on land, except that the characteristics of the filter are changed from profile to profile.

EXPERIMENTAL DATA ON THE NATURE OF WATER REVERBERATION RECORDS

In Figure 10 a textbook-type example of the water reverberation problem in the Persian Gulf is provided. Water depth is 195 ft according to fathometer data. Successive arrivals with a constant time interval are very apparent. However, the time interval indicates a trap depth of 210 ft. Note that successive reflections are inverted due to the inversion at the free surface and lack of inversion at the water bottom. Velocities computed from normal moveout and the delta-T due to dip across the record are shown.

Normal Moveout Data

The results of a conventional normal movement analysis of 15 profiles from the Persian Gulf (Figure 19) are shown in Figure 11. Average velocities were computed by using straight ray-path formulae, and the vertical lines in Figure 11 show the standard deviation. The computed velocity ranges from about 6000 to 8000 ft per sec. There is no evidence for average velocity values of 5000 ft per sec which should be present if purely water-confined energy were dominant on any part of the record. The actual velocity as a function of depth is not well established for this area, but it is expected to be of the order of 7000 ft per sec to a depth of 1000 to 1500 ft, and interval velocity below that depth should increase with depth from an initial value of about 10,000 ft per sec. On the basis of Figure 11, it may be inferred that energy associated with the very shallow section (reverberations set up by primaries and section multiples) is dominant over the first 1.0 to 1.5 sec, with deeper penetration dominating in the interval from 2 to 3 sec.

An additional suite of more than 200 profiles from an entirely different water-reverberation area were similarly analyzed with similar results. The measured normal-movement velocities were consistently intermediate between expected values for subsurface primaries and purely water-confined energy.

It is concluded on the basis of normal moveout that the dominant data on nearly all of the water-reverberation records we have studied represents water reverberations connected with reflections from the subsurface and that the

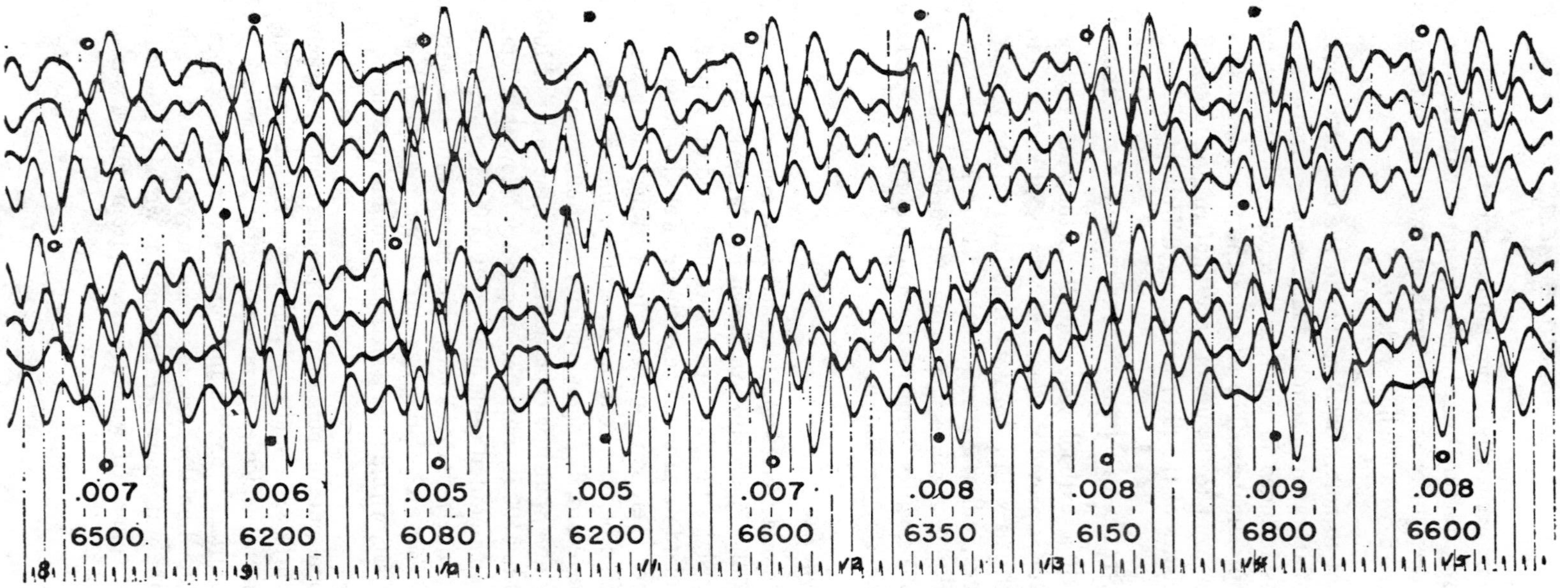

FIG. 10. Textbook example of a Persian Gulf water reverberation record.

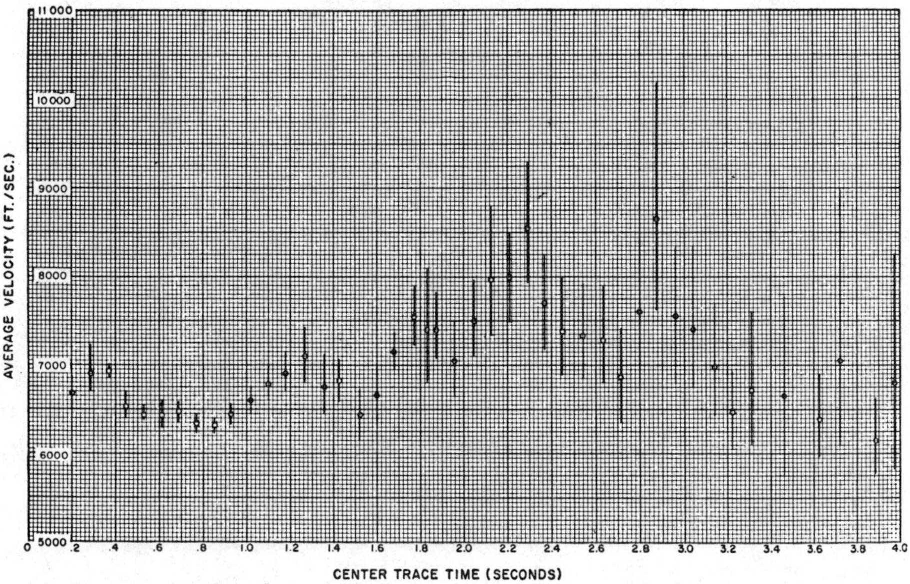

FIG. 11. Average velocity versus center trace time from a normal moveout analysis of the profiles shown in Figure 19.

travel path is made up of the ordinary subsurface travel path plus a path representing a large number of successive reflections within the water layer .

A more striking example of the normal moveout criterion is supplied in Figure 12 in which the singing is not nearly as obvious as in the previous examples. This record was dynamically corrected for the expected normal moveout for primary events. Up to about 0.5 sec the events have quite excessive normal moveout, in this case actually close to that for a 5000-ft-per-sec average velocity. At about 0.6 sec, an event appears with nearly the proper normal moveout, and successive reverberations connected with this subsurface reflection may be seen following. This sort of behavior is very common in areas where a moderate singing record problem is present.

The utility of dynamically corrected record sections is immediately apparent for this type of work where normal moveout constitutes a significant criterion in the identification of events.

Frequency Analysis

The Fourier transform of four segments of the fourth trace of the profile illustrated in Figure 10 was computed. The amplitude as a function of frequency is displayed in Figure 13. Amplitude coefficients were computed at 1-cps intervals over the range 45 cps to 63 cps, and at 3-cps intervals over the rest of the range from 6 to 120 cps. The pass band through which the data had been filtered previous to analysis was about 20 to 120 cps including the pressure transducer

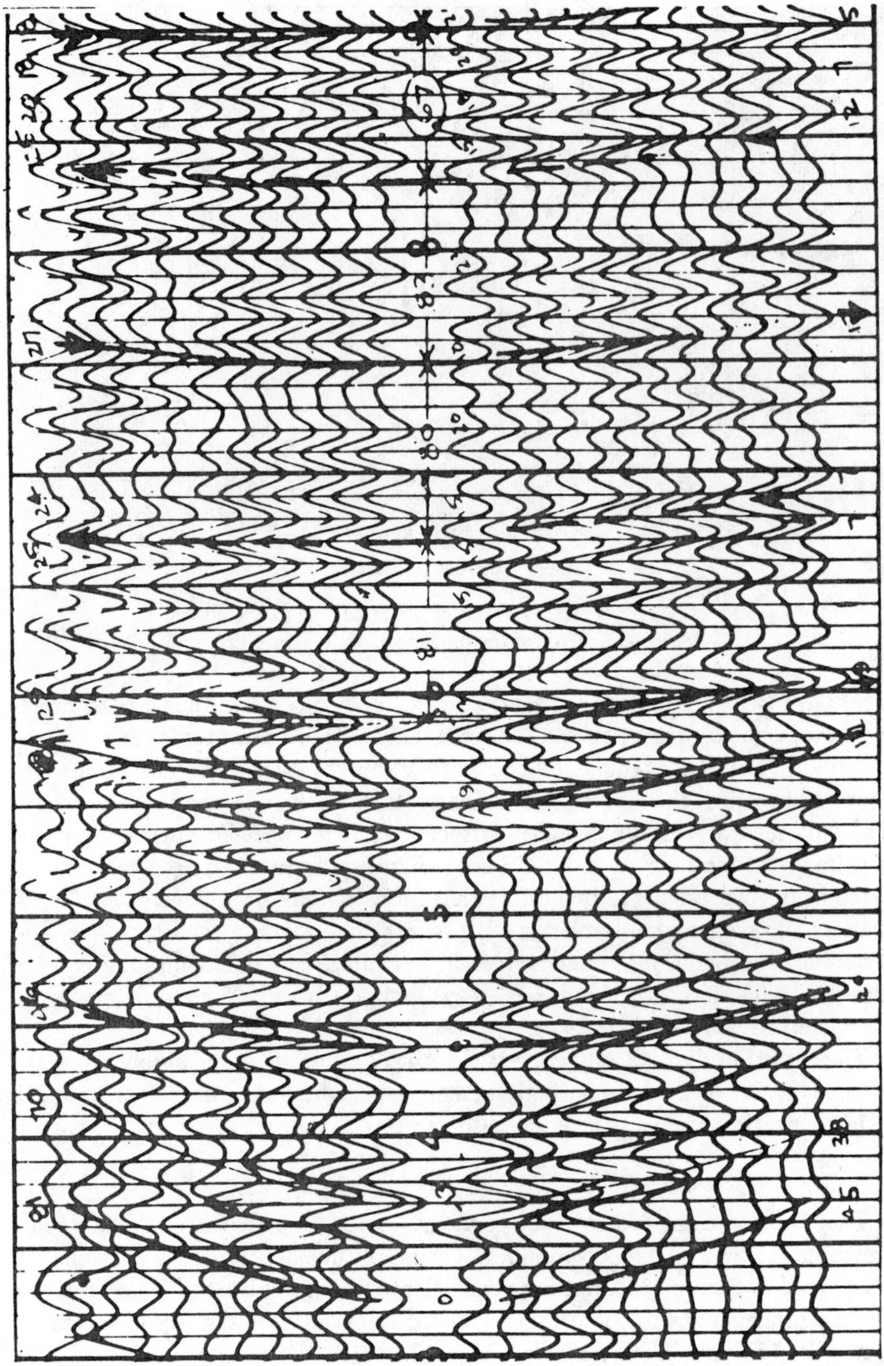

FIG. 12. A moderate water reverberation record dynamically corrected for expected true section velocity.

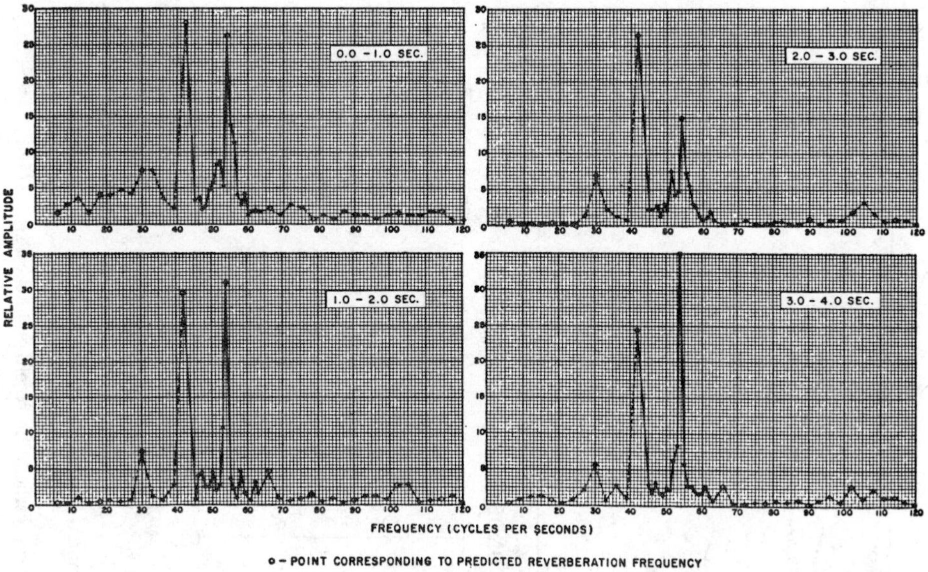

FIG. 13. Frequency analysis of four segments of Trace 4, Figure 10.

response. Seismometer depth was 25 ft, and the effective response of this filter is shown in Figure 4. Water depth was 195 ft. Accepting 42 cps and 54 cps as resonant frequencies of the water-layer filter, we predict resonances at 6 cps and all odd harmonics plotted as open circles in Figure 13. The sharply peaked response, particularly at 42 and 54 cps, is quite apparent on all segments of the record. The data fits the concept of regarding the water layer as a filter if we accept a trap depth different from that indicated by the fathometer. This difference has been found generally to be quite common in the Persian Gulf and has been attributed to the presence of a thin, soft layer overlying the hard bottom. The high frequency fathometer sees the top of this thin layer as the first dominant reflector, while the lower seismic frequencies are most affected by the hard bottom interface.

In Figure 13, the complete dominance of the harmonics at 42 and 54 cps cannot be explained at all on the basis of instrumental filters and can only be partially explained on the basis of seismometer depth. This dominance is attributed to the relative enrichment of frequencies from 30 to 60 cps in the energy returning from the subsurface. The relative complexity of the spectra is additional evidence indicating the importance of the energy returning from the subsurface.

Amplitude Decay Data

In Figure 14, several traces from a constant-gain recording in the Persian Gulf, and amplitude-versus-time measurements taken from that recording are shown. A plot of amplitude decay with the effect of spherical divergence approxi-

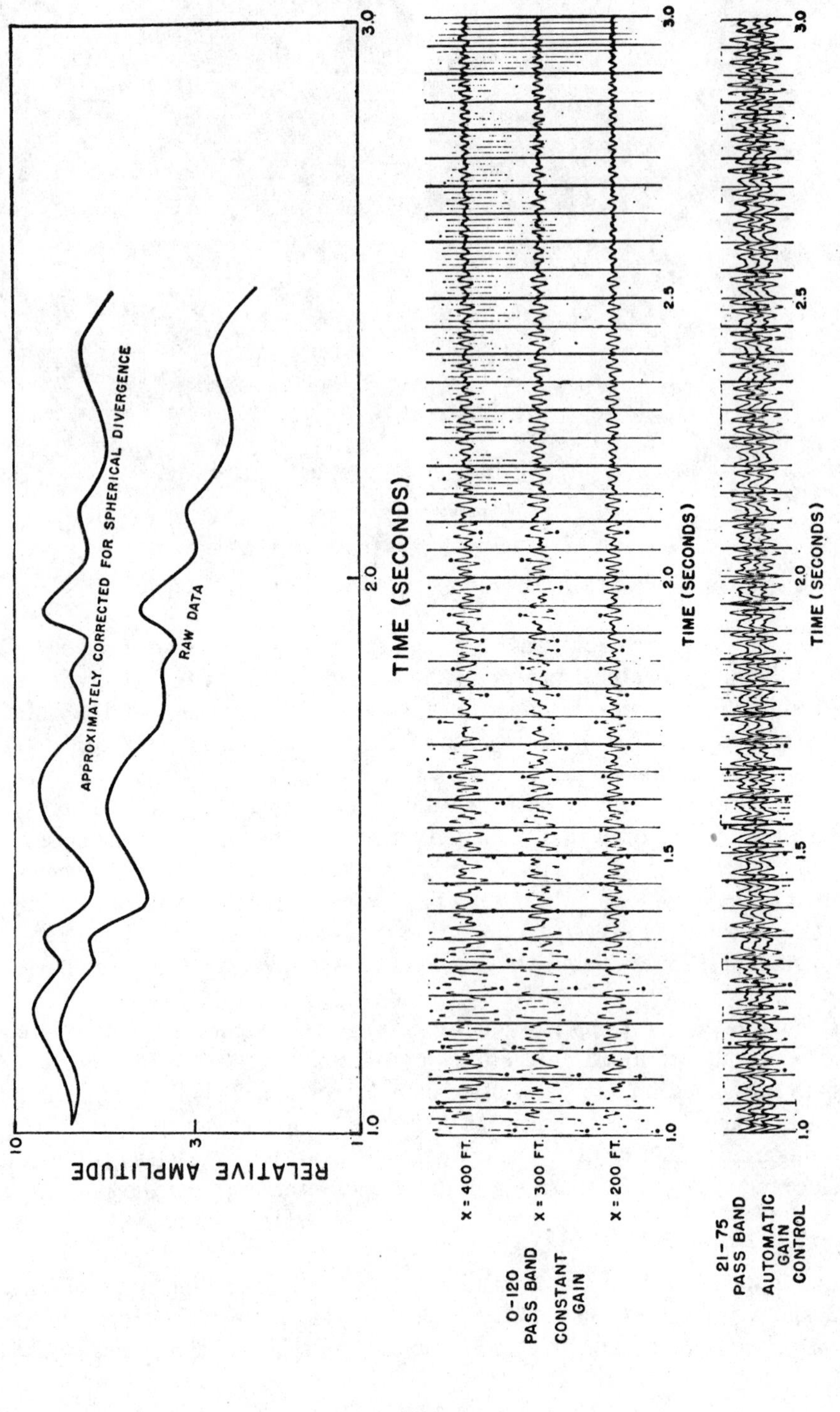

FIG. 14. Example of the amplitude decay as a function of time on a water reverberation record from the Persian Gulf.

mately removed is also shown. By comparison with Figures 7–9, it may be seen that Figure 14 cannot be explained on the basis of water-confined energy, but is reasonably consistent with the concept of dominant subsurface excited reverberations.

Structural Evidence

Purely water-confined reverberation energy should show structure corresponding approximately to an exaggeration of trap-bottom structure, the extent of the exaggeration increasing with time. On a particular profile, apparent dip should increase regularly with time. Reverberation energy excited by subsurface reflections, on the other hand, should be biased by water-bottom structure but should also contain subsurface structure. On a particular profile, dip-versus-time should be made up of segments showing local change as a function of time controlled by water-bottom structure, but it should be displaced in absolute value from the dip predicted from water-bottom structure. However, a study of these factors has not led to a definitive conclusion based on structural evidence. Structural behavior may be examined in the Persian Gulf record section (Figure 19) in which the fathometer and trap depth data are shown.

Conclusions

The existence of the water reverberation problem is characterized by an apparent singing or dominant repetition interval on the records, by a sharply peaked frequency spectrum, by normal moveout intermediate between that expected for the true section velocity and that expected for water velocity, and by an abnormally slow energy decay. Experimental data confirms the prediction that the dominant energy is connected with subsurface reflected energy and is consistent with the concept of regarding the water-layer effect approximately as a linear filtering effect.

REMOVAL OF THE WATER REVERBERATION EFFECT

Referring to Figure 15, the water-reverberation effect can be regarded as a linear filtering effect rather than noise. The desired primary reflection data passes through two sections of the water-layer filter, with transfer function

$$K(\omega) = \left(\frac{1}{1 + Re^{-j\omega\tau_w}}\right)^2. \tag{18}$$

If no non-linear processes are involved in recording the signal, by passing the recorded data through the inverse of the filter in equation (18), the desired primary reflection data may be recovered in its original form. The frequency response of the required inverse filter, $H(\omega)$, is defined by

$$[H(\omega)]\left[\frac{1}{1 + Re^{-j\omega\tau_w}}\right]^2 = 1$$

or

$$H(\omega) = (1 + Re^{-i\omega\tau_w})^2. \tag{19}$$

The difference between this filtering concept and the usual seismic filtering concept requires emphasis. In general, we record a desired seismic signal, $f(t, x)$, to which is added noise, $N(t, x)$. The characteristics of f and N are examined in terms of frequency and wave-length. We then use a filter with a pass band limited to the region where $F(\omega)/N(\omega)$ is large, or we design a multiple-seismometer array with a pass band where $F(k)/N(k)$ is large. We filter out those frequencies or wave lengths where the signal-to-noise ratio is low and accept the resulting degradation in the signal. An equivalent approach to singing records would be to regard the water reverberations as noise, and use a narrow band-pass filter

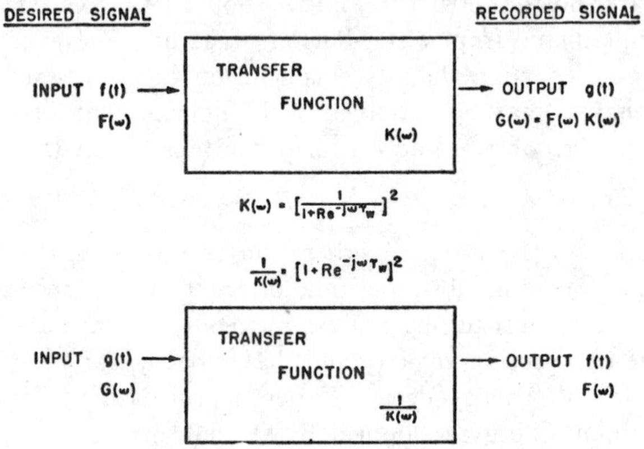

WATER REVERBERATIONS—
AN INVERSE FILTERING PROBLEM

DESIRED SIGNAL	SEISMIC NOISE	RECORDED SIGNAL	
f(t)	N(t)	f(t) + N(t)	time
F(ω)	N(ω)	F(ω) + N(ω)	frequency
F(k)	N(k)	F(k) + N(k)	wave length

SEISMIC NOISE – A SELECTIVE FILTERING PROBLEM

FIG. 15. Block diagram contrasting the water reverberation and the seismic noise problems.

fitted between two resonant peaks of the water-layer filter, or to use an arbitrary comb-notching filter to eliminate the reverberation frequencies. In the case of the water-reverberation problem, this sort of approach would often result in excessive degradation of the desired signal. By using the inverse filter, on the other hand, the primary reflection events should be recovered with the wave form which would have been obtained if the water layer had been removed.

Section-multiple reflections are not removed by this inverse water-layer filtering, nor are the reverberations from section multiples completely eliminated. However, the reverberating energy from section multiples is considerably attenuated, while valid initial arrivals from primary reflections are not attenuated. The compounding effect of the water layer on section multiples is reduced considerably. The effect of the inverse filter in equation (19) on the n, j, k term in equation (14) is a reduction by the ratio,

$$\frac{(j-k)(j-k-1)}{(n+j-k)(n+j-k-1)},$$

which represents also the increase in the ratio of primary reflection amplitude to the amplitude of energy connected with section multiples. For example, in Figure 9, the amplitude of the reverberations due to the first section multiple would be attenuated by 22 db at $(t/\tau)=18$, and by 33 db at $(t/\tau)=25$.

To obtain the impulse response of the required inverse filter, we transform (19) to the time domain, obtaining

$$h(t) = \delta(t) + 2R\delta(t-\tau) + R^2\delta(t-2\tau). \tag{20}$$

The frequency response of the required filter is shown in Figure 16 for a water depth of about 200 ft and for a bottom coefficient of 1.0 and 0.5. The phase response of the filter is particularly interesting. In applying inverse filtering to a set of records it is necessary to determine whether the dominant reverberation problem is due to an acoustically hard or soft bottom, and it is necessary to determine the effective trap depth and the water-bottom coefficient for each profile.

The seismometer-depth filtering effect is of a similar form to the required inverse filter. Thus if one recorded, in the case of an acoustically hard bottom, with velocity seises located at the bottom, the filtering effect is of the form

$$P(\omega) = (1 + e^{-j\omega\tau_w}) \tag{21}$$

which constitutes a notching filter with notches at the water-reverberation resonant frequencies. The same effect could be achieved in the case of an acoustically soft bottom by the use of pressure seismometers at the trap bottom. There are serious drawbacks to this approach, in addition to the operational difficulties. First, particularly in hard bottom areas, the significant lower trap boundary is often deeper than the physical bottom. Thus dragging the seises on the physical bottom does not provide the desired result. Second, the use of seismometer depth to attenuate the water reverberations is a noise-filtering approach, and would

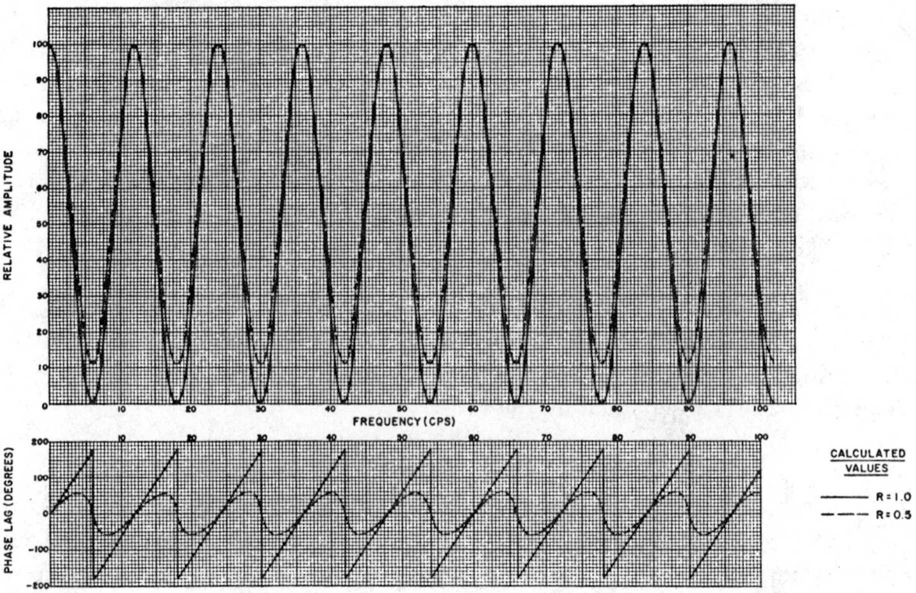

FIG. 16. Frequency response of the two-section inverse filter.

often, particularly in deep water, be expected to produce excessive degradation of the desired signal.

Discussion of Approximations

The examination of the problem has been confined to the plane-wave case. On the later portion of the record, this analysis should nearly apply directly, since the reflected events at that time behave essentially as plane waves. On the early part of the record, and in the behavior of the water confined energy, the essential deviations of the point-source problem from the plane-wave problem are the following:

(1) time differences due to normal moveout,
(2) amplitude variations due to spherical divergence,
(3) variations in the effective reflection coefficient due to the spherical wave front.

Due to normal moveout, on the outer traces during the early part of the record, the interval between successive water reverberations should be less than that on the center traces. The interval should then increase with time, tending to the interval observed on the center traces. However, normal moveout effects may be removed from the outer traces by a number of available dynamic correction devices. If the normal moveout is removed in a continuous manner, the point-source record may be changed as far as time relationships are concerned to the approximate equivalent of the plane-wave case, except for some interference effects.

Due to spherical divergence, the amplitude decays as $1/r$, where r is the length of the travel path. This factor could be approximately compensated for in data processing. Also the seismic amplitude suffers non-geometrical attenuation in the subsurface, whereas it suffers practically no non-geometric attenuation in the water. In Figures 7–9 these factors would increase the importance of terms with small j relative to those with large j. These factors have a very significant effect on the results in Figures 7–9, but they do not alter the basic conclusions which have been drawn.

For wave lengths large relative to the distance between the source and the ocean bottom, the ocean bottom has an effective reflection coefficient different from that of the plane-wave case and varying with frequency. The filtering error thus introduced would be greatest at low frequencies and in shallow water. For example, using a case discussed by Pekeris (1948, p. 49), the amplitude difference between the plane wave and spherical wave for the first reflection from the bottom would be about 12 percent at 20 cps for a 200-ft water depth. Because of this factor, one of the sections of the water-layer "filter" through which primary reflections pass deviates from the plane-wave case at low frequencies.

The problem has been examined for flat, parallel boundaries. Slight dip or structure in the ocean bottom (which is greatly magnified on the reverberating record) reduces slightly the effect of the inverse filtering technique. Severe dip or structure in the trap bottom combined with a very high bottom-reflection coefficient results in considerable complications and the problem is not amenable to inverse filtering techniques.

A number of additional deviations of the actual case from the theoretical cases discussed are readily apparent.

Examples

An example of the approximate application of this inverse filtering technique to a textbook-type record is shown in Figure 17. The data were recorded with conventional automatic gain control, which constitutes a non-linear process, and R was set somewhat arbitrarily to one, using only a single section of the inverse water-layer filter. This record is from the Persian Gulf; water depth is about 200 ft. These are 75–21 cps playbacks of a wide band recording. No dynamic corrections were applied for playback.

In Figure 18, an example from Lake Maracaibo is provided. Note that the dominant frequencies on the unprocessed record are 30 cps and 58 cps; this indicates that the dominant reverberation problem was due to an acoustically soft bottom.

Figures 19 and 20 are variable density record sections of a suite of records from the Persian Gulf, illustrating the approximate application of the inverse filtering technique. The suite consists of 15 profiles providing continuous subsurface coverage of about 3.5 miles.

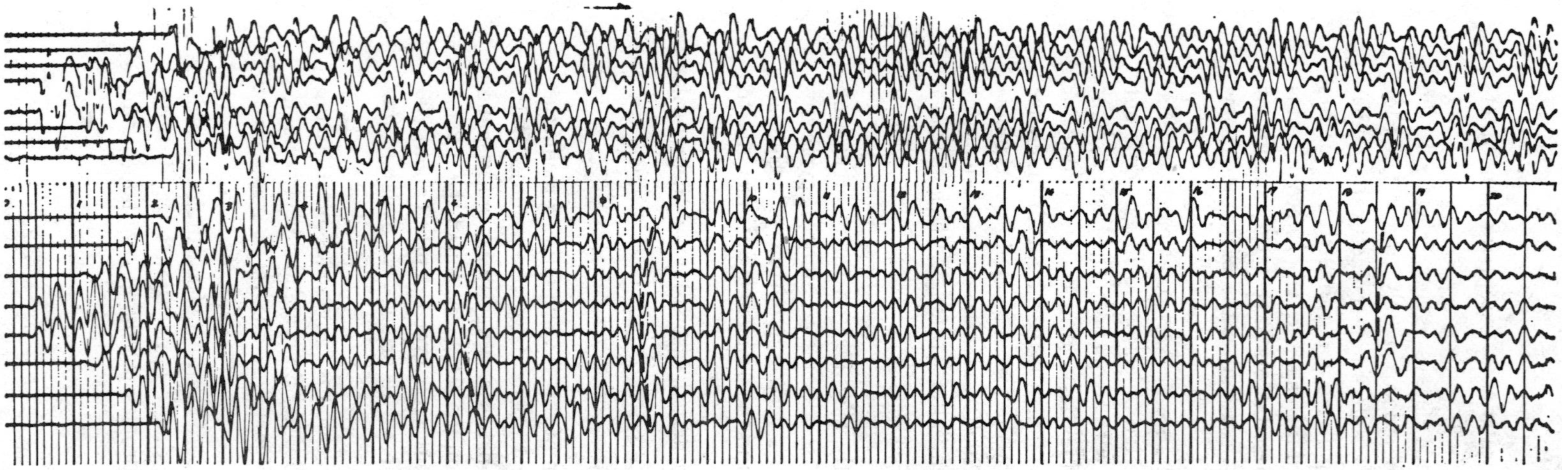

Fig. 17. An illustration of approximate inverse filtering applied to a Persian Gulf record. The upper record is a conventional playback (75–21 cps pass band) of a reverberation record from the Persian Gulf. The lower record is a playback of the same record with approximate inverse filtering applied. This illustrates the effectiveness of inverse filtering on conventionally recorded data taken in deep water (200 ft) for a hard bottom.

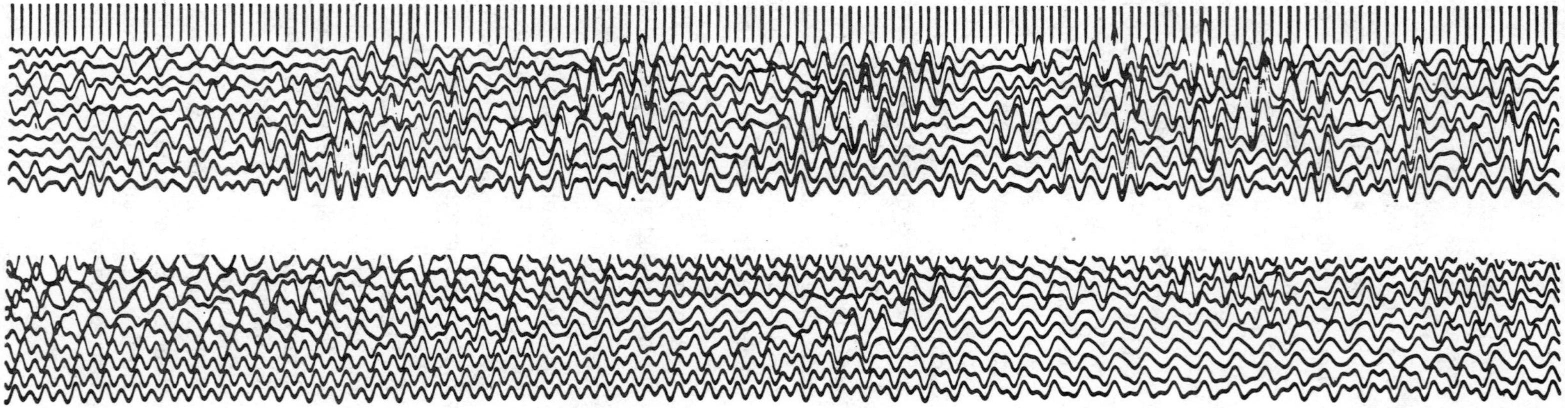

Fig. 18. The lower record is a conventional playback of an end-on profile from Lake Maracaibo. Water depth was about 80 ft, and the two dominant frequencies on the record are 30 cps and 58 cps indicating that the effective trap bottom is acoustically soft. The upper record is a playback of the same data with inverse filtering applied.

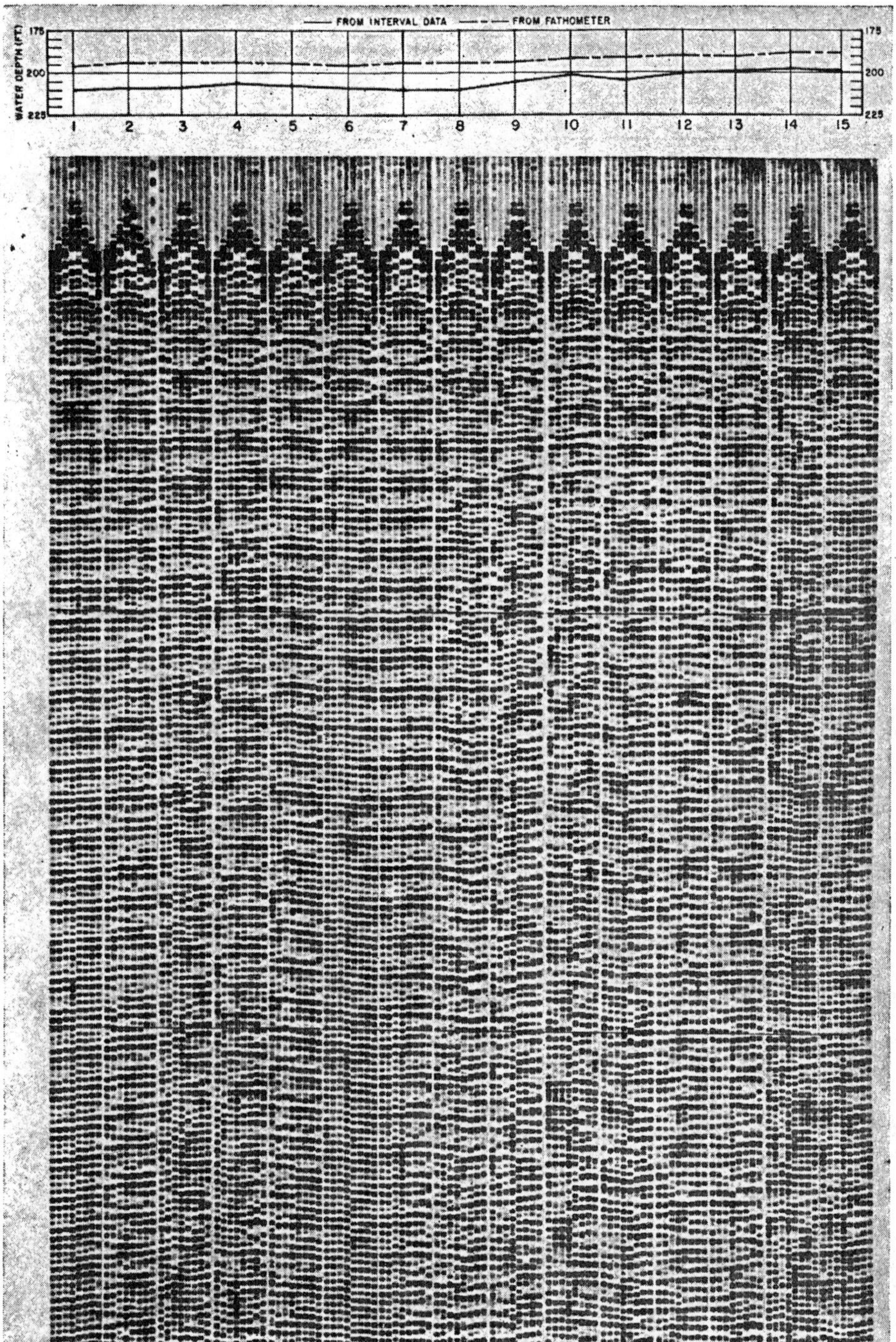

FIG. 19. A record section from the Persian Gulf. A 21–75-cps pass band was used in playback, and NMO corrections were applied to compensate for *observed* normal moveout. Profiles 1–7 represent severe water reverberation records while profiles 8–15 constitute a less obvious example of the problem. In the upper chart the solid line represents calculated water trap depth based on the apparent repetition interval and frequency content of the records.

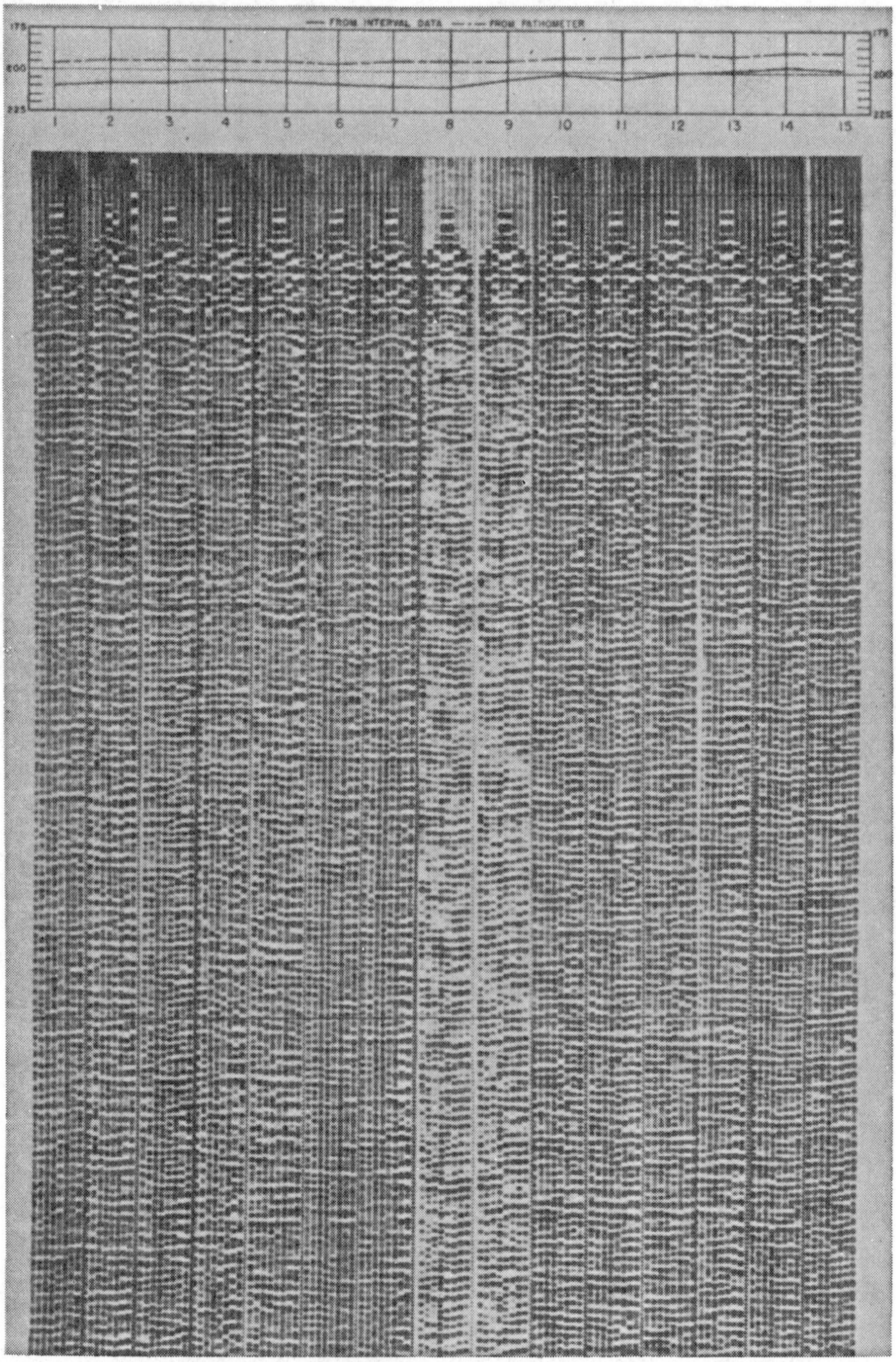

FIG. 20. This record section is identical to Figure 19 except that inverse filtering has also been approximately applied. Any valid events should show reverse normal-moveout.

The data were recorded on magneDISCs from linear arrays of flat-response piezoelectric pressure transducers at a depth of 25 ft, using a 15–120-cps pass band and conventional automatic gain control. Figure 19 is a 75–21-cps playback corrected for *observed* normal moveout. That is, the dynamic corrections applied were larger than those for expected section velocity. The processed section (Figure 20) had the same dynamic corrections applied; hence valid reflection events should show reverse normal moveout. The inverse filtering technique was only approximately applied in the same sense as indicated for Figure 17.

Discussion of Examples

An examination of Figures 17–20 shows that approximate inverse filtering has removed the dominant singing effect or periodicity on the records and has brought out some apparent reflection events which were masked by the ringing on the unprocessed records. These uncovered events in general have less normal moveout than the dominant energy appearing at the same record time on the unprocessed records. This is particularly apparent in the reverse normal moveout of the events in the processed Persian Gulf record section. It means that the processed data at a given time represents deeper subsurface penetration. Some of the events on the processed records are probably initial primary reflection arrivals, and others are initial section-multiple arrivals. Insufficient velocity control is available to establish clearly which events are primary reflections. The quality of the reflections on the processed data is poor, and this is due in part to the fact that only an approximate application of the inverse filtering principal has been made. These comparisons illustrate what can be done in the data processing of conventionally recorded marine seismic data.

For the optimum application of this technique special high fidelity recording is necessary. The non-linear AGC must be eliminated or optimally minimized. The amount of signal attenuation in inverse filtering may be as high as 30 db in extreme cases, for example, in the "dead" areas on the processed record in Figure 17. Thus this data-processing technique truly requires the full dynamic range available on today's magnetic recorders.

CONCLUSIONS

Experimental and theoretical evidence indicates that on the majority of singing records the dominant energy represents water reverberations excited by reflections from the subsurface. Singing records in untreated form thus contain some structural information about the subsurface.

The effect of the water layer may be approximately treated as a linear filtering effect and thus may be removed or attenuated by applying the principle of inverse filtering during data processing. The application of inverse filtering constitutes a practical production technique which, under ideal conditions, approximately results in the elimination of the dominant apparent ringing on marine records, the recovery of primary reflection events with the wave form which they would have had if the water-rock interface had been removed, and a substantial increase in the ratio of valid primary reflection events to section multiple energy.

ACKNOWLEDGMENTS

The author thanks Willis S. Shelton who did much of the computational and data-processing work involved and offered many valuable comments on the problem. The author thanks Geophysical Service Incorporated for the permission to publish this paper.

REFERENCES

Bortfeld, R., 1956, Multiple Reflexionen in Nordwestdeutschland: Geophys. Prosp., v. 4, p. 394–423.

Burg, K. E., Ewing, Maurice, Press, Frank, and Stulken, E. J., 1951, A seismic wave guide phenomenon: Geophysics, v. 16, p. 594–612.

Cole, R. H., 1948, Underwater explosions: Princeton, Princeton University Press.

Jones, J. L., et al., 1958, Acoustic characteristics of a lake bottom: Acoustical Society America Jour., v. 30, p. 142–145.

Pekeris, C. L., 1948, Theory of propagation of explosive sound in shallow water: Geol. Soc. America Mem. 27.

Poulter, T. C., 1950, Geophysical studies of the Antarctic: Palo Alto, Stanford Research Institute.

Sarrafian, G. P., 1956, Marine seismic model: Geophysics, v. 21, p. 320–336.

Smith, W. O., 1958, Recent underwater surveys using low frequency sound to locate shallow bedrock: Geol. Soc. America Bull., v. 69, p. 69–98.

Reprinted from Geophysics v. 25, no. 1, p. 130-140

ELIMINATION OF SEISMIC GHOST REFLECTIONS BY MEANS OF A LINEAR FILTER*

J. P. LINDSEY†

ABSTRACT

A technique is described for elimination of ghost reflections on magnetically recorded seismograph records by means of a linear filter. The application of this filter does not alter the character of primary reflections although eliminating the ghost reflections. The principal assumption made in the development of the technique is that the effect of AGC in altering the amplitude ratio of primary and ghost reflections is uniform for all record time.

A realization of the required filter is given and a measurement technique is outlined for detecting the existence of ghost reflections based on the autocorrelation function of the seismograph trace.

The existence of a large velocity discontinuity above a seismic shot has long been recognized as the source of "ghost" reflections appearing on the seismogram In such instances, the downgoing wavefront set up by the shot is characterized by energy moving directly downward from the shotpoint followed in space and time by energy reflected from the overlying discontinuity. When detected, this downgoing wavefront appears as two wavelets displaced in time by approximately twice the travel time from the shot to the discontinuity and with possible differences in shape. Any difference in shape may be attributed to tuning effects of the ground between the shot and the discontinuity and the sphericity of the incident wavefront at the discontinuity. These relationships are illustrated in Figure 1.

The wavelet associated with the incident component of the downgoing wavefront will be designated as $b(t)$, a time function having a LaPlace transform $B(s)$ (Gardner and Barnes, 1942; Goldman, 1955). The wavelet associated with the reflected component of the downgoing wavefront will have the LaPlace transform $\mathcal{R}H(s)B(s)e^{-s\tau}$ where $\mathcal{R}$ is the reflection coefficient at the overlying boundary, $H(s)$ represents the tuning effects of the two passes through the strata between the shotpoint and the boundary, and $e^{-s\tau}$ is the LaPlace time delay operator having τ seconds delay (Gardner and Barnes, 1942; Goldman, 1955). The symbol τ represents the time lag of the ghost wavefront. Any pure loss of the overlying strata will be included in the factor $\mathcal{R}$. Having made these definitions, the LaPlace transform of the total wavelet representing the downgoing wavefront is,

$$B(s) - \mathcal{R}H(s)B(s)e^{-s\tau}. \qquad \text{(a)}$$

When an impulse is presumed to excite the subsurface layering, the response wavelet received by the surface detector will be denoted by the time function

* Presented at Twelfth Annual Midwestern Meeting, El Paso, Texas, April 28, 1959. Manuscript received by the Editor July 14, 1959.

† Phillips Petroleum Company—Research Division, Bartlesville, Okla.

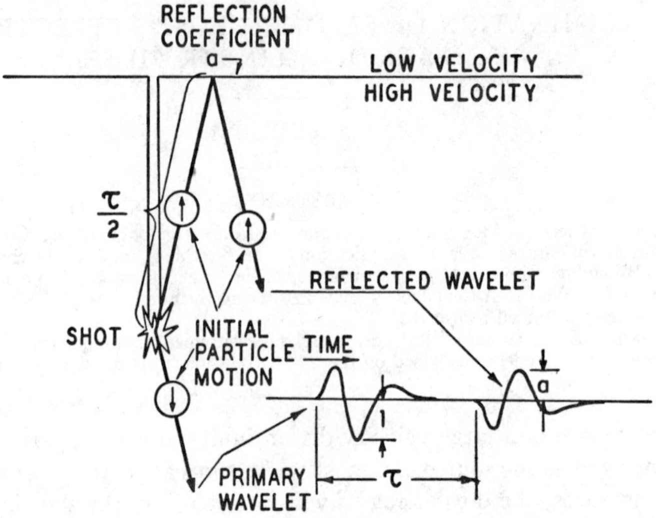

FIG. 1. Illustration of the relation between primary and ghost wavefronts generated by the seismic shot.

$r(t)$ having LaPlace transform $R(s)$. Therefore, for the excitation represented by expression (a), the surface response will be,

$$R(s)\left[B(s) - \mathcal{A}H(s)B(s)e^{-s\tau}\right]. \tag{b}$$

The response desired at the surface is that generated by only the incident wavefront set up by the shot. This component of the surface response is given by the LaPlace expression,

$$R(s)B(s). \tag{c}$$

Now the filter required to eliminate ghost reflections may be defined as one that converts the total response at the surface to the primary reflections alone. This conversion process will be specified as the operation of linear filtering (Robinson, 1954). This filter, designated $F(s)$, may be mathematically described as,

$$F(s)R(s)\left[B(s) - \mathcal{A}H(s)B(s)e^{-s\tau}\right] = R(s)B(s),$$

$$F(s) = \frac{R(s)B(s)}{R(s)\left[B(s) - \mathcal{A}H(s)B(s)e^{-s\tau}\right]},$$

$$F(s) = \frac{1}{1 - \mathcal{A}H(s)e^{-s\tau}}. \tag{1}$$

Realization of this filter is made quite easy by observation of the similarity between the transfer function as given by (1), and the general expression for the

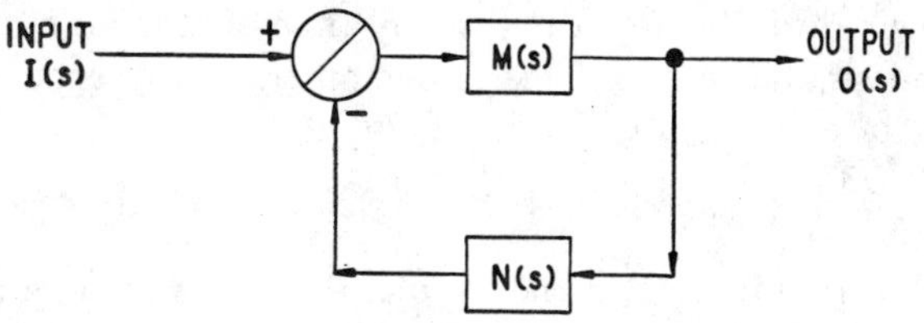

$$\frac{O(s)}{I(s)} = \frac{M(s)}{1 + M(s)N(s)}$$

FIG. 2. Generalized single loop feedback system employing negative feedback.

transfer function of a system with feedback. The generalized single-loop feedback system and its transfer function are shown in Figure 2. The numerator of the transfer function represents the transfer function of the forward loop and the second term of the denominator represents the product of all transfer functions around the loop. By analogy, the required filter, when represented in the configuration of a single-loop feedback system, will have a forward loop transfer function of unity and a transfer function around the loop of $\mathcal{a}H(s)e^{-s\tau}$. The resulting feedback system is indicated in Figure 3.

This filter converts a primary reflection plus ghost reflection at the input into a primary reflection only at the output. Likewise a series of primary reflections plus their respective ghosts is transformed into the primary reflection series only. This action is more easily seen by considering a single primary-ghost wavelet sequence. Upon entering the filter, a primary wavelet appears at the output with no delay and simultaneously is stored in the delay unit. As its ghost wavelet enters the filter τ seconds later, the stored primary wavelet is transformed by $H(s)$ to have the same shape as the ghost wavelet, and thus cancels the ghost wavelet at the output of the filter.

It is important to point out here that only the portions of the seismic record subject to the ghosting conditions should be admitted to the input of the ghost elimination filter. Any event such as first-break wavelets, which have no ghost

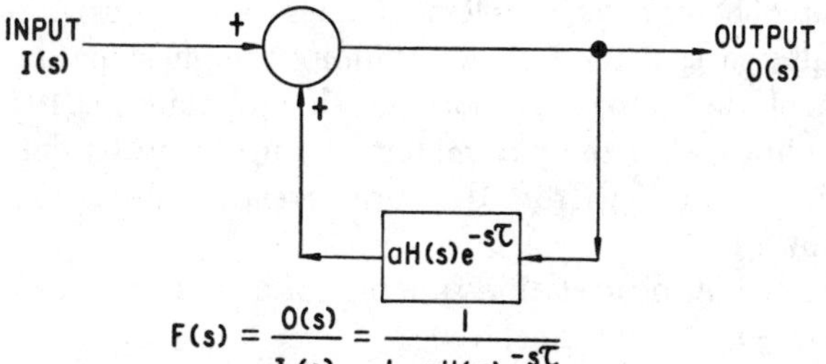

$$F(s) = \frac{O(s)}{I(s)} = \frac{1}{1 - aH(s)e^{-s\tau}}$$

FIG. 3. Feedback system which realizes the transfer function of equation (1). (Note that feedback must be positive).

wavelet following, will be duplicated many times at the output of the filter with obviously undesirable results. This is demonstrated by expanding the transfer function $F(s)$ into an infinite series:

$$F(s) = 1 + \mathcal{A}H(s)e^{-s\tau} + \mathcal{A}^2H^2(s)e^{-2s\tau} + \mathcal{A}^3H^3(s)e^{-3s\tau} + \cdots \tag{2}$$

When the earth tuning, $H(s)$, is negligible, this series reduces to,

$$F(s) = 1 + \mathcal{A}e^{-s\tau} + \mathcal{A}^2e^{-2s\tau} + \mathcal{A}^3e^{-3s\tau} + \cdots \tag{3}$$

This is the response of the ghost-elimination filter to an impulse input, and consists of a series of impulses decaying in amplitude at the rate $\mathcal{A}$ per time τ, and spaced τ seconds apart. Likewise, if any single event enters the filter, the output is an infinite repetition of the event every τ seconds and decaying exponentially in amplitude. This problem is usually resolved by electronically eliminating early arrivals on the record before passing the signals through the filter.

The "earth" filter, represented by $H(s)$, will of course, take on many forms depending upon the nature of the strata overlying the shot. If the base of the weathering is the ghost-producing boundary and there is no inhomogeniety between the shot and the weathering, there may be no need for the earth filter at all. If the surface is the ghost-producing boundary and no high contrast layering is present between the shot and the surface, a relatively simple filter that emphasizes the low frequencies with about a six db per octave attenuation will suffice. The filter should be phase corrected and have its three db attenuation point at a frequency appropriate for the ground characteristic. A large degree of inhomogeneity or complicated layering between the shot and the ghosting boundary generally washes out the ghost energy sufficiently that ghosts are not the primary problem. In the majority of cases where elimination of ghost reflections is required there is no need for the earth filter.

The obvious problem in setting up the ghost-elimination filter is to determine the parameters $\mathcal{A}$, $H(s)$ (if required), and τ. If field procedure permits, these may be obtained from a down-hole seismometer recording with the seismometer placed below the shot level. Aside from the problems associated with obtaining this trace, it often happens that magnetically recorded data is already in hand and the field work completed before the problem of ghost reflections is recognized. Consequently an alternate method of determining the ghost parameters is desirable. Examination of the autocorrelation function (Robinson, 1954) of the seismic record will indicate that the parameters $\mathcal{A}$ and τ may be determined, although accurate definition of $H(s)$ from the autocorrelation function is seldom possible, as will be shown later.

The truncated autocorrelation function of a seismic trace measured over the time interval $t_2 - t_1$ is defined as,

$$A(\sigma) = \frac{1}{t_2 - t_1}\int_{t_1}^{t_2} X(t)X(\sigma + t)dt. \tag{4}$$

The interval $t_2 - t_1$ will be chosen so that only events subject to the ghosting conditions are included. When ghosts are present on the trace, the function $X(t)$ may be expressed as

$$X(t) = Y(t) - \mathcal{A}Y(t - \tau), \tag{5}$$

if the effects of earth tuning are neglected. In this notation, $Y(t)$ represents the primary reflection series.

Substituting expression (5) into (4) yields,

$$A_{p+g}(\sigma) = \frac{1}{t_2 - t_1}\int_{t_1}^{t_2} [Y(t) - \mathcal{A}Y(t - \tau)][Y(\sigma + t) - \mathcal{A}Y(\sigma + t - \tau)]dt,$$

and,

$$A_{p+g}(\sigma) = \frac{1}{t_2 - t_1}\left[\int_{t_1}^{t_2} Y(t)Y(\sigma + t) + \mathcal{A}^2\int_{t_1}^{t_2} Y(t - \tau)Y(\sigma + t - \tau)dt - \mathcal{A}\int_{t_1}^{t_2} Y(t)Y(\sigma + t - \tau)dt - \mathcal{A}\int_{t_1}^{t_2} Y(\sigma + t)Y(t - \tau)dt\right]. \tag{6}$$

The autocorrelation function of the primary reflection series alone is,

$$A_p(\sigma) = \frac{1}{t_2 - t_1}\int_{t_1}^{t_2} Y(t)Y(\sigma + t)dt. \tag{7}$$

Comparison of (6) and (7) gives a relation between the autocorrelation function of primary plus ghost reflections to the autocorrelation function of primary reflections alone:

$$A_{p+g}(\sigma) = (1 + \mathcal{A}^2)A_p(\sigma) - \mathcal{A}A_p(\sigma + \tau) - \mathcal{A}A_p(\sigma - \tau). \tag{8}$$

Thus, the autocorrelation function of a trace having primary and ghost reflections is composed of the sum of three autocorrelation functions of the primary reflection series. These three autocorrelation "wavelets" are separated in time by τ seconds, and the ratio of the amplitude of the center wavelet to that of the adjacent two wavelets is $(\mathcal{A}^2+1)/\mathcal{A}$. Figure 4 illustrates these features. Two of these autocorrelation wavelets exist only because of the ghosting conditions. That is, if the quantity $\mathcal{A}$ in (8) were reduced to zero, which would be the case when no ghost is present, only one autocorrelation wavelet would exist. Therefore, the presence of ghost reflections on a record is indicated by the three-wavelet appearance of the autocorrelation function. When analyzing the autocorrelation function for ghost parameters, it is convenient to compute $\mathcal{A}$ from the amplitude of the ghost-indicating autocorrelation wavelet normalized to the autocorrelation value at zero stepout. This amplitude, denoted by m, is,

$$m = \frac{\mathcal{A}}{\mathcal{A}^2 + 1}. \tag{9}$$

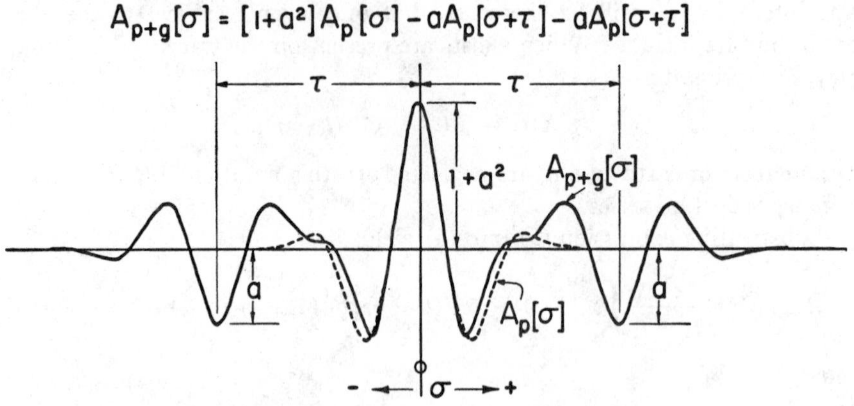

FIG. 4. Illustration of the relation between the autocorrelation function of a seismic trace with ghost reflections and the autocorrelation function of the same trace with only primary reflections. (Note the "tails" produced by the ghost reflections.)

Solution for $\mathfrak{a}$ may be made in terms of m.

$$\mathfrak{a} = \frac{1}{2m} - \sqrt{\frac{1}{4m^2} - 1}. \tag{10}$$

The relation (10) is plotted in Figure 5. The maximum value of m is 0.5 when the reflection coefficient at the ghosting boundary is unity.

It is possible in theory to determine the earth tuning from the characteristics of the autocorrelation function. If there is significant tuning of the earth, it is recognized in the autocorrelation function by a difference in character between the center wavelet shape and the adjacent wavelet shapes at stepout, $\pm\tau$. By relating the spectra of these two wavelet components of the autocorrelation function, the amplitude and phase characteristic of the earth filtering may be determined. In practice this technique usually fails because the detail characteristic of the two wavelet shapes is lost in the "noise" of the autocorrelation function which results from a finite measurement interval, and thus an imperfect statistical sampling.

The ability of the autocorrelation function to portray the character of the basic seismic wavelet generating the trace is directly related to the degree of randomness of the time of arrival and amplitude of the myriad of these basic wavelets. The autocorrelation function of a stationary random series of identically shaped wavelets is the autocorrelation function of the wavelet (Robinson, 1954). Experience indicates that as little as 0.5 sec samples of a seismic trace reasonably well approximates these conditions of randomness. The large number of appreciable reflection coefficients obtained by uniformly sampling the interval velocity log lends support to this viewpoint.

The ghost delay time (τ) is readily measured from the autocorrelation function. Comparison of this time to twice the uphole time as indicated by the shothole seismometer always shows the uphole time to be less than half the ghost time. A possible explanation for this discrepancy lies in the method of picking the uphole time. It is customary to make the pick on first-break energy which gives the travel time for high frequency components. The autocorrelation measure-

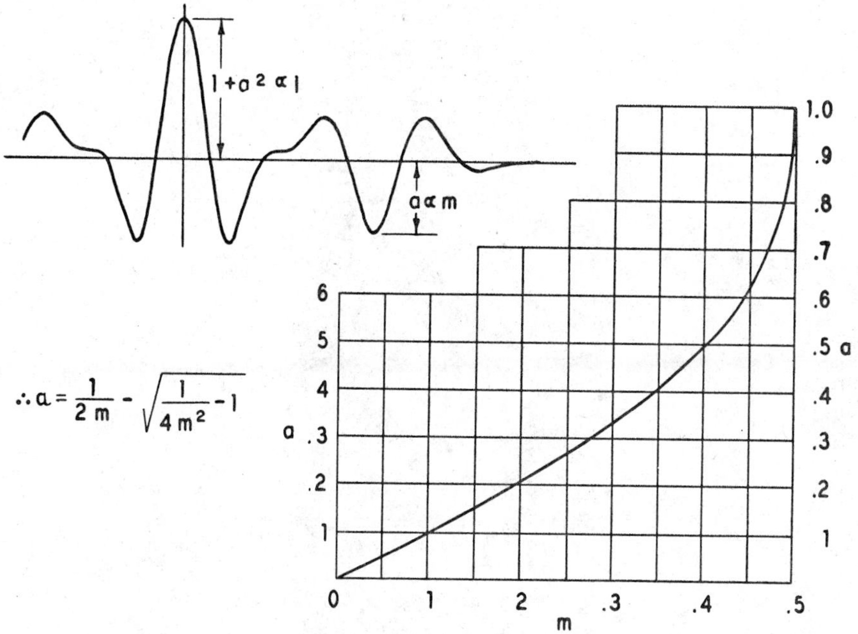

FIG. 5. A plot of the relation between reflection coefficient $\mathfrak{A}$ and the the ratio m obtained from the autocorrelation function.

ment indicates a ghost travel time that applies to the frequency of major content on the record. If there is any dispersion in the surface layering these two travel times will be different.

In Figure 6 the significant part of a field record is shown in which ghost reflections are prevalent. With only the record to judge by, it is difficult to be certain of the existence of ghosts. The autocorrelation function of each trace of this record was measured over a time interval of approximately 3 sec including the 800 millisec shown. These are shown in Figure 7. For better averaging of the extraneous "noise" of the autocorrelation functions, the sum of the twelve autocorrelation functions was used to determine the ghost parameters. In this case the value of τ is 0.109 sec and the value of m is 0.47. Referring to Figure 5, $\mathfrak{A}$ is determined to be approximately 0.7. No earth tuning is apparent since the component wavelet shapes are very nearly the same.

After elimination of the ghost reflections by means of the filter, the record

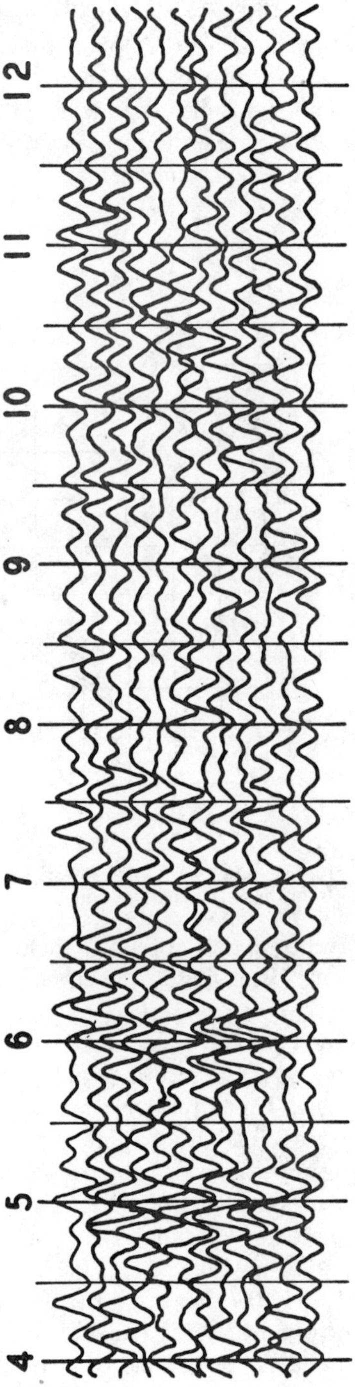

Fig. 6. A portion of a field record containing ghost reflections.

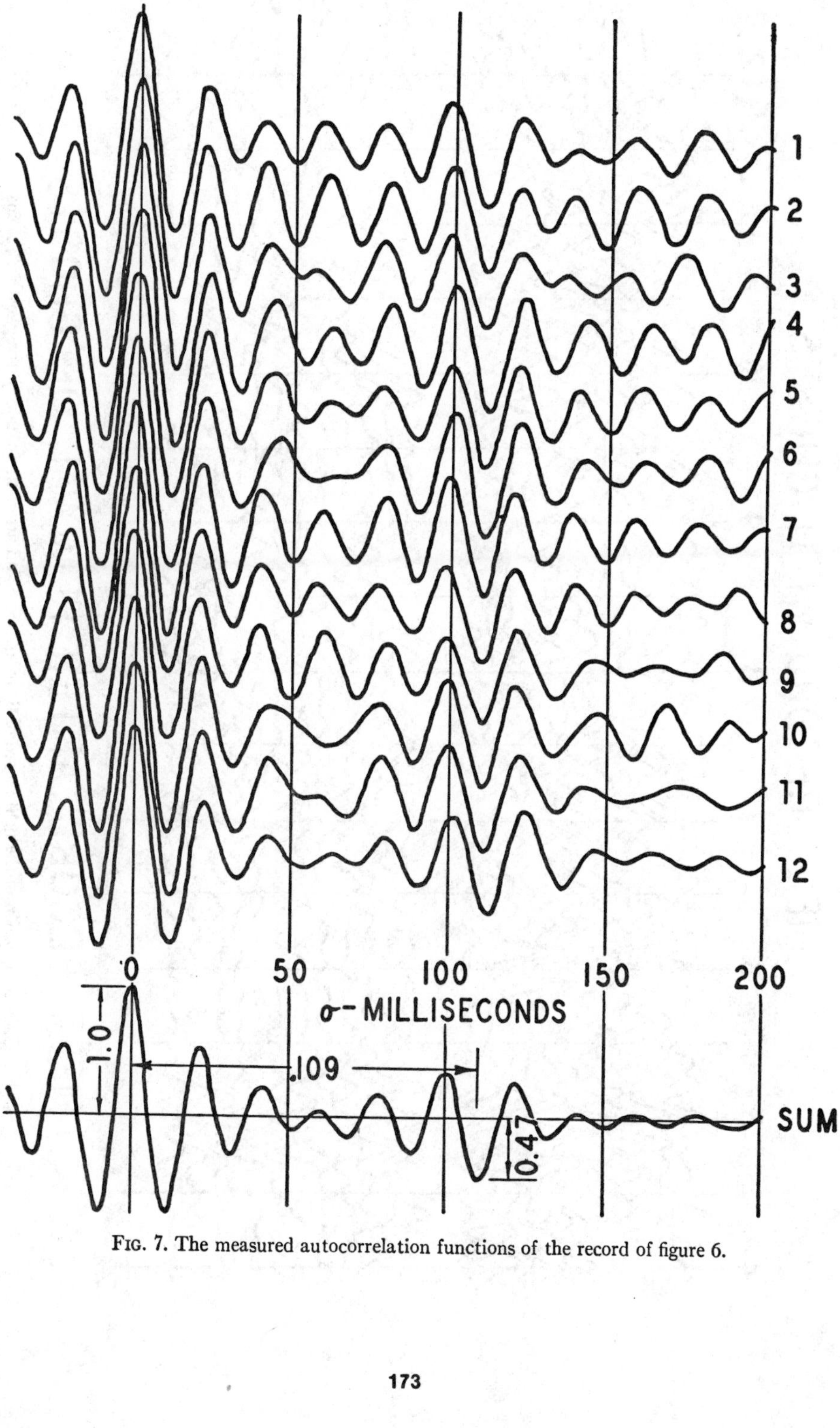

FIG. 7. The measured autocorrelation functions of the record of figure 6.

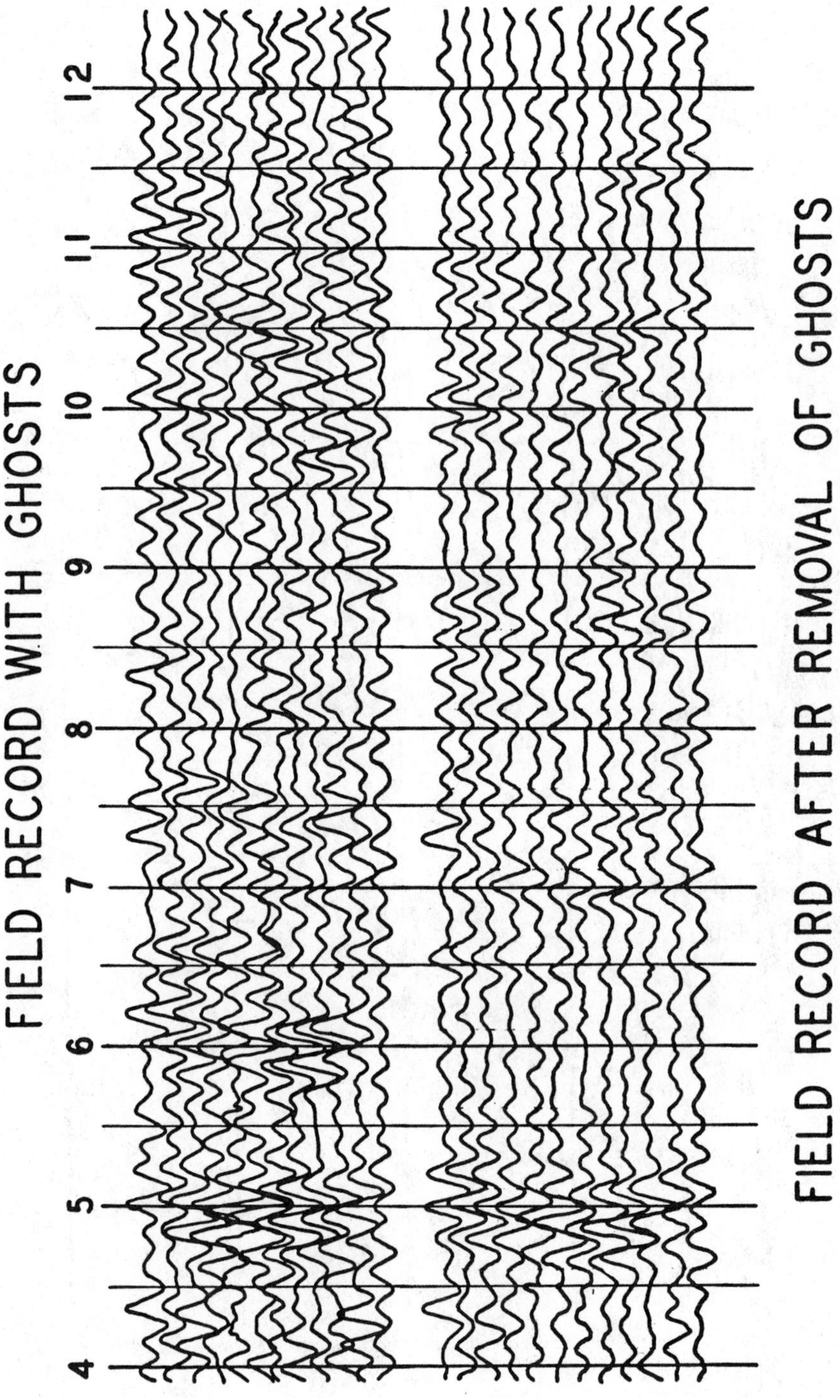

FIG. 8. A comparison of the field record of figure 6 before and after elimination of the ghost reflections by means of an appropriate filter.

appearance is somewhat different as indicated in Figure 8. Note the primary reflection at 0.860 sec that had been obscured by a ghost arrival.

The effects of AGC have been ignored in this treatment although it is possible for rapidly changing gain to alter the amplitude ratio between a primary and its ghost along the record. A slow AGC usually minimizes this problem so that it is negligible. Very rapid AGC should be avoided if ghost elimination is to be attempted in this manner.

REFERENCES

Gardner, M. F., and Barnes, John L., 1942, Transients in linear systems: New York, John Wiley and Sons, Inc.

Goldman, Stanford, 1955, Transformation calculus and electrical transients: Englewood Cliffs, Prentice-Hall, Inc.

Robinson, E. A., 1954, Predictive Decomposition of time series with applications to seismic exploration: MIT Geophysical Analysis Group Report No. 7.

III. DIGITAL INVERSE FILTERING

Much of the discussion of deconvolution has been centered on minimum-phase wavelets; that is, one-sided stable wavelets which have one-sided stable inverses. R. B. Rice presents an analysis emphasizing symmetric (zero-phase) inverse filters for symmetric wavelets. Ricker wavelets are used as examples. He finds that (as a rule of thumb) with filters of comparable length to the wavelets, the center trough of Ricker wavelets can be contracted to about 50 percent of the original breadth. High-frequency noise, as might be expected, has a disastrous effect. Rice points out that the biggest problem in designing inverse convolution filters is determining the shape of the wavelet. Obtaining a satisfactory estimate of the source pulse remains a current problem and the subject of research.

J. F. Claerbout and E. A. Robinson in a short note give the conditions under which the error in least-squares inverse filtering will go to zero as the length of the operator tends to infinity.

W. T. Ford and J. H. Hearne present a lucid discussion of inverse filtering, including an explanation of the mysterious process of introducing "white noise" into the system. The student will also find worked examples of Z-transforms (negative exponent), convolution, and the factorization of autocorrelations. Unfortunately, the details of the concluding seismic example are less clearly presented.

The paper by W. R. Burns examines the important issue of just how high in frequency one should go in deconvolution, or "How weak must a component be for an attempted expansion of its amplitude to do more harm than good?" A subjective answer is often given: One chooses a post-deconvolution filter which gives results that are pleasing to the eye. In this paper, a statistical answer is given: The autocorrelation on which the operator design is based is a sample autocorrelation; the deconvolution operator is thus a random variable; and one should select an output spectrum that minimizes the sum of a fixed and a statistical error. The fixed error is due to the difference between the desired and selected spectrum; the statistical error arises from the expected rms difference between the selected spectrum and the sample output spectrum. It is briefly noted that an additional noise term can be included in the analysis, and surely the noise spectrum must be part of a comprehensive answer to the above question.

M. R. Foster et al present a fascinating investigation of optimum design parameters for deconvolution. Among other conclusions, they find that it is good practice to use a rectangular truncator on the sample autocorrelation at a lag of about 10 percent of the data window; this has often been done in practice as a purely empirical rule. Also, where the data are severely distorted by reverberation, improved results can be achieved by prewhitening.

K. L. Peacock and Sven Treitel present a generalized approach to predictive deconvolution: By extending the prediction distance beyond one sample, one can control the degree of resolution or wavelet contraction. The relation of prediction operators and prediction error operators is carefully developed, and the fact is noted that prediction error operators can be designed without a scaling factor, so that the amplitude variations in the input data can be preserved if desired. It is shown that the output autocorrelation tends to vanish beyond the prediction distance for an interval equal to the length of the operator. As a good compromise between wavelet contraction and low signal-to-noise ratio in the output trace, a prediction distance roughly equal to the second zero crossing of the autocorrelation is recommended. A final very practical note: For the removal of long-period reverberation, a relatively long prediction distance and short operator length give a substantial saving in computer time. This fine paper was awarded recognition as the Outstanding Paper in GEOPHYSICS during 1969.

J. N. Galbraith points out that the Kolmogorov spectrum factorization can be used to obtain the final value of the mean-square-error in prediction. This figure can in turn be used to normalize the error for finite length operators and thereby judge their effectiveness.

The final paper of this section, by R. J. Wang and S. Treitel, deals with computational aspects of deconvolution filter design. A development of gradient methods for the solution of the normal equations is presented which includes instructive illustrations by means of very simple numerical examples. In a trial on seismic data, a satisfactory approximate solution was developed in only one-sixth the time required for the frequently used Levinson algorithm. The authors find that the gradient methods are particularly well suited for implementation on special-purpose floating-point multiply-and-add hardware.

Reprinted from Geophysics v. 27, no. 1, p. 4-18

INVERSE CONVOLUTION FILTERS*

R. B. RICE†

The difficult problem of trying to locate stratigraphic traps with the reflection seismograph would be simplified (at least in good record areas) if it were possible to perform the inverse of the reflection process, i.e., to "divide out" the reflection wavelet of which the record is composed, leaving only the impulses representing the reflection coefficients. This process has been discussed by Robinson under the title "predictive decomposition," but his approach requires that the basic composition wavelet be a one-sided, damped, minimum-phase time function. Most seismic wavelets which we observe or are accustomed to working with (e.g. the symmetric Ricker wavelet) are not of this class. The purpose of this paper is to discuss a digital computer approach to the problem. Finite, bounded inverse filter functions are obtained which will reduce seismic wave forms to best approximations to the unit impulse in the least squares sense. The degree of approximation obtained depends upon the time length of the inverse filter. Inverse filter functions of moderate length produce approximate unit impulses whose breadths are 50 percent or less than those of the original wavelets. Hence, these filters will increase resolution well beyond the practical limits of instrumental filters. Their effectiveness is more or less sensitive to variations in the peak frequency and shape of the composition wavelet, and to interference, depending upon individual conditions. Although this sensitivity problem can be solved to some extent through the proper design of the inverse filter, it is aggravated by the usual lack of knowledge about the form of the composition wavelet.

INTRODUCTION

The problem of trying to locate stratigraphic traps with the reflection seismograph is a difficult one. One of the reasons for the difficulty is the lack of detail or resolution on the seismic record. Hence, the problem would be simplified (at least in good record areas) if it were possible to perform the inverse of the reflection process, i.e., to "divide out" the reflection wavelet of which the record is composed, leaving only the impulses representing the reflection coefficients. From another point of view, this process consists of applying a filter which contracts each reflection wavelet to a spike representing the arrival time and amplitude of that reflection.

In his paper dealing in part with "wavelet contraction," Ricker (1953) discusses this problem from an instrumental point of view. He was able to design electronic filters which will reduce conventional seismic wavelets to 70 or 80 percent of their original breadth. The application of such filters to seismograms only achieves a partial transformation back to the reflection coefficient function, but a considerable improvement in resolution is effected.

Robinson (1957) has treated the inverse filtering problem under the title "predictive decomposition." Starting with a seismic trace or portion thereof, he gives theoretical and statistical methods for computing (1) one form of the seismic wavelet composing the trace and (2) the "prediction operator" or inverse wavelet for effecting the contraction to a sequence of spikes. His techniques are based on discrete or digitized time functions which are amenable to treatment on digital computers. This approach has the advantage that completely arbitrary wavelet shapes or impulsive responses, which may be quite expensive if not impossible to simulate electronically, are handled with ease.

Robinson restricts his attention to cases in which the basic composition wavelet is a one-sided, damped, minimum-phase time function, a mathematical definition of which will be given later. The reason for imposing this restriction is that these are the only wavelets for which bounded, one-sided inverses exist. However, most seismic wavelets which we observe or are accustomed to working with (e.g. the theoretical Ricker wavelet) are not of this restricted class.

* Presented at the 30th Annual SEG Meeting, Galveston, Texas, November 9, 1960. Manuscript received by the Editor June 9, 1961.

† The Ohio Oil Company, Littleton, Colorado.

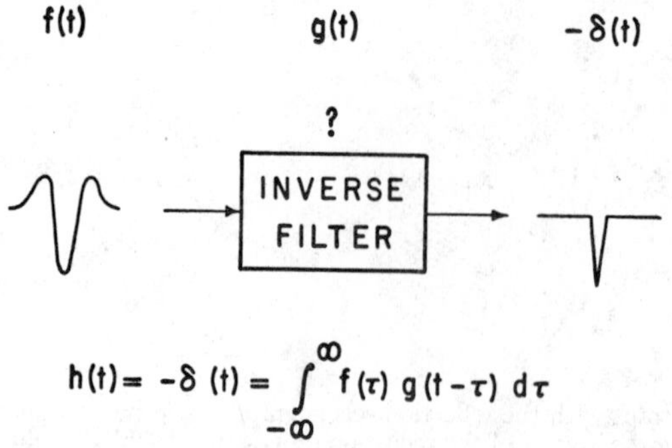

FIG. 1. The problem of determining an inverse filter function, $g(t)$, which will transform a given wavelet, $f(t)$, into a negative unit impulse, $-\delta(t)$.

The purpose of this paper is to present methods for computing approximate inverse filter functions for arbitrary, non-minimum-phase seismic wavelets, and to discuss some of the properties of these inverses which may affect their application to actual seismic records.

STATEMENT OF THE PROBLEM

The problem to be treated here is indicated in Figure 1. Given some arbitrary seismic wave form, $f(t)$, which is assumed to be the basic reflection wavelet composing a seismogram, find, by digital means, an inverse filter function, $g(t)$, which will transform $f(t)$ into a unit impulse, or the best approximation thereto in some sense. In Figure 1, and in what follows, the negative unit impulse, or delta function, is used so that the trough of the input waveform will correspond to a trough of the output function.

It is assumed that we are dealing with linear systems, so that the filter process is defined mathematically by the familiar convolution integral:

$$h(t) = \int_{-\infty}^{\infty} f(\tau)g(t - \tau)d\tau, \qquad (1)$$

where $h(t)$, the output function, is equal to the negative delta function, $-\delta(t)$, only when the problem has an exact solution in terms of a finite, bounded $g(t)$. As will be seen later, given a wave form $f(t)$, there is in general no finite, bounded $g(t)$ which will produce the exact negative delta function for $h(t)$. In these cases $f(t)$ and $-\delta(t)$ may be called "incompatible." The chief concern of this study is to investigate the effectiveness of filter functions, $g(t)$, which produce the best approximations to $-\delta(t)$ in some sense.

It should be mentioned that electronic circuit theoreticians have been concerned with the inverse convolution filtering problem for some time from the point of view of synthesizing electronically the impulsive response of a "black box" when one knows the input and output wave forms (e.g., Ba Hli (1954), and Kautz (1954)). However, they rarely deal with delta function outputs, being more concerned with situations in which $f(t)$ and $h(t)$ are "compatible" than with "incompatible" ones. For "compatible" cases, simple methods are available for determining $g(t)$. Hence, their principal problem is to obtain rational fraction approximations to the system function, i.e. the Laplace transform of $g(t)$, which lead to optimum realizable networks.

MATHEMATICAL BACKGROUND

In actual cases, $f(t)$ and $g(t)$ must be taken to be zero everywhere except on some finite time interval. Hence, relation (1) may be written as

$$h(t) = \int_{0}^{t} f(\tau)g(t - \tau)d\tau. \qquad (1')$$

There are several possible approaches to the problem of obtaining $g(t)$ when $f(t)$ and $h(t)$ are known. One method is to take the Laplace transform, or the more general LaPlace-Stieltjes transform, of both sides which yields the well-known result (e.g., Gardner and Barnes (1942), p. 228),

$$H(s) = F(s)\,G(s), \qquad (2)$$

where $H(s)$, $F(s)$, and $G(s)$ are the Laplace transforms of $h(t)$, $f(t)$, and $g(t)$, respectively. Since our interest is in discrete representations, the Laplace transforms may be replaced by the Dirichlet series representations:

$$F(s) = \sum_{\nu=1}^{m} A_\nu e^{-\nu(\Delta t)s},$$
$$G(s) = \sum_{\nu=1}^{N} X_\nu e^{-\nu(\Delta t)s}, \text{ and} \qquad (3)$$
$$H(s) = \sum_{\nu=1}^{n} B_\nu e^{-\nu(\Delta t)s},$$

where A_ν, X_ν, and B_ν represent the areas under the curves $f(t)$, $g(t)$, and $h(t)$, respectively, for the νth equal interval Δt, and where it is assumed that

$A_\nu=0$ for $\nu>m$, $X_\nu=0$ for $\nu>N$, and $B_\nu=0$ for $\nu>n$. Letting a_ν, x_ν, and b_ν represent midpoint amplitude values of $f(t)$, $g(t)$, and $h(t)$, respectively, and taking $\Delta t=1$, the above relations reduce to the approximate expressions:

$$F(s) = \sum_{\nu=1}^{m} a_\nu e^{-\nu s},$$

$$G(s) = \sum_{\nu=1}^{N} x_\nu e^{-\nu s}, \text{ and} \tag{4}$$

$$H(s) = \sum_{\nu=1}^{n} b_\nu e^{-\nu s}.$$

Since these are polynomials in $e^{-\nu s}$, it is evident that the unknown amplitudes, x_ν, can be obtained by synthetic division of $H(s)$ by $F(s)$.

Another method is to represent the time functions in terms of the socalled "generating functions" of Laplace (e.g., Jordan (1947) p 21). The generating function, $F(u)$, of the function, $f(t)$, can be written as

$$F(u) = \sum_{\nu=1}^{m} a_\nu u^\nu, \tag{5}$$

where a_ν has the same meaning as before; and similarly for $G(u)$ and $H(u)$. It has been shown by Piety (1951) that, under the proper conditions, the generating function of the convolution integral (1) can be approximated accurately by

$$H(u) = F(u)\, G(u). \tag{6}$$

Thus, the unknown amplitudes, x_ν, can again be obtained through polynomial division of $H(u)$ by $F(u)$.

The accuracy of any of these approximations will depend upon the size of the interval used in sampling the original functions. It is known that, if the waveforms have a cut-off frequency of f_c, then $1/2f_c$ is an adequate interpolation interval. In the absence of a realistic cut-off frequency, an arbitrary cut-off may be made at that point at which the amplitude spectrum is down, say 40 db, or 1/100 of the peak value.

In all the studies described in this paper, a sampling interval of 2 ms has been used, with a corresponding cut-off frequency of 250 cps.

Having defined the inverse filter function in terms of a polynomial division process, the next step is to find out what happens when this division

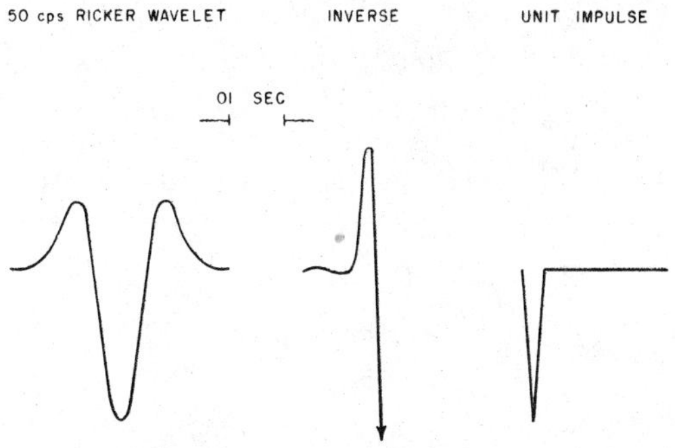

FIG. 2. The exact inverse of a 50 cps Ricker wavelet obtained by polynomial division.

is attempted. For $h(t)$ equal to the negative unit impulse, its polynomial approximation reduces to a single term, $-u$. When this is divided by the polynomial representation of the wavelet $f(t)$, usually the quotient or inverse coefficients oscillate and increase more and more rapidly until they are exceeding all reasonable bounds. This is illustrated in Figure 2 which shows the first portion of the inverse resulting for a symmetric 50 cps Ricker wavelet.

From the point of view of complex variable theory, this is the case for which $F(u)$, with u interpreted as a complex variable, has at least one root inside the unit circle $|u|=1$. Then $G(u)$ cannot have a convergent power series expansion on and within the unit circle and hence no bounded inverse exists. In the special case in which $F(u)$ has no roots on the unit circle, but has roots both interior and exterior to it, $G(u)$ has a Laurent series expansion

$$G(u) = \sum_{s=-\infty}^{\infty} g_s u^s,$$

which converges on the unit circle. Such an expansion is called "two-sided" in contrast to a "one-sided" power series expansion. It is possible in this instance, if the convergence is rapid enough, to use a reasonable number of terms of this two-sided expansion as the inverse wavelet, but we shall not be concerned with this approach.

If $F(u)$ has all its roots within the unit circle, then the Laurent expansion reduces to the singular part

$$G(u) = \sum_{s=-\infty}^{0} g_s u^s,$$

which converges on the unit circle but which is strictly "nonrealizable" because it has no finite starting time.[1] Again, if the convergence is rapid enough, it is possible to use a finite number of terms of this expansion as the inverse filter. If, in addition to roots within the unit circle, $F(u)$ also has one or more roots on the unit circle, then the above expansion for $G(u)$ will not converge on the unit circle, and hence it cannot in any way be used as an inverse operator.

In any of the above cases, $F(u)$ is "non-minimum-phase" (at least one root inside the unit circle), and an attempted one-sided polynomial division of $H(u)$ by $F(u)$ will produce an unbounded inverse.

If, on the other hand, $F(u)$ has no roots on or within the unit circle, it is called "minimum-phase." Then $G(u)$ does have a convergent power series expansion for $|u| \leq 1$, which can be used for the inverse filter. This is the case treated by Robinson (1957), as mentioned in the introduction.

In summary then, there is usually no one-sided bounded inverse which will transform observed or theoretical seismic wavelets into the unit impulse.

This problem may be examined from another point of view which may be more lucid and which will form the basis for later remarks about the solution. Again, if convolution is represented as polynomial multiplication and coefficients of like terms on either side of the equality are equated, the following series of linear simultaneous equations in the unknown inverse amplitudes, x_i, is obtained:

$$
\begin{array}{llr}
a_1 x_1 & = b_1 & \\
a_2 x_1 + a_1 x_2 & = b_2 & \\
\vdots & \vdots & \\
a_m x_1 + a_{m-1} x_2 + \cdots + a_1 x_m & = b_m & \\
\quad a_m x_2 + \cdots + a_2 x_m + a_1 x_{m+1} & = b_{m+1} & (7) \\
\quad\quad \ddots & \vdots & \\
\quad\quad a_m x_m + x_{m-1} x_{m+1} + \cdots a_1 x_{n-m+1} & = b_{n-m+1} & \\
\quad\quad\quad \ddots & \vdots & \\
\quad\quad\quad a_m x_{n-m+1} + \cdots & = b_n &
\end{array}
$$

For $h(t)$ equal to the negative unit impulse, $b_1 = -1$ and $b_v = 0$ for $v > 1$. Hence, the polynomial division process is equivalent to solving the first equation for x_1, substituting this in the second equation with $b_2 = 0$, and solving for x_2, etc. Since this is a set of infinitely many equations in infinitely many unknowns, the solution will never terminate unless the values of x_i obtained in the solution of the first $n-m+1$ equations exactly satisfy the last $m-1$ equations. This is the condition for "compatibility" mentioned earlier.

LEAST SQUARES COMPUTATION OF INVERSE FILTERS

Since, in general, there is no exact solution to the problem, methods for obtaining approximate solutions are called for. There are several possibilities. One may relax requirements on the wavelet $f(t)$, on the unit impulse, or on both. Then one can ask for x_v's which will give the best approximation to these modified functions in the Tchebycheff sense (minimize the maximum deviation), in the least squares sense (minimizing the sum of the squares of the deviations), or in some other sense.

A number of approaches have been tried. It appears that the most useful solution is obtained in terms of the best least squares approximation to the unit impulse. That is to say, assuming all the x_v's in equation 7 are zero for $v > N$, we ask for the unique set of x's which will minimize the sum of the squares of the deviations of the calculated b_v's from the desired ones, $b_1 = -1$, and $b_v = 0$ for $v > 1$.

[1] Filters, whether electronic or digital, are usually termed "realizable" only when they permit one to work in "real time." However, if the filtering is done with respect to "nominal time" (e.g. the time scale on a recorded seismogram), then "nonrealizable" filters can be used. Such "nonrealizable" filters are in fact filters with large time-delays. Usually, electronic filters are used for real time applications and digital computers for nominal time applications, although there are numerous exceptions.

The derivation of the least squares solution is most conveniently given in terms of matrix algebra. Let A represent the $n\times N$ matrix of coefficients, X the Nth order solution column vector, B the nth order column vector of right-hand elements, and E the nth order column vector whose elements are the deviations of the calculated b_ν's from the desired ones. Then,

$$AX - B = E, \tag{8}$$

and the sum of the squares of the deviations is

$$\begin{aligned} E'E &= (AX - B)'(AX - B) \\ &= (X'A' - B')(AX - B) \\ &= X'A'AX - X'A'B - B'AX + B'B, \end{aligned} \tag{9}$$

where the primes denote transposes.

To obtain the normal equations, take the partial derivatives of (9) with respect to each of the elements of X and set the result equal to zero. We then obtain the normal equations,

$$X'A'AU_k + U'A'AX - U_k'A'B - B'AU_k = 0 \text{ for all } k, \tag{10}$$

where U_k is the Nth order unit column vector with unity in the kth position. Rearrange (10) so that

$$(X'A'A - B'A)U_k = U_k'(-A'AX + A'B). \tag{10'}$$

Now, $A'A$ is an Nth order square matrix. Hence $X'A'A$ and $(X'A'A'-B'A)$ are Nth order row vectors. On the other hand, $(-A'AX+A'B)$ is an Nth order column vector. Thus, for any k, the left-hand side of (10′) is the element in the kth position of the row vector $(X'A'A-B'A)$ and the right-hand side is the element of the kth position of the column vector $(-A'AX+A'B)$. Since the equality is valid for all k, we must have

$$(X'A'A - B'A)' = -A'AX + A'B$$

or

$$A'AX = A'B \tag{11}$$

which gives

$$X = (A'A)^{-1}A'B \tag{12}$$

and

$$X' = B'A(A'A)^{-1}. \tag{13}$$

Substituting these quantities in (9), the least squares error reduces to

$$E'E = B'B - B'AX. \tag{14}$$

It is easily verified from these relations that X satisfying $AX=B$ is both a necessary and sufficient condition for $E'E=0$. As indicated earlier, this can only happen in special situations for finite X, X', and B, which, in matrix language, are those cases for which the rank of the augmented matrix AB is equal to the rank of A. However, when B, by adding 0's, A, and X are allowed to become infinite, A approaches a square matrix, and there is an exact, though useless, X which satisfies $AX=B$. This is the same solution obtained by the successive substitution or polynomial division processes mentioned earlier.

From these heuristic considerations and more detailed analyses of the form of $E'E$ for low-order systems, the following theorem is conjectured to be true, although a general proof is not yet at hand:

Theorem: The least squares error $E'E$ decreases monotonically as the number of nonzero terms in the inverse filter function, X, is allowed to increase.

If true, this theorem indicates that a wavelet may be reduced to as accurate an approximation to the unit impulse as one desires by an appropriate choice of the length of the inverse filter function. This is an important consideration for applications and will be illustrated later.

Before considering some specific inverse filters computed by the least squares method, it is of interest to look at the form of the normal equations (11) in more detail. The matrix of coefficients $A'A$ has the symmetric form,

$$A'A = \begin{pmatrix} \alpha_0 & \alpha_1 & \cdots & \alpha_{n-m} \\ \alpha_1 & \alpha_0 & \cdots & \alpha_{n-m-1} \\ \cdot & \cdot & & \cdot \\ \cdot & \cdot & & \cdot \\ \cdot & \cdot & & \cdot \\ \alpha_{n-m} & \alpha_{n-m-1} & \cdots & \alpha_0 \end{pmatrix}, \tag{15}$$

where

$$\alpha_k = \sum_{i=1}^{m-k} a_i a_{i+k}, \qquad k = 0, 1, \cdots, n-m,$$

with $\alpha_k=0$ for $k>m-1$. Now the α_k are the amplitudes of the autocorrelation of the original wavelet, with

$$\alpha_0 = \sum_{i=1}^{m} a_i^2$$

being the maximum center value, α_1 the next value on either side of the center, etc.

The right-hand member of (11) is

$$A'B = \begin{pmatrix} \sum_{i=1}^{m} a_i b_i \\ \sum_{i=1}^{m} a_i b_{i+1} \\ \cdot \\ \cdot \\ \cdot \\ \sum_{i-1}^{m} a_i b_{i+n-m} \end{pmatrix}. \quad (16)$$

For the case when the negative unit impulse is placed at the mth point ($b_m = -1$, $b_v = 0$ for $v \neq m$), this vector reduces to

$$\begin{pmatrix} -a_m \\ -a_{m-1} \\ \cdot \\ \cdot \\ \cdot \\ -a_{2m-m} \end{pmatrix}, \quad (16')$$

the elements of which are the negative amplitudes of the original wavelet in reverse order down to the a_{2m-n}th one.

The above observations show that the normal equations are easily formed from the amplitudes of the original wavelet and its autocorrelation. Hence, with a polynomial multiplication routine to calculate the autocorrelation, least squares inverse filters can be computed using a standard program for solving systems of linear equations.

A special case of interest is the one for which the original wavelet is symmetric; m, n, and $N(=n-m+1)$ are odd; and the negative unit impulse is placed at the $\frac{1}{2}(n+1)$st point. Then, for $n > 2m-1$, the right-hand vector (16) will be symmetric around the center value $-a_{\frac{1}{2}(m+1)}$, with zeros at the top and bottom, and the inverse filter function will be symmetric. If the inverse filter is taken to be the same length as the original wavelet ($N=m$, $n=2m-1$, $b_m=-1$), there are no zeros at the upper right- and lower left-hand corners of $A'A$ (15). Also the right-hand members (16) consist precisely of the negative amplitudes of the original wavelet with no additional zeros.

The case of symmetric wavelets and symmetric inverses is the one dealt with most frequently in the illustrations to follow. For this case, the m normal equations (11) can be reduced to $\frac{1}{2}(m+1)$ independent equations, thereby greatly reducing the computation time and the amount of computer storage required.

ILLUSTRATIVE EXAMPLES

Figure 3 shows the least squares inverses and resulting approximations to the symmetric negative unit impulse obtained for symmetric Ricker wavelets peaked at 75 and 37.5 cps, respectively

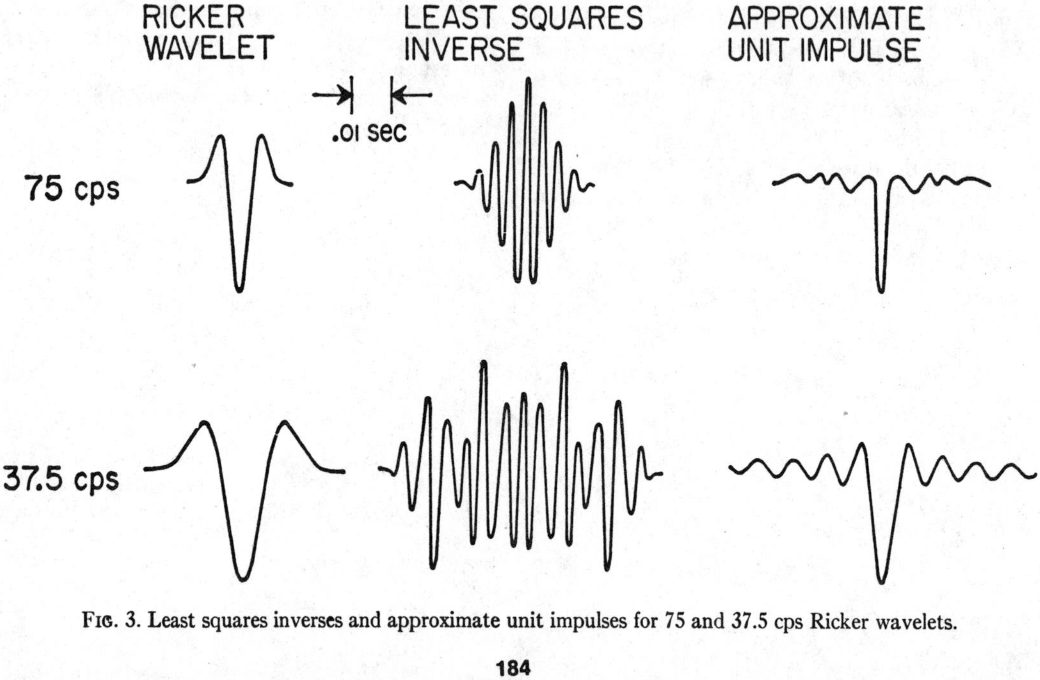

FIG. 3. Least squares inverses and approximate unit impulses for 75 and 37.5 cps Ricker wavelets.

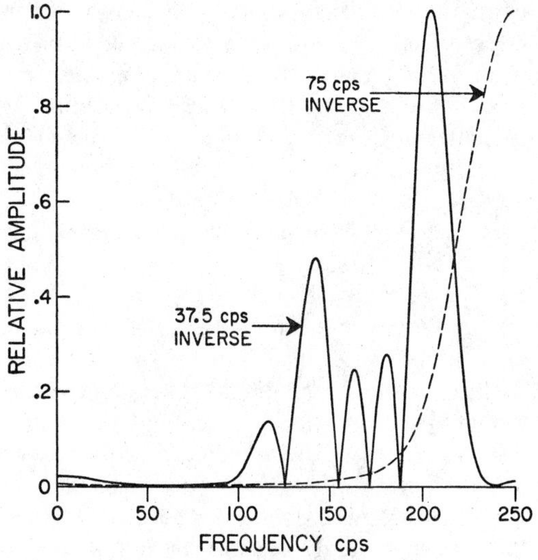

FIG. 4. Amplitude spectra of 75 and 37.5 cps inverses of Figure 3.

The tails of the Ricker wavelets have been cut off at a level of 0.002 percent of the maximum amplitude. In each case, the number of points in the inverse is the same as the number defining the Ricker wavelet.

The resulting approximation to the negative unit impulse for the 75 cps case is narrower and has less ripple on either side of the center trough than the one for the 37.5 cps case. However, it should be noted that in each instance the center trough of the Ricker wavelet has been reduced to about 50 percent of its original breadth. This rule-of-thumb holds for any frequency when the inverse and the wavelet are of the same length.

Note that the two inverses are not similar in shape as one might expect. The reason is that the same interpolation interval (2ms) has been used in both cases, so that the frequency characteristics of the wavelets and inverses have different relationships to the cut-off frequency (250 cps).

Figure 4 shows the amplitude spectra for the 75 and 37.5 cps inverse filters. The spectrum of the 75 cps inverse rises smoothly to a peak at 250 cps, whereas the 37.5 cps inverse spectrum has a local maximum at 140 cps and the major peak at about 200 cps. Note the small amount of low-frequency content in both cases which, as will be seen later, can be troublesome. The phase spectra (not shown) are linear with the slope of course depending upon the choice of zero time.

The corresponding inverses for 50 and 25 cps symmetric Ricker wavelets are exhibited in Figure 5. Again, the wavelets are reduced by the inverses to about 50 percent of their original breadths.

Figure 6 shows the amplitude spectra for the 50 and 25 cps inverses. The spectrum of the 25 cps inverse rises smoothly to a peak frequency of about 100 cps, then falls off erratically. On the other hand, the 50 cps spectrum peaks locally at 170 cps and has its major peak at about 215 cps.

The theorem conjectured above is illustrated in Figure 7 which shows the inverses and approximate unit impulses for the 75 cps Ricker wavelet

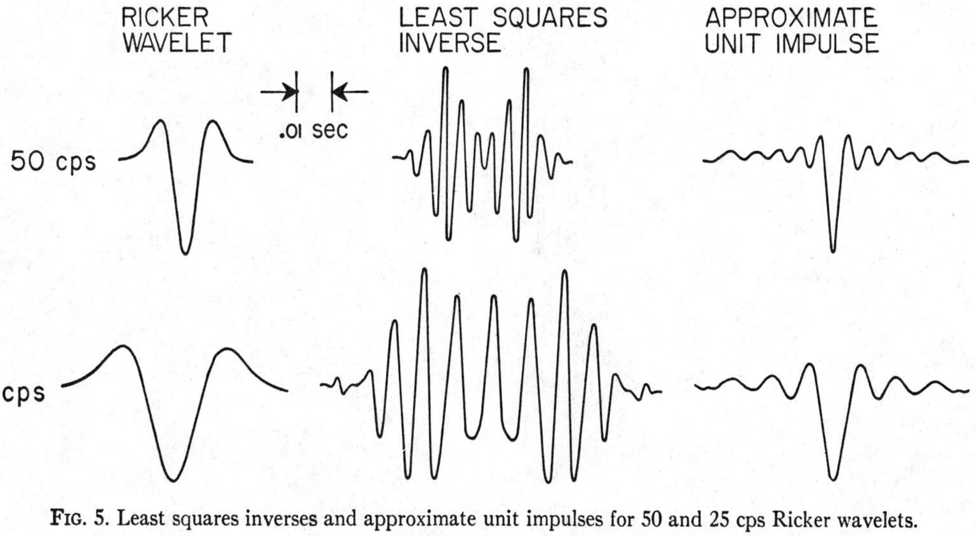

FIG. 5. Least squares inverses and approximate unit impulses for 50 and 25 cps Ricker wavelets.

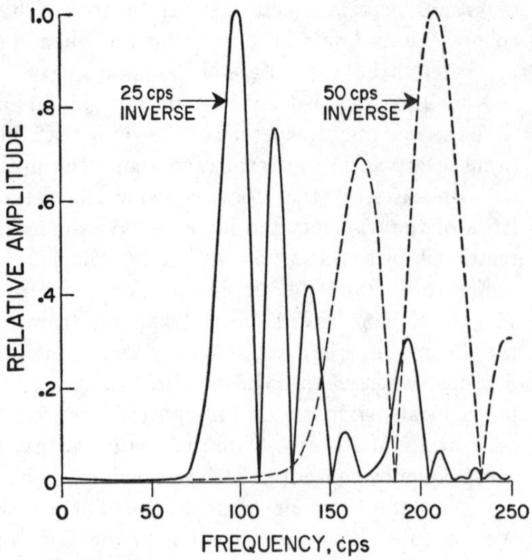

FIG. 6. Amplitude spectra of 50 and 25 cps inverses of Figure 5.

when the number of points in the inverse is increased from 17 to 25, then to 41. The improvement in the shape of the unit impulse approximation is quite striking although the breadth, which is limited by the cut-off frequency, remains essentially constant. The spectra for these inverses are presented in Figure 8. Note that, as the inverse increases in length, the spectrum exhibits an increasingly steeper slope starting at higher and higher frequencies.

These results indicate that the least squares inverse filters can be made as nearly perfect as one desires, except for the limitations imposed on the size of systems of normal equations that can be solved on today's digital computers. Experience to date indicates that double-precision arithmetic (18 or 20 decimal digits) is required for about 20th order or larger systems. Round-off error is more troublesome than in most cases because there are few or no zeros present in the coefficient matrix.

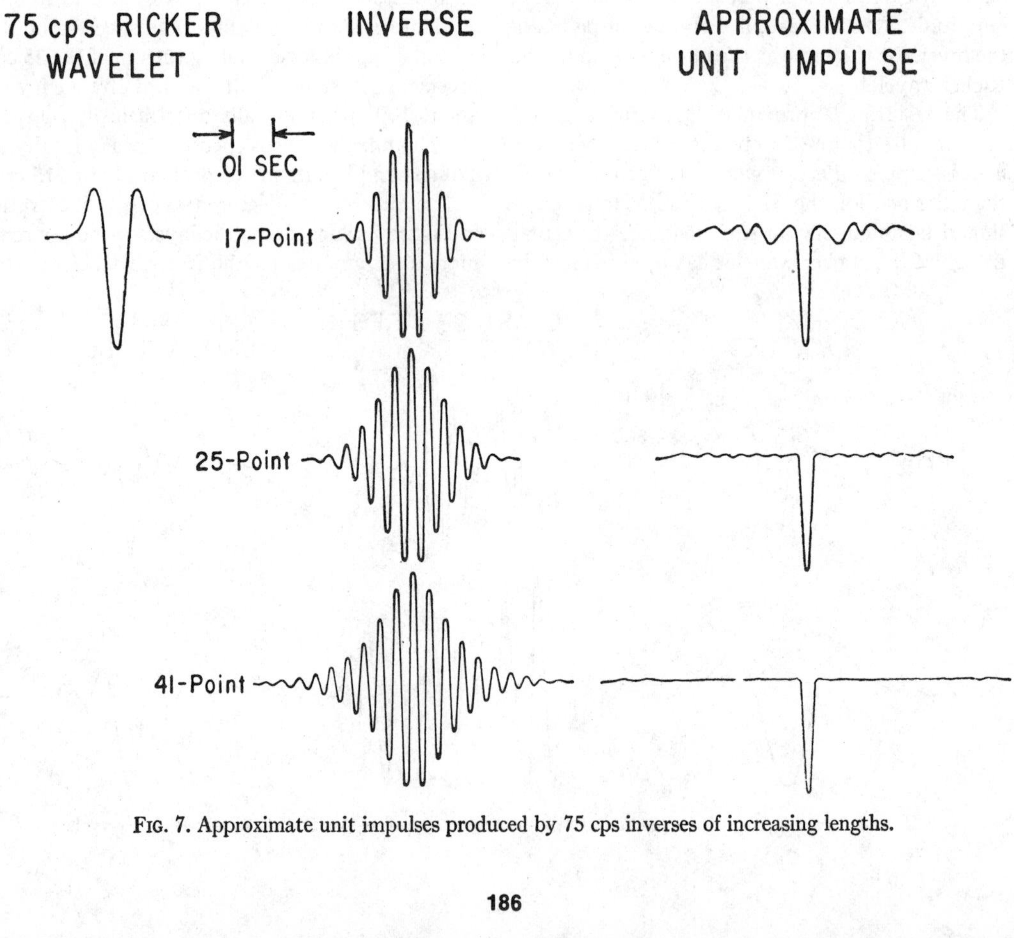

FIG. 7. Approximate unit impulses produced by 75 cps inverses of increasing lengths.

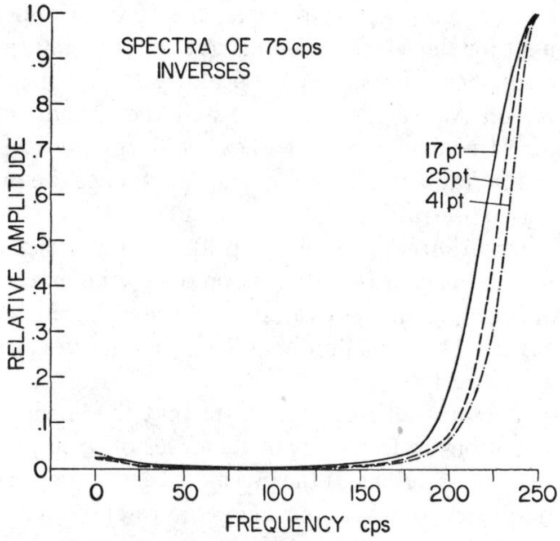

FIG. 8. Amplitude spectra of 75 cps inverses of Figure 7

Next, it is of interest to see how effective these inverse filters are in transforming synthetic seismograms back to the reflection coefficient functions from which they were derived. The results will be indicative of possibilities under ideal condictions; i.e., assuming that the basic composition wavelet is known, is invariant with respect to time, and that there is no interference present.

Trace (c) of Figure 9 represents a reflection coefficient function obtained from an actual continuous velocity log on a Nebraska well, assuming constant density. Trace (a) is the synthetic seismogram obtained by convolving the 75 cps symmetric Ricker wavelet of Figure 3 with Trace (c), and Trace (b) is the result of applying the 75 cps inverse of Figure 3 to Trace (a). The detailed agreement between Traces (b) and (c) is excellent. If one were able to do this well on an actual siesmogram, certainly there would be much less difficulty in making detailed, accurate stratigraphic and lithologic interpretations from seismic records However, in practice there are many complicating factors which will be discussed below.

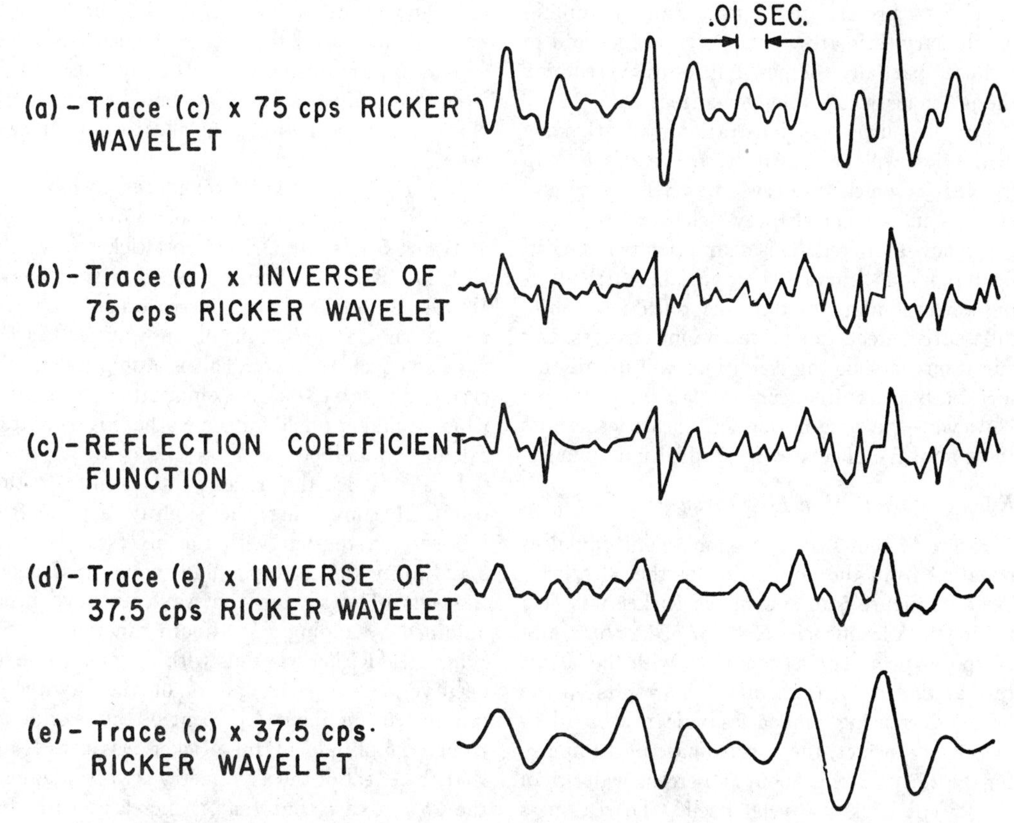

FIG. 9. Results of applying 75 and 37.5 cps inverses of Figure 3 to synthetic seismograms computed from a CVL on a Nebraska well.

Trace (e) of Figure 9 is the synthetic seismogram obtained from the same reflection coefficient function (c) using the 37.5 cps symmetric Ricker wavelet of Figure 3 , and Trace (d) is the result of filtering this trace with the 37.5 cps inverse. In this case, some of the detail has been lost, but the agreement with Trace (c) is still fairly good.

In Figure 10, similar results, based on the same reflection coefficient (Trace (c)), are shown for the cases of the 50 and 25 cps Ricker wavelets and their inverses of Figure 5. The 50 cps inverse produces quite good agreement (Trace (b)) with the original reflection coefficient function, but the 25 cps inverse does not restore much of the detail. Better results can be obtained in any case by using a longer inverse.

In the application of the inverse filtering technique to actual seismograms, there are a number of factors which may significantly affect the results. These include variations in the frequency and character of the composition wavelet, and the presence of interference. Some of these effects have been investigated synthetically to obtain preliminary estimates of their significance and to evaluate partially the possibilities of overcoming the problems which they generate.

For the purpose of illustrations, an arbitrary distinction will be drawn between variations in frequency which preserve wavelet form and changes in wavelet shape which leave the peak frequency unaltered. This distinction may be difficult to find in practice where variable earth filtering corresponding to different reflection times, differential effects due to weathering changes, and variations in shooting conditions will usually affect both peak frequency and wavelet shape. However, the assumption of linearity permits these additive effects to be studied individually.

Effect of Variations in Peak Frequency

Figure 11 shows the approximate unit impulses resulting from the application of the 50 cps inverse of Figure 5 to symmetric Ricker wavelets having peak frequencies of 44, 37.5, 25, 56.25, and 75 cps, respectively, as compared with the 50 cps case at the top of the figure. The results on the left for decreasing frequencies indicate a broadening effect. In fact, the approximate unit impulse for the 25 cps case is about 30 percent wider than the 25 cps Ricker wavelet itself. The reason is that the inverse amplifies the low frequencies in the wavelet relative to the intermediate frequencies (Figure 6). For example, the 10 cps component for the 50 cps inverse is five times the 50 cps component. Hence, as the peak frequency of the Ricker wavelet is decreased and the amount of high-frequency content above 100 cps becomes insignificant, the low-frequency portion of the inverse spectrum becomes dominant.

On the other hand, if the peak frequency of the wavelet is greater than that on which the inverse is based, as in the two cases on the right side of Figure 11, the additional high frequencies are amplified too much.

The over-all results indicate that the effect of variations in frequency of the order of ± 10 percent is not serious. If the inverse is based on a peak frequency which is too high, the resolution will suffer proportionately. If the peak frequency is too low, the amount of ripple will increase.

Effect of Variations in Wavelet Shape

Next consider variations in wavelet shape corresponding to changes in the form of the amplitude and/or phase spectra of the composition wavelet which leave the peak frequency unaltered Obviously, there is no end to the number of different kinds of variations that could be considered However, space permits the illustration of two types.

Figure 12 shows the approximate unit impulses obtained by applying the 17-point, 75 cps inverse of Figure 3 to symmetric 75 cps Ricker wavelets with 17, 13, and 11 nonzero values, respectively. It will be noted that this increase in the sharpness of cut-off of the tails of the wavelet has no adverse effect whatever. This is a desirable property of the least squares computation technique which would not hold for some other inverse filter calculation methods.

In Figure 13, the traces on the right are the result of convolving the slightly asymmetric wavelets on the left with the inverses (Figures 3 and 5) for the corresponding symmetric Ricker wavelets. The asymmetric wavelets have been obtained by adding a saw-tooth function to the symmetric Ricker wavelets with the amplitude of positive and negative peaks of the saw-tooth function being about 4 percent of the maximum trough amplitude of the Ricker wavelet. Note that the effect of the asymmetry is very slight in the 75 cps case, but that it increases with decreasing frequency, becoming completely overriding in the 25 cps case. This is explained by the

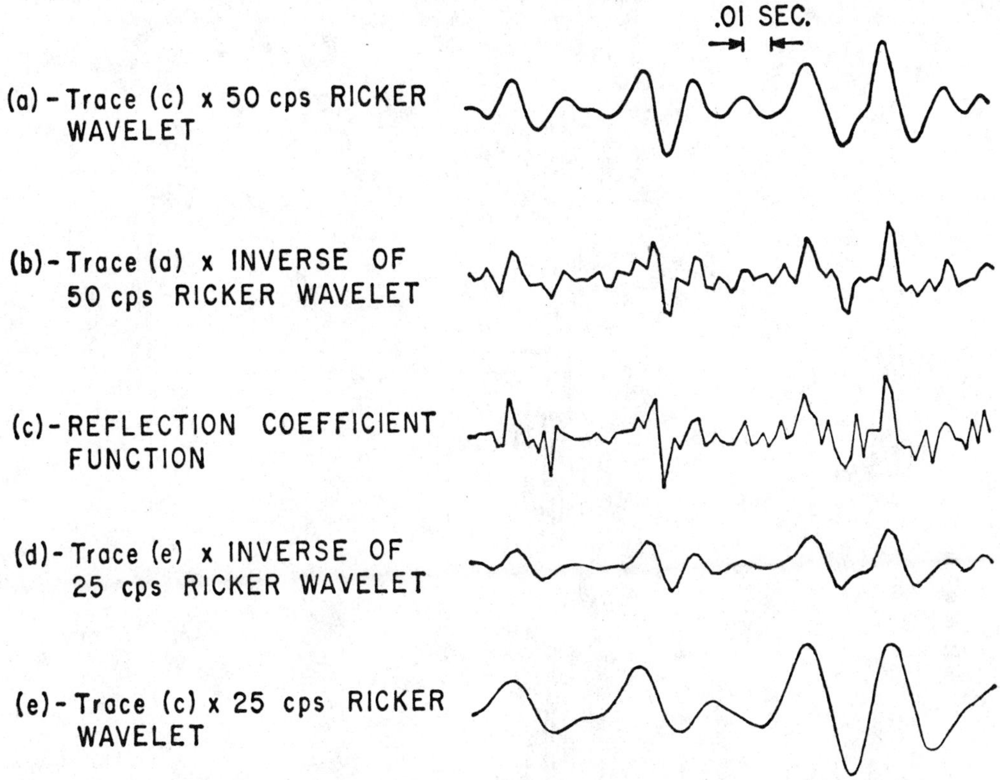

FIG. 10. Results of applying 50 and 25 cps inverses of Figure 5 to synthetic seismograms computed from a CVL on a Nebraska well.

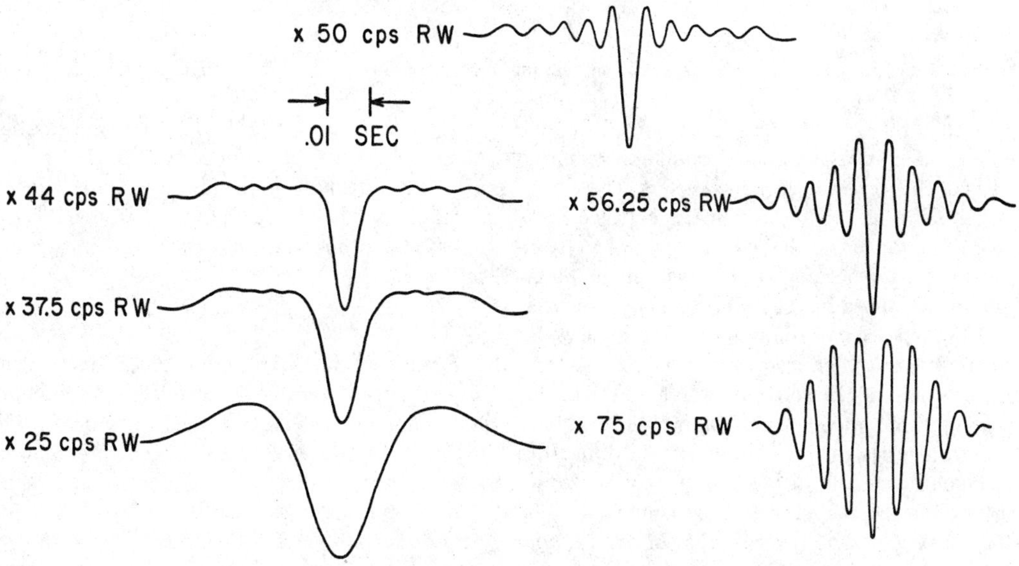

FIG. 11. Convolutions of 50 cps inverse of Figure 5 with Ricker wavelets of different peak frequencies.

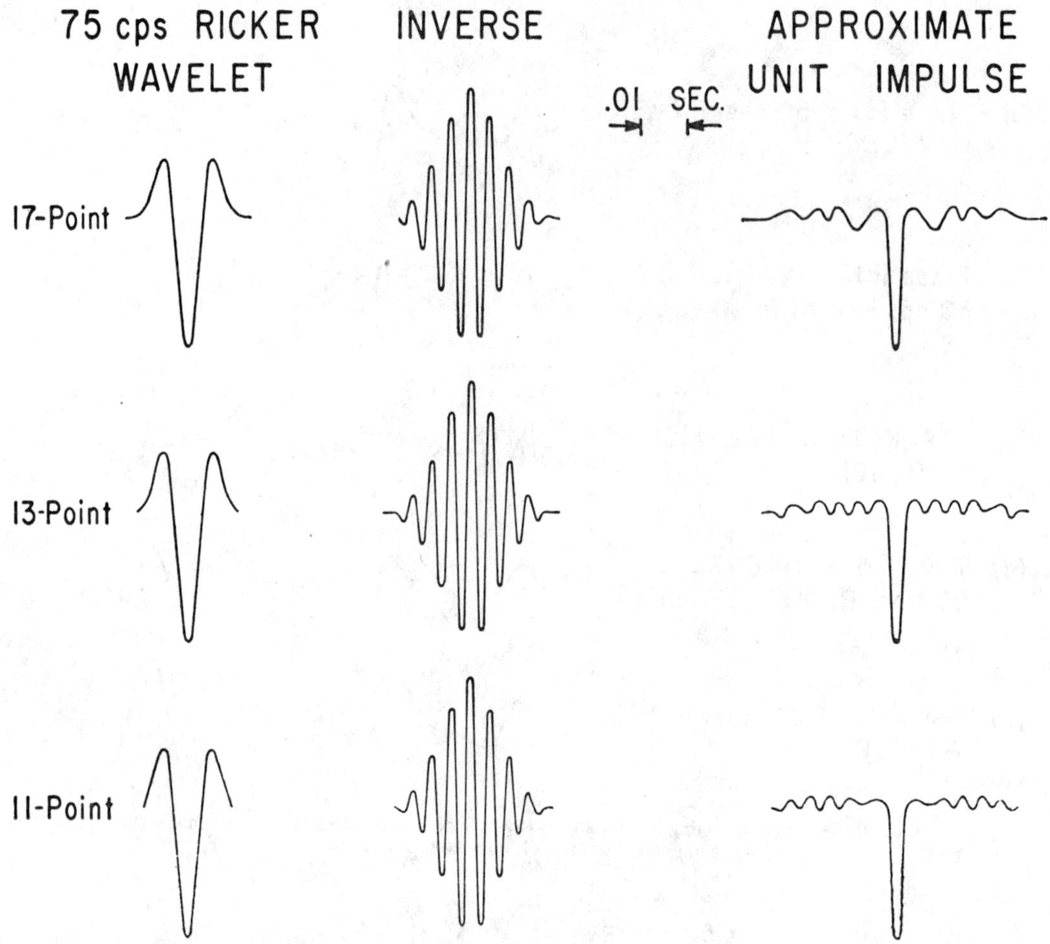

FIG. 12. Effect of approximate unit impulse for 75 cps case of cutting off tails of Ricker wavelet.

fact that the saw-tooth function added in each case contains a larger proportion of high-frequency components than the Ricker wavelets. These components are not significant in the extreme high-frequency range where the peak of the 75 cps inverse spectrum occurs, but are increasingly differentially amplified by the lower frequency inverses.

Most types of variations in wavelet shape would not add such high-frequency components and hence would not be so troublesome. If the varied forms of the wavelet are known, the inverse filter can be designed to circumvent the difficulties as much as possible, although the amount of resolution effected may have to be comprised. The biggest problem, however, is that of determining the form of the composition wavelet. The standard technique based on the power spectrum of the autocorrelation of the assumed stationary time series (e.g., Robinson (1957)) gives no information about the phase spectrum of the wavelet, which, of course, strongly influences the shape. Hence, reliance must be placed on theoretical or empirical knowledge other than that contained in the seismogram.

Effect of Random Noise

Finally, a few synthetic studies have been carried out to investigate the effect of random interference on the inverse filtering process. The digital computer was used to generate random noise spikes with an assigned density and maximum amplitude in a prescribed interval. The upper left-hand Trace I of Figure 14 shows such a set of random noise spikes superimposed on the latter portion of the 50 cps synthetic seismogram (Trace (a)) of Figure 10. The maximum amplitude

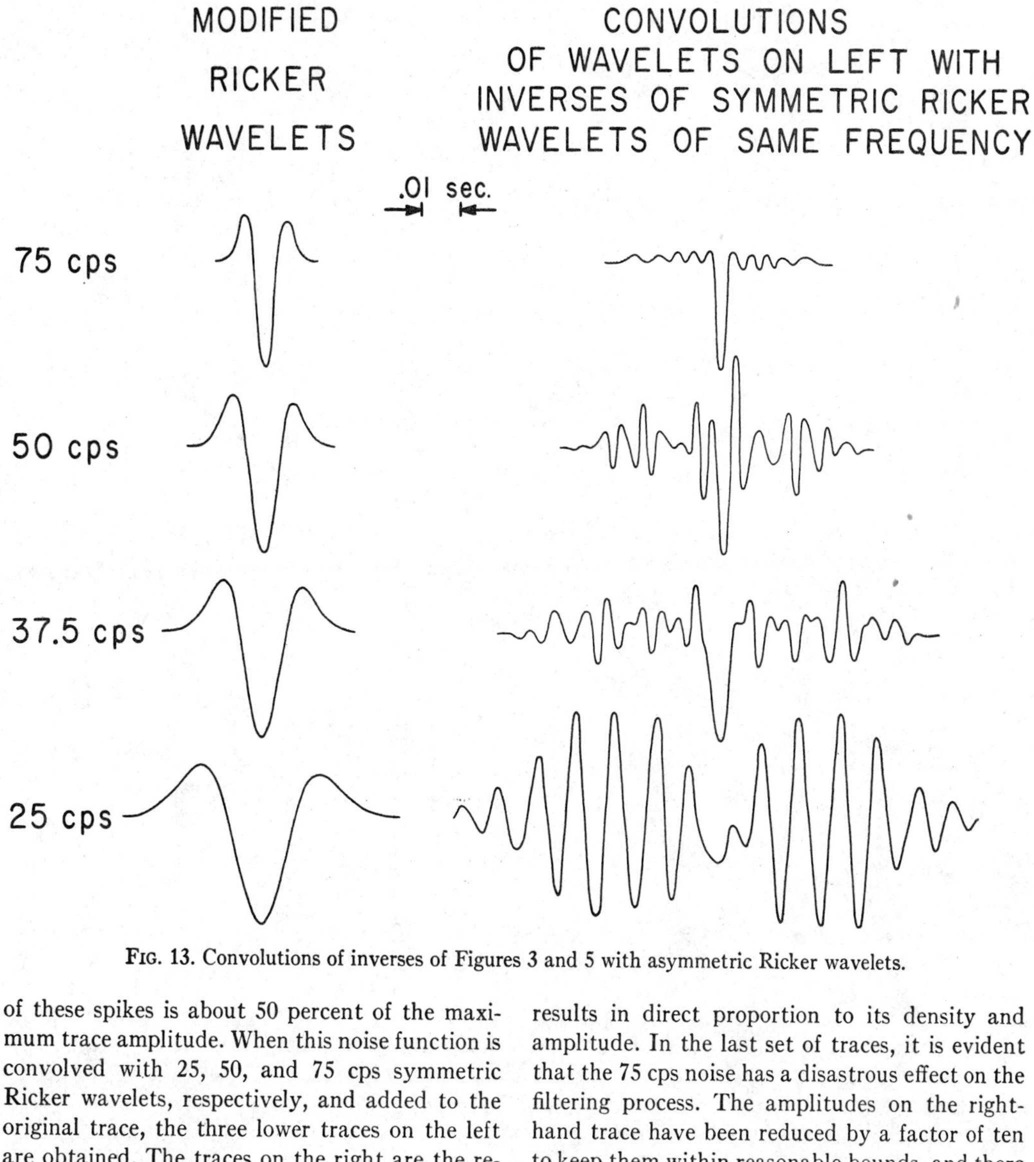

FIG. 13. Convolutions of inverses of Figures 3 and 5 with asymmetric Ricker wavelets.

of these spikes is about 50 percent of the maximum trace amplitude. When this noise function is convolved with 25, 50, and 75 cps symmetric Ricker wavelets, respectively, and added to the original trace, the three lower traces on the left are obtained. The traces on the right are the result of applying the 50 cps inverse filter of Figure 5 to the left-hand traces.

In comparing the second trace on the right with the upper right-hand trace, it is observed that the details have been fairly well restored in the presence of the 25 cps noise, but that there is a low-frequency component remaining. For the case when the interference is made the same frequency as the signal (third set of traces), the inverse filter cannot distinguish between the signal and the noise. Hence, the interference affects the results in direct proportion to its density and amplitude. In the last set of traces, it is evident that the 75 cps noise has a disastrous effect on the filtering process. The amplitudes on the right-hand trace have been reduced by a factor of ten to keep them within reasonable bounds, and there is no resemblance to the top trace.

These examples indicate that random interference may not be too troublesome if it is of substantially lower frequency than the reflections, although the very low frequencies will be amplified by the inverse filter to some extent as mentioned earlier in the discussion of Figure 11. The requirement is that the interference must not have significant frequency content beyond the point at which the spectrum of the inverse filter starts to rise rapidly. If the cut-off frequency can be made

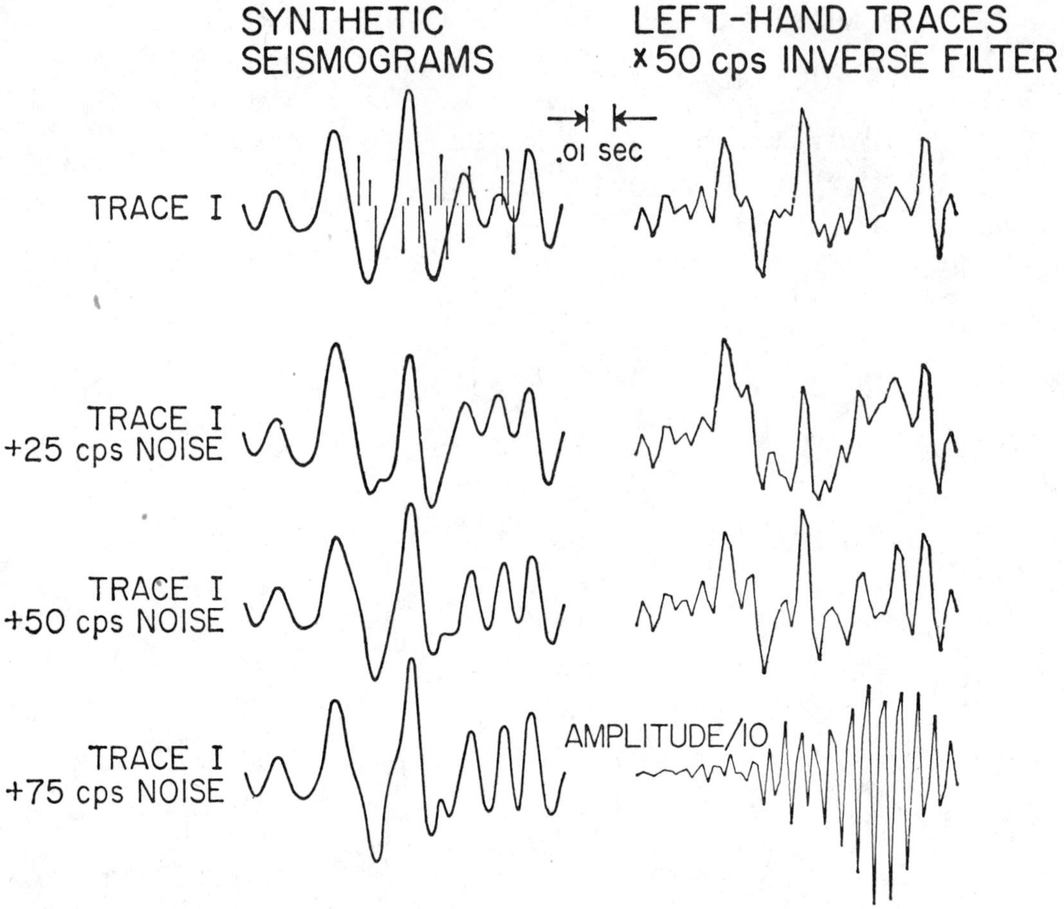

FIG. 14. Effect of random noise of different frequencies on results of applying 50 cps inverse of Figure 5 to 50 cps synthetic seismogram.

high enough and the inverse filter long enough, this requirement can usually be fulfilled. If this is not accomplished, the results, as in the lower right-hand trace, will be self-explanatory.

It should be noted that significant reading errors introduced in digitizing seismograms will have the same effect as high-frequency interference. Hence, it is necessary to develop some procedure for digitizing traces with sufficient accuracy. Commercial oscillographic readers are usually employed.

SUMMARY AND CONCLUSIONS

In summary, it has been demonstrated that it is possible to compute inverse filter functions by digital techniques which, under ideal conditions, will increase the resolution of reflection effects on seismograms well beyond the limits that are practical with instrumental filters. It has been shown that the effectiveness of these filters is more or less sensitive to variations in the peak frequency and shape of the wavelet composing the seismogram, and to interference. This sensitivity can be overcome to some extent by designing the filter so that it will not amplify unwanted frequencies. In many instances, the amount of resolution effected will have to be compromised with the sensitivity problem, which is aggravated by the usual lack of knowledge about the form of the composition wavelet. Nevertheless, the method appears to hold enough promise as a new seismic interpretation tool in the search for stratigraphic traps to warrant further study of its application to field records.

ACKNOWLEDGMENTS

The author is greatly indebted to his assistant, Mr. R. L. Massey, for his help in most phases of the work on this paper. He would also like to thank Dr. E. A. Robinson for his helpful reading of the manuscript.

REFERENCES

Ba Hli, F., 1954, A general method for time domain network synthesis: Transactions of the IRE Professional Group on Circuit Theory, v. CT-1, n. 3, p. 21–28.

Gardner, M. F., and Barnes, J. L., 1942, Transients in linear systems, v. 1: New York, John Wiley and Sons, Inc.

Jordan, Charles, 1947, Calculus of finite differences, 2nd edition: New York, Chelsea Publishing Co.

Kautz, W. H., 1954, Transient synthesis in the time domain: Transactions of the IRE Professional Group on Circuit Theory, v. CT-1, n. 3, p. 29–38.

Piety, R. G., 1951, A linear operational calculus of empirical functions: Phillips Petroleum Co. Research Division Report 106–12-51R.

Ricker, Norman, 1953, Wavelet contraction, wavelet expansion, and the control of seismic resolution: Geophysics, v. 18, p. 769–792.

Robinson, E. A., 1957, Predictive decomposition of seismic traces: Geophysics, v. 22, p. 767–778.

Reprinted from Geophysics v. 29, no. 1, p. 118-120

THE ERROR IN LEAST-SQUARES INVERSE FILTERING*

J. F. CLAERBOUT† AND E. A. ROBINSON†

Least-squares inverse filtering always involves consideration of the error. Under certain conditions the error will go to zero as the length of the filter tends to infinity. It is shown that the error will go to zero if either: (1) the waveform being inverted is minimum-phase, or (2) if the output is chosen to come after a sufficiently long time delay. If the waveform being inverted is not minimum-phase and if in addition the output is not chosen to be delayed, then the error will be finite and may be large.

INTRODUCTION

The equations for least-squares inverse filtering are well known (Rice, 1962; Simpson et al, 1961–1963; Claerbout, 1963); however, the conditions under which the least-squares error goes to zero do not seem to be so well known. The following conditions will be shown: (1) If the filter is one-sided, i.e., if it is a zero-delay inverse, then the error will tend to zero if, and only if, the wavelet being inverted is minimum-phase; (2) if the inverse filter is two-sided, i.e., the output is not just inverted, but also sufficiently delayed, then the energy in the error will tend to zero as $1/m$, where m is the number of filter coefficients.

ERROR FOR ZERO-DELAY FILTER

First, we see that if the waveform to be inverted is minimum-phase (its spectrum has no zeros inside the unit circle $z=e^{i\omega}$), then the error from a one-sided least-squares inverse will tend to zero as the filter length tends to infinity. This follows since it is well known (Robinson, 1954, 1957) that if the wavelet to be inverted is minimum-phase, an exact inverse can be found by the method of polynomial division (a Taylor series for the inverse of the spectrum converges on the unit circle). Since this inverse has zero error, a least-squares error method must also give zero error.

Next, we show that if the wavelet, $\mathbf{b}$, is not minimum-phase, then the error for a zero-delay inverse will not go to zero. Now it will be necessary to derive the least-squares equations.

Let $\mathbf{b}=(b_0,b_1, \cdots, b_n)$ be the given wavelet to be inverted, and let $\mathbf{a}=(a_0,a_1, \cdots ,a_m)$ be the filter coefficients which are to do this inversion. Then the desired output, $\mathbf{d}$, of the convolution of $\mathbf{a}$ with $\mathbf{b}$ is a column vector with elements $\mathbf{d}=(d_0,d_1, \cdots ,d_{m+n})=(1,0,0, \cdots ,0)$.

Convolution or filtering can be defined by the matrix multiplication

$$\begin{bmatrix} c_0 \\ c_1 \\ \cdot \\ \cdot \\ \cdot \\ \cdot \\ \cdot \\ \cdot \\ c_{m+n} \end{bmatrix} = \begin{bmatrix} b_0 & & & \text{zeros} \\ b_1 & b_0 & & \\ b_2 & b_1 & \ddots & \\ \cdot & b_2 & & b_0 \\ \cdot & & & b_1 \\ b_n & & & \cdot \\ & b_n & & \cdot \\ & & \ddots & \cdot \\ \text{zeros} & & & b_n \end{bmatrix} \begin{bmatrix} a_0 \\ a_1 \\ \cdot \\ \cdot \\ \cdot \\ a_m \end{bmatrix}$$

or

$$\mathbf{c} = \mathbf{Ba}.$$

The error in the output of the filter is $\mathbf{c}-\mathbf{d}$. Letting T denote transpose, the sum-squared error is

$$E = (\mathbf{c} - \mathbf{d})^T(\mathbf{c} - \mathbf{d}).$$

Setting the derivative of E with respect to each of the filter coefficients equal to zero, one arrives at simultaneous equations called the normal

* Work done under Contract AF 19(604)7378 at the Massachusetts Institute of Technology sponsored by The Advanced Research Projects Agency (Project VELA UNIFORM). Manuscript received by the Editor June 11, 1963.

† Now at the University Institute of Statistics, Uppsala, Sweden.

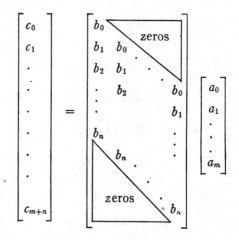

equations (for more of the algebraic steps see Rice (1962), Simpson et al (1961–1963), or Claerbout (1963)):

$$\mathbf{B}^T\mathbf{B}\mathbf{a} = \mathbf{B}^T\mathbf{d}.$$

The matrix $\mathbf{B}^T\mathbf{B}$ has rows and columns which are the autocorrelation of the wavelet **b,** the zero lag of the autocorrelation being on the main diagonal. $\mathbf{B}^T\mathbf{B}$ has no information about the phase spectrum of **b.** If $\mathbf{d}=(1,0,0,\cdots)$, then the vector $\mathbf{B}^T\mathbf{d}$ is $\mathbf{B}^T\mathbf{d}=(b_0,0,0,\cdots)$. A scale factor, b_0, in the inhomogeneous part (right side) of the normal equations can only affect the solution filter by a scale factor. Thus, there is no information about the phase spectrum of **b** in the normal equations. Any member of a class of wavelets with a given correlation will produce the same normal equations as any other member within a scale factor. Hence, the solution filters are all the same except for a scale factor. It is well known (Robinson, 1962) that the wavelet, **b**, can be represented as

$$\mathbf{b} = \mathbf{p} * \mathbf{w} \text{ (letting ``*'' denote convolution),}$$

where **p** is an all-pass phase delaying filter and **w** is a minimum-phase wavelet with the same autocorrelation function as **b.** An all-pass phase delaying filter is a filter which does not attenuate any frequency, but may delay its phase. Now suppose we convolve **b** with $(a_0,a_1,\cdots,a_m)$. As m tends to infinity we have noted that **a** tends to be proportional to a zero error inverse of **w.** Thus,

$$\mathbf{b} * \mathbf{a} = \mathbf{p} * \mathbf{w} * \mathbf{a} \sim \mathbf{p}.$$

This can equal $(1,0,0,\cdots)$ only if **p** is the trivial phase-shift filter (no phase shift). Hence, we have the conclusion that the error cannot tend to zero unless **b** is minimum-phase.

ERROR FOR FILTER WITH DELAY

Finally, we show the hardest part of the theorem that if the filter, **a,** is supposed to put out a delayed impulse, $\mathbf{d}=(\cdots,0,0,1,0,0,\cdots)$, then the sum-squared error $(\mathbf{Ba}-\mathbf{d})^T(\mathbf{Ba}-\mathbf{d})$ will tend to zero regardless of the phase characteristic of **b.** We consider the class of least-squares inverse filters with output at all possible different lags. There are $n+m+1$ possible desired outputs; arranged in a square matrix they are

$$\mathbf{D} = \left[(\mathbf{d}^0)(\mathbf{d}^1)\cdots(\mathbf{d}^{m+n})\right] = \begin{bmatrix} 1 & 0 & 0 & \cdots & 0 \\ 0 & 1 & 0 & & \\ 0 & 0 & 1 & & \\ \vdots & & & \ddots & \\ 0 & & & & 1 \end{bmatrix}.$$

Now consider the sum-squared error for any one of the desired outputs. It is

$$E(\mathbf{d}) = (\mathbf{Ba} - \mathbf{d})^T(\mathbf{Ba} - \mathbf{d}).$$

Solving the normal equations for **a,**

$$\mathbf{a} = (\mathbf{B}^T\mathbf{B})^{-1}\mathbf{B}^T\mathbf{d},$$

and inserting into the expression for error we get

$$E(\mathbf{d}) = (\mathbf{B}(\mathbf{B}^T\mathbf{B})^{-1}\mathbf{B}^T\mathbf{d}-\mathbf{d})^T (\mathbf{B}(\mathbf{B}^T\mathbf{B})^{-1}\mathbf{B}^T\mathbf{d}-\mathbf{d}).$$

Now if we consider the sum of the sum-squared errors for all of the possible desired outputs **d,** we obtain

$$E_{\text{total}} = E(\mathbf{d}^0) + E(\mathbf{d}^1) + \cdots + E(\mathbf{d}^{m+n})$$
$$= \text{trace}\,(\mathbf{B}(\mathbf{B}^T\mathbf{B})^{-1}\mathbf{B}^T\mathbf{D}-\mathbf{D})^T (\mathbf{B}(\mathbf{B}^T\mathbf{B})^{-1}\mathbf{B}^T\mathbf{D}-\mathbf{D}).$$

The trace of a matrix is just the sum of the diagonal elements. The kth diagonal element is just the error for $\mathbf{d}^k$. We will proceed to simplify this expression and come to the remarkable conclusion that it is independent of the filter length m.

First we note that **D** is an identity matrix.

$$E_{\text{total}} = \text{trace}\,(\mathbf{B}(\mathbf{B}^T\mathbf{B})^{-1}\mathbf{B}^T-\mathbf{D})^T (\mathbf{B}(\mathbf{B}^T\mathbf{B})^{-1}\mathbf{B}^T-\mathbf{D})$$

and since $(\mathbf{B}^T\mathbf{B})^{-1}$ is symmetric,

$$\begin{aligned} E_{\text{total}} &= \text{trace}\,(\mathbf{B}(\mathbf{B}^T\mathbf{B})^{-1}\mathbf{B}^T-\mathbf{D})(\mathbf{B}(\mathbf{B}^T\mathbf{B})^{-1}\mathbf{B}^T-\mathbf{D}) \\ &= \text{trace}\,(\mathbf{B}(\mathbf{B}^T\mathbf{B})^{-1}\mathbf{B}^T\mathbf{B}(\mathbf{B}^T\mathbf{B})^{-1}\mathbf{B}^T \\ &\quad -2\mathbf{B}(\mathbf{B}^T\mathbf{B})^{-1}\mathbf{B}^T+\mathbf{D}) \\ &= \text{trace}\,(D-\mathbf{B}(\mathbf{B}^T\mathbf{B})^{-1}\mathbf{B}^T) \\ &= \text{trace}\,\mathbf{D}-\text{trace}\,\mathbf{B}(\mathbf{B}^T\mathbf{B})^{-1}\mathbf{B}^T. \end{aligned}$$

Now we use the fact that the trace of a product of two matrices does not depend on the order of the product, i.e., trace $\mathbf{M}_1\mathbf{M}_2 =$ trace $\mathbf{M}_2\mathbf{M}_1$. This is true even if the matrices are not square.

$$\begin{aligned} E_{\text{total}} &= \text{trace}\,\mathbf{D} - \text{trace}\,(\mathbf{B}^T\mathbf{B})(\mathbf{B}^T\mathbf{B})^{-1} \\ &= (m+n+1) - (m+1) \\ &= n. \end{aligned}$$

Hence, we have the conclusion that the sum of the sum-squared errors for inverse filters at all possible lags is independent of the filter length m. In fact, it is just equal to n, the number of coefficients minus one of the wavelet to be inverted.[1]

As the length, m, of the filter goes to infinity the total error $E_{\text{total}} = n$ is spread out over a larger and larger interval (of size $n+m+1$). Hence, given n, if we let $E_s(m)$ be the smallest error, for some delay, for a given m, then we must have

$$E_s(m) \leq \frac{n}{m+n+1}\,.$$

Thus, as $m \to \infty$, $E_s \to 0$.

DISCUSSION OF FINITE ERROR

If the sum-squared error were the same for all possible delays, it would be $n/(m+n+1)$ for any particular inverse filter. This need not be the case, however. It may be considerably greater for small and/or large delays. For example, when one tries to invert a nonminimum-phase wavelet with zero delay, one obtains a finite and possibly large error.

The largest possible sum-squared error for any inverse filter is unity. This is because the zero filter, $\mathbf{a} = (0,0,\cdots,0)$, would have a sum-squared error of unity and any least-squares error inverse would have no more than this error. The maximum error would actually be obtained if one were to try to invert without delay a wavelet which had undergone a pure delay, say $\mathbf{b} = (0, b_1, b_2, \cdots, b_n)$. In this case the right side of the normal equations is zero and hence the filter is zero and no good at all.

[1] R. A. Wiggins pointed out that n is just the number of zeros in the Z-transform of $\mathbf{b}$. This is also the number of poles above (or below) the real frequency axis of the inverse energy density spectrum of $\mathbf{b}$. It seems as if one should be able to base a proof using contour integration upon this fact.

ACKNOWLEDGMENTS

We would like to thank A. H. Booker and R. A. Wiggins for helpful comments and discussions.

REFERENCES

Claerbout, J. F., 1963, Digital filters and applications to seismic detection and discrimination: M.S. thesis, Massachusetts Institute of Technology.

Rice, R. B., 1962, Inverse convolution filters: Geophysics v. 27, pp. 4–18.

Robinson, E. A., 1954, Predictive decomposition of time series with applications to seismic exploration: Ph.D. thesis, Massachusetts Institute of Technology.

———, 1957, Predictive decomposition of seismic traces: Geophysics, v. 22, pp. 767–778.

———, 1962, Random wavelets and cybernetic systems: London, Griffin and New York, Stechert-Hafner.

Simpson, S. M., Robinson, E. A., Claerbout, J. F., Galbraith, J. N., Wiggins, R., Clark, J., Ross, W. P., and Pan, C., 1961–63, Reports to Advanced Research Projects Agency: Project VELA UNIFORM.

Reprinted from Geophysics v. 31, no. 5, p. 917-926

LEAST-SQUARES INVERSE FILTERING†

WAYNE T. FORD* AND JAMES H. HEARNE**

Suppose we are given the autocorrelation function of a certain unknown sampled signal. Although a number of different signals might produce the given autocorrelation function, only one of these is minimum-delay. Denoting this minimum-delay unknown signal by the matrix **K**, the given autocorrelation may be written in the form **K′K** where **K′** is the transpose of **K**.

It is desired to determine approximately the inverse of this unknown signal **K**; that is, we wish to determine a vector **X** so that **KX** is as close to $\mathbf{B}' = (1, 0, 0, \cdots, 0)$ as possible in a least-squares sense. If **K** were known, Rice shows that

$$\mathbf{X} = (\mathbf{K'K})^{-1}\mathbf{K'B}.$$

At first glance, the above formula appears useless as **K′** is unknown. However, although **K′** is indeed unknown, **K′B** has the form $\mathbf{K'B} = (c, 0, 0, \cdots, 0)'$ where the scalar c simply plays the role of a scale factor. Thus, we determine **X** by simply selecting a convenient multiple of the first column of the inverse of the known matrix **K′K**.

Although we do not present the details of the computer programming involved in the above calculation, we do present some simple examples to illustrate the process.

INTRODUCTION

A number of authors (e.g., Rice, 1962; Treitel, 1964; Wiggins, 1965) have treated various aspects of the problem of determining digital filters that satisfy certain given conditions. Rice (1962) studies the problem of determining an inverse filter function, $g(t)$, which transforms a known function into a unit impulse, or some "best" approximation to a unit impulse. Suppose we are given the autocorrelation function of an unknown minimum-phase function, $f(t)$. We shall be interested in determining the minimum-phase inverse filter function, $g(t)$, which would transform this unknown function into a unit impulse, or some "best" approximation to a unit impulse. We remark that Rice's (1962) methods would allow one to estimate the original unknown function, $f(t)$, from its inverse, $g(t)$. Moreover, since $f(t)$ and $g(t)$ are inverses of each other, their respective autocorrelation functions are inverses of each other.

Various methods of solution of our problem are possible. The essential aspects of solution using Fourier transforms are known (Papoulis, 1962, and Ford, 1966). Also, a Z-transform approach is available (Sakrison, 1965). The object of the present paper is to discuss and illustrate a matrix algebra approach to the problem. However, we shall employ Z-transforms to clarify ideas and to develop proofs.

Although we shall present some new results, we shall carefully discuss the entire problem in the hope that we may reduce the obscurity which now seems to surround some aspects of the problem. Since we at first found ourselves easily confused at certain points in the problem, we shall give an abstract presentation rather than an intuitive one. We hope to lighten the load of abstraction by the inclusion of detailed numerical examples.

DEFINITIONS

Vectors and sequences

It is natural to denote a *row-vector* of dimension $N+1$ by $(h_0, h_1, \cdots, h_N)$. The conjugate transpose of this row-vector will be denoted by $(h_0, h_1, \cdots, h_N)'$. This is a *column vector* containing the components $\bar{h}_0, \bar{h}_1, \cdots, \bar{h}_N$, where $\bar{h}_n$ represents the complex conjugate of h_n. However, we shall assume vectors to have real components unless otherwise indicated.

It will often be necessary to identify the com-

† Presented at the 35th Annual SEG Meeting in Dallas, Texas, November 17, 1965. Manuscript received by the Editor January 3, 1966; revised manuscript received June 8, 1966.

* Mathematics Department, University of Houston; Consultant to Ray Geophysical Division, Mandrel Industries, Inc.

** Ray Geophysical Division, Mandrel Industries, Inc.

ponents of a vector with a certain list of consecutive "sample times." We shall indicate this situation by placing a subscript and superscript on a vector to denote the initial and final "sample time," respectively. Thus, $(1, 2, 3)_5^7$ represents an infinite sequence $(\cdots, h_0, h_1, \cdots)_{-\infty}^{\infty}$ of numbers such that $h_5=1$, $h_6=2$, $h_7=3$, and $h_n=0$ whenever $n<5$ or $n>7$. We shall use $(h_n)_{-\infty}^{\infty}$ as an abbreviation for $(\cdots, h_0, h_1, \cdots)$.

Z-transforms

Suppose we are given an infinite sequence of real numbers $\mathbf{h}=(h_n)_{-\infty}^{\infty}$. The Z-transform (Jury, 1958 and 1964) of h is defined by

$$Z(\mathbf{h}) = H(z) = \sum_{n=-\infty}^{\infty} h_n z^{-n}. \quad (1)$$

We may expect this series to converge in some ring in the complex z-plane. That is, $H(z)$ is defined for $r<|z|<R$. We define $\mathbf{h}^*=(h_n^*)_{-\infty}^{\infty}=(h_{-n})_{-\infty}^{\infty}$, and

$$H^*(z) = \sum_{n=-\infty}^{\infty} h_n^* z^{-n} = \sum_{m=-\infty}^{\infty} h_{-m}^* z^m$$

$$= \sum_{m=-\infty}^{\infty} h_m z^m = H(z^{-1}). \quad (2)$$

We must have $r<1<R$ in order that $H^*(z)$ be well defined. However, many applications involve finite sequences, and there is no question of convergence.

Example. Let $\mathbf{h}=(4, -2, 3, 1)_{-1}^2$. We have $\mathbf{h}^*=(1, 3, -2, 4)_{-2}^1$, $H(z)=4z-2+3z^{-1}+z^{-2}$, $H^*(z)=z^2+3z-2+4z^{-1}$, $r=0$, and $R=\infty$.

Minimum-phase operators

We shall say that $\mathbf{h}$ is a minimum-phase digital operator, H is minimum phase, or $\mathbf{h}$ is minimum phase, if and only if $H(z)$ is finite and nonzero whenever $|z|\geq 1$. There are a number of equivalent definitions. One of these is that H is minimum phase if and only if $H^*(z)$ is finite and nonzero whenever $|z|\leq 1$. The similarity of these two definitions can lead to some confusion. We shall say that H is quasi-minimum phase if $H(z)$ is finite and nonzero whenever $|z|>1$. That is, a quasi-minimum phase operator may have a pole and/or a root on the unit circle. We remark that "minimum delay" is often used as a synonym for "minimum phase" (Treitel and Robinson, 1964). If either $\mathbf{h}$ or $\mathbf{h}^*$ is minimum phase, the other is said to be maximum phase (or maximum delay), except that a time shift must be applied to $\mathbf{h}^*$ in order to obtain a causal operator.

We shall say that $\mathbf{h}$ is causal if $\mathbf{h}=(h_n)_0^{\infty}$. For $\mathbf{h}$ to be minimum phase, it is clearly necessary (but not sufficient) that $\mathbf{h}$ is causal and h_0 is nonzero.

Examples. $(1, -5, 6)_{-1}^1$ is not minimum phase since it is not causal. $(1, -5, 6)_0^2$ is not minimum phase since $1-5z^{-1}+6z^{-2}=(1-2z^{-1})(1-3z^{-1})$ is zero if $z=2$ or $z=3$. However, $(6, -5, 1)_0^2$ is minimum phase since $6-5z^{-1}+z^{-2}=(2-z^{-1})(3-z^{-1})$ is zero only if $z=1/2$ or $z=1/3$, and is defined for all nonzero z.

Inverses. If H is minimum phase, it has a minimum-phase inverse. The Z-transform of the inverse can be obtained by the formal division operation $1/H(z)$. If H is merely quasi-minimum phase, it is useless to seek an inverse for $H(z)$.

Example. We have seen that $(6, -5, 1)_0^2$ is minimum phase. Its inverse, $(C_0, C_1, \cdots)$, is minimum phase and given by $C_0=1/6$, $C_1=5/36$, $C_{n+2}=(5C_{n+1}-C_n)/6$ for $n=0, 1, 2, 3, \cdots$.

Convolution and autocorrelation

Let $\mathbf{h}$ and $\mathbf{k}$ be two sequences of the kind discussed above. The Z-transform of the convolution of $\mathbf{h}$ and $\mathbf{k}$ is the product of $H(z)$ and $K(z)$. The autocorrelation of $\mathbf{h}$ is the sequence with Z-transform given by the product of $H(z)$ and $H^*(z)$.

Examples. Let $\mathbf{h}=(1, -5, 6)_m^{m+2}$ and $k=(1, 2, 3)_0^2$. The Z-transform of the convolution of $\mathbf{h}$ and $\mathbf{k}$ is

$$\begin{aligned} &H(z)K(z) \\ &= (z^{-m} - 5z^{-m-1} + 6z^{-m-2})(1 + 2z^{-1} + 3z^{-2}) \\ &= z^{-m} - 3z^{-m-1} - z^{-m-2} - 3z^{-m-3} \\ &\quad + 18z^{-m-4} \end{aligned}$$

and the convolution is $(1, -3, -1, -3, 18)_m^{m+4}$. The autocorrelation of $\mathbf{h}$ has the Z-transform

$$\begin{aligned} &H(z)H^*(z) \\ &= (z^{-m}-5z^{-m-1}+6z^{-m-2})(6z^{m+2}-5z^{m+1}+z^m) \\ &= 6z^2-35z+62-35z^{-1}+6z^{-2}. \end{aligned}$$

Time-invariance. We express the above situation in the remark that autocorrelation is time-invariant. That is, a shift of subscripts in a sequence has no effect on the autocorrelation of the sequence.

FACTORIZATION OF AUTOCORRELATIONS

Root patterns

Let $\mathbf{h}$ represent a sequence of real numbers. Since $H(z)$ has real coefficients, $H(z_0)=0$ if and only if $H(\bar{z}_0)=0$. That is, the nonreal roots of $H(z)$ occur in conjugate pairs. Also, $H(z_0)=0$ if and only if $H^*(z_0^{-1})=0$. That is, the roots of $H(z)$ and $H^*(z)$ are reciprocal to each other in pairs. Thus, if z_0 is nonreal and $H(z_0)=0$, it follows that $H(z)\,H^*(z)$ has at least the four roots, z_0, $\bar{z}_0$, $(z_0)^{-1}$, and $(\bar{z}_0)^{-1}$. In particular, if $|z_0|=1$ and $H(z_0)=0$, it follows that z_0 has even multiplicity as a root of $H(z)\,H^*(z)$.

Finite autocorrelations

Let $\mathbf{h}=(h_n)_0{}^n$ and assume that $h_0h_n\neq 0$. $H(z)$ will have precisely N roots in the complex z-plane, and we may write

$$H(z) = h_N \prod_{n=1}^{N} (z^{-1} - z_n^{-1})$$

$$= h_0 z^{-N} \prod_{n=1}^{N} (z - z_n) \qquad (3)$$

since

$$H(\infty) = h_0 = h_N \prod_{n=1}^{N} (-z_n^{-1}).$$

Also, we have

$$H^*(z) = h_0 z^N \prod_{n=1}^{N} (z^{-1} - z_n)$$

$$= h_0 \prod_{n=1}^{N} (1 - z_n z) \qquad (4)$$

and

$$H(z)H^*(z)$$

$$= h_0{}^2 z^{-N} \prod_{n=1}^{N} (z - z_n)(1 - z_n z). \qquad (5)$$

Although it is likely that neither $H(z)$ nor $H^*(z)$ is minimum phase, we shall see that there exists $\mathbf{k}=(k_n)_0{}^N$ which is quasi-minimum phase and such that

$$H(z)H^*(z) = K(z)K^*(z). \qquad (6)$$

We construct $K(z)$ by an examination of the expression $(z-z_n)\,(1-z_nz)$ for $n=1, 2, \cdots, N$. If $|z_n|<1$, include $(z-z_n)$ as a factor of $K(z)$ and $(1-z_nz)$ as a factor of $K^*(z)$. If $|z_n|>1$, include $(1-z_nz)$ as a factor of $K(z)$ and $(z-z_n)$ as a factor of $K^*(z)$. If $z_n=\pm 1$, use either of the above rules. If $z_n=e^{i\theta}, \theta\neq 0, H(z)\,H^*(z)$ includes the expression

$$z(-e^{i\theta})(1-e^{i\theta}z)(z-e^{-i\theta})(1-e^{-i\theta}z)$$

$$=\left[(z-e^{i\theta})(z-e^{-i\theta})\right]^2$$

as a factor. In this case, include $(z\_e^{i\theta})\,(z-e^{i\theta})$ as a factor of both $K(z)$ and $K^*(z)$. Next, include the z^{-N} in equation (5) as a factor of $K(z)$. Finally multiply both $K(z)$ and $K^*(z)$ by $\pm h_0$ where we choose the sign so that k_0 is positive. Thus, $K(z)$ is quasi-minimum phase, and equation (6) is satisfied. Clearly, $K(z)$ is uniquely determined except for the sign choice on k_0. Also, while we have a constructive proof of the existence of $K(z)$, this constructive method is useless for actual computation unless N is quite small.

Example. Let $\mathbf{h}=(12, 31, 15, 2)_0{}^3$. The autocorrelation of $\mathbf{h}$ is $(24, 242, 867, 1334, 867, 242, 24)_{-3}{}^3$, and we have

$$H(z)H^*(z)$$

$$=24z^3+242z^2+867z+1334+867z^{-1}+242z^{-2}+24z^{-3}$$

$$=z^{-3}(2+z)(1+2z)(3+z)(1+3z)(4+z)(1+4z).$$

There are eight distinct sequences having this same autocorrelation. Selecting convenient time origin in each case, we may write the Z-transforms of these eight sequences in the form

$$24 + 26z^{-1} + 9z^{-2} + z^{-3} = (2 + z^{-1})(3 + z^{-1})(4 + z^{-1})$$

$$z^3 + 9z^2 + 26z + 24 = (z + 2)(z + 3)(z + 4)$$

$$12 + 31z^{-1} + 15z^{-2} + 2z^{-3} = (1 + 2z^{-1})(3 + z^{-1})(4 + z^{-1})$$

$$2z^3 + 15z^2 + 31z + 12 = (2z + 1)(z + 3)(z + 4)$$

$$8 + 30z^{-1} + 19z^{-2} + 3z^{-3} = (2 + z^{-1})(1 + 3z^{-1})(4 + z^{-1})$$

$$3z^3 + 19z^2 + 30z + 8$$

$$= (z + 2)(3z + 1)(z + 4)$$

$$6 + 29z^{-1} + 21z^{-2} + 4z^{-3}$$
$$= (2 + z^{-1})(3 + z^{-1})(1 + 4z^{-1})$$

$$4z^3 + 21z^2 + 29z + 6$$
$$= (z + 2)(z + 3)(4z + 1)$$

where we have paired the possibilities to indicate that the second member of a pair would have been $H^*(z)$ if the first member of the same pair had been $H(z)$. Only one of these eight Z-transforms is minimum phase. That is,

$$K(z) = 24 + 26z^{-1} + 9z^{-2} + z^{-3}$$

in this case.

Infinite autocorrelations

There are at most 2^N distinct sequences having a given finite autocorrelation, $\mathbf{a} = (an)_{-N}{}^N$. As we have seen, precisely one of these will be minimum phase. Although we shall be primarily interested in finite autocorrelations, it seems desirable to include some remarks on the situation resulting from $N = \infty$.

Let $\mathbf{a} = (a_n)_{-\infty}{}^{\infty}$ be an autocorrelation of some sequence, $\mathbf{h} = (h_n)_0{}^{\infty}$. Then

$$H(z) = \sum_{n=0}^{\infty} h_n z^{-n}, \quad |z| > r \tag{7}$$

where the assumption that $r < 1$ is necessary for the existence of the autocorrelation of $\mathbf{h}$. Now, we have

$$H^*(z) = \sum_{n=0}^{\infty} h_n z^n, \quad |z| < r^{-1}. \tag{8}$$

A quasi-minimum-phase $K(z)$ exists, and the first few values of k_n can be approximated. Sakrison (1965) gives an effective method for accomplishing this.

LEAST-SQUARES INVERSE FACTORIZATION

Matrix notation

Let $\mathbf{h} = (h_n)_0{}^M$ and $\mathbf{x} = (x_n)_0{}^N$. We express the convolution of $\mathbf{h}$ and $\mathbf{x}$ as $\mathbf{y} = (y_n)_0{}^{M+N}$. We wish to express $\mathbf{y}$ as the result of a matrix multiplication. To do this, we use the elements of $\mathbf{h}$ to construct the matrix (Claerbout and Robinson, 1964)

$$\mathbf{H} = \begin{bmatrix} h_0 & & & & \\ h_1 & h_0 & & \text{zeros} & \\ h_2 & h_1 & h_0 & & \\ \cdot & h_2 & h_1 & \ddots & \\ \cdot & \cdot & h_2 & \ddots & \\ \vdots & \vdots & \vdots & & \\ h_M & \cdot & \cdot & & \\ & h_M & \cdot & & h_0 \\ & & h_M & & h_1 \\ & & & & h_2 \\ & \text{zeros} & & \ddots & \vdots \\ & & & & h_M \end{bmatrix} \tag{9}$$

having $M+N+1$ rows and $N+1$ columns. Then, we have

$$\mathbf{y}' = \mathbf{H}\mathbf{x}'. \tag{10}$$

Minimization

Let $\mathbf{d} = (d_n)_0{}^{M+N}$ be given. Suppose we want to find a vector $\mathbf{x}$ such that the length of $(\mathbf{Hx}' - \mathbf{d}')$ is as small as possible. That is, we want to choose $\mathbf{x}$ such that

$$\sum_{n=0}^{M+N} (y_n - d_n)^2 = (\mathbf{Hx}' - \mathbf{d}')(\mathbf{Hx}' - \mathbf{d}') \tag{11}$$

is minimized. The solution (Rice, 1962) is given by

$$\mathbf{x}' = (\mathbf{H}'\mathbf{H})^{-1}\mathbf{H}'\mathbf{d}'. \tag{12}$$

The expression $(\mathbf{H}'\mathbf{H})^{-1}\,\mathbf{H}'$ is the so-called generalized inverse of the matrix, $\mathbf{H}$. Cline (1965) gives several references to works on generalized inverses.

We note that $\mathbf{H}'\mathbf{H}$ is a square matrix $(N+1)$ by $(N+1)$ of Toeplitz type (Grenander et al., 1958). Denoting the autocorrelation of $\mathbf{h}$ by $\mathbf{a} = (a_n)_{-M}{}^M$, we have

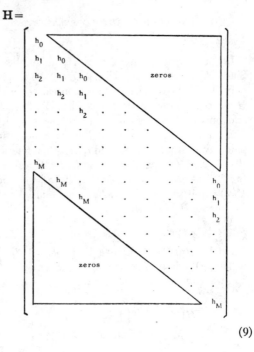

$$
\mathbf{H'H} = \begin{bmatrix} a_0 & a_1 & \cdot & \cdot & \cdot & a_{N-1} & a_N \\ a_1 & a_0 & \cdot & \cdot & \cdot & a_N & a_{N-1} \\ \cdot & \cdot & \cdot & \cdot & \cdot & \cdot & \cdot \\ \cdot & \cdot & \cdot & \cdot & \cdot & \cdot & \cdot \\ \cdot & \cdot & \cdot & \cdot & \cdot & \cdot & \cdot \\ \cdot & \cdot & \cdot & \cdot & \cdot & a_0 & a_1 \\ a_N & a_{N-1} & \cdot & \cdot & \cdot & a_1 & a_0 \end{bmatrix} \tag{13}
$$

Methods of calculation of $(\mathbf{H'H})^{-1}$ are available (Levinson, 1947) although special care is required for large N. If we assume h_0 is nonzero, it follows that $\mathbf{H'H}$ is nonsingular. To see this, suppose $\mathbf{H'H}$ is singular. Then there exists a nonzero $\boldsymbol{\lambda}=(\lambda_n)_0^N$ such that $\mathbf{H'H}\boldsymbol{\lambda}'=\mathbf{0}$. Thus, $(\mathbf{H}\boldsymbol{\lambda}')'(\mathbf{H}\boldsymbol{\lambda}')=\boldsymbol{\lambda}\mathbf{H'H}\boldsymbol{\lambda}'=\mathbf{0}$.
That is,

$$
\begin{aligned}
0 &= h_0\lambda_0, \\
0 &= h_1\lambda_0 + h_0\lambda_1, \\
0 &= h_2\lambda_0 + h_1\lambda_1 + h_0\lambda_2, \\
&\;\;\vdots \\
\text{and}\quad 0 &= h_N\lambda_0 + h_{N-1}\lambda_1 + \cdots + h_0\lambda_N.
\end{aligned}
$$

It is clear that h_0 nonzero will imply that each λ_n is zero, and $\mathbf{H'H}$ is nonsingular. We are certainly justified in assuming h_0 is nonzero because of the time-invariance of autocorrelations.

Factorization

Suppose we are given an autocorrelation, $\mathbf{a}=(a_n)_{-M}^M$, of some $\mathbf{h}=(h_n)_0^M$. There is a unique (except for a factor of ± 1) minimum-phase operator $\mathbf{k}$ having the same autocorrelation providing we assume that $H(z)$ has no root on the unit circle. We postpone consideration of the case where $H(z)$ has such a root.

We want to find an operator $\mathbf{g}_N=(g_{n,N})_0^N$ such that the length of $\mathbf{Kg}_N'-\mathbf{d}_N'$ is as small as possible, where $\mathbf{d}_N=(1, 0, \cdots, 0)_0^{M+N}$. We use relation (12) to write

$$
\mathbf{g}_N' = (\mathbf{K'K})^{-1}\mathbf{K'd}_N' = (\mathbf{H'H})^{-1}\mathbf{K'd}_N', \tag{14}
$$

where the dimensionality of $\mathbf{K}$ and $\mathbf{H}$ depends on N in the manner shown in relation (9). At first glance, this relation appears useless since $\mathbf{K}'$ is an unknown matrix. However, although $\mathbf{K}'$ is indeed unknown, we have

$$
\mathbf{K'd}_N' = (k_0, 0, 0, \cdots, 0)'. \tag{15}
$$

Thus $\mathbf{g}_N$ is the product of the scalar, k_0, and the first column of the inverse of $(\mathbf{H'H})$. We may take $k_0=1$ without loss of generality since this is simply a choice of scale.

Phase considerations

It is known (Claerbout, 1964 and Robinson, 1963) that the sequence of vectors $\mathbf{g}_2, \mathbf{g}_3, \cdots, \mathbf{g}_N, \cdots$ converges to the minimum-phase inverse of the minimum-phase operator $\mathbf{k}$. Moreover, it is known (Robinson, 1963) that each vector in the sequence of approximations is also minimum phase.

COMPUTING CONSIDERATIONS

Quasi-minimum-phase operators

We have been assuming that $H(z)$ has no zeros on the circumference of the unit circle. If $H(z)$ has zeros on the unit circle, we should expect difficulties since $K(z)$ is merely quasi-minimum phase.

Example. Suppose $H(z)=1+z^{-1}$. Formal division gives

$$
H^{-1}(z) = 1 - z^{-1} + z^{-2} - z^{-3} + \cdots, \tag{16}
$$

which is undefined at $z=-1$. Although we can calculate $\mathbf{g}_N$, it is clearly not useful to do so.

Center-point adjustment

If $\mathbf{h}$ is a physical quantity such as a portion of a seismic trace, one does not expect $H(z)$ to have a root which falls precisely on the unit circle. However, the measured approximation to $H(z)$ may have roots on the unit circle and very probably will have roots that are very near the unit circle.

We may correct this situation by arbitrarily increasing the center-point, a_0, of the autocorrelation. This may be thought of as an adjustment of the various components of **h** in such a way that a_0 is slightly increased while the remaining coefficients in the autocorrelation are unchanged. Such a process is often referred to as the introduction of some "white noise" into the system.

Cosine transforms

For z on the unit circle, the transform of the autocorrelation of **h** has the form

$$H(e^{i\theta})H(e^{-i\theta})$$

$$= a_0 + \sum_{n=1}^{M} a_n(e^{in\theta} + e^{-in\theta})$$

$$= \left| H(e^{i\theta}) \right|^2 = a_0 + 2\sum_{n=1}^{M} a_n \cos n\theta. \quad (17)$$

It is clear that this transform is never negative. Also, if the transform is zero for some value of θ, this will not be so when a_0 has been slightly increased.

We have found that replacement of a_0 by approximately 1.05 a_0 is effective in the seismic application to be discussed below. Some theoretical work remains to be done in this area.

DECONVOLUTION

Let **h** represent a portion of a seismic trace. The so-called deconvolution of the trace can be thought of as the application of two filters in sequence. One of these filters is the minimum-phase inverse of $H(z)$. The other is a convenient so-called "diagnostic wavelet" which we also take to be minimum-phase. Perhaps we might choose a band-pass filter as our "diagnostic wavelet." However, this choice would vary according to geographical area, seismic interpretation, etc.

In any case, let $\mathbf{w} = (w_n)_0^k$ represent the appropriate minimum-phase "diagnostic wavelet." Then, we can develop a single filter from **w** and $\mathbf{g}_N$ by convolution. This filter, $\mathbf{w} * \mathbf{g}_N$, would have the effect of applying **w** and $\mathbf{g}_N$ in sequence.

Z-transforms and Fourier transforms

As we have mentioned above, our filter can be developed using Z-transform or Fourier transform methods. Sakrison (1965) shows how one can approximate $\mathbf{w} * \mathbf{g}_N$ by development of the minimum-phase operator whose autocorrelation has the Z-transform

$$\frac{W(z)W^*(z)}{H(z)H^*(z)}. \quad (18)$$

Also, Fourier transform methods may be applied to the ratio of the power spectra of **w** and **h** to yield similar results.

Compatibility

We have successfully applied all three of these approaches to the deconvolution of seismic traces, and we have obtained comparable results.

However, the matrix approach reported in this paper seems to be somewhat more satisfactory than the other two methods for two reasons. First, the matrix approach is the only one of the three which involves an attempt to minimize the errors introduced as a result of the necessity of using a filter of finite length. Second, the computer calculations in the matrix approach are simpler and cheaper than those in the other two methods.

It should be emphasized that the practical geophysical advantage, ignoring computing costs, of the matrix approach over the other two methods is slight.

A SEISMIC EXAMPLE

Since our interest in these processes originated in the problem of reducing the distortion of seismic data due to near surface reverberatory layers, we include a seismic example. The data in this example is taken from three adjacent traces of a typical marine seismic record.

Conventions of data display

We note that the vertical scale on the attached plots of power spectra is linear rather than logarithmic. Also, the attached autocorrelations are plotted with a time-varying vertical scale which maintains a constant envelope to ease visual observation of the periodicities in the autocorrelation. Of course, these aspects of display are not used in the details of the computer calculations.

Discussion of the figures

Our three-trace example is typical of the data which we have processed by all three methods of filter design.

We show the original traces and the results of the three deconvolution processes in Figure 3.

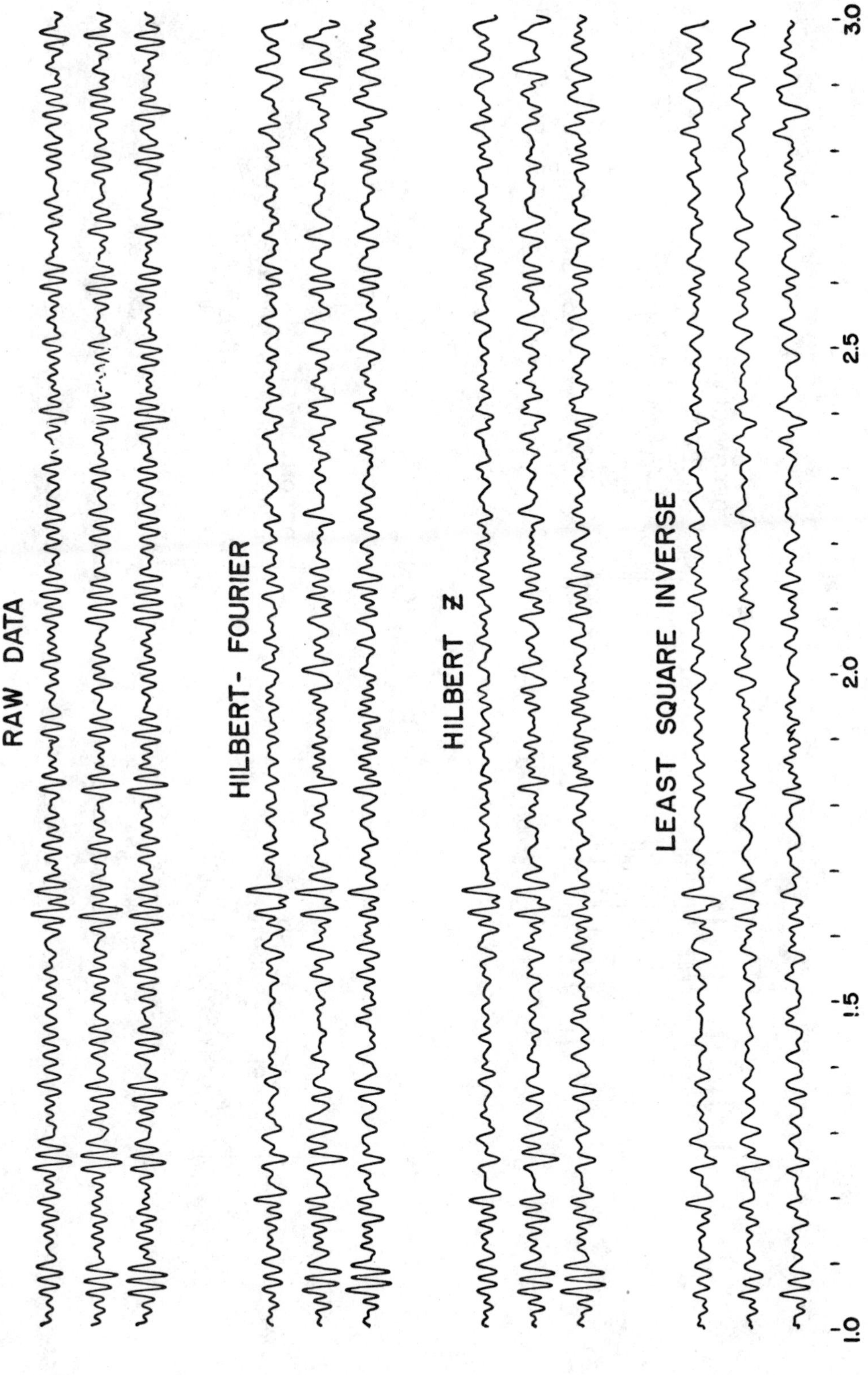

Fig. 1. Original traces together with the corresponding traces resulting from three deconvolution processes.

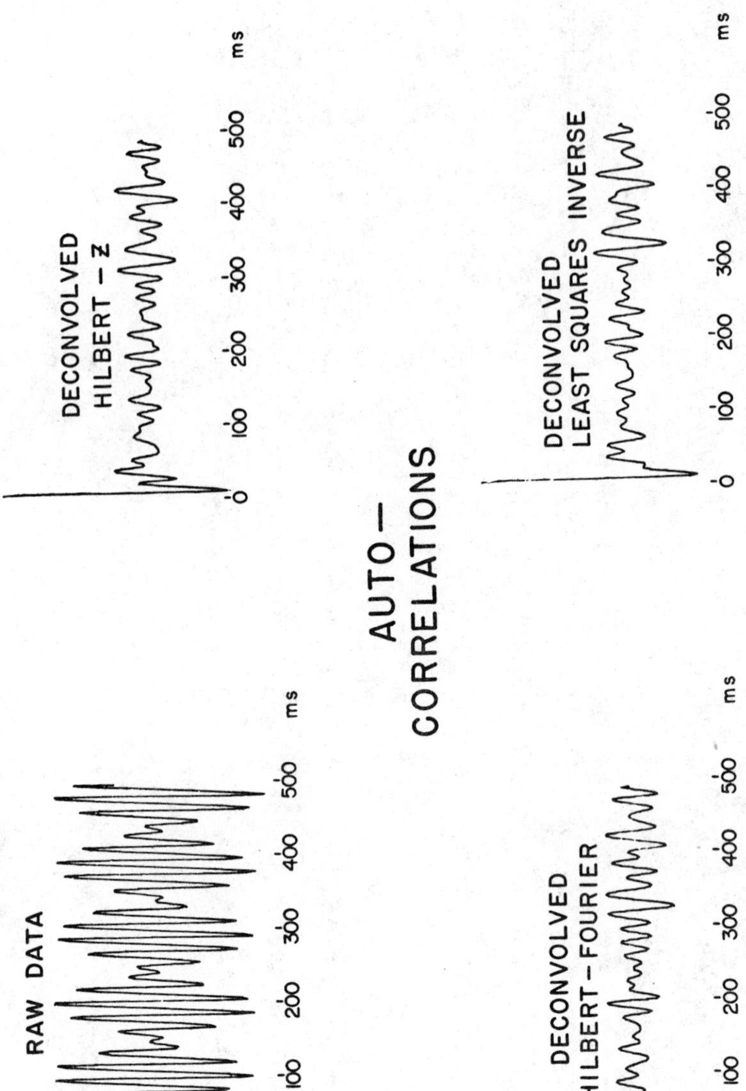

FIG. 2. Autocorrelations of the center traces from each of the four sets of traces in Figure 1.

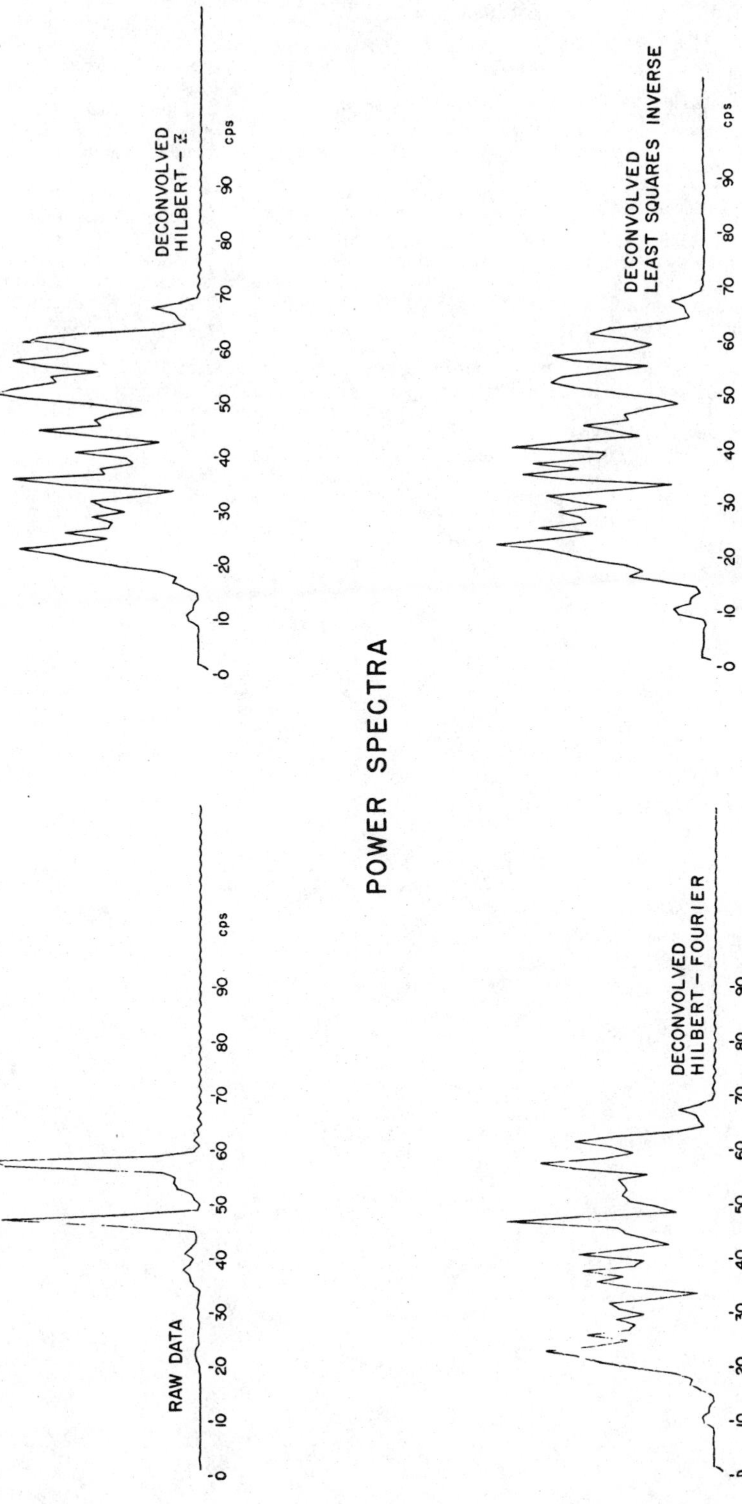

FIG. 3. Power spectra of the center traces from each of the four sets of traces in Figure 1.

We have selected the center trace of this three-trace example for a more detailed study. We show the autocorrelations of the center trace of the original data and the deconvolved data in Figure 2 and the corresponding power spectra in Figure 3.

REFERENCES

Claerbout, J. F., and Robinson, E. A., 1964, The error in least-squares inverse filtering: Geophysics, v. 29, p. 118–120.

Cline, R. E., 1965, Representations for the generalized inverse of sums of matrices: J. SIAM Numer. Anal., v. 2, p. 99–114.

Ford, Wayne T., 1966, Estimation of a minimum-phase operator from a portion of its amplitude spectrum: In preparation.

Grenander, Ulf and Szego, Gabor, 1958, Toeplitz forms and their applications: Berkeley, University of California.

Jury, E. I., 1958, Sampled-data control systems: New York, John Wiley & Sons.

——— 1964, Theory and application of the Z-transform method: New York, John Wiley & Sons.

Levinson, Norman, 1947, The Wiener RMS (root mean square) error criterion in filter design and prediction: Jour. Math & Physics, v. 25, p. 261–278.

Papoulis, A., 1962, The Fourier integral and its application: New York, McGraw-Hill Book Co., Inc.

——— 1965, Probability, random variables, and stochastic processes: New York, McGraw-Hill Book Co., Inc.

Rice, R. B., 1962, Inverse convolution filters: Geophysics, v. 27, p. 4–18.

Robinson, E. A., 1963, Structural properties of stationary stochastic processes with applications: in Time Series Analysis edited by M. Rosenblatt, New York, John Wiley & Sons, p. 170–196.

Sakrison, David J., 1965, Relation of the real and imaginary parts of the *Z*-transform of a realizable time function: In preparation.

Treitel, S. and Robinson, E. A., 1964, The stability of digital filters: IEEE Transactions on Geoscience Electronics, v. 2, p. 6–18.

Wiggins, Ralph A. and Robinson, E. A., 1965, Recursive solution to the multichannel filtering problem: Jour. Geophys. Res., v. 70, p. 1885–1891.

Reprinted from Geophysics v. 33, no. 2, p. 255-263

A STATISTICALLY OPTIMIZED DECONVOLUTION†

W. R. BURNS*

There are many innate errors involved in the process of deconvolution. Of interest here is an analysis of one of these. In particular, the errors resulting from the pseudo randomness of the seismic records are calculated. These errors are present even assuming that the many assumptions involved in the deconvolution process are correct. Such assumptions are in themselves dangerous. The errors calculated thus represent the most optimistic conditions.

A technique to calculate the useful amount of deconvolution is derived using the error information. The technique is applied, example results given, and the practical difficulties discussed. The technique is found to be of primary interest as a learning device and research tool. Its complexities at present limit its use in routine processing. It is found, however, to yield information which cannot otherwise be objectively obtained.

INTRODUCTION

The purpose of this paper is to determine the extent to which deconvolution can be usefully applied. Deconvolution involves applying a linear filter to a seismic record, the power response of the filter being designed to remove the smoothing effects caused by the original convolution process. If one analyzes the seismic record in terms of frequency components, the process is equivalent to increasing the amplitude of the weaker components. The process had been frequently described in the literature; Rice (1962), D'Hoeraene (1962). Physically, the question answered in this paper is "How weak must a component be for an attempted expansion of its amplitude to do more harm than good?" The present solution, in fact, goes further to indicate exactly how much the weak component amplitude should be increased. If one assumes a decrease in amplitude with frequency in the high frequency portion of the spectrum, the solution gives the added information of how high in frequency one should deconvolve.

The particular interest of this paper is a presentation of the mathematics necessary to answer the above questions. Although examples of results will be given, a discussion of the interpretation of information gained from the results is not of primary interest here.

PROBLEM ANALYSIS

If one were given the true power spectrum of the seismic record, there would be no problem. The limit on the deconvolution would be set by the limit of one's interest in high frequency content. However, this is not the case. The power spectrum must be obtained by a Fourier analysis of the autocorrelation of the record. The autocorrelation can only be estimated on the basis of a finite time average. An infinite time average would be required to obtain the true autocorrelation. The time limit is imposed by two factors: (1) the record is of finite length, and (2) the true power response changes as a function of time. As a result of the finite time average, the power spectrum estimate of the record is, in fact, a random variable. The deconvolution operator, obtained from this power spectrum estimate, is then also a random variable.

The present approach may be illustrated as follows. Assume the convolution process forming the seismic record described by the equation:

$$s(t) = r(t) * p(t)$$

where

- $s(t)$ denotes the seismic record as a function of time,
- $p(t)$ denotes the seismic pulse,
- $r(t)$ denotes the reflection coefficient series

Letting $P(\omega)$, $S(\omega)$, and $R(\omega)$ denote the power spectrum of the pulse, record, and reflection coefficient series respectively, one has:

$$S(\omega) = R(\omega)P(\omega)$$

† Manuscript received by the Editor 16 March 1967; revised manuscript received 7 July 1967.

* Gulf Research & Development Company, Pittsburgh, Pennsylvania. Now with the National Radio Astronomy Observatory, Charlottesville, Virginia.

In the present analysis, the power spectrum $R(\omega)$ is assumed a constant. This is the usual assumption that the reflection coefficient log has a "white" spectrum. Since the various spectra shall be assumed normalized, the spectrum $R(\omega)$ is taken as unity. One then has:

$$S(\omega) = P(\omega)$$

Let the power spectrum of the deconvolution operator be denoted by $I(\omega)$ and the spectrum of the deconvolution output be denoted by $O(\omega)$. The deconvolved power response is then given by:

$$O(\omega) = I(\omega)P(\omega)$$

If the output power response is desired to be that of the reflection coefficient series, it follows that:

$$I(\omega) = 1/P(\omega)$$

This is well known and follows a simple substitution of terms. In the present approach, a weighting function $C(\omega)$ is assigned to the process such that:

$$I(\omega) = C(\omega)/P(\omega)$$

$C(\omega)$ is, in fact, the desired solution. The criterion for obtaining $C(\omega)$ shall be a minimization of the square of a fixed and a statistical error. The fixed error is defined as the difference between the results one would like from deconvolution and the results with which one must be content. Hence:

Fixed error $(\omega) = R(\omega) - I(\omega)P(\omega)$.

The statistical error arises from the fact that the deconvolution operator is a random variable. Suppose one were to take several different sections of a seismic record, and for each section calculate an operator and deconvolve the entire record. A different deconvolved power spectrum would be obtained for each deconvolution. This is true even assuming perfect stationarity. Let $O(\omega)$ represent the average deconvolved power spectrum. Let $O^*(\omega)$ denote the output power spectrum for any one of the cases examined. The statistical error is defined as the root-mean square of the difference between $O(\omega)$ and $O^*(\omega)$, averaged over all possible record section samples.

Graphically, the optimization under discussion is shown in Figure 1. In the graphical representation, the pulse power spectrum $P(\omega)$ is assumed to have a maximum at $\omega = 0$. This was done only for clarity. In the problem solution, $P(\omega)$ is normalized to its peak value which may occur at any frequency.

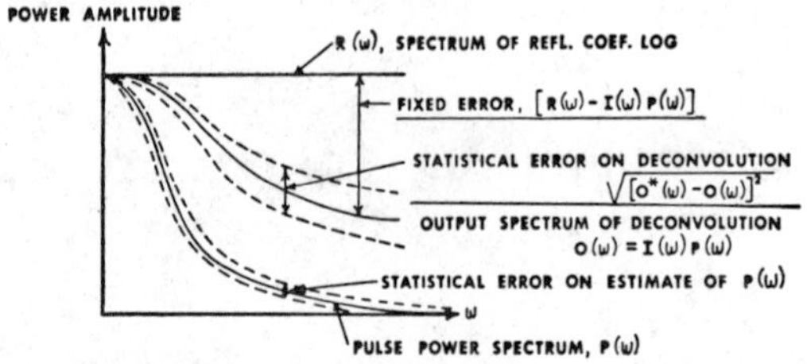

FIG. 1. Diagrammatic representation of the power spectrum and estimation errors before and after deconvolution.

PROBLEM SOLUTION

A systematic problem description is illustrated in block diagram form in Figure 2. The following description parallels the block diagram. The discussion describes only the use of the various steps and refers to the appendices for a detailed description.

The pulse power spectrum is first taken from a portion of the seismic record. The sample portion, often referred to as the window, is shown in Figure 2 to be of length T. The autocorrelation estimate is denoted by $\rho^*(\tau)$ and the true value by $\rho(\tau)$. The calculation of both $\rho^*(\tau)$ and $\rho(\tau)$ is shown in Appendix I and, in addition, a calculation of the covariance of the statistical error associated with $\rho^*(\tau)$. It is well known that the estimate $\rho^*(\tau)$ is biased; that is, an average of many estimates $\rho^*(\tau)$ does not yield the true autocorrelation $\rho(\tau)$. The bias is calculated in Appendix I and is assumed removed from the estimate so as to yield an unbiased estimate.

The pulse power spectrum estimate $P^*(\omega)$ is obtained by a Fourier transform of the autocorrelation estimate. The pulse spectrum estimate is not biased if the autocorrelation bias was corrected; that is, the average of a large number of estimates does yield the true spectrum $P(\omega)$. A weighting function $W(\tau)$ is introduced in the calculation of $P^*(\omega)$. This is done by necessity and not for convenience. The autocorrelation estimate is defined only for regions of τ less than T, T being the record window length. By introducing the weighting function, one can assume the autocorrelation estimate $\rho^*(\tau)$ known for all τ, and then let $W(\tau)$ be zero for those regions of τ for which $\rho^*(\tau)$ is now known. $W(\tau)$ in effect causes

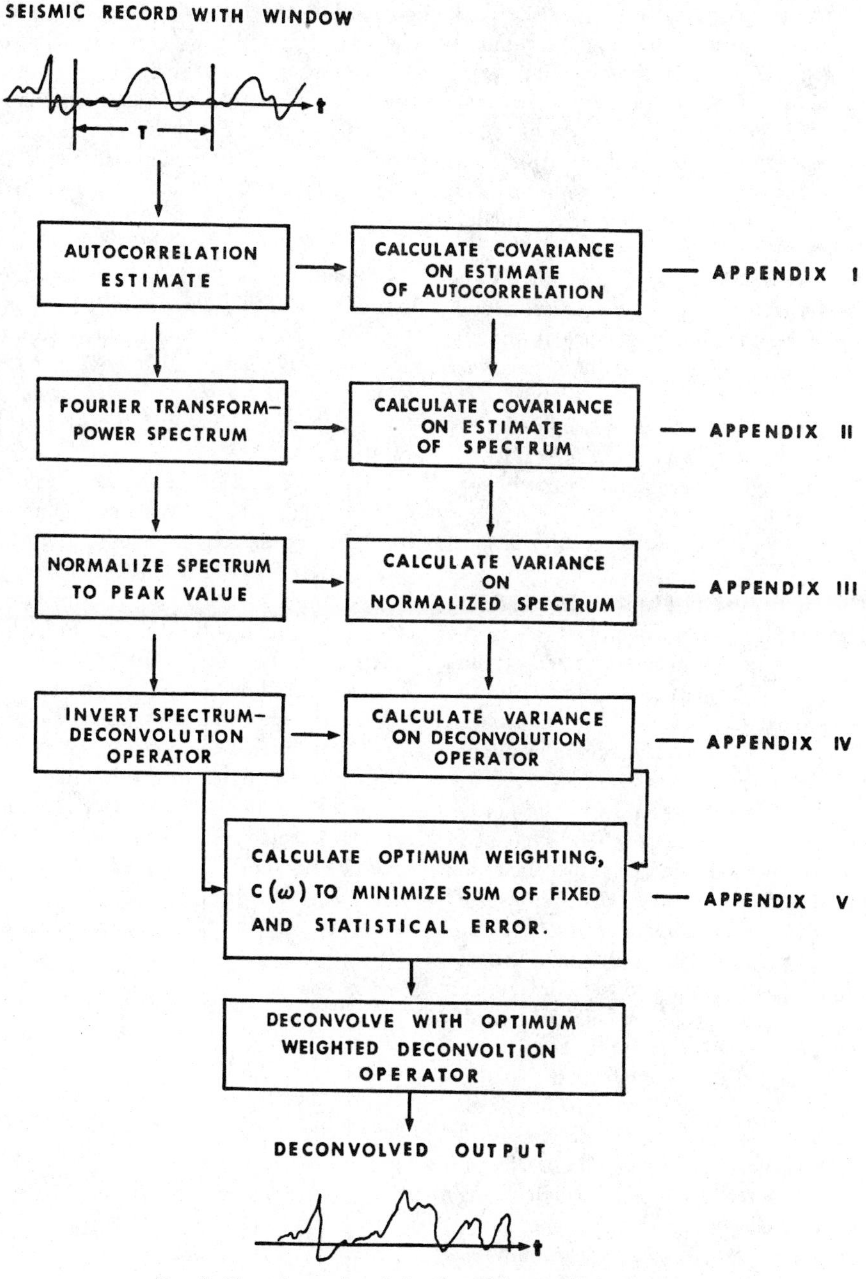

FIG. 2. Flow chart of optimization of deconvolution operator.

the phenomenon commonly referred to as the power spectrum estimate. It follows that one has some choice in the smoothing by letting $W(\tau)$ take on values other than just unity and zero. A detailed technical description of $P^*(\omega)$ is given in Appendix II. The covariance of the statistical error associated with $P^*(\omega)$ is also shown in the same appendix.

The next step, as shown in the block diagram, is the normalization of $P^*(\omega)$. This is done because one is interested only in the shape and not the absolute magnitude of the spectrum. The normalization is done with respect to the peak value and involves merely dividing the estimate $P^*(\omega)$ for each value of ω by the maximum value of $P^*(\omega)$ for all ω. The normalization, unfortunately, causes some difficulty in transferring the information concerning the statistical error from the unnormalized to the normalized estimate. The normalized estimate is denoted by $\hat{P}^*(\omega)$, the hat indicating $P^*(\omega)$ has been normalized. In fact, only the variance on $\hat{P}^*(\omega)$ is required to solve the problem under consideration. The normalization, however, requires that the covariance on the unnormalized estimate be known. That is, one must know how the error on $P^*(\omega)$ is related to the error on the value used for the normalization. A detailed description is given in Appendix III.

The estimate of the deconvolution or inverse operator is given by:

$$I^*(\omega) = C(\omega)/\hat{P}^*(\omega)$$

Here $C(\omega)$ is the assigned statistical weighting function previously described and $\hat{P}^*(\omega)$ is the normalized pulse power spectrum previously discussed. In effect, $C(\omega)$ tells one how much to deconvolve at a particular value of ω. The above definition is used only when the normalized pulse spectrum estimate is larger than its expected error. Otherwise, $I^*(\omega)$ is defined as zero. The mathematics of the formulation of $I^*(\omega)$ is given in Appendix IV as is also the information concerning its statistical variance. A note might be inserted concerning the variance transfer from the estimate $\hat{P}^*(\omega)$ to the estimate $I^*(\omega)$. In the transfer the variable $\hat{P}^*(\omega)$ was assumed to have Gaussian statistics. The discussion at the end of Appendix II lends support to such an assumption.

The calculation of $C(\omega)$ completes the solution and effects the desired optimization. The calculation is given in Appendix V. As previously described, $C(\omega)$ is chosen so as to minimize the sum of the square of the fixed and statistical error associated with the deconvolution. The sum of the squares as opposed to a linear sum was chosen primarily for mathematical convenience. If a linear sum is used, one must also insure that $C(\omega)$ be positive. A minimization of both a linear sum and weighted sums remains to be investigated. The errors were graphically shown in Figure 1. The solution of $C(\omega)$ is shown in Appendix V and is seen to have a rather simple form. Also, the solution varies in a way that would be physically expected. When there is no error as in the case of infinite window length, a complete deconvolution is indicated, $C(\omega)=1$ for all ω. As the window length is decreased causing a greater statistical error, $C(\omega)$ decreases. Similarly, as pulse energy at a particular value of ω decreases, so does $C(\omega)$.

EXAMPLES AND DISCUSSION

This paper was written to describe the mathematics involved in the optimized deconvolution and not to discuss interpretation of results. A few examples, however, shall be given and, in addition, a few problems indicated which are encountered in carrying out the proposed deconvolution.

Figures 3 and 4 give two examples of the proposed deconvolution. The figures do not show deconvolved traces but instead give the pulse spectrum and optimized deconvolved spectrum. Both examples were obtained using a procedure close to that described. The procedure used differed slightly in that: (1) numerical approximations were used instead of analytic functions, and (2) the variance transformation from $P^*(\omega)$ to $I^*(\omega)$ was further simplified. The results, however, correctly represent the described procedure.

The two examples illustrated represent pulse spectra selected arbitrarily from a profile or several field tapes. The profile was deconvolved both using the present and the more classical technique. In the latter, a post-deconvolution

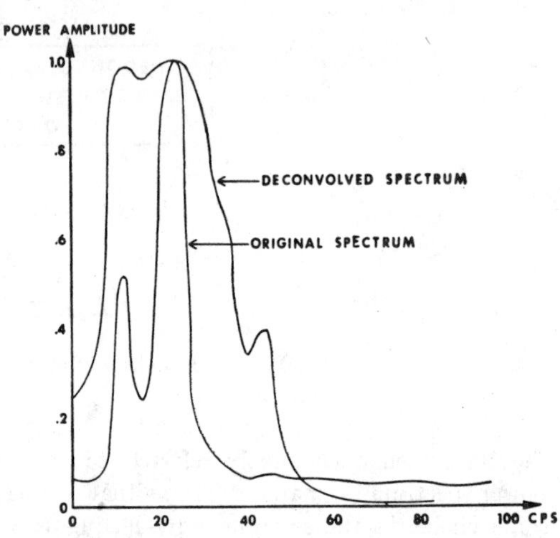

FIG. 3. Power spectrum before and after optimized deconvolution of seismic record from Area *A*.

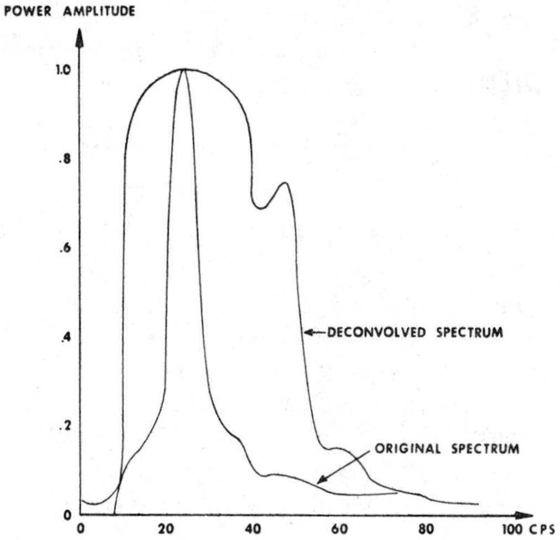

FIG. 4. Power spectrum before and after optimized deconvolution of seismic record from Area *B*.

filter is selected by eye. A visual comparison was then made between the two. To evaluate by such a comparison is difficult because neither member may be assumed "correct." One can only examine the difference between the two; the difference cannot meaningfully be attributed to "error" on one or the other. Synthetic seismograms were also made for the area represented by the profile and the respective deconvolutions compared to these. Evaluation by this type of comparison is also difficult. In the first place, the idea that the records represent a convolution process is only an assumption. As a result of this and many other assumptions involved, similarity between deconvolutions and synthetic seismograms is seldom what one would like. It would, therefore, be unfair to use such similarity to show whether improvement had been made. Secondly, the comparison is even further complicated by the fact that two different pulses are required in the construction of the synthetic seismograms. This occurs because the classical and proposed deconvolutions may have substantially different spectral content. The synthetic seismograms, used for comparison in each case, then must also contain substantially different spectral content.

PRACTICAL DIFFICULTIES

In applying the procedure to an arbitrary seismic record, $P(\omega)$ must be known. That is, the true pulse power spectrum $P(\omega)$ is necessary to estimate the error on the estimate $P^*(\omega)$. In a practical application, one can only assume the estimate $P^*(\omega)$ equal to $P(\omega)$ for the error calculations. The assumption is valid since the calculated errors are assumed good only to the first order.

In the described analysis, the errors under study were assumed to be caused by the inherent statistical fluctuation in the reflection coefficient log. If desired, an additional noise term may be added. The effect is to increase the variance on the estimate $P^*(\omega)$. Such an addition would seem extremely reasonable for the case where the calculated estimate $P^*(\omega)$ does not tend to zero as ω increases.

Additional difficulties occur when a whole tape rather than an individual seismic trace is deconvolved. One may choose to calculate the weighting function $C(\omega)$ from only a few traces and to use the calculated inverse operators for the whole tape. One may, for example, calculate the inverse operator on only every third trace. In such a case, the power spectrum of the trace used for calculation is assumed to represent the spectrum of each trace deconvolved with the calculated operator. The present technique causes this assumption to be much more critical than in a normal deconvolution, because the optimum inverse operator may have holes in the spectral domain. One of these holes may fall in an interval over which the true spectrum has significant content. The hole may then cause the oscillations usually associated with a heavy discontinuity in the frequency domain.

Problems may also occur if one chooses to calculate a separate optimum operator for each trace. This is especially true if a minimum-phase function is used for the deconvolution. In such a case, the phase-frequency function of the deconvolution operator is set by the power spectrum of the operator. Since this spectrum is the result of the present optimization, the optimization, in fact, sets the phase function of the deconvolution. Frequently, the power spectrum of traces forming a record differ strongly from trace-to-trace in the high frequency portion. This fact, along with the phase-amplitude relation discussed, may combine to produce an undesirable result. What is commonly referred to as "high frequency correlation" on the deconvolution may be destroyed. The same tendency exists, but to a lesser extent, in classical deconvolution.

CONCLUSIONS

The errors resulting from the innate pseudo randomness of a seismic record have been calculated. A technique to calculate the extent to which deconvolution may be usefully applied has been described. Two examples have been given and an evaluation of their results briefly discussed. Practical difficulties have been indicated and discussed.

The difficulties indicated should not be so construed as to strongly interfere with the use of the technique. It is the author's belief, however, that its principal value at present lies as a research tool, as opposed to use in routine data processing. The technique as given is too complex for use in routine processing. Unless practical simplifications are obtained, one must be content with using it as a learning device. The value of the device lies in the fact that at present the resulting information cannot be objectively obtained another way. Comparative schemes are largely subjective in that one chooses the inverse weighting function "by eye." This is done by selecting a post-deconvolution filter. Such an approach is both practical and useful but has the inherent disadvantage of subjectivity.

APPENDIX I

AUTOCORRELATION ESTIMATE AND ERROR

It is assumed that $s(t)$ is stationary and ergodic. Its autocorrelation is defined as the infinite time average:[1]

$$\rho(\tau) = \overline{s(t)s(t+\tau)}^{\,\infty},$$

Expressing $\rho(\tau)$ in integral form, we obtain

$$\rho(\tau) = \lim_{T\to\infty} \frac{1}{T}\int_0^T s(t)s(t+\tau)dt$$

In forming the autocorrelation estimate, the total information available is the section of $s(t)$ taken from the seismic window. The estimate is defined as the finite time average:

$$\rho^*(\tau) = \overline{s(t)s(t+\tau)}^{\,T},$$

[1] Some authors use a t and Σ to represent the time and ensemble averages, respectively. In the present case, one would write

$$\overline{s(t)s(t+\tau)}^{\,T}t,\quad \overline{s(t)s(t+\tau)}^{\,\infty}t,\quad \text{and}\quad \overline{\rho^*(\tau)}\,\Sigma.$$

Here the window length T is used to indicate that the average is formed on a section of $s(t)$ of length T. The estimate in integral form is

$$\rho^*(\tau) = \frac{1}{T}\int_0^{T-\tau} s(t)s(t+\tau)dt$$

$\overline{\rho^*(\tau)}$ = infinite ensemble average of $\rho^*(\tau)$; Lee (1960). That is, $\overline{\rho^*(\tau)}$ represents the average of an infinite number of calculations of $\rho^*(\tau)$. Note that $\rho(\tau)$ is a constant in the ensemble sense, i.e., each calculation gives the same result. Because $\rho^*(\tau)$ involves only a finite time average, it is not a constant in the ensemble sense; i.e., each calculation of $\rho^*(\tau)$ gives a new result

$$\overline{\rho^*(\tau)} = \operatorname*{Lim}_{N\to\infty} \frac{1}{N}\sum_{K=1}^{N} \rho_K{}^*(\tau)$$

where $\rho_K{}^*(\tau)$ indicates the Kth calculation of

$$\overline{s(t)s(t+\tau)}^{\,T}$$

Covariance on

$$\begin{aligned}\rho^*(\tau) &= \sigma^2(\tau_1, \tau_2)\\ &= \overline{[\rho^*(\tau_1) - \overline{\rho^*(\tau_1)}][\rho^*(\tau_2) - \overline{\rho^*(\tau_2)}]}\end{aligned}$$

or

$$\sigma^2(\tau_1, \tau_2) = \overline{\rho^*(\tau_1)\rho^*(\tau_2)} - \overline{\rho^*(\tau_1)}\;\overline{\rho^*(\tau_2)}$$

$\overline{\rho^*(\tau)}$ represents an infinite ensemble average of $\rho^*(\tau)$, the autocorrelation estimate. The relation between this ensemble average and the true autocorrelation is well known

$$\rho(\tau) = \frac{T}{T-|\tau|}\,\rho^*(\tau)$$

The estimate $\rho^*(\tau)$ is assumed corrected, if necessary, before a transform is taken to obtain the power spectrum estimate. The covariance is also assumed altered for the correction

$$\rho^*(\tau)\text{ corrected} = \frac{T}{T-|\tau|}\,\rho^*(\tau)$$

One then has an unbiased estimate in that, $\overline{\rho^*(\tau)}$ corrected $=\rho(\tau)$ true autocorrelation.

Assuming ergodicity, and interchanging time and ensemble averages, one has:

$$\sigma^2(\tau_1, \tau_2) = \frac{1}{T^2}\int_{\mu=0}^{+T}\int_{v=0}^{+T} \overline{s(\mu)s(\mu+\tau_1)s(v)s(v+\tau_2)}\, d\mu dv - \rho(\tau_1)\rho(\tau_2)$$

Letting

$$M(a, b, c) = \overline{s(t)s(t+a)s(t+b)s(t+c)}$$

and

$$\alpha = v - u, \qquad \beta = u,$$

one has:

$$\overline{s(u)s(u+\tau_1)s(v)s(v+\tau_2)} = \overline{s(u)s(u+\tau_1)s(u+\alpha)s(u+\alpha+\tau_2)} = M(\tau_1, \alpha, \alpha+\tau_2)$$

$$\sigma^2(\tau_1, \tau_2) = \frac{1}{T^2}\int_0^T\int_0^{T-\alpha} M(\tau_1, \alpha, \alpha+\tau_2)d\beta d\alpha + \frac{1}{T^2}\int_{-T}^0\int_{-\alpha}^T M(\tau_1, \alpha, \alpha+\tau_2)d\beta d\alpha$$
$$- \rho(\tau_1)\rho(\tau_2)$$

$$\sigma^2(\tau_1, \tau_2) = \frac{1}{T^2}\int_{\alpha=-T}^{+T}\int_{\beta=|\alpha|}^{+T} M(\tau_1, \alpha, \alpha+\tau_2)d\beta d\alpha - \rho(\tau_1)\rho(\tau_2)$$

Integrating out β and using the assumption $M(\tau_1, \alpha, \alpha+\tau_2)$ is nonzero only for α/T small, one has:

$$\sigma^2(\tau_1, \tau_2) = \frac{1}{T}\int_{-T}^{+T} [M(\tau_1, \alpha, \alpha+\tau_2) - \rho(\tau_1)\rho(\tau_2)]d\alpha$$

For similar treatment, see Bendat (1958). Bendat considers the special case $\tau_1 = \tau_2$. In such case $M(\tau_1, \alpha, \alpha+\tau_2)$ is even in α. This is not true in the present case and therefore his results do not directly apply.

$M(a,b,c)$ is called the 4th order correlation function of $s(t)$. $s(t)$ was assumed to be the result of an impulse series $r(t)$ convolved with a pulse $p(t)$. $M(a, b, c)$ may be obtained if $r(t)$ (impulse series or reflection coefficient log) is assumed represented by shot noise (Rice, 1944, 1945). In the assumed representation, the impulses occur randomly in time and have an unspecified amplitude distribution. This is the representation used by Robinson (1957). Under this assumption, the fourth order correlation function may be written as follows (Rice, 1944, 1945; Leneman, 1965);

$$M(\tau_1, \alpha, \alpha+\tau_2) = \mu^2(\overline{r^2})^2 [\rho_P(\alpha+\tau_1-\tau_2)\,\rho_P(\alpha) + \rho_P(\alpha+\tau_1)\rho_P(\alpha-\tau_2)]$$
$$+ \rho(\tau_1)\rho(\tau_2) + \mu\,\overline{r^4}\int_{n=0}^{\infty} P(n)P(n+\tau_1)P(n+\alpha)P(n+\alpha+\tau_2)dn$$
$$\tau_1, \tau_2 \geq 0$$

where $\rho_P(\tau)$ denotes the pulse autocorrelation, μ is the density of impulses forming $r(t)$, and $\overline{r^2}$ and $\overline{r^4}$ are the second and fourth moments of the series $r(t)$. The first and third order moments of the series have been assumed zero. Under the same shot noise representation, the record autocorrelation $\rho(\tau)$ is related to the pulse autocorrelation by $\rho(\tau) = \mu\,\overline{r^2}\,\rho_P(\tau)$. A substitution of M into the previous integral equation would yield the desired error estimate. The expressions, however, shall be left in this form as they shall shortly be expressed in the Fourier domain.

APPENDIX II

PULSE POWER SPECTRUM ESTIMATE AND ERROR

The pulse power spectrum estimate is the Fourier transform of the autocorrelation estimate. A weighting function $W(\tau)$ is included in the transform

$$P^*(\omega) = \frac{1}{2\pi}\int_{-\infty}^{+\infty} W(\tau)\rho^*(\tau)\cos(\omega\tau)d\tau$$

$\rho^*(\tau)$ is defined only for $\tau \leq T$, T being the averaging time in the estimate of $\rho^*(\tau)$. By introducing the weighting function, one can assume $\rho^*(\tau)$ known for all τ and let $W(\tau)$ force the truncation. The covariance on the power spectrum estimate is defined by:

$$\sigma^2(\omega_1, \omega_2) = \overline{[P^*(\omega_1) - P(\omega_1)][P^*(\omega_2) - P(\omega_2)]}$$

Note: Here $\rho^*(\tau)$ is assumed corrected for its bias so

$$\overline{P^*(\omega)} = P(\omega)$$

$$\sigma^2(\omega_1, \omega_2) = \left(\frac{1}{2\pi}\right)^2 \int_{\tau_1}\int_{\tau_2} \sigma^2(\tau_1, \tau_2) W(\tau_1) W(\tau_2) \cdot e^{-j(\omega_1\tau_1 - \omega_2\tau_2)} d\tau_1 d\tau_2$$

One has from Appendix I:

$$\sigma^2(\tau_1, \tau_2) = \frac{\mu^2 (\overline{r^2})^2}{T} \int_{-T}^{+T} [\rho_p(\alpha + \tau_1 - \tau_2)\rho_p(\alpha) + \rho_p(\alpha + \tau_1)\rho_p(\alpha - \tau_2)] d\alpha + \frac{\mu \overline{r^4}}{T} \int_{-T}^{+T}\int_{t=0}^{\infty} P(n) P(n + \tau_1) P(n + \alpha) \cdot P(n + \alpha + \tau_2) dn d\alpha$$

The contribution to the covariance in the frequency domain from the first integral in the above expression is by Weinreb (1963).

$$\frac{\mu^2(\overline{r^2})^2}{2\pi T} \int_{-\infty}^{+\infty} P^2(\omega) W(\omega + \omega_1) \cdot [W(\omega + \omega_2) + W(\omega - \omega_2)] d\omega$$

Note: $W(\omega)$ denotes the Fourier transform of $W(\tau)$.

The contribution from the second integral in the above expression may be obtained as follows: For pulse lengths, nonzero portion of $P(t)$, much less than T, the limits $-T$, $+T$, may be extended to $-\infty$, $+\infty$. An integration followed by a double transform then yields:

$$\frac{\mu \overline{r^4}}{T} \int_\omega P(\omega) W(\omega + \omega_1) d\omega \int_\omega P(\omega) W(\omega + \omega_2) d\omega$$

Combining the terms yields the covariance in the frequency domain:

$$\sigma^2(\omega_1, \omega_2) = \frac{\mu^2(\overline{r^2})^2}{2\pi T} \int_{-\infty}^{+\infty} P^2(\omega) W(\omega + \omega_1) [W(\omega + \omega_2) + W(\omega - \omega_2)] d\omega + \frac{\mu \overline{r^4}}{T} \int_\omega P(\omega) W(\omega + \omega_1) d\omega \int_\omega P(\omega) W(\omega + \omega_2) d\omega$$

In the above expression, the first term represents the contribution to the covariance resulting from the Gaussian part of the reflection coefficient record. That is, if the reflection coefficient amplitude statistics are described by a Gaussian and a non-Gaussian part, the first term in the above expression follows from the Gaussian part. $(1/\mu)$ represents the average time interval between impulses in $r(t)$. If the pulse length of $p(t)$ is much larger than the pulse interval $(1/\mu)$ of $r(t)$, the first term in the above expression dominates. In other words, the statistics of $s(t)$ become Gaussian. The second term in the above expression may, in such a case, be neglected.

APPENDIX III

NORMALIZATION OF POWER SPECTRUM ESTIMATE AND VARIANCE OF ERROR

The normalization of the power spectrum is done by dividing each point of the spectrum by the maximum value. Letting $\hat{P}^*(\omega)$ represent the normalized spectral estimate, one has:

$$\hat{P}^*(\omega) = P^*(\omega)/P^*(\omega)_{max}$$

The variance on the estimate $\hat{P}^*(\omega)$ is defined by:

$$\hat{\sigma}^2(\omega) = \overline{[\hat{P}^*(\omega) - \overline{\hat{P}^*(\omega)}]^2}$$

If there exists a good signal-to-noise ratio on the estimate $P^*(\omega)_{max}$, $\hat{\sigma}^2(\omega)$ can be approximated in terms of the covariance of the unnormalized estimate, Weinreb (1963).

$$\hat{\sigma}^2(\omega) = \frac{1}{[P^*(\omega)_{max}]^2} [\sigma^2(\omega, \omega) - 2\gamma\sigma^2(\omega, \omega_0) + \gamma^2\sigma^2(\omega_0, \omega_0)]$$

$$\gamma = P^*(\omega)/P^*(\omega)_{max}$$

Here ω_0 denotes the value ω at which $P^*(\omega)$ is maximum.

APPENDIX IV

INVERSE OPERATOR I (W) AND ERROR CALCULATION

Define the inverse operator estimate by

$$I^*(\omega) = C(\omega)/P^*(\omega) \quad \text{for } P^*(\omega) \geq \sqrt{\hat{\sigma}^2(\omega)}$$

$$I^*(\omega) = 0 \quad \text{for } P^*(\omega) < \sqrt{\hat{\sigma}^2(\omega)}$$

Let $\sigma_I^2(\omega)$ denote the variance on the estimate $I^*(\omega)$ and $\Delta_I(\omega)$ denote the bias.

$$\Delta_I(\omega) = \overline{[I^*(\omega) - I(\omega)]}$$

$$\sigma_I^2(\omega) = \overline{[I^*(\omega) - \overline{I^*(\omega)}]^2}$$

where $I^*(\omega)$ is defined as above and $I(\omega) = 1/P(\omega)$. In calculating both $\Delta_I(\omega)$ and $\sigma_I^2(\omega)$, Gaussian statistics are assumed on the estimate $\hat{P}^*(\omega)$. The previous normalization makes it difficult to give a rigorous argument for the assumption as Gaussian statistics do not necessarily transfer through a nonlinear process. The assumption is only critical for ω at which $\hat{P}^*(\omega)$ has a poor signal-to-noise ratio. For these values of ω, $P^*(\omega)_{\max}$ appears to be constant, i.e., has low variance compared to $P^*(\omega)$. In such a case, the Gaussianness of $\hat{P}^*(\omega)$ follows from that of $P^*(\omega)$ which, in turn, follows from the closing argument of Appendix II.

For each value of ω, letting $X = P^*(\omega)$, $X_0 = P(\omega)$, and $\sigma = \sqrt{\sigma^2(\omega)}$ one has:

$$\overline{I^*(\omega)} = \frac{C(\omega)}{\sqrt{2\pi}\sigma} \int_\sigma^\infty \frac{1}{X} \cdot \exp\left\{-\frac{(X - X_0)^2}{2\sigma^2}\right\} dx$$

$$\overline{[I^*(\omega)]^2} = \frac{C^2(\omega)}{\sqrt{2\pi}\sigma} \int_\sigma^\infty \frac{1}{X^2} \cdot \exp\left\{-\frac{(X - X_0)^2}{2\sigma^2}\right\} dx$$

The above integrals cannot be analytically integrated. It is not clear whether a solution can be easily obtained with the aid of tables. A numerical solution is recommended. Such a solution is rapid because of the strong convergence. Note that $C(\omega)$ remains an unknown during the calculation. After the above calculation, the estimate $I^*(\omega)$ should be redefined by $I^*(\omega)_{\text{corrected}} = I^*(\omega) - \Delta I(\omega)$, thus, removing the bias. $\sigma_I^2(\omega)$ shall be written as $\sigma_I^2(\omega) = C^2(\omega) f(\omega)$, where $f(\omega)$ is used to denote the solution to the variance integral.

It should be noted that the solution indicated in the present appendix was not, in fact, used in the case of the discussed examples. Instead, the variance calculation on $I^*(\omega)$ was divided into several different regions of X_0/σ. Different approximations were then made in each region. The segmenting did not lead to the simplifications one would hope. If segmenting is to be used, it should be done instead on the integrals given in this appendix.

APPENDIX V

SOLUTION FOR OPTIMUM WEIGHTING FUNCTION

The weighting function $c(\omega)$ is obtained by choosing a compromise between the fixed and statistical errors. In present case the sum of the mean squares is minimized.

(Fixed error)$^2 = [R(\omega) - I(\omega)P(\omega)]^2$
(Statistical error)$^2 = \sigma_I^2(\omega)$
ϵ = (Fixed error)^2 + (Statistical error)^2
$= [R(\omega) - I(\omega)P(\omega)]^2 + \sigma_I^2(\omega)$
Noting the $R(\omega)$ = unity, $I(\omega) = C(\omega)/P(\omega)$, and $\sigma_I^2(\omega) = C^2(\omega) f(\omega)$

where $f(\omega)$ is that solution of the integral for the variance on $I^*(\omega)$, Appendix IV, one has:

$$\epsilon = (1 - C(\omega))^2 + C^2(\omega) f(\omega)$$

Obtaining $C(\omega)$ to minimize the total error yields

$$C(\omega) = \frac{1}{1 + f(\omega)}$$

REFERENCES

Bendat, J. S., 1958, Principles and applications of random noise theory: New York, John Wiley & Sons, p. 260.

D'Hoeraene, J., 1962, Deconvolution de traces réeles: Geophys. Prosp., v. 10, p. 68–83; presented at the 21st Meeting of the European Association of Exploration Geophysicists in Trieste, 13–15 Dec. 1961.

Lee, Y. W., 1960, Statistical theory of communication: New York, John Wiley & Sons, Inc.

Leneman, O. A. Z., 1965, On some new results in shot noise: Proc. of I.E.E.E., December.

Rice, R. B., 1962, Inverse convolution filters: Geophysics, v. 27, p. 4–18.

Rice, S. O., 1944, 1945, Mathematical analysis of random noise: Bell System Technical Journal; Pt. I, v. 23, Jul '44, p. 282–332; Pt. II, v. 24, Jan. '45, p. 46–108.

Robinson, E. A., 1957, Predictive decomposition of seismic traces: Geophysics, v. 22, p. 767.

Weinreb, S., 1963, A digital spectral analysis technique and its application to radio astronomy: M.I.T. Tech. Report, no. 412.

Reprinted from Geophysics v. 33, no. 6, p. 945-949

USE OF MONTE CARLO TECHNIQUES IN OPTIMUM DESIGN OF THE DECONVOLUTION PROCESS†

M. R. FOSTER,* R. L. SENGBUSH,* AND R. J. WATSON*

The deconvolution process is widely used to enhance seismic data by suppressing distortions of the shot pulse caused by such things as reverberations and ghosts. The process consists of estimating the correlation function from the data, determining the inverse filter using the Levinson algorithm, and applying the inverse filter to the data.

This paper is concerned with the estimation problem. Certain conclusions about the estimation problem are suggested by the theory of power spectra developed by Tukey and others. By means of a Monte Carlo simulation of the deconvolution process, we have tested these conclusions:

(1) Severely distorted data should be prewhitened.
(2) Truncators (lag windows) with the same number of degrees of freedom yield the same error.
(3) There is an optimum number of degrees of freedom for a fixed data window.
(4) Due to time variance in the data, there is an optimum length of data window.

Monte Carlo simulation can be used to estimate the optimum values (3) and (4) and so improve the performance of the deconvolution process.

The history of the deconvolution method in seismic exploration has been reviewed by Enders Robinson (1966). We will here treat an aspect of the deconvolution process which was not discussed in Robinson's paper: the problem of estimating the required correlation function.

The deconvolution process is a highly nonlinear one, and a direct mathematical analysis of the effect of errors in the estimation of the correlation function is difficult. For this reason, we have made Monte Carlo experiments to study the effect of the design parameters in the estimation procedure. In these Monte Carlo experiments we assume that the data-generation process can be described as shown in Figure 1. We generate a reflectivity function by means of a reflectivity ensemble which is Gaussian and white. The reflectivity function is fed through a filter B which simulates the seismic wavelet, and then through a second filter G which simulates the distortion. Various types of distortion can be simulated in this way, including ghosts and reverberations. The output of the distortion filter is mixed with white Gaussian noise generated by noise ensemble to produce the simulated data $X(t)$. The data $X(t)$ is then fed into the deconvolution operation which can be described by means of the diagram in Figure 2.

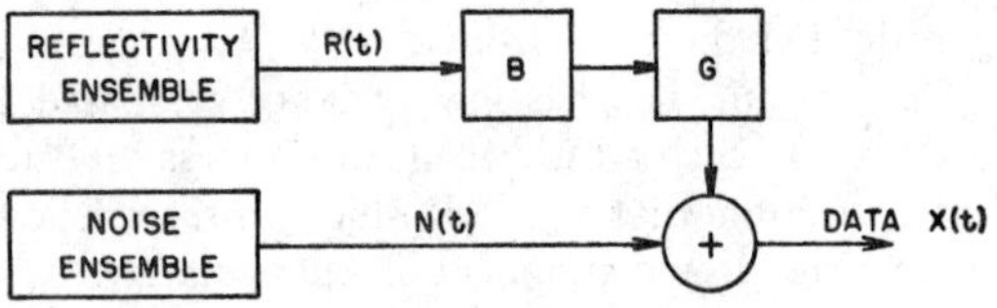

FIG. 1. Mathematical model of data-generation process.

The first step in the deconvolution operation is to estimate the correlation function denoted by $\hat{\rho}$, which is then fed into the well-known Levinson algorithm to compute the coefficients of the inverse filter (Levinson, 1947). The inverse filter is then applied to the raw data to produce an estimate of the reflectivity function up to an unknown amplitude factor denoted by α in this diagram. The estimated reflectivity function is then fed through a filter denoted by D to produce the synthetic seismogram from which the distortion has hopefully been removed.

Our concern in this paper is with the box labeled "Estimation of the Correlation Function." We will suppose that the data has been sampled and that the samples are $X_1, X_2, \cdots, X_N$. (We leave aside here the problem of choosing an

† Presented at the 36th Annual International SEG Meeting in Houston, Texas November 9, 1966. Manuscript received by the Editor March 1, 1968; revised manuscript received July 23, 1968.

* Mobil Research and Development Corporation, Dallas, Texas.

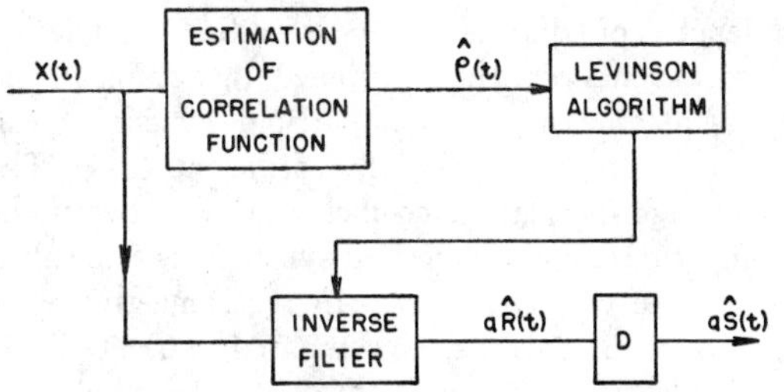

FIG. 2. Block diagram of deconvolution procedure.

appropriate sampling interval since this is generally well understood.) The number N of samples is commonly called the length of the data window. The formula for the estimation of the correlation function is then:

$$\hat{\rho}(k) = W_M(k)\frac{1}{N}\sum_{j=1}^{N-k} X_j X_{j+k}. \qquad (1)$$

The weights denoted $W_M(k)$ are referred to as the lag window or truncator. They are required to have the properties (Amos and Koopmans, 1963):

$$\begin{aligned} W_M(k) &= W_M(-k), \\ 0 \le W_M(k) &\le W_M(0) = 1, \qquad (2) \\ W_M(k) &= 0 \quad \text{for } |k| > M. \end{aligned}$$

A number of such truncators have been proposed by statisticians. Some of them are shown in Table 1.

As can be seen from equation (1), the estimation procedure is completely specified when the lag window $W_M(k)$ and the length N of the data window are given. Thus the role of the designer of the estimation procedure is to furnish these quantities.

Results from the theory of power spectral analysis

In the statistical literature, the estimation problem has been mainly treated in connection with power spectral analysis. It is of interest to review the conclusions from this work.

The power spectrum is simply estimated as the Fourier transform of the estimated correlation function:

$$\hat{f}(\lambda) = \frac{1}{2\pi}\sum_{k=-M}^{M}\hat{\rho}(k)e^{-ik\lambda}. \qquad (3)$$

From formulas due originally to Tukey (Blackman and Tukey, 1958), it is possible to show that the expected mean square error is given approximately by the sum of two terms:

$$E|\hat{f}(\lambda) - f(\lambda)|^2 = \frac{f(\lambda)^2}{n} + \frac{f''(\lambda)^2\pi^4 n^4}{36N^4}. \qquad (4)$$

(For a derivation of equation (4) see the Appendix.) Here $f(\lambda)$ is the true power spectrum and n is the number of "degrees of freedom" associated with the lag window. In Table 1, n is tabulated for the common lag windows.

It will be observed that the lag window enters into equation (4) only through the number of degrees of freedom n. The first term decreases with n and the second increases with n. Thus as a function of n, the error should have a single minimum. Finally the second term decreases with N while the first is independent of N.

Formula (4) is an approximation which is valid only under the assumption that the power spectrum is sufficiently smooth. As Blackman and Tukey (1958) have emphasized, this assumption can and should be made to hold by "prewhitening" the data. In this process one passes the data through the prewhitening filter to reduce peaks and troughs in the power spectrum.

We conclude from spectral analysis:

(1) Lag windows with the same number of degrees of freedom yield the same error.
(2) The error has a single minimum as a

Table 1. Degrees of freedom for common lag windows from Amos and Koopmans (1963)

Lag window	$W_M(k)$	n
Rectangular	$1, \quad \lvert k\rvert \le M$	$0.5\ N/M$
Triangular	$1 - \frac{\lvert k\rvert}{M}, \quad \lvert k\rvert \le M$	$1.5\ N/M$
Hamming	$0.54+0.46\cos\frac{\pi k}{M}, \quad \lvert k\rvert \le M$	$1.25\ N/M$
Hanning	$0.50+0.50\cos\frac{\pi k}{M}, \quad \lvert k\rvert \le M$	$1.33\ N/M$
Parzen	$\begin{cases} 1-6\left\lvert\frac{k}{M}\right\rvert^2+6\left\lvert\frac{k}{M}\right\rvert^3, & \lvert k\rvert \le \frac{M}{2} \\ 2\left(1-\left\lvert\frac{k}{M}\right\rvert\right)^3, & \frac{M}{2} \le \lvert k\rvert \le M \end{cases}$	$1.8\ N/M$
Daniell	$\frac{\sin kh}{kh}, \quad 0 \le \lvert k\rvert \le N$	$\frac{Nh}{\pi}$

function of the number of degrees of freedom.

(3) For a fixed number of degrees of freedom, the error decreases with the length of the data window N.

The above conclusions can only be expected to hold if the data has been properly prewhitened.

Monte Carlo experiments

In order to determine whether the conclusions from power spectral analysis carry over to deconvolution, we carried out the Monte Carlo simulation diagramed in Figure 3. The reflectivity function $R(t)$ is generated by a white noise ensemble as previously described and passed through the data-generation process. The data-generation process includes the effect of the seismic wavelet, the distortion operation, and additive white noise. The resulting data are then presented to the deconvolution process along with a data window N and a lag window W_M. The output of the deconvolution process is an estimate of the desired synthetic seismogram up to an unknown amplitude factor. The reflectivity function is passed directly through the D filter to produce the true synthetic seismogram which is then compared with the estimated synthetic seismogram. For the comparison it is necessary to use an amplitude independent measure. The correlation coefficient defined by

$$C_{S\hat{S}} = \frac{\sum_i (S_i - \bar{S})(\hat{S}_i - \bar{\hat{S}})}{[\sum_i (S_i - \bar{S})^2 \sum_i (\hat{S}_i - \bar{\hat{S}})^2]^{1/2}} \qquad (5)$$

is suitable for this purpose.

This process is repeated a large number of times with the same lag window W_M and the

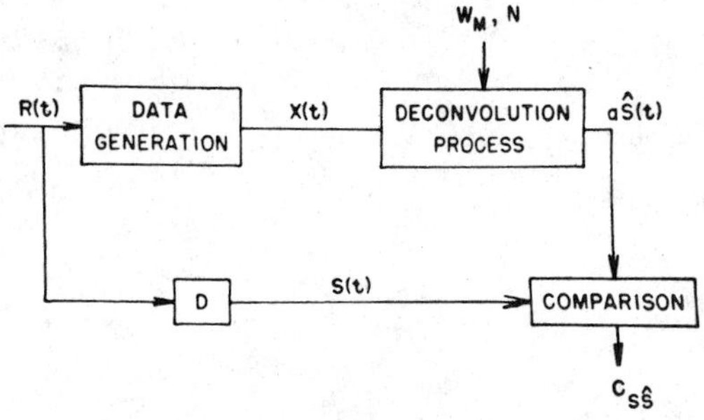

FIG. 3. Block diagram of Monte Carlo simulation.

length of data N. The resulting correlation coefficients are then averaged to obtain an ensemble measure of comparison, which we call the average correlation coefficient $\bar{C}_{S\hat{S}}$. The average correlation coefficient $\bar{C}_{SX}$ between the input data and the desired synthetic is also calculated after restricting X to the same frequency band as S. This may be compared to $\bar{C}_{S\hat{S}}$ to judge the degree of improvement due to deconvolution.

In Table 2 we show the results of a test of conclusion (1) from the theory of spectral analysis. In this test the distortion was a simulated reverberation corresponding to a two-way water time T_W of 40 ms, and a water-bottom reflection coefficient R_W of 0.6. The length of the data window was 250 with a sampling interval of 4 ms. The number of degrees of freedom for each of the truncators was 4.8. As can be seen from the table, there is a negligible difference between the results for different lag windows.

We have repeated this test for many choices of the parameters and also for ghosting distortions. So long as the distortion is not too severe, the conclusion holds in all cases; there is no significant difference between truncators having the same number of degrees of freedom.

For severe distortion, this is no longer the case without prewhitening. For example, with a water-bottom reflection coefficient of 0.8 and otherwise the same parameters as in Table 2, the average correlation coefficients were 0.77 and 0.55 with rectangular and triangular truncators respectively. With prewhitening and with a crude three-point operator (Backus, 1959), these were 0.84 and 0.83 respectively. Thus, even for severe distortion, the conclusion appears to be valid if the data are first prewhitened.

The second conclusion from spectral analysis is that the error should have a single minimum as a function of the number of degrees of freedom. Thus the average correlation coefficient should have a single maximum. A test of this conjecture is shown in Figure 4. As we would expect, the average correlation coefficient does have a single maximum within the limits of statistical fluctuation.

Finally the third conclusion from spectral analysis is that for a fixed number of degrees of freedom, the error should decrease with the length of the data window. We see indeed that in Figure 5 the average correlation coefficient increases as the length of the data window increases.

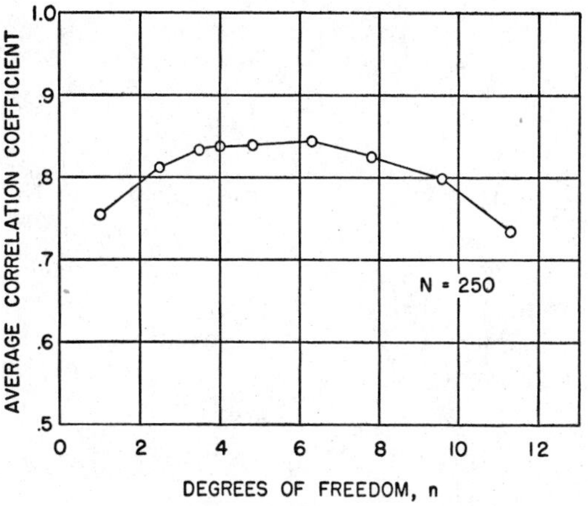

FIG. 4. Test of conclusion (2): The error has a single minimum as a function of n. Parameters, $R_w=0.6$, $T_w=40$ ms, $\Delta t=4$ ms, noise-to-signal ratio$=0.01$.

This latter conclusion depends, of course, upon the assumption that the data-generation process is time-invariant. For actual seismic data, this assumption is only approximately true. To investigate this effect, we simulated time variance in the data by changing the seismic wavelet with time.

McDonal et al (1958) and others have shown from field measurements that to a first order of approximation, the amplitude spectrum of the wavelet decreases exponentially as the first power of the frequency-distance product. Our model of time variance was chosen to conform reasonably to these observations and is stated in equation (6).

$$B(\tau, t) = \frac{1}{2\pi}\int_{-\infty}^{\infty} B(\omega)e^{-\alpha|\omega|\tau}e^{i\omega(t-\tau)}d\omega, \quad (6)$$

where $B(\tau, t)$ is the seismic wavelet at one-way traveltime τ, $B(\omega)$ is the Fourier transform of $B(\tau, t)$ at $\tau=0$, and α is the attenuation coefficient. Equation (6) is the same as the McDonal model if the velocity of the medium is constant. No attempt was made to simulate the phase shift which must accompany frequency-dependent attentuation because this effect would have had little bearing on the results of the experiment.

Attentuation results in a general decay in the amplitude of the generated data with time. Such decay is usually compensated in the field by expander or AVC recording, and these operations were simulated in the computer experiments.

We show the results of the test of conclusion (3) for the time-variant case in Figure 6. The results are for $\alpha=0.0015$. This value was assumed to be reasonably typical, since it was estimated from actual seismograms and obtained independently by averaging measured attentuation constants. The figure shows that the average correlation coefficient decreases for data window lengths beyond a certain optimum value. Different values of α gave similar results except, of course, that the maxima in the correlation coefficient versus N curves varies with α.

CONCLUSIONS

We conclude from this study:

(1) Severely distorted data should be prewhitened. For reverberations this may be done by using the three-point operator technique (Backus, 1959) or by other well-known methods.

(2) The choice of truncator is unimportant except for its number of degrees of freedom n. Referring to Table 1, we see that the rectangular truncator has the smallest M for a given n and so

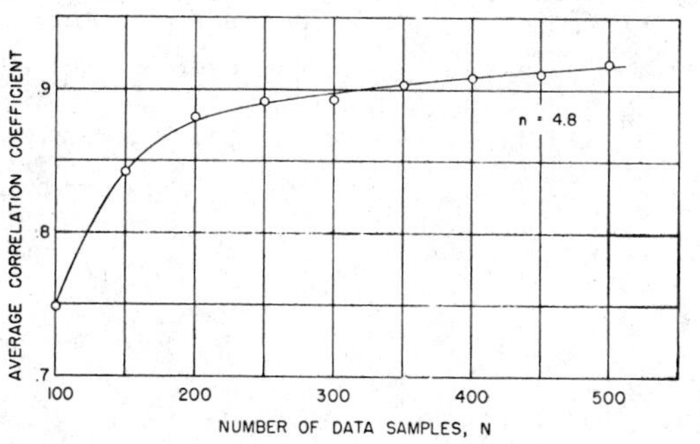

FIG. 5. Test of conclusion (3): For n fixed, the error decreases with increasing N. Parameters, $R_w=0.6$, $T_w=40$ ms, $\Delta t=4$ ms, noise-to-signal ratio$=0.001$.

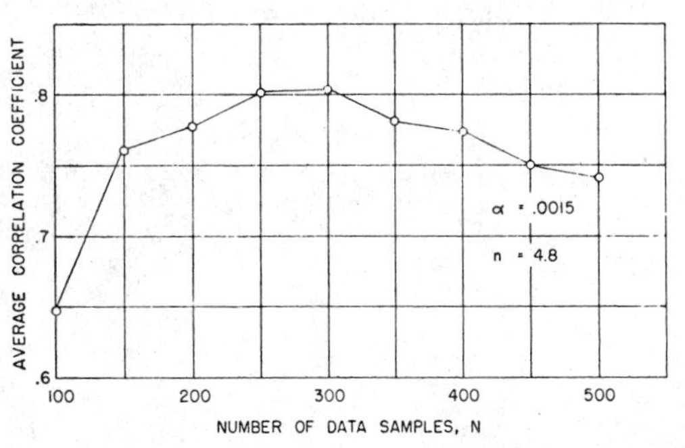

FIG. 6. Failure of conclusion (3) when the seismic wavelet is time-variant. Parameters, $R_w=0.6$, $T_w=40$ ms, $\Delta t=4$ ms, noise-to-signal ratio$=0.001$.

Table 2. Test of conclusion 1: lag windows with same n yield, same error

Lag window	Average correlation coefficient $C_{S\hat{S}}$
Rectangular	0.836
Triangular	0.821
Hamming	0.815
Parzen	0.831
Daniell	0.822
Sinc squared	0.839
Noise/Signal=0.01	R_w=0.6
N=250	T_w=40 ms
n=4.8	Δt=4 ms
C_{SX}=0.29	

requires the least number of samples in the correlation function. It would probably be preferred on this basis.[1]

(3) There is an optimum value of n. If the distortion can be approximately estimated, this could be determined by running the simulation. For the reverberation case we illustrate in Table 2 and Figure 4, it lies between 4 and 7. Its exact value is not critical.

(4) Due to time variance in the data there is an optimum data length N. Again this can be approximately determined by running the simulation. In our particular study it was approximately 300, corresponding to 1.2 sec of data.

APPENDIX

The expected mean square error can be written:

$$E|\hat{f}(\lambda) - f(\lambda)|^2 = E|\hat{f}(\lambda) - E\hat{f}(\lambda)|^2 + |E\hat{f}(\lambda) - f(\lambda)|^2.$$

The first of these is the variance of the estimator, and the second is the square of the bias error. Using approximate formulas originally due to Tukey (Amos and Koopmans, p. 25), we have for the variance:

$$V[\hat{f}(\lambda)] = \frac{2\pi}{N}\int_{-\pi}^{\pi} W_M{}^2(\mu - \lambda)f(\mu)^2 d\mu$$

and for the bias error:

$$b[\hat{f}(\lambda)] = \int_{-\pi}^{\pi} W_M(\mu - \lambda)f(\mu)d\mu - f(\lambda).$$

Here $W_M(\lambda)$ is the spectral window corresponding to the lag window $W_M(k)$. If we make a Taylor's expansion of $f(\mu)$ about the point λ and retain only leading terms,

$$V[\hat{f}(\lambda)] = \frac{2\pi}{N}f(\lambda)\int_{-\pi}^{\pi} W_M{}^2(\lambda)d\lambda = \frac{f(\lambda)}{n},$$

$$b[\hat{f}(\lambda)] = \frac{f''(\lambda)}{2}\int_{-\pi}^{\pi} W_M(\mu)\mu^2 d\mu.$$

This last term may be approximated by introducing the equivalent rectangular window (Amos and Koopmans, p. 24) of width $h=2\pi n/N$ and height $1/2h$. Using this approximation. we obtain

$$b[\hat{f}(\lambda)] = \frac{\pi^2 n^2 f''(\lambda)}{6N^2},$$

and formula (4) follows.

[1] Since rectangular truncators may lead to nonpositive-definite covariance matrices, it is common practice to add a small multiple of the identity matrix before applying the Levinson algorithm (Foster, 1961).

REFERENCES

Amos, D. E., and Koopmans, L. H., 1963, Tables for distribution of coefficient of coherence, Sandia Corp. Monograph SCR-483.

Backus, Milo M., 1959, Water reverberations—their nature and elimination, Geophysics, v. 24, no. 2, p. 233–261.

Blackman, R. B., and Tukey, J. W., 1958, The measurement of Power Spectra: New York, Dover, 190 p.

Foster, Manus, 1961, An application of the Wiener Kolmogorov Smoothing Theory to matrix inversion: J. Soc. Indust. Appl. Math., v. 9, no. 3, p. 387–392.

Levinson, Norman, 1947, The Wiener RMS error criterion in filter design and prediction: J. Math. & Physics, v. 25, p. 261–278 (reprinted as Appendix B to Wiener, 1949).

McDonal, F. J., Angona, F. A., Mills, R. L., Sengbush, R. L., Van Nostrand, R. G., and White, J. E., 1958, Attenuation of shear and compressional waves in Pierre Shale: Geophysics, v. 23, no. 3, p. 421–439.

Robinson, Enders A., 1966, Multichannel z-transforms and minimum-delay: Geophysics, v. 31, no. 3, p. 482–500.

Reprinted from Geophysics v. 34, no. 2, p. 155-169

PREDICTIVE DECONVOLUTION: THEORY AND PRACTICE†

K. L. PEACOCK* AND SVEN TREITEL*

Least-squares inverse filters have found widespread use in the deconvolution of seismograms. The least-squares prediction filter with unit prediction distance is equivalent within a scale factor to the least-squares, zero-lag inverse filter. The use of least-squares prediction filters with prediction distances greater than unity leads to the method of predictive deconvolution which represents a more generalized approach to this subject. The predictive technique allows one to control the length of the desired output wavelet, and hence to specify the desired degree of resolution. Events which are periodic within given repetition ranges can be attenuated selectively. The method is thus effective in the suppression of rather complex reverberation patterns.

INTRODUCTION

The Wiener filter is one of the most effective tools for the digital reduction of seismic traces. It constitutes the keystone of many current deconvolution methods. In one realization this filter is used to deconvolve a reverberating pulse train into an approximation of a zero-delay unit impulse. More generally it is possible to arrive at Wiener filters which remove repetitive events having specified periodicities. In this context the Wiener filter is better viewed as a predictor of coherent energy than merely as a spiker of "leggy" wave trains.

The prediction filter used in this treatment gives rise to the method of *predictive deconvolution*. We remark that Robinson's Ph.D. thesis (1954), if written today, would be entitled "*Predictive Deconvolution* of Time Series with Applications to Seismic Exploration," since the older term *decomposition* has given way to the newer term *deconvolution*. The method of predictive deconvolution has been described in a paper by Robinson (1966), in which the author advocates a prediction distance greater than unity. A discussion of the general properties of the digital Wiener filter has been given by Robinson and Treitel (1967).

BASIC CONCEPTS

The digital filtering process is described by the discrete convolution formula

$$y_\tau = \Delta t \sum_t x_t a_{\tau-t},$$

where x_t is the input, a_t is the filter, y_τ is the output, and Δt is the sampling increment. No loss of generality will result if we assume Δt to be unity. In the sequel t and τ are discrete time variables and $\Delta t=1$ unless otherwise specified.

If a_t is a *prediction* operator with prediction distance α, the output y_τ will be an estimate of the input x_t at some future time $t+\alpha$. We thus write

$$y_\tau = \sum_t x_t a_{\tau-t} = \hat{x}_{t+\alpha}, \tag{1}$$

where $\hat{x}_{t+\alpha}$ is an estimate of $x_{t+\alpha}$.

An error series may be defined as the difference between the true value $x_{t+\alpha}$ and the estimated or predicted value $\hat{x}_{t+\alpha}$,

$$\epsilon_{t+\alpha} = x_{t+\alpha} - \hat{x}_{t+\alpha}. \tag{2}$$

Thus ϵ_t is an output series which represents the nonpredictable part of x_t.

Replacement of the term $\hat{x}_{t+\alpha}$ in equation (2) with its equivalent as defined by equation (1) results in

† Presented at the 38th Annual International SEG Meeting in Denver, Colorado, October 1, 1968. Manuscript received by the Editor October 10, 1968.

* Pan American Petroleum Corp., Research Center, Tulsa, Okla.

$$\epsilon_{t+\alpha} = x_{t+\alpha} - \sum_t x_t a_{\tau-t}. \quad (3)$$

The z-transform of equation (3) is

$$z^{-\alpha}E(z) = z^{-\alpha}X(z) - X(z)A(z). \quad (4)$$

Multiplication of both sides of equation (4) by z^α yields

$$E(z) = X(z) - z^\alpha X(z)A(z) = X(z)[1 - z^\alpha A(z)]. \quad (5)$$

The quantity $[1-z^\alpha A(z)]$ is the z-transform of the so-called *prediction error operator*. It is seen to be the difference between the zero-delay unit spike and the prediction operator $A(z)$ delayed by the prediction distance α. Thus one may calculate the error series ϵ_t by computing $\hat{x}_{t+\alpha}$ from equation (1), and follow with a subtraction as defined by equation (2). Alternatively, one may compute the error series in a single step by use of the prediction error operator. Let us assume that a seismic trace is represented by the convolution of an uncorrelated reflection coefficient series with a reverberating pulse train, which by its nature is rich in repetitive energy (see Appendix A). The prediction error filter will then remove the predictable portion of such a trace, which to a good approximation will be given by the repetitive energy in the reverberations. The output of this filtering operation is the error series ϵ_t, which within the framework of this model constitutes the estimate of the reflection coefficient series of the layered subsurface.

Suppose that the prediction operator is given by the n-length series

$$a_t = a_0, a_1, \cdots, a_{n-1}.$$

Then the corresponding prediction error operator with prediction distance α is

$$f_t = 1, \overbrace{0, 0, \cdots, 0}^{\alpha - 1 \text{ zeros}}, -a_0, -a_1, \cdots, -a_{n-1}.$$

We must now deal with the explicit design of the prediction operator, a task which will be accomplished in the next section.

THE LEAST-SQUARES PREDICTIVE FILTERING MODEL

A general least-squares filter model involves the three signals illustrated in Figure 1, namely (1) the input signal x_t, (2) the desired output signal z_t, and (3) the actual output signal y_t. Minimization of the energy existing in the difference between the desired output z_t and the actual output y_t, i.e., minimization of the expression

$$I = \sum_t (z_t - y_t)^2,$$

results in the least-squares, or Wiener filter described by Robinson and Treitel (1967).

The n-length Wiener filter results from the solution of the normal equations with matrix representation,

$$\begin{bmatrix} r_0 & r_1 & \cdots & r_{n-1} \\ r_1 & r_0 & \cdots & r_{n-2} \\ & & \vdots & \\ r_{n-1} & r_{n-2} & \cdots & r_0 \end{bmatrix} \begin{bmatrix} f_0 \\ f_1 \\ \vdots \\ f_{n-1} \end{bmatrix} = \begin{bmatrix} g_0 \\ g_1 \\ \vdots \\ g_{n-1} \end{bmatrix}, \quad (6)$$

where r_t is the autocorrelation of the input, g_t is the crosscorrelation between the desired output and the input, and f_t is the Wiener filter.

We have seen in the previous section that the prediction operator is that filter which acts on an input trace up to time t and estimates the trace amplitude at some future time $t+\alpha$. Thus it is reasonable to define the desired output for the predictive filter as a time-advanced version of the input x_t. We can now express the prediction filter in terms of a particular Wiener filter, namely the one for which the desired output trace is simply a time-advanced version of the input trace.

In order to solve equation (6) we must know

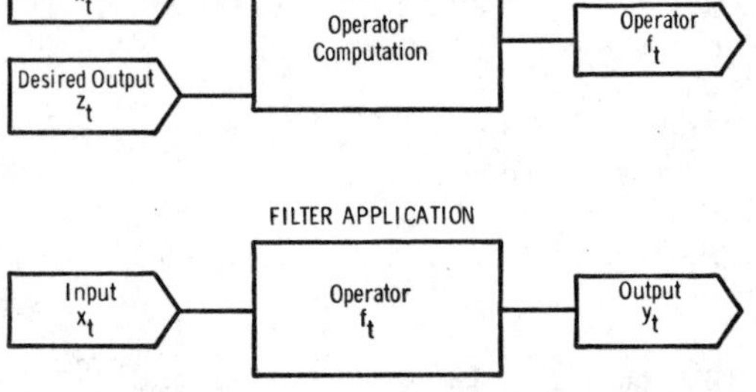

FIG. 1. A general model which illustrates the design and application of the Wiener filter.

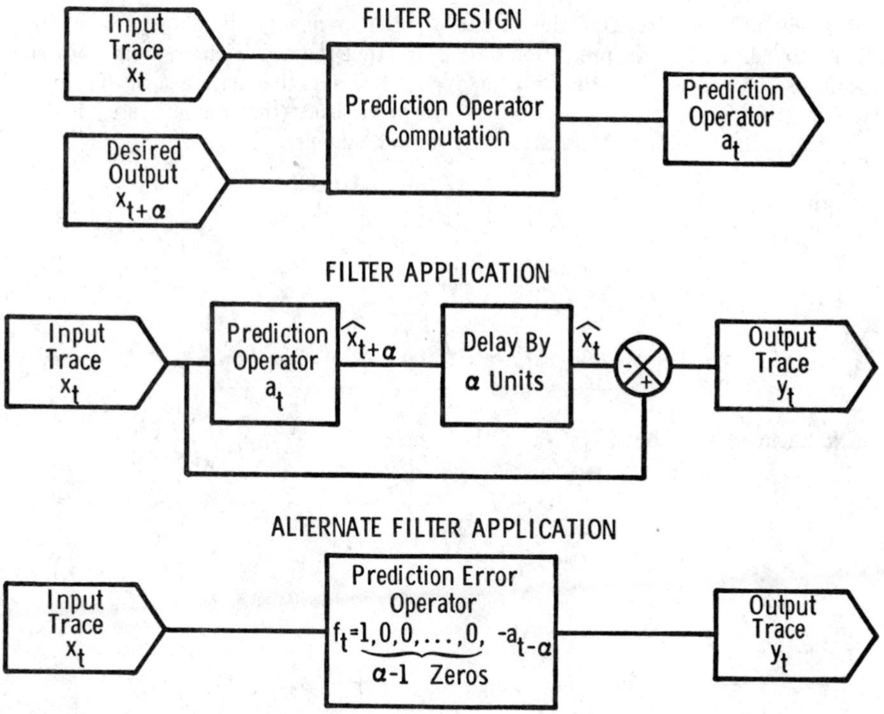

FIG. 2. The predictive filter model which illustrates prediction operator design (upper diagram) and application (center diagram). Alternately, the prediction error operator may be formed and utilized as illustrated in the lower diagram.

the autocorrelation of the input and the positive lag coefficients of the crosscorrelation between the desired output and the input. The autocorrelation of the input trace is given by

$$r_\tau = \sum_t x_t x_{t-\tau},$$

while the crosscorrelation between the desired output and input traces is

$$g_\tau = \sum_t z_t x_{t-\tau}. \tag{7}$$

Since the desired output for the prediction operator is a time-advanced version of the input, i.e., since

$$z_t = x_{t+\alpha},$$

equation (7) becomes

$$g_\tau = \sum_t x_{t+\alpha} x_{t-\tau} = \sum_t x_t x_{t-(\tau+\alpha)} = r_{\tau+\alpha}.$$

Thus, the crosscorrelation between the desired output and the input is by definition equal to the autocorrelation of the input for lags $\geq \alpha$. The normal equations (6) become

$$\begin{bmatrix} r_0 & r_1 & \cdots & r_{n-1} \\ r_1 & r_0 & \cdots & r_{n-2} \\ & & \vdots & \\ r_{n-1} & r_{n-2} & \cdots & r_0 \end{bmatrix} \begin{bmatrix} a_0 \\ a_1 \\ \vdots \\ a_{n-1} \end{bmatrix} = \begin{bmatrix} r_\alpha \\ r_{\alpha+1} \\ \vdots \\ r_{\alpha+n-1} \end{bmatrix}. \tag{8}$$

The solution to the above matrix equation yields the prediction operator illustrated in the upper diagram of Figure 2. This prediction operator can be utilized as indicated in the center diagram of Figure 2. Alternately, the corresponding prediction error operator can be formed from the prediction operator and utilized as indicated in the lower diagram of Figure 2.

We have thus shown how one can use the Wiener least squares error criterion to generate the least squares prediction operator, and how the prediction error operator is derived from its corresponding prediction operator. Our aim in the next section is to indicate how the prediction *error* operator can also be expressed in the form of a particular Wiener filter.

PREDICTIVE FILTERING AND DECONVOLUTION

We shall demonstrate that the least-squares deconvolution filter which ideally transforms an unknown signal to an impulse at zero delay is equivalent to the prediction *error* filter for which the prediction distance α is unity. The matrix relation for the prediction operator a_t with prediction distance unity ($\alpha=1$) and length n is obtained by setting $\alpha=1$ in equation (8),

$$\begin{bmatrix} r_0 & r_1 & \cdots & r_{n-1} \\ r_1 & r_0 & \cdots & r_{n-2} \\ & & \vdots & \\ r_{n-1} & r_{n-2} & \cdots & r_0 \end{bmatrix} \begin{bmatrix} a_0 \\ a_1 \\ \vdots \\ a_{n-1} \end{bmatrix} = \begin{bmatrix} r_1 \\ r_2 \\ \vdots \\ r_n \end{bmatrix}. \quad (9)$$

The above system may be written in the form of the n simultaneous linear equations,

$$\begin{aligned} r_0a_0 + r_1a_1 + \cdots + r_{n-1}a_{n-1} &= r_1 \\ r_1a_0 + r_0a_1 + \cdots + r_{n-2}a_{n-1} &= r_2 \\ &\vdots \\ r_{n-1}a_0 + r_{n-2}a_1 + \cdots + r_0a_{n-1} &= r_n. \end{aligned}$$

Let us modify the above system by first subtracting the coefficient r_i from both sides of the ith row of each equation such that the right-hand side vanishes (the original system is shown within the rectangle),

$$\begin{aligned} -r_1 + \quad & \boxed{r_0a_0 + r_1a_1 + \cdots + r_{n-1}a_{n-1} = r_1} \quad -r_1 \\ -r_2 + \quad & r_1a_0 + r_0a_1 + \cdots + r_{n-2}a_{n-1} = r_2 \quad -r_2 \\ & \vdots \\ -r_n + \quad & r_{n-1}a_0 + r_{n-2}a_1 + \cdots + r_0a_{n-1} = r_n \quad -r_n. \end{aligned}$$

Let us next augment the above system in the form,

$$\begin{aligned} -r_0 + \quad & r_1a_0 + r_2a_1 + \cdots + r_na_{n-1} = -\beta \\ -r_1 + \quad & \boxed{r_0a_0 + r_1a_1 + \cdots + r_{n-1}a_{n-1} = r_1} \quad -r_1 \\ -r_2 + \quad & r_1a_0 + r_0a_1 + \cdots + r_{n-2}a_{n-1} = r_2 \quad -r_2 \\ & \vdots \\ -r_n + \quad & r_{n-1}a_0 + r_{n-2}a_1 + \cdots + r_0a_{n-1} = r_n \quad -r_n. \end{aligned}$$

This system may be written

$$\begin{aligned} r_0 - r_1a_0 - r_2a_1 - \cdots - r_na_{n-1} &= \beta \\ r_1 - r_0a_0 - r_1a_1 - \cdots - r_{n-1}a_{n-1} &= 0 \\ &\vdots \\ r_n - r_{n-1}a_0 - r_{n-2}a_1 - \cdots - r_0a_{n-1} &= 0, \end{aligned}$$

for which the associated matrix equation is

$$\begin{bmatrix} r_0 & r_1 & \cdots & r_n \\ r_1 & r_0 & \cdots & r_{n-1} \\ & & \vdots & \\ r_n & r_{n-1} & \cdots & r_0 \end{bmatrix} \begin{bmatrix} 1 \\ -a_0 \\ \vdots \\ -a_{n-1} \end{bmatrix} = \begin{bmatrix} \beta \\ 0 \\ \vdots \\ 0 \end{bmatrix}. \quad (10)$$

We now see that the Wiener filter of equation (10) can be identified as the unit prediction *error* operator associated with the prediction operator of equation (9). Let us rewrite equation (10) in the form

$$\begin{bmatrix} r_0 & r_1 & \cdots & r_n \\ r_1 & r_0 & \cdots & r_{n-1} \\ & & \vdots & \\ r_n & r_{n-1} & \cdots & r_0 \end{bmatrix} \begin{bmatrix} b_0 \\ b_1 \\ \vdots \\ b_n \end{bmatrix} = \begin{bmatrix} \beta \\ 0 \\ \vdots \\ 0 \end{bmatrix}, \qquad (11)$$

where

$$b_0 = 1$$

$$b_i = -a_{i-1}, \qquad i = 1, \cdots, n$$

$$\beta = \sum_{i=0}^{n} b_i r_i.$$

In Appendix B we describe the standard deconvolution method, which is based on the use of the least-squares, zero-delay inverse filter. We note that the system of normal equations for the inverse filter given by equation (B-1) is identical to the system of equations (11), except for a scale factor β. Thus the $(n+1)$-length prediction error operator with prediction distance unity is identical to the zero-delay inverse filter of length $(n+1)$, except for a scale factor.

We shall now show that the predictive filter with prediction distance greater than unity can also serve as a deconvolution operator, and thus it turns out that the predictive filtering technique constitutes a more generalized approach to deconvolution. We remark that under certain assumptions described in Appendix A, the autocorrelation of an input seismic signal can be identified with the autocorrelation of the source wavelet[1].

The inverse filter described in Appendix **B** shapes the unknown source wavelet to an impulse at zero lag time. We will show here that the predictive filter shapes the unknown source wavelet of length $\alpha+n$ to another unknown wavelet of length α. Thus, by having control of the desired output wavelet length, one may specify the desired degree of resolution.

The predictive filter matrix equation for filter length n and prediction distance α is given by equation (8),

$$\begin{bmatrix} r_0 & r_1 & \cdots & r_{n-1} \\ r_1 & r_0 & \cdots & r_{n-2} \\ & & \vdots & \\ r_{n-1} & r_{n-2} & \cdots & r_0 \end{bmatrix} \begin{bmatrix} a_0 \\ a_1 \\ \vdots \\ a_{n-1} \end{bmatrix} = \begin{bmatrix} r_\alpha \\ r_{\alpha+1} \\ \vdots \\ r_{\alpha+n-1} \end{bmatrix},$$

or

$$\begin{aligned} r_0 a_0 + r_1 a_1 + \cdots + r_{n-1} a_{n-1} &= r_\alpha \\ r_1 a_0 + r_0 a_1 + \cdots + r_{n-2} a_{n-1} &= r_{\alpha+1} \\ &\vdots \\ r_{n-1} a_0 + r_{n-2} a_1 + \cdots + r_0 a_{n-1} &= r_{\alpha+n-1}. \end{aligned} \qquad (12)$$

The above system can be augmented in such a way that the prediction operator is converted into its corresponding prediction error operator. This is accomplished by the addition of suitable terms to both sides of the equations (12). Proceeding as in the case of the unit prediction error filter (equations (9) et seq.), one obtains,

$$\begin{aligned}
&-r_0 1 - r_1 0 - \cdots - r_{\alpha-1} 0 + r_\alpha a_0 + r_{\alpha+1} a_1 + \cdots + r_{\alpha+n-1} a_{n-1} = -\rho_0 \\
&-r_1 1 - r_0 0 - \cdots - r_{\alpha-2} 0 + r_{\alpha-1} a_0 + r_\alpha a_1 + \cdots + r_{\alpha+n-2} a_{n-1} = -\rho_1 \\
&\qquad \vdots \\
&-r_{\alpha-1} 1 - r_{\alpha-2} 0 - \cdots - r_0 0 + r_1 a_0 + r_2 a_1 + \cdots + r_n a_{n-1} = -\rho_{\alpha-1} \\
&\qquad \vdots \\
&-r_\alpha 1 - r_{\alpha-1} 0 - \cdots - r_1 0 + \boxed{r_0 a_0 + r_1 a_1 + \cdots + r_{n-1} a_{n-1} = r_\alpha} - r_\alpha \\
&-r_{\alpha+1} 1 - r_\alpha 0 - \cdots - r_2 0 + \boxed{r_1 a_0 + r_0 a_1 + \cdots + r_{n-2} a_{n-1} = r_{\alpha+1}} - r_{\alpha+1} \\
&\qquad \vdots \\
&-r_{\alpha+n-1} 1 - r_{\alpha+n-2} 0 - \cdots - r_n 0 + \boxed{r_{n-1} a_0 + r_{n-2} a_1 + \cdots + r_0 a_{n-1} = r_{\alpha+n-1}} - r_{\alpha+n-1}
\end{aligned}$$

where the original set (12) is enclosed by the rectangle. The associated matrix equation is

[1] The source wavelet is here meant to be the shot pulse modified by near-surface reverberations.

$$
\begin{bmatrix}
r_0 & r_1 & \cdots & r_{\alpha+n-1} \\
r_1 & r_0 & \cdots & r_{\alpha+n-2} \\
 & & \vdots & \\
r_{\alpha-1} & r_{\alpha-2} & \cdots & r_n \\
r_\alpha & r_{\alpha-1} & \cdots & r_{n-1} \\
 & & \vdots & \\
r_{\alpha+n-1} & r_{\alpha+n-2} & \cdots & r_0
\end{bmatrix}
\begin{bmatrix}
1 \\ 0 \\ \vdots \\ 0 \\ -a_0 \\ \vdots \\ -a_{n-1}
\end{bmatrix}
=
\begin{bmatrix}
\rho_0 \\ \rho_1 \\ \vdots \\ \rho_{\alpha-1} \\ 0 \\ \vdots \\ 0
\end{bmatrix}, \quad (13)
$$

where

$$\rho_0 = r_0 - (r_\alpha a_0 + r_{\alpha+1} a_1 + \cdots + r_{\alpha+n-1} a_{n-1})$$
$$\rho_1 = r_1 - (r_{\alpha-1} a_0 + r_\alpha a_1 + \cdots + r_{\alpha+n-2} a_{n-1})$$
$$\vdots$$
$$\rho_{\alpha-1} = r_{\alpha-1} - (r_1 a_0 + r_2 a_1 + \cdots + r_n a_{n-1}).$$

The solution of the above matrix equation yields the prediction error operator with prediction distance α. Let us interpret this equation in terms of the Wiener filter model, where the left-hand matrix is the input autocorrelation matrix, and where the elements of the right-hand column vector constitute the positive lag values of the crosscorrelation between the desired output and the input. Subject to the assumptions given in Appendix A, the autocorrelation function $r_0, r_1, \cdots, r_{\alpha+n-1}$ can be identified with the autocorrelation of a source wavelet of length $\alpha+n$. However, we still require an interpretation of the crosscorrelation,

$$g_\tau = \underbrace{\rho_0, \rho_1, \cdots, \rho_{\alpha-1},}_{\alpha \text{ terms}} \; \underbrace{0, \cdots, 0.}_{n \text{ zeros}}$$

Although the crosscorrelation function is complicated, we can make one important observation. Since the crosscorrelation vanishes for lags greater than $\alpha-1$, the length of the implied desired output wavelet cannot be greater than α. In other words, the input wavelet is of length $\alpha+n$, while the implied desired output wavelet is of length α, and hence the prediction error operator shortens an input wavelet of length $\alpha+n$ to an output wavelet of length α. Since α is an independent variable, we are free to select whatever length we choose for the desired output wavelet. We conclude that the predictive filter leads to a more generalized approach to deconvolution, in which one may control the desired degree of resolution or wavelet contraction.

We have shown earlier in this section that the zero-delay least-squares, inverse filter is equal within a scale factor to the prediction error filter with prediction distance unity ($\alpha=1$). Experience has taught us that the output from these filters cannot in general be interpreted with ease. This is due to the presence of high-frequency components in the deconvolved trace, which result from the fact that this kind of deconvolution makes use of inverse, or "spiking" filters. One improves this condition by passing the raw deconvolved trace through suitable low-pass filters, by smoothing the autocorrelation function, or by other related means. We suggest that the use of prediction error filters with arbitrary prediction distance α leads to a deconvolution method in which one has more effective control on the desired degree of resolution. It is also significant to note that the inverse filter deconvolution method requires the insertion of an arbitrary scaling factor into the right side of the normal equations (i.e., the element β of equation (11) is arbitrary). No such scaling factor is needed in the predictive filter model, and hence the trace-to-trace amplitude variation which occurs in the input data can be preserved if so desired. In addition, no time need be spent by the computer in analyzing the output data to determine the scaling factor.

It is of some interest to establish how the predictive filters presented in this section perform on an idealized, noise-free reverberation model. These matters are discussed in Appendix C, where we also show that under appropriate simplifying conditions the prediction error filter becomes identical to the 3-point filter of Backus (1959).

APPLICATIONS OF PREDICTIVE DECONVOLUTION

The concept which permits resolution control by means of the prediction distance parameter has been introduced in the previous section. We have seen that the crosscorrelation between the desired output and the input is zero between lag positions α and $\alpha+n-1$, and we were thus able to deduce that the implied desired output pulse cannot be of length greater than α. We note from the autocorrelation matrix of equation (13) that the predictive filter does not utilize any autocorrelation coefficient beyond lag position $\alpha+n-1$. Our model thus implies that the source wavelet is of length $\alpha+n$.

Since this filter attempts to shape the input into some desired output, we can argue that the autocorrelation of the actual output data will tend to vanish between lag positions α and $\alpha+n-1$. This is because the autocorrelation of the implied desired output wavelet vanishes for lags greater than $\alpha-1$. The predictive filter will thus modify the input in such a way that the autocorrelation of the actual output will tend to vanish between α and $\alpha+n-1$. We cannot expect the autocorrelation to be 0 everywhere beyond lag$=\alpha+n-1$, since the filter computation makes use of no autocorrelation coefficients for lags greater than $\alpha+n-1$.

Anstey (1966) describes how the autocorrelation can be an interpretative aid in the analysis of reverberatory problems. When we consider that the predictive filter is designed only from knowledge of the input autocorrelation and that the magnitude of this input autocorrelation at a particular lag is an indication of the degree of predictability at that lag, we see that the autocorrelogram is a very important entity to gauge the effectiveness of dereverberation by means of predictive deconvolution.

Thus we set our parameters α (the prediction distance) and n (the prediction operator length) such that predictable (i.e., repetitive) energy having periods between α and $\alpha+n-1$ time units will tend to be removed, and hence the autocorrelation of the output will tend to vanish between lags α and $\alpha+n-1$.

Another means to measure the effectiveness of the predictive deconvolution process has been given by Wadsworth et al (1953). These authors point out that the reduction in energy content of the output trace relative to the input trace gives a measure of the predictable energy removed by the filtering operation in the range $t=\alpha$ to $t=\alpha+n-1$.

A given reverberation may be characterized as either "short-period" or "long-period." Long-period reverberations appear on a correlogram as distinct waveforms which are separated by quiet

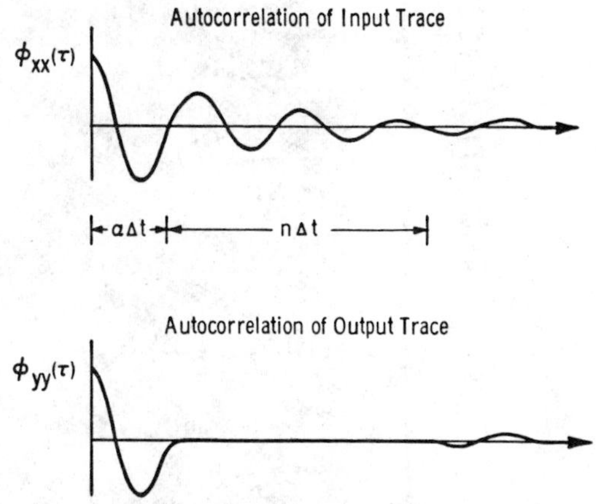

FIG. 3. A typical autocorrelation of an input trace with a moderate amount of short-period reverberation is illustrated by the upper diagram. The lower diagram illustrates the appearance of the output autocorrelation after application of the prediction error operator with prediction distance α and length n, as established in the upper diagram. If the reverberating pattern on the autocorrelation is highly regular, one need set n such that only the first full cycle is spanned. If the pattern is irregular, the significant portion of the reverberation should be spanned.

zones. Short-period reverberations appear on a correlogram in the form of decaying waveforms which are not separated by any noticeable quiet intervals.

The upper diagram of Figure 3 illustrates the autocorrelation of a typical trace which exhibits a moderate degree of short-period reverberation. The prediction distance α is chosen to specify the degree of wavelet contraction desired. As α approaches unity, more contraction and consequently more high-frequency noise is introduced. Thus we choose α so that we may obtain a compromise between wavelet contraction and signal-to-noise ratio in the output trace. Preliminary studies indicate that α should be set roughly equal to the lag that corresponds to the second zero crossing of the autocorrelation function. The lower diagram of Figure 3 illustrates the fact that the autocorrelation of the output signal trace tends to zero between lags α and $\alpha+n-1$.

Figure 4 shows three different predictive deconvolution runs on offshore traces having reverberations with characteristics somewhat in-between our definitions of the "short-period" and "long-period" types. The product $\alpha\Delta t$ has been given the values 32, 16, and 4 ms. Since this data has a sampling interval of 4 ms, the third run actually corresponds to deconvolution by the

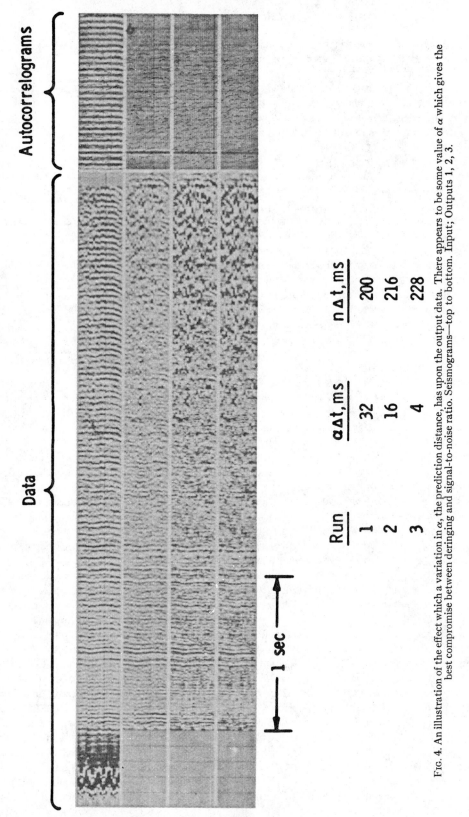

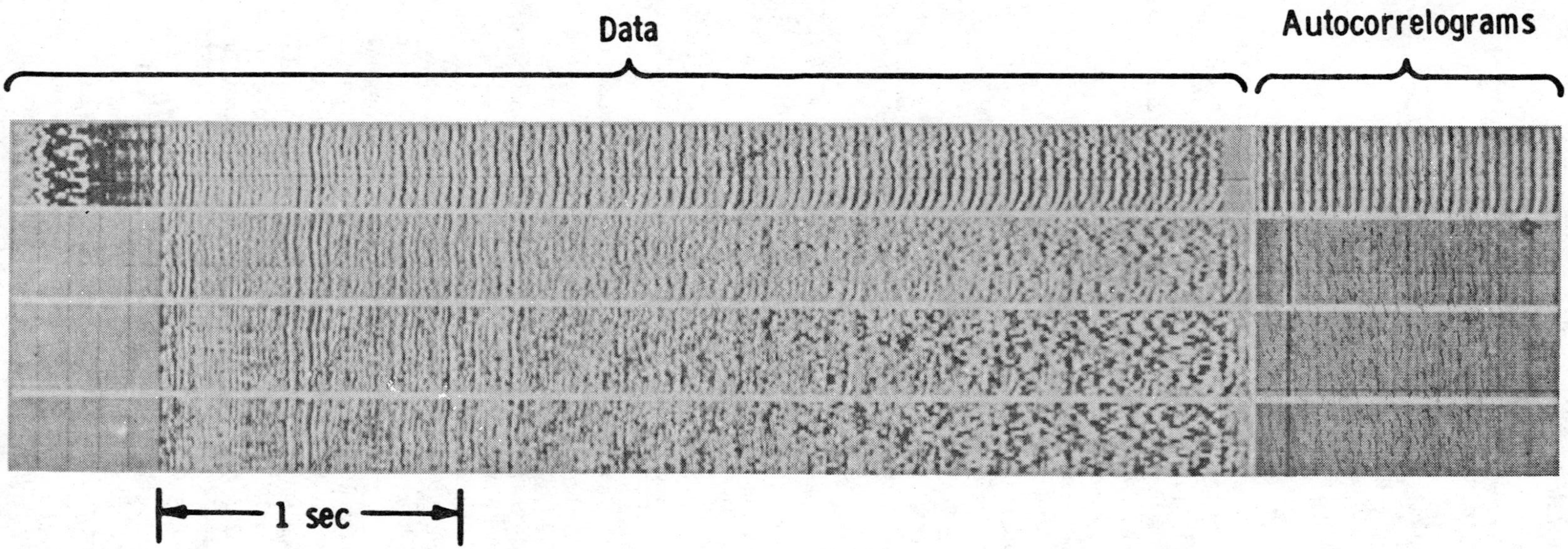

Run	$\alpha\Delta t$, ms	$n\Delta t$, ms
1	32	200
2	16	216
3	4	228

FIG. 4. An illustration of the effect which a variation in α, the prediction distance, has upon the output data. There appears to be some value of α which gives the best compromise between deringing and signal-to-noise ratio. Seismograms—top to bottom. Input; Outputs 1, 2, 3.

zero-delay least-squares inverse filter. We see that for $\alpha\Delta t = 32$ ms ($\alpha = 8$), we obtain a good dereverberation which does not exhibit the noise build-up associated with the smaller prediction distances.

The upper diagram of Figure 5 illustrates the autocorrelation of a typical trace with long-period reverberations. We define the appropriate prediction distance α such that the window to be deleted on the autocorrelogram begins just before the onset of the first multiple indication. Depending upon the nature and period of the reverberation, we define the filter length n such that the window to be deleted spans one, two, or more orders of the multiple pattern. The autocorrelation of the resulting output trace will show very little energy between lags α and $\alpha + n - 1$. In addition, further repetitions of the waveform centered at multiples of $\alpha + n/2$ will be attenuated.

We note that the predictive filter enables us to suppress selected waveform portions of the autocorrelation function. This is highly advantageous since some waveforms on the autocorrelation might be due to accidentally strong correlations between certain primary reflections, and in this case we would choose not to suppress them. We may indeed avoid their suppression by the proper selection of the prediction distance α, and we then concentrate on those waveforms associated with reverberations. We remark that we can often successfully remove the long-period multiple indication from the autocorrelation by means of the predictive filter. Even so, the net change on the section itself is not always significant. Perhaps more study will reveal better ways of selecting the parameters such that long-period reverberations will be better attenuated by the predictive filter.

Figure 6 illustrates a predictive deconvolution run on a record with short-period reverberations. The data and associated autocorrelograms show the pulse compression which has been obtained.

Figure 7 depicts a predictive deconvolution run on marine data which exhibits long-period reverberations. We note that the prediction distance in this case is 150 ms and that the filter length is only 60 ms. If one were to deconvolve these traces with the unit prediction error filter, it would be necessary to make the filter length at least equal to 210 ms. This would require much more time to process the data. Kunetz and Fourmann (1968) have reached similar conclusions by a somewhat different line of reasoning. We note

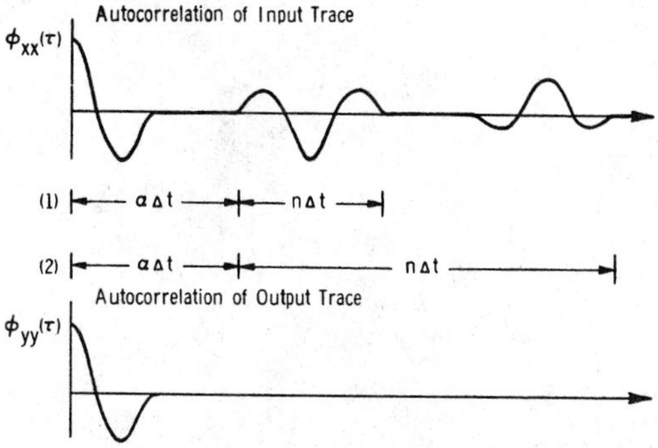

FIG. 5. A typical autocorrelation of an input trace with a moderate amount of long-period reverberation is illustrated by the upper diagram. The lower diagram illustrates the appearance of the output autocorrelation after application of the prediction error operator with prediction distance α and length n as established in the upper diagram. The parameters should be defined as indicated in (1) if the ringing is of a first-order nature, or as indicated in (2) if the ringing is of a second-order nature.

that we have achieved a successful attenuation of the reverberations on the autocorrelograms and a moderately successful dereverberation of the data itself.

CONCLUSIONS

The predictive filter is a very flexible tool for the deconvolution of seismic traces. The ability to specify the prediction distance implies the ability to control output resolution, and this means that a broad range of complex reverberatory problems can be successfully attacked with the present methods. Repetitive waveforms of a particular period can be selectively attenuated, and this is accomplishable without any significant disturbance of waveforms which one may wish to retain. The autocorrelogram is a valuable interpretative device for reverberation analysis and should be used on a routine basis. It would be desirable to have still better criteria for the determination of optimum values of such filter parameters as the prediction distance α and the filter length n. More research and evaluation of the methods presented in this paper are therefore in order.

ACKNOWLEDGMENTS

The authors wish to express their thanks to Mr. C. W. Frasier for the use of some of his unpublished results and to the Pan American Petroleum Corporation for permission to publish this paper.

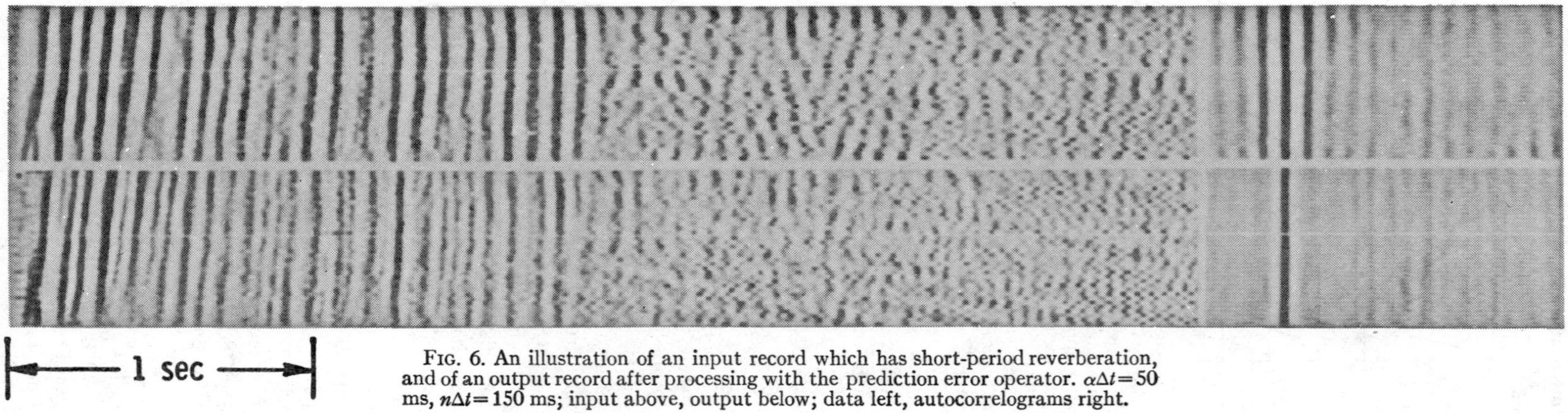

FIG. 6. An illustration of an input record which has short-period reverberation, and of an output record after processing with the prediction error operator. $\alpha\Delta t = 50$ ms, $n\Delta t = 150$ ms; input above, output below; data left, autocorrelograms right.

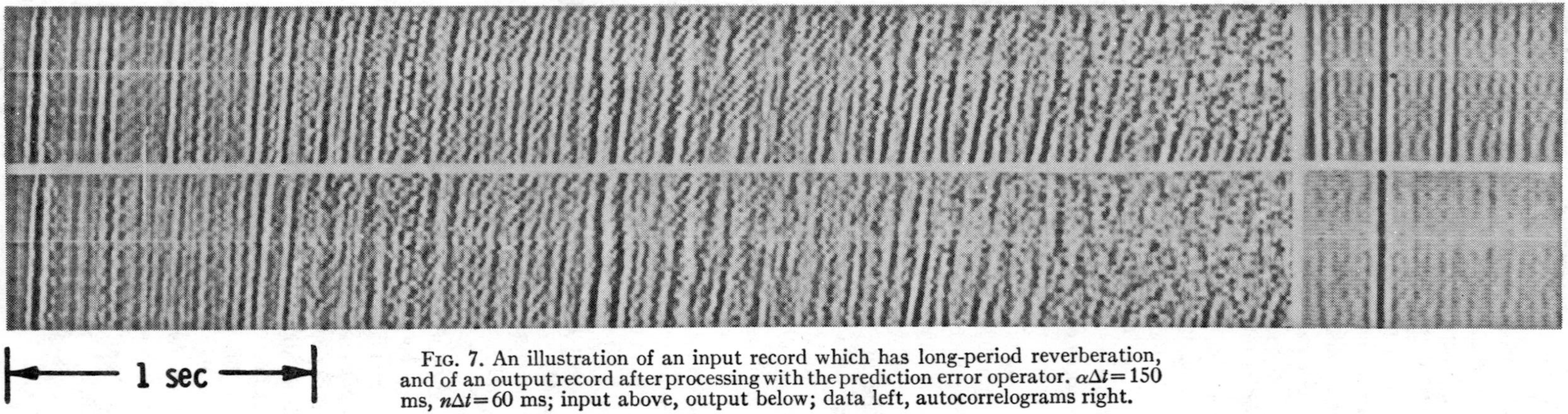

FIG. 7. An illustration of an input record which has long-period reverberation, and of an output record after processing with the prediction error operator. $\alpha\Delta t = 150$ ms, $n\Delta t = 60$ ms; input above, output below; data left, autocorrelograms right.

REFERENCES

Anstey, N. A., 1966, The sectional autocorrelogram and the sectional retrocorrelogram: Geophys. Prosp., v. 14, p. 389–426.

Backus, M. M., 1959, Water reverberations—Their nature and elimination: Geophysics, v. 24, p. 233–261.

Jenkins, G. M., 1961, General considerations in the analysis of spectra: Technometrics, v. 3, no. 2, p. 133–166.

Kunetz, G., and Fourmann, J. M., 1968, Efficient deconvolution of marine seismic records: Geophysics, v. 33, p. 412–423.

Robinson, E. A., 1954, Predictive decomposition of time series with applications to seismic exploration: Ph.D. thesis, MIT, Cambridge, Mass.

——— 1966, Multichannel z-transforms and minimum-delay: Geophysics, v. 31, p. 482–500.

——— and Treitel, S., 1967, Principles of digital Wiener filtering: Geophys. Prosp., v. 15, p. 311–333.

Wadsworth, G. P., Robinson, E. A., Bryan, J. G., and Hurley, P. M., 1953, Detection of reflections on seismic records by linear operators: Geophysics, v. 18, p. 539–586.

APPENDIX A

THE AUTOCORRELATION OF A SEISMIC TRACE

Under the proper assumptions the autocorrelation of a seismic trace is an estimate of the autocorrelation of the "basic" seismic wavelet.[2] The derivation presented here is similar to one given by Robinson and Treitel (1967).

Suppose we have a signal x_t which results from the convolution of a basic wavelet p_t with an uncorrelated series n_t, where we assume that n_t can be identified with the reflection coefficient series of a layered medium (Robinson, 1954), that is,

$$x_t = p_t * n_t.$$

The z-transform of the autocorrelation of x_t is given by

$$\Phi_{xx}(z) = [P(z)N(z)][P(1/z)N(1/z)].$$

The above equation can be rewritten in the form,

$$\Phi_{xx}(z) = [P(z)P(1/z)][N(z)N(1/z)],$$

which is the z-transform of

$$\phi_{xx}(\tau) = \phi_{pp}(\tau) * \phi_{nn}(\tau). \qquad \text{(A-1)}$$

Therefore the autocorrelation of x_t is equal to the convolution of the autocorrelation of p_t with the autocorrelation of n_t. Since n_t is an uncorrelated series, we obtain

$$\phi_{nn}(\tau) = E_n \qquad \text{for } \tau = 0$$

and

$$\phi_{nn}(\tau) = 0 \qquad \text{for } \tau \neq 0,$$

where E_n is the energy in n_t. Thus equation (A-1) reduces to

$$\phi_{xx}(\tau) = \sum_t \phi_{nn}(t)\phi_{pp}(\tau - t) = E_n\phi_{pp}(\tau),$$

and we see that the autocorrelation of x_t is simply a scaled version of the autocorrelation of p_t. This means that subject to the above assumptions, we can obtain an estimate of ϕ_{pp} even though we do not know p_t itself. The consistency of the autocorrelation estimates can be improved through use of suitable weighting functions. A good discussion of these matters is given by Jenkins (1961).

APPENDIX B

THE INVERSE FILTER MODEL

The Wiener filter model requires that the autocorrelation of the input and the positive lag values of crosscorrelation between the desired output and the input be known. The basic seismic wavelet is generally unknown; however, we can calculate its autocorrelation and the required crosscorrelation if we make the proper assumptions.

Appendix A shows that an estimate of the basic wavelet autocorrelation can be obtained from the input trace. If we assume the desired output to be an impulse at zero lag time, the crosscorrelation between desired output and input also becomes an impulse at zero lag time. In other words, since the crosscorrelation is given by

$$\phi_{dp}(\tau) = \sum_t d_t p_{t-\tau} \text{ for } \tau = 0, 1, \cdots, n-1,$$

where $d_t = 1, 0, 0, \cdots$ is the desired output signal and $p_t = p_0, p_1, p_2, \cdots$ is the basic wavelet or input signal, we see that

$$\phi_{dp}(\tau) = p_0, 0, 0, \cdots; \qquad \tau = 0, 1, \cdots, n-1,$$

which can be scaled in the form,

$$\phi_{dp}(\tau) = 1, 0, 0, \cdots.$$

The matrix equation for the Wiener filter (Robinson and Treitel, 1967) then becomes,

[2] The basic seismic wavelet is assumed to be either the initial shot pulse, or the initial shot pulse modified by near-surface reverberations.

$$\begin{bmatrix} r_0 & r_1 & \cdots & r_{n-1} \\ r_1 & r_0 & \cdots & r_{n-2} \\ & & \vdots & \\ r_{n-1} & r_{n-2} & \cdots & r_0 \end{bmatrix} \begin{bmatrix} f_0 \\ f_1 \\ \vdots \\ f_{n-1} \end{bmatrix} = \begin{bmatrix} 1 \\ 0 \\ \vdots \\ 0 \end{bmatrix}, \qquad \text{(B-1)}$$

where the f_t are the n coefficients which shape the basic wavelet p_t to an approximation of the impulse at zero lag time.

We have thus assumed a model of the form

$$x_t = p_t * n_t,$$

where x_t is the signal trace and n_t is an uncorrelated series which represents the reflection coefficients of the layered subsurface. Since the desired output is an impulse at zero lag time, we see that the model requires a filter f_t such that

$$f_t * p_t \doteq 1$$

or

$$f_t \doteq \overset{-1}{p_t},$$

where the symbol $\doteq$ means "approximately equal to." The filter f_t then deconvolves the input trace as follows:

$$y_t \doteq x_t * f_t \doteq p_t * \overset{-1}{p_t} * n_t$$

$$y_t \doteq \delta_t * n_t = n_t,$$

where δ_t is the unit impulse function, and in this sense the output trace tends to approximate the subsurface reflection coefficient series.

APPENDIX C

A STUDY OF A TWO-LAYER MARINE REVERBERATION MODEL

Impulse response of first-order component

Figure C-1 represents an idealized noise-free model of an offshore seismic situation. Reflector 1 is the water surface, reflector 2 is the water bottom, reflector 3 is some strong interface beneath the water bottom, and S is the source location just beneath the water surface. The associated normal incidence reflection coefficients are 1, c_1 and c_2, respectively, while the transmission coefficient across reflector 2 is t_1. If c is the downward reflection coefficient, the corresponding upward reflection coefficient is $-c$. From physical considerations, we know that the magnitudes of all re-

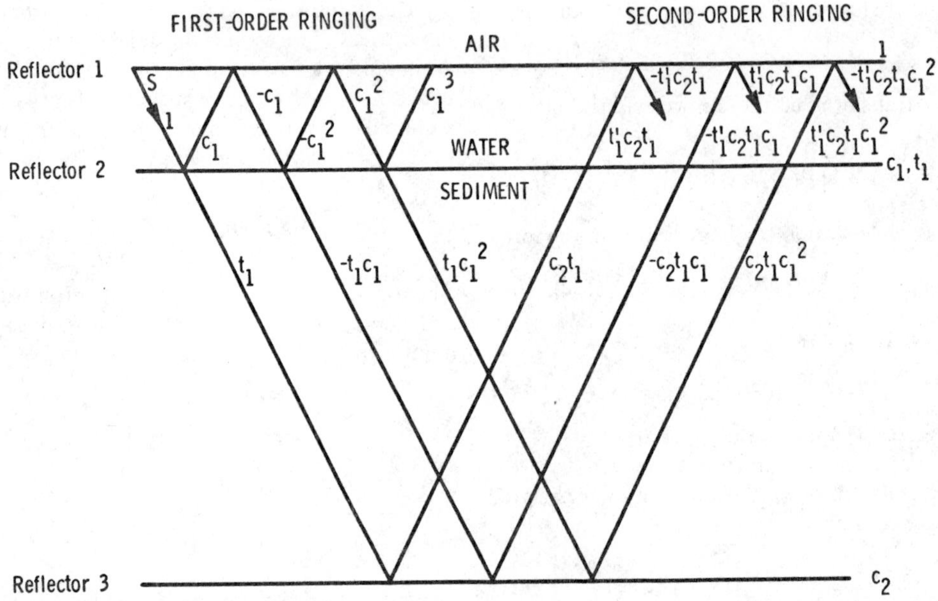

FIG. C-1. First- and second-order ringing in a 2-layer marine model.

flection coefficients are less than unity. The two-way traveltime through the water layer is τ_1.

Let us compute the first-order reverberation portion of the two-layer impulse response as indicated in Figure C-1. This response can be expressed in terms of the following z-transform:

$$R_1(z) = 1 - c_1 z^{\tau_1} + c_1^2 z^{2\tau_1} + \cdots . \qquad \text{(C-1)}$$

Multiplication of equation (C-1) by $c_1 z^{\tau_1}$ produces

$$c_1 z^{\tau_1} R_1(z) = c_1 z^{\tau_1} - c_1^2 z^{2\tau_1} + c_1^3 z^{3\tau_1} + \cdots , \qquad \text{(C-2)}$$

while addition of equations (C-1) and (C-2) yields the expression

$$R_1(z) = \frac{1}{1 + c_1 z^{\tau_1}} . \qquad \text{(C-3)}$$

We have thus obtained the z-transform of the first-order reverberation portion of the two-layer impulse response.

(C.1) *Impulse response of second-order component*

Figure C-1 indicates the pattern of all raypaths which contribute to second-order reverberations. A component from the original impulse is reflected from reflector 3 and re-enters the water layer to be reflected from the surface. At this point its amplitude is $-t_1' c_2 t_1$, where t_1' is the upward transmission coefficient across reflector 2. Hence an impulse of amplitude $-t_1' c_2 t_1$ is introduced into the first layer, which again will generate the associated first-order ringing already given by equation (C-3). However, the onset of this ringing occurs with a time delay of $\tau_1 + \tau_2$, where τ_2 is the two-way traveltime in the second layer. The z-transform of this component is

$$R_{2,1}(z) = z^{\tau_1+\tau_2}(-t_1' c_2 t_1) \frac{1}{1 + c_1 z^{\tau_1}} ,$$

where the subscripts of $R(z)$ denote response order and associated components, respectively. Likewise, the next pulse entering the water layer in the above manner generates the second component of the second-order response,

$$R_{2,2}(z) = z^{2\tau_1+\tau_2}(t_1' c_2 t_1 c_1) \frac{1}{1 + c_1 z^{\tau_1}} .$$

The third component of the second-order response is

$$R_{2,3}(z) = z^{3\tau_1+\tau_2} (-t_1' c_2 t_1 c_1^2) \frac{1}{1 + c_1 z^{\tau_1}} .$$

One can continue this analysis up to any number of additional components. Summation of the above series of equations produces the complete second-order response,

$$\begin{aligned} R_2(z) &= R_{2,1}(z) + R_{2,2}(z) + R_{2,3}(z) + \cdots \\ &= z^{\tau_1+\tau_2} (-t_1' c_2 t_1) \frac{1}{1 + c_1 z^{\tau_1}} \cdot [1 - c_1 z^{\tau_1} + c_1^2 z^{2\tau_1} + \cdots] \\ &= z^{\tau_1+\tau_2}(-t_1' c_2 t_1) \frac{1}{(1 + c_1 z^{\tau_1})^2} \end{aligned}$$

where we recall that $|c_1| < 1$. Since $z^{\tau_1+\tau_2}$ is simply a delay factor and $-t_1' c_2 t_1$ is a constant, we may shift the time origin and normalize the second-order response. This yields

$$R_2(z) = \frac{1}{(1 + c_1 z^{\tau_1})^2} , \qquad \text{(C-4)}$$

an expression which we see to be the square of the first-order response given by equation (C-3).

Removal of first-order ringing

Let us incorporate the first-order impulse response into the predictive deconvolution model. We will assume that the two-way traveltime through the water layer is τ_1 sample units. Thus our impulse response becomes,

$$x_1(t) = 1, \underbrace{0, 0, \cdots, 0,}_{\tau_1 - 1 \text{ zeros}} -c_1, \underbrace{0, 0, \cdots, 0,}_{\tau_1 - 1 \text{ zeros}} c_1^2, \cdots .$$

In order to compute the predictive filter, we require the autocorrelation of $x_1(t)$, which is

$$\begin{aligned} r_\tau &= 1 + c_1^2 + c_1^4 + \cdots , \qquad \tau = 0. \\ r_\tau &= 0, \qquad 0 < \tau < \tau_1. \\ r_\tau &= -c_1(1 + c_1^2 + c_1^4 + \cdots) \\ &= -c_1 r_0, \qquad \tau = \tau_1, \end{aligned}$$

and so on. Thus the autocorrelation of $x_1(t)$ can

be written

$$r_\tau = E_x, \underbrace{0, 0, \cdots, 0}_{\tau_1 - 1 \text{ zeros}}, -cE_x, \cdots,$$

where E_x is the energy in $x_1(t)$. Let the filter length n be *less* than τ_1, and let the prediction distance be $\alpha = \tau_1$. Then the normal equations become

$$\begin{bmatrix} r_0 & 0 & \cdots & 0 \\ 0 & r_0 & \cdots & 0 \\ & & \vdots & \\ 0 & 0 & \cdots & r_0 \end{bmatrix} \begin{bmatrix} a_0 \\ a_1 \\ \vdots \\ a_{n-1} \end{bmatrix} = \begin{bmatrix} r_{\tau_1} \\ 0 \\ \vdots \\ 0 \end{bmatrix}.$$

The only member of this system whose right side does not vanish is,

$$r_0 a_0 = r_{\tau_1},$$

and thus,

$$a_0 = r_{\tau_1}/r_0 = -c_1 r_0/r_0 = -c_1.$$

The associated prediction error operator is

$$f_2(t) = 1, \underbrace{0, 0, \cdots, 0}_{\tau_1 - 1 \text{ zeros}}, c_1. \qquad \text{(C-5)}$$

In practice it is not necessary to set the prediction distance α exactly to τ_1. The present model permits α to take on any value as long as it is less than or equal to τ_1. Furthermore, the filter length must be such that the inequality $\alpha + n > \tau_1$ holds true.

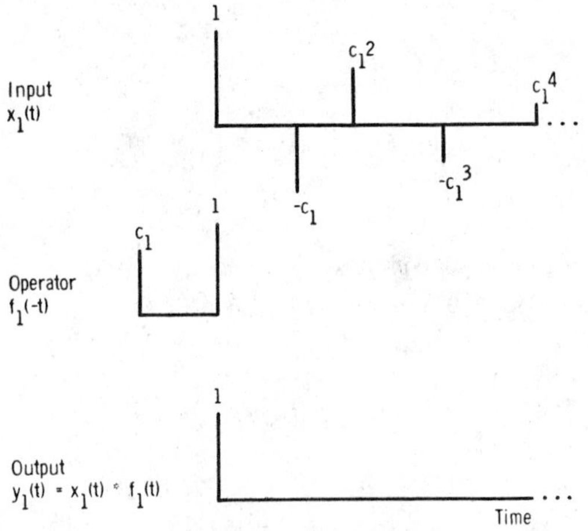

FIG. C-2. Deconvolution of a first-order ringing system. The operator is shown in time-reversed form.

We note that the z-transform of the prediction error operator of equation (C-5) is $1 + c_1 z^{\tau_1}$ which is the inverse of the first-order impulse response given by equation (C-3). If this first-order impulse response is convolved with the above prediction error filter, the output will be 1; in other words, we will have deconvolved the ringing signal. Figure C-2 illustrates the input signal $x_1(t)$, the prediction error operator $f_1(t)$, and the output signal $y_1(t)$ for this situation.

Removal of second-order ringing

Let us use the predictive deconvolution method on the second-order portion of the two-layer impulse response given by equation (C-4). This response can be written,

$$x_2(t) = 1, \underbrace{0, 0, \cdots, 0}_{\tau_1 - 1 \text{ zeros}}, \quad -2c_1, \underbrace{0, 0, \cdots, 0}_{\tau_1 - 1 \text{ zeros}},$$

$$3c_1^2, \underbrace{0, 0, \cdots, 0}_{\tau_1 - 1 \text{ zeros}}, \quad -4c_1^3, \cdots.$$

The autocorrelation of $x_2(t)$ is

$$r_\tau = 1 + 4c_1^2 + 9c_1^4 + 16c_1^9 + \cdots$$

$$= \frac{1 + c_1^2}{(1 - c_1{}^2)^3}, \qquad \tau = 0.$$

$$r_\tau = 0, \qquad 0 < \tau < \tau_1.$$

$$r_\tau = -2c_1 - 6c_1^3 - 12c_1^5 - 20c_1^7 + \cdots$$

$$= \frac{-2c_1}{(1 - c_1{}^2)^3}, \qquad \tau = \tau_1.$$

$$r_\tau = 0, \qquad \tau_1 < \tau < 2\tau_1.$$

$$r_\tau = 3c_1^2 + 8c_1^4 + 15c_1^6 + 24c_1^8 + \cdots$$

$$= \frac{-c_1^4 + 3c_1^3}{(1 - c_1^2)^3}, \qquad \tau = 2\tau_1,$$

and so on. Thus the normalized autocorrelation of x_t becomes

$$r_\tau = 1 + c_1^2, \underbrace{0, 0, \cdots, 0}_{\tau_1 - 1 \text{ zeros}},$$

$$-2c_1, \underbrace{0, 0, \cdots, 0}_{\tau_1 - 1 \text{ zeros}}, \quad 3c_1^2 - c_1^4, \cdots.$$

If the filter length is $n=\tau_1+1$ and the prediction distance is $\alpha=\tau_1$, the normal equations become,

$$\begin{array}{c}\\ \\ \tau_1-1 \\ \text{rows}\end{array}\left\{\begin{bmatrix} 1+c_1^2 & \overbrace{0 \quad 0 \quad \cdots \quad 0}^{\tau_1-1\ \text{columns}} & -2c_1 \\ 0 & 0 \quad 0 \quad \cdots \quad 0 & 0 \\ & \vdots & \\ 0 & 0 \quad 0 \quad \cdots \quad 0 & 0 \\ -2c_1 & 0 \quad 0 \quad \cdots \quad 0 & 1+c_1^2 \end{bmatrix}\right. \begin{bmatrix} a_0 \\ a_1 \\ \vdots \\ a_{\tau_1-1} \\ a_{\tau_1} \end{bmatrix} = \begin{bmatrix} -2c_1 \\ 0 \\ \vdots \\ 0 \\ 3c_1^2 - c_1^4 \end{bmatrix}.$$

The two nonvanishing equations of this system yield the solution

$$a_0 = -2c_1$$

and

$$a_{\tau_1} = -c_1^2.$$

Hence the associated prediction error operator is

$$f_2(t) = 1, \underbrace{0, 0, \cdots, 0,}_{\tau_1-1\ \text{zeros}}$$

$$2c_1, \underbrace{0, 0, \cdots, 0,}_{\tau_1-1\ \text{zeros}} c_1^2. \qquad \text{(C-6)}$$

This particular prediction error operator is identical to the three-point filter of Backus (1959). We thus see that in the noise-free case the present predictive deconvolution model yields the classical results obtained on the basis of strictly deterministic considerations. The predictive deconvolution scheme allows a more general attack on the dereverberation problem, as the present treatment has sought to demonstrate.

It is not necessary to set the prediction distance exactly equal to τ_1, nor is it necessary to set the filter length exactly equal to τ_1+1. However, these parameters must be set such that $\alpha \leq \tau_1$, $n \geq \tau_1$, and $\alpha + n \geq 2\tau_1$.

The z-transform of equation (C-6) is

$$F_2(z) = 1 + 2c_1 z^{\tau_1} + c_1^2 z^{2\tau_1} = (1 + c_1 z^{\tau_1})^2,$$

which is the inverse of the z-transform of the second-order impulse response given by equation (C-4). Thus, convolution of the second-order impulse response $x_2(t)$ with the prediction error operator $f_2(t)$ produces a zero delay spike (Figure C-3). In other words, the second-order ringing system has been deconvolved by means of the prediction error operator.

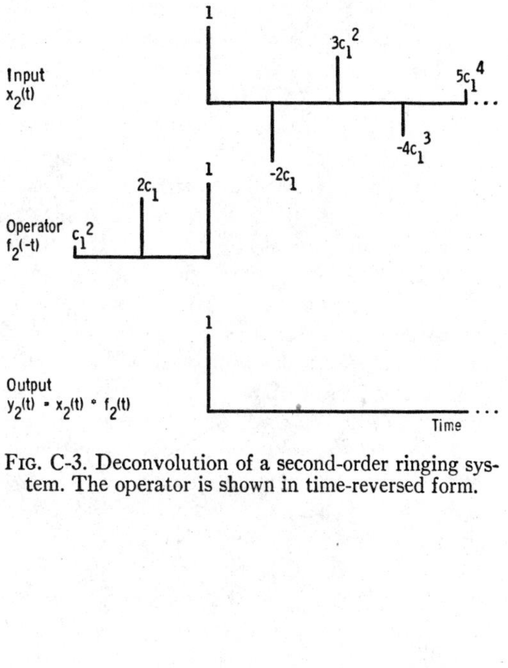

FIG. C-3. Deconvolution of a second-order ringing system. The operator is shown in time-reversed form.

Reprinted from Geophysics v. 36, no. 2, p. 261-265

PREDICTION ERROR AS A CRITERION FOR OPERATOR LENGTH†

JAMES N. GALBRAITH*

Prediction error filtering has been widely used for deconvolution. The mean squared error in prediction is a monotonically nonincreasing function of operator length, and the value of the error is readily available from the Wiener-Levinson algorithm. In general, the value of this error for the infinitely long operator is not known a priori. It is shown that the final value of the error can be obtained by considering the Kolmogorov spectrum factorization. Simple criteria can then be established for operator effectiveness and length.

Prediction error filtering, which has been used for the deconvolution of seismic records for several years, has proved to be one of the most successful techniques of deconvolution; the Wiener-Levinson algorithm is an efficient way of computing the filter. A by-product of this algorithm is the value of the prediction error at each operator length as the operator is extended. The mean squared error is the variance of the filtered output and is a monotonically nonincreasing function of operator length. A function of this sort conceivably could provide a means of determining a suitable operator length. One need only extend the operator until the error falls to an acceptable level and then stop. Unfortunately, the normalization of the error in general is not known, so that one does not know a priori to what value it should converge. The difficulty becomes apparent when we consider a case of the type shown in Figure 1. For a unit length operator, the prediction error starts at r_0, the variance of the input. It then drops down and levels off before falling to its final value, ϵ_∞. If one simply monitors the prediction error without knowing the final value, one might conclude that the error had converged to ϵ_∞ when it first leveled off. If the final value of the error were known, no such false conclusion would be drawn.

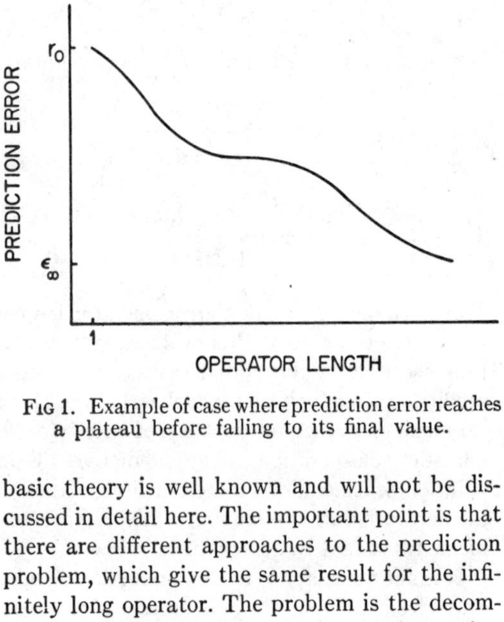

FIG 1. Example of case where prediction error reaches a plateau before falling to its final value.

This paper discusses a way to determine ϵ_∞. The basic theory is well known and will not be discussed in detail here. The important point is that there are different approaches to the prediction problem, which give the same result for the infinitely long operator. The problem is the decomposition or factorization of stationary time series. Possible approaches to factorization are the least squares method with the Wiener-Levinson algorithm and the Kolmogorov spectrum factorization.

Let us now discuss a model for a stationary time series and the basic relationships which exist between the model, the prediction operator, and

† Presented at the 39th Annual International SEG Meeting, Calgary, Alberta, September 16, 1969. Manuscript received by the Editor October 30, 1969; revised manuscript received July 27, 1970.

* Geoscience Inc.; now with Mobil Research and Development Corporation, Dallas, Texas 75234.

the prediction error operator. Our model for the time series is a white light (uncorrelated) series convolved with a minimum-phase wavelet. In the seismic case, this model corresponds to an unreverberated seismic trace convolved with the reverberation operator. The reverberated trace can be written, in z-transform notation (using the z-transform convention of Robinson and Treitel, 1964), as

$$s(z) = w(z)\xi(z), \tag{1}$$

where $s(z)$ is the reverberated trace or time series, $w(z)$ is the wavelet or reverberation operator, and $\xi(z)$ is the white light series. This decomposition of the time series is unique in the sense that $w(z)$, the minimum-phase wavelet, is unique to within a scale factor; that is, the decomposition

$$s(z) = w'(z)\xi'(z), \tag{2}$$

where $w'(z) = c\, w(z)$, $\xi'(z) = (1/c)\xi(z)$, and c is a constant, is equally valid. Minimum phase decomposition is possible only for stationary time series that meet certain conditions (for example, an absolutely continuous spectral distribution) which were given by Kolmogorov and are summarized by Robinson (1954, 1967) and Galbraith (1963). In practice, we generally assume that the conditions are met. Other factorizations of the time series with nonminimum-phase wavelets are possible. However, we know that the reverberation operator is minimum phase (Sherwood and Trorey, 1965); hence, the minimum-phase decomposition is the one of interest.

Some other relationships between wavelets, prediction operators, and prediction error operators are basic to the concept of the factorization. The optimum prediction operator $p(z)$ is that operator which best (least squares) predicts the next value of the time series. It operates only on past values to predict the present value of the series and may be expressed as a polynomial in z,

$$p(z) = \sum_{i=1}^{\infty} p_i z^i. \tag{3}$$

The prediction error operator $p_e(z)$ computes the difference between the actual and the predicted time series. The output of the filtering operation is an error series, and its mean squared value is the mean squared error. Hence,

$$\begin{aligned} p_e(z) &= 1 - p(z) = 1 - \sum_{i=1}^{\infty} p_i z^i \\ &= \sum_{i=0}^{\infty} p_{e_i} z^i. \end{aligned} \tag{4}$$

The prediction error operator is minimum phase and is proportional to the inverse of the minimum-phase wavelet,

$$\begin{aligned} w(z) &= \sum_{i=0}^{\infty} w_i z^i = \frac{w_0}{p_e(z)} \\ &= \frac{w_0}{1 - \sum_{i=1}^{\infty} p_i z^i}. \end{aligned} \tag{5}$$

The prediction error operator and the wavelet which we show here are infinitely long. The prediction error operator obtained by time domain methods, such as the Wiener-Levinson algorithm, is a finite length operator which is minimum phase and is a least squares approximation to the infinitely long operator. As the length goes to infinity, the operator coefficients converge to the ones for the infinitely long case.

The formulation of the least squares prediction error problem follows easily from the prediction problem. For prediction, we determine the operator or filter such that mean square difference between the input and the output is a minimum. To determine the prediction error operator, we may use equation (4) above after $p(z)$ is determined or, equivalently, we can apply this requirement to the normal equations for the prediction operator. This procedure leads to the familiar normal equations for $p_e'(z)$,

$$\begin{bmatrix} r_0 & r_1 & r_2 & \cdots & r_n \\ r_1 & r_0 & r_1 & & \cdot \\ r_2 & & r_0 & & \cdot \\ \cdot & & & \cdot & \\ \cdot & & & \cdot & r_1 \\ r_n & \cdots & & r_1 & r_0 \end{bmatrix} \begin{bmatrix} p'_{e_0} \\ p'_{e_1} \\ p'_{e_2} \\ \cdot \\ \cdot \\ p'_{e_n} \end{bmatrix} = \begin{bmatrix} \epsilon_n \\ 0 \\ 0 \\ \cdot \\ \cdot \\ 0 \end{bmatrix}, \tag{6}$$

where r_i is the ith lag of the autocorrelation, p'_{e_i} is the least squares estimate of the prediction error operator, and ϵ_n is the expected error for a filter of length $n+1$. In this form of the normal equations, $p'_{e_0} = 1$. The p_{e_i} have been primed because they are the n-length least squares approximation to p_{e_i} and are not necessarily equal to p_{e_i} as long as n is finite. As the Wiener-Levinson iteration proceeds (Wiggins and Robinson, 1965), the next value of p'_{e_i} is determined; all previous values of p'_{e_i} for $i \geq 0$ may change; and ϵ_n decreases or remains constant.

The expected error ϵ_n is the variance of the output of the $(n+1)$-length prediction error operator; it is the final value of this, ϵ_∞, which we would like to know. The difficulty encountered in finding ϵ_∞ with the Wiener-Levinson or any other time domain algorithm for the solution of the least squares problem can be seen by considering equations (1) and (5). The wavelet shape w_i is unknown, since it is proportional to the inverse of the prediction error operator p_{e_i}; and only a least squares approximation of the first $n+1$ terms of p_{e_i} is known from the solution of equation (6). The variance of ξ_i is also unknown. Therefore, the mean squared value of the output of the infinitely long operator cannot be computed. The basic ambiguity in the factorization as shown in equation (2) is not particularly important in determining ϵ_∞. Knowledge of $w(z)$ and $\xi(z)$ for one particular factorization is sufficient.

As we have just seen, this knowledge is not readily available from the least squares formulation and solution to the problem. There are other approaches to the problem of finding the prediction error operator or minimum-phase wavelet which do lead to relatively simple methods of finding ϵ_∞. One such approach is the Kolmogorov spectrum factorization. The Kolmogorov technique can be used to find the minimum-phase wavelet from the power spectrum of the process. This method, a frequency domain method, permits one to compute the first n terms of the complete wavelet. The Wiener-Levinson algorithm, a time-domain method, permits one to compute an n-term least squares approximation to the prediction error operator. In the limit as n goes to infinity, the methods are equivalent, since the Kolmogorov method yields $w(z)$ and the Wiener-Levinson yields $p_e(z)$, and these are related by equation (5). One could use the Hilbert z-transform instead of the Kolmogorov factorization, as was shown by Sakrison et al (1967), or the Wold decomposition (e.g., Robinson, 1954). Details of the Kolmogorov factorization have been described in the literature (Kolmogorov, 1939; Robinson, 1954, 1967; Galbraith, 1963) and will not be repeated here. However, a brief description of the mathematics is in order. The problem is to find a minimum-phase time function given the power spectrum $\Phi(\omega)$ of the wavelet. Since the time series is described by equation (1), $\Phi(\omega)$ is the power spectrum of $s(z)$, the seismic trace. The amplitude of the Fourier transform of the wavelet is, therefore, the square root of the power spectrum. The phase is unknown and must be determined according to the minimum-phase condition. We represent the wavelet by $B(\omega)$,

$$B(\omega) = A(\omega)e^{i\phi(\omega)} = \sqrt{\Phi(\omega)}e^{i\phi(\omega)}, \quad (7)$$

where $A(\omega)$ is the amplitude and is known and $\phi(\omega)$ is the phase and is unknown. We now can take the logarithm of $B(\omega)$ and obtain an expression which involves $\phi(\omega)$ as a linear term. The minimum-phase condition is that there are no poles or zeros inside the unit circle; hence, the function is analytic in this region and may be expanded in a Taylor series. We can then find the coefficients of the series, since we know the value of the function on the unit circle. This is just a boundary value problem where there are no sources or sinks inside the boundary and the value of the function $[\frac{1}{2}\log\Phi(\omega)]$ is given on the boundary (unit circle). The existence and uniqueness theorems which apply to boundary value problems apply here, so that existence and uniqueness are established.

The result of the factorization is an expression for $k(z)$, the minimum-phase wavelet. This expression is

$$k_0 = \exp\left[\frac{1}{2\pi}\int_0^\pi \log\Phi(\omega)d\omega\right] \quad (8)$$

for the first term, and

$$k(z) = k_0 \exp\left[\frac{1}{\pi}\sum_{m-1}^{\infty} z^m \int_0^\pi \cdot\log\Phi(\omega)\cos m\omega\, d\omega\right] \quad (9)$$

for the entire wavelet. The time series is then described by

$$s(z) = k(z)\eta(z),$$

where $\eta(z)$ is a white light series with unit variance. As we shall see, the fact that $\eta(z)$ has unit variance is extremely important.

Let us now compare two minimum-phase

factorizations of the time series $s(z)$:

$$s(z) = w(z)\xi(z); \tag{10}$$

$$s(z) = k(z)\eta(z), \tag{11}$$

where $w(z)$ and $k(z)$ are proportional to one another, since both are minimum phase with power spectra which are proportional. The mean squared error in prediction is simply the variance of the output of the prediction error operation. Therefore, we can operate on $s(z)$ with the infinitely long prediction error operator $p_e(z)$ and take the variance of the result to obtain ϵ_∞. Thus,

$$\begin{aligned}\epsilon_\infty &= \text{var}\,[p_e(z)s(z)] \\ &= \text{var}\,[p_e(z)k(z)\eta(z)].\end{aligned} \tag{12}$$

Using equation (5), we have $p_e(z)k(z)=k_0$; and, since $\text{var}[\eta(z)]=1$,

$$\epsilon_\infty = k_0^2. \tag{13}$$

Similarly, from equation (10),

$$\epsilon_\infty = w_0^2 \text{ var } [\xi(z)]; \tag{14}$$

that is, the expected error for the Wiener-Levinson algorithm approaches k_0^2 as $n \to \infty$. Once the value to which the error converges is known, it is relatively easy to set up criteria for operator length and measures of filter effectiveness. We can compute k_0 from equation (8) and then have

$$\frac{\epsilon_n}{k_0} \underset{n\to\infty}{\to} 1. \tag{15}$$

The quantity ϵ_n is the expected error for the nth iteration $[(n+1)$-length filter$]$ and can be monitored as the filter is extended. Other measures are also useful. The zero lag of the autocorrelation r_0 is the variance of the input series and is a measure of the total energy. The prediction error for the nth iteration ϵ_n is the variance or energy of the output; and the output is the unpredicted amount. The minimum value of ϵ_n is ϵ_∞. The ratio

$$\frac{r_0 - \epsilon_\infty}{r_0} \tag{16}$$

is then the fraction of the original energy which is predictable, and, for a filter of length $n+1$, the ratio

$$\frac{r_0 - \epsilon_n}{r_0} \tag{17}$$

is the fraction of energy actually predicted. When expressed as a percent, expression (17) is called the percent reduction (Wadsworth et al, 1953). The ratio of the two quantities (16) and (17),

$$\frac{r_0 - \epsilon_n}{r_0 - \epsilon_\infty}, \tag{18}$$

is then the fraction of predictable energy which is actually predicted. This is a quantitative measure of operator effectiveness as a function of filter length, which can be used as a criterion for filter length.

The ratio given in expression (17) is also useful by itself as a measure of operator effectiveness. For example, if n is large so that one is fairly sure that $\epsilon_n \simeq \epsilon_\infty$, and if $(r_0-\epsilon_n)/r_0$ is small compared to 1, the amount of predictable energy is small and deconvolution is probably a waste of time and money. Similarly, whenever $(r_0-\epsilon_n)/r_0$ is small, deconvolution is ineffective with that length filter.

Another way of deriving the expression for the final value of the prediction error once the wavelet $k(z)$ is known leads to a more general expression, which is useful when gapped deconvolution is used. By gapped deconvolution, we mean prediction error filtering with a prediction distance greater than one. We have the z-transform notation for $s(z)$ as given in equation (11) and the equivalent time domain expression,

$$s_t = \sum_{i=0}^{\infty} k_i \eta_{t-i}, \tag{19}$$

where η_t is the unit-variance, zero-mean series and k_i is the Kolmogorov wavelet. If we wish to predict s_t at some time $t+m$ in the future, that is, wish to predict s_{t+m}, we can use only the information on hand at time t. This information, as can be seen from relation (19), is all of k_i and η_t up to time t. The predicted value of s_{t+m}, $\hat{s}_{t+m}$, can be written from equation (19) as

$$\hat{s}_{t+m} = \sum_{i=0}^{\infty} k_i \eta_{t+m-i}.$$

However, η_{t+1} to η_{t+m} are not known; and the best estimate of them is their average, which is zero. Hence,

$$\hat{s}_{t+m} = \sum_{i=m}^{\infty} k_i \eta_{t+m-i}.$$

The mean squared error in prediction ϵ is then the average value of the squared difference between the actual and predicted values of the time series.

$$\begin{aligned} \epsilon &= E(s_{t+m} - \hat{s}_{t+m})^2 \\ &= E\left[\sum_{i=0}^{\infty} k_{i+m}\eta_{t-i+m} - \sum_{i=m}^{\infty} k_i \eta_{t-i+m}\right]^2 \\ &= E\left[\sum_{i=0}^{m-1} k_i \eta_{t+m-i}\right]^2 = \sum_{i=0}^{m-1} k_i^2 E(\eta_t)^2 \qquad (20) \\ &= \sum_{i=0}^{m-1} k_i^2. \end{aligned}$$

This reduces to k_0^2 for unit prediction distance ($m=1$) as we saw before. If we wish to compute the final value of the expected error for $m>1$, we need to compute m terms of the wavelet using equation (9), and the computation expense increases. If simple deconvolution ($m=1$) is of interest, relation (8) can be used.

It is probable that one can gain a reasonably good idea concerning the convergence of the error to its final value by computing k_0^2 for only a few traces in a section, so that some quality control can be gained at modest expense, particularly if fast Fourier transform (FFT) hardware is available for the computation of $\Phi(\omega)$.

We have seen that the final value of the expected error in prediction can be computed from the power spectrum of the time series, and that several useful quantitative measures can be defined easily. The most practical measure in terms of cost is the simple ratio

$$R = \frac{r_0 - \epsilon_n}{r_0},$$

where $0 \le R \le 1$. This is the effectiveness of an $(n+1)$-length filter. It is useful even when ϵ_∞ is not computed, since it is the fraction of energy predicted by that filter. If R is small compared to 1, the filter will not be effective. If it is close to one, the filter will be effective. If one wants to know how effective the most effective filter can be, then ϵ_∞ must be computed.

REFERENCES

Galbraith, J. N., 1963, Computer studies of microseism statistics with applications to prediction and detection: Ph.D. Thesis, Mass. Inst. of Tech.

Kolmogorov, A., 1939, Sur l'interpolation et extrapolation des suites stationaires: C. R. Acad. Sci. Paris, v. 208, p. 2043–2045.

Robinson, E. A., 1954, Predictive decomposition of time series with applications to seismic exploration: Ph.D. Thesis, Mass. Inst. of Tech.

——— 1967, Predictive decomposition of time series with applications to seismic exploration: Geophysics, v. 32, p. 428–484.

Robinson, E. A., and Treitel, S., 1964, Principles of digital filtering: Geophysics, v. 29, p. 395–404.

Sakrison, D. J., Ford, W. T., and Hearne, J. H., The z transform of a realizable time function: IEEE Trans. Geo. Elect., v. GE-5, no. 2.

Sherwood, J. W. C., and Trorey, A. W., 1965, Minimum phase and related properties of the response of a horizontally stratified absorptive earth to plane acoustic waves: Geophysics, v. 30, p. 191–197.

Wadsworth, G. P., Robinson, E. A., Bryan, J. G., and Hurley, P. M., 1953, Detection of reflections on seismic records by linear operators: Geophysics, v. 18, p. 539–586.

Wiggins, R. A., and Robinson, E. A., 1965, Recursive solution to the multichannel filtering problem: J. Geophys. Res., v. 70, p. 1885–1891.

Reprinted from Geophysics v. 38, no. 2, p. 310-326

THE DETERMINATION OF DIGITAL WIENER FILTERS BY MEANS OF GRADIENT METHODS†

R. J. WANG* AND S. TREITEL*

The normal equations for the discrete Wiener filter are conventionally solved with Levinson's algorithm. The resultant solutions are exact except for numerical roundoff. In many instances, approximate rather than exact solutions satisfy seismologists' requirements. The so-called "gradient" or "steepest descent" iteration techniques can be used to produce approximate filters at computing speeds significantly higher than those achievable with Levinson's method. Moreover, gradient schemes are well suited for implementation on a digital computer provided with a floating-point array processor (i.e., a high-speed peripheral device designed to carry out a specific set of multiply-and-add operations). Levinson's method (1947) cannot be programmed efficiently for such special-purpose hardware, and this consideration renders the use of gradient schemes even more attractive.

It is, of course, advisable to utilize a gradient algorithm which generally provides rapid convergence to the true solution. The "conjugate-gradient" method of Hestenes (1956) is one of a family of algorithms having this property. Experimental calculations performed with real seismic data indicate that adequate filter approximations are obtainable at a fraction of the computer cost required for use of Levinson's algorithm.

INTRODUCTION

The standard formulation of the discrete single-channel Wiener filter problem (Treitel and Robinson, 1966) leads to a system of normal equations, which we shall write in the form:

$$\mathbf{AU} = \mathbf{B}.$$

Here $\mathbf{A}$ is an $n \times n$ input autocorrelation matrix, $\mathbf{U}$ is an $n \times 1$ column vector of the n unknown filter coefficients $(u_1, u_2, \cdots, u_n)$, and $\mathbf{B}$ is an $n \times 1$ column vector of crosscorrelation coefficients. This system is commonly solved by means of an algorithm due to Levinson (1947), which has been described in simplified terms by Treitel and Robinson (1966, Appendix III). Levinson's solution is exact except for numerical roundoff, and the computation time is proportional to the square of the number of filter coefficients, namely n^2. Quite often, one will be satisfied with only an approximate solution; in such cases one has recourse to the so-called "gradient" schemes, which produce acceptable estimates of the exact solution in substantially shorter computation times. Most gradient methods have the further advantage that they are particularly suited for implementation on special-purpose floating-point multiply-and-add hardware. On the other hand, the structure of the Levinson algorithm does not lend itself to an efficient utilization of such equipment, and this circumstance results in additional benefits from the gradient schemes.

A gradient[1] method is based on the principle that an original guess of the solution vector $\mathbf{U}$ can be iteratively adjusted in accordance with the progressive diminution in magnitude of a specified error function. At every step of such an iteration, one strives to make this magnitude decrease

[1] The name "gradient method" is not unique in the literature. These schemes are variously called the "method of steepest descent," the "hill-climbing method," and the "method of stochastic approximation."

† Manuscript received by the Editor June 13, 1972; revised manuscript received August 23, 1972.

* Amoco Production Company, Tulsa, Oklahoma 74102.

as much as possible, for in this manner, a good approximation to **U** will be obtained in a minimum number of steps.

The successive estimates $\mathbf{X}_i$ of the true solution **U** can be conveniently described by the relation

$$\mathbf{X}_i = \mathbf{X}_{i-1} + \alpha_{i-1}\mathbf{P}_{i-1}, \qquad i = 1, 2, \cdots,$$

where X_i is an $n\times 1$ column vector which constitutes the ith estimated solution of the system $\mathbf{AU}=\mathbf{B}$. The $n\times 1$ vectors $\mathbf{P}_i$ are called "direction vectors," whose explicit form depends on a particular choice of the error function, while the α_i's are suitable positive real numbers, which may or may not depend on the estimates $\mathbf{X}_i$. The iteration always commences with a selection of the initial approximant $\mathbf{X}_0$. In general, the number of iterations required to reduce the error to a predetermined level is small when the deviation of $\mathbf{X}_0$ from the true solution **U** is also small. Since the computing time is directly proportional to the number of iterations, such a choice of $\mathbf{X}_0$ is obviously appropriate.

Sometimes, it is desirable to solve a new set of normal equations of like order for each successive trace of a seismic record. Alternatively, time-adaptive Wiener filtering requires the solution of two or more systems of normal equations of like order for successive gates of a given trace (Wang, 1969). In either event one can frequently assume that the Wiener filters do not differ drastically from trace to trace or from gate to gate. For such situations, one might begin by using Levinson's method on the first trace or the first gate, as the case may be. Then, once the exact filter has been found, the coefficients of this solution can be chosen as the elements of the initial approximant $\mathbf{X}_0$ for the second trace, or gate. The approximate solution obtained in this manner is next identified with the initial approximant for the third trace or gate, and this process continues until all input data have been so treated.

Many variants of the gradient technique have been proposed in the literature. Here we shall describe the gradient method of Forsythe and Wasow (1960) and the so-called "conjugate-gradient" method of Hestenes (1956), which are particularly suited for the solution of the normal equations. In the sequel, these two techniques will be identified, respectively, as the *g*-method and the *c-g*-method. Their application to Wiener filtering problems will be illustrated by means of very simple numerical examples, as well as with actual seismic data.

THE GRADIENT METHOD OF FORSYTHE AND WASOW

We first describe the *g*-method of Forsythe and Wasow (1960, p. 224 ff). This scheme is actually less efficient than the alternate *c-g*-method of Hestenes (1956), but it serves as a very convenient introduction to the general subject of gradient techniques.

Consider the system of normal equations

$$\mathbf{AU} = \mathbf{B}, \tag{1}$$

where **A** is an $n\times n$ symmetric and positive definite autocorrelation matrix (Robinson, 1967, p. 43). Let $\mathbf{X}_i$ be the ith estimate of the solution vector **U**, and let the ith residual vector $\mathbf{R}_i$ be given by

$$\mathbf{R}_i = \mathbf{B} - \mathbf{AX}_i. \tag{2}$$

When $\mathbf{X}_i=\mathbf{U}$, we have, by (1),

$$\mathbf{R}_i = \mathbf{B} - \mathbf{AU} = \boldsymbol{\theta},$$

where $\boldsymbol{\theta}$ is the null vector. Hence, the energy, or the sum of the squares of the elements of $\mathbf{R}_i$ is zero. In general, $\mathbf{X}_i\neq\mathbf{U}$, and the objective of any gradient technique is to reduce the energy of $\mathbf{R}_i$, or of a function which is related to this energy, as rapidly as possible to zero. The energy of $\mathbf{R}_i$ can be expressed in the form

$$\mathbf{R}_i^T\mathbf{R}_i,$$

where the superscript $(^T)$ denotes transposition. But (2) can be rewritten as

$$\mathbf{R}_i = \mathbf{A}(\mathbf{A}^{-1}\mathbf{B} - \mathbf{X}_i), \tag{3}$$

where $\mathbf{A}^{-1}$ is the inverse of **A**. As a result, we have

$$\mathbf{R}_i^T\mathbf{R}_i = (\mathbf{A}^{-1}\mathbf{B} - \mathbf{X}_i)^T\mathbf{A}^T\mathbf{A}(\mathbf{A}^{-1}\mathbf{B} - \mathbf{X}_i),$$

or

$$\mathbf{R}_i^T\mathbf{R}_i = (\mathbf{A}^{-1}\mathbf{B} - \mathbf{X}_i)^T\mathbf{A}^2(\mathbf{A}^{-1}\mathbf{B} - \mathbf{X}_i),$$

since **A** is symmetric, that is $\mathbf{A}^T=\mathbf{A}$. Although it is possible to derive a gradient method which progressively diminishes the magnitude of $\mathbf{R}_i^T\mathbf{R}_i$ as i increases, a computationally more attractive scheme is based on the progressive reduction in magnitude of the function

$$\mathbf{E}(\mathbf{X}_i) = \mathbf{R}_i^T\mathbf{A}^{-1}\mathbf{R}_i.$$

The scalar-valued quantity $\mathbf{E}(\mathbf{X}_i)$ is called the "error function," which becomes, by virtue of (3):

$$\mathbf{E}(\mathbf{X}_i) = (\mathbf{A}^{-1}\mathbf{B} - \mathbf{X}_i)^T\mathbf{A}(\mathbf{A}^{-1}\mathbf{B} - \mathbf{X}_i). \quad (4)$$

Since $\mathbf{A}$ is positive definite, its inverse $\mathbf{A}^{-1}$ is also positive definite, and, thus, $\mathbf{E}(\mathbf{X}_i)$ is a real quadratic form whose minimum value of zero occurs when $\mathbf{X}_i=\mathbf{A}^{-1}\mathbf{B}$, that is, when $\mathbf{X}_i=\mathbf{U}$ (see Hildebrand, 1965, p. 48 and p. 70).

Expanding (4), we obtain

$$\mathbf{E}(\mathbf{X}_i) = \mathbf{X}_i^T\mathbf{A}\mathbf{X}_i - 2\mathbf{X}_i^T\mathbf{B} + \mathbf{B}^T\mathbf{A}^{-1}\mathbf{B}, \quad (5)$$

where we have made use of the fact that $\mathbf{A}^{-1}$ is symmetric since $\mathbf{A}$ is symmetric (Hildebrand, 1965, p. 57). The direction along which $\mathbf{E}(\mathbf{X}_i)$ diminishes most rapidly with respect to a given approximant $\mathbf{X}_i$ is the so-called "direction of steepest descent," which is obtainable from the negative, or "downhill" *gradient*,

$$
\begin{aligned}
-\frac{\partial}{\partial \mathbf{X}_i}[\mathbf{E}(\mathbf{X}_i)] &= -\frac{\partial}{\partial \mathbf{X}_i}[\mathbf{X}_i^T\mathbf{A}\mathbf{X}_i - 2\mathbf{X}_i^T\mathbf{B} + \mathbf{B}^T\mathbf{A}^{-1}\mathbf{B}] \\
&= 2(\mathbf{B} - \mathbf{A}\mathbf{X}_i) \\
&= 2\mathbf{R}_i.
\end{aligned}
$$

Since $\mathbf{E}(\mathbf{X}_i)$ thus diminishes most rapidly in the direction of the residual vector $\mathbf{R}_i$, Forsythe and Wasow choose the iteration

$$\mathbf{X}_i = \mathbf{X}_{i-1} + \alpha_{i-1}\mathbf{R}_{i-1}, \quad (6)$$

where α_{i-1} is a real and positive number yet to be found. We thus see that, for this particular choice of the error function $\mathbf{E}(\mathbf{X}_i)$, the direction vector $\mathbf{P}_i$ introduced in the previous section is made identical to the residual vector $\mathbf{R}_i$.

We will want to determine α_{i-1} in such a way that the error function $\mathbf{E}(\mathbf{X}_i)$ is minimized for each i. This can be done by first substituting the iteration formula (6) into equation (5),

$$
\begin{aligned}
E(\mathbf{X}_i) = {} & \mathbf{X}_{i-1}^T\mathbf{A}\mathbf{X}_{i-1} + 2\alpha_{i-1}\mathbf{R}_{i-1}^T\mathbf{A}\mathbf{X}_{i-1} \\
& + \alpha_{i-1}^2\mathbf{R}_{i-1}^T\mathbf{A}\mathbf{R}_{i-1} - 2\mathbf{X}_{i-1}^T\mathbf{B} \\
& - 2\alpha_{i-1}\mathbf{R}_{i-1}^T\mathbf{B} + \mathbf{B}^T\mathbf{A}^{-1}\mathbf{B}.
\end{aligned}
$$

The value of α_{i-1} which minimizes $\mathbf{E}(\mathbf{X}_i)$ is obtained by differentiating this error function with respect to α_{i-1} and equating the result to zero,

$$
\begin{aligned}
\frac{\partial \mathbf{E}(\mathbf{X}_i)}{\partial \alpha_{i-1}} = {} & 2\mathbf{R}_{i-1}^T\mathbf{A}\mathbf{X}_{i-1} + 2\alpha_{i-1}\mathbf{R}_{i-1}^T A\mathbf{R}_{i-1} \\
& - 2\mathbf{R}_{i-1}^T\mathbf{A} = 0.
\end{aligned}
$$

Furthermore,

$$\frac{\partial^2 \mathbf{E}(\mathbf{X}_i)}{\partial \alpha_{i-1}^2} = 2\mathbf{R}_{i-1}^T\mathbf{A}\mathbf{R}_{i-1}.$$

But $\mathbf{A}$ is positive definite, and, hence, $\mathbf{R}_{i-1}^T\mathbf{A}\mathbf{R}_{i-1} > 0$. This means that the value obtained by solving the equation $\partial\mathbf{E}(\mathbf{X}_i)/\partial\alpha_{i-1}=0$ for α_{i-1} indeed minimizes $\mathbf{E}(\mathbf{X}_i)$. This solution is

$$\alpha_{i-1} = \frac{\mathbf{R}_{i-1}^T\mathbf{R}_{i-1}}{\mathbf{R}_{i-1}^T\mathbf{A}\mathbf{R}_{i-1}}. \quad (7)$$

The algorithm for the *g*-method can thus be presented in the form of the three relations

$$
\left.
\begin{aligned}
&\mathbf{R}_{i-1} = \mathbf{B} - \mathbf{A}\mathbf{X}_{i-1}, \\
&\alpha_{i-1} = \frac{\mathbf{R}_{i-1}^T\mathbf{R}_{i-1}}{\mathbf{R}_{i-1}^T\mathbf{A}\mathbf{R}_{i-1}}, \\
&\mathbf{X}_i = \mathbf{X}_{i-1} + \alpha_{i-1}\mathbf{R}_{i-1},
\end{aligned}
\right\} \quad i = 1, 2, \cdots, \quad (8)
$$

where $\mathbf{X}_0$ is a suitable initial approximant to the solution $\mathbf{U}$. Appendix A demonstrates that this algorithm converges with increasing i; namely, that

$$\lim_{i\to\infty} \mathbf{X}_i = \mathbf{U}.$$

It is evident from equation (7) that the parameter α_{i-1} is a function of the residual, or direction vector $\mathbf{R}_{i-1}$, and that this parameter must be recomputed for every step of the iteration. On the other hand, Forsythe and Wasow show that α_{i-1} can also be determined in such a manner that its value remains constant throughout the iteration. In this case, one has

$$\alpha_{i-1} = \alpha = \frac{2}{\lambda_{\min} + \lambda_{\max}}, \quad (9)$$

where $\lambda_{\min}$ and $\lambda_{\max}$ are, respectively, the smallest and largest real and positive eigenvalues of the positive definite matrix **A**. Convergence of the algorithm (8) can also be demonstrated in this case. The simplicity of (9) provides a temptation for its use in preference to (7), but the calculation of λ_{min} and λ_{max} is itself not a trivial problem, particularly if the order n of the matrix **A** is large. Furthermore, use of (9) does not guarantee that $\mathbf{E}(\mathbf{X}_i)$ be as small as possible for every i, and this in turn implies that convergence of the algorithm is slower than if (7) is used.

Let us return to the algorithm of equation (8). We see that approximately $(2n^2+3n)$MAD's[2] are required for each iteration step, where n is the order of the matrix **A**. We shall now show that this required number of MAD's can be reduced by a simple rearrangement of the algorithm. If we substitute equation (6) into the expression for the residual

$$\mathbf{R}_i = \mathbf{B} - \mathbf{A}\mathbf{X}_i, \tag{10}$$

we obtain

$$\begin{aligned} \mathbf{R}_i &= \mathbf{B} - \mathbf{A}(\mathbf{X}_{i-1} + \alpha_{i-1}\mathbf{R}_{i-1}) \\ &= (\mathbf{B} - \mathbf{A}\mathbf{X}_{i-1}) - \alpha_{i-1}\mathbf{A}\mathbf{R}_{i-1} \\ &= \mathbf{R}_{i-1} - \alpha_{i-1}\mathbf{A}\mathbf{R}_{i-1}. \end{aligned} \tag{11}$$

Setting $\mathbf{Q}_{i-1}=\mathbf{A}\mathbf{R}_{i-1}$, we can rewrite equations (7) and (8) as

$$\alpha_{i-1} = \frac{\mathbf{R}_{i-1}^T\mathbf{R}_{i-1}}{\mathbf{R}_{i-1}^T\mathbf{Q}_{i-1}}, \tag{12}$$

where

$$\mathbf{R}_{i-1} = \mathbf{R}_{i-2} - \alpha_{i-2}\mathbf{Q}_{i-2}, \tag{13}$$

and

$$\mathbf{X}_i = \mathbf{X}_{i-1} + \alpha_{i-1}\mathbf{R}_{i-1}. \tag{14}$$

Equations (12), (13), and (14), together with $\mathbf{R}_0=\mathbf{B}-\mathbf{A}\mathbf{X}_0$ and $\mathbf{Q}_{i-1}=\mathbf{A}\mathbf{R}_{i-1}$, constitute the shortened algorithm for the g-method. We see that this algorithm now requires only (n_2+4n)MAD's for each step. For large n, this algorithm will thus take only half as much computing time as the scheme of equation (8).

THE CONJUGATE-GRADIENT METHOD OF HESTENES

The g-method described in the preceding section does not yield the exact solution within a finite number of iterations. Hestenes (1956) has shown that a gradient method can be made to yield the exact solution of the system $\mathbf{AU}=\mathbf{B}$. This technique is called the conjugate-gradient method (c-g-method), and its iteration procedure terminates with an *exact* solution in m steps, where m is less than or equal to n, the order of the matrix **A**. In the c-g-method, the direction vectors $\mathbf{P}_i$, $i=0, 1, \cdots$, are a set of **N**-conjugate vectors,[3] and one of these vectors must be generated and stored to be used for each step of the iteration.

Let $\mathbf{P}_i$ and $\mathbf{g}_i$, $i=0, 1, \cdots$, be two sets of nonnull vectors satisfying the relations (Hestenes, 1956, equations 4:2, p. 88)

$$\left.\begin{aligned} \mathbf{P}_i^T\mathbf{N}\mathbf{P}_j &= 0 \quad \text{if } i \neq j \\ &\neq 0 \quad \text{if } i = j \end{aligned}\right\}, \tag{15}$$

and

$$\left.\begin{aligned} \mathbf{g}_i^T\mathbf{K}\mathbf{g}_j &= 0 \quad \text{if } i \neq j \\ &\neq 0 \quad \text{if } i = j \end{aligned}\right\}, \tag{16}$$

where **N** and **K** are both positive definite Hermitian matrices of order n. Hestenes then shows that the **N**-conjugate vectors $\mathbf{P}_i$, $i=0, 1, \cdots$, and the **K**-conjugate vectors $\mathbf{g}_i$, $i=0, 1, \cdots$, may be generated by the following algorithm (Hestenes, 1956, equations 4:1a–d, pp. 87–88):

$$a_i = \mathbf{g}_i^T\mathbf{P}_i/\mathbf{P}_i^T\mathbf{N}\mathbf{P}_i, \tag{17}$$

$$b_i = -(\mathbf{N}\mathbf{P}_i)^T(\mathbf{K}\mathbf{g}_{i+1})/\mathbf{P}_i^T\mathbf{N}\mathbf{P}_i, \tag{18}$$

$$\mathbf{g}_{i+1} = \mathbf{g}_i - a_i\mathbf{N}\mathbf{P}_i, \tag{19}$$

and

$$\mathbf{P}_{i+1} = \mathbf{K}\mathbf{g}_{i+1} + b_i\mathbf{P}_i, \tag{20}$$

where $\mathbf{P}_0=\mathbf{K}\mathbf{g}_0$. The above algorithm to generate $\mathbf{P}_i$ and $\mathbf{g}_i$, $i=0, 1, \cdots$, terminates after m steps, where $m \leq n$. For the last, or mth step, the vector $\mathbf{g}_m$ reduces to the null vector.

We shall now return to the problem of solving the system

$$\mathbf{AU} = \mathbf{B} \tag{1}$$

by means of the c-g-method. **A** is an $n\times n$ positive definite Hermitian matrix, and **B** is an $n\times 1$ col-

[2] The term "MAD" denotes a multiplication followed by an addition.

[3] The vectors $\mathbf{P}_0, \mathbf{P}_1, \cdots$ are said to be **N**-conjugate (or **N**-orthogonal) if $\mathbf{P}_i^T\mathbf{N}\mathbf{P}_j=0$, $i\neq j$, where **N** is a positive definite Hermitian matrix. A matrix is Hermitian if its elements a_{ij} satisfy the relation $a_{ij}^*=a_{ji}$, where the superscript (*) denotes the complex conjugate.

umn vector. We again denote the residual error at the ith step by $\mathbf{R}_i=\mathbf{B}-\mathbf{A}\mathbf{X}_i$, where the n-dimensional vector $\mathbf{X}_i$ is the ith estimate of the unknown vector $\mathbf{U}$. Furthermore, Hestenes chooses an error function of the form:

$$\mathbf{E}(\mathbf{X}_i) = \overset{T}{\mathbf{R}_i}\mathbf{H}\mathbf{R}_i, \tag{21}$$

where $\mathbf{H}$ is a predetermined positive definite Hermitian matrix of order n. Then it can be shown (see Appendix B) that the following equations, together with the algorithm described by equations (17)–(20), solve (1) in m steps, where $m\le n$ (Hestenes 1956, equations 3:2, p. 85, and equations 3:5, p. 86):

$$\mathbf{g}_i = \mathbf{A}^*\mathbf{H}\mathbf{R}_i, \tag{22}$$

$$\mathbf{N} = \mathbf{A}^*\mathbf{H}\mathbf{A}, \tag{23}$$

$$c_i = \overset{T}{\mathbf{g}_i}\,\mathbf{P}_i, \tag{24}$$

$$d_i = \overset{T}{\mathbf{P}_i}\mathbf{N}\mathbf{P}_i, \tag{25}$$

$$a_i = c_i/d_i, \tag{26}$$

and

$$\mathbf{X}_{i+1} = \mathbf{X}_i + a_i\mathbf{P}_i. \tag{27}$$

In equations (22) and (23), the superscript (*) denotes the complex conjugate transpose. Let us next specialize these results for the case of the normal equations. Then $\mathbf{A}$ is positive definite and symmetric, i.e., $\mathbf{A}=\mathbf{A}^*$. Let $\mathbf{H}=\mathbf{A}^{-1}$, $\mathbf{N}=\mathbf{A}$, and $\mathbf{K}=\mathbf{I}$, where $\mathbf{I}$ is the identity matrix. Since $\mathbf{A}$ is positive definite, so are $\mathbf{H}$ and $\mathbf{N}$. Equations (21), (22), (23), and (25) can thus be rewritten in the form:

$$\mathbf{E}(\mathbf{X}_i) = \overset{T}{\mathbf{R}_i}\,\mathbf{A}^{-1}\,\mathbf{R}_i, \tag{28}$$

$$\mathbf{g}_i = \mathbf{R}_i, \tag{29}$$

$$\mathbf{N} = \mathbf{A}, \tag{30}$$

and

$$d_i = \overset{T}{\mathbf{P}_i}\mathbf{A}\mathbf{P}_i. \tag{31}$$

Comparison of equations (4) and (28) reveals that both the g-method of Forsythe and Wasow and the present version of the c-g-method of Hestenes are based on identical error functions.

Some further algebraic manipulations described in Appendix C produce the algorithm (Hestenes, 1956, equations 8:4a–g, p. 100):

$$\left.\begin{aligned} c_{-1} &= 1,\ \mathbf{P}_{-1} = \boldsymbol{\theta},\\ \mathbf{R}_i &= \mathbf{B} - \mathbf{A}\mathbf{X}_i,\\ c_i &= \overset{T}{\mathbf{R}_i}\mathbf{R}_i,\\ b_{i-1} &= c_i/c_{i-1}, \qquad i = 0, 1, \cdots, m,\\ d_i &= \overset{T}{\mathbf{P}_i}\mathbf{A}\mathbf{P}_i, \qquad m \le n\\ a_i &= c_i/d_i,\\ \mathbf{P}_i &= \mathbf{R}_i + b_{i-1}\mathbf{P}_{i-1},^{4}\\ \mathbf{X}_{i+1} &= \mathbf{X}_i + a_i\mathbf{P}_i, \end{aligned}\right\} \tag{32}$$

where $\boldsymbol{\theta}$ is the $n\times 1$ null vector. Instead of using $\mathbf{R}_i=\mathbf{B}-\mathbf{A}\mathbf{X}_i$, the residual error vector can also be generated by means of the alternate relation

$$\mathbf{R}_{i+1} = \mathbf{R}_i - a_i\mathbf{A}\mathbf{P}_i, \tag{33}$$

where $\mathbf{R}_0=\mathbf{B}-\mathbf{A}\mathbf{X}_0$. Equation (33) may be derived by substituting (27) into the expression $\mathbf{R}_{i+1}=\mathbf{B}-\mathbf{A}\mathbf{X}_{i+1}$ and then setting $\mathbf{R}_i=\mathbf{B}-\mathbf{A}\mathbf{X}_i$.

We see from the algorithm (32) that the required number of MAD's for each step is approximately $(2n^2+4n)$, where n is the order of $\mathbf{A}$. If we replace the relation $\mathbf{R}_i=\mathbf{B}-\mathbf{A}\mathbf{X}_i$ by equation (33) with $\mathbf{R}_0=\mathbf{B}-\mathbf{A}\mathbf{X}_0$, and if we set $\mathbf{Q}_i=\mathbf{A}\mathbf{P}_i$, then the algorithm for the c-g-method becomes:

$$\left.\begin{aligned} c_{-1} &= 1,\ \mathbf{P}_{-1} = \boldsymbol{\theta},\\ \mathbf{R}_0 &= \mathbf{B} - \mathbf{A}\mathbf{X}_0,\\ c_i &= \overset{T}{\mathbf{R}_i}\mathbf{R}_i,\\ b_{i-1} &= c_i/c_{i-1},\\ \mathbf{P}_i &= \mathbf{R}_i + b_{i-1}\mathbf{P}_{i-1}, \qquad i = 0, 1, 2, \cdots, m,\\ \mathbf{Q}_i &= \mathbf{A}\mathbf{P}_i, \qquad m \le n\\ d_i &= \overset{T}{\mathbf{P}_i}\mathbf{Q}_i,\\ a_i &= c_i/d_i,\\ \mathbf{X}_{i+1} &= \mathbf{X}_i + a_i\mathbf{P}_i,\\ \mathbf{R}_{i+1} &= \mathbf{R}_i - a_i\mathbf{Q}_i. \end{aligned}\right\} \tag{34}$$

The algorithm (34) requires only (n^2+5n)MAD's for each step, and when n is large, the computing time is thus approximately halved with respect to the prior algorithm (32).

Comparison of the algorithms (12)–(14) and

[4] For $i=0$, we have that $\mathbf{P}_0=\mathbf{R}_0+b_{-1}\mathbf{P}_{-1}$, and hence we *define* $\mathbf{P}_{-1}$ to be the null vector $\boldsymbol{\theta}$.

(34) shows that the required computing times for both the *g*-method and the *c-g*-method are quite alike for large n. The main disadvantage of the *c-g*-method in comparison to the *g*-method is that an extra vector $\mathbf{P}_i$ needs to be generated and stored at each step. More importantly, however, the residual error norm[5] for the *c-g*-method generally decreases at a faster rate than is the case for the *g*-method. This fact will be illustrated in the next section, where both methods are demonstrated by means of very simple numerical examples.

NUMERICAL ILLUSTRATIONS OF THE *g*- AND *c-g*-ALGORITHMS

We shall here demonstrate how the algorithm (12)–(14) for the *g*-method and the algorithm (34) for the *c-g*-method can be applied to the solution of the matrix equation (1). This will be done with simple numerical examples. Let us set

$$\mathbf{A} = \begin{bmatrix} 4 & -1 \\ -1 & 4 \end{bmatrix} \quad \text{and} \quad \mathbf{B} = \begin{bmatrix} 2 \\ 0 \end{bmatrix}.$$

The exact solution is

$$\mathbf{U} = \begin{bmatrix} \frac{8}{15} \\ \frac{2}{15} \end{bmatrix}.$$

We first work this problem by means of the *g*-method, and use the algorithm given by equations (12)–(14) for this purpose. We first make the arbitrary choice for the initial approximant X_0 in the form

$$\mathbf{X}_0 = \begin{bmatrix} 0 \\ 0 \end{bmatrix}.$$

Then

$$\mathbf{R}_0 = \mathbf{B} - \mathbf{A}\mathbf{X}_0 = \mathbf{B} = \begin{bmatrix} 2 \\ 0 \end{bmatrix},$$

and the residual error norm[6] is $\|\mathbf{R}_0\| = (\mathbf{R}_0^T\mathbf{R}_0)^{1/2} = 2$.

Now, since

$$\mathbf{Q}_0 = \mathbf{A}\mathbf{R}_0 = \begin{bmatrix} 4 & -1 \\ -1 & 4 \end{bmatrix}\begin{bmatrix} 2 \\ 0 \end{bmatrix} = \begin{bmatrix} 8 \\ -2 \end{bmatrix},$$

it follows from (12) that

$$\alpha_0 = \frac{\mathbf{R}_0^T\mathbf{R}_0}{\mathbf{R}_0^T\mathbf{Q}_0} = \frac{4}{16} = \frac{1}{4}.$$

From (14), we have

$$\mathbf{X}_1 = \mathbf{X}_0 + \alpha_0\mathbf{R}_0 = \begin{bmatrix} 0 \\ 0 \end{bmatrix} + \frac{1}{4}\begin{bmatrix} 2 \\ 0 \end{bmatrix} = \begin{bmatrix} 1/2 \\ 0 \end{bmatrix},$$

and from (13),

$$\mathbf{R}_1 = \mathbf{R}_0 - \alpha_0\mathbf{Q}_0 = \begin{bmatrix} 2 \\ 0 \end{bmatrix} - \frac{1}{4}\begin{bmatrix} 8 \\ -2 \end{bmatrix} = \begin{bmatrix} 0 \\ 1/2 \end{bmatrix},$$

and

$$\|\mathbf{R}_1\| = (\mathbf{R}_1^T\mathbf{R}_1)^{1/2} = \frac{1}{2}.$$

Again, since

$$\mathbf{Q}_1 = \mathbf{A}\mathbf{R}_1 = \begin{bmatrix} 4 & -1 \\ -1 & 4 \end{bmatrix}\begin{bmatrix} 0 \\ 1/2 \end{bmatrix} = \begin{bmatrix} -1/2 \\ 2 \end{bmatrix},$$

then

$$\alpha_1 = \frac{\mathbf{R}_1^T\mathbf{R}_1}{\mathbf{R}_1^T\mathbf{Q}_1} = \frac{1}{4}.$$

We now have

$$\mathbf{X}_2 = \mathbf{X}_1 + \alpha_1\mathbf{R}_1 = \begin{bmatrix} 1/2 \\ 0 \end{bmatrix} + \frac{1}{4}\begin{bmatrix} 0 \\ 1/2 \end{bmatrix} = \begin{bmatrix} 1/2 \\ 1/8 \end{bmatrix}.$$

The residual error is

$$\mathbf{R}_2 = \mathbf{R}_1 - \alpha_1\mathbf{Q}_1 = \begin{bmatrix} 0 \\ 1/2 \end{bmatrix} - \frac{1}{4}\begin{bmatrix} -1/2 \\ 2 \end{bmatrix} = \begin{bmatrix} 1/8 \\ 0 \end{bmatrix},$$

[5] This norm is defined in the following section.

[6] The residual error norm is defined to be equal to the square root of the sum of the squares of the elements of the residual vector $\mathbf{R}_i$. The notation $\|\mathbf{R}_i\|$ for this norm is standard.

and the error norm is

$$\|R_2\| = (R_2^T R_2)^{1/2} = 1/8.$$

At this point, it is comforting to note that the error norms $\|R_0\|$, $\|R_1\|$, and $\|R_2\|$ form a monotonically decreasing sequence. Let us proceed with one more step of the iteration. Having determined R_2, Q_2 and α_2 become

$$Q_2 = AR_2 = \begin{bmatrix} 4 & -1 \\ -1 & 4 \end{bmatrix} \begin{bmatrix} 1/8 \\ 0 \end{bmatrix}$$

$$= \begin{bmatrix} 1/2 \\ -1/8 \end{bmatrix},$$

and

$$\alpha_2 = \frac{R_2^T R_2}{R_2^T Q_2} = \frac{1}{4}.$$

From (14), we have

$$X_3 = X_2 + \alpha_2 R_2$$

$$= \begin{bmatrix} 1/2 \\ 1/8 \end{bmatrix} + \frac{1}{4} \begin{bmatrix} 1/8 \\ 0 \end{bmatrix} = \begin{bmatrix} 17/32 \\ 1/8 \end{bmatrix}.$$

Finally, from (13), the residual error is

$$R_3 = R_2 - \alpha_2 Q_2$$

$$= \begin{bmatrix} 1/8 \\ 0 \end{bmatrix} - \frac{1}{4} \begin{bmatrix} 1/2 \\ -1/8 \end{bmatrix} = \begin{bmatrix} 0 \\ 1/32 \end{bmatrix},$$

and the error norm becomes

$$\|R_3\| = (R_3^T R_3)^{1/2} = 1/32.$$

After the third step, the approximate g-method solution to the matrix equation (1) is thus found to be

$$X_3 = \begin{bmatrix} 17/32 \\ 1/8 \end{bmatrix},$$

with an associated error norm of $\|R_3\| = 1/32$.

Let us next work this problem by means of the version of the c-g-method described by equations (34). We again choose

$$X_0 = \begin{bmatrix} 0 \\ 0 \end{bmatrix}$$

together with

$$P_{-1} = \begin{bmatrix} 0 \\ 0 \end{bmatrix}, \quad c_{-1} = 1.$$

From algorithm (34), we have, for $i=0$,

$$R_0 = B - AX_0 = \begin{bmatrix} 2 \\ 0 \end{bmatrix},$$

$$c_0 = R_0^T R_0 = 4,$$

$$b_{-1} = c_0/c_{-1} = 4,$$

$$P_0 = R_0 + b_{-1} P_{-1} = \begin{bmatrix} 2 \\ 0 \end{bmatrix},$$

$$Q_0 = AP_0 = \begin{bmatrix} 4 & -1 \\ -1 & 4 \end{bmatrix} \begin{bmatrix} 2 \\ 0 \end{bmatrix} = \begin{bmatrix} 8 \\ -2 \end{bmatrix},$$

$$d_0 = P_0^T Q_0 = 16,$$

$$a_0 = c_0/d_0 = 1/4,$$

$$X_1 = X_0 + a_0 P_0 = \begin{bmatrix} 1/2 \\ 0 \end{bmatrix},$$

and

$$R_1 = R_0 - a_0 Q_0 = \begin{bmatrix} 2 \\ 0 \end{bmatrix} - \frac{1}{4} \begin{bmatrix} 8 \\ -2 \end{bmatrix}$$

$$= \begin{bmatrix} 0 \\ 1/2 \end{bmatrix}.$$

This completes the first step. For the next step, we let $i=1$ in the algorithm (34),

$$c_1 = R_1^T R_1 = 1/4,$$

$$b_0 = c_1/c_0 = 1/16,$$

$$P_1 = R_1 + b_0 P_0 = \begin{bmatrix} 0 \\ 1/2 \end{bmatrix} + \frac{1}{16} \begin{bmatrix} 2 \\ 0 \end{bmatrix}$$

$$= \begin{bmatrix} 1/8 \\ 1/2 \end{bmatrix},$$

$$Q_1 = AP_1 = \begin{bmatrix} 4 & -1 \\ -1 & 4 \end{bmatrix} \begin{bmatrix} 1/8 \\ 1/2 \end{bmatrix}$$

$$= \begin{bmatrix} 0 \\ 15/8 \end{bmatrix},$$

$$d_1 = \overset{T}{\mathbf{P}}_1\mathbf{Q}_1 = 15/16,$$

$$a_1 = c_1/d_1 = 4/15,$$

and

$$\mathbf{X}_2 = \mathbf{X}_1 + a_1\mathbf{P}_1 = \begin{bmatrix} 1/2 \\ 0 \end{bmatrix} + \frac{4}{15}\begin{bmatrix} 1/8 \\ 1/2 \end{bmatrix}$$

$$= \begin{bmatrix} 8/15 \\ 2/15 \end{bmatrix},$$

and the residual vector R_2 becomes

$$\mathbf{R}_2 = \mathbf{R}_1 - a_1\mathbf{Q}_1 = \begin{bmatrix} 0 \\ 1/2 \end{bmatrix} - \frac{4}{15}\begin{bmatrix} 0 \\ 15/8 \end{bmatrix}$$

$$= \begin{bmatrix} 0 \\ 0 \end{bmatrix}.$$

Since the order of the matrix **A** is $n=2$, the *c-g*-method yields the *exact* solution in $m=n=2$ steps. The error norm sequences for the *g*-method and the *c-g*-method are, respectively,

$$(2,\ \ 1/2,\ \ 1/8,\ \ 1/32),$$

and

$$(2,\ \ 1/2,\ \ 0).$$

The above examples thus illustrate the generally faster convergence rate of the *c-g*-method.

The performance of either algorithm is also expressible in terms of the normalized square error (NSE),

$$(\mathrm{NSE})_i = \frac{(\mathbf{X}_i - \mathbf{U})^T(\mathbf{X}_i - \mathbf{U})}{\mathbf{U}^T\mathbf{U}}, \quad (35)$$

where the notation $(\mathrm{NSE})_i$ indicates that this error measure is computed at the end of the ith iteration. We recognize the numerator and denominator of (35) to constitute the energy in the difference vector $\mathbf{X}_i - \mathbf{U}$ and in the exact solution vector **U**, respectively.

In the above simple example, the *c-g*-method yields for $i=0$,

$$\mathbf{X}_0 - \mathbf{U} = -\mathbf{U}, \quad \text{since} \quad \mathbf{X}_0 = \begin{bmatrix} 0 \\ 0 \end{bmatrix}.$$

Therefore,

$$(\mathrm{NSE})_0 = \frac{(-\mathbf{U})^T(-\mathbf{U})}{\mathbf{U}^T\mathbf{U}} = 1.$$

For $i=1$,

$$\mathbf{X}_1 - \mathbf{U} = \begin{bmatrix} 1/2 \\ 0 \end{bmatrix} - \begin{bmatrix} 8/15 \\ 2/15 \end{bmatrix}$$

$$= \begin{bmatrix} -1/30 \\ -2/15 \end{bmatrix},$$

and, hence,

$$(\mathrm{NSE})_1 = \frac{\left(-\frac{1}{30}\right)^2 + \left(-\frac{2}{15}\right)^2}{\left(\frac{8}{15}\right)^2 + \left(\frac{2}{15}\right)^2} = \frac{1}{16}.$$

For $i=2$, the solution is exact so that $\mathbf{X}_2 = \mathbf{U}$, and, thus,

$$(\mathrm{NSE})_2 = 0.$$

The sequence of normalized square errors is, accordingly,

$$(1,\ \ 1/16,\ \ 0),$$

and we observe that the NSE and residual error norm sequences differ numerically, even though both are valid criteria for the determination of convergence rates.

The behavior of the NSE criterion for particular systems of larger order is shown in Figure 1, where the NSE has been plotted as a function of the number of iterations m for four given systems of orders $n=30$, 60, 120, and 240. The solid and the dashed curves indicate this error behavior for the *g*-method and for the *c-g*-method, respectively. We note that for a given value of m and n, the *c-g*-algorithm always yields a smaller NSE than is the case for the *g*-algorithm.

Both the residual error norm $\|\mathbf{R}_i\|$ as well as the NSE serve as convenient performance parameters, but from (35) we deduce that the NSE criterion *directly* measures the deviation between the true and the approximate solutions. This criterion should be used whenever the true solution **U** is known, but in most practical situations this is naturally not the case, and one then falls back on the residual error norm $\|\mathbf{R}_i\|$,

$$\|\mathbf{R}_i\| = (\overset{T}{\mathbf{R}}_i\mathbf{R}_i)^{1/2},$$

where $\mathbf{R}_i$ is the residual vector given by equation (2),

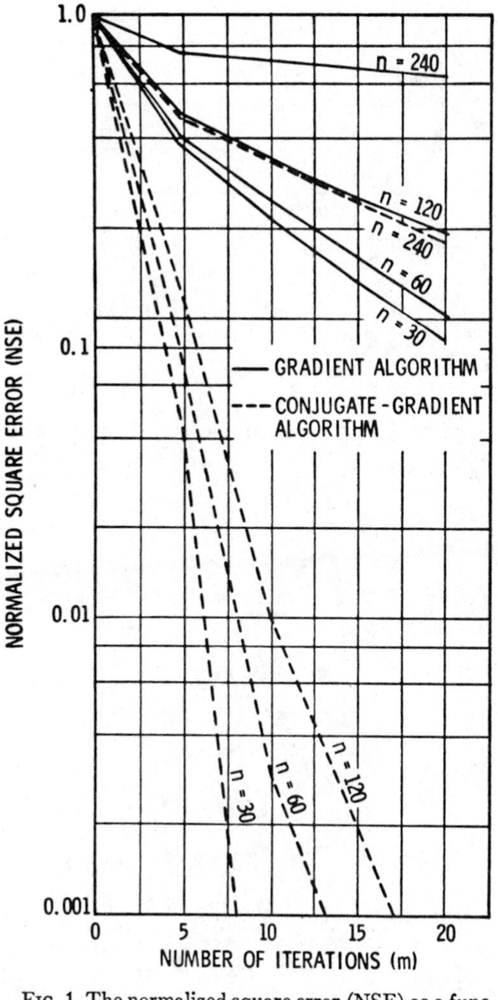

FIG. 1. The normalized square error (NSE) as a function of iteration length (m) for four given systems of order $n=30, 60, 120$, and 240.

$$\mathbf{R}_i = \mathbf{B} - \mathbf{A}\mathbf{X}_i. \tag{2}$$

COMPUTATION TIME COMPARISONS BETWEEN THE *c-g* AND THE LEVINSON ALGORITHM

We have already indicated the generally more rapid convergence of the *c-g*-method relative to the *g*-method; in the sequel we accordingly restrict our discussion to seismic implementations of the *c-g*-algorithm alone. Since the normal equations are usually solved by means of Levinson's method, we shall present an analysis of the savings in computer time that can be achieved by use of the *c-g*-algorithm in preference to the more conventional algorithm of Levinson.

It has been stated in the section on Hestenes' method that the *c-g*-algorithm (34) requires (n^2+5n)MAD's for each iteration step. Of this total, n^2 MAD's are needed for the evaluation of the product of the nth order matrix $\mathbf{A}$ into the n-rowed column vector $\mathbf{P}_i$,

$$\mathbf{Q}_i = \mathbf{A}\mathbf{P}_i, \tag{36}$$

where $\mathbf{A}$ is given by

$$\begin{bmatrix} \phi_1 \; \phi_2 & \cdots \; \phi_n \\ \phi_2 \; \phi_1 & \cdots \; \phi_{n-1} \\ \vdots & \vdots \\ \phi_n \; \phi_{n-1} & \cdots \; \phi_1 \end{bmatrix}.$$

The real-valued sequence $(\phi_1, \phi_2, \cdots, \phi_n)$ represents the center point and right half of a discrete autocorrelation function. The matrix $\mathbf{A}$ is symmetric and Toeplitz, i.e., all elements situated along any diagonal are identical (Treitel and Robinson, 1966, Appendix III).

Consider the first column and the first row oı $\mathbf{A}$. If we start at the bottom of the first column and proceed upwards to the top-most element, and then proceed toward the right to the right-most element of the first row, we generate the $(2n-1)$-length sequence,

$$(\phi_n, \phi_{n-1}, \cdots, \phi_2, \phi_1, \phi_2, \cdots, \phi_{n-1}, \phi_n).$$

This sequence is symmetric about the value ϕ_1. Let us rewrite it in the form

$$(v_1, v_2, \cdots, v_{n-1}, v_n, v_{n+1}, \cdots, v_{2n-2}, v_{2n-1}),$$

so that the new sequence is symmetric about the value v_n. If we denote the jth elements of the column vectors $\mathbf{P}_i$ and $\mathbf{Q}_i$ by the symbols $p_j^{(i)}$ and $q_j^{(i)}$, respectively, we find that the product

$$\mathbf{Q}_i = \mathbf{A}\mathbf{P}_i$$

can be written

$$q_j^{(i)} = \sum_{k=1}^{n} p_k^{(i)} v_{(n+j)-k}, \quad j = 1, 2, \cdots, n.$$

This expression can be evaluated at very high speeds by recourse to special-purpose floating point convolution hardware. Therefore each matrix product of the type (36) is computable on such a device, and most of the arithmetic required by the *c-g*-algorithm can be carried out in this efficient manner.

The calculations to be described below were carried out on an IBM 360/75 computer. In particular, the n^2 MAD's required per evaluation of the matrix product (36) were performed on the IBM 2938 array processor. This device is peripheral for the IBM 360/75 system. Although the MAD time for this device is about 0.22 μsec,[7] the time required to access the array processor from the computer's main frame is approximately 1250 μsec. Therefore, an algorithm that requires frequent exchange of digital data between the main frame and the array processor tends to become inefficient. It turns out that the Levinson algorithm requires several such exchanges per iteration step. Furthermore the Levinson scheme involves the calculation of vector dot products of progressively increasing length. This sequence of operations cannot be efficiently carried out by means of the above "convolutional" formulation. Hence, little would be gained by implementing this algorithm on the array processor.

In order to derive *theoretical* computing time formulas for both algorithms, we shall find it convenient to let:

t_1 = main frame MAD time;
t_2 = array processor MAD time;
t_3 = array processor access time;
T_1 = total computing time for m steps of the *c-g*-algorithm, $(m \leq n)$; and
T_2 = total computing time for Levinson's algorithm.

Since the *c-g*-algorithm (34) requires n^2+5n MAD's per step, we have

$$T_1 \approx m(n^2 t_2 + 5nt_1 + t_3). \tag{37}$$

It can also be shown (see Treitel and Robinson, 1966) that, for Levinson's algorithm,

$$T_2 \approx (2.5n^2 + 4.5n)t_1. \tag{38}$$

The above expressions for T_1 and T_2 are only approximate due to neglect of times required for logical branching and various arithmetic operations involving only scalars. Whenever $T_1 = T_2$, both algorithms require the same computing time; whenever $T_1 < T_2$, the *c-g*-algorithm becomes more efficient. In particular, the *c-g*-method remains computationally faster as long as m, the

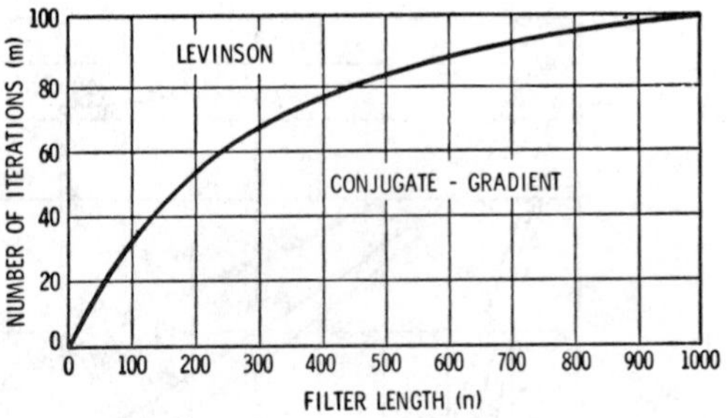

FIG. 2. The separation curve dividing the (m, n)-plane into regions of relative efficiency for the *c-g* and Levinson algorithms.

number of *c-g*-iteration steps, satisfies the inequality

$$m < \frac{(2.5n^2 + 4.5n)t_1}{(n^2 t_2 + 5nt_1 + t_3)}.$$

For the IBM 360/75—IBM 2938 configuration, we have[8]

$$t_1 \approx 11 \ \mu\text{sec},$$

$$t_2 \approx 0.22 \ \mu\text{sec},$$

and

$$t_3 \approx 1250 \ \mu\text{sec},$$

and, thus, m must satisfy

$$m < \frac{(27.5n^2 + 49.5n)}{(0.22n^2 + 55n + 1250)}$$

if the *c-g*-method is to have the advantage of speed.

Figure 2 shows a plot of the above relation when m is identically equal to the right-hand side of the above expression. The area below this curve is the region in which the *c-g*-method is computationally more efficient than Levinson's algorithm. Excluding the effects of numerical round-off, the *c-g*-algorithm produces an exact solution in at most $m = n$ steps (see Appendix B). Such a use of the method would in general be ill-advised, since the curve of Figure 2 clearly indicates that Levinson's algorithm is far more efficient as the number of iterations m approaches the filter length n. The crucial advantage of the *c-g*-algorithm results

[7] This figure was obtained by averaging MAD times for 500 trials of a convolution of two 250-length sequences.

[8] These figures are approximate and were determined empirically by means of computer "benchmark" tests (Lee, 1970).

from the fact that excellent approximations to the exact solution can usually be obtained for values of m small enough such that they fall substantially below the curve of Figure 2. This will certainly be the case when the original estimate $\mathbf{X}_0$ is already a close approximation of the exact solution $\mathbf{U}$. An example for which such conditions hold will be described in the next section.

We illustrate the above considerations with the particular system for $n=120$, for which the NSE-versus-m curve is given in Figure 1. Suppose that an NSE of 0.01 is acceptable, and that the system is to be solved by the *c-g*-method. From the dashed curve labeled $n=120$, we see that this NSE value is achievable for $m=10$ steps. We then proceed to Figure 2, where we observe that the point $m=10$, $n=120$ falls well below the m-versus-n curve, and, thus, the *c-g*-method can be expected to yield significant savings in computer time in comparison to Levinson's algorithm. Rough estimates of the respective times required for this problem on the IBM 360/75—IBM 2938 configuration can be obtained with the aid of expressions (37) and (38). These give

$$T_1 \approx 110 \text{ msec } (c\text{-}g),$$

and

$$T_2 \approx 400 \text{ msec (Levinson)},$$

and, hence, for the present example, the method of conjugate-gradients is about four times faster than Levinson's technique.

THE APPLICATION OF THE *c-g*-METHOD TO PREDICTIVE DECONVOLUTION

In this section we demonstrate the effective use of the *c-g*-algorithm for the solution of systems of normal equations associated with single-channel predictive deconvolution operators. The pertinent theory has been discussed elsewhere (Peacock and Treitel, 1969); here we merely indicate how the computation times can be shortened by recourse to the *c-g*-algorithm.

Figure 3(A) shows the input to a 36-trace, single-channel predictive deconvolution experiment. The uniform sampling increment for this data is 4 msec. Throughout the example the prediction filter length was held constant at $n=60$ weighting coefficients, and all computations were carried out for a prediction distance of 38 msec.

The filter for the first (topmost) trace was determined by Levinson's algorithm. This filter was then used as the initial estimate for the *c-g*-solution of the normal equations corresponding to the second trace. The solution obtained in this manner was in turn used as the initial estimate for the filter to be computed for the third trace, and so on. For all traces the number of iterations was held fixed at $m=3$ steps.

In order to establish a basis for comparison, the entire set of 36 filters was also computed with Levinson's algorithm. The deconvolved record resulting from the application of these *exact* operators is shown in Figure 3(B). On the other hand, the *approximate* operators obtained with the *c-g*-algorithm yield the deconvolved record pictured in Figure 3(C). Both outputs are remarkably alike in appearance. At the time these calculations were performed the plotting facilities for the IBM 360/75—IBM 2938 configuration were not operational, and therefore the present experiment was carried out on a CDC 3300 computer. However, use of the timing formulas (37) and (38) leads to the conclusion that a 60-length filter ($n=60$) requiring 3 *c-g*-iterations ($m=3$) can be evaluated on the IBM 360/75—IBM 2938 system in about 16 msec; on the other hand, the exact filter of this length determined only by use of the IBM 360/75's main frame requires roughly 100 msec. A sixfold savings in computer time is thus achievable in this manner, and the output resulting from the approximate *c-g*-filters does not result in a visually significant degradation of the deconvolved seismogram.

It is of interest to compare the memory functions of the exact prediction filters with their corresponding *c-g*-approximations. Figure 4 shows three memory functions. The first, labeled "EXACT, TR 1," is a plot of the exact filter computed by Levinson's algorithm for the first trace of the test record of Figure 3(A). The curves labeled "EXACT, TR 2" and "C-G, TR 2" are plots of the exact and conjugate-gradient versions, respectively, of the prediction filter for the second trace. We observe that these latter two filters differ in shape, even though their corresponding outputs are quite similar in appearance. The reason for this paradox becomes evident when one examines the amplitude and phase spectra of the filters, given by Figures 5 and 6, respectively. Here we note that the exact and *c-g*-filters for the second trace differ relatively little in their amplitude and phase characteristics for frequencies below about 0.5 f_N where f_N is the Nyquist, or

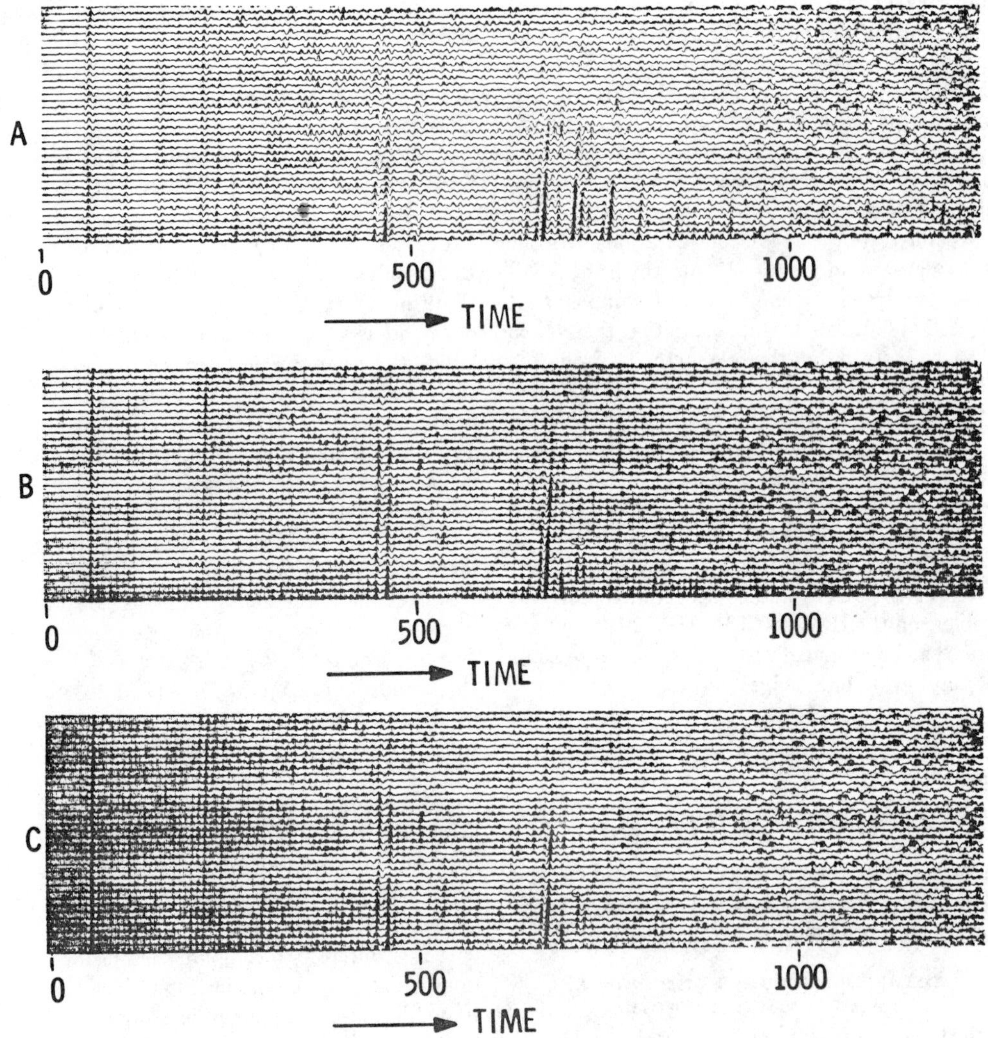

FIG. 3. Predictive deconvolution (time in msec). (A) Input field data. (B) Deconvolution with Levinson's algorithm. (C) Deconvolution with conjugate—gradient algorithm ($m=3$ iterations).

folding frequency. Since most of the spectral content of the test record traces is concentrated in the lower frequency bands, one sees qualitatively why both filters produce such similar-looking output.

From Figures 4, 5, and 6 we deduce also that the exact filter solutions for the first and second traces do not differ very much from each other. This observation tends to support the strategy which identifies the filter obtained for a given trace with the initial estimate of the filter to be determined for the following trace. In this particular instance three *c-g*-iterations per filter were found to be sufficient to produce quite satisfactory approximations to the exact solutions.

DISCUSSION AND CONCLUSIONS

The normal equations associated with the discrete Wiener filter can be solved with very high efficiency by means of the conjugate-gradient algorithm of Hestenes (1956). The implementation of this scheme on a computer system provided with a floating-point array processor leads to processing times which are significantly less

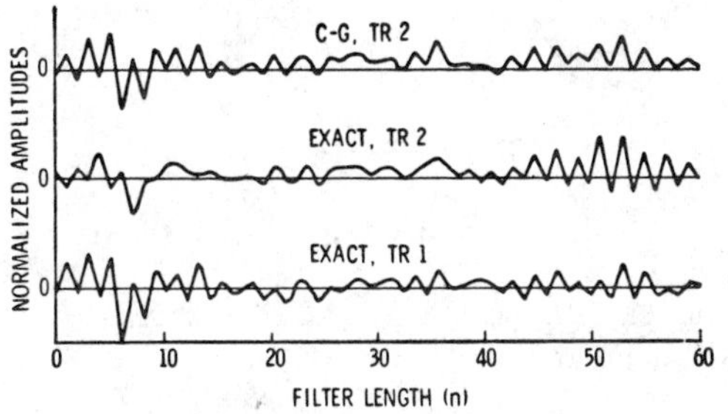

FIG. 4. Memory functions of deconvolution filters for traces 1 and 2 of the example of Figure 3.

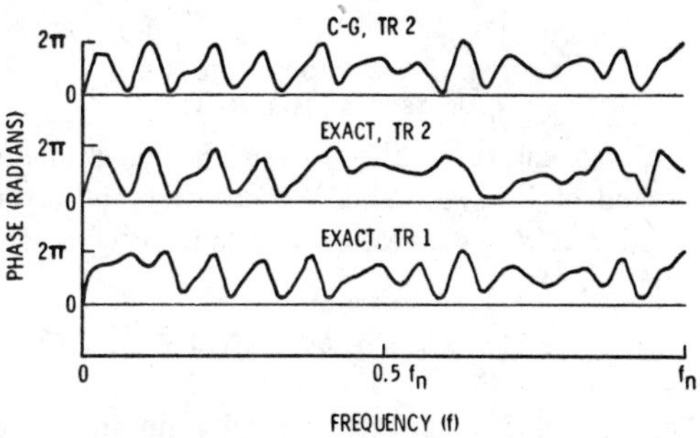

FIG. 6. Phase spectra of deconvolution filters for traces 1 and 2 of the example of Figure 3.

than those achievable with the more conventional Levinson algorithm. In contrast to this latter scheme, the method of conjugate gradients yields only approximate solutions but our analysis definitely suggests that such approximations are adequate for most seismic applications. The present deconvolution experiment reduced the computer time to one sixth of that required by the conventional Levinson approach.

The above field example can also be viewed as an instance of a "space-adaptive" deconvolution procedure, in which the filter for the $(k+1)$th trace is iteratively determined from an initial guess identified with the filter previously found for the kth trace. It would also be possible to use the *c-g*-algorithm in an efficient implementation of a "time-adaptive" deconvolution scheme. In such a case one would divide a given trace into two or more gates according to some optimizing criterion (Wang, 1969). The filter for the first gate could be computed exactly with the aid of Levinson's method, and this filter would constitute the original approximant for the solution of the normal equations associated with the second gate. This process would continue until filters for all gates of a given trace have been so determined.

It has been remarked previously that the Levinson algorithm yields an exact solution provided that numerical roundoff remains bounded throughout the calculation. Gradient methods are generally designed in such a way that an error function is minimized at each iteration step, and it often turns out that this procedure enables one to obtain solutions of systems too ill-conditioned for attack by Levinson's scheme.

Finally, the method of conjugate gradients can also be used for the rapid approximate solution of the normal equations associated with the multichannel Wiener filter (Treitel, 1970).

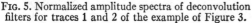

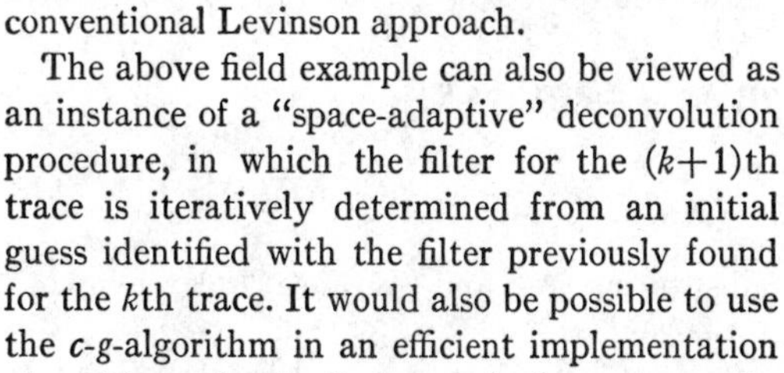

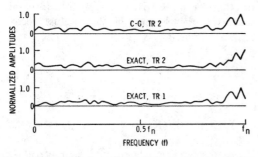

FIG. 5. Normalized amplitude spectra of deconvolution filters for traces 1 and 2 of the example of Figure 3.

REFERENCES

Forsythe, G. E., and Wasow, W. R., 1960, Finite difference methods for partial differential equations: New York, John Wiley and Sons.

Hestenes, M. R., 1956, The conjugate-gradient method for solving linear systems, *in* Proceedings of symposia in applied mathematics, v. VI: New York, McGraw-Hill Book Co., Inc.

Hildebrand, F. B., 1965, Methods of applied mathematics, second edition: New York, Prentice-Hall, Inc.

Lee, E. (1970), IBM Corp., personal communication.

Levinson, H., 1947, The Wiener RMS error criterion in filter design and prediction: J. Math. and Physics, v. 25, p. 261–278.

Peacock, K. L., and Treitel, S., 1969, Predictive deconvolution: theory and practice: Geophysics, v. 34, p. 155–169.

Robinson, E. A., 1967, Multichannel time series analysis with digital computer programs: San Francisco, Holden-Day, Inc.

Treitel, S., 1970, Principles of digital multichannel filtering: Geophysics, v. 35, p. 785–811.

Treitel, S., and Robinson, E. A., 1966, The design of high-resolution digital filters: IEEE Trans. on Geosci. Electron., v. GE-4, p. 25–38.

Wang, R. J., 1969, The determination of optimum gate lengths for time-varying Wiener filtering: Geophysics, v. 34, p. 683–695.

APPENDIX A

CONVERGENCE PROOF FOR THE GRADIENT ALGORITHM

It was shown in the section on the gradient method of Forsythe and Wasow that the error function $\mathbf{E}(\mathbf{X}_i)$ is a real quadratic form given by equation (4),

$$\mathbf{E}(\mathbf{X}_i)=(\mathbf{A}^{-1}\mathbf{B}-\mathbf{X}_i)^T\mathbf{A}(\mathbf{A}^{-1}\mathbf{B}-\mathbf{X}_i). \qquad \text{(A1)}$$

This quadratic form takes on its minimum value of zero when the ith approximant $\mathbf{X}_i$ becomes identically equal to the true solution $\mathbf{U}$, i.e., when $\mathbf{X}_i=\mathbf{A}^{-1}\mathbf{B}=\mathbf{U}$. Whenever $\mathbf{X}_i\neq\mathbf{U}$, there follows

$$\mathbf{E}(\mathbf{X}_i) > 0,$$

and the convergence of the g-algorithm is assured if the condition,

$$\mathbf{E}(\mathbf{X}_i) < \mathbf{E}(\mathbf{X}_{i-1})$$

is obeyed for all $i\geq 1$. It was also demonstrated that the symmetry of the matrix $\mathbf{A}$ permits us to express $\mathbf{E}(\mathbf{X}_i)$ in the form given by equation (5),

$$\mathbf{E}(\mathbf{X}_i) = \mathbf{X}_i^T\mathbf{A}\mathbf{X}_i - 2\mathbf{X}_i^T\mathbf{B} + \mathbf{B}^T\mathbf{A}^{-1}\mathbf{B}.$$

The third term of this expression is real and positive. Since this term is independent of $\mathbf{X}_i$, convergence of the g-algorithm is still assured as long as

$$\mathbf{E}'(\mathbf{X}_i) < \mathbf{E}'(\mathbf{X}_{i-1}),$$

where

$$\mathbf{E}'(\mathbf{X}_i) = \mathbf{X}_i^T\mathbf{A}\mathbf{X}_i - 2\mathbf{X}_i^T\mathbf{B}. \qquad \text{(A2)}$$

For convenience we reorder the algorithm (8) in the form,

$$\mathbf{X}_i = \mathbf{X}_{i-1} + \alpha_{i-1}\mathbf{R}_{i-1}, \qquad \text{(A3)}$$

$$\mathbf{R}_{i-1} = \mathbf{B} - \mathbf{A}\mathbf{X}_{i-1}, \qquad \text{(A4)}$$

and

$$\alpha_{i-1} = \frac{\mathbf{R}_{i-1}^T\mathbf{R}_{i-1}}{\mathbf{R}_{i-1}^T\mathbf{A}\mathbf{R}_{i-1}}\cdot \qquad \text{(A5)}$$

From (A2), we have,

$$\begin{aligned}\mathbf{E}'(\mathbf{X}_i) &= {}_i\mathbf{X}^T\mathbf{A}\mathbf{X}_i-2\mathbf{B}^T\mathbf{X}_i\\ &= (\mathbf{X}_{i-1}+\alpha_{i-1}\mathbf{R}_{i-1})^T\mathbf{A}(\mathbf{X}_{i-1}+\alpha_{i-1}\mathbf{R}_{i-1})\\ &\quad -2\mathbf{B}^T(\mathbf{X}_{i-1}+\alpha_{i-1}\mathbf{R}_{i-1})\\ &= (\mathbf{X}_{i-1}^T\mathbf{A}\mathbf{X}_{i-1}-2\mathbf{B}^T\mathbf{X}_{i-1})\\ &\quad +2\alpha_{i-1}\mathbf{R}_{i-1}^T\mathbf{A}\mathbf{X}_{i-1}\\ &\quad +\alpha_{i-1}^2\mathbf{R}_{i-1}^T\mathbf{A}\mathbf{R}_{i-1}-2\alpha_{i-1}\mathbf{B}^T\mathbf{R}_{i-1}.\end{aligned}$$

But since $\mathbf{E}'(\mathbf{X}_{i-1})=\mathbf{X}_{i-1}^T\mathbf{A}\mathbf{X}_{i-1}-2\mathbf{B}^T\mathbf{X}_{i-1}$, there follows that

$$\begin{aligned}\mathbf{E}'(\mathbf{X}_i) &= \mathbf{E}'(\mathbf{X}_{i-1}) + \alpha_{i-1}(2\mathbf{R}_{i-1}^T\mathbf{A}\mathbf{X}_{i-1}\\ &\quad + \alpha_{i-1}\mathbf{R}_{i-1}^T\mathbf{A}\mathbf{R}_{i-1} - 2\mathbf{B}^T\mathbf{R}_{i-1}).\end{aligned}$$

For convergence, $\mathbf{E}'(\mathbf{X}_{i-1})>\mathbf{E}'(\mathbf{X}_i)$, or

$$\begin{aligned}\mathbf{E}'(\mathbf{X}_{i-1}) - \mathbf{E}'(\mathbf{X}_i) &= \alpha_{i-1}(2\mathbf{B}^T\mathbf{R}_{i-1}\\ &\quad - 2\mathbf{R}_{i-1}^T\mathbf{A}\mathbf{X}_{i-1} - \alpha_{i-1}\mathbf{R}_{i-1}^T\mathbf{A}\mathbf{R}_{i-1}) > 0.\end{aligned}$$

Since $\alpha_{i-1}>0$, whenever $\mathbf{R}_{i-1}$ is not null, convergence requires that

$$\begin{aligned}&2\mathbf{B}^T\mathbf{R}_{i-1} - 2\mathbf{R}_{i-1}^T\mathbf{A}\mathbf{X}_{i-1}\\ &\qquad - \alpha_{i-1}\mathbf{R}_{i-1}^T\mathbf{A}\mathbf{R}_{i-1} > 0. \qquad \text{(A6)}\end{aligned}$$

Substituting (A4) and (A5) into (A6), we have

$$\begin{aligned}&2\mathbf{B}^T(\mathbf{B} - \mathbf{A}\mathbf{X}_{i-1}) - 2(\mathbf{B} - \mathbf{A}\mathbf{X}_{i-1})^T\mathbf{A}\mathbf{X}_{i-1}\\ &\qquad - (\mathbf{B} - \mathbf{A}\mathbf{X}_{i-1})^T(\mathbf{B} - \mathbf{A}\mathbf{X}_{i-1})\\ &\qquad = \mathbf{B}^T\mathbf{B} - 2\mathbf{B}^T\mathbf{A}\mathbf{X}_{i-1} + (\mathbf{A}\mathbf{X}_{i-1})^T(\mathbf{A}\mathbf{X}_{i-1})\\ &\qquad = (\mathbf{B} - \mathbf{A}\mathbf{X}_{i-1})^T(\mathbf{B} - \mathbf{A}\mathbf{X}_{i-1})\\ &\qquad = \mathbf{R}_{i-1}^T\mathbf{R}_{i-1}.\end{aligned}$$

But the quantity $\mathbf{R}_{i-1}^T\mathbf{R}_{i-1}$ is merely the sum of the squares of the elements of the residual vector $\mathbf{R}_{i-1}$, and, thus,

$$\mathbf{R}_{i-1}^T\mathbf{R}_{i-1} > 0.$$

This condition, in turn, implies that

$$\mathbf{E}'(\mathbf{X}_i) < \mathbf{E}'(\mathbf{X}_{i-1}), \qquad i = 1, 2, \cdots.$$

Hence the gradient algorithm (8) converges as i increases.

APPENDIX B

CONVERGENCE PROOF FOR THE *c-g* ALGORITHM (22)-(27)

Let us consider an iterative process described by

$$\mathbf{X}_{i+1} = \mathbf{X}_i + \alpha_i \mathbf{P}_i \qquad \text{(B1)}$$

which solves the matrix equation $\mathbf{AU}=\mathbf{B}$, where $\mathbf{A}$ is an $n \times n$ symmetric Toeplitz matrix. We furthermore choose $\mathbf{E}(\mathbf{X}_i) = \mathbf{R}_i^T \mathbf{H} \mathbf{R}_i$, where $\mathbf{H}$ is a positive definite Hermitian matrix of order n, and $\mathbf{R}_i$ is the residual vector $\mathbf{R}_i = \mathbf{B} - \mathbf{A}\mathbf{X}_i$. The vectors $\mathbf{P}_i$, $i=0, 1, \cdots, n-1$, are direction vectors yet to be determined.

Equations (22)–(26) are here rewritten in the rearranged form,

$$\mathbf{g}_i = \mathbf{A}^*\mathbf{H}\mathbf{R}_i, \qquad \text{(B2)}$$

$$\mathbf{N} = \mathbf{A}^*\mathbf{H}\mathbf{A}, \qquad \text{(B3)}$$

$$a_i = c_i/d_i, \qquad \text{(B4)}$$

$$c_i = \mathbf{g}_i^T \mathbf{P}_i, \qquad \text{(B5)}$$

$$d_i = \mathbf{P}_i^T \mathbf{N} \mathbf{P}_i, \qquad \text{(B6)}$$

where $\mathbf{g}_i$ differs from the gradient of $\mathbf{E}(\mathbf{X}_i)$ only by a factor of $-\frac{1}{2}$, and where $\mathbf{A}^*$ is the conjugate transpose of the matrix $\mathbf{A}$. Then it can be shown (Hestenes, 1956, equation 3:3) that

$$\mathbf{E}(\mathbf{X}_i) - \mathbf{E}(\mathbf{X}_{i+1}) = \{|a_i|^2 - |\alpha_i - a_i|^2\} d_i. \qquad \text{(B7)}$$

For the iterative process (B1) to converge, (B7) must be nonnegative. Since $\mathbf{H}$ is positive definite, so is $\mathbf{N}$, and $d_i > 0$ when $\mathbf{P}_i \neq 0$. (B7) is nonnegative if and only if α_i is of the form

$$\alpha_i = a_i(1 + \epsilon_i), \qquad \text{(B8)}$$

where $\epsilon_i \leq 1$. From (B7) and (B8) one deduces that (B7) is maximum when $\alpha_i = a_i$, and, then,

$$\mathbf{E}(\mathbf{X}_i) - \mathbf{E}(\mathbf{X}_{i+1}) = |a_i|^2 d_i = |c_i|^2/d_i \qquad \text{(B9)}$$

(Hestenes, 1956, p. 86, equation 3:4). Setting $\alpha_i = a_i$ in (B1), we have

$$\mathbf{X}_{i+1} = \mathbf{X}_i + a_i \mathbf{P}_i, \qquad \text{(B10)}$$

and substitution of (B10) into the expression

$$\mathbf{R}_{i+1} = \mathbf{B} - \mathbf{A}\mathbf{X}_{i+1}$$

yields

$$\mathbf{R}_{i+1} = \mathbf{R}_i - a_i \mathbf{Q}_i. \qquad \text{(B11)}$$

where $\mathbf{Q}_i = \mathbf{A}\mathbf{P}_i$, and $\mathbf{R}_0 = \mathbf{B} - \mathbf{A}\mathbf{X}_0$. Now, from (B2), (B3), and (B11),

$$\begin{aligned} \mathbf{g}_{i+1} &= \mathbf{A}^*\mathbf{H}\mathbf{R}_{i+1} = \mathbf{A}^*\mathbf{H}(\mathbf{R}_i - a_i\mathbf{Q}_i) \\ &= \mathbf{g}_i - a_i \mathbf{N}\mathbf{P}_i. \end{aligned} \qquad \text{(B12)}$$

The vector dot product of $\mathbf{g}_{i+1}$ and $\mathbf{P}_j$ is

$$\mathbf{g}_{i+1}^T \mathbf{P}_j = \mathbf{g}_i^T \mathbf{P}_j - a_i \mathbf{P}_i^T \mathbf{N}\mathbf{P}_j. \qquad \text{(B13)}$$

But since the vectors $\mathbf{P}_i$ are $\mathbf{N}$-conjugate,[9] we have

$$\mathbf{g}_i^T \mathbf{N}\mathbf{P}_j = 0,$$

or,

$$\mathbf{g}_{i+1}^T \mathbf{P}_j = \mathbf{g}_i^T \mathbf{P}_j, \qquad i \neq j. \qquad \text{(B14)}$$

if $i=j$, (B13) becomes

$$\mathbf{g}_{i+1}^T \mathbf{P}_i = \mathbf{g}_i^T \mathbf{P}_i - a_i \mathbf{P}_i^T \mathbf{N}\mathbf{P}_i.$$

But from equation (17), we have

$$a_i = \mathbf{g}_i^T \mathbf{P}_i / \mathbf{P}_i^T \mathbf{N}\mathbf{P}_i,$$

and, therefore,

$$\mathbf{g}_{i+1}^T \mathbf{P}_i = 0. \qquad \text{(B15)}$$

Next let us replace i by $j+k$ in (B14),

$$\mathbf{g}_{j+k+1}^T \mathbf{P}_j = \mathbf{g}_{j+k}^T \mathbf{P}_j.$$

In particular, this yields for $k=1$,

$$\mathbf{g}_{j+2}^T \mathbf{P}_j = \mathbf{g}_{j+1}^T \mathbf{P}_j,$$

a quantity which vanishes by virtue of (B15). Using induction, one then obtains the condition,

$$\mathbf{g}_i^T \mathbf{P}_j = 0 \quad \text{for} \quad i > j \qquad \text{(B16)}$$

and, in particular,

$$\mathbf{g}_n^T \mathbf{P}_i = 0 \text{ for } i = 0, 1, \cdots, n-1. \qquad \text{(B17)}$$

This set of relations can also be written

$$\begin{bmatrix} p_1^{(1)} & p_1^{(2)} & \cdots & p_1^{(n)} \\ p_2^{(1)} & p_2^{(2)} & \cdots & \\ \cdot & \cdot & \cdots & \cdot \\ p_{n-1}^{(1)} & p_{n-1}^{(2)} & \cdots & p_{n-1}^{(n)} \end{bmatrix} \begin{bmatrix} g_n^{(1)} \\ g_n^{(2)} \\ \cdot \\ \cdot \\ g_n^{(n)} \end{bmatrix} = \begin{bmatrix} 0 \\ 0 \\ \cdot \\ \cdot \\ 0 \end{bmatrix},$$

[9] See footnote [3] p. 313.

where $p_i^{(j)}$ and $g_n^{(j)}$ are elements of the vectors $\mathbf{P}_i$ and $\mathbf{g}_n$, respectively. Since the direction vectors $\mathbf{P}_i$ are linearly independent (Hestenes, 1956, Lemma 3.2), the above square matrix is nonsingular, and, thus, the vector $\mathbf{g}_n$ is itself null. Then, (B12) yields

$$\begin{aligned}\mathbf{g}_n &= \mathbf{A}^*\mathbf{H}\mathbf{R}_n \\ &= \mathbf{A}^*\mathbf{H}(\mathbf{B} - \mathbf{A}\mathbf{X}_n) = \boldsymbol{\theta},\end{aligned}$$

where $\boldsymbol{\theta}$ is the null vector. Therefore, we must have

$$\mathbf{A}\mathbf{X}_n = \mathbf{B},$$

and we conclude that the algorithm (22–27) in conjunction with relations (17–20) solves the system $\mathbf{AU}=\mathbf{B}$ exactly in at most n steps.

APPENDIX C

PROOF THAT THE *c-g*-ALGORITHM (32) IS EQUIVALENT TO THE ALGORITHM (22)–(27)

It has been shown in Appendix B that the algorithm (22)–(27) with equations (17)–(20) solves the system $\mathbf{AU}=\mathbf{B}$ in at most n steps, where n is the order of $\mathbf{A}$. We shall here demonstrate that the algorithm (32) is equivalent to the algorithm (22)–(27) when $\mathbf{H}=\mathbf{A}^{-1}$, $\mathbf{N}=\mathbf{A}$, and $\mathbf{K}=\mathbf{I}$, where $\mathbf{I}$ is the identity matrix. Under these conditions, we have shown that equations (21), (22), (23), and (25) can be expressed in the form (28), (29), (30), and (31), respectively.

Substituting equation (29) into equation (17) and using $\mathbf{N}=\mathbf{A}$ we obtain

$$a_i = \mathbf{R}_i^T\mathbf{P}_i/\mathbf{P}_i^T\mathbf{A}\mathbf{P}_i.$$

From equation (20), we have

$$\mathbf{P}_i = \mathbf{R}_i + b_{i-1}\mathbf{P}_{i-1}, \qquad (\mathbf{P}_0 = \mathbf{R}_0),$$

and substituting this expression into the numerator of a_i yields

$$a_i = \mathbf{R}_i^T(\mathbf{R}_i + b_{i-1}\mathbf{P}_{i-1})/P_i^T\mathbf{A}\mathbf{P}_i.$$

But, from (B15), we deduce that

$$\mathbf{g}_{i+1}^T\mathbf{P}_i = \mathbf{R}_{i+1}^T\mathbf{P}_i = \mathbf{R}_i^T\mathbf{P}_{i-1} = 0,$$

and, thus,

$$a_i = \mathbf{R}_i^T\mathbf{R}_i/\mathbf{P}_i^T\mathbf{A}\mathbf{P}_i. \tag{C1}$$

Next let us use equations (29) and (30) together with $\mathbf{N}=\mathbf{A}$ and $\mathbf{K}=\mathbf{I}$ in equations (18)–(20). This gives

$$b_i = -\mathbf{P}_i^T\mathbf{A}\mathbf{R}_{i+1}/\mathbf{P}_i^T\mathbf{A}\mathbf{P}_i, \tag{C2}$$

$$\mathbf{R}_{i+1} = \mathbf{R}_i - a_i\mathbf{A}\mathbf{P}_i, \tag{C3}$$

and

$$\mathbf{P}_{i+1} = \mathbf{R}_{i+1} + b_i\mathbf{P}_i, \tag{C4}$$

where $\mathbf{P}_0=\mathbf{R}_0$. From (C3), we obtain

$$-\mathbf{A}\mathbf{P}_i = (\mathbf{R}_{i+1} - \mathbf{R}_i)/a_i, \tag{C5}$$

and substitution of (C5) into (C2) yields

$$b_i = (\mathbf{R}_{i+1} - \mathbf{R}_i)^T\mathbf{R}_{i+1}/a_i\mathbf{P}_i^T\mathbf{A}\mathbf{P}_i. \tag{C6}$$

From the definition (16) and equation (29) it follows that

$$\mathbf{R}_i^T\mathbf{R}_{i+1} = 0,$$

and use of (C1) then produces

$$\begin{aligned}b_i &= \mathbf{R}_{i+1}^T\mathbf{R}_{i+1}/a_i\mathbf{P}_i^T\mathbf{A}\mathbf{P}_i \\ &= \mathbf{R}_{i+1}^T\mathbf{R}_{i+1}/\mathbf{R}_i^T\mathbf{R}_i.\end{aligned} \tag{C7}$$

Now let

$$c_i = \mathbf{R}_i^T\mathbf{R}_i, \tag{C8}$$

where

$$\mathbf{R}_i = \mathbf{B} - \mathbf{A}\mathbf{X}_i. \tag{C9}$$

Then, equations (C1) and (C7) become

$$a_i = c_i/d_i, \tag{C10}$$

$$b_i = c_{i+1}/c_i, \tag{C11}$$

where $d_i=\mathbf{P}_i^T\mathbf{N}\mathbf{P}_i$. Since $\mathbf{P}_0=\mathbf{R}_0$, (C4) can be rewritten

$$\mathbf{P}_i = \mathbf{R}_i + b_{i-1}\mathbf{P}_{i-1}, \tag{C12}$$

where

$$\mathbf{P}_{-1} = 0.$$

Also, if we choose $c_{-1}=1$, we may write (C11) as

$$b_{i-1} = c_i/c_{i-1}. \tag{C13}$$

Let us use $\mathbf{N}=\mathbf{A}$ and rearrange equations (C8)—(C13) and equations (25) and (27) in the form

$$c_{-1} = 1, \qquad \mathbf{P}_{-1} = \boldsymbol{\theta},$$

$$\mathbf{R}_i = \mathbf{B} - \mathbf{A}\mathbf{X}_i,$$

$$c_i = \mathbf{R}_i^T\mathbf{R}_i,$$

$$b_{i-1} = c_i/c_{i-1},$$

$$d_i = \mathbf{P}_i^T\mathbf{A}\mathbf{P}_i,$$

$$a_i = c_i/d_i, \tag{C14}$$

$$\mathbf{P}_i = \mathbf{R}_i + b_{i-1}\mathbf{P}_{i-1},$$

$$\mathbf{X}_{i+1} = \mathbf{X}_i + a_i\mathbf{P}_i,$$

where $\mathbf{\theta}$ is the $n\times 1$ null vector.

This set of relations constitutes the final *c-g* algorithm (32). It is thus seen to be a specialization of the earlier algorithm (22)–(27) for the case when $\mathbf{H}=\mathbf{A}^{-1}$, $\mathbf{N}=\mathbf{A}$, and $\mathbf{K}=\mathbf{I}$.

IV. APPLICATIONS TO WELL LOGGING

In the application of deconvolution to seismic data, it is usually necessary to design the filter from an analysis of the data. Well-logging applications, however, permit a design from a priori information. M. R. Foster et al investigate the use of optimum inverse filters to shorten the spacing of velocity logs. The data can be visualized as a moving average of the traveltimes for short increments of depth; inverse filtering gives an estimate of the traveltimes for these shorter intervals. The velocity logs appear to be troubled with "noise spikes" (a non-Gaussian noise) which must be removed before inverse filtering. This noise is caused by the instrument banging against the wall of the hole. Examples are given of finite memory inverse filters which are optimum for various residual noise levels; the results obtained using these inverse filters compare quite favorably with actual short spacing field data.

C. F. George, Jr. et al discuss the application of inverse filtering to induction log analysis. Here again, the vertical investigation characteristic of the log, corresponding to an impulse response, is given a priori. Indeed, the authors state that for a given set of logging devices the geometric factors and required filters could be computed once and stored in the computer for future use. Noise does not appear to be a significant problem. An interesting feature of this application is the use of a ratio of computed and measured logs of the same frequency response; this ratio, normally unity, departs from unity in invaded zones.

Reprinted from Geophysics v. 27, no. 3, p. 317-326

OPTIMUM INVERSE FILTERS WHICH SHORTEN THE SPACING OF VELOCITY LOGS*

M. R. FOSTER,† W. G. HICKS,‡ AND J. T. NIPPER†

A long-spacing velocity log contains almost the same information as an ideal short-spacing log, but in a distorted form with added noise. The distortion can be thought of as a moving average or smoothing filter. Its inverse, called a "sharpening" filter by astronomers, amplifies noise. If the inverse is to be useful, it must be designed with a balance between errors due to noise amplification and those due to incomplete sharpening. The Wiener optimum filter theory gives a prescription for achieving this balance. The result is called an *optimum inverse filter*.

We have calculated finite-memory optimum inverse filters using the IBM 704. We have applied them to actual data, digitized in the field, to produce synthetic short-spacing velocity logs. These we have compared with their field counterparts. The synthetic logs have less calibration error and are free from noise spikes. The general agreement is good.

INTRODUCTION

The problem of correcting observational data for instrument distortion is common to many fields of science. Robinson (1957), Kunetz (1961), and Rice (1962) discuss the problem for seismic traces. Fellgett and Schmeidler (1952) describe an astronomical problem in which the distorting operation is a "smoothing" filter. In this case they have termed the correction process a "sharpening" of observational data.

Sharpening data requires a surprising amount of finesse. The difficulty arises from the fact that a sharpening filter tends to amplify noise in the data. Unless steps are taken to prevent it, the amplified noise can completely dominate everything of interest. For this reason, a balance must be struck between errors due to magnification of the noise and errors due to incomplete sharpening.

The modern filter design theory which stems from the work of Wiener (1949) gives a prescription for achieving this balance. The sharpening operation prescribed by this theory is a filter which is an optimum compromise in a certain definite sense. This leads us to refer to it as the *optimum inverse filter*.

Our purpose in this paper is to describe the theory of the finite-memory optimum inverse filter in the context of a relatively simple geophysical problem. We have chosen the problem of producing synthetic short-spacing velocity logs from long-spacing field logs. In this application we limit ourselves to finite-memory filters for which pure time-domain methods are most appropriate. In the absence of noise, these reduce to least squares inverse filters of the type described by Rice (1962). Infinite memory inverse filters are discussed fully by Kunetz (1961), Robinson (1957), and Wiener (1949).

To avoid misunderstanding, we would like to state at this point that we believe firmly that information cannot be increased by processing data (Kullback, 1959). If there is an apparent paradox in producing seemingly more detailed short-spacing logs from those of longer spacing, it is only apparent. The information content of the log, in the sense of Shannon (1949), is determined by the ratio of logging speed to pulse repetition rate. This ratio is normally the same for both short- and long-spacing field logs, about two inches.

DISTORTION PRODUCED BY THE INSTRUMENT

A velocity logging instrument measures the travel time of a compressional pulse over a distance called the *spacing*. Stripling (1958) shows that for single-detector instruments, the spacing is approximately the separation of source and de-

* Presented at the 31st Annual International SEG Meeting, November 9, 1961. Manuscript received by the Editor August 28, 1961.

† Socony Mobil Oil Co., Inc., Dallas, Texas.

‡ Deceased July 11, 1961.

tector. For double-detector instruments, it is the separation of the detectors.

Let the spacing be Mh when h is a small depth interval and M is an integer. Let t_i be the time required for a compressional pulse to travel from depth ih to depth $(i+1)h$. If h is sufficiently small, then h/t_i is the local compressional velocity at depth ih. Thus, the t_i are the data we wish to measure.

However, an ideal instrument with spacing Mh will measure instead the travel time $t_i+t_{i+1}+\cdots+t_{i+M-1}$ over M of these intervals, or it can be calibrated to measure the "interval" travel time:

$$T_i = \frac{t_i + t_{i+1} + \cdots + t_{i+M-1}}{M}.$$

This can be written in the form of a digital filtering operation:

$$T_i = \sum_{j=-\infty}^{\infty} a_j t_{i-j}, \qquad (1)$$

provided the filter coefficients are defined by:

$$a_j = \begin{cases} 1/M \text{ for } j = -M+1, -M+2, \cdots, 0 \\ 0 \quad \text{for all other } j\text{'s.} \end{cases} \qquad (2)$$

The operation specified by equations (1) and (2) is called a moving average by statisticians. It is frequently used to smooth data and suppress random errors (noise). Thus, the distortion due to an ideal instrument is that of a smoothing filter. By regarding the coefficients (2) as samples of a continuous curve, the filter response can be visualized graphically as a rectangular pulse of height $1/M$ and width M.

An instrument operating in a field environment will not produce exactly the T_i of formula (1). The actual data will be:

$$T_i' = T_i + N_i. \qquad (3)$$

Here the error, N_i, is due to various causes which we will lump together and refer to as noise. A certain type of noise caused by the instrument banging against the wall of the hole can be surgically removed from the data (at least for the long spacing logs). An illustration of this surgery by digital computer is shown in Figure 1. We will suppose that this type of noise has been removed and is not included in N_i.

INVERSE FILTERS

Exact inverse filters

While not of direct use, the exact inverse is a limiting case of the optimum inverse and serves to round out the discussion. By an exact inverse is meant a digital filter which inverts the operation (1):

$$t_i = \sum_{j=-\infty}^{\infty} b_j T_{i-j}. \qquad (4)$$

Using (1) and (2) in (4), it is found that the response of an exact inverse filter must satisfy:

$$\frac{1}{M}\sum_{j=m}^{m+M-1} b_j = \begin{cases} 1 \text{ if } m = 0 \\ 0 \text{ if } m \neq 0. \end{cases}$$

It is not difficult to solve this system and, in fact, it has many solutions. One of them is shown in Figure 2 by employing the device mentioned above of regarding the filter coefficients as samples of a continuous curve.[1]

If an exact inverse is applied to T_i' rather than to T_i, then the result is t_i plus an error term due to the noise, N_i. There is, in practice, an additional error due to the infinite memory of the exact inverse and the finite domain of the data. The error terms completely dominate in all cases we have tried. Thus, the exact inverse is of no direct practical value and we are forced to an explicit consideration of the noise. This will lead to finite-memory optimum inverse filters.

The Wiener criterion

We seek an operation which, when applied to the data, T_i', will give as good an estimate of the t's as possible. If we denote the coefficients of this operation by $b_{-L}, b_{-L+1}, \cdots, b_{-L+K-1}$, then the result of applying it to the data is:

$$\hat{t}_m = \sum_{j=-L}^{-L+K-1} b_j(T_{m-j} + N_{m-j}). \qquad (5)$$

Here K is the memory "length" of the inverse filter and L is the filter "lag." Wiener's criterion for choosing the coefficients is:

$$\langle \sum (t_m - \hat{t}_m)^2 \rangle_{\text{av}} = \text{minimum}, \qquad (6)$$

[1] Brillouin (1956) describes how digital filters can be treated completely in continuous terms by employing band-limited functions.

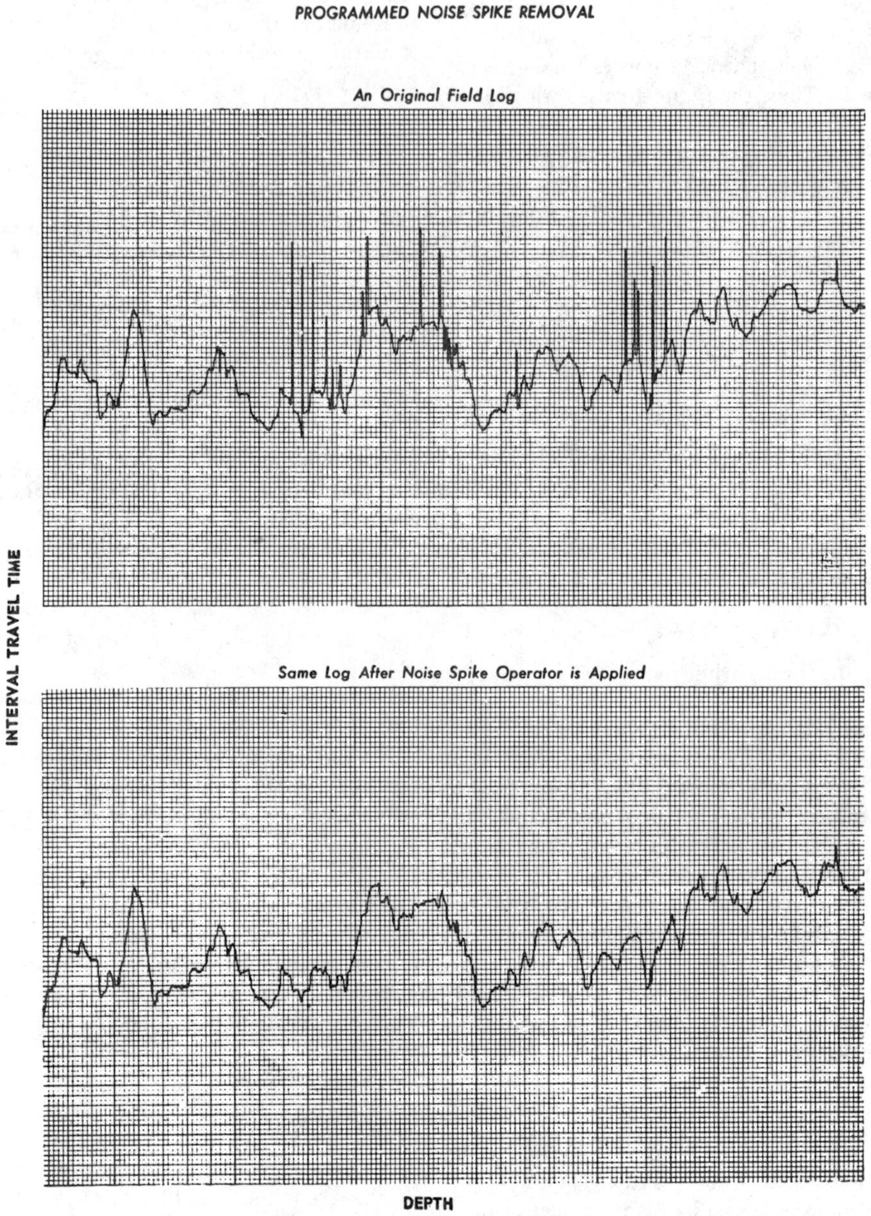

FIG. 1. Noise spike removal with a digital computer.

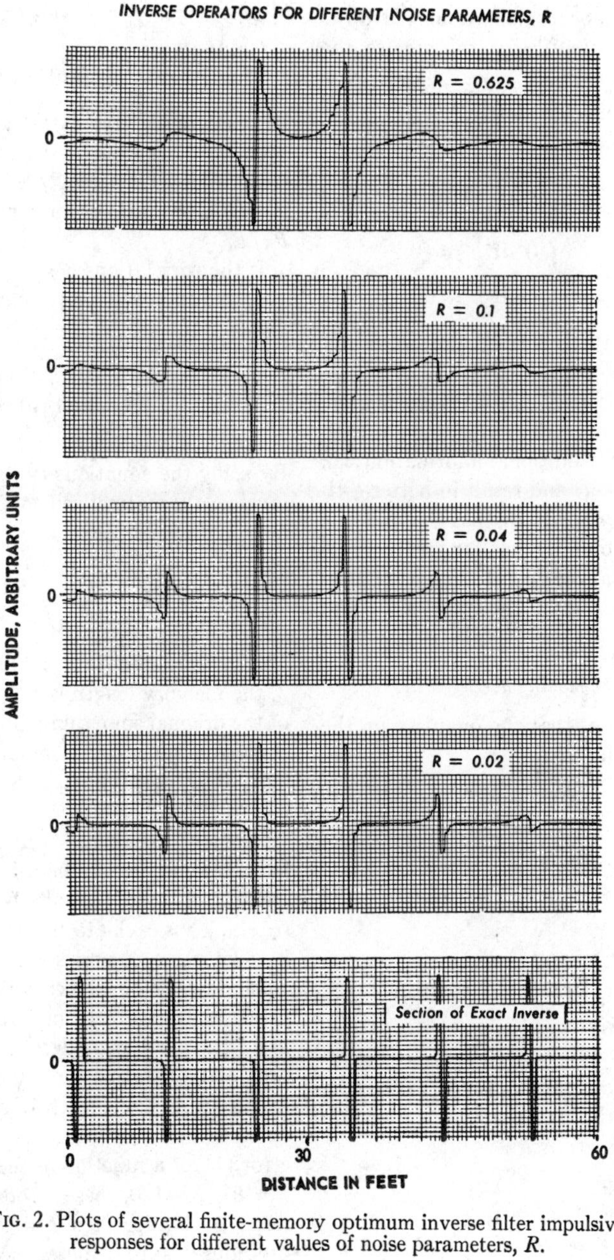

FIG. 2. Plots of several finite-memory optimum inverse filter impulsive responses for different values of noise parameters, *R*.

where the average is over all possible T's and N's. A filter (b_i) which satisfies the criterion is termed an *optimum inverse filter*. In calculating the average, we will employ the so-called "minimum information" or "maximum entropy" ensemble (Shannon, 1949). For this choice the correlations are very simple[2]:

$$\langle t_m t_{m+k}\rangle_{av} = \begin{cases} 0 & \text{if } k \neq 0 \\ P_S & \text{if } k = 0 \end{cases},$$

$$\langle t_m N_{m+k}\rangle_{av} = 0, \tag{7}$$

$$\langle N_m N_{m+k}\rangle_{av} = \begin{cases} 0 & \text{if } k \neq 0 \\ P_N & \text{if } k = 0 \end{cases}.$$

Here P_N and P_S may be termed the noise and signal power, respectively.

It should be emphasized that if more than the "minimum information," P_N/P_S, is known, then it should be used. The additional information will alter the correlations (7) and result in better estimates of the t's. To properly discuss the role of a priori information would lead us far afield. It requires a more general viewpoint than that provided by the Wiener filter theory (Raiffa and Schlaifer, 1961).

Optimum inverses to the moving average filter

Necessary conditions that the quantity on the left of (6) be a minimum are:

$$\frac{\partial}{\partial b_i}\langle \textstyle\sum_m (t_m - \hat{t}_m)^2\rangle_{av} = 0,$$

for $i=-L, -L+1, \cdots, -L+K-1$. This is the same as:

$$\textstyle\sum_m \left\langle (t_m - \hat{t}_m)\dfrac{\partial \hat{t}_m}{\partial b_i}\right\rangle_{av} = 0, \tag{8}$$

for $i=-L, -L+1, \cdots, -L+K-1$. Upon inserting the expression (1) for T_i in the expression (5) for $\hat{t}_m$, and then this expression in the system (8), we obtain the result:

$$\sum_{K=-L}^{K-L-1} A_{jk} b_k = a_{-j}$$

$$j = -L, \cdots, K-L-1, \tag{9}$$

[2] It is convenient to remove the mean travel-time both from the data and the t's before application of the optimum inverse. We assume in what follows that this has been done.

where

$$A_{jk} = \begin{cases} (1+R)/M & \text{if } j=k \\ \dfrac{M-|j-k|}{M^2} & \text{if } |j-k| \leq M \text{ and } j \neq k \\ 0 & \text{otherwise.} \end{cases} \tag{10}$$

In deriving (9) from (8), the correlations (7) have been used, as well as the definition (2) of the smoothing filter. The R/M factor which appears in (10) is the ratio of noise power to signal power, P_N/P_S.

If the correlations (7) are inserted into (6), the expected estimation error can be written:

$$\langle \textstyle\sum_m (t_m - \hat{t}_m)^2\rangle = \text{constant}\cdot\left[1 - \sum_{i,j} b_i b_j A_{ij}\right].$$

By using the equations (9) defining the optimum inverse, the normalized minimum expected estimation error is then:

$$\epsilon^2 = 1 - \sum_{j=-L}^{K-L-1} b_j a_{-j}. \tag{11}$$

For the optimum inverse filters shown in Figure 2, the memory length is 120 (six times the length of the original smoothing filter), the lag is 62, and the errors given by formula (11) are 0.37, 0.47, 0.61, 0.82, respectively, for the R values 0.02, 0.04, 0.10, and 0.625.

Strictly speaking, the optimum memory length, K, is the number of data points. For this choice only one t would be estimated, namely t_{K-L-1}. Thus each t to be estimated would require its own operator and specific lag, L. In practice we find this to be unnecessary. For noise-to-signal ratios appropriate to our application, the minimum expected error decreases rapidly to an approximately stationary value for K equal to five or six times the length of the moving average operator. This determines our choice of K. The error (11) is a function of the lag, L. For the case $R=0.1$, $K=120$, these errors are 0.62, 0.61, 0.62, respectively, for lags $L=29$, 62, and 110. Our procedure is to choose the operator with the minimum lag unless parts of the sharpened log which cannot be obtained by this operator are required. The relation between the lag and the recovered section of sharpened log is shown in Figure 6.

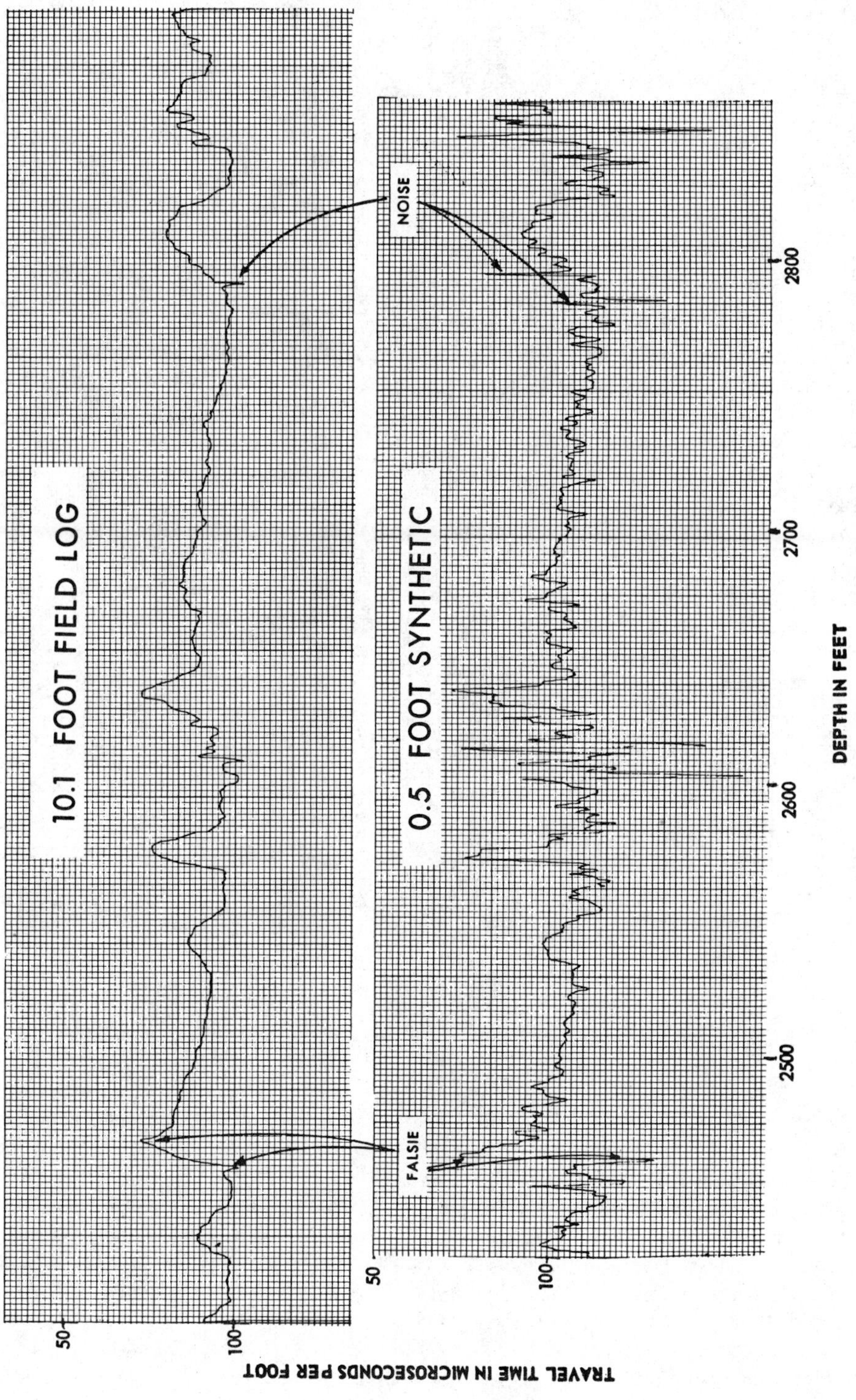

FIG. 3. Ten and one-tenth-ft tool spacing field log and the 0.5-ft synthetic obtained from it by applying an optimum inverse filter.

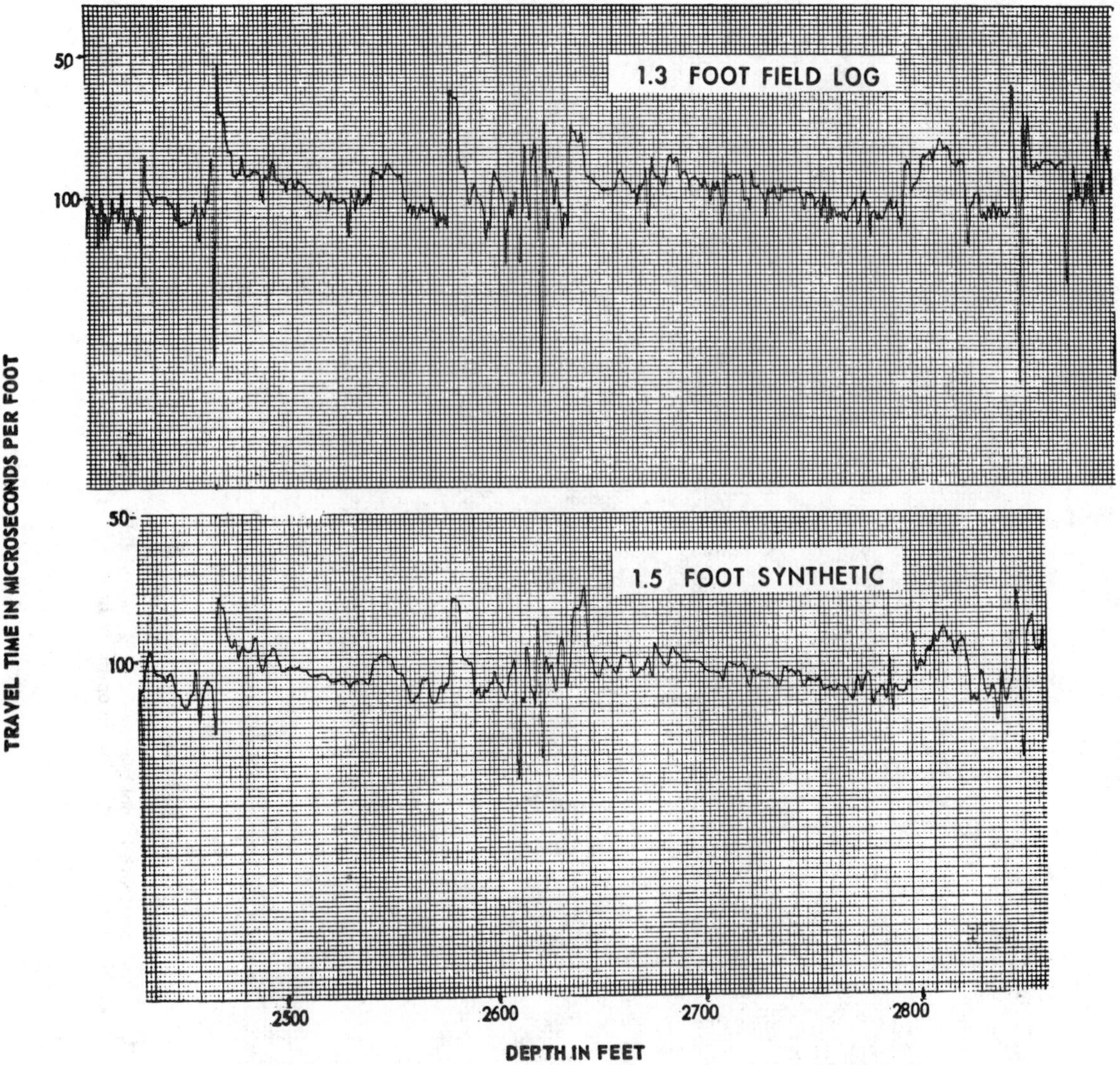

FIG. 4. Comparison of 1.3-ft field log and 1.5-ft synthetic. The 1.5-ft synthetic was obtained from the 0.5-ft synthetic (Figure 3) by applying (1) with $h=0.5$ and $M=3.0$.

APPLICATION TO FIELD LOGS

To test the theory, velocity logs were recorded at 0.5-ft intervals on punched paper tape in the field. They were obtained for spacings of 1.3, 2.8, 6.1, and 10.1 ft, using a double-detector instrument in the Nels Benson No. 1 well, Jack County, Texas. In what follows, these will be referred to as "field logs" to distinguish them from the "synthetic logs" produced by the computer.

The operator corresponding to $R=0.1$ in Figure 2 was applied to the 10.1-ft field log to produce a synthetic log with a spacing of 0.5 ft. Both the field log and the short spacing synthetic are shown in Figure 3. The total length of these logs is about 440 ft.

A noise spike of the type which would normally be removed has been left near the right side of the log. This noise is amplified slightly by the inverse operator. On the left-hand side of the log is a false excursion (called a "falsie") of a type discussed by Stripling (1958). It is due to fluid-filled cavities in the borehole rather than to velocity variations in the adjacent formations. This, too, is amplified slightly by the inverse operation. Other "falsies"

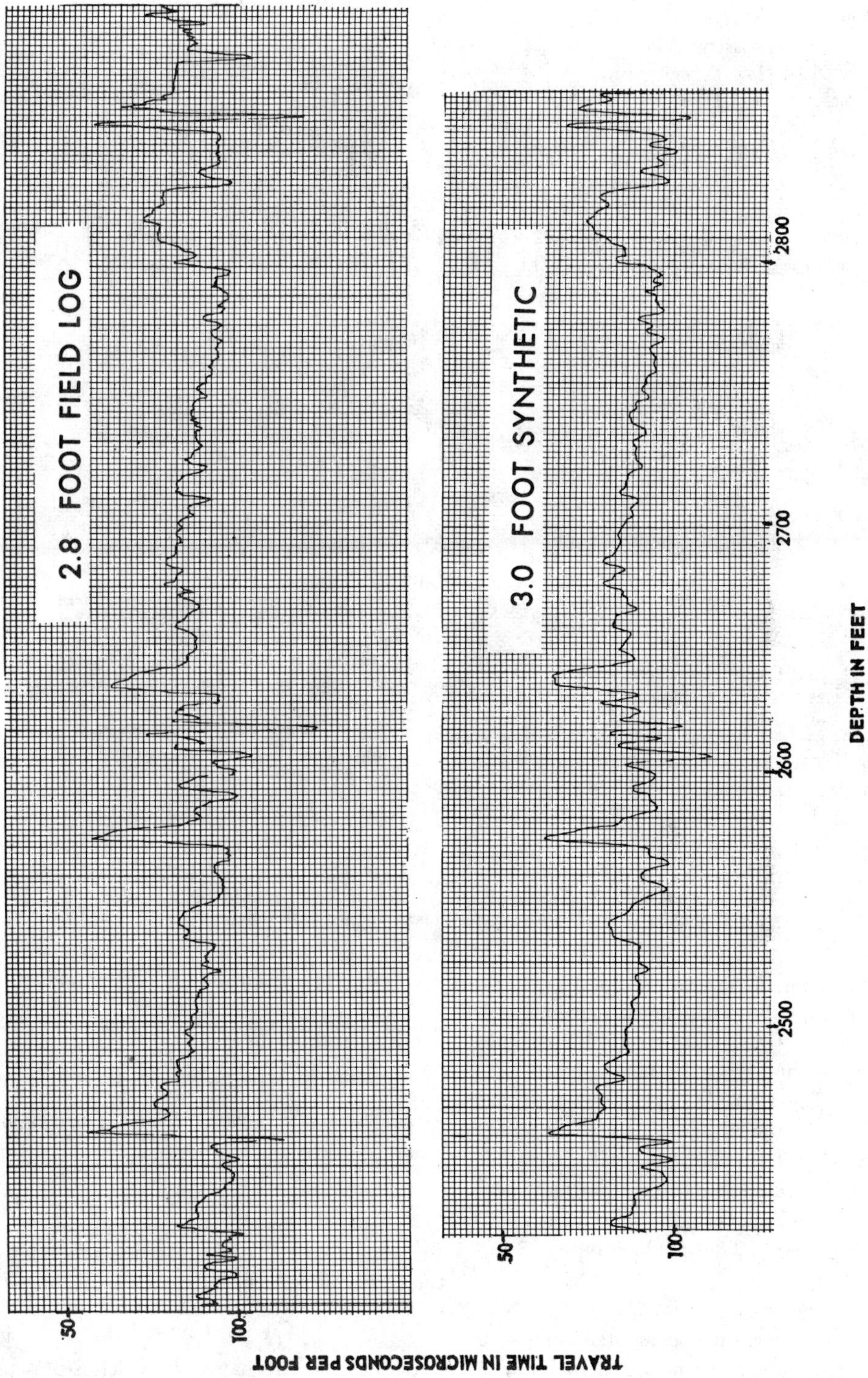

FIG. 5. Plots of 2.8-ft field log and 3.0-ft synthetic. The 3.0-ft synthetic was obtained from the 0.5-ft synthetic by applying (1) with $h=0.5$ and $M=6.0$.

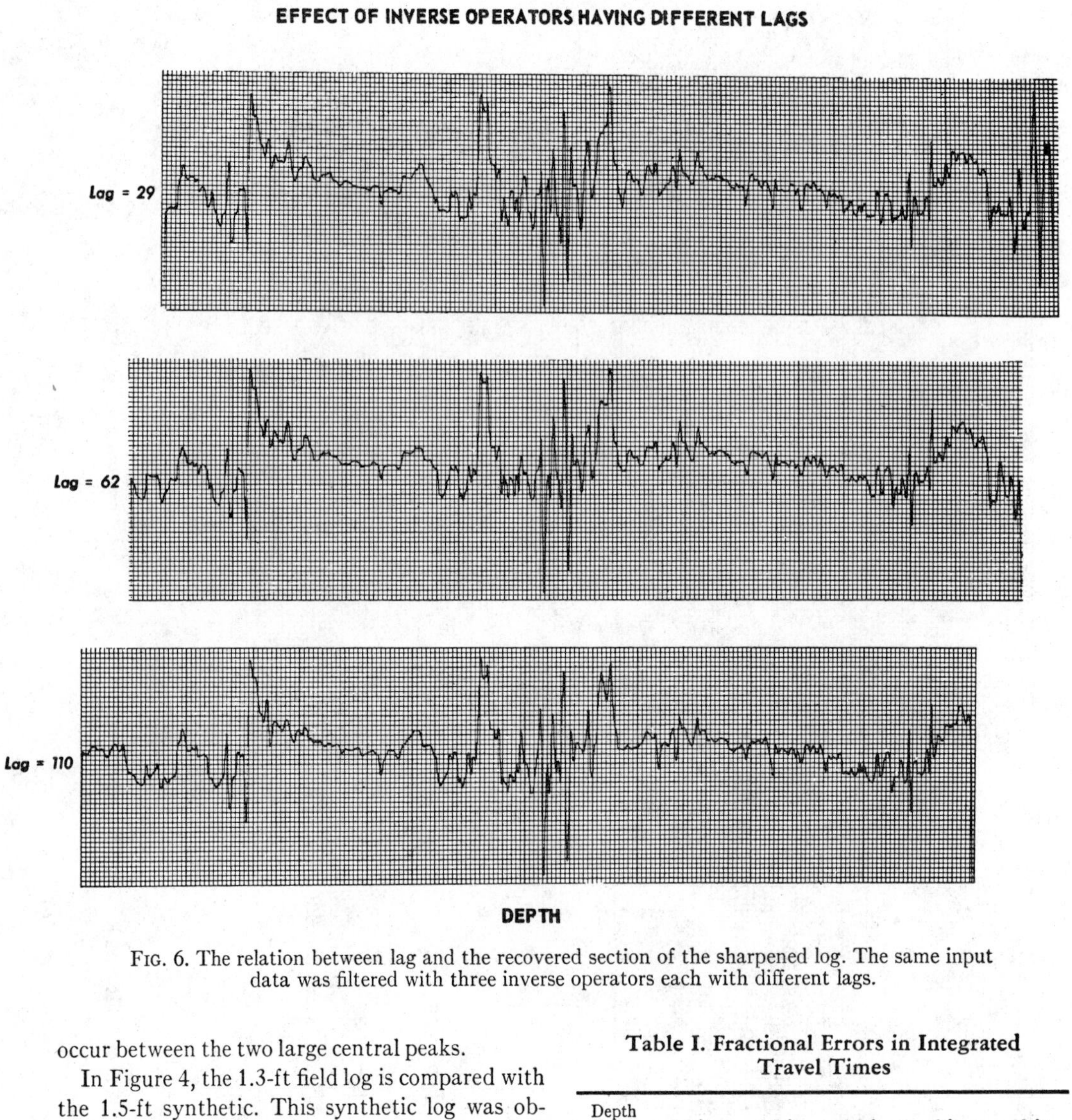

FIG. 6. The relation between lag and the recovered section of the sharpened log. The same input data was filtered with three inverse operators each with different lags.

occur between the two large central peaks.

In Figure 4, the 1.3-ft field log is compared with the 1.5-ft synthetic. This synthetic log was obtained by applying the smoothing operation (1) with $M=3$. The most striking difference is that the amplitude of the "falsies" is much less on the synthetic. Apart from this, and small effects due to the spacing difference, the agreement is generally good. It is even better in the comparison of Figure 5 between the 2.8-ft field log and the 3.0-ft synthetic.

One test of inner consistency is shown in Table I. Integrated travel times computed from synthetic logs of different spacing are compared with the integrated travel times computed from the 10.1-ft field log. The agreement is good even for the very short spacing logs.

Table I. Fractional Errors in Integrated Travel Times

Depth From Top of Log (ft)	0.5-ft Synthetic	1.5-ft Synthetic	3.0-ft Synthetic	6-ft Synthetic	10-ft Synthetic
40	−0.0007	0.0009	0.0042	0.0121	0.0250
120	−0.0028	−0.0013	−0.0001	0.0037	0.0087
200	−0.0037	−0.0034	−0.0030	−0.0016	0.0010
280	−0.0031	−0.0029	−0.0024	−0.0010	0.0012
360	−0.0027	−0.0026	−0.0022	−0.0007	0.0009

CONCLUSIONS

We have shown that detail which is present in short spacing logs is contained in distorted form in the long-spacing log as well. How well it can be recovered depends upon the noise level of the long-spacing log. The optimum inverse filter is, in a particular definite sense, the best answer to this recovery problem.

There are some reasons for preferring the synthetic short-spacing logs to their field counterparts. As was shown in Table I, short-spacing synthetic logs give good values of the integrated travel time. This indicates that calibration error is much reduced over that of short-spacing field logs, which often show errors of 5 percent or higher. Such error reduction can be important if the velocity logs are to be used for porosity estimation (Berry, 1959). Another advantage is that noise "spikes" can be removed from the source data automatically as was shown in Figure 1. For short-spacing field logs it is not always possible to separate these "spikes" from legitimate detail.

Improved field digital recording techniques (Schlumberger, 1960) and existing computers are adequate tools for the practical application of the optimum inverse filter. However, the application will become more attractive if the signal-to-noise ratio of the long-spacing field logs can be increased, for example by the use of more powerful transducers.

The optimum inverse, while developed here in the context of velocity logging, is not limited to this application. Any type of geophysical measurement, for which instrument distortion can be described by a filtering operation, is a possible candidate.

ACKNOWLEDGMENTS

We are indebted to A. A. Stripling for suggesting the application to velocity logs and for many useful comments. The aid of D. R. Miller and R. L. Atkins in preparing the figures is gratefully acknowledged. We thank R. L. Caldwell, L. Massé, and F. J. McDonal for their support and encouragement, and the management of Socony Mobil Oil Company, Inc. for permission to publish this paper.

REFERENCES

Berry, J. E., 1959, Acoustic velocity in porous media: J. Pet. Tech., v. 11, pp. 262–270.

Brillouin, L., 1956, Science and information theory, Academic Press, N. Y.

Fellgett, P. B., and Schmeidler, F. B., 1952, On the sharpening of observational data with special application to the darkening of the solar limb: Royal Astr. Soc. Notices, v. 112, p. 445.

Kullback, S., 1959, Information theory and statistics, John Wiley and Sons, New York, pp. 18–22.

Kunetz, G., 1961, Essai d'analyse de traces sismiques: Geophysical Prospecting, v. 9, p. 317.

Raiffa, H., and Schlaifer, R., 1961, Applied statistical decision theory, Harvard, Boston.

Rice, R. B., 1962, Inverse convolution filters: Geophysics, v. 27, pp. 4–18.

Robinson, E. A., 1957, Predictive decomposition of seismic traces: Geophysics, v. 22, p. 767.

Shannon, C. E. and Weaver, W., 1949, The Mathematical theory of communication, Univ. of Illinois Press, Urbana.

Shlumberger Well Surveying Corporation, Memorandum to API Subcommittee on Data Processing, November 8, 1960.

Stripling, A. A., 1958, Velocity log characteristics: Petroleum Transactions, AIME, v. 213, pp. 207–212.

Wiener, N., 1949, Extrapolation, interpolation and smoothing of stationary time series, John Wiley and Sons, New York.

Reprinted from Geophysics v. 29, no. 1, p. 93-104

APPLICATION OF INVERSE FILTERS TO INDUCTION LOG ANALYSIS*

C. F. GEORGE, JR.,† H. W. SMITH,‡ AND F. X. BOSTICK, JR.‡

The general availability of digital computing equipment has motivated the investigation of the application of filtering techniques of communication theory to the analysis of electric and induction well logs.

It will be shown that, with certain assumptions, the induction log may be represented as a linear filter with an impulse response and system function. A response equalization filter, which filters a shallow investigation log for comparison with a deep investigation log, may be used to scan pairs of induction, or induction and electric logs to detect invaded formations. An iterative process may be used to obtain an approximate solution for the depth of invasion from three appropriately chosen logs. Inverse convolution filtering may be used to obtain approximate resistivity values from apparent resistivity values indicated by the logs in radial zones concentric with the borehole for induction logs with different coil spacings.

INTRODUCTION

A resistivity logging device indicates only an apparent formation resistivity, which represents the weighted effect of the volume of earth within its depth of investigation. The availability of digital computing equipment has opened the door to new approaches to the problem of log analysis. This paper shows that the mathematics of linear filter theory may be applied to logs to investigate the formation resistivity profile in both the vertical and radial directions.

REPRESENTATION OF THE INDUCTION LOG AS A LINEAR FILTER

The vertical investigation characteristic of the two-coil induction log, $g(z)$, given by equation (1) below and derived from the geometric factor, serves as a simple introductory example for one type of analysis that may be applied to log data in an effort to specify more accurately true conductivity values from apparent values (Doll, 1949; Dakhnov, 1962, pp. 273–286).

$$\begin{aligned} g(z) &= 1/2L, \; -L/2 < z < L/2 \\ &= L/8z^2, \; z < -L/2, \; z > +L/2, \end{aligned} \tag{1}$$

where

z = downhole distance relative to center of tool,
L = spacing of coils.

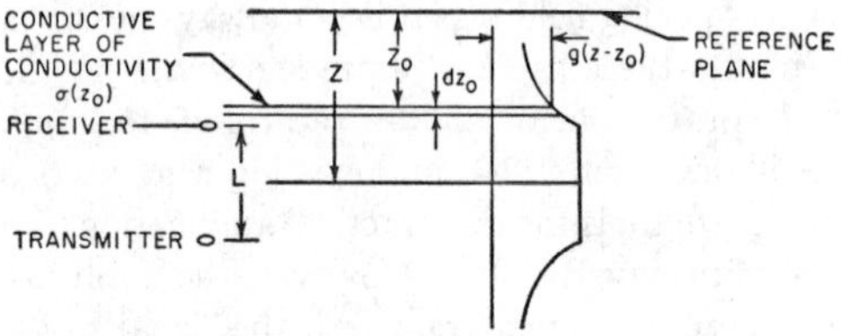

FIG. 1. Representation of induction log as a linear filter.

The "geometric factor" concept neglects propagation effects which have been analyzed by Duesterhoeft (1961a, 1961b, 1962). This is a reasonable approximation for high resistivities or short coil spacings. The application of inverse convolution filtering techniques to induction log analysis was first suggested to the authors by W. C. Duesterhoeft (George, 1963, pp. 58–61). Inverse filters have also been applied to the analysis of velocity logs (Foster et al, 1962).

The differential conductive layer dz_0 of conductivity $\sigma(z_0)$ in Figure 1 contributes a differential quantity $dv(z)$ to the total log reading observed at position z:

$$dv(z) = \sigma(z_0)g(z - z_0)dz_0. \tag{2}$$

The total log reading at position z is the integral of equation (2) over all z_0:

$$v(z) = \int_{-\infty}^{\infty} \sigma(z_0)g(z - z_0)dz_0. \tag{3}$$

* Presented at session seven of SWIRECO in Dallas, Texas, on April 17, 1963. Based on thesis submitted in partial fulfillment of Ph.D. requirements at the University of Texas, May, 1963.

† The Atlantic Refining Company, Dallas, Texas. Formerly with the Department of Electrical Engineering, The University of Texas.

‡ Department of Electrical Engineering, The University of Texas, and Consultants to Lane Wells, Inc.

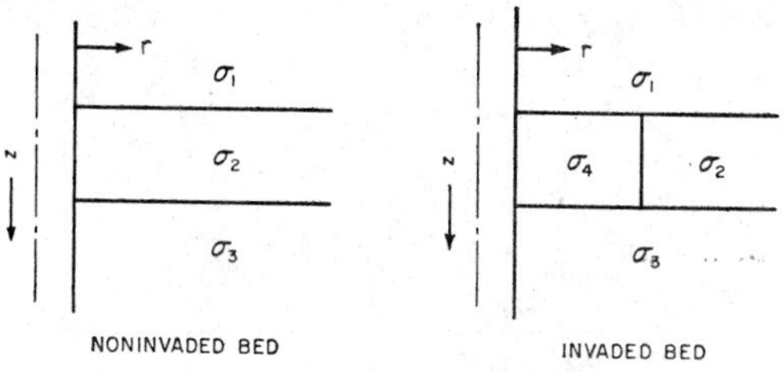

FIG. 2. Invaded and noninvaded beds.

Equation (3) is recognized as being of the form of the convolution integral of a linear time-invariant filter. The Fourier transform of equation (3) may be taken, giving, where frequency has the units reciprocal distance,

$$V(\omega) = S(\omega)G(\omega). \qquad (4)$$

It may be observed from equation (1) that as the coil spacing of the log is reduced, the geometric factor is "sharpened" and the vertical response of the device is increased. This broadens the frequency bandwidth of $G(\omega)$ as a result of the inverse time bandwidth relationship of a linear filter. However, as the coil spacing is reduced, the depth of penetration of the tool is reduced, and the increased vertical resolution is accompanied by decreased area of investigation.

The system function, $G(\omega)$, for a two-coil induction log with a coil spacing of two, is obtained by taking the Fourier transform of equation (1) (George, 1963, pp. 103–104):

$$G(\omega) = \frac{1}{2}\left\{\frac{\sin\omega}{\omega} + \cos\omega - \frac{\omega\pi}{2} + \omega S_i(\omega)\right\}. \qquad (5)$$

RESPONSE EQUALIZATION FILTERING

The response of two induction logs with different coil spacings to both a uniform bed and an invaded bed, indicated in Figure 2, may be considered. The quanity σ_{A1} shall represent the apparent conductivity indicated by the shorter coil-spacing log, and σ_{A2} shall represent the reading of the log with the longer coil spacing. The log with the short coil spacing will be more responsive to thin beds than the log with the long coil spacing as a result of its wider frequency bandwidth. Therefore, in response to a thin bed, σ_{A1} is not usually equal to σ_{A2}. Let the signal σ_{A1} be filtered by a filter which makes the frequency response of log 1 equal to the frequency response of log 2:

$$\bar{\sigma}_{A1}(z) = \int_{-\infty}^{\infty} \sigma_{A1}(z_0) h_1(z - z_0) dz_0, \qquad (6)$$

where $h_1(z)$ is a filter which makes the frequency response of log 1 equal to the frequency response of log 2.

An approximation to the filter $h_1(z)$ described may be obtained in the following manner.

Let

$$H_1(\omega) = \frac{G_2(\omega)}{G_1(\omega)}, \qquad \omega \leq A,$$
$$= 0, \qquad \omega \geq A, \qquad (7)$$

where:

A = that value of ω for which $G_2(\omega)$ first goes to zero,
$G_2(\omega)$ = Fourier transform of the total vertical investigation geometric factor of log 2,
$G_1(\omega)$ = Fourier transform of the total vertical investigation geometric factor of log 1.

A truncation error is introduced by the finite limits of $H_1(\omega)$, but this error is generally small enough that acceptable results may be obtained. Figure 3 shows the relative shape of $G_1(\omega)$, $G_2(\omega)$, and $H_1(\omega)$ for a typical case in which the normalized coil spacing of log 2 is one and that of log 1 is one-half.

The filtered log 1 response, $\bar{\sigma}_{A1}$, will equal the response of log 2, σ_{A2}, for noninvaded uniform beds. However, for invaded beds $\bar{\sigma}_{A1}$ will usually not equal σ_{A2}. An appreciable difference between $\bar{\sigma}_{A1}$ and σ_{A2} indicates that the conductivities of the formations surrounding the well bore are not uniform radially.

A guard electrode log may be filtered to give

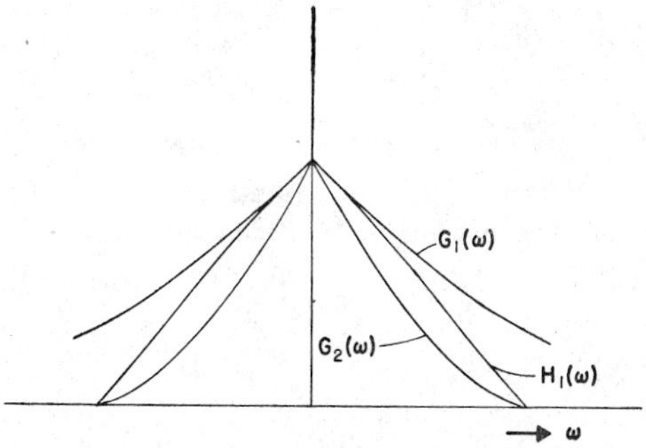

FIG. 3. Transfer function of response equalization filter.

the same response as log 2 by letting $h_1(z)$ become equal to $g_2(z)$, where $g_2(z)$ represents the vertical investigation characteristic of log 2. This assumes the vertical response of the guard log to be infinite in comparison with the induction log. Although the guard log does not have a geometrical factor, the highly idealized approach of the Owens-Greer method of analysis of the guard log in prolate spheroidal coordinates gives an approximate expression for the radial response of the device (Owen and Greer, 1951; Dakhnov, 1962, pp. 233–268). The filtered guard log then has the same vertical response as induction log 2. In terms of frequency this is equivalent to considering the shaded portion of Figure 4 of the "flat" vertical frequency characteristic of the guard log.

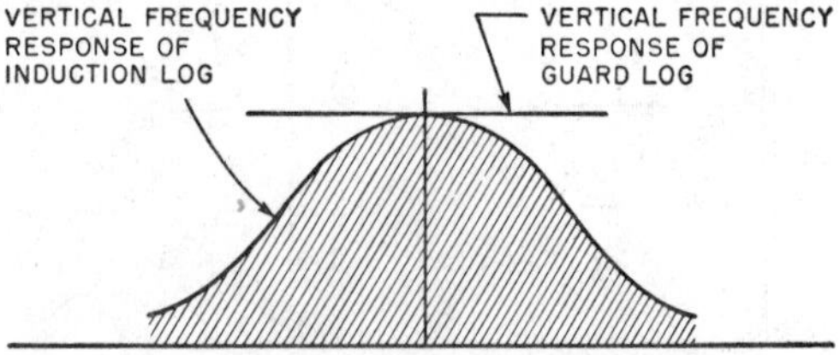

FIG. 4. Vertical response characteristics of guard and induction logs.

INVERSE CONVOLUTION FILTERING

Inverse convolution filters may be used to restore a signal which has been distorted in passing through a system. An inverse convolution filter is essentially an inverse filter in the band of interest, that is, its transfer function approximates the inverse of the transfer function of the distorting system. It may, however, be designed to be useful in the presence of appreciable noise (George, 1963, pp. 40–42).

An approximate value, $\bar{\sigma}(z)$, of formation conductivity, σ_0, of a radially uniform bed in equation (3) may be obtained by filtering the apparent conductivity $v(z)$ by an inverse convolution filter. An inverse convolution filter function $h(z)$, may be derived by treating the vertical investigation characteristic of the induction log, $g(z)$, as the impulse response of a linear filter (George et al, 1962). The approximation, $\bar{\sigma}(z)$, to formation conductivity may be obtained by convolving $h(z)$ with the apparent conductivity $v(z)$ indicated by the device:

$$\bar{\sigma}(z) = \int_{-\infty}^{\infty} h(z_0)v(z - z_0)dz_0. \quad (8)$$

MULTIPLE INDUCTION LOG ENHANCEMENT

The previous section suggested a means of increasing the vertical resolution of an induction log passing through beds in which the resistivity was uniform in the radial direction. This type of analysis will now be extended to a system of two, two-coil induction logs traversing beds containing invaded zones. This analysis was first given by Bostick (George, 1963, pp. 61–67). The coils of logs will be located symmetrically about a common reference point to eliminate the need for shifting one log relative to the other. The subscript "1" will refer to the log with the shorter coil spacing. This example is chosen for its simplicity to illustrate a procedure which may be extended to the more interesting cases of focused systems and many coiled induction logs. The procedure may also be extended to include investigation of more than two radial conductivity zones.

Equation (3) may be generalized to include radial variations, thus giving as the response of log 1 at point z,

$$\sigma_{A1}(z) = \iint_{\text{All Space}} \sigma(r, z_0)g_1(r, z - z_0)dz_0dr, \quad (9)$$

where σ_{A1} is the apparent conductivity indicated by log 1 and g_1 is the geometrical factor for log 1.

Similarly, for log 2,

$$\sigma_{A2}(z) = \iint_{\text{All Space}} \sigma(r, z_0)g_2(r, z - z_0)dz_0dr. \quad (10)$$

The response of the system to two cylindrical conductivity zones concentric about the borehole, indicated in Figure 5, will be considered. This would correspond to an invaded zone and an uncontaminated zone. Within each zone the conductivity will be considered only as a function of z. Inverse convolution filter functions may be derived which, when operating on σ_{A1} and σ_{A2} in the manner indicated below by equation (11), yield approximate values for the true conductivity of zones 1 and 2. The derivation of the inverse convolution filter functions, $h_{ij}(z)$, is given in Appendix A.

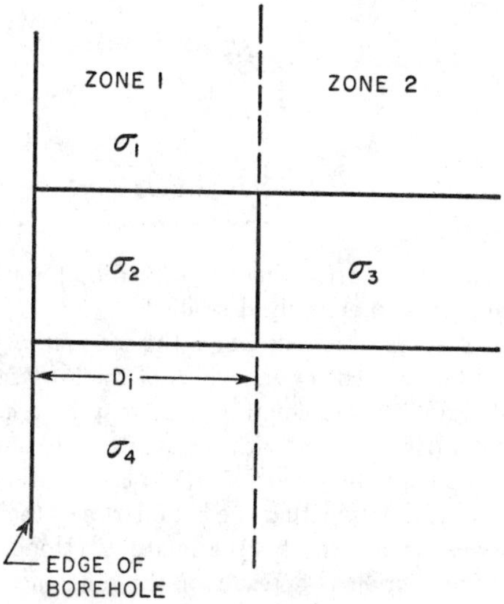

FIG. 5. Two-zone invaded bed.

$$\bar{\sigma}_1(z) = \int_{-\infty}^{\infty} \sigma_{A1}(z_0) h_{11}(z - z_0) dz_0 + \int_{-\infty}^{\infty} \sigma_{A2}(z_0) h_{12}(z - z_0) dz_0,$$

$$\bar{\sigma}_2(z) = \int_{-\infty}^{\infty} \sigma_{A1}(z_0) h_{21}(z - z_0) dz_0 + \int_{-\infty}^{\infty} \sigma_{A2}(z_0) h_{22}(z - z_0) dz_0. \quad (11)$$

Figure 6 illustrates the shape of $h_{11}(z)$, $h_{12}(z)$, $h_{21}(z)$, and $h_{22}(z)$ for the case of normalized coil spacings of unity and one-half and an invaded zone depth of 0.625.

SOLUTION FOR THE DEPTH OF INVASION

The multiple-induction-log enhancement analysis assumed knowledge of the depthof invasion, D_i. An iterative solution, first suggested by H. W. Smith, will be illustrated by considering two induction logs with different coil spacings and a guard electrode log traversing the two-zone bed of Figure 5 (George, 1963, pp. 75–83).

If the width of the uniform bed of Figure 5 is wide enough that the contributions to the log reading from the adjacent beds are small, then σ_1 and σ_2 are constant over that range of z for which the partial geometric factors $g_{ik}(z)$ have a nonnegligible value. The integrations may then be executed, giving

$$\sigma_{A1} = \sigma_1 g_{11}(D_i) + \sigma_2 g_{12}(D_i),$$
$$\sigma_{A2} = \sigma_1 g_{21}(D_i) + g_2 g_{22}(D_i), \quad (12)$$

where

$$g_{11}(D_i) + g_{12}(D_i) = 1.0,$$
$$g_{21}(D_i) + g_{22}(D_i) = 1.0.$$

The $g_{ik}(D_1)$ is an integrated partial geometric factor:

$$g_{i1}(D_i) = \int_{-\infty}^{\infty} \int_{0}^{D_i} g_i(r, z) dr dz,$$

$$g_{i2}(D_i) = \int_{-\infty}^{\infty} \int_{D_i}^{\infty} g_i(r, z) dr dz.$$

A value for D_i may be assumed, the integrated geometric factors computed or selected from previous calculations of these quantities, and the equations solved for σ_1 and σ_2 to give

$$\sigma_1 = \frac{\sigma_{A1} g_{22}(D_i) - \sigma_{A2} g_{12}(D_i)}{g_{11}(D_i) g_{22}(D_i) - g_{21}(D_i) g_{12}(D_i)},$$
$$\sigma_2 = \frac{\sigma_{A2} g_{11}(D_i) - \sigma_{A1} g_{21}(D_i)}{g_{11}(D_i) g_{22}(D_i) - g_{21}(D_i) g_{12}(D_i)}. \quad (13)$$

From the calculated conductivity values and the assumed depth of invasion, the reading of the guard electrode log may be computed. The computed value of the guard log may then be compared with the actual reading. If the two values are not equal, then the assumed D_i may be incremented by an amount ΔD_i, and the process repeated until the difference between the computed and actual guard log readings is less than a predetermined quantity ϵ which may be expressed

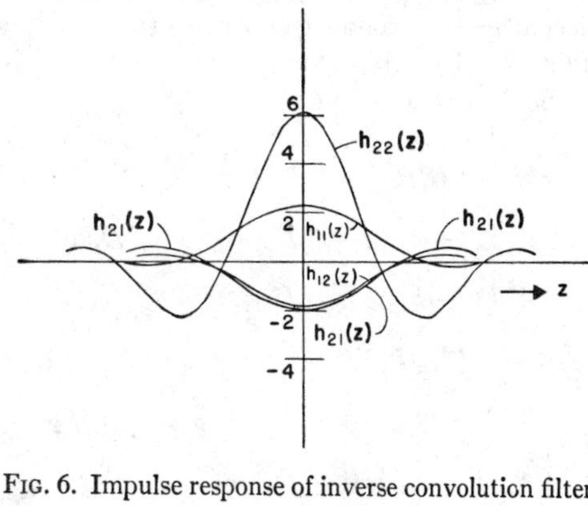

FIG. 6. Impulse response of inverse convolution filters.

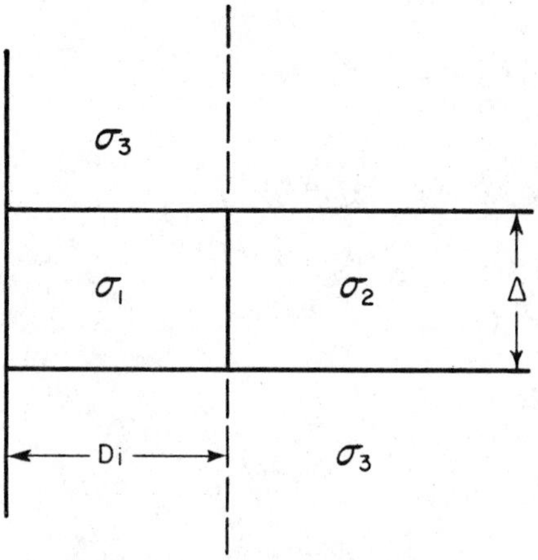

FIG. 7. Thin invaded bed of width Δ.

as a percent of the log reading. The procedure is applicable to log combinations other than the one given above.

The process described gives accurate results only for wide invaded beds. The partial geometric factors $g_{ik}(z)$ of an induction log actually have a nonnegligible value over a range several coil-spacings wide (George, 1963, pp. 75–83). The geometric contribution, however, of a bed of known width to the total log reading may be calculated from a knowledge of the geometric factor. This knowledge may be used to extend the iterative solution for D_i to the case of thin beds by using center bed values.

A simple example may be selected to illustrate the extension of the solution for D_i to the case of the thin bed of Figure 7 of width Δ. The quantity k_{ik} shall represent the percent contribution to the partial geometric factor, $g_{ik}(D_i)$, of a bed of width Δ. The contribution of σ_1 in Figure 7 to the log reading is $\sigma_1 g_{11}(D_i)k_{11}$. The contribution of the adjacent beds of conductivity σ_3 to the total log reading is $(1-k_{11})g_{11}(D_i)\sigma_3$. Equation (12) may then be rewritten as

$$\begin{aligned}\sigma_{A1} &= \sigma_1 g_{11}(D_i)k_{11} \\ &\quad + (1 - k_{11})\sigma_3 g_{11}(D_i) + \sigma_2 g_{12}(D_i)k_{12} \\ &\quad + (1 - k_{12})\sigma_3 g_{12}(D_i), \\ \sigma_{A2} &= \sigma_1 g_{21}(D_i)k_{21} \qquad (14) \\ &\quad + (1 - k_{21})\sigma_3 g_{21}(D_i) + \sigma_2 g_{22}(D_i)k_{22} \\ &\quad + (1 - k_{22})\sigma_3 g_{22}(D_i),\end{aligned}$$

where

$$k_{ij} = \frac{\int_{-\Delta/2}^{\Delta/2} g_{ij}(z)dz}{\int_{-\infty}^{\infty} g_{ij}(z)dz}.$$

The above equations may be solved by the iterative process previously described.

In cases of thin beds where the conductivity of the adjacent beds or the width of the interesting bed is not known, equation (12) may be used in the iterative solution with the guard log to obtain an approximate value of D_i. This solution is meaningful only within the boundaries of the bed in question as seen by the guard electrode log. This restriction is a result of the difference between the assumed infinite vertical resolution characteristics of the unfiltered guard log and the vertical resolution characteristics of the induction log.

EXAMPLE OF APPLICATION OF TECHNIQUES

The techniques previously described will now be applied to log data from a system of two induction logs of coil spacings 1.0 and 0.5 and a guard log of length 0.5. Log 1 will refer to the induction log of coil spacing 0.5, and log 2 will refer to the induction log of coil spacing 1.0. Distances, both down the well bore and radially, are given in terms of long coil spacing, the long coil spacing being normalized to unity. The response of this system to the formations of Figure 8 will be investigated. Resistivity, depth of invasion, and formation factor are shown for each bed. The response of the two induction logs may be computed by use of partial geometric factors for four cylindrical zones concentric with the borehole. Log readings are displayed on a logarithmic scale.

The first step in the analysis of the log data is to scan the log and determine the interesting (invaded) beds. Figure 9 gives the apparent resistivity indicated by log 1 and log 2 and the value of log 1 filtered in such a way as to have the frequency response of log 2. The invaded beds are identified by the separation between log 2 and filtered log 1. A ratio log, the ratio of filtered log 1 to log 2, is also shown. The ratio log, normally unity in noninvaded regions, departs from unity in invaded zones. Figure 10 gives the re-

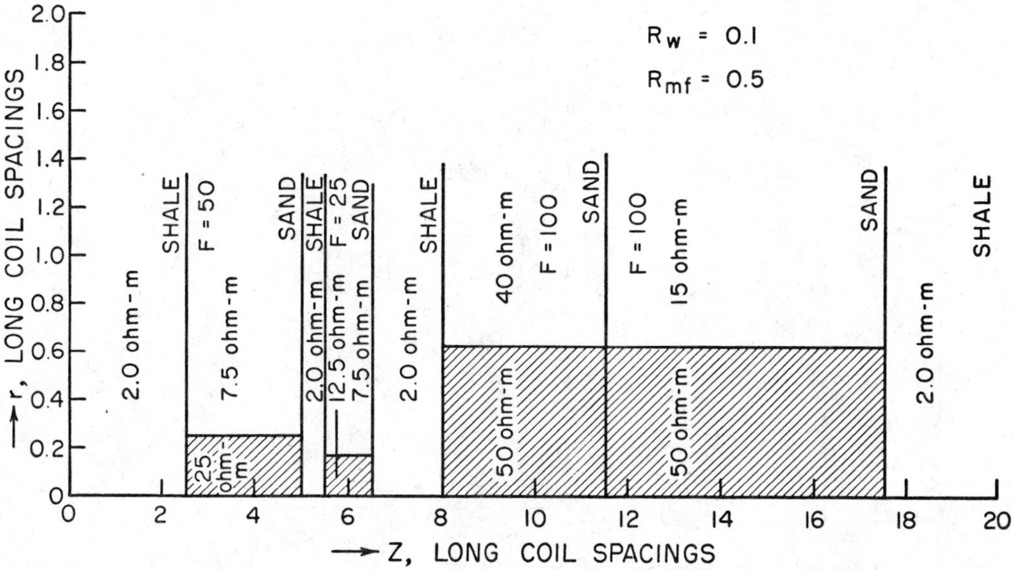

FIG. 8. Bed formations.

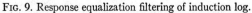

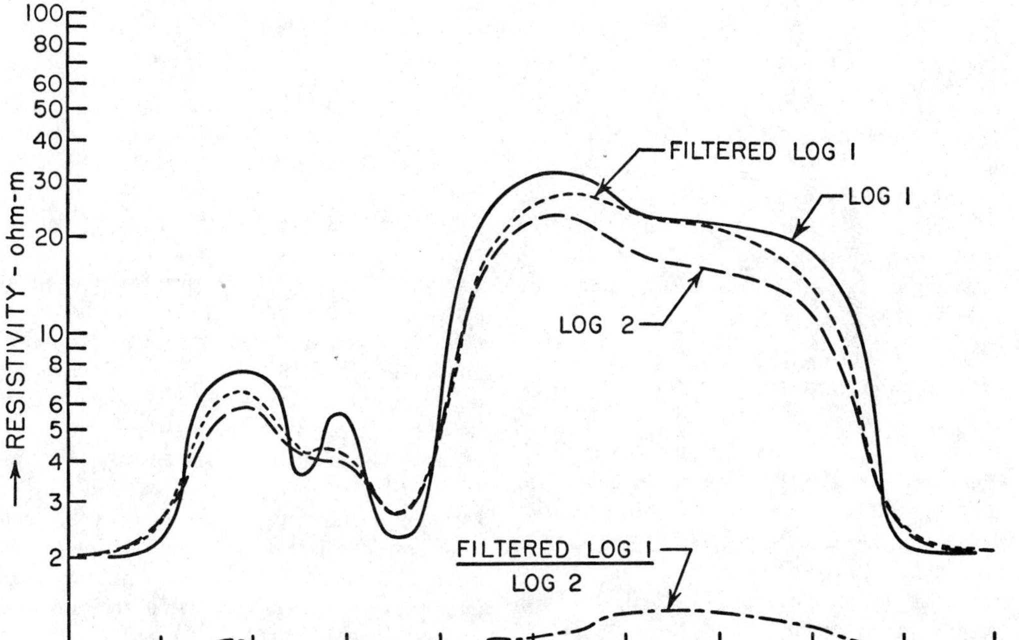

FIG. 9. Response equalization filtering of induction log.

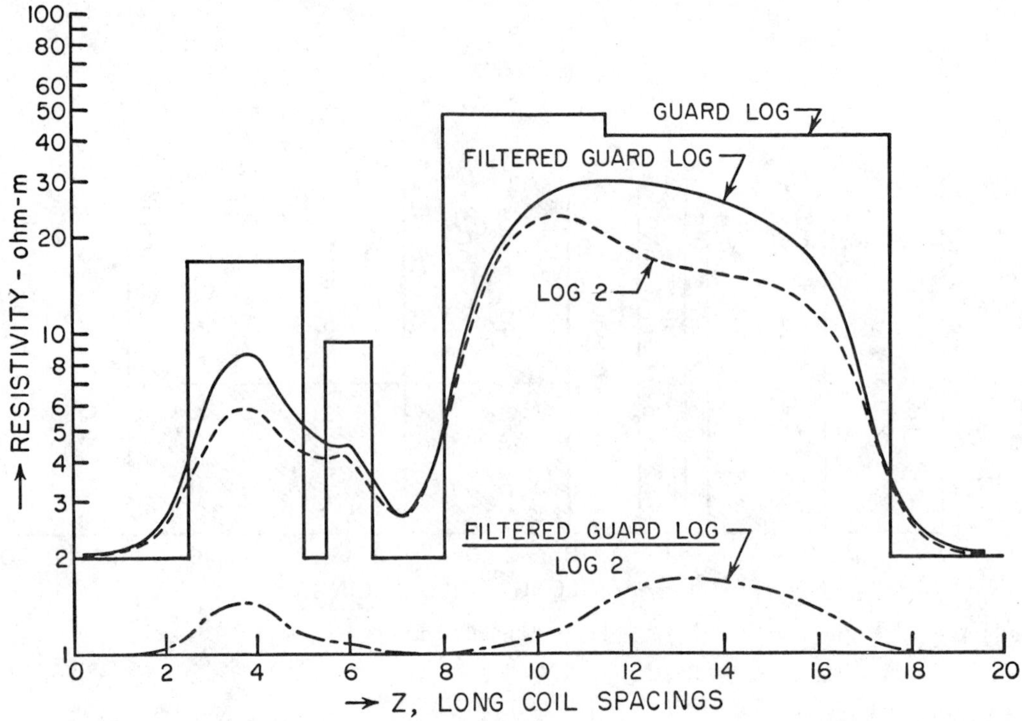

FIG. 10. Response equalization filtering of guard log.

sponse of log 2 and the guard log, the filtered guard log, and the ratio log for the filtered guard log and log 2.

Figure 11 shows the actual D_i and D_i computed from the iterative solution. In regions between invaded beds where no value for the computed D_i is given, the solution did not converge. The quantity

$$\left(\frac{\text{filtered log 1}}{\text{log 2}} - 1\right)$$

is also shown on the linear scale of Figure 11. Figure 12 gives the three log readings and the iterative solution for the resistivity of the inside zone 1 and the outside zone 2, corresponding to the values of D_i given in Figure 11. The interesting formations have now been located, the depth of invasion determined, and the approximate values of resistivity obtained.

Having obtained a value for D_i, inverse convolution filters, based on the value of D_i, may be convolved with the log readings to obtain approximate conductivity values in two radial zones all down the entire length of the well bore traversed by the logging devices.[1] Figure 13 shows the approximate zone conductivity values for an assumed depth of invasion of 0.625 which is the value of D_i for the two wide beds of Figure 8. Figure 14 shows the results of the same calculation for D_i equal to 0.5. Figure 15 shows the results of this calculation based on the correct value of D_i for each section of the log. In this example the boundaries of the interesting beds are defined by changes in the guard log reading. More sophisticated techniques for determination of the bed boundaries may sometimes be required. In practice, inflection points in the self-potential curve may indicate boundaries between shales and porous formations. The necessary information has now been obtained from the logs to construct an approximate resistivity map of the formations

[1] Filtered values may be obtained for points at distances equal to or greater than half the "width" of the filtering functions from the end of the log. Filtered values may be obtained for the entire length of the log by assuming the log is bounded by uniform beds half the width of the filter-function width and of value equal to the last value of the actual log.

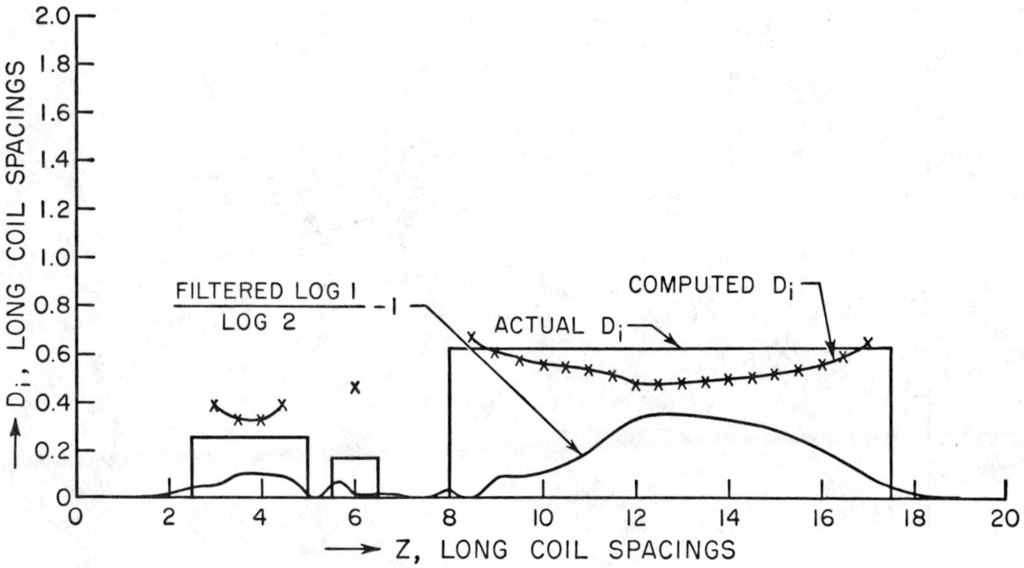

FIG. 11. Solution for depth of invasion.

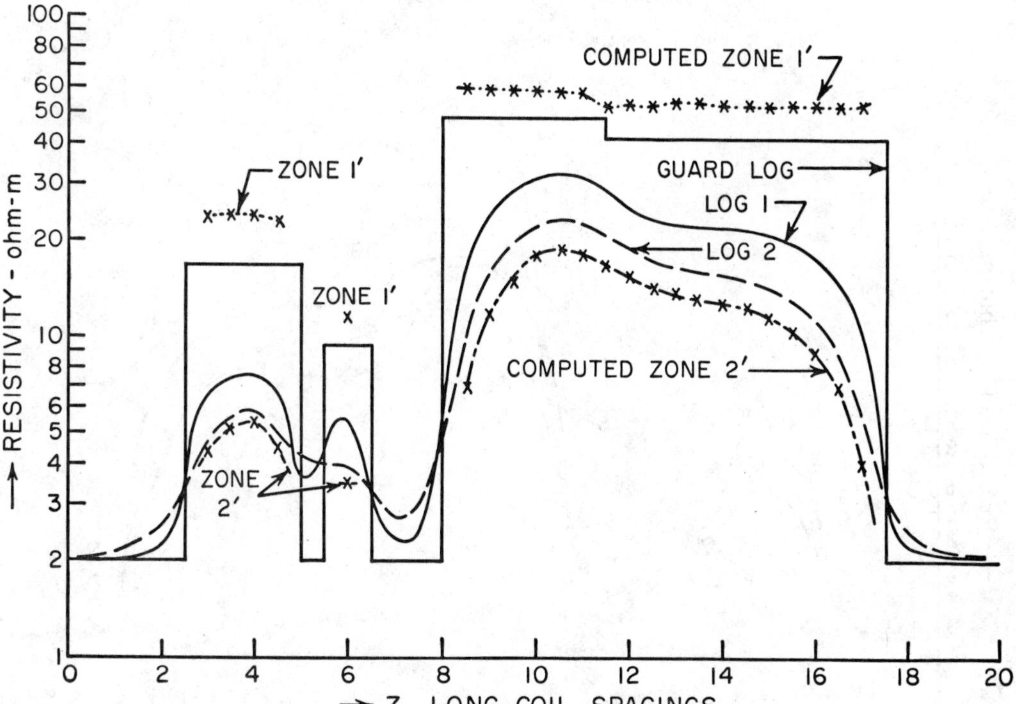

FIG. 12. Iterative solution for zone conductivities.

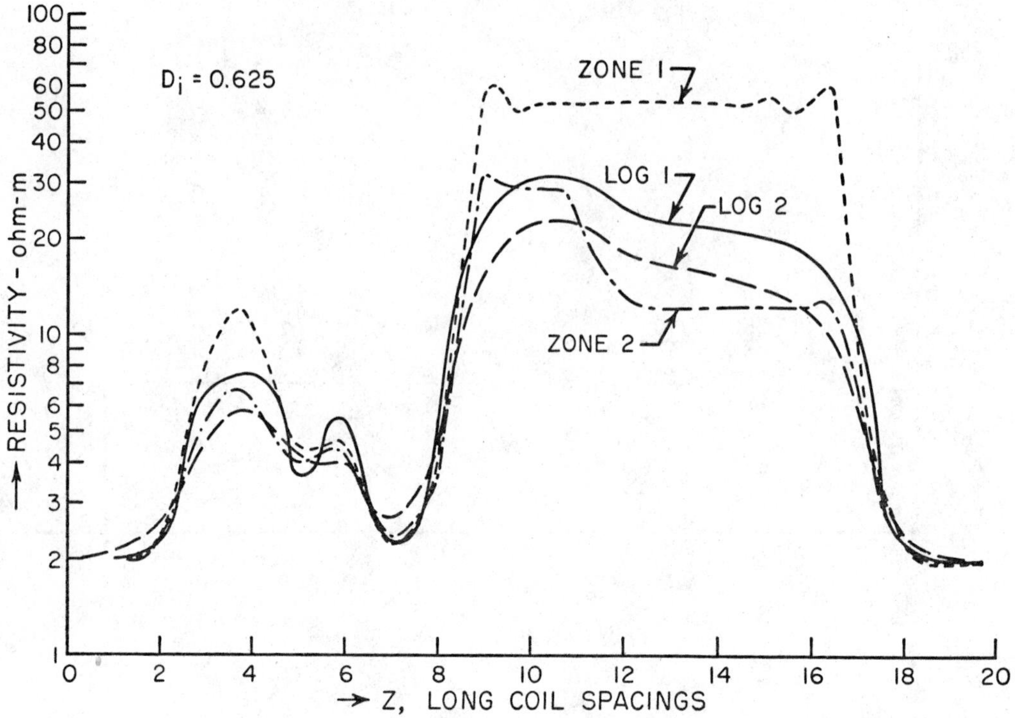

FIG. 13. Zone conductivities for assumed $D_i = 0.625$.

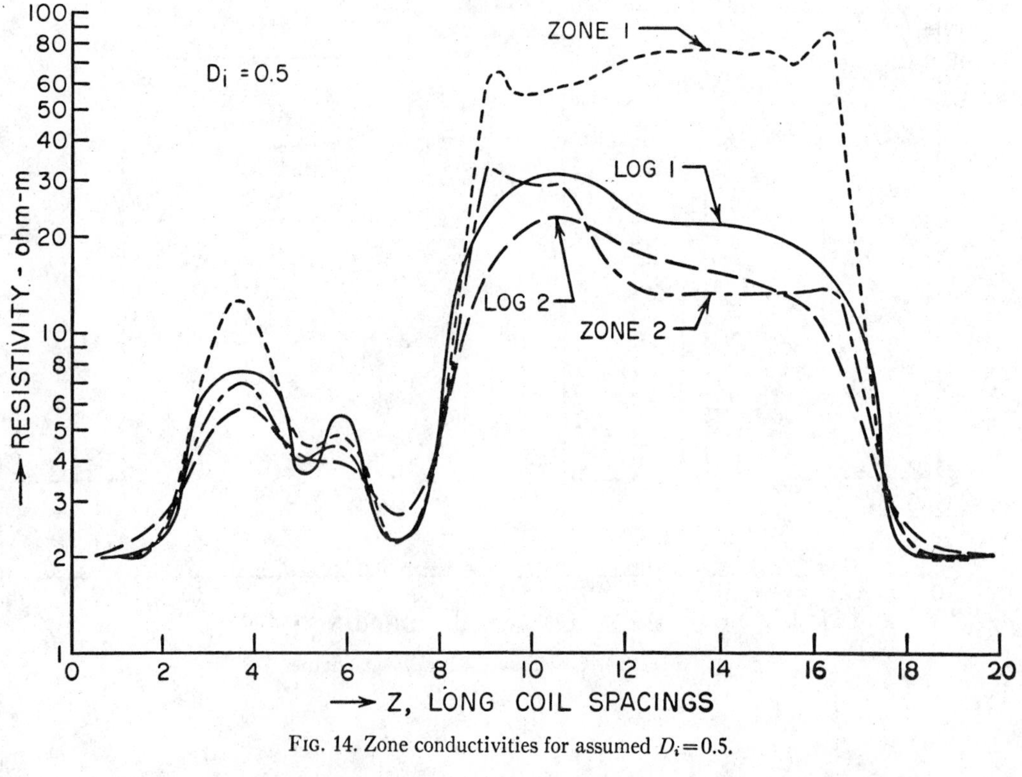

FIG. 14. Zone conductivities for assumed $D_i = 0.5$.

surrounding the borehole for this section of the distance traversed by the logging instruments.

CONCLUSION

The procedures previously described are examples of techniques applicable to automation. They would not be practical for hand calculation. The example of the previous section illustrates the possibilities presented by automatic interpretation of logs. All results which have been presented were computed on a CDC 1604 digital computer. For a given set of logging devices, the geometric factors, response equalization filter functions, and inverse convolution filter functions could be computed once and stored in the memory of the computer. The log analyst may then interpret the results of the computer program. Functions programmed for a digital computer are specified only for finite ranges of their arguments, but effects of time limiting and band limiting are usually small (George, 1963, pp. 94–97).

The examples given above have been simplified to illustrate more easily the application of the theory and are not suggestive of its limitations. Although mud column effects have been neglected in the example, these could be incorporated in the analysis or the basic instrument could be compensated for these effects. With the addition of more logs with different investigation characteristics, more detailed analysis of radial variations of resistivity could be made. For example, with certain assumptions, three appropriately chosen induction logs and a guard log could be used to obtain approximate conductivities of the invaded zone, the annulus region, and the undisturbed region. The type of analysis described in this study may lead not only to improved log interpretation, but also to better selection of logging tools with differing depths of investigation. Experience with actual field data will determine the practical limitations of these techniques.

ACKNOWLEDGMENTS

The authors wish to thank Lane Wells, Inc., for whom the study was conducted, for permission to publish the material.

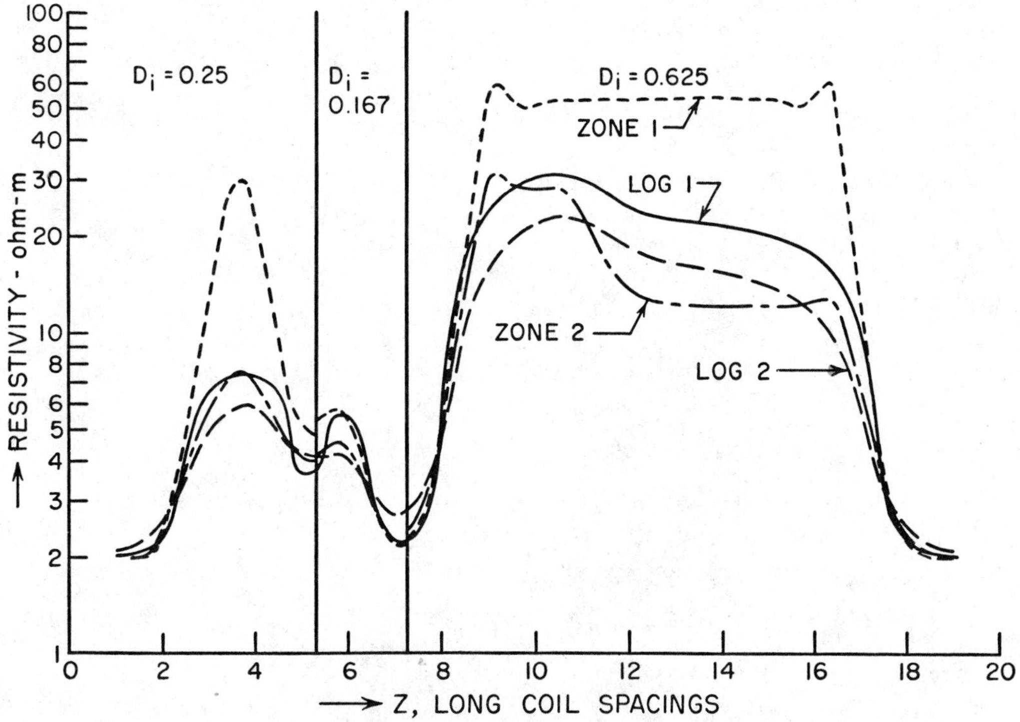

FIG. 15. Zone conductivities for correct values of D_i.

APPENDIX A—DERIVATION OF INVERSE CONVOLUTION FILTER FUNCTIONS FOR MULTIPLE INDUCTION LOG ENHANCEMENT

The response of two induction logs to the two-zone bed of Figure (4), where D_i represents the depth of invasion, or radius of zone 1, is given by:

$$\sigma_{A1}(z) = \int_{-\infty}^{\infty} \sigma_1(z_0)\left[\int_0^{D_i} g_1(r, z - z_0)dr\right]dz_0 + \int_{-\infty}^{\infty} \sigma_2(z_0)\left[\int_{D_i}^{\infty} g_1(r, z - z_0)dr\right]dz_0,$$

$$\sigma_{A2}(z) = \int_{-\infty}^{\infty} \sigma_1(z_0)\left[\int_0^{D_i} g_2(r, z - z_0)dr\right]dz_0 + \int_{-\infty}^{\infty} \sigma_2(z_0)\left[\int_{D_i}^{\infty} g_2(r, z - z_0)dr\right]dz_0. \quad \text{(A1)}$$

Let

$$g_{11}(z - z_0) = \int_0^{D_i} g_1(r, z - z_0)dr,$$

$$g_{12}(z - z_0) = \int_{D_i}^{\infty} g_1(r, z - z_0)dr,$$

$$g_{21}(z - z_0) = \int_0^{D_i} g_2(r, z - z_0)dr,$$

$$g_{22}(z - z_0) = \int_{D_i}^{\infty} g_2(r, z - z_0)dr, \quad \text{(A2)}$$

where g_{ik} is the partial geometric factor of log i in zone k; that is, the integral of the product of g_{ik} and σ_k gives the contribution of the kth zone to the log reading. Equation (A1) may be rewritten as

$$\sigma_{A1}(z) = \int_{-\infty}^{\infty} \sigma_1(z_0)g_{11}(z - z_0)dz_0 + \int_{-\infty}^{\infty} \sigma_2(z_0)g_{12}(z - z_0)dz_0,$$

$$\sigma_{A2}(z) = \int_{-\infty}^{\infty} \sigma_1(z)g_{21}(z - z_0)dz_0 + \int_{-\infty}^{\infty} \sigma_2(z_0)g_{22}(z - z_0)dz_0, \quad \text{(A3)}$$

or

$$\sigma_{A1}(z) = g_{11}(z) * \sigma_1(z) + g_{12}(z) * \sigma_2(z),$$
$$\sigma_{A2}(z) = g_{21}(z) * \sigma_1(z) + g_{22}(z) * \sigma_2(z). \quad \text{(A4)}$$

Equations (A4) are reduced to algebraic equations by the Fourier transformation:

$$S_{A1}(\omega) = G_{11}(\omega)S_1(\omega) + G_{12}(\omega)S_2(\omega),$$
$$S_{A2}(\omega) = G_{21}(\omega)S_1(\omega) + G_{22}(\omega)S_2(\omega). \quad \text{(A5)}$$

The inverse convolution filter motivates an approximate solution of the following form:

$$\overline{S}_1(\omega) = H_{11}(\omega)S_{A1}(\omega) + H_{12}(\omega)S_{A2}(\omega),$$
$$\overline{S}_2(\omega) = H_{21}(\omega)S_{A1}(\omega) + H_{22}(\omega)S_{A2}(\omega), \quad \text{(A6)}$$

where the $\overline{S}$ quantities approximate the S quantities. Substituting equation (A5) into equation (A6) gives

$$\overline{S}_1(\omega) = F_{11}(\omega)S_1(\omega) + F_{12}(\omega)S_2(\omega),$$
$$\overline{S}_2(\omega) = F_{21}(\omega)S_1(\omega) + F_{22}(\omega)S_2(\omega), \quad \text{(A7)}$$

where

$$F_{11}(\omega) = H_{11}(\omega)G_{11}(\omega) + H_{12}(\omega)G_{21}(\omega),$$
$$F_{12}(\omega) = H_{11}(\omega)G_{12}(\omega) + H_{12}(\omega)G_{22}(\omega),$$
$$F_{21}(\omega) = H_{21}(\omega)G_{11}(\omega) + H_{22}(\omega)G_{21}(\omega),$$
$$F_{22}(\omega) = H_{21}(\omega)G_{12}(\omega) + H_{22}(\omega)G_{22}(\omega). \quad \text{(A8)}$$

To eliminate the contribution of S_2 to $\overline{S}_1$ and the contribution of S_1 to $\overline{S}_2$, F_{12} and F_{21} are set equal to zero:

$$F_{12} = F_{21} = 0.$$

Equation (A8) may be solved for the H's to obtain

$$H_{11}(\omega) = \left[\frac{G_{22}(\omega)}{G_{11}(\omega)G_{22}(\omega) - G_{12}(\omega)G_{12}(\omega)}\right]F_{11}(\omega),$$

$$H_{12}(\omega) = -\left[\frac{G_{12}(\omega)}{G_{11}(\omega)G_{22}(\omega) - G_{12}(\omega)G_{21}(\omega)}\right]F_{11}(\omega),$$

$$\text{(A9)}$$

$$H_{22}(\omega) = \left[\frac{G_{11}(\omega)}{G_{11}(\omega)G_{22}(\omega) - G_{12}(\omega)G_{21}(\omega)}\right]F_{22}(\omega),$$

$$H_{21}(\omega) = -\left[\frac{G_{21}(\omega)}{G_{11}(\omega)G_{22}(\omega) - G_{12}(\omega)G_{21}(\omega)}\right]F_{22}(\omega).$$

The criterion of maximum possible response improvement with minimum overshoot again suggests the selection of the "hanning" function for $F(\omega)$. Therefore, since the denominators of all the $H_{ik}(\omega)$'s are the same, let

$$\begin{aligned} F_{11}(\omega) &= F_{22}(\omega) \\ &= 1/2 + 1/2 \cos K\omega, \ \omega \leq \pi/K \\ &= 0, \ \omega \geq \pi/K, \end{aligned} \tag{A10}$$

where K is some number such that $F_{ij}(\omega)$ becomes zero before the first zero of the denominator of equation (A9). The inverse transform of the $H(\omega)s$ may be taken to obtain a set of functions $h_{ik}(z)$ which when convolved with the log readings give approximate zone conductivity values in the manner indicated by the equations:

$$\bar{\sigma}_1(z) = \int_{-\infty}^{\infty} \sigma_{A1}(z_0) h_{11}(z - z_0) dz_0 + \int_{-\infty}^{\infty} \sigma_{A2}(z_0) h_{12}(z - z_0) dz_0,$$

$$\bar{\sigma}_2(z) = \int_{-\infty}^{\infty} \sigma_{A1}(z_0) h_{21}(z - z_0) dz_0 + \int_{-\infty}^{\infty} \sigma_{A2}(z_0) h_{22}(z - z_0) dz_0. \tag{A11}$$

REFERENCES

Dakhnov, V. N., 1962, Geophysical well logging: Golden, Colorado School of Mines.

Doll, H. G., 1949, Introduction to induction logging and application to logging of wells drilled with oil base mud: Jour. of Pet. Tech., v. 1, pp. 148–162.

Duesterhoeft, Jr., W. C., 1961, Propagation effects in induction logging: Geophysics, v. 26, pp. 192–204.

———, Hartline, Ralph E., and Thomsen, H. Sandoe, 1961, The effect of coil design on the performance of the induction log: Jour. of Pet. Tech., v. 13, pp. 1137–50.

———, and Smith, H. W., 1962, Propagation effects on radial response in induction logging: Geophysics, v. 27, pp. 463–469.

Foster, M. R., Hicks, W. G., and Nipper, J. T., 1962, Optimum inverse filters which shorten the spacing of velocity logs: Geophysics, v. 27, pp. 317–326.

George, Jr., C. F., 1963, Application of special filtering techniques to well log analysis: doctoral dissertation, The University of Texas, Austin.

———, Smith, H. W., and Bostick, Jr., F. X., 1962, The application of inverse convolution techniques to improve signal response of recorded geophysical data: IRE Proceedings, v. 50, pp. 2313–2319.

Owen, J. E., and Greer, W. J., 1951, The guard electrode logging system: Petroleum Transactions, AIME, v. 192, pp. 347–356.

V. MULTICHANNEL DECONVOLUTION

W. A. Schneider et al present a two-channel technique for the elimination of ghost arrivals. To be sure, some data improvement can be obtained by simply aligning the primaries and (vertically) stacking, but the authors show that ghost energy can be rejected over a much wider band of frequencies by filtering each trace before summation. An example is presented where such filters are designed on the basis of flat spectrum primary power, flat spectrum ghost power, and flat spectrum random noise (at 0.1 the level of primary and ghost power). The only remaining design parameter is the ghost reflection moveout time, and this can be estimated from an uphole-time measurement; the filter design thus does not require an analysis of reflections. Very impressive deghosting is achieved using field data from an uphole survey taken near Sherman, Texas. The effectiveness of the processing is further demonstrated by applying it so as to enhance ghost energy and reject primaries. The same experimental data are also used to construct a two-interface model of the near-surface. An autocorrelation analysis demonstrates that the base of the weathering and the surface both produce substantial ghosts, 0.6 being the estimated reflection coefficient for each interface.

In a related paper, Schneider, E. R. Prince, Jr., and B. F. Giles develop three-channel optimum filters for multiple removal. Again, a substantial improvement over simple stacking is achieved on synthetic records. This problem is more complex than ghost elimination, since different filters are required for each seismometer group and for different time segments of the record. In processing, the filters are stored in core and brought in at the appropriate time. An interesting feature of the filter design is an allowance for imperfect alignment of the primary energy.

E. B. Davies and E. J. Mercado report on an investigation of multichannel deconvolution with emphasis on field-recorded data. While they find that multichannel deconvolution results in somewhat better ghost suppression, they conclude that on field-recorded common-depth-point data the multichannel deconvolution processing results are not significantly better than those obtained by stacking and single-channel deconvolution.

The tutorial paper by Sven Treitel should be invaluable to those who wish to learn the basics of digital multichannel filtering. A brief quote: ". . . few elementary treatments outlining the principles of this subject have been published. The present contribution represents an effort in this direction and, hopefully, provides the necessary background needed to follow more advanced material." The reader will find careful definitions (note the positive exponent in Z-transforms) and many worked-out examples. A peculiarity of interest is the zero error that can be achieved with multichannel Wiener filters for finite-length inputs.

Reprinted from Geophysics v. 29, no. 5, p. 783-805

A NEW DATA-PROCESSING TECHNIQUE FOR THE ELIMINATION OF GHOST ARRIVALS ON REFLECTION SEISMOGRAMS†

WILLIAM A. SCHNEIDER,* KENNETH L. LARNER,‡ J. P. BURG,* AND MILO M. BACKUS*

A new data-processing technique is presented for the separation of initially up-traveling (ghost) energy from initially down-traveling (primary) energy on reflection seismograms. The method combines records from two or more shot depths after prefiltering each record with a different filter. The filters are designed on a least-mean-square-error criterion to extract primary reflections in the presence of ghost reflections and random noise. Filter design is dependent only on the difference in uphole time between shots, and is independent of the details of near-surface layering. The method achieves wide-band separation of primary and ghost energy, which results in 10–15 db greater attenuation of ghost reflections than can be achieved with conventional two- or three-shot stacking (no prefiltering) for ghost elimination.

The technique is illustrated in terms of both synthetic and field examples. The deghosted field data are used to study the near-surface reflection response by computing the optimum linear filter to transform the deghosted trace back into the original ghosted trace. The impulse response of this filter embodies the effects of the near-surface on the reflection seismogram, i.e. the cause of the ghosting. Analysis of these filters reveals that the ghosting mechanism in the field test area consists of both a surface- and base-of-weathering layer reflector.

INTRODUCTION

In exploration seismology when an explosive source is detonated below the free surface, the initial up-traveling energy is either absorbed or returned to the subsurface by the near-surface reflection complex. If the reflected energy forms a coherent wavefront, then we may expect to record both primary and ghost arrivals from every reflector at depth, separated in time by approximately twice the traveltime from the shot to the near-surface reflection complex. This "double image" view of the subsurface which results from ghosting is clearly undesirable for the following reasons:

(1) Doubling the number of reflection events on the record compounds the subsurface interpretation.

(2) Ghosts and primaries may interfere destructively, thereby masking subsurface information.

(3) The ghost-reflection interference may vary considerably within a prospect, causing poor character correlation.

The latter is perhaps the most undesirable feature of ghosting. Since the ghost reflections are sensitive to near-surface conditions in the vicinity of the shotpoint, such as thickness of the weathering layer, roughness of surface and base-of-weathering reflector, depth of the shot, position of water table, to mention a few, we might well expect ghosting to vary significantly within short lateral distances. This variation in the ghosting can destroy meaningful fine structure in the primary reflections which may convey stratigraphic trap information. For these as well as other reasons, ghost arrivals are classified as seismic noise along with surface waves, multiple reflections, scattered energy, wind noise, etc.

The elimination of ghost energy from seismic records may be accomplished by exploiting the space-time relationship between primary and ghost arrivals on a single or multichannel basis. An example of the former is provided in Lindsey's (1960) article whereby a simple positive feedback loop is employed to effectively cancel ghosts from a single near-surface reflector against

† Presented at the 32nd Annual International SEG Meeting, Calgary, Alberta, Canada, September 19, 1962; Pacific Section AAPG, SEG, and SEPM, Los Angeles, April 25, 1963. Manuscript received by the Editor July 3, 1963.

* Texas Instruments Incorporated, Dallas, Texas.

‡ Graduate Student, Massachusetts Institute of Technology, Department of Geology and Geophysics, Cambridge, Massachusetts.

primaries. In Lindsey's notation, the filter to apply to each trace is given by

$$\frac{1}{(1 - aH(s)e^{-s\tau})} \qquad (1)$$

a = near-surface reflection coefficient.
$H(s)$ = near-surface filter to account for differences in primary and ghost wavelet.
τ = delay between primary and ghost.
s = Laplace-transform variable.

As is evident from equation (1) this method of ghost rejection requires specific and detailed information about the ghosting mechanism, which, in all but the simplest cases, is difficult to obtain from the field records. As Lindsey demonstrates, the autocorrelation function of the input data may in principle furnish the parameters a, $H(s)$, and (τ) with sufficient accuracy for the feedback technique to be employed. This requires, however, (1) a high signal-to-noise ratio, (2) wide-band signal information, and (3) a simple total seismic system wavelet. Any significant amplitude in the correlation function due to the primaries themselves at other than zero lag would seriously hamper isolating that part of the correlation due to ghosting, and the specification of the parameters for equation (1). This will become evident in the following.

The multichannel techniques currently available for ghost elimination consist of vertical shot stacking (Hammond, 1962), and the use of distributed charges (Musgrave, Ehlert, and Nash, 1958; Sparks and Silverman, 1953). These, of course, differ only in degree and not kind as they both represent vertical stacking. The philosophy of vertical shot stacking, for ghost elimination, finds its basis in the time relationship that exists between primaries and ghosts from different shot depths. If we have several shots at different depths recorded at the same seismometer position, and static correct these traces to a common shot depth, the primary reflections will time tie and the ghosts, of course, will not. By stacking the time-shifted traces, it is clear that the primary energy will be enhanced relative to the ghost energy. The stacking may be accomplished in the playback center with the individually recorded shots, or in the field with delayed charges or distributed charges whose detonation velocity is matched to the propagation velocity of the surrounding medium. This is a brute-force ghost-cancellation technique which works the ghost reflections against themselves. It is attractive in that it does not require knowledge of the ghosting mechanism or the primary-ghost relationship. The major shortcoming of the conventional shot-stacking scheme results from its narrow-band rejection capability when stacking only a few shots. This limitation is serious for two reasons: (1) There is an industry-wide movement to achieve wide-band seismic reflection data for fuller utilization of the available signal spectrum; therefore, narrow-band techniques are decidedly undesirable if alternatives exist; (2) The narrow-band rejection capability makes the shot separation a critical parameter; the shot spacing must be very nearly $\frac{1}{4}$ wavelength for the peak frequency in the spectrum of the ghost wavelet in order to deghost successfully even with narrow-band input.

These considerations lead us to the conclusion that the ideal ghost-elimination processor should require: (1) both wide-band ghost rejection and signal preservation capabilities, and (2) minimum knowledge of the ghosting mechanism and the primary-ghost relationship. These requirements can be well met by allowing for individually filtering the several shots before stacking as is demonstrated in the following.

METHOD

This multichannel-processing technique is schematically characterized below.

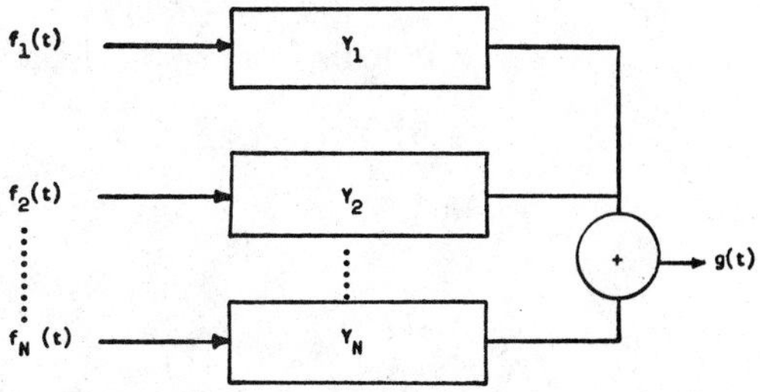

Where $f_i(t)$ is the input trace from the ith shot at a given seismometer location, the Y_i's are filters applied to the traces before summation, and $g(t)$ is the output of the processor.

The complex transfer functions $Y_i(f)$ are designed, on a least-mean-square-error criterion, to extract primary events from random noise and ghost events, on the basis of the differential time dependence of primaries and ghosts on hole depth. The least-squares-filter design is presented in Appendix A.

Although any number of shots can be stacked by this process, two shots usually provide ample attenuation of the ghost energy. Therefore, we will restrict our attenuation to two-channel processes.

In the specification of the signal and noise model it is assumed that the signal and noise are statistically independent. This assumption insures that the resulting filters $Y_1(f)$ and $Y_2(f)$ will not require knowledge of the ghosting mechanism for their specification. The fact that the primaries and ghosts are in fact correlated does not invalidate the process, but rather reflects that we are not using all the information in the data available to us. By utilizing this latter correlation we could obtain a somewhat more effective deghosting processor. However, the difficulty in estimating the primary-ghost correlation and the impracticality of having to redesign the filters for each new ghosting situation presently outweighs any potential gain. Before proceeding to examine specific examples of the use of the processor, it is instructive to compare its theoretical response with that of the conventional shot-stacking technique. The parameter of interest here is the ratio of (ghost/primary) amplitude response as a function of frequency. Clearly, we desire this to be as small as possible consistent with signal preservation and random noise rejection at all frequencies. If we define the signal and noise for the two shots as

$$S_1(t) = P(t)$$

(primary arrivals from shallow shot)

$$S_2(t) = P(t)$$

(primary arrivals from deep shot)

and

$$N_1(t) = G(t)$$

(ghost arrivals from shallow shot)

$$N_2(t) = G(t - \Delta t)$$

(ghost arrival from deep shot)

where Δt is the ghost reflection moveout between the two shots after static correcting the traces to a common depth, and equals twice the difference in uphole ($2\Delta t_{uh}$) time; then the responses for primaries and ghosts for the conventional stack are, respectively,

$$S_1(t) + S_2(t) = 2P(t)$$
$$N_1(t) + N_2(t) = G(t) + G(t - \Delta t). \quad (2)$$

Transforming equation (2) to the frequency domain we have

response for primaries

$$= 2\bar{P}(f)$$

response for ghosts

$$= \bar{G}(f)(1 + e^{-i2\pi f\Delta t}), \quad (3)$$

where

$$\bar{x}(f) = \int_{-\infty}^{\infty} x(t)e^{-i2\pi ft}dt$$

and

$$x(t) = \int_{-\infty}^{\infty} \bar{x}(f)e^{i2\pi ft}df$$

are Fourier-transform pairs.

From equation (3), the transfer function of the conventional stack process for primary events is the constant 2, and for ghosts $(1+e^{-i2\pi f\Delta t})$. The amplitude of the ratio of ghost-to-primary-transfer functions is clearly

$$\left| \cos \pi f\Delta t \right| , \quad (4)$$

a well-known result.

Similarly, the responses for primaries and ghosts for the present process, which we call "Optimum Wide-Band Stack" process, are, respectively,

response for primaries

$$= \bar{P}(f)(Y_1(f) + Y_2(f))$$

response for ghosts

$$= \bar{G}(f)(Y_1(f) + Y_2(f)e^{-i2\pi f\Delta t})$$

and the amplitude of the ratio of ghost-to-primary-transfer functions is

$$\frac{\left| Y_1(f) + Y_2(f) \right|}{\left| Y_1(f) + Y_2(f)e^{-i2\pi f\Delta t} \right|} \cdot \quad (5)$$

Expressions (4) and (5) are plotted in Figure 1. The difference in the breadth of the reject bands of the two processes is apparent. In particular, for a ΔT of 10 msec, the -20 db reject range for the conventional two-shot stack is only 47 to 53 cps, whereas the present process provides a 20–80 cps range. The latter increased the rejection bandwidth by a factor of 10 over the conventional two-channel stack. This difference, as well as the general effectiveness of the method, is best illustrated in the following deghosting examples using both synthetic and field examples.

RESULTS

The first example is a highly idealized synthetic case, but it serves to put the results of Figure 1 in the more familiar light of wavelets that might be expected to occur on a reflection seismogram.

Figure 2 shows the idealized cross section including shot and recording geometry used to combat ghost reflections on a multichannel basis. Two or more shots are recorded at each seismometer group for stacking purposes. The shot separation is determined by a rough knowledge of the velocity in the shot interval, and the center frequency of the signal spectrum.

Figure 3 depicts a possible primary, and ghost reflection resulting from the shots A and B recorded at one seismometer group. The traces A and B have been time-shifted to a common shot depth, thereby lining up primary events. The ghost reflections in this case show a 10-msec moveout or a difference in uphole time $\Delta T_{uh}=5$ msec. The figure also shows a comparison of the conventional stack and the present system for ghost rejection. The former method simply adds the two traces with a resulting attenuation of ghost amplitude of about -7 db, while the latter filters each trace with different filters and then stacks, achieving in this instance better than -25 db rejection of the ghost energy.

The primary and ghost wavelets in this example are identical but reversed in polarity. Their amplitude spectrum is given by

$$\overline{P(f)} = \left(1 - \cos\frac{2\pi f}{100}\right) \qquad 0 < f \leq 100$$

$$\overline{P(f)} = 0 \qquad f > 100 \text{ cps},$$

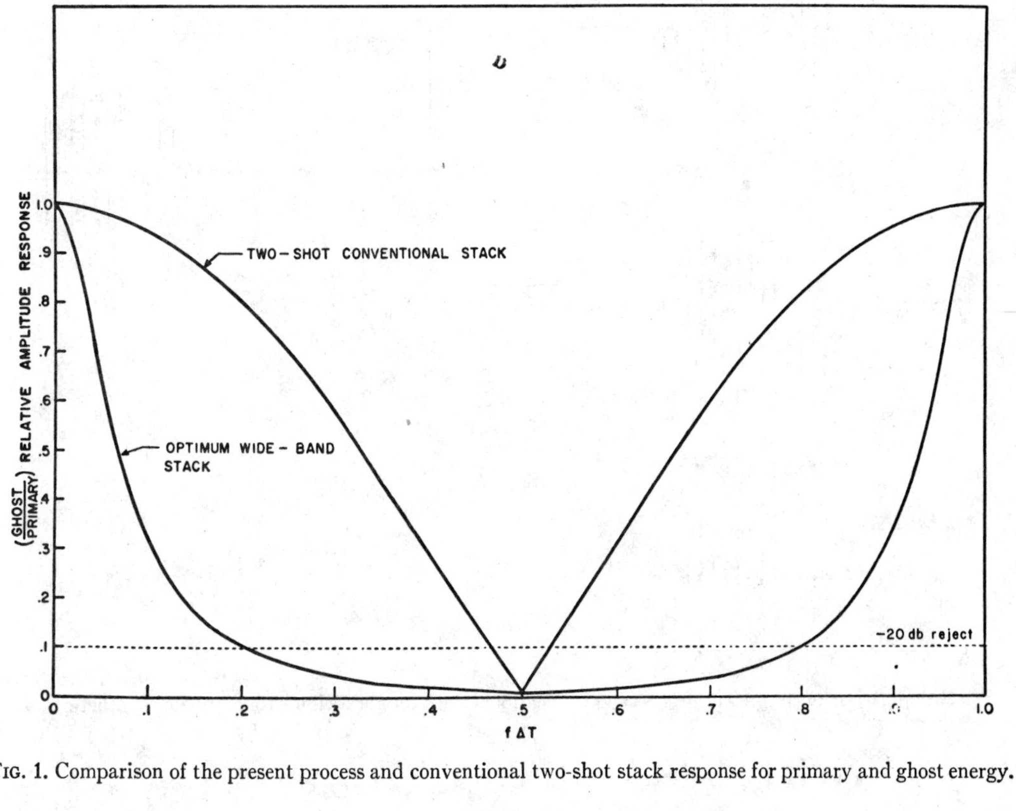

FIG. 1. Comparison of the present process and conventional two-shot stack response for primary and ghost energy.

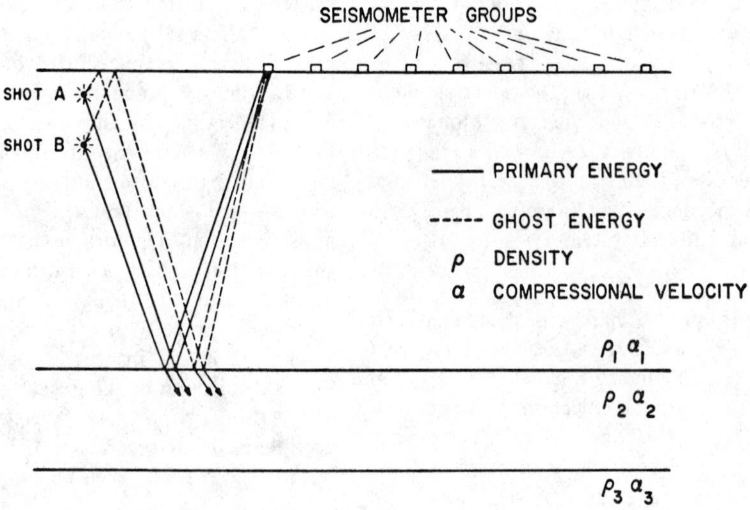

FIG. 2. Idealized ghost problem in reflection seismology.

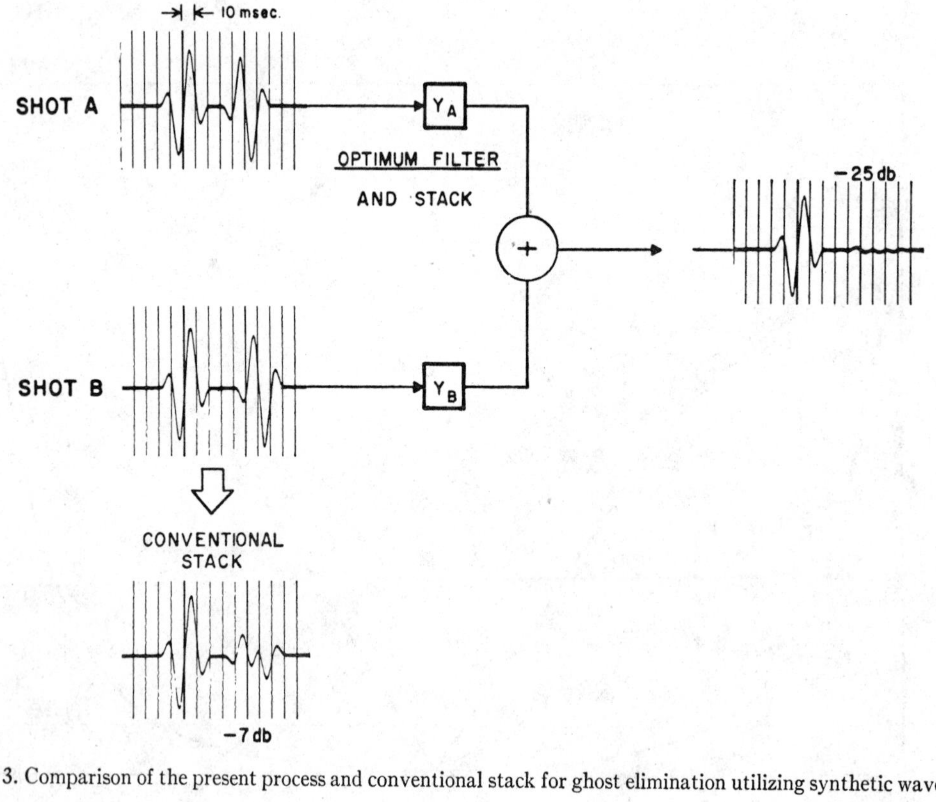

FIG. 3. Comparison of the present process and conventional stack for ghost elimination utilizing synthetic wavelets.

which is peaked at 50 cps. With a $\Delta T = 10$ msec for the ghost wavelets, it is evident that this spectrum would have its peak frequency centered at $f_{\text{peak}} \cdot \Delta T = .5$ in Figure 1. Thus, the results in Figure 3 represent the maximum rejection possible for either process using this wavelet. The advantages of the wide-band rejection and signal-preservation characteristics of the process are self-evident. In addition, no specific knowledge of the primary-ghost relationship or the spectrum of the wavelets, other than the $\Delta T = 2\Delta T_{uh}$, is required to implement the filters Y_A and Y_B. Fortunately, the difference in uphole time can be obtained accurately in the field with little difficulty if reasonable care is taken to record an uphole pulse with each shot. Other data which can provide an independent check on the ΔT are: (1) accurate knowledge of the shot separation and vertical velocity in that depth increment, (2) correlation functions, (3) difference of arrival times of both primaries and ghost from shots at different depths.

The second example uses field data taken from an uphole survey near Sherman, Texas.

The spread geometry and details of the uphole survey are shown in Figure 4. Two holes were shot from 180 ft to 40 ft in 20-ft increments using 5-lb charges. The recording array was designed to reject the prominent surface waves, based on previous noise studies in the area. Only 12 of the available 24 channels were used on the surface spread; the remaining 12 were employed with a vertical array of seismometers cemented in the ground as part of an auxiliary study. Programmed gain control and a recording passband of 10–350 cps were employed.

The upper half of Figure 5 shows the center four traces obtained from SP 2 at the 180-ft and 160-ft shot depths. They have been corrected for normal moveout and static corrected to the deeper shot. Several strong reflection complexes are evident from .2–.4 sec and commencing again at about 1.3 sec. The shallow marker at .31 sec, and the first part of the reflection complex at about 1.4 sec, correlate well between records from different shot depths indicating these events are primaries (solid circles); however, the remainder of the reflections do not have good character correlation nor do they necessarily time-tie. This is particularly evident for the events between .34–.40 sec. They lag the primary at .31 sec by less than or equal to $2T_{uh}$, or 82 and 76 msec for the 180-ft and 160-ft shots, respectively. As will be demonstrated, these events are ghost reflections from the base of the weathering (open circles) and surface reflectors.

The lower half of Figure 5 shows the results of conventional and optimum wide-band shot stacking of the 180-ft and 160-ft records. The results of the conventional stack indicate fair improvement in ghost reduction over the input traces directly above. This is predictable from the two-shot-stack response for ghost energy in Figure 1. The reflection peak frequency of the records varies from about 50 cps for the early events to about 30 cps for the later arrivals. The ghost moveout for the 20-ft shot interval is about 6 msec as can be seen from the records. Therefore, the product $F_{\text{peak}}\Delta T$ in Figure 1 ranges between .18 and .30 giving predicted ghost rejection of about 3–6 db. The actual achieved rejection appears to be slightly less than this. The optimum wide-band stack results shown in the last 12 traces indicate that ghost energy has been attenuated of the order of 10–20 db, which is in general agreement with the response curves of Figure 1 for the frequencies and moveout involved. In particular, it is noted that the low-frequency "leggy" portion of the deep reflector at 1.5 sec is, in fact, ghost energy. This is unattenuated by the conventional stack, but is well rejected by the present process. Both shot-stacking processes represented in Figure 5 preserve signal (primary) amplitude and waveform, but they differ markedly in their rejection capabilities. Furthermore, the only information necessary for the implementation of the present process was the ghost moveout, or twice the difference in uphole times = 6 msec.

The very nature of ghosting makes its *positive* identification on the basis of one shot depth almost impossible, even though the ghost reflections may be strong. It is felt this is the primary reason that the importance of ghost reflections has been a somewhat controversial subject in reflection seismology. If, however, the seismogram is viewed as a function of shot depth, the primary and ghost energy are clearly separable on the basis of their arrival-time relationship. To illustrate this, a representative trace from each shot depth at SP 2 was gathered onto one record to facilitate the shot depth comparison. Rather than select a particular trace from each depth, however, a velocity filter (Embree, Burg, and

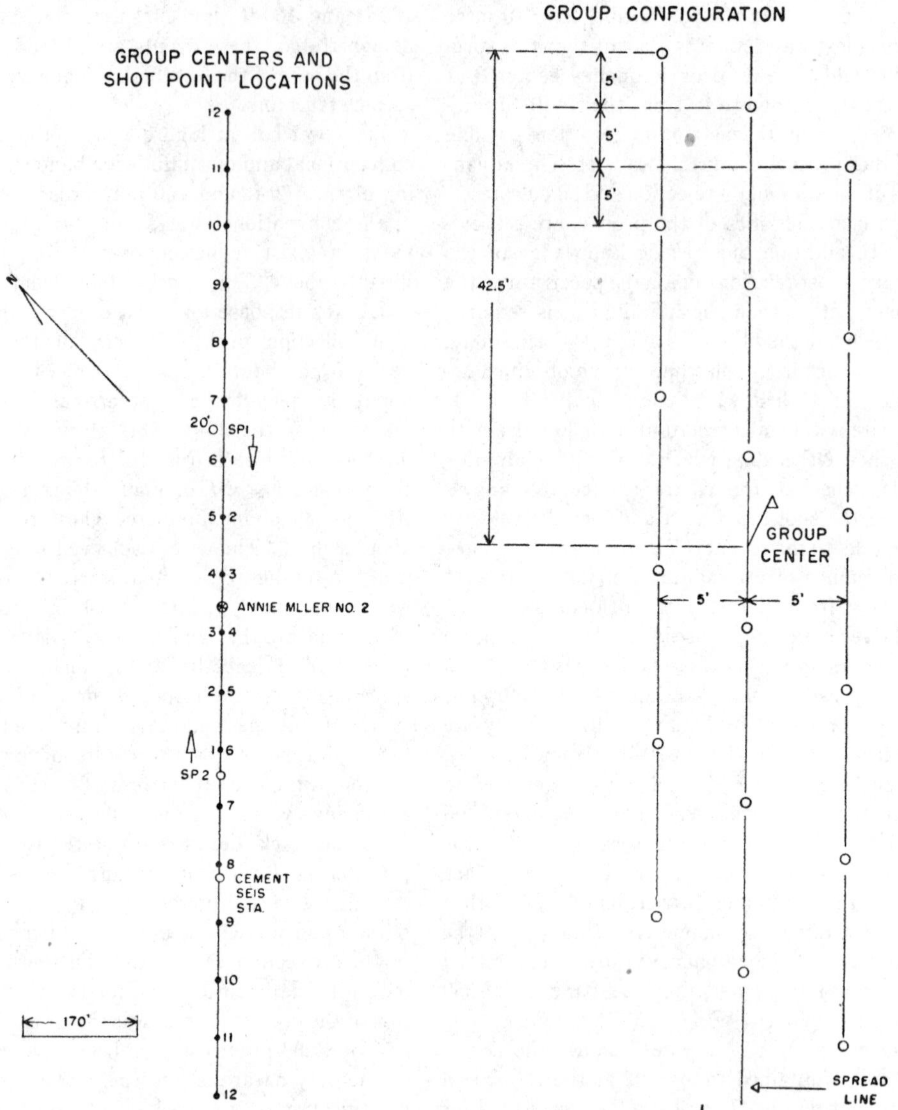

FIG. 4. Spread configuration for uphole survey taken near Sherman, Texas.

Backus, 1963) was applied to each 12-trace record to yield a single-trace representation of each shot depth. The application of the velocity filter here achieved additional rejection of the horizontally travelling shot-generated noise that was not cut out by the seismometer groups. Thus, each single trace representation is essentially free of horizontally traveling energy. What remains is composed of primaries, multiples, ghosts, and any scattered energy with less than two msec per trace moveout across the spread.

The gathered traces are shown in Figure 6 as the top eight traces. They have not been static corrected to a common depth. Therefore, primaries will arrive earlier from the deep than from the shallow shots while the opposite is true for ghosts. The presence of both primary and ghost moveout is evident with the formation of the characteristic "*V*" pattern due to generically related primaries and ghosts. Several of these can be seen at record times of .34 sec, 1.45 sec, 1.7 sec, and 1.86 sec. There can be no doubt on this

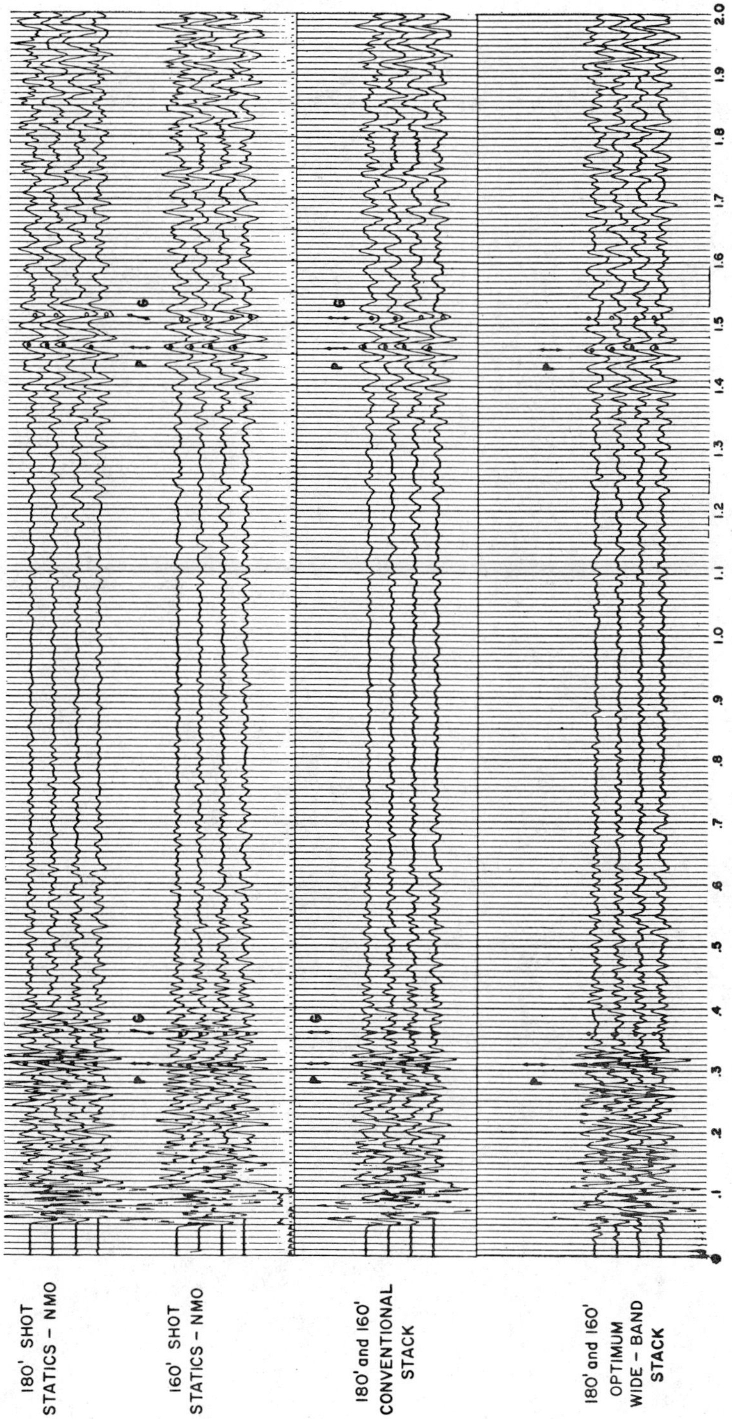

FIG. 5. 180-ft and 160-ft shot records before and after stacking.

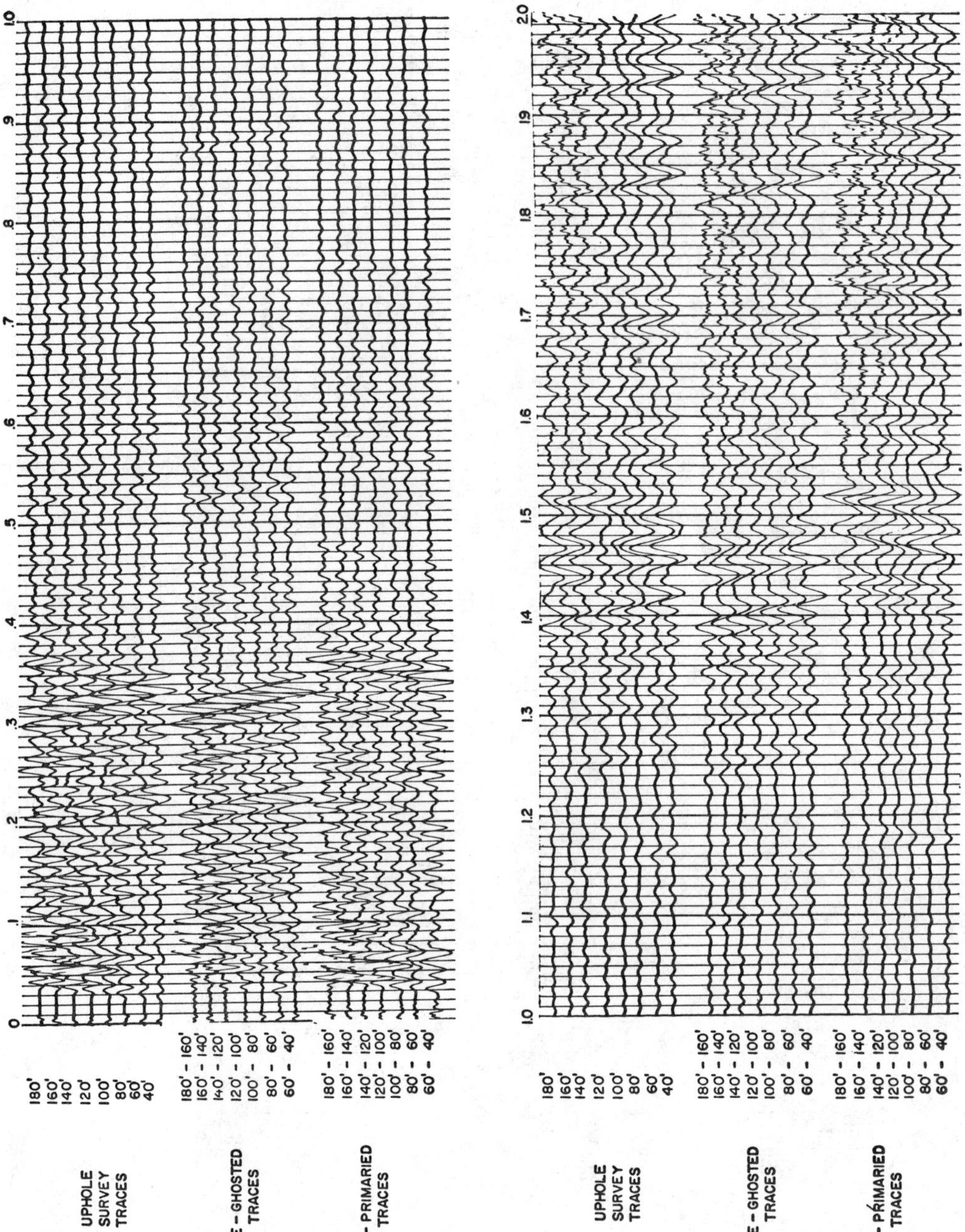

FIG. 6. Gathered uphole traces, processed to remove ghosts and primaries using the present process.

type of a display that ghost energy is both present and strong, whereas single depth information would not be nearly as revealing. The effect of shot depth on character correlation when ghost energy is present can best be seen in the reflection complex at about 1.45 sec. Notice the drastic changes in waveform across the 140 feet of shot depth resulting from primary-ghost interference. Similar horizontal variations also can be expected in the primary-ghost relationship for reasons previously mentioned. The removal of this ghost interference is mandatory before meaningful significance can be attached to reflection waveforms. The present process was applied to adjacent pairs of uphole traces resulting in the deghosted traces immediately below them in Figure 6. The first deghosted trace results from "optimum stacking" the 180-ft and 160-ft traces, etc. The most striking result is that the deghosted traces show only primary moveout. The obvious ghost-picks on the input traces have completely vanished for visual purposes, which implies about −20 db rejection. Other significant features are: (1) Several shallow reflections have been enhanced. (2) A number of weak reflectors in the quiet zone .35–1.2 sec have become pickable, (3) Character correlation is excellent across the 140 ft of shot depth. The latter is particularly noticeable at about 1.4 sec where the correlation is equally good for the high-frequency reflection sequence onset as it is for the low-frequency tail. This again illustrates the wideband capability of the processor. The high degree of correlation is in no way generated by the process of stacking. In fact, alternate deghosted traces have no common data between them. This degree of correlation is expected if the following conditions are fulfilled by the data:

(1) constant charge size,

(2) elimination of horizontally traveling energy (surface waves, etc.),

(3) receiver position constant for all shots,

(4) medium in which the shots occur is homogeneous,

(5) ghost reflections eliminated.

Conditions (1) and (3) were accomplished in the field, (2) was achieved by the recording array and velocity filtering, (4) implies that the same wavelet was put into the ground for each shot which was substantiated by monitoring the uphole wavelet, and a velocity survey, showing constant velocity between 30 ft and 260 ft, and (5) was accomplished by the present process.

To further evaluate the ghost problem and to understand its relationship to the near-surface, the process was applied to the data to enhance ghost energy and reject primaries. This processing is shown as the last seven traces in Figure 6. The amplitude of the deprimaried traces has been boosted by about 6 db to bring them up to the level of the input and deghosted traces. Notice there is only ghost moveout present. Furthermore, the ghost reflections appear to be simply related to the primaries. For example, the strong ghost at about .35 sec is the mirror image of its primary, and has the correct delay for a ghost off the base of the weathering. There is a second ghost following the aforementioned by 30 msec, which would correspond to a ghost off the surface. Thus, it appears that the ghosting mechanism in this area consists of two specular reflectors, one at the base of the weathering layer and the other at the surface. These observations may be confirmed in a quantitative manner through the use of both correlation functions, and the estimation of the impulse response of the near-surface obtained from computing the optimum linear operator for transforming the deghosted traces back into the original ghosted traces.

GHOST ANALYSIS USING CORRELATION FUNCTIONS

The use of the autocorrelation function, and its interpretation for a single ghosting horizon has been discussed by Lindsey (1960). His analysis can be readily extended to more complex ghosting situations. Of particular interest here is the predicted autocorrelation function for a double ghost mechanism.

Let the trace represented by primary energy only be $p(t)$. The ghosted trace resulting from two near-surface reflectors (above the shot) is approximately given by

$$g(t) = p(t) - \alpha_1 p(t - \tau_1) - \beta p(t - \tau_2) \quad (6)$$

τ_1 = two-way traveltime from shot to the first ghost generator.

τ_2 = two-way traveltime from shot to the second ghost generator.

α_1 = reflection coefficient of the first reflector, and

β = apparent reflection coefficient of the second reflector.

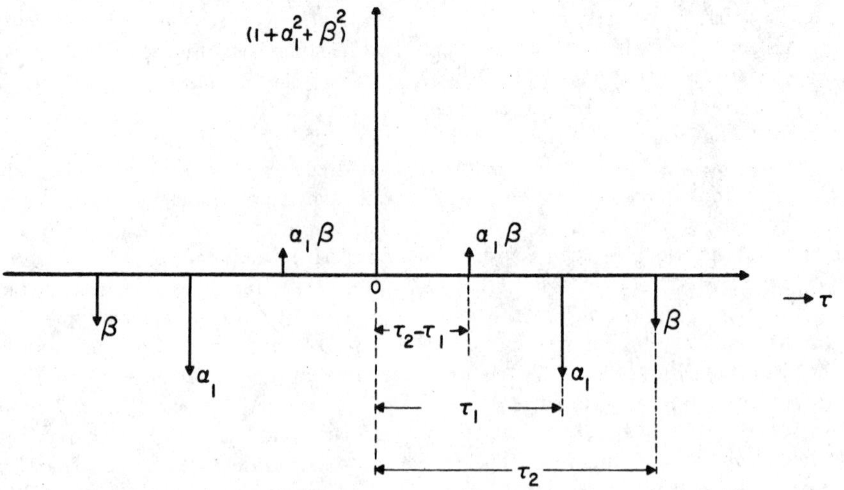

FIG. 7. Impulse autocorrelation function for primary plus double-ghost problem.

The true reflection coefficient of the second reflector, α_2, is related to β by $\alpha_2=\beta/(1-\alpha_1^2)$. In the above, multiple reflections between the two reflectors have been neglected.

The correlation function $\phi(\tau)$ for the ghosted trace is given by

$$\phi_{p+g}(\tau) = \frac{1}{T_2 - T_1}\int_{T_1}^{T_2} g(t)g(t+\tau)dt.$$

Substituting in the above for $g(t)$ given by equation (6), expanding and collecting like terms, the ghosted autocorrelation function may be written as a sum of seven terms:

$$\begin{aligned}\phi_{p+g}(\tau) = {} & (1+\alpha_1^2+\beta^2)\phi_p(\tau) \\ & + \alpha_1\beta\phi_p(\tau-\tau_2+\tau_1) \\ & + \alpha_1\beta\phi_p(\tau+\tau_2-\tau_1) \\ & - \alpha_1\phi_p(\tau-\tau_1) - \alpha_1\phi_p(\tau+\tau_1) \\ & - \beta\phi_p(\tau-\tau_2) - \beta\phi_p(\tau+\tau_2) \qquad (7)\end{aligned}$$

where

$$\phi_p(\tau) = \frac{1}{T_2 - T_1}\int_{T_1}^{T_2} p(t)p(t+\tau)dt$$

is the autocorrelation function of the primary reflection sequence. Expression (7) may also be written as the convolution of $\phi_p(\tau)$ with a sequence of impulses as

$$\phi_{p+g}(\tau) = \int_{-\infty}^{\infty} \phi_p(t)h(\tau - t)dt,$$

with

$$\begin{aligned}h(\tau) = {} & (1+\alpha_1^2+\beta^2)\delta(\tau) + \alpha_1\beta\delta(\tau-\tau_2+\tau_1) \\ & + \alpha_1\beta\delta(\tau+\tau_2-\tau_1) - \alpha_1\delta(\tau-\tau_1) \\ & - \alpha_1\delta(\tau+\tau_1) - \beta\delta(\tau-\tau_2) \\ & - \beta\delta(\tau-\tau_2). \qquad (8)\end{aligned}$$

It is more instructive to examine $h(\tau)$, the impulse autocorrelation, than $\phi_{p+g}(\tau)$ since the latter contains the smearing effects of the reflection wavelet.

The impulse autocorrelation function $h(\tau)$ for primary and double-ghost situation is shown in Figure 7. Note that if $\beta\rightarrow 0$, the autocorrelation function reduces to that given by Lindsey (1960) after convolution with the primary autocorrelation function $\phi_p(\tau)$. The addition of a second ghost reflector results in two additional features in the structure of the correlation function as seen in Figure 7. The negative spike of amplitude β at $\pm\tau_2$ sec results from the correlation of primary and ghost reflections separated in time by τ_2 sec, that is, the second set of ghosts; while the positive feature of amplitude $\alpha_1\beta$ at $\pm(\tau_2-\tau_1)$ seconds is due to the correlation between the two ghosts themselves. The degree to which these features may be seen on an autocorrelation function depends, of course, upon the resolving power of the primary correlation $\phi_p(\tau)$. If the latter is very nearly a delta function itself, then our picture in Figure 7 would remain essentially unchanged. If, however, $\phi_p(\tau)$ is slowly decaying away from zero lag because of reverberations or narrow-band signal information, then our impulse representation would be severely distorted after convolution, and our ability to identify particular wiggles in the correlation function with ghosting would be practically nil.

The autocorrelation functions were computed for the input traces in Figure 6 and are shown in Figure 8 arranged in order of decreasing hole depth. Only the positive half of the correlation function is shown, it being symmetric about zero lag or $\tau=0$. Twice the uphole time, $2T_{uh}$, as determined from the uphole pulse is indicated by an arrow below each correlation function. There is generally good agreement between this time and a trough in the correlation functions predicted for the primary-surface ghost correlation. A second stronger trough can be seen preceding the surface ghost trough by about 30 msec, and

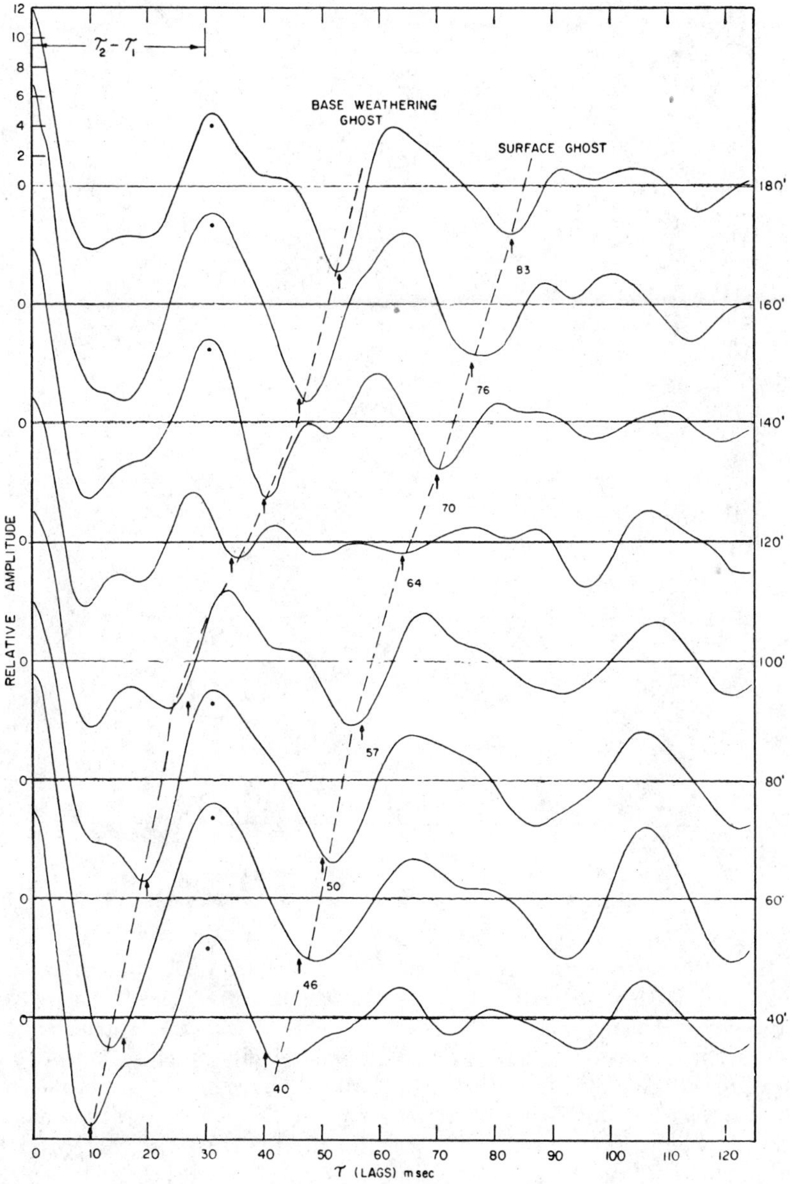

FIG. 8. Autocorrelation functions of uphole traces.

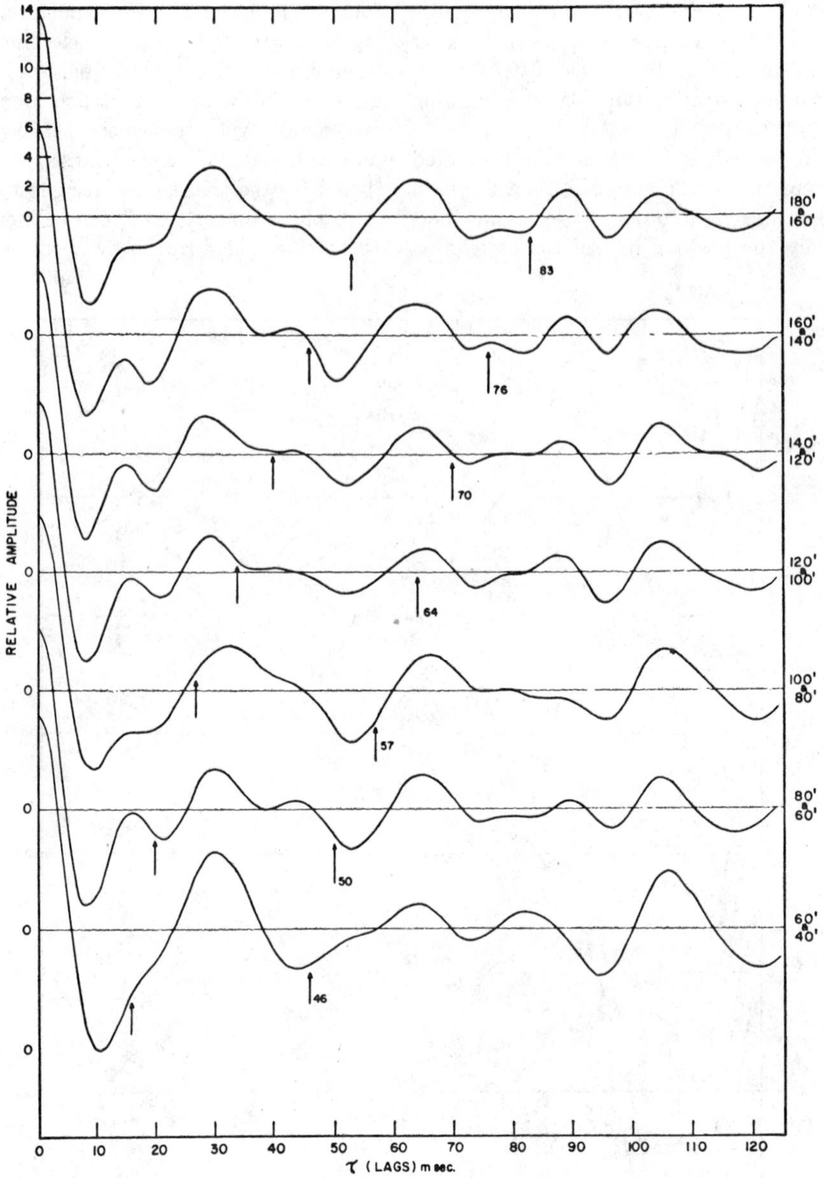

FIG. 9. Autocorrelation functions of deghosted traces.

paralleling the latter's moveout between shot depths. This trough is identified with a base-of-the-weathering ghost reflection, which is the most prominent ghost on the records. There is also a positive correlation at about 30 msec independent of depth, that is, predicted by Figure 7 for two ghosts separated by 30 msec.

The correlation functions of the deghosted traces of Figure 6 are shown in Figure 9. The two troughs associated with ghosting which move across the correlations in Figure 8 are no longer evident. In fact, the residual wiggles show good reproducibility from shot depth to shot depth with the exception of the shallower shots. They represent the structure in the autocorrelation of the effective seismic wavelet, $\phi_p(\tau)$, which is independent of shot depth for reasonable depths.

Several additional features become clear with

the before and after correlations of Figures 8 and 9. The 120-ft ghosted correlation function shows anomalous lack of correlation for τ greater than 30 msec. At this depth, however, the ghost troughs coincide with peaks in the primary correlation $\phi_p(\tau)$ causing destructive interference. The primary autocorrelation functions in Figure 9 do have an intrinsic peak at about 30-msec lag, however, this peak is reduced from Figure 8 indicating that the positive correlation of the two ghosts contributed to its amplitude.

It becomes evident what difficulties might be encountered in attempting to use amplitude information from the ghosted correlation functions to estimate reflection coefficients and near-surface filtering effects on the ghost reflections. Not having to specify these parameters makes shot-stacking a particularly attractive deghosting technique.

ESTIMATE OF NEAR-SURFACE REFLECTION RESPONSE

Once we have obtained a ghost-free estimate of the subsurface reflection sequence, we can work backwards and achieve a quantitative description of the near-surface reflection complex. This is accomplished by computing the operator or filter which transforms a deghosted trace into one of the ghosted traces from whence it came. The impulse response or time-domain representation of this operator should embody the effects of the near-surface on the reflection seismograms, e.g., the cause of the ghosting.

Mathematically, we seek the operator $h(\tau)$ which will transform a ghost-free trace $p(t)$ back into a ghosted trace $g(t)$ or which satisfies the relation,

$$g(t) = \int_{-\infty}^{\infty} p(\tau)h(t-\tau)d\tau. \qquad (9)$$

At this point the reader may wonder if he is not caught in a vicious circle of first removing ghosts and then replacing them in the data. To expel any such fears a word of explanation is in order. Recall that the present process by design only requires knowledge of the difference in uphole times for its implementation. No other information about the ghosts is required such as: reflection coefficients, primary-ghost time lag, near-surface filtering effects, or details of the ghosting mechanism. Therefore, these parameters which are interesting from the standpoint of studying the near-surface and its effect on the reflection seismogram do not automatically drop out of the processing. They may be inferred, of course, by visually examining the traces before and after deghosting, and computing correlation functions and the like. The integral equation (9) however, provides a quantitative measure of these near-surface parameters in terms of the unknown function $h(t)$.

The least-squares estimate of $\hat{h}(t)$ may be obtained from $g(t)$ and the deghosted estimate of the primary reflection sequence $\tilde{p}(t)$. That is, we seek the filter $\hat{h}(t)$ which minimizes the mean-square error,

$$\overline{e(t)^2} = \frac{1}{2T}\int_{-T}^{T}[g(t) - \hat{h}(t) * \tilde{p}(t)]^2 dt \qquad (10)$$

where $*$ denotes convolution. Solutions of equation (10) for discrete data are given by Levinson, 1947, and more recently by Levin, 1960, and Robinson, 1963.

In matrix form, the least-squares estimate of $\hat{h}(t)$ is,

$$\begin{bmatrix} \phi_p(0) & \phi_p(1) & \phi_p(2) & \cdots & \phi_p(N) \\ \phi_p(1) & \phi_p(0) & \phi_p(1) & \cdots & \\ \vdots & & & & \\ \phi_p(N) & & \cdots & & \phi_p(0) \end{bmatrix} \begin{bmatrix} \hat{h}(0) \\ \hat{h}(1) \\ \hat{h}(2) \\ \vdots \\ \hat{h}(N) \end{bmatrix} = \begin{bmatrix} \phi_{pg}(0) \\ \phi_{pg}(1) \\ \phi_{pg}(2) \\ \vdots \\ \phi_{pg}(N) \end{bmatrix} \qquad (11)$$

where $\phi_p(i)$ is the ith lag autocorrelation of $\tilde{p}(t)$, and $\phi_{pg}(i)$ is the ith lag crosscorrelation of $\tilde{p}(t)$, and $g(t)$. The special symmetry properties of the correlation matrix in (11) allow an iterative solution for the filter weights $\hat{h}(i)$. That is, the solution vector $[\hat{h}(i)]^T$ can be obtained by operations on the previous vector $[\hat{h}(i-1)]^T$ without actually inverting the $i\times i$ correlation matrix, as demonstrated by Levinson, 1947.

Figure 10 shows the operator $\hat{h}(t)$ calculated for several different shot depths at SP 2.

At each depth represented, the operator was

computed from a deghosted trace and the ghosted trace corresponding to that depth. The spike in the operators at $t=0$ simply reproduces the primary events when convolved with $p(t)$. The first strong negative trough produces the base-of-weathering ghosts, and the second shallow, broad trough correlating with twice the uphole time gives rise to the surface ghost. As also observed on the autocorrelation functions, this double set of troughs is separated by 30 msec. The base-of-the-weathering trough remains constant up the hole in both amplitude and character, whereas the shallow surface ghost trough exhibits sizeable waveform fluctuations from shot to shot. This may be indicative of the noise level of the data.

The reflection coefficients α_1 and β defined in equation (6) were estimated from the operators in Figure 10 by taking the ratio of ghost trough to primary peak at $t=0$. Admittedly, this may be a crude approximation to the true reflection coefficients in view of the difference in frequency content between the base-of-weathering and surface ghost trough. Clearly β and α_1 are frequency dependent, or at least our simple model of two sharp ghosting horizons forces them to be. Since we have not allowed for frequency dependent absorption within the weathering layer, any effects due to absorption are necessarily lumped in the parameter β. Keeping these facts in mind, the average reflection coefficients obtained from the near-surface responses in Figure 10 are:

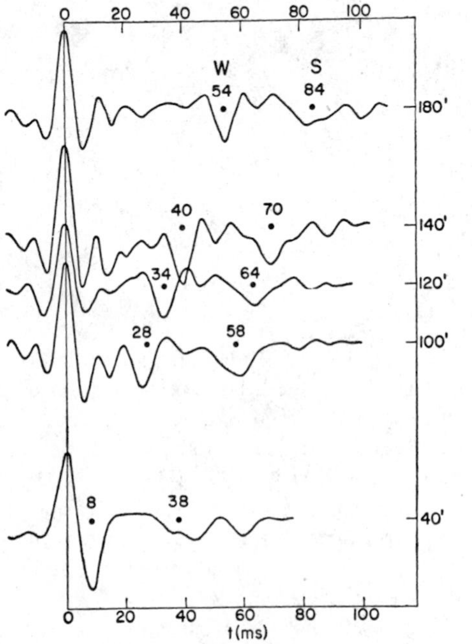

FIG. 10. Near-surface response estimate obtained for several shot depths.

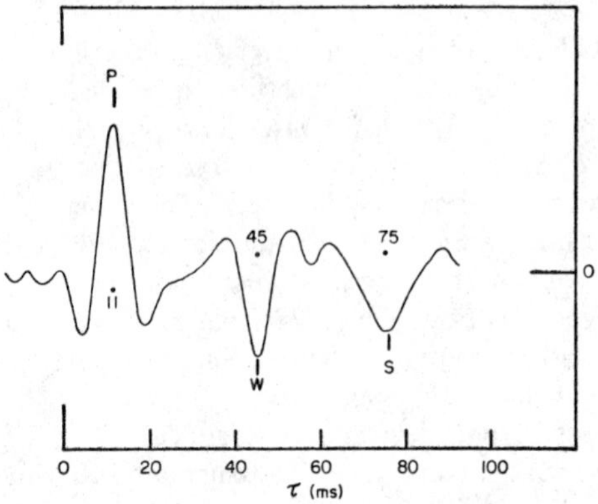

FIG. 11. Near-surface response estimate obtained from transforming the 180-ft deghosted trace into the 120-ft original trace.

$$\overline{\alpha_1} = .47,$$

$$\overline{\beta} = .28,$$

giving for the surface reflection coefficient

$$\overline{\alpha_2} = \frac{\beta}{1 - \alpha_1^2} = \frac{.28}{.78} = .36.$$

These estimates provide a lower limit for the reflection coefficients because of the imperfect rejection of ghost energy and because of noise correlation between deghosted and original traces.

To reduce the effect of residual ghost and noise correlation, the optimum filter to transform the 180-ft deghosted trace into the 120-ft original trace was computed. The filter, limited to about 70 cps on the high side, is shown in Figure 11. The reduction of "primary" correlation relative to ghost correlation is apparent. Crudely compensating for the low-frequency drift in the operator, the values,

$$\overline{\alpha_1} \quad .6$$

$$\overline{\beta} \quad .4$$

$$\overline{\alpha_2} \quad .6$$

are obtained. This yields a model indicated in Figure 12.

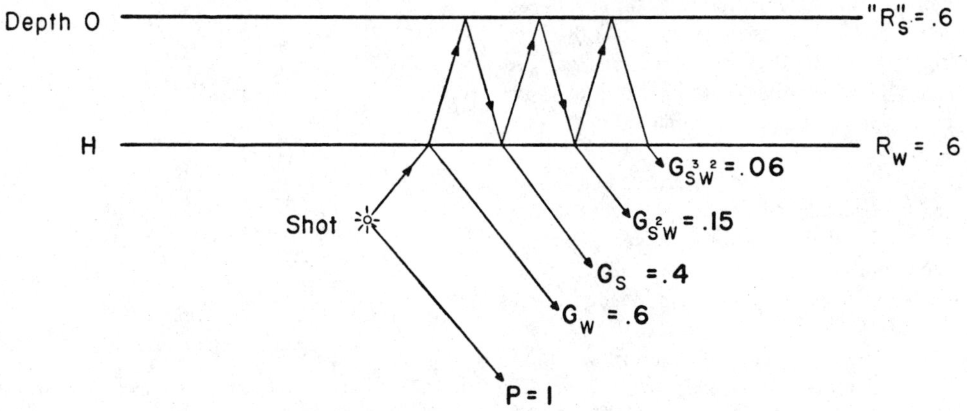

FIG. 12. Two-interface model for near-surface.

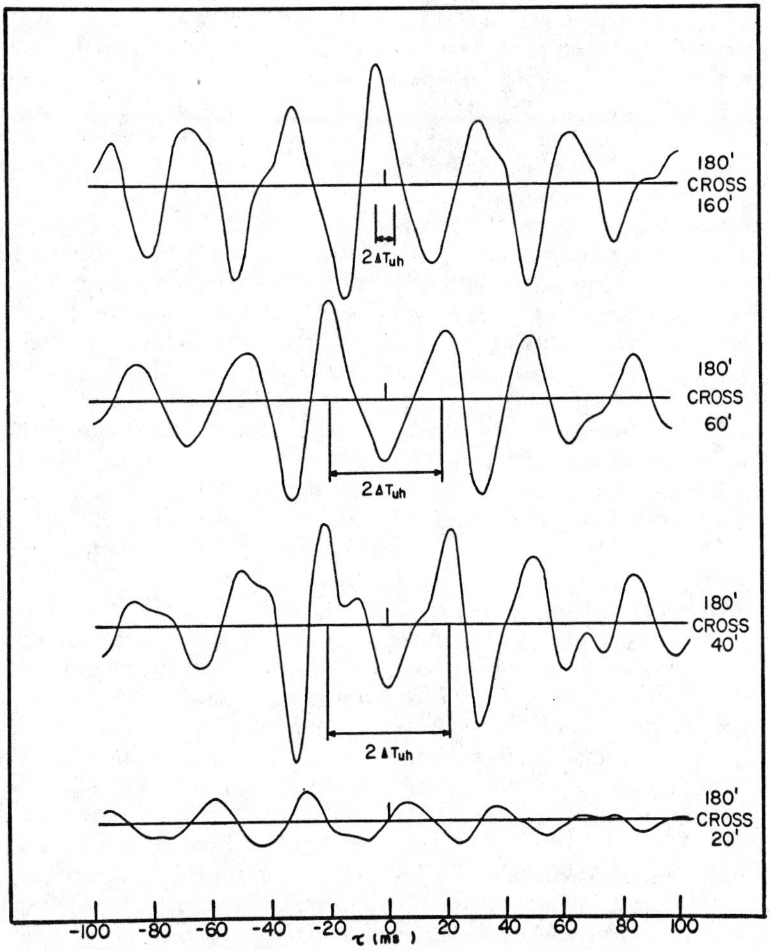

FIG. 13. Crosscorrelation functions computed between different shotpoints for the original velocity-filtered uphole traces.

The fact that the effective reflection coefficient of the surface is less than one (i.e. $R_s \approx .6$) is presumably due to energy dissipation in the weathering layer. This is further supported by the apparent lack of high frequency (above 50 cps) in the weathering ghost.

The simple two-interface discrete model shown in Figure 12 yields a result that the total energy in the ghost complex should be about 54 percent of the total initially down-traveling wave, $(.6^2+.4^2+.15^2+.06^2=.54)$. The adequacy of this simple model can be further tested by examination of the crosscorrelation function computed between different shot depths for the original velocity filtered traces, shown in Figure 13. As exemplified in the crosscorrelation between the 180-ft shot and the 40-ft shot, all of the primary energy should correlate at a lag $\tau=-\Delta T_{uh}$, and all of the initially up-traveling energy should correlate at a lag $\tau=+\Delta T_{uh}$, no matter how complex the ghosting mechanism is. The 180–160 ft correlation is uninterpretable because of the small ΔT_{uh} involved. The 20-ft shot is in the weathering and yields a very unreliable correlation. The two center correlations in Figure 13 suggest that the initially up-going energy contributes about 80 percent as much energy to the reflection record as the initially down-going energy. This compares with a 54 percent figure accounted for by analysis on the basis of the simple two-reflector model in Figure 12. The 80 percent figure suggests that about 20 percent of the initially up-going energy in the reflection-record frequency range is dissipated in the weathering layer or scattered by undulation in the boundaries of the upper reflectors.

The major portion of the analysis was focused on SP 2. However, the SP 1 data yielded similar findings, with the exception that the ghosting appeared somewhat more complex. This was evident on the uphole traces as well as the autocorrelation functions, which were less interpretable from a ghosting standpoint than those in Figure 8 from SP 2. After deghosting, the SP 1 traces correlated very well with the SP 2 deghosted traces.

The estimate of the near-surface reflection response at SP 1 obtained from transforming deghosted into ghosted traces revealed differences in the near-surface response between SP 1 and SP 2. Figure 14 shows a comparison of the near-surface

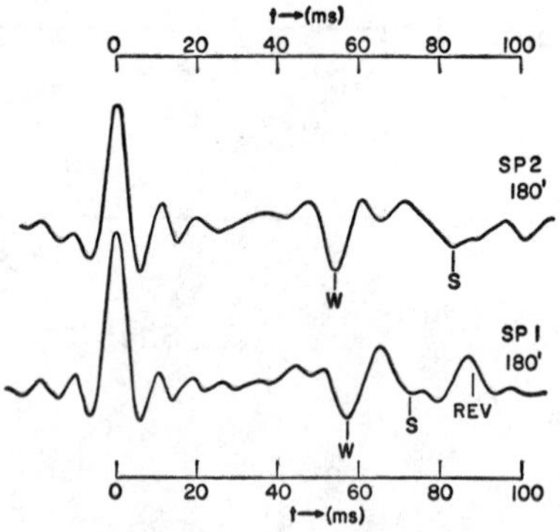

FIG. 14. Comparison of near-surface response estimates obtained at SP 1 and SP 2 for the 180-ft shot.

response estimates obtained at the two shotpoints for the 180-ft shot depth. The portions of the operators relating to primaries are almost identical. Twice the uphole times indicated by S are significantly different at the two shotpoints, yet the base-of-the-weathering reflector remains essentially fixed, therefore, the traveltime through the weathering at SP 1 is about 7 msec shorter than at SP 2. The base-of-the-weathering ghost trough at SP 1 is of comparable strength to the SP 2 trough, but does not possess the symmetry of the latter. Furthermore, $2T_{uh}=S$ at SP 1 does not correspond to a well-defined ghost trough; and finally an additional strong positive feature appears at the first reverberation time within the weathering layer. Therefore, it is likely that the total interval from about 50 to 90 msec at SP 1 represents ghosting, the details of which are not resolved as simple troughs or peaks but rather form a complex interference.

CONCLUSIONS

A new deghosting technique is presented which utilizes two shots at different depths. It differs from conventional stacking in that separate filters are applied to the two shots before stacking. The filters are designed in an optimum manner to enhance initially down-traveling energy and reject initially up-traveling energy. The resulting process has wide-band ghost reject and signal-preservation characteristics. No specific knowledge of the ghost mechanism is required for its implementation other than, (1) determining the difference in uphole times, (2) assuming a spectral content for signal and noise, usually "white" over

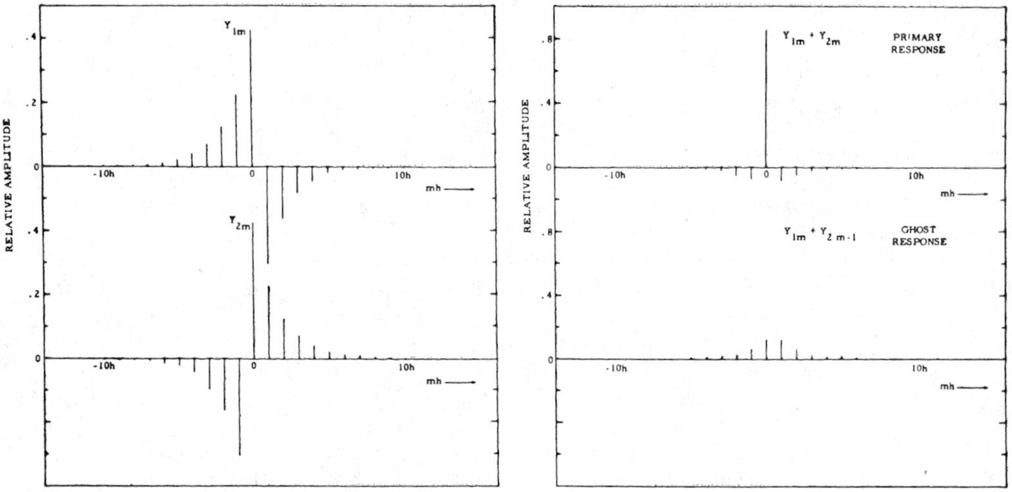

FIG. 15. Left half of the figure shows the sampled impulse response of the deghosting filters, and the right half shows the impulse response for primaries and ghosts obtained by suitably adding the filter responses.

the frequency range of interest, and (3) insuring that both shots are below the ghosting mechanism. The method will handle ghosting from a reflection complex or ghosts from a single reflector equally well.

The degree of primary and ghost energy separation achievable with the present process permits the quantitative investigation of the near-surface reflection response by computing the operator which transforms the deghosted into the ghosted trace.

ACKNOWLEDGMENTS

We wish to acknowledge the assistance of Earl C. Wisler in early experimental applications of this technique. We also thank Texas Instruments Incorporated and Geophysical Service Inc. for permission to publish this paper.

REFERENCES

Burg, J. P., Three-dimensional filtering with an array of seismometers: Geophysics, v. 29, p. 693–713.

Embree, P., Burg, J. P., and Backus, M. M., 1963, Wide-band velocity filtering—the Pie-Slice process: Geophysics, v. 28, p. 948–974.

Foster, M. R., and Sengbush, R. L., (to be published) Design of sub-optimum filter systems for multi-trace seismic data processing: Geophysical Prospecting.

Hammond, J. W., 1962, Ghost elimination from reflection records: Geophysics, v. 27, p. 48–60.

Levin, M. J., 1960, Optimum estimations of impulse response in the presence of noise: IRE Transactions C.T., v. CT-7, p. 50–56.

Levinson, N., 1947, A heuristic exposition of Wiener's mathematical theory of prediction and filtering: Journal of Math. and Physics, v. 26, p. 110–119.

Lindsey, J. P., 1960, Elimination of seismic ghost reflections by means of a linear filter: Geophysics, v. 25, p. 130–140.

Musgrave, A. W., Ehlert, G. W., and Nash, D. M., Jr., 1958, Directivity effect of elongated changes: Geophysics, v. 23, p. 81–96.

Robinson, E. A., 1963, Mathematical development of discrete filters for the detection of nuclear blasts: Jour. Geoph. Res., v. 63, p. 5559–5567.

Sparks, N. R., and Silverman, D., 1953, Pressure field around distributed charges: Seventh Annual Midwestern Meeting of the Society of Exploration Geophysicists, Dallas, Texas, November 12–13.

Spieker, L. J., Burg, J. P., Backus, M. M., and Strickland, L., 1961, Seismometer array and data processing system: Final Report Phase I, AFTAC contract AF 33(600)-41840, Project VT/077.

Wiener, N., 1949, Extrapolation interpolation and smoothing of stationary time series: New York, John Wiley and Sons, p. 104–116.

APPENDIX A

LEAST-SQUARES TWO-CHANNEL FILTER DESIGN

The classical least-squares technique is applied to two-channel sampled time functions for optimum-filter design by an obvious extension of Levinson's (1947) single-channel treatment.

Consider the two-channel time functions $f_1(t)$ and $f_2(t)$ to be represented by their sampled values at points $t=kh$, where h is the sample period and k is an index. Let the sampled functions $f_1(kh)$ and $f_2(kh)$ be regarded as the sequences b_{1k} and b_{2k} representing signal and noise, that is:

$$b_{1k} = S_{1k} + N_{1k},$$

$$b_{2k} = S_{2k} + N_{2k},$$

where

$$S_{1k} = \text{signal on channel 1,}$$
$$S_{2k} = \text{signal on channel 2,}$$
$$N_{1k} = \text{noise on channel 1,}$$
$$N_{2k} = \text{noise on channel 2.}$$

We wish to determine the pair of linear filters which, when convolved, respectively, with the input sequences b_{1k} and b_{2k} and summed, will give an output as similar as possible to the desired signal on one of the input channels, say S_{1k}. The filter sequences Y_{1k} and Y_{2k} are designed such that the total power in the error signal

$$E_k \equiv S_{1k+p} - \sum_{n=0}^{M} Y_{1n} b_{1k-n} - \sum_{n=0}^{M} Y_{2n} b_{2k-n} \qquad \text{(A-1)}$$

is minimized. The power in the error signal is given by the sum of the squares of the E_k's, or stated in formula

$$\begin{aligned} I &= \lim_{N\to\infty} \frac{1}{2N+1} \sum_{k=-N}^{N} E_k^2 \\ &= \lim_{N\to\infty} \frac{1}{2N+1} \sum_{k=-N}^{N} (S_{1k+p})^2 - 2 \sum_{n=0}^{M} Y_{1n} \lim_{N\to\infty} \frac{1}{2N+1} \sum_{k=-N}^{N} S_{1k+p} b_{1k-n} \\ &\quad - 2 \sum_{n=0}^{M} Y_{2n} \lim_{N\to\infty} \frac{1}{2N+1} \sum_{k=-N}^{N} S_{1k+p} b_{2k-n} \\ &\quad + 2 \sum_{n=0}^{M} \sum_{m=0}^{M} Y_{1n} Y_{2m} \lim_{N\to\infty} \frac{1}{2N+1} \sum_{k=-N}^{N} b_{1k-n} b_{2k-m} \\ &\quad + \sum_{n=0}^{M} \sum_{m=0}^{M} Y_{1n} Y_{1m} \lim_{N\to\infty} \frac{1}{2N+1} \sum_{k=-N}^{N} b_{1k-n} b_{1k-m} \\ &\quad + \sum_{n=0}^{M} \sum_{m=0}^{M} Y_{2n} Y_{2m} \lim_{N\to\infty} \frac{1}{2N+1} \sum_{k=-N}^{N} b_{2k-n} b_{2k-m}, \end{aligned} \qquad \text{(A-2)}$$

which is to be minimized with respect to the Y_{1k}'s and Y_{2k}'s.

Equation (A-2) may be simplified by introducing the correlation functions.

The autocorrelation function for a sampled time function is defined as,

$$R_a(k) = \lim_{N\to\infty} \frac{1}{2N+1} \sum_{e=-N}^{N} a_e a_{e-k},$$

and the crosscorrelation function between two sampled time functions as

$$R_{ba}(k) = \lim_{N\to\infty} \frac{1}{2N+1} \sum_{e=-N}^{N} (a_e b_{e-k}) = R_{ab}(-k).$$

Substituting the correlation functions into equation (A-2) we have:

$$\begin{aligned} I = R_{s_1}(0) &- 2 \sum_{n=0}^{M} Y_{1n} R_{b_1 s_1}(n+p) \\ &- 2 \sum_{n=0}^{M} Y_{2n} R_{b_2 s_1}(n+p) \\ &+ 2 \sum_{n=0}^{M} \sum_{m=0}^{M} Y_{1n} Y_{2m} R_{b_2 b_1}(m-n) \\ &+ \sum_{n=0}^{M} \sum_{m=0}^{M} Y_{1n} Y_{1m} R_{b_1}(m-n) \\ &+ \sum_{n=0}^{M} \sum_{m=0}^{M} Y_{2n} Y_{2m} R_{b_2}(m-n). \end{aligned} \qquad \text{(A-3)}$$

For I to be a minimum with respect to the Y_{1k}'s and Y_{2k}'s we must have

$$\frac{\partial I}{\partial Y_{1k}} = 0 \qquad k = 0, 1, 2 \cdots M$$

$$\frac{\partial I}{\partial Y_{2k}} = 0$$

or

$$\frac{\partial I}{\partial Y_{1k}} = -2R_{b_1s_1}(k+p) + 2\sum_{m=0}^{M} Y_{2m}R_{b_2b_1}(m-k) + 2\sum_{m=0}^{M} Y_{1m}R_{b_1}(k-m) = 0$$

$$\frac{\partial I}{\partial Y_{2k}} = -2R_{b_2s_1}(k+p) + 2\sum_{m=0}^{M} Y_{1m}R_{b_2b_1}(k-m) + 2\sum_{m=0}^{M} Y_{2m}R_{b_2}(k-m) = 0 \quad \text{(A-4)}$$

This system of equations may be written in matrix form as

$$\begin{bmatrix} r_0 & r_1 & r_2 \cdots r_M \\ r_{-1} & r_0 & r_1 \\ r_{-2} & \cdot & \\ & \cdot & \\ & & \cdot \\ r_{-M} & & r_0 \end{bmatrix} \begin{bmatrix} Y_0 \\ Y_1 \\ \cdot \\ \cdot \\ \cdot \\ Y_M \end{bmatrix} = \begin{bmatrix} r_p \\ \gamma_{1+p} \\ \gamma_{2+p} \\ \\ \\ \gamma_{M+p} \end{bmatrix} \quad \text{(A-5)}$$

where each of the r_i's, Y_i's, and γ_j's are sub-matrices defined as follows

$$r_{m-k} = \begin{bmatrix} R_{b_1}(m-k) & R_{b_2b_1}(m-k) \\ R_{b_1b_2}(m-k) & R_{b_2}(m-k) \end{bmatrix}$$

$$Y_m = \begin{bmatrix} Y_{1m} \\ Y_{2m} \end{bmatrix}$$

$$\gamma_{k+p} = \begin{bmatrix} R_{b_1s_1}(k+p) \\ R_{b_2s_1}(k+p) \end{bmatrix}.$$

Notice also that $r_{m-k} = r^{T}_{k-m}$.

An iterative solution to (A-5) is also possible (Robinson, 1963) as in the single-channel case. The parameter p in R_k and equation (A-1) relates to whether or not one desires to do filtering, filtering and prediction, or interpolation. If $p=0$, we attempt to extract S_k from knowledge of the b's up to time $t=hk$. If $p>0$, we attempt to predict S_{k+p} ahead in time by $t=hp$ sec, and if $p<0$, we use values of the b's in future time to interpolate the value of S_{k-p}.

Up until now the derivation has been completely general without regard to the deghosting problem. In order to design the deghosting filters on a theoretical model the various correlation functions in (A-4) must be specified.

The signal and noise description for the ghost-reflection problem on a two-channel basis may be adequately specified as:

Shallow shot

$$f_1(t) = P_1(t) + G_1(t) + N_1(t)$$

Deep shot

$$f_2(t) = P_2(t) + G_2(t) + N_2(t)$$

where

$P_{1,2}(t)$ = primary reflection sequence on channel 1, 2,

$G_{1,2}(t)$ = ghost reflection sequence on channel 1, 2,

$N_{1,2}(t)$ = random noise on channel 1, 2.

If the traces $f_1(t)$ and $f_2(t)$ are static-shifted by the difference in uphole time $=\Delta T_{uh}$ then $P_1(t) = P_2(t) = P(t)$, and $G_1(t) = G(t)$, $G_2(t) = G(t-2\Delta T_{uh})$, based on the assumption that both shots are in the same medium below the ghosting horizon. We further assume that the random noise is white with power density level N_0, and is uncorrelated with primaries or ghosts. The final assumption concerns the crosscorrelation of primaries and ghosts. Of course, physically they are correlated as Figure 8 shows, however, the measurement of this correlation is difficult even with good data. Furthermore, the primary ghost correlation can be expected to change drastically in short horizontal distances due to variable near-surface conditions as well as variations in shot depth. From the practical consideration of filter design we neglect the primary ghost correlation and accept the slight loss in filter effectiveness.

With these assumptions, the various auto- and crosscorrelations become in the continuous case:

$$\begin{aligned} R_{b_1}(\tau) &= \phi_p(\tau) + \phi_g(\tau) + N_0\delta(\tau), \\ R_{b_2}(\tau) &= \phi_p(\tau) + \phi_g(\tau) + N_0\delta(\tau), \\ R_{b_1b_2}(\tau) &= \phi_p(\tau) + \phi_g(\tau - 2\Delta T_{uh}), \\ R_{b_2b_1}(\tau) &= \phi_p(\tau) + \phi_g(\tau + 2\Delta T_{uh}), \\ R_{b_1s_1}(\tau) &= \phi_p(\tau), \\ R_{b_2s_1}(\tau) &= \phi_p(\tau). \end{aligned} \quad \text{(A-6)}$$

In order that the filters have a wide-band capability for separating primaries and ghosts we assume that the spectral content of the primaries and ghosts is white over the frequency range of interest. That is, we are not introducing any frequency difference between signal and noise, but rather exploiting only the spatial correlation differences. Thus, the correlations become

$$\phi_p(\tau) = P\delta(\tau),$$

$$\phi_g(\tau) = G\delta(\tau), \qquad \text{(A-7)}$$

where P and G are the signal and ghost-power levels and $\delta(\tau)$ is the delta function.

Figure 15 shows the impulse response of the filters Y_{1m} and Y_{2m} designed by means of equation (A-5), using correlations defined by (A-6) and (A-7) for the following parameters:

$P = 1.0$ primary power level,

$G = 1.0$ ghost power level,

$N_o = .1$ random noise power level,

$2\Delta T_{uh} = 1.0$ times the sample period $= h$,

$M = 20$ number of filter points minus one,

$p = -10.$

The figure also shows the impulse response for primaries obtained by adding $Y_{1m}+Y_{2m}$, and the impulse response for ghosts obtained by shifting and adding $Y_{1m}+Y_{2m-1}$. Note that the response for primaries is very nearly a delta function with maximum side-lobe levels of less than 10 percent of the main peak, whereas the impulse response for ghosts is uniformly small.

The time-domain formulation of the least-square filter problem presented here is clearly superior to a frequency-domain formulation if the resulting filters are to be used as finite digital convolution operators. The reason being that the former provides an optimum finite length operator, whereas the latter yields an optimum amplitude and phase spectrum which must be Fourier transformed and in general truncated in the time domain. In the limit of infinite-length operators, the time and frequency domain results are identical. Inasmuch as filter or operator length is an important economic factor in the implementation of this and other related techniques it is desirable to achieve the maximum effectiveness of the filter with the least number of points. The time-domain solution insures that each of the N-filter points used is optimum. In addition, by examining the mean-square error as a function of number of filter points, it is possible in most problems to select a minimum filter length beyond which the mean-square error essentially levels off. That is, the addition of more points to the filter does not significantly improve its performance in the mean-square-error sense.

The frequency domain on the other hand, if not best suited for digital filter design, does provide physical insight and interpretation that is lacking in equations (A-5) and (A-6).

The frequency-domain solution to multiple time series least-mean-square-error filter problems is contained in Wiener's (1949) original work. Certain important simplifications in the mathematics result when both the past and future of the time series are available for filter design, as is the case of magnetic-tape-recorded seismic data. The solution for the optimum filters $Y_j(f)$ operating on both past and future is (Spieker, Burg, Backus, and Strickland, 1961; Foster and Sengbush, 1964; and Burg, 1964) given by the matrix equation,

$$[S_{ij}(f) + N_{ij}(f)][Y_j(f)] = [S_{ik}(f)]$$

$$i = 1, 2, \cdots, N,$$

$$j = 1, 2, \cdots, N, \qquad \text{(A-8)}$$

where $S_{ij}(f)$ and $N_{ij}(f)$ are the cross-spectral densities between channels i and j for signal and noise, respectively. The column vector $[S_{ik}(f)]$ consists of cross-spectral densities between signal in channel i and the desired signal to be estimated at channel k.

The $Y_j(f)$ are, of course, the unknown filters to be applied to each channel in the multichannel signal-extraction process. Equation (A-8) implies that signal and noise are uncorrelated. An additional term in the equation would allow for the latter if it exists.

Using the signal and noise model for the ghosting problem defined by equation (A-6), and transforming to the frequency domain, we have a 2×2 matrix inversion problem for the filters $Y_1(f)$ and $Y_2(f)$;

$$\begin{bmatrix} \Phi_p(f) + \Phi_g(f) + N_0 & \Phi_p(f) + \Phi_g(f)e^{-i2\pi f 2\Delta T_{uh}} \\ \Phi_p(f) + \Phi_g(f)e^{i2\pi f 2\Delta T_{uh}} & \Phi_p(f) + \Phi_g(f) + N_0 \end{bmatrix} \begin{bmatrix} Y_1(f) \\ Y_2(f) \end{bmatrix} = \begin{bmatrix} \Phi_p(f) \\ \Phi_p(f) \end{bmatrix} \quad \text{(A-9)}$$

where

$$\Phi_p(f) = \int_{-\infty}^{\infty} \phi_p(\tau)e^{-i2\pi f\tau}d\tau$$

$$\Phi_g(f) = \int_{-\infty}^{\infty} \phi_g(\tau)e^{-i2\pi f\tau}d\tau.$$

The solutions of (A-9) are,

$$Y_1(f) = \frac{\gamma_R + \gamma_c(1 - e^{-i2\pi f 2\Delta T_{uh}})}{\gamma_R(\gamma_R + 2\gamma_c + 2) + 2\gamma_c(1 - \cos 2\pi f 2\Delta T_{uh})}$$

$$Y_2(f) = Y_1{}^*(f) \qquad * = \text{complex conjugate.} \quad \text{(A-10)}$$

The γ's are the noise-to-signal-power ratios, that is,

$$\gamma_R = \frac{N_o}{\Phi_p}$$

$$= \text{random-noise-to-signal-power ratio}$$

$$\gamma_c = \frac{\Phi_g}{\Phi_p}$$

$$= \text{coherent-noise-to-signal-power ratio.}$$

As previously mentioned, the response for primaries is obtained by adding $Y_1(f)$ and $Y_2(f)$ which yields,

$$Y_1(f) + Y_2(f) = \frac{2\gamma_R + 2\gamma_c(1 - \cos 2\pi f 2\Delta T_{uh})}{\gamma_R(\gamma_R + 2\gamma_c + 2) + 2\gamma_c(1 - \cos 2\pi f 2\Delta T_{uh})}. \quad \text{(A-11)}$$

For $\gamma_R \ll \gamma_c$ the response for primary energy is very nearly equal to unity for all frequencies except at and near the set $f_n = n/2\Delta T_{uh}$, $n = 0, 1, 2 \cdots$ where the response becomes,

$$Y_1(f_n) + Y_2(f_n) = \frac{2}{(\gamma_R + 2\gamma_c + 2)}. \quad \text{(A-12)}$$

The f_n are alias frequencies at which the stack (either conventional or the present process) cannot distinguish between primary and ghost energy. The associated wavelengths $\lambda_n = 2d/n$ are submultiples of twice the vertical shot separation. The denominator of (A-11) which is common to both $Y_1(f)$ and $Y_2(f)$ is essentially a compensation filter to undo the distortion introduced in the primaries as a consequence of ghost attenuation.

The response for ghosts may be obtained in a similar fashion by adding $Y_1(f)$ and $Y_2(f)e^{-i2\pi f 2\Delta T_{uh}}$ giving,

[*See* A-13 *at top of next page*]

The latter is small at all frequencies except the set f_n, at which the primary and ghost responses are equal as evidenced by (A-13). At $f_m = m/4\Delta T_{uh}$, $m = 1, 3, 5, 7, \cdots$ equation (A-13) has a zero irrespective of the value of γ_R and γ_c. At these frequencies the ghost energy is 180° out of phase between the two shots and cancellation is effected by simply adding the two traces. This frequency corresponds to the node in the response curves presented in Figure 1. The latter were obtained from equations (A-11) and (A-13) with $\gamma_R = .1$ and $\gamma_c = 1.0$.

It would seem that by allowing the random-noise-to-signal-power ratio γ_R to become arbitrarily small one could effect perfect undistorted separation of primaries and ghosts. That is, equations (A-11) and (A-13) tend to unity and zero, respectively, as $\gamma_R \to 0$. The difficulty, of course, is that the amplitude density of the filters (A-10) becomes infinite at $f = f_n$ for $\gamma_R = 0$. This implies that the impulse response of the filters is not square integrable over the time interval $(-\infty, \infty)$. Indeed, the impulse response of the filters for $\gamma_R = 0$ can be obtained by expanding the periodic frequency spectra (A-10) in a Fourier series and equating the coefficients a_m to delta function amplitudes at $t = m2\Delta T_{uh}$. That is,

$$V_1(f) + V_2(f)e^{-i2\pi f2\Delta T_{uh}} = e^{-i2\pi f\Delta T_{uh}} \frac{2\gamma_R \cos 2\pi f\Delta T_{uh}}{\gamma_R(\gamma_R + 2\gamma_c + 2) + 2\gamma_c(1 - \cos 2\pi f2\Delta T_{uh})}. \quad \text{(A-13)}$$

$$h_1(t) = \sum_{m=-\infty}^{\infty} a_{1m}\delta(t - m2\Delta T_{uh})$$

$$h_2(t) = \sum_{m=-\infty}^{\infty} a_{2m}\delta(t - m2\Delta T_{uh}) \quad \text{(A-14)}$$

where

$$a_{1m} = \begin{cases} +\frac{1}{2} \text{ for } m \leq 0 \\ -\frac{1}{2} \text{ for } m > 0 \end{cases}$$

$$a_{2m} = \begin{cases} -\frac{1}{2} \text{ for } m < 0 \\ +\frac{1}{2} \text{ for } m \geq 0 \end{cases}$$

clearly,

$$h_1(t) + h_2(t) = \text{response for primaries} = \delta(t),$$

$$h_1(t) + h_2(t - 2\Delta T_{uh}) = \text{response for ghosts} = 0.$$

These filters do not decay with increasing or decreasing time, and are obviously not practical for digital application. The role of random noise in (A-10) clearly provides stability and convergence for the impulse responses such that they may be truncated after a finite number of samples without degradation of the filters. The time-domain formulation circumvents the problem of having to truncate the filters at all by solving (A-5) for optimum N-point filters. The latter will, of course, depend upon γ_R as well as the other parameters used, however, they are the best N-point operators in the mean-square-error sense.

Reprinted from Geophysics v. 30, no. 3, p. 348-362

A NEW DATA-PROCESSING TECHNIQUE FOR MULTIPLE ATTENUATION EXPLOITING DIFFERENTIAL NORMAL MOVEOUT†

WILLIAM A. SCHNEIDER*, E. R. PRINCE, JR.*, AND BEN F. GILES*

A new data-processing technique is presented which utilizes optimum multichannel digital filtering in conjunction with common subsurface horizontal stacking for the efficient rejection of multiple reflections. The method exploits the differential normal moveout between primary and multiple reflections that results from an increase in average velocity with depth.

Triple subsurface coverage is obtained in the field; the common subsurface traces are individually prefiltered with different filters and stacked. The digital filters are designed on the least-mean-square-error criteria to preserve primaries (signal) in the presence of multiples (noise) of predictable normal moveout, and random noise.

The method achieves wide-band separation of primary and multiple energy with only a three-point stack; it can work effectively with small normal moveout differences eliminating the need for long offsets and the attendant signal degradation due to wide-angle reflections; it does not require equal multiple moveout on the triplet of traces stacked; and finally the method is not sensitive to small errors in statics or predicted normal moveout.

The technique is illustrated in terms of synthetic examples selected to encompass realistic field situations, and the parameter specification necessary for the multichannel filter design.

INTRODUCTION

The identification and elimination of multiples has been one of the most perplexing reflection seismic noise problems facing the exploration geophysicist.

Reverberations or "ringing" records encountered in marine shooting, due to the water-layer energy trap, may be largely overcome by inverse filtering techniques (Backus, 1959). Deep-section multiples present a different problem. Their effect is not to stretch out individual reflections as does the reverberant water layer, but rather to add significantly to the total number of recorded events and greatly complicate their interpretation. If there is an increase in the average velocity with depth, then primaries and multiples may be distinguished on the basis of their normal moveout differences. We would like not only to distinguish between the latter, but also to attenuate multiples which may be interfering with or masking valid primaries. Conventionally, the normal moveout difference is exploited for multiple attenuation by shooting multiple subsurface coverage, applying static and dynamic corrections, and stacking the common depth traces with different shot-recorded distances (Mayne, 1962; Shock, 1963). The excess multiple normal moveout results in partial cancellation of the multiple energy, while primaries add in-phase.

The new method presented in this paper (Optimum Horizontal Stack[1]) advances the concept of horizontal common-depth-point stacking a step further. With the present process, optimum utilization of the excess multiple moveout for multiple attenuation is achieved through the application of special filters prior to stacking. This results in drastic reduction of the multiple energy as will be illustrated in the following examples.

The application of the present process, as well as any common-depth-point method, requires shooting multiple subsurface coverage. In most instances, triple coverage provides sufficient multiple attenuation with the present process. Thus, the results to follow pertain to a threefold optimum stack; however, the method may be applied to any degree of multiple subsurface coverage. Figure 1 shows a proposed shooting and recording procedure to obtain triple coverage. Of course, the latter may be obtained by any number of other shot-recording configurations, the relative merits of which are not discussed here. The top row of circles in the figure schematically indi-

[1] A GSI Service Mark.

† Paper presented at the 33rd Annual International SEG Meeting, New Orleans, Louisiana, October 23, 1963. Manuscript received by the Editor August 18, 1964.

* Geophysical Service Inc., Dallas, Texas.

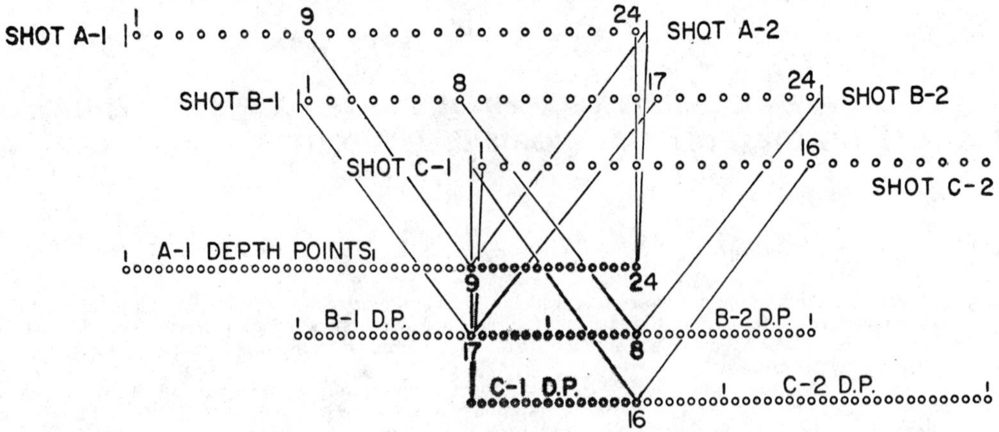

FIG. 1. Proposed spread for three-fold horizontal stack.

cates the group positions for shots A-1 and A-2, shot one-half a group interval off the end of the spread. The corresponding depth points are indicated below. The spread is picked up and moved one-third its length with group positions for shots B-1 and B-2, as shown, and their corresponding depth points below. Shots C-1 and C-2 are handled similarly to produce 300 percent subsurface coverage, as indicated in the cross-hatched area. For instance, group 1 from shot C-1, group 17 from B-1, and group 9 from A-2 have the same depth point. Similarly, one can proceed on through and specify the remaining triplets with common subsurface coverage. After eight such groups have been specified, the pattern repeats itself. Thus, for this particular shooting geometry, there are only eight different combinations of triple subsurface coverage.

The manner in which the present process or the conventional stack exploits the excess multiple normal moveout is illustrated in Figure 2. Traces X_1, X_2, and X_3 represent three different shot-receiver separations having common subsurface coverage. The events labeled P represent primary energy on the three channels in the stack as it would appear after static and dynamic corrections have been applied. That is, the primary energy is in-phase on the three traces in the stack.

Events labeled M represent the appearance of multiple energy on the three common-depth-point traces as they would appear at a given record time, T. In this particular case, we are employing a velocity function which is appropriate to the Louisiana Gulf Coast. The single-bounce multiple moveouts as seen are representative of that velocity function, for $X_1=110$ ft, $X_2=3{,}410$ ft, $X_3=3{,}630$ ft, and the record time $T=$three sec. The conventional method of suppressing multiples consists of stacking or adding these three channels, the output of which is shown as the fourth trace in Figure 2. It has been normalized by one-third so that primary energy comes through with the same amplitude as on the individual traces. The primary has been perfectly preserved, but the multiple has been attenuated only 5.4 db by the three-channel stack. This illustrates the need in general for stacking 600 or 1,200 percent subsurface coverage in the conventional processing. Without adding a large number of traces, one cannot achieve very significant rejection of multiple energy because of the differential moveout relationships which are normally encountered.

The sole difference between the conventional stack and the present process is the insertion of specially designed digital filters Y_1, Y_2, and Y_3, and then stacking. The resulting output, shown in the extreme right of the figure, preserves primary energy with no visible amplitude or waveform distortion, yet the multiple is attenuated on the order of 19 db. The filters are designed on the least-mean-square-error criteria to minimize the error between the desired signal (primary) and the signal plus multiple energy on the input channels X_1, X_2, and X_3. Viewed as a multichannel process, the signal-plus-noise description of the process is essentially embodied in the figure. That is, the primary or signal energy arrives simultaneously on all three channels with approximately equal amplitude and waveform, while the multiple or noise energy arrives with a given set of delays. The noise is also assumed to have approximately equal amplitude and waveform on the three traces. The multiple moveout relationships are, of course, dependent on the multiple velocity function, the spread geometry, the particular group of three traces being stacked, and the record time gate. Given a time gate on the record during which it is desired to suppress the multiple reflections, one can then put limits on the range of multiple moveouts to be expected in that time gate. For the example in Figure 2, the multiple moveout between channels 1 and 2 is 15±2 ms, and 17±2 ms between channels 1 and 3. This applies to the record time gate, 2.5 to 4.0 sec. We make the further assumption that the multiples are uniformly distributed within that range of moveout. This, in essence, provides the mathematical model for the least-mean-square-error filter design, the details of which are given in Appendix A.

It is recalled that there are eight different sets of three traces; consequently, filters Y_1, Y_2, and Y_3 must be designed for eight different groups giving a total of 24 different filters for this particular shooting scheme. In general, this set of 24 filters can be used throughout a prospect, providing that the primary or multiple velocity functions do not change drastically within the prospect.

Figure 3 shows the impulse response of the filters for channels 1, 2, and 3 appropriate to one of the eight groups. For digital application as a convolution, sampled versions of the latter are employed. Viewed on an individual basis, they are not particularly revealing. That is, it is not intuitively obvious why this set of filters is optimum for the particular signal and noise problem previously defined. Nonetheless, when they are applied to the three traces in the stack and summed, their combined action is such that they will pick out that part of the energy on the three traces which has signal or primary relationships, and reject that part of the energy on the three traces which has multiple relationships. The information needed to design the filters comes from the primary and multiple velocity functions as obtained from well velocity data, or an X^2, T^2 analysis of an expanding spread velocity survey (Musgrave, 1962; Dix, 1955).

Figure 4 relates how this latter information is used to provide the necessary input to the filter design program. The residual multiple moveout differences are shown for groups 1, 4, and 8 as a function of record time in seconds. In each instance, the top curve represents the difference in residual multiple moveout between channels 3 and 1, while the bottom curve is the difference in residual multiple moveout between channels 2

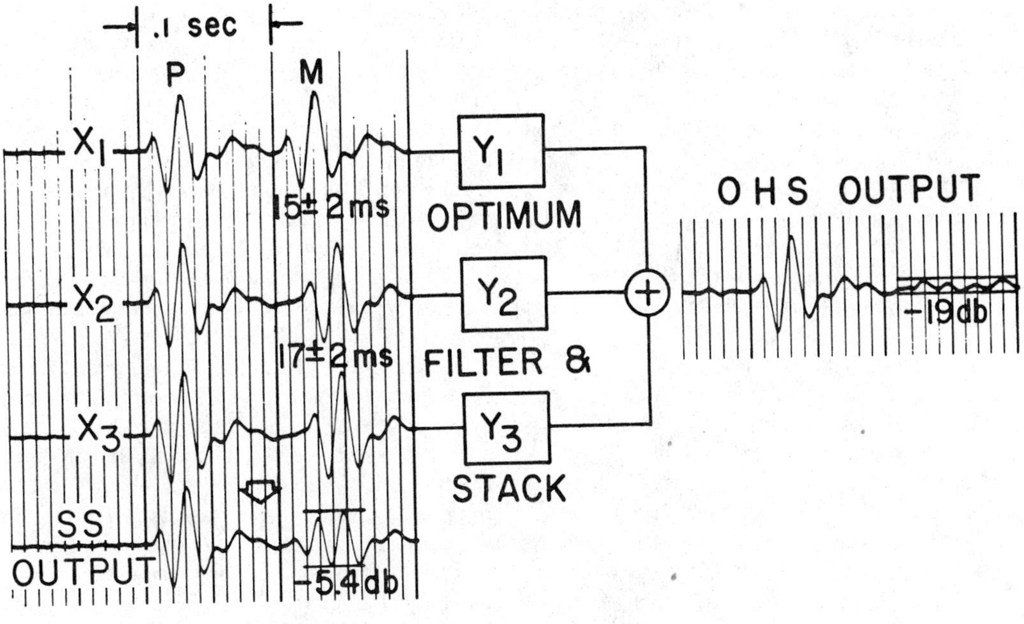

FIG. 2. Comparison of present process and conventional three-fold horizontal stack.

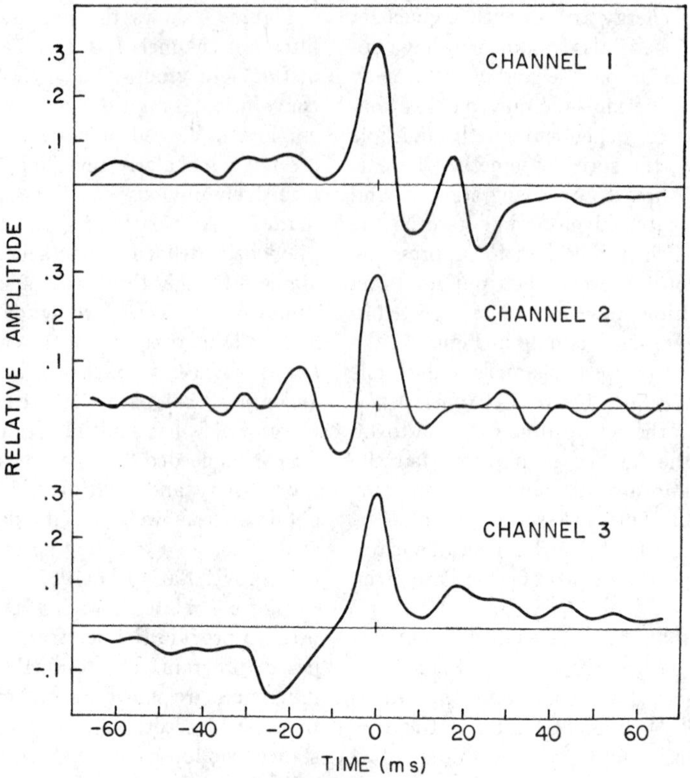

FIG. 3. Impulse responses of optimum horizontal stack filters.

and 1 as a function of record time. These curves again serve to illustrate the fact that the multiple moveout is a function, not only of the group, but also of record time. Once a time gate of interest is selected, then the range of multiple moveout to be expected in that time gate may be read off the curves. In this example, 2.5 to 4 sec is indicated as the gate of interest. Thus, on group 4, the expected multiple moveout between channels 1 and 3 is 23 ± 2 ms, and 9 ± 1 ms between channels 1 and 2. In a similar manner, the expected multiple moveout for the remainder of the groups can be read off of curves of this type.

EXAMPLES

In the first synthetic example, a comparison is given between the conventional straight stack and the present process as a function of group for multiples at approximately three sec of record time. In Figure 5 traces X_1, X_2, and X_3 represent the offset position for the traces having common subsurface coverage. The first event, "*P*," is a primary as it would appear on any of the triple subsurface coverage traces for reference purposes. The first event indicated under group 1 represents the appearance of a multiple at approximately three sec of record time. The residual moveout relationships are 15 ms between 1 and 2, and 17 ms between 1 and 3. The multiple moveout relationships continuously change as we proceed from group 1 to group 8, wherein there is just one ms of moveout between channels 1 and 2, and 30 sec of moveout between channels 1 and 3. Again, these residual moveouts are characteristic of velocity functions in the Louisiana Gulf Coast region.

The straight stack output obtained by adding the three common subsurface traces directly above is shown and labeled as trace *SS*. Primary energy comes through undistorted, and the multiples receive varying degrees of attenuation from -12 db in the best case to -4 db in the worst case. The straight stack achieves best attenuation when the multiple moveouts are almost equal as in group 3 with 11 ms and 21 ms between the three channels. In situations where

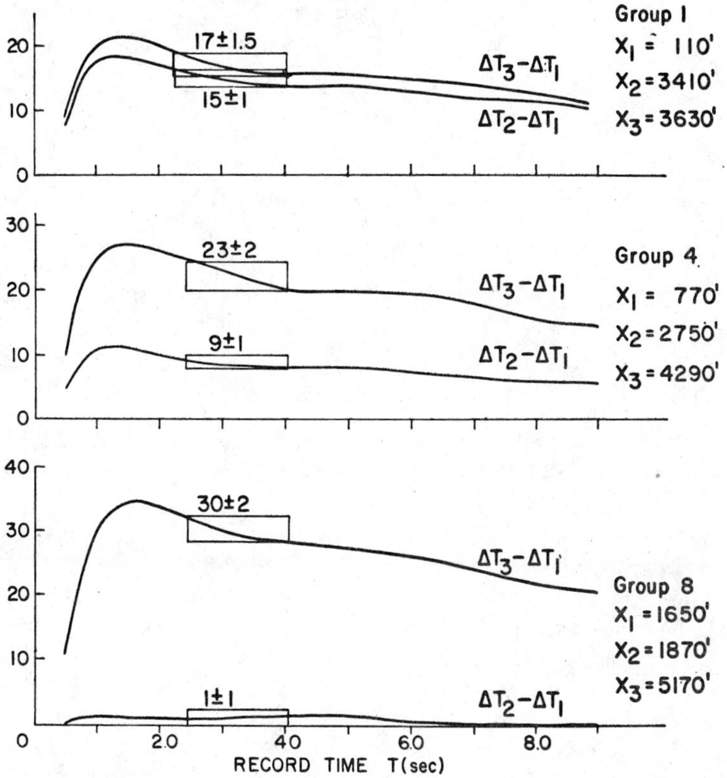

FIG. 4. Residual multiple normal moveout differences for design of stacking filters.

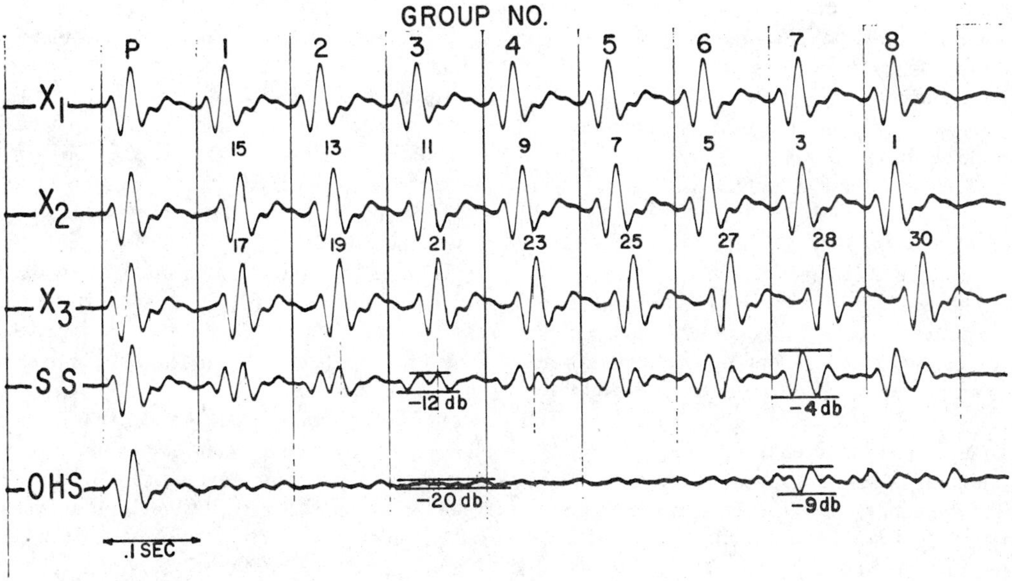

FIG. 5. Comparison of present process and conventional horizontal stack for different groups of three-fold coverage.

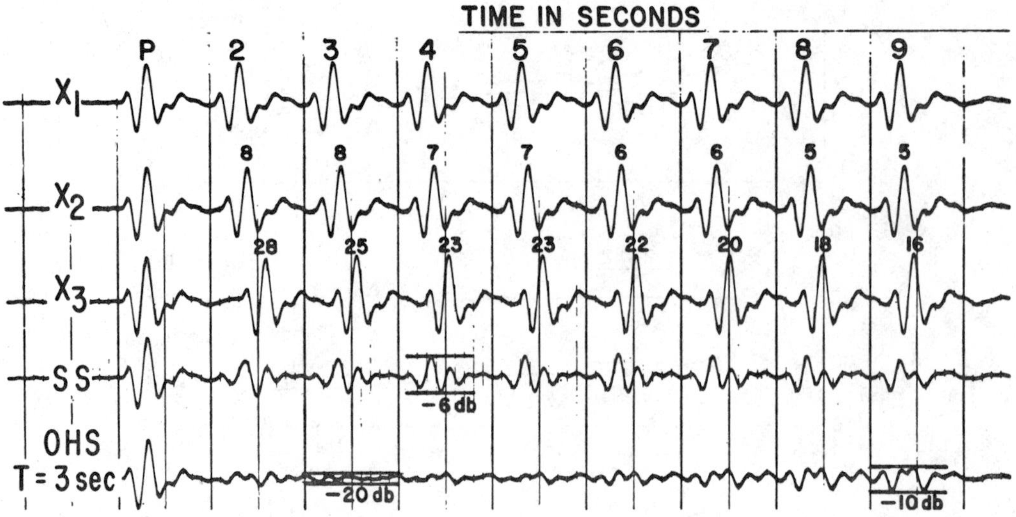

FIG. 6. Comparison of present process and conventional horizontal stack for multiples at different record times

two of the multiples are almost in-phase, as for groups 7 and 8, the straight stack is not able to achieve any significant degree of attenuation on a three-channel basis.

The output of the present process is shown as the last trace. It was obtained by applying the appropriate filters to each of the groups 1 through 8, and stacking. The primary also comes through with no visible distortion in waveform or amplitude; however, for groups 1 through 6, the process achieves on the order of 20 db attenuation of the multiple energy. For groups 7 and 8, the attenuation is somewhat less, about 9 or 10 db. For the particular multiple relationships on groups 7 and 8 even the optimum stack process has difficulty in attenuating this energy. The latter corresponds to space aliasing. These circumstances can be somewhat minimized in the field by suitably designing the spread geometry and offset, having some a priori knowledge of the multiple velocity function.

The next synthetic example shown in Figure 6 illustrates a comparison of the conventional straight stack and the optimum horizontal stack as a function of record time. In this case, traces X_1, X_2, and X_3 apply to the particular offsets for group 5. The first event, "*P*," is again a primary reflection as it would appear on the three traces. The remainder of the events are multiples as they would appear on group 5 as a function of record time from 2 through 9 sec, for the particular velocity function used. At two sec record time, the multiple moveout relationships show 8 ms between channels 1 and 2, and 28 between 1 and 3. At 9 sec record time, there are only 5 ms between channels 1 and 2, and 16 between 1 and 3, which is to be expected. The straight stack output *SS* was obtained by adding the three channels and normalizing by one-third. The primary comes through with no visible distortion as before, and the conventional stack achieves about 6 db attenuation of the multiple energy nearly independent of record time.

The output of the present process shown as the last trace in the figure was obtained by designing filters at and about three sec record time, applying them, and stacking to achieve the output trace. The primary comes through undistorted, and in the vicinity of 2 to 5.5 sec record time, the process achieves 20 db attenuation of the multiple energy. For multiples coming later than 6 sec, the process begins to lose its attenuating ability, and multiple rejection drops off to 10 db in the worst case. Although these multiples clearly fall outside the filter design specifications, they are still being attenuated no worse than with straight stack.

For this example it would appear that 6 sec of useable record would be most adequate; however, this is purely a function of the multiple velocity. In another area, the multiple moveout relationships may vary as much as indicated in Figure 6 over one or two sec of record time, in which case the present process would have a very narrow time window over which it would be operating at optimum efficiency. Under these circumstances, a time-varying process is employed. The design parameters for the latter are shown in Figure 7. Here the differences in multiple moveout for one of the groups, group 5, are shown as a function of record time. If it is desirable to have the process work over the entire record time, the multiple moveout may be segmented as indicated into four different time gates: 1 to 2.5, 2.5 to 4, 4 to 7, and 7 to 9 sec. In each of these time gates, a different filter set is designed based on the expected multiple moveouts as indicated. In processing, the various sets are stored in the computer's core and brought into the convolution sequentially at the appropriate record time.

An example of the application of the time-varying horizontal stack process is shown in Figure 8. Here we have the identical input as for the previous example, that is, the multiple relationships as a function of time for group number 5, as well as the straight stack output, have been repeated for comparison purposes. (The time-varying present process output is shown as the last trace.) The input traces X_1, X_2, and X_3 have been filtered differently in the four time gates, with the appropriate filters and stacked. In this case, a uniform 20 db rejection of the multiple energy has been achieved over the entire record.

The preceding examples have served to illustrate the theoretical effectiveness of the present process. It is, of course, of major interest to know the sensitivity of the process to static errors—errors in specifying the primary multiple velocity—as well as to amplitude fluctuation between primary and multiple energy on the three traces stacked. Partial answers to these questions are supplied in the following.

Figure 9 shows the effect of having small static errors in the traces of a particular group. The top three traces indicate again the three traces in a particular stack with both primary and multiple energy on the trace at 3, 3.5, and 4 sec of record time. The horizontal stack filters designed for multiples at T=three sec were applied, the output of which is shown as the first trace in the bottom half of the record. Primary energy is preserved and the multiples are attenuated as before. The same input was then static shifted by the amount of zero ms on X_1, -2 ms on the X_2, and -1 ms on X_3, shown as the second set of three traces in the top half of the record. The same horizontal stacking filters were applied, the output of which is shown as the last trace in the record. The primary amplitude and waveform comes through essentially undistorted, and there has been only a slight loss of attenuation of the multiple energy as the result of static errors. In general, as the static errors are increased until they become a significant fraction of the multiple

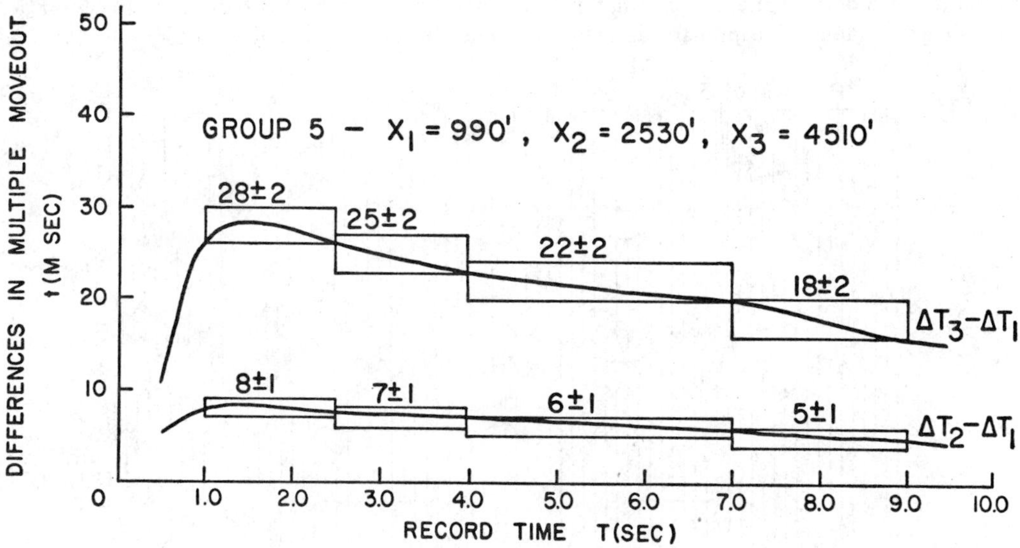

FIG. 7. Time-varying filter design specification.

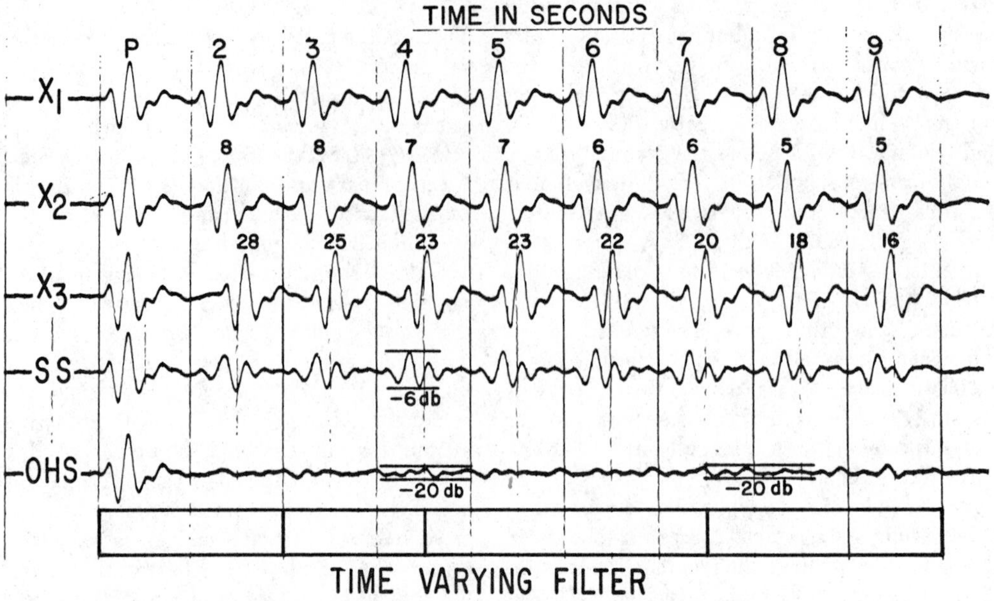

FIG. 8. Application of time-varying horizontal stack filters.

moveout, the attenuation of the process drops off to a point where it is comparable with a straight stack.

Figure 10 illustrates the effect of amplitude variation on the present process. The top three traces represent three traces in the stack with both primary and multiple energy shown. The appropriate horizontal stack filters were applied, the output of which is shown by the first trace in the bottom half of the figure, at somewhat higher gain than the input. The primary as before is undistorted, and 22 db attenuation of the multiple energy is achieved. Amplitude variations were introduced in the second set of three traces with −3, 0, and +3 db variations placed on both the primary and multiple energy. The same horizontal stack filters were applied and the output is shown as the last trace in the figure. The primary energy appears to be well preserved, with a loss of attenuation of about 6 db in the multiple. To minimize this effect, trace-to-trace equalization is employed before filtering and

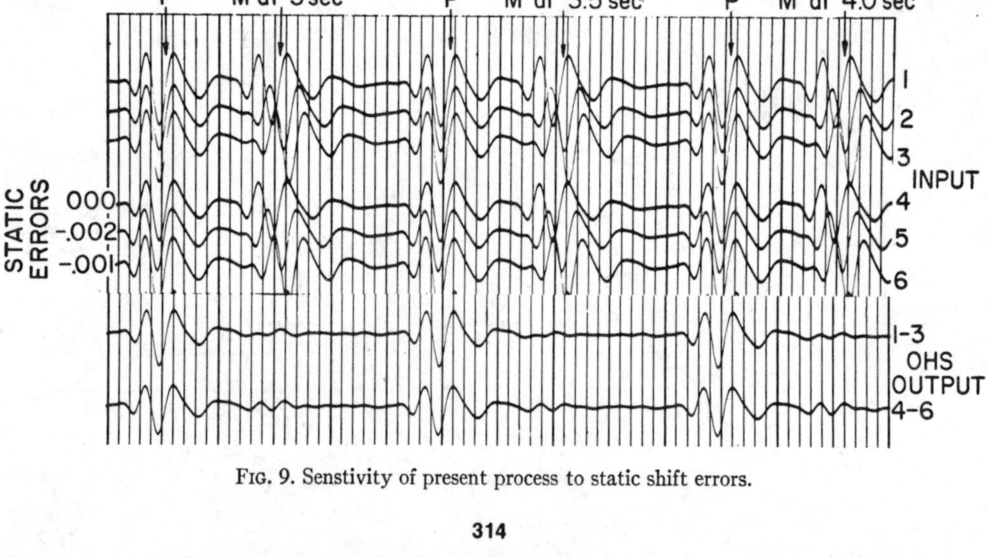

FIG. 9. Senstivity of present process to static shift errors.

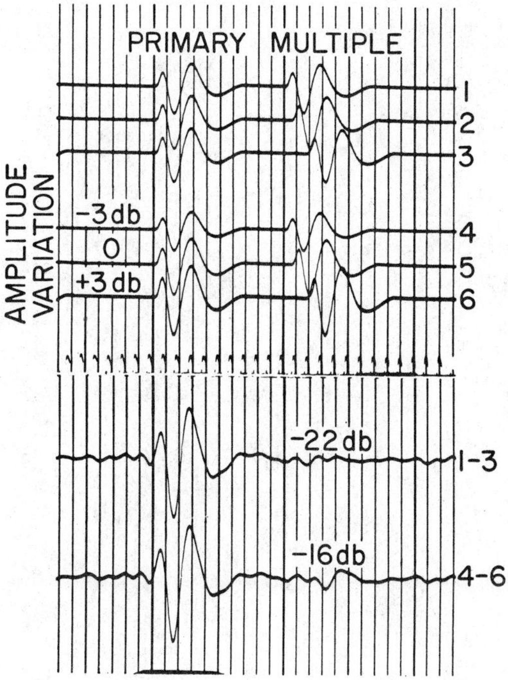

FIG. 10. Sensitivity of present process to amplitude errors.

stacking so that less than ±3 db variation in amplitude is maintained.

The final synthetic example illustrates the application of the present process to a record section on which primary and multiples are superimposed to simulate a more realistic field situation.

Figure 11 shows a portion of synthetic record section generated from a single synthetic seismogram which was reproduced on all 24 channels. At about 1.1 sec an "unconformity" was introduced showing dipping beds for deeper structure. The shallow beds are flat lying, and the section is multiple free. Multiple reflections were generated by introducing residual multiple moveout into the flat-lying shallow section. This was increased in amplitude relative to the primary section by approximately 6 db, shifted down over the deeper section, and superimposed on it.

Figure 12 shows the appearance of the primary-plus-multiple record section as it would appear recorded on end-overs. Notice that the multiple residual moveout clearly shows through, and it is, in fact, the most predominant type of energy visible. The flat-lying primary section is not

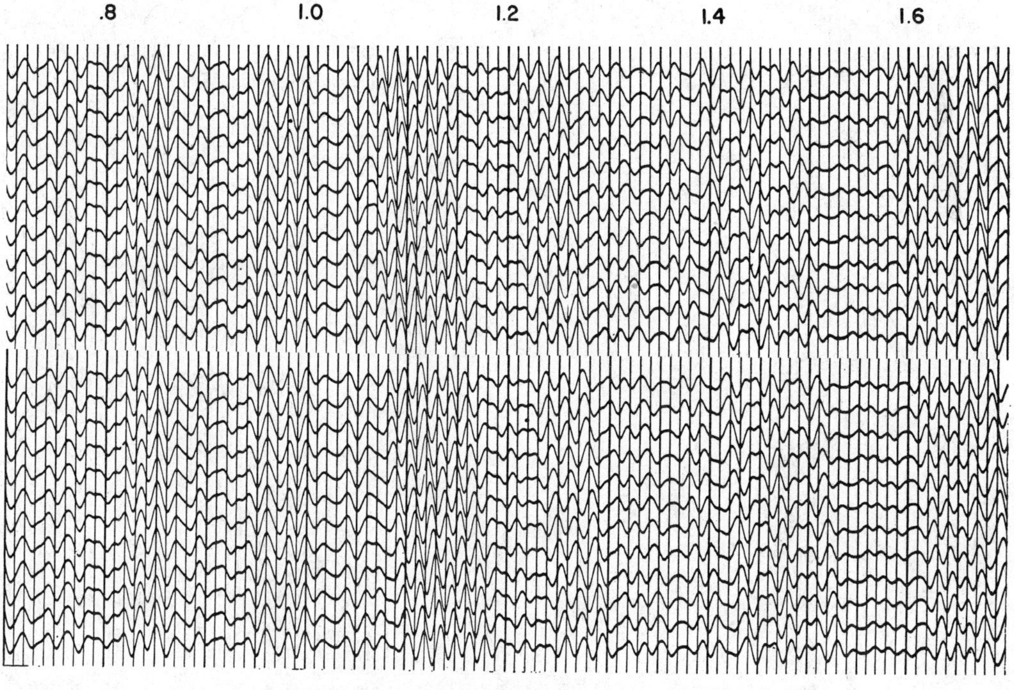

FIG. 11. Synthetic record section without multiples.

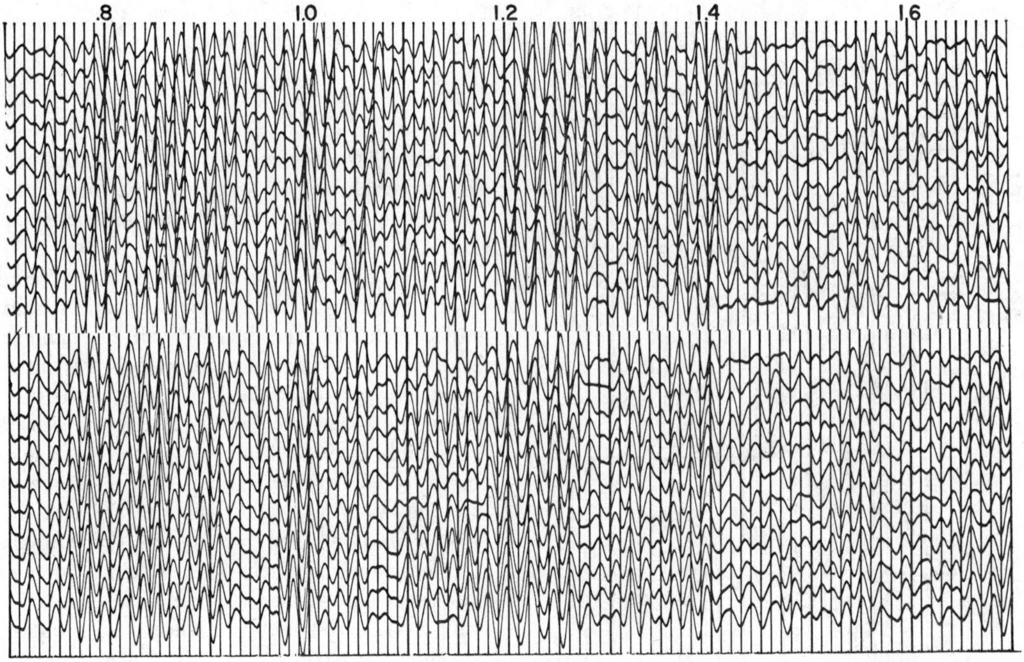

FIG. 12. Input record section with multiples (end-over).

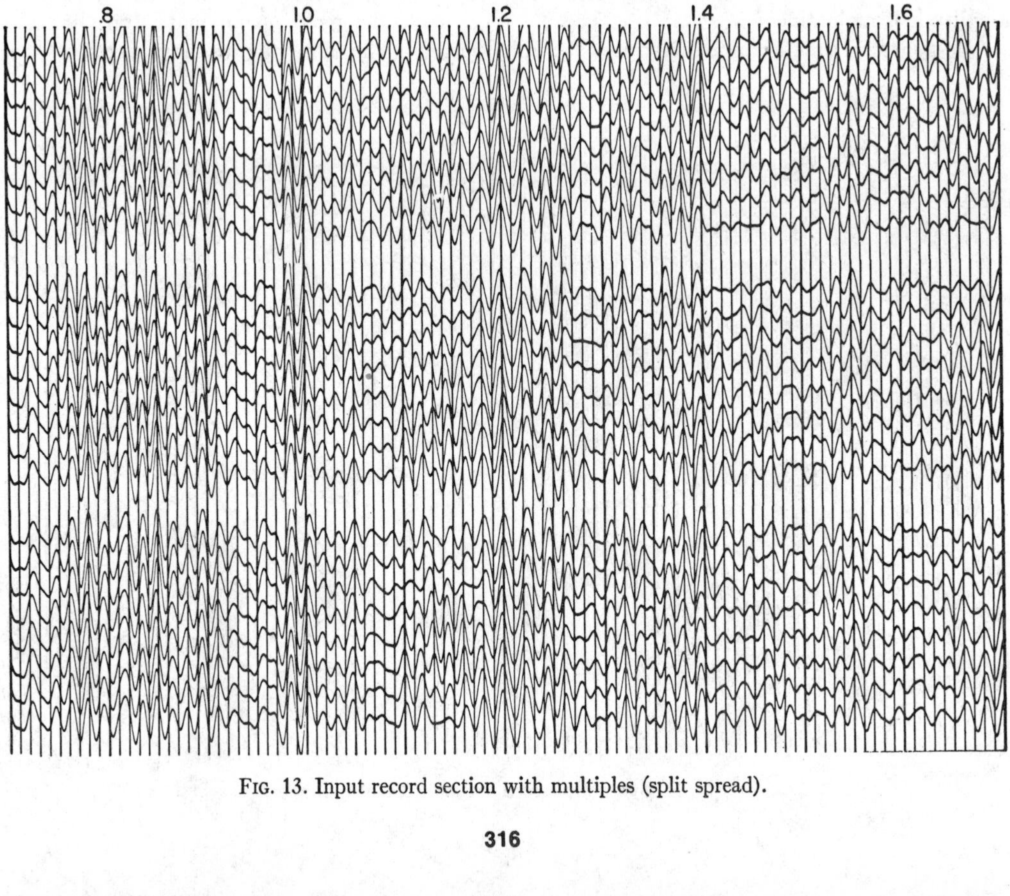

FIG. 13. Input record section with multiples (split spread).

readily discernible nor is the dipping unconformable structure on the original primary section; both are masked by the multiple moveout on the end-over spread.

On a split or short spread used to record the section in Figure 13, the residual multiple moveout is very small. Consequently, the entire section appears to have flat dip. Since the primary input section did have unconformable dip below 1.1 sec, it is evident this is masked by multiple energy at a 2:1 multiple-to-primary ratio.

Triple subsurface coverage was generated, using a Gulf Coast velocity function, and the spread geometry shown in Figure 1. The resulting triple subsurface coverage was filtered and stacked using the present process. A section of the output is shown in Figure 14.

Both the shallow flat dip and the deeper unconformable dip are evident. The process has not achieved perfect trace-to-trace reproduction, but this is the result of residual multiple energy. Multiple attenuation of the order of 18 to 20 db has been achieved.

A conventional stack of the same input data is shown in Figure 15. The output of the conventional stack is not as good as the present process; in particular, the familiar roof-topping effect of residual multiple energy can be seen at about .85 sec and 1.25 sec. Some of the unconformable dip in the primary section is visible at about 1.1 sec, and approximately 1.4 sec; however, there is considerable amplitude variation in the primaries that do come through. We can expect, on a basis of previous synthetic examples, that the multiples have been attenuated by about 6 db in the conventional stack. Thus, the output of the conventional stack should show a signal-to-multiple ratio of 1 to 1, which appears to be consistent with Figure 15.

In a more detailed comparison of the input, present process, and conventional stack sections (Figure 16), the major energy bands on the input come through the present process with good fidelity in both waveform, amplitude, and dip; whereas, the same bands on the conventional stack are highly variable due to interfering residual multiple energy.

In summary, the present process is an effective

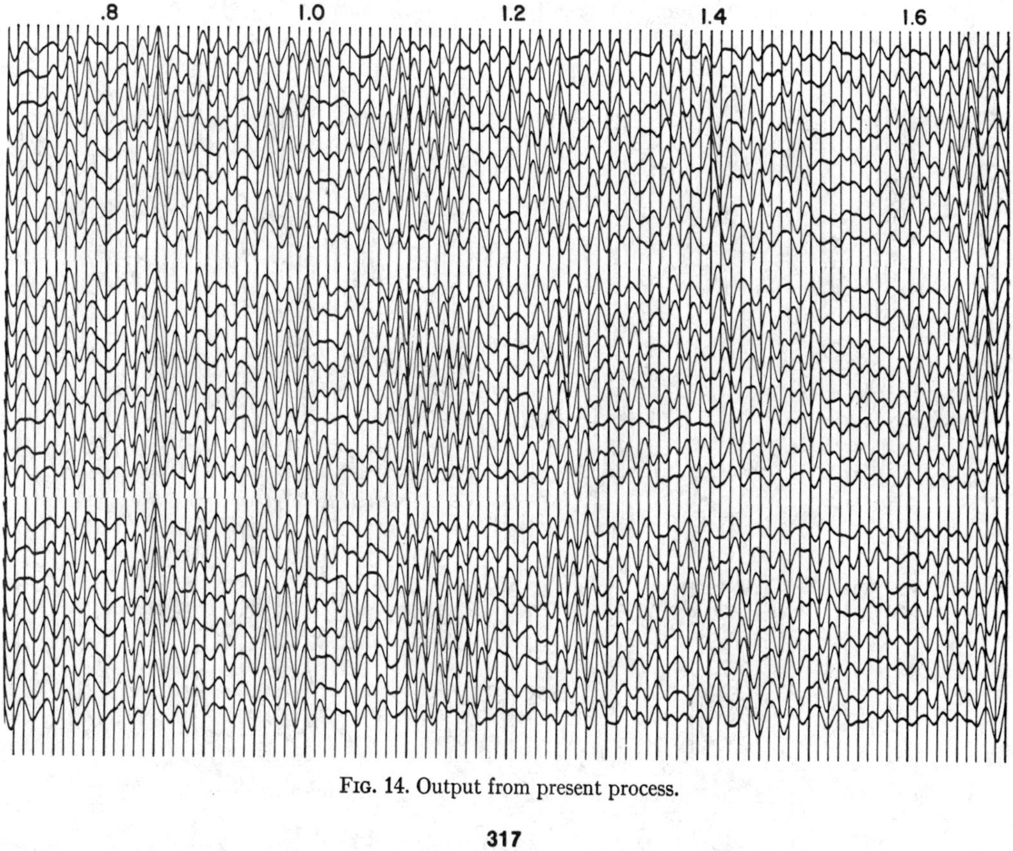

FIG. 14. Output from present process.

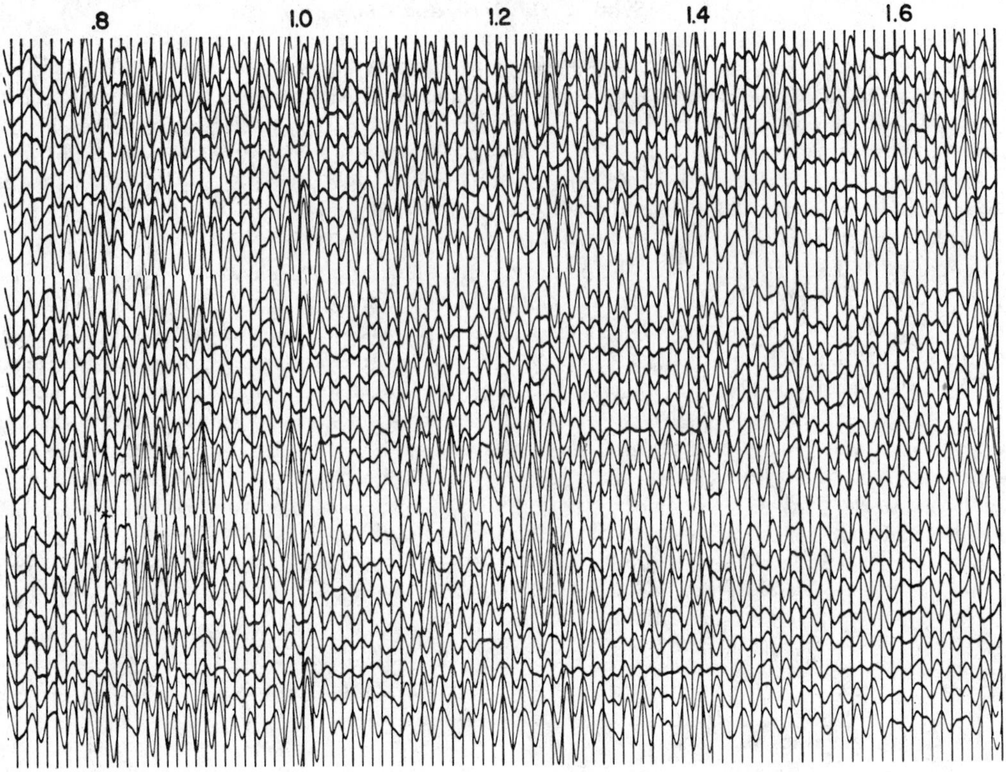

FIG. 15. Output from conventional stack.

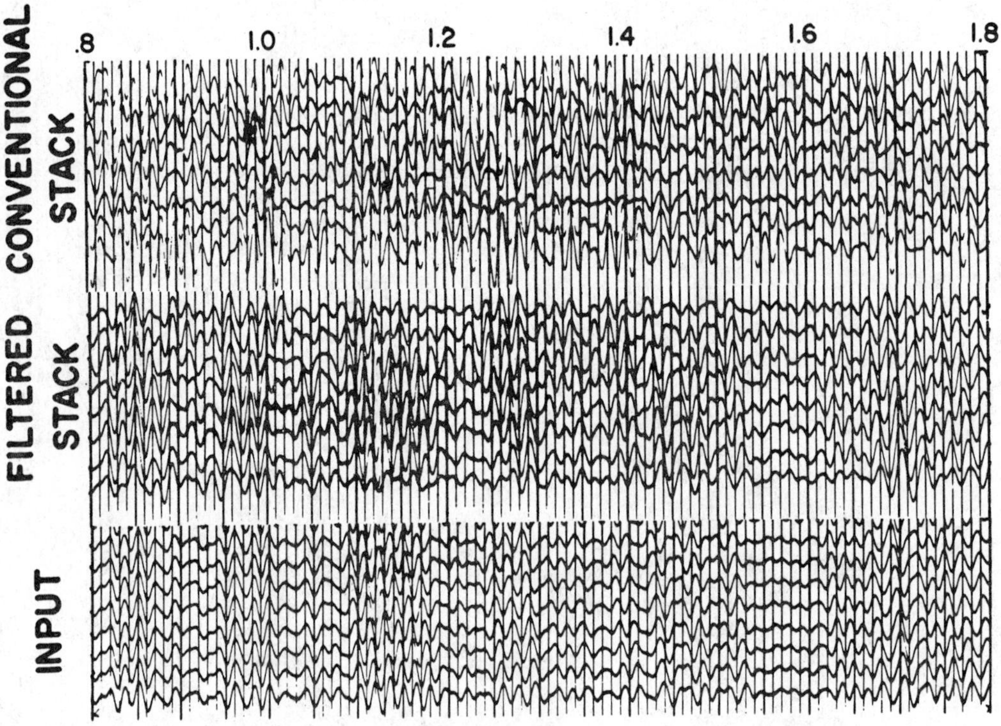

FIG. 16. Comparison of input, present process, and conventional stack.

tool for the elimination of deep-section multiples differentiated from primaries on a basis of normal moveout. The method combines the concept of horizontal common-depth-point stacking with optimum multichannel filter theory to design a three-channel processor for the extraction of primaries in the presence of multiples and random noise. Its major advantages over the conventional straight stack are:

1) It provides greater multiple attenuation with fewer traces over a wide frequency band.
2) It provides for more effective utilization of the excess multiple moveout permitting use of shorter spreads with less signal degradation due to offset.
3) It permits the preservation of high-frequency signal data which is generally lost in heavy mixes involving 600 to 1,200 percent subsurface coverage incurred in conventional horizontal stacking.

ACKNOWLEDGMENTS

The authors wish to express their thanks to Texas Instruments Incorporated and Geophysical Service Inc. for permission to publish this paper.

REFERENCES

Backus, Milo, 1959, Water reverberations—Their nature and elimination: Geophysics, v. 24, p. 233–261.
Burg, J. P., 1964, Three-dimensional filtering with an array of seismometers: Geophysics, v. 29, p. 693–713.
Dix, C. H., 1955, Seismic velocities from subsurface measurements: Geophysics, v. 20, p. 68–86.
Foster, M. R., Sengbush, R. L., and Watson, R. J., 1964, Design of suboptimum filter systems for multitrace seismic data processing: Geophys. Prosp., v. 12, p. 173–191.
Mayne, W. H., 1962, Common reflection point horizontal data stacking techniques: Geophysics, v. 27, no. 6, Part II, p. 927–938.
Musgrave, A. W., 1962, Applications of the expending reflection spread: Geophysics, v. 27, p. 981–993.
Schneider, W. A., Larner, K. L., Burg, J. P., and Backus, M. M., 1964, A new data-processing technique for the elimination of ghost arrivals on reflection seismograms: Geophysics, v. 29, p. 783–805.
Shock, Lorenz, 1963, Roll-along and drop-along seismic techniques: Geophysics, v. 28, p. 831–841.
Spieker, L. J., Burg, J. P., Backus, M. M., and Strickland, L., 1961, Seismometer array and data processing system: Final Report Phase I, AFTAC Contract AF 33(600)-41840, Project VT/077.
Wiener, Norbert, 1950, Extrapolation, interpolation, and smoothing of stationary time series: Jointly published by The Technology Press of MIT, and New York, John Wiley and Sons.

APPENDIX A

LEAST-MEAN-SQUARE-ERROR FILTER DESIGN FOR OPTIMUM HORIZONTAL STACK

The multichannel filter representing the present process may be characterized as three input terminals to a filter and summation network as shown below:

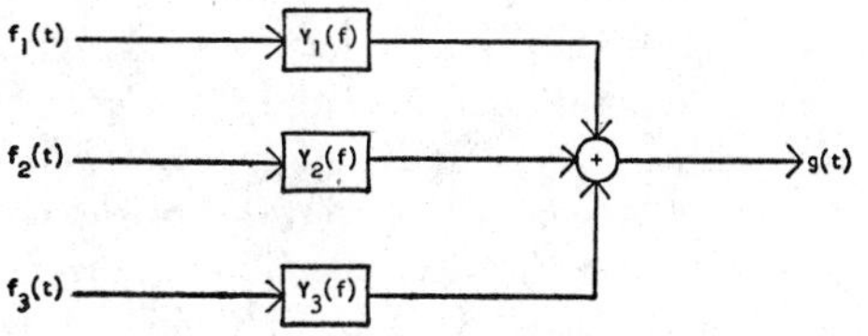

FIG. A-1

In the present problem, the inputs $f_i(t)$, $i=1, 2, 3$, are assumed to consist of signal and noise defined as follows:

$$f_i(t) = s_i(t) + n_i(t) \qquad i = 1, 2, 3$$
$$s_i(t) = s(t) \text{ (primary reflection sequence)} \qquad i = 1, 2, 3$$
$$n_i(t) = m_i(t) + r_i(t). \qquad \text{(A-1)}$$

The noise component $n_i(t)$ is divided into two parts; a coherent component $m_i(t)$ representing multiple energy on channel i, and a random component $r_i(t)$ uncorrelated between the three channels $i=1, 2, 3$.

Our criterion of goodness for the process, being least-mean-square error, requires that the average squared difference between the output $g(t)$ and the desired output $s(t)$ be a minimum, or mathematically:

$$\text{min. } \overline{(s(t) - g(t))^2}$$
$$= \text{min.} \frac{1}{2T} \int_{-T}^{T} \left[s(t) - \sum_{i=1}^{3} f_i(t) * h_i(t) \right]^2 dt, \qquad \text{(A-2)}$$

where $*$ denotes convolution, and $h_i(t)$ are the filter impulse responses.

The frequency domain solution of (A-2) is contained in Wiener's 1950 work. Certain important simplifications result in the mathematics when both past and future of the time series are available for digital filtering, as is the case with magnetic-tape-recorded seismic data. That is, the usual physical realizability constraints may be relaxed. The frequency domain solution of (A-2) for filters $Y_i(f)$ operating on both past and future time (Spieker, Burg, Backus, and Strickland, 1961; Foster, Sengbush, and Watson, 1964; Burg, 1964), is given by the matrix equation:

$$[S_{ij}^*(t)+N_{ij}^*(f)]\{Y_j(f)\}=\{S_{io}^*(f)\}, \quad \text{(A-3)}$$

where $S_{ij}(t)$ and $N_{ij}(t)$ are the cross-spectral densities between channels i and j for signals and noise, respectively. The column vector $\{S_{io}^*(f)\}$ consists of the complex conjugates of the cross-spectral densities between signal in channel i and the desired output signal. Equation (A-3) is inverted for the solution vector $\{Y_j(f)\}$, $j=1, 2, 3$, at the frequencies $f_0, f_1, f_2 \cdots f_{\text{max}}$, spanning the band of interest. The latter is approximately 10–100 cps in the exploration seismic problem. The frequency filters $Y_i(f)$ are inverted to the time domain by inverse Fourier transform and applied as digital convolution operators. An application of (A-3) to the deghosting problem is discussed by Schneider, Larner, Burg, and Backus (1964).

The matrix entries in equation (A-3) are specified in terms of the signal and noise model implied in Figure 2 and equations (A-1).

The signal correlation statistics may be modeled as follows. The signal autocorrelation on channel i is:

$$\phi_{ii}^s(\tau)=\frac{1}{2T}\int_{-T}^{T} s_i(t)s_i(t-\tau)dt \quad \text{(A-4)}$$

using (A-1),

$$=\frac{1}{2T}\int_{-T}^{T} s(t)s(t-\tau)dt$$

$$=\phi_{ss}(\tau)$$

for $i=1, 2, 3$.

The signal crosscorrelation between channels i and j would be the same using (A-1); however, we now allow for imperfect alignment of primary energy between channels in the stack due to incorrect statics, normal moveout, etc. The signal crosscorrelation may be written as:

$$\phi_{ij}^s(\tau)=\frac{1}{2T}\int_{-T}^{T} s_i(t)s_j(t-\tau)dt$$

$$=\phi_{ss}(\tau+\alpha_j-\alpha_i), \quad \text{(A-5)}$$

where α_i and α_j are the time alignment errors of primaries on channels i and j, respectively. These are considered as random variables governed by the joint probability density function $p(\alpha_i, \alpha_j)$. The expectation value of the signal crosscorrelation becomes,

$$E[\phi_{ij}^s(\tau)]$$

$$=\int_{-\infty}^{\infty}\int_{-\infty}^{\infty} p(\alpha_i, \alpha_j)\phi_{ss}(\tau+\alpha_j-\alpha_i)d\alpha_i d\alpha_j. \quad \text{(A-6)}$$

We make the assumption that static errors, etc., are statistically independent between channels $p(\alpha_i, \alpha_j)=p(\alpha_i)p(\alpha_j)$, and that the probability density functions for delays are uniform;

$$p(\alpha_i)=\begin{cases}\dfrac{1}{2\Delta S_i} & \text{if } |\alpha_i| \le \Delta S_i \text{ sec}\\ 0 & \text{if } |\alpha_i| > \Delta S_i \text{ sec}\end{cases} \quad \text{(A-7)}$$

$$i=1, 2, 3,$$

where $\pm\Delta S_i$ is the expected range of signal misalignment on channel i. Using (A-7) in equation (A-6) we see that the signal crosscorrelation expectation value is a cascade filtration or double convolution of the signal autocorrelation with two "box-car" functions,

$$E[\phi_{ij}^s(\tau)]=\int_{-\infty}^{\infty} d\alpha_j p(\alpha_j)$$

$$\cdot\int_{-\infty}^{\infty} d\alpha_i p(\alpha_i)\phi_{ss}(\tau+\alpha_j-\alpha_i). \quad \text{(A-8)}$$

The signal crosspower density $S_{ij}(f)$ may be obtained by Fourier transformation of (A-8); or simply recalling that convolution in the time domain corresponds to multiplication in the frequency domain, we have,

$$S_{ij}(f)=\Phi_{ss}(f)\frac{\sin 2\pi f\Delta S_i}{2\pi f\Delta S_i}\,\frac{\sin 2\pi f\Delta S_j}{2\pi f\Delta S_j}, \quad \text{(A-9)}$$

where

$$\Phi_{ss}(f)=\int_{-\infty}^{\infty}\phi_{ss}(\tau)e^{-i2\pi f\tau}d\tau,$$

and

$$\frac{\sin 2\pi f\Delta S_i}{2\pi f\Delta S_i}=\int_{-\infty}^{\infty}\rho(\alpha_i)e^{-i2\pi f\alpha_i}d\alpha_i$$

for $i \neq j$. For $i=j$, the diagonal signal elements of matrix equations (A-3) are,

$$S_{ii}(f) = \Phi_{ss}(f) \qquad \text{(A-10)}$$

by (A-4).

The noise matrix elements are derived in a similar fashion; however, a somewhat different interpretation of the delay probability densities $p(\alpha_i)$, $i=1, 2, 3$, are required. In the signal case, the latter accounted for random fluctuations in signal alignment. For the multiple component of the noise $m_i(t)$, the delay probability densities must incorporate the expected range of multiple residual normal moveout in the time gate of interest, as well as static fluctuations to be expected.

Taking channel 1 as reference, the probability densities for delay are given by,

$$p(\alpha_1) = \begin{cases} \dfrac{1}{2\Delta M_1} \text{ if } \alpha_1 \leq \Delta M_1 \\ 0 \text{ if } |\alpha_1| > \Delta M_1 \end{cases} \qquad \text{(A-11)}$$

$$p(\alpha_i) = \begin{cases} \dfrac{1}{2\Delta M_i} \text{ if } M_i - \Delta M_i \leq \alpha_i \leq M_i + \Delta M_i \\ 0 \text{ if } M_i - \Delta M_i > \alpha_i > M_i + \Delta M_i \end{cases}$$

for $i=2, 3$.

The average differences in residual moveout in the time gate of interest, between multiples on channels 1 and 2, and 1 and 3, are M_2 and M_3, respectively. These correspond to 15 and 17 ms, respectively, in Figure 2. The range in moveout difference between 1 and 2, and 1 and 3 is determined by the parameter ΔM_i, $i=1, 2, 3$.

Using (A-11) the off-diagonal, coherent-noise, cross-spectral densities are

$$N_{ij}(f) = \Phi_{mm}(f) \frac{\sin 2\pi f \Delta M_i}{2\pi f \Delta M_i} \frac{\sin 2\pi f \Delta M_j}{2\pi f \Delta M_j} \cdot e^{-i2\pi f (M_i - M_j)}, \qquad \text{(A-12)}$$

where

$$\Phi_{mm}(f) = \int_{-\infty}^{\infty} \phi_{mm}(\tau) e^{-i2\pi f \tau} \, d\tau$$

is the multiple autopower spectrum, and $M_1=0$.

The diagonal noise elements consist of the multiple and random noise autopower spectra,

$$N_{ii}(f) = \Phi_{mm}(f) + \Phi_{rr}(f). \qquad \text{(A-13)}$$

Finally, the elements of column vector on the right-hand side of equation (A-3) may be specified as follows. The desired output signal $s_o(t)$ is identical with the signal (primary reflection sequence) on each of the three channels in the stack, except that we assume no uncertainty in statics or arrival time. Thus the delay probability densities are

$$p(\alpha_o) = \delta(\alpha_o), \qquad \text{(A-14)}$$

where α_o is the output static referenced to itself, and the $p(\alpha_i)$, $i=1, 2, 3$, are given by (A-7). Taking the Fourier transform we have

$$S_{io}(t) = \Phi_{ss}(f) \frac{\sin 2\pi f \Delta S_i}{2\pi f \Delta S_i} \qquad \text{(A-15)}$$

for $i=1, 2, 3$.

This completely specifies the matrix equation. The functions (A-9), (A-10), (A-12), (A-13), and (A-15) may be readily programmed as part of the inversion routine. The autospectra $\Phi_{ss}(f)$, $\Phi_{mm}(f)$, and $\Phi_{rr}(f)$ must either be estimated, or assumed "white" with different levels over the frequency range of interest. In the absence of measured spectra, the "whiteness" assumption forces the optimum stack to work equally hard at all frequencies exploiting only the phase relationships of signal and noise. Since the multichannel phase relationships provide the basic signal enhancement leverage in this case, amplitude assumptions are relatively unimportant.

Reprinted from Geophysics v. 33, no. 5, p. 711-722

GEOPHYSICS

MULTICHANNEL DECONVOLUTION FILTERING OF FIELD RECORDED SEISMIC DATA†

E. B. DAVIES AND E. J. MERCADO*

Several writers have proposed the use of multichannel filters for the elimination of coherent noise on seismic records. One filter of this type which can be constructed is a multichannel Wiener filter which has a multichannel input and a single channel output. In this form, it is applicable to data collected for vertical or horizontal common-depth-point stack processing.

The choice of desired output characteristics for this Wiener filter is flexible and, for example, can be tuned to correspond to multichannel deconvolution.

The results of the application of filters of this type to field and synthetic data, in general, show little if any advantage over single-channel deconvolution. This failure appears to be connected with the low cross coherence of both noise and reflection signal on field-recorded, common-depth-point traces.

INTRODUCTION

The seismic data processing technique known as deconvolution or inverse filtering has proven to be a useful process in improving the quality of seismic data. The deconvolution method, which is based on least-mean-square-error filtering can be extended to multichannel systems. It has been suggested by Robinson (1966) that this concept may be of use in seismic data processing. In this paper, we discuss a multichannel input, single-channel output process of this type which is called multichannel deconvolution.

Such a process should apply to common-depth-point (CDP) data. This is due to the fact that on a group of CDP traces, direct reflection events should be correlatable with zero moveout while other events should correlate with differential moveout from trace to trace. The multichannel process utilizes both auto- and crosscorrelation functions of the input channels. It would thus seem that this process should have some advantage over combinations of stacking and single-channel deconvolution in increasing the resolution of reflection records.

Theory of the method

The multichannel input, single-channel output, least-mean-square-error (Wiener) filter is shown schematically in Figure 1. The notation in this figure is:

$x_k(t)$ = input trace on the kth channel

$f_k(t)$ = impulse response of the kth channel filter

$g_k(t)$ = output trace on the kth channel resulting from the convolution of $x_k(t)$ with $f_k(t)$

$g(t) = \sum_{k=1}^{L} g_k(t)$ = single-channel output trace

$Z(t)$ = desired single-channel output trace

$e(t)$ = error signal which results from the difference between the actual single-channel output and the desired single-channel output.

If the input data is in discrete, uniformly sampled form, the notation is:

$$x_k(t) = [x_{k0}, x_{k1}, x_{k2}, \cdots, x_{kN}]$$
$$f_k(t) = [f_{k0}, f_{k1}, f_{k2}, \cdots, f_{kM}]$$
$$g_k(t) = [g_{k0}, g_{k1}, g_{k2}, \cdots, g_{kN+M}]$$
$$g(t) = [g_0, g_1, g_2, \cdots, g_{N+M}]$$
$$Z(t) = [Z_0, Z_1, Z_2, \cdots, Z_{N+M}]$$
$$e(t) = [e_0, e_1, e_2, \cdots, e_{N+M}].$$

With data of this type the equations which describe the least-squares filter can be written in matrix form.

† Manuscript received by the Editor January 26, 1968.

* Gulf Research and Development Company, Pittsburgh, Pennsylvania.

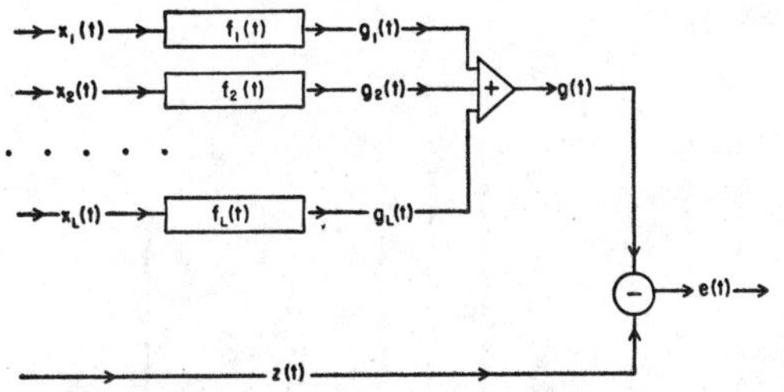

FIG. 1. Multichannel input, single-channel output, least-squares filter.

The equation which represents the convolution of the input traces and the summation which yields $g(t)$ can be written as

$$\begin{vmatrix} X_0 & 0 & \cdot & 0 \\ X_1 & X_0 & \cdot & \cdot \\ \cdot & X_1 & \cdot & 0 \\ X_N & \cdot & \cdot & X_0 \\ 0 & X_N & 0 & X_1 \\ \cdot & \cdot & \cdot & \cdot \\ 0 & 0 & \cdot & X_N \end{vmatrix} \begin{vmatrix} F_0 \\ F_1 \\ \cdot \\ F_M \end{vmatrix} = \begin{vmatrix} g_0 \\ g_1 \\ \cdot \\ \cdot \\ \cdot \\ \cdot \\ g_{N+M} \end{vmatrix}, \qquad (1)$$

where $X_k=[x_{1k}, x_{2k}, \cdot\ x_{Lk}]$, a 1 by L matrix and

$$F_k = \begin{vmatrix} f_{1k} \\ f_{2k} \\ \cdot \\ f_{Lk} \end{vmatrix}, \text{ an } L \text{ by 1 matrix.}$$

Equation 1 can be written symbolically as

$$XF = G, \qquad (2)$$

where

X is an $N+M+1$ by $(M+1)L$ matrix,
F is a vector of dimension $(M+1)L$,

and

G is a vector of dimension $N+M+1$.

The desired output is a vector Z of dimension $N+M+1$. By applying the least-mean-square error criterion, the filter F obtained by solving equation (2) minimizes the quantity $(Z\text{-}XF)^T$-$(Z\text{-}XF)$ as a function of F.

It can be shown that the desired filter F is the solution of the equation

$$X^TXF = X^TZ. \qquad (3)$$

This equation can be written in the form

$$\Phi F = \Psi, \qquad (4)$$

where

Φ is an $(M+1)L$ matrix and is the multichannel autocorrelation matrix of the input traces,

and

Ψ is a vector of dimension $(M+1)L$ and shows the crosscorrelation between the input traces and the desired output.

Equation (4) can be written in the expanded form

$$\begin{vmatrix} \phi(0) & \phi(-1) & \cdot & \phi(-M) \\ \phi(1) & \phi(0) & \cdot & \cdot \\ \cdot & \cdot & \cdot & \cdot \\ \phi(M) & \cdot & \cdot & \phi(0) \end{vmatrix} \cdot \begin{vmatrix} F_0 \\ F_1 \\ \cdot \\ F_M \end{vmatrix} = \begin{vmatrix} \psi(0) \\ \psi(1) \\ \cdot \\ \psi(M) \end{vmatrix}, \qquad (5)$$

where

$$\phi(k) = \begin{vmatrix} \phi_{11}(k) & \phi_{12}(-k) & \cdot & \phi_{1L}(-k) \\ \phi_{21}(k) & \phi_{22}(k) & \cdot & \cdot \\ \cdot & \cdot & \cdot & \cdot \\ \phi_{L1}(k) & \cdot & \cdot & \phi_{LL}(k) \end{vmatrix}$$

and

$$\psi(k) = \begin{vmatrix} \psi_{1Z}(k) \\ \psi_{2Z}(k) \\ \cdot \\ \psi_{LZ}(k) \end{vmatrix}.$$

The desired digital filter is found in the vector F and this can be found from (4) by calculating the inverse of the matrix Φ and solving the equation

$$F = \Phi^{-1}\Psi. \qquad (6)$$

The main difficulty lies in finding the inverse of this large Φ matrix. The development of recursion methods by Robinson (1963) and Wiggins and Robinson (1965) have made possible the solution of equation (6) for filters with impulse responses of meaningful length.

Equation (6) can be used to determine special-purpose multichannel filters either by presetting the correlation matrices or by specifying idealized

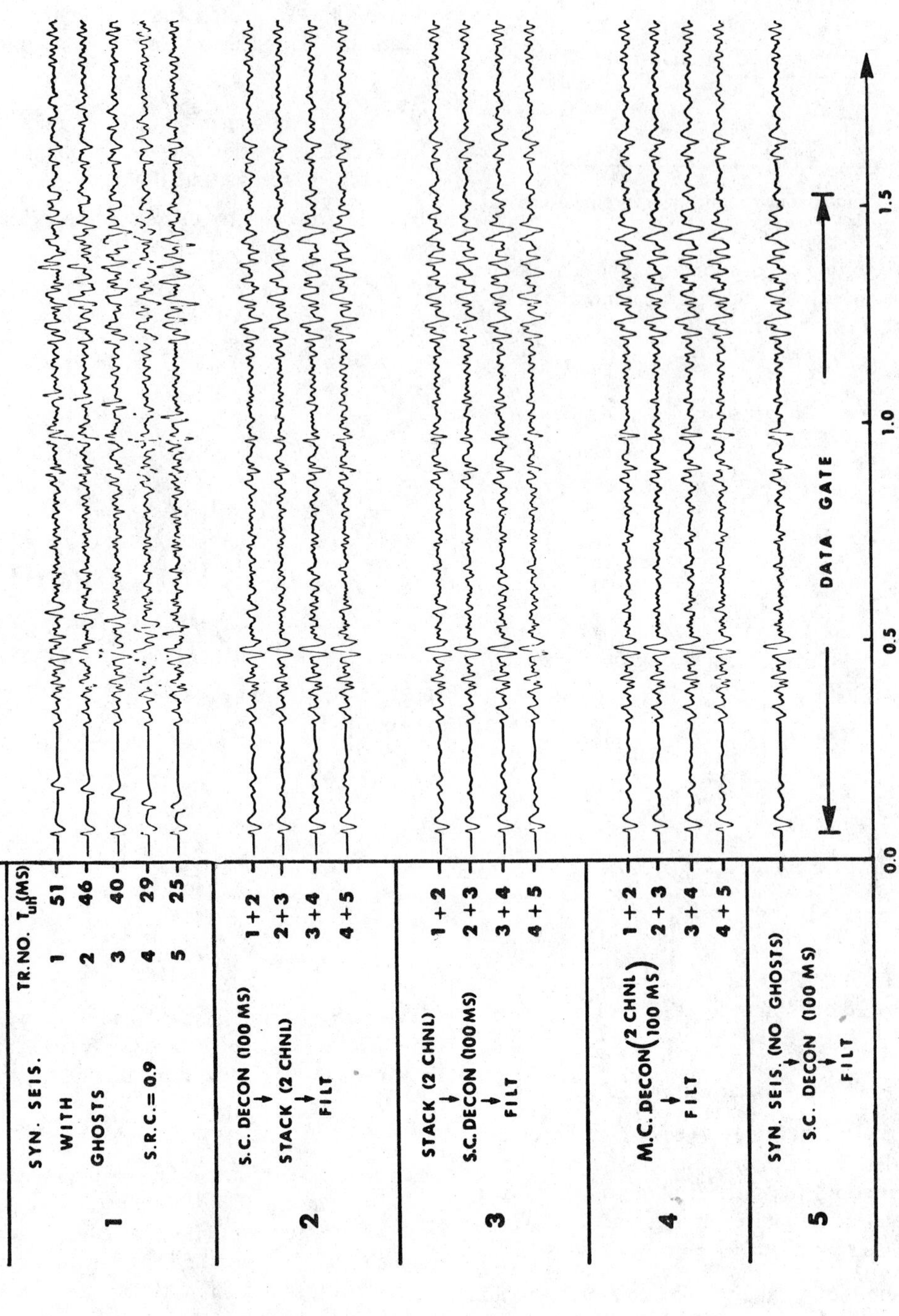

FIG. 2. Synthetic, vertical, CDP data

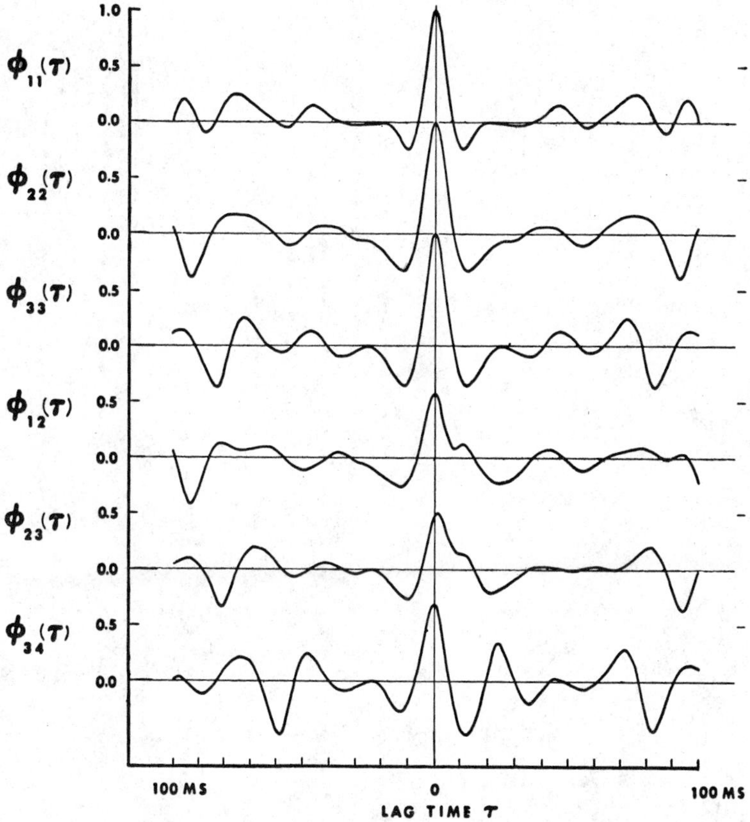

FIG. 3. Correlation functions: synthetic, vertical, CDP data.

input and output data. Examples of filters of this type are ghost suppression filters (Schneider et al, 1964) and optimum horizontal stack filters (Schneider, Prince, and Giles, 1965).

The approach taken for the work described here was to apply the multichannel filter method directly to field recorded data. It can be seen from equation (5) that the filter values are partly determined by a matrix which contains both the autocorrelation functions of each input trace and the crosscorrelation functions between input traces. Thus, if an event is common to each trace it should create significant crosscorrelation peaks while random events should show a low crosscorrelation value. Specific types of correlatable noise should be separable from correlatable signal by means of correlation peaks appearing at different lag values.

Application of method to common-depth-point data

Data from common-depth-point shooting, either vertical or horizontal, is adaptable to a multichannel input, single-channel output filter. Consider an ideal set of CDP traces which have been perfectly time corrected. The reflection events (signal) should yield crosscorrelation peaks at zero lag while correlatable or organized noise such as ghosts or multiples should produce crosscorrelation peaks at other lag values.

In theory, the characteristics of the desired output $Z(t)$ can be specified in many ways, but such specification should have a reasonable physical basis. In the case of CDP traces we assume that the reflection energy on each trace results from the same set of reflection coefficients and the same seismic pulse. The crosscorrelation functions between the input traces and the desired output $Z(t)$ are set to be nonzero only at zero lag values. This corresponds to setting $Z(t)$ to be the set of reflection coefficients or a zero-delay spike. This is the specification for multichannel deconvolution. For field data, there is a good chance that the above assumptions are not satisfied.

In processing common-depth-point field data,

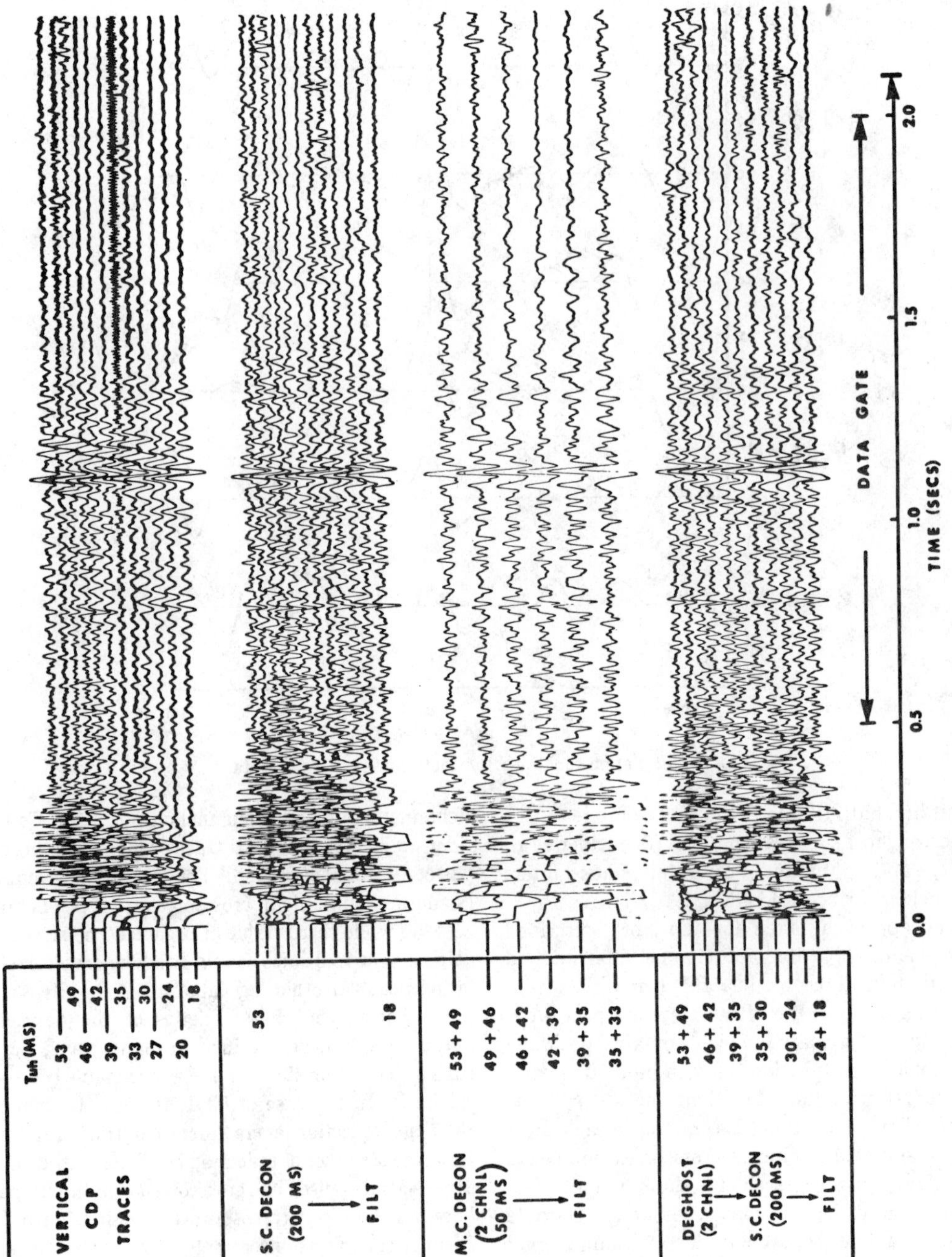

Fig. 4. Field, vertical, CDP data.

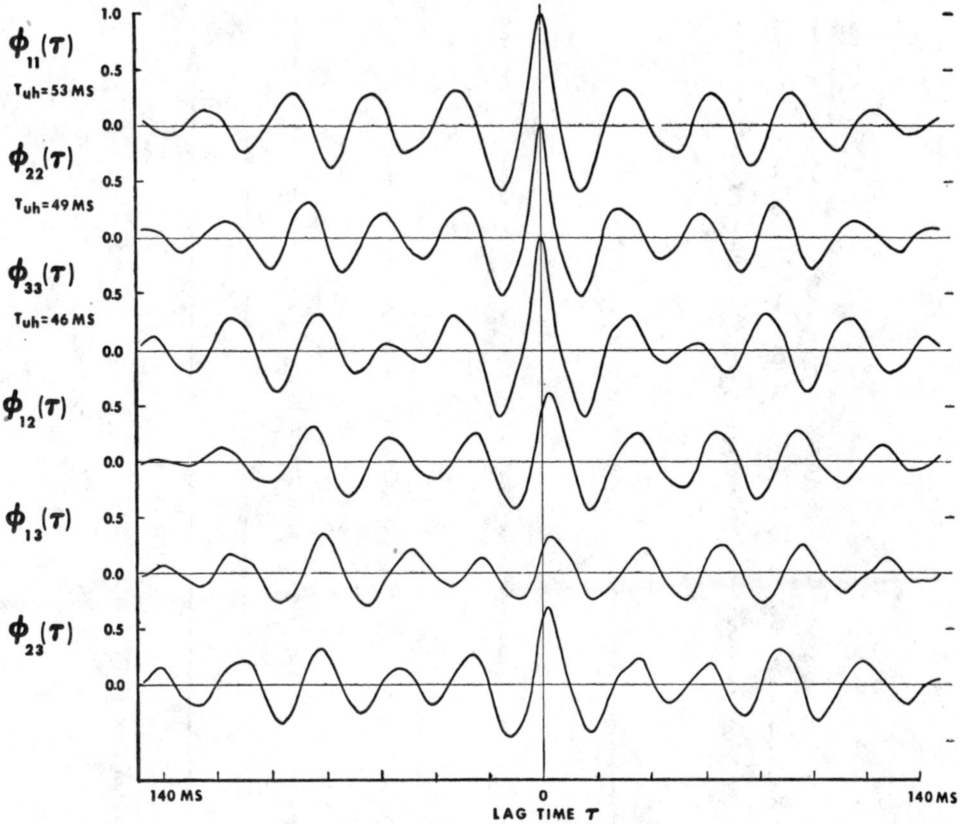

FIG. 5. Correlation functions: field, vertical, CDP data.

the following options are possible:

(1) stacking followed by single-channel deconvolution
(2) single-channel deconvolution followed by stacking
(3) multichannel deconvolution.

The question arises as to the equivalence of these processes. It is obvious that the deconvolve-stack process (2) is different from the other two since it does not use crosscorrelation terms in the input-correlation matrices. However, in the special case where there is essentially zero crosscorrelation between input traces, the multichannel input process and the deconvolve-stack process are essentially the same. The stack-deconvolve process (1) contains basically the same correlation information as the multichannel process (3), but the operators determined by the least-square process and their manner of application to the data differ in the two methods.

RESULTS

In this section, the multichannel deconvolution process is evaluated by a study of the results obtained using both synthetic and field data as input. The field data tested contained examples of horizontal and vertical CDP records plus marine, horizontal CDP control obtained both with dynamite and nondynamite sources.

1) Synthetic, vertical, common-depth-point data

The synthetic data in this case consist of a set of ghosted seismograms arranged so as to simulate an uphole survey. This simulation is shown as the first group of traces on Figure 2. Examples of the correlation functions for some of these input traces are given on Figure 3. Trace groups 2, 3, and 4 show the results of the three types of deconvolution processes which were discussed in the previous section. Group 5 shows a deconvolved ghost-free seismogram.

Each type of deconvolution filtering should

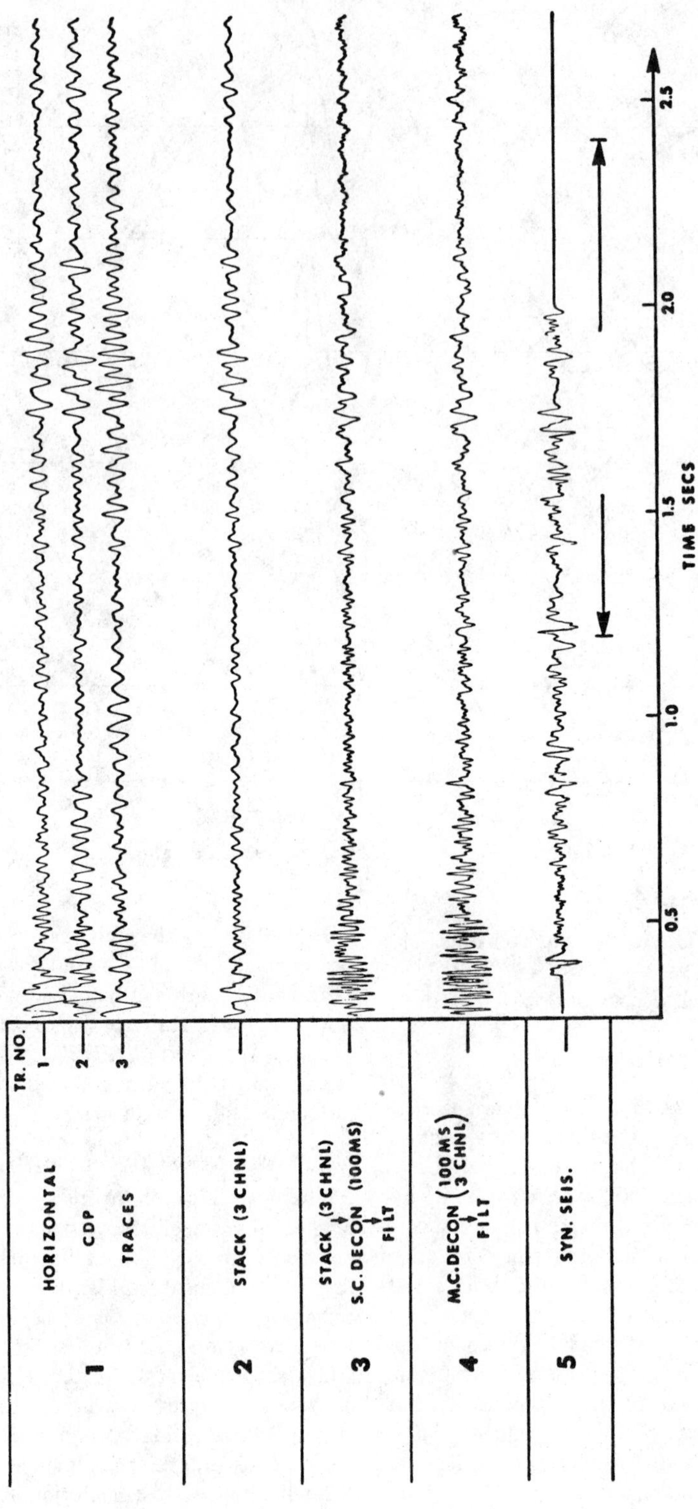

FIG. 6. Field, horizontal, CDP data.

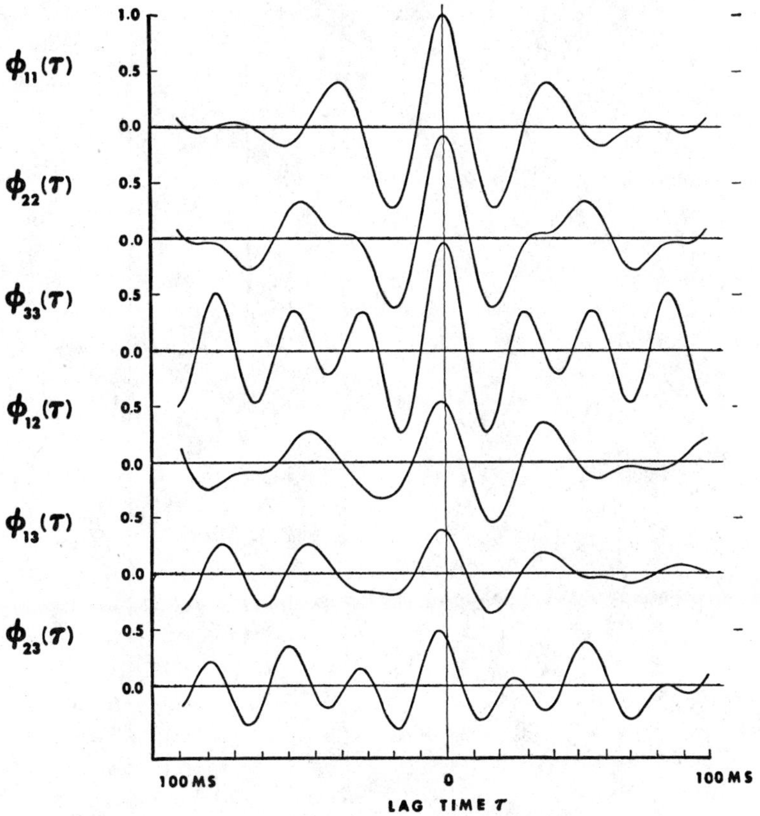

FIG. 7. Correlation functions: field, horizontal, CDP data.

tend to suppress the ghost event to some extent. In general, the results for all three processes are quite similar but small differences do exist. There is remarkable similarity between deconvolution-stack and stack-deconvolution for traces with the longer uphole times. This is mainly because the ghost delay is greater than the operator length, and hence, the crosscorrelation terms for the ghost event are not completely entered into the correlation matrices (see Figure 3). On the other hand, for traces with shorter uphole times, the correlation functions contain more ghost information, and the differences between the traces processed by the two sequences become more pronounced. The multichannel deconvolution process appears to make more efficient use of the ghost correlation information. The output traces are fairly similar regardless of the input pair used, and these traces show a fairly good fit to the deconvolved, ghost-free seismogram.

2) Field-recorded, vertical, common-depth-point data

The first group of traces on Figure 4 shows data from a field-recorded uphole survey. There is evidence that ghost events are present. Figure 5 shows typical correlation functions for these traces; it is obvious that the effect of ghosts on these functions is much less than for the synthetic example of Figure 3.

In this example, we have chosen to show the results of processes other than the stack-deconvolve or deconvolve-stack sequences used on the synthetic ghosted data. Trace groups 2, 3, and 4 of Figure 4 show the results.

The traces of Group 2 were produced by single-channel deconvolution using a fairly long operator. Ghost events are still visible on the output traces, and trace-to-trace similarity is poor. The multichannel deconvolution results (Group 3) show somewhat better ghost suppression and more evidence of pulse broadening than do the

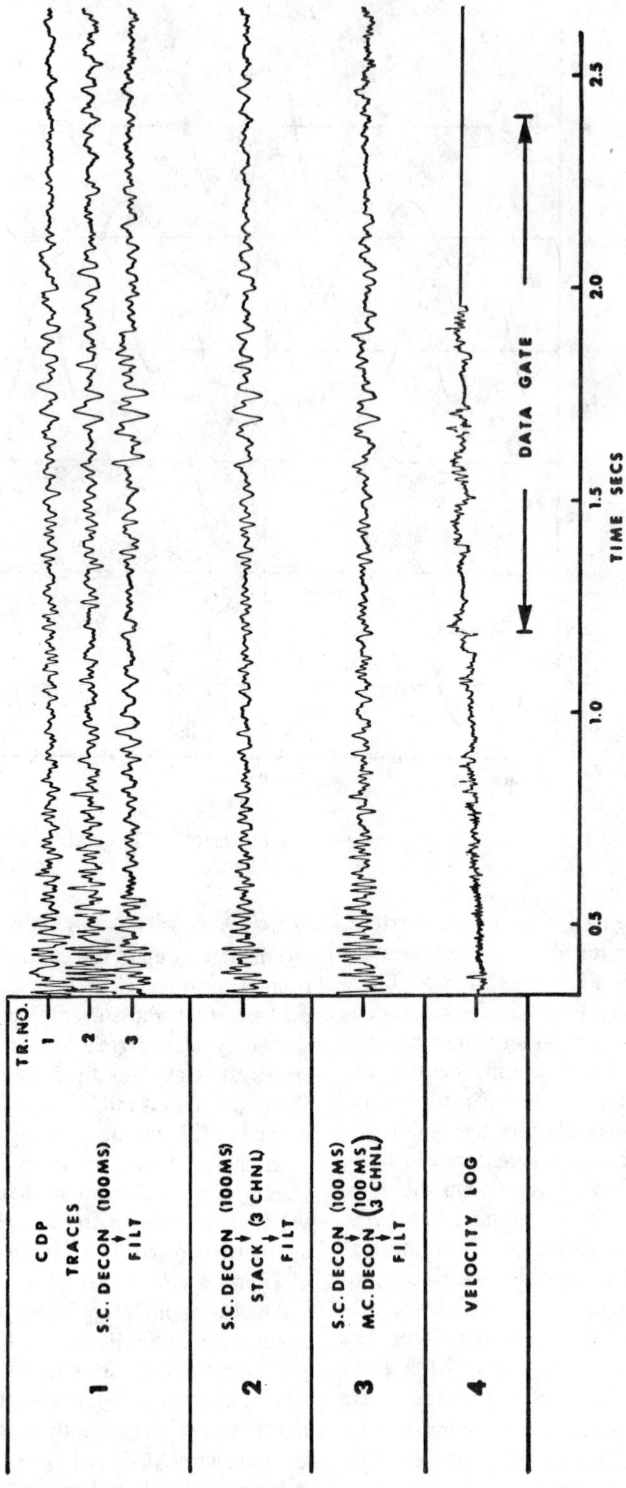

FIG. 8. Field, horizontal, CDP data.

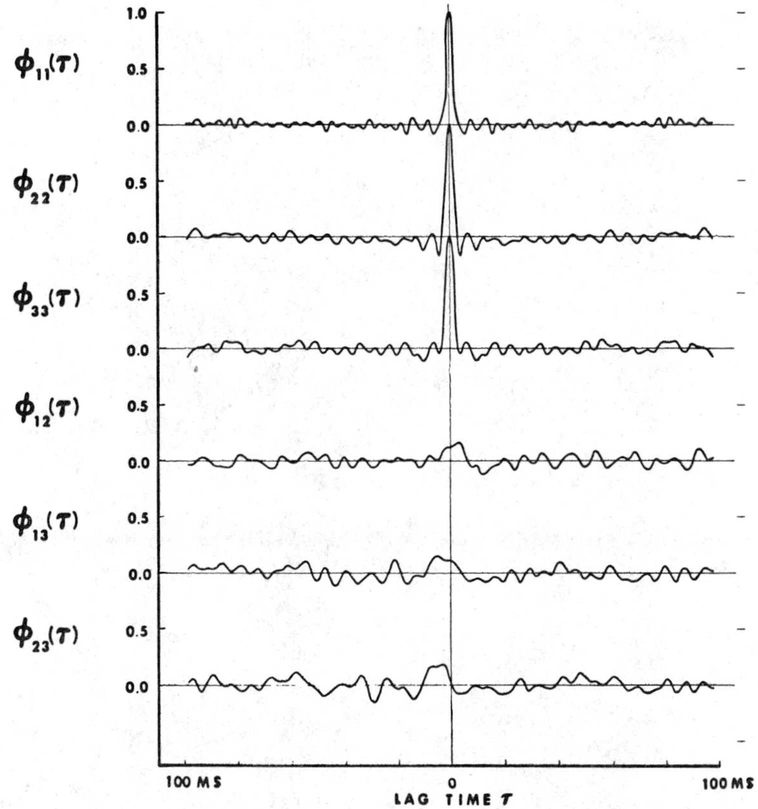

FIG. 9. Correlation functions: field, horizontal, CDP data (deconvolved).

single-channel deconvolution traces. However, trace-to-trace similarity is poor. The lack of distinctive ghost information in the correlation functions reduced the efficiency of the multichannel process in this case.

Probably the best ghost suppression has been obtained by the use of a two-channel, deghost filter. The subsequent application of a fairly long single-channel deconvolution filter has not achieved significant pulse shortening (group 4).

3) Field-recorded, horizontal, common-depth-point data

The first group of traces on Figure 6 shows a trace-collect of three horizontal common-depth-point traces obtained from a field survey. Single-shot records in the area indicate that record quality is fair, but the traces shown indicated little in the way of common reflection events. The correlation functions for these traces are given in Figure 7. The autocorrelation functions also reflect the differences between traces while the crosscorrelation functions appear to carry little useful information.

Traces 2, 3, and 4 of Figure 6 show the results of stack, stack-single-channel deconvolution, and multichannel deconvolution, respectively. In this case, there is little difference between the results produced by the two processes. Also shown in Figure 6 is a synthetic seismogram prepared from a velocity log located within a few hundred feet of the common-depth-point. Agreement between the processed traces and the synthetic seismogram is not good.

The first three traces of Figure 8 show a filtered version of the single-channel, deconvolved, CDP traces of Figure 6. The corresponding, unfiltered, CDP traces were then stacked and filtered, and the output trace is shown as trace 2. This trace thus results from the single-channel deconvolution-stack sequence and is comparable with traces 3 and 4 of Figure 6. These three traces are quite

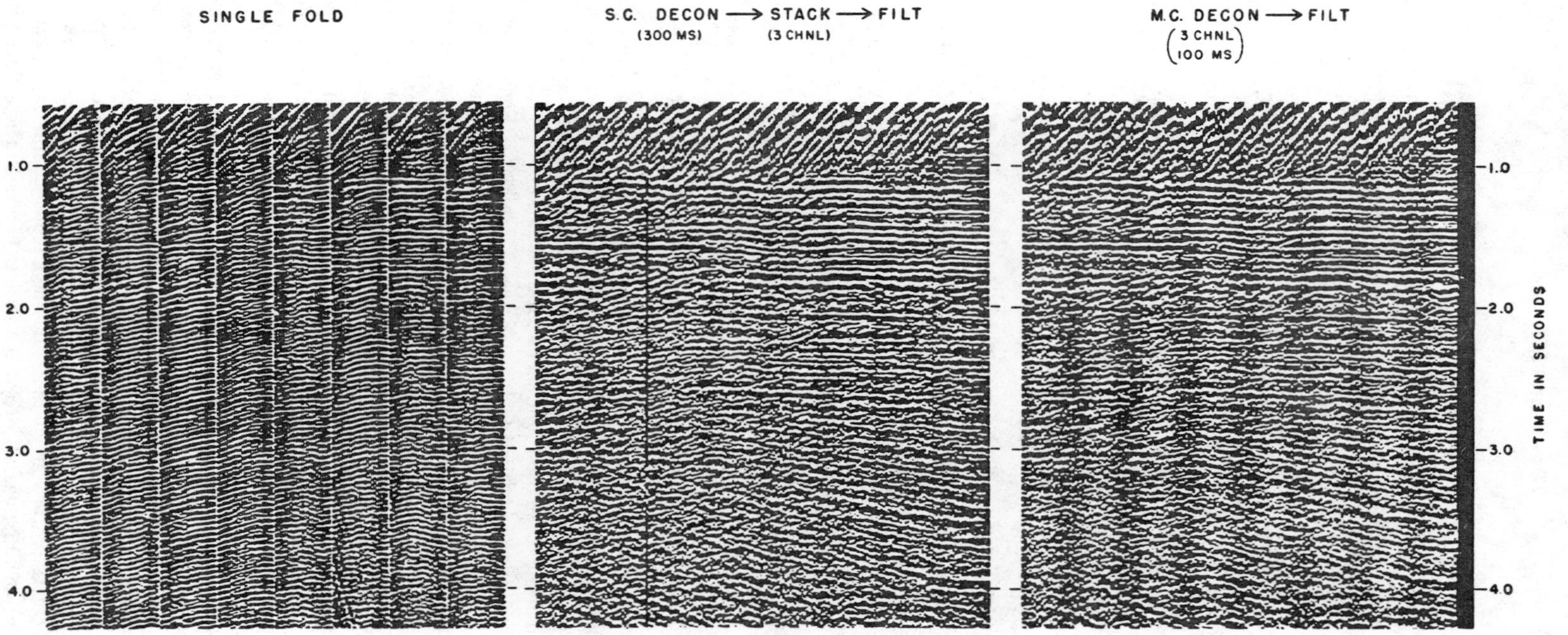

FIG. 10. Processed sections; marine profile.

similar and show that the three deconvolution processes yield essentially the same results on this set of field data.

As a final step, multichannel deconvolution was applied to the deconvolved, common-depth-point traces. The output trace from this procedure (trace 3, Figure 8) is not much different from the other output traces, but does appear to show more pulse broadening. This result demonstrates that single-channel deconvolution of the original CDP traces has not improved the information content of the crosscorrelation functions. This is also illustrated by the nature of the correlation functions of Figure 9.

4) *Marine profile*

The data presented so far has dealt only with isolated sets of common-depth-point traces. In these cases, the multichannel deconvolution process has shown only marginal improvement over other deconvolution sequences. Another method of evaluation is to compare processes on a profile of records. The marine profile of Figure 10 presents an example of this type.

It can be seen from the single-shot section that the records are moderately ringing and that deep reflections are obscured. Both the single-channel deconvolution-stack process and multichannel deconvolution are equally efficient in suppressing ringing and recovering deep events. However, the latter process seems to produce better lateral correlation of events and more effective pulse shortening.

CONCLUSIONS

The main conclusion that can be drawn from this study is that the application of the multichannel deconvolution process to field-recorded, common-depth-point data yields results which are not significantly better than those obtained by combinations of stacking and single-channel deconvolution methods.

Lack of improvement in the case of field data seems to be due to the lack of significant information in the crosscorrelation functions for common-depth-point traces.

ACKNOWLEDGMENT

The authors' wish to thank Dr. W. H. Guilinger and Dr. E. K. Darby for their valuable assistance in this study. We would also like to thank the Gulf Research & Development Company for permission to publish this report.

REFERENCES

Robinson, E. A., 1963, Mathematical development of discrete filters for the detection of nuclear explosions: J. Geophys. Res., v. 68, p. 5559–5567.

——— 1966, Multichannel *z*-transforms and minimum delay: Geophysics, v. 31, p. 482–500.

Schneider, W. A., Larner, K. L., Burg, J. P., Backus, M. M., 1964, A new data-processing technique for the elimination of ghost arrivals on reflection seismograms: Geophysics, v. 29, p. 783–805.

Schneider, W. A., Prince, E. R., Giles, B. F., 1965, A new data-processing technique for multiple attenuation exploiting differential normal moveout: Geophysics, v. 30, p. 348–362.

Wiggins, R. A., and Robinson, E. A., 1965, Recursive solution to the multichannel filtering problem: J. Geophys. Res., v. 70, p. 1885–1891.

Reprinted from Geophysics v. 35, no. 5, p. 785-811

PRINCIPLES OF DIGITAL MULTICHANNEL FILTERING†

SVEN TREITEL*

The transition from single-channel to multichannel data processing systems requires substantial modifications of the simpler single-channel model. While the response function of a single-channel digital filter can be specified in terms of scalar-valued weighting coefficients, the corresponding response function of a multichannel filter is more conveniently described by matrix-valued weighting coefficients. Correlation coefficients, which are scalars in the single-channel case, now become matrices.

Multichannel sampled data are manipulated with greater ease by recourse to multichannel z-transform theory. Exact inverse filters are calculable by a matrix inversion technique which is the counterpart to the computation of exact single-channel inverse operators by polynomial division. The delay properties of the original filter govern the stability of its inverse. This inverse is expressible in the form of a two-stage cascaded system, whose first stage is a single-channel recursive filter.

Optimum multichannel filtering systems result from a generalization of the single-channel least squares error criterion. The corresponding correlation matrices are now functions of coefficients which are themselves matrices. The system of normal matrix-valued equations that is obtained in this manner can be solved by means of Robinson's generalization of the Wiener-Levinson algorithm. Inverse multichannel filters are designed by specifying the desired output to be an identity matrix rather than a unit spike; if this matrix occurs at zero lag, the least squares filter is minimum-delay. Simple numerical examples serve to illustrate the design principles involved and to indicate the types of problems that can be attacked with multichannel least squares processors.

"The good 'geophysicist' should beware of mathematicians and all those who make empty prophecies. The danger already exists that mathematicians have made a covenant with the devil to darken the spirit and confine man in the bonds of Hell."

Paraphrased from St. Augustine

1.0 INTRODUCTION

Single-channel time series theory has by now found widespread application in various geophysical data processing systems. Both the fundamental concepts of this theory as well as innumerable examples of its implementation in seismic analysis have been extensively discussed in the recent literature. The field of multichannel time series has received comparatively less attention; indeed, few elementary treatments outlining the principles of this subject have been published. The present contribution represents an effort in this direction and, hopefully, provides the necessary background needed to follow more advanced material.

Our primary concern in applied work is with discrete processes; that is, the time series are sampled at some convenient uniform rate. Therefore, the derivations are carried out entirely in discrete terms. Such an approach not only yields results that are easily programmable for the digital computer, but it also avoids many of the mathematical complexities that we encounter in the attempt to discretize results previously obtained in continuous time.

A sampled seismic trace is an example of a discrete single-channel time series, while an assemblage of two or more such traces constitutes an example of a discrete multichannel time series. One sampled value per time increment is needed

† Received by the Editor June 11, 1970.

* Research Center, Pan American Petroleum Corp., Tulsa, Oklahoma 74102.

to describe a single channel, or simple time series, while k sampled values per time increment are required to describe a multichannel, or multiple time series consisting of k simple time series. The algebra of multichannel processes has much in common with the algebra of single-channel processes. However, whereas the latter involves operations with scalars, the former involves the more complicated operations with matrices. Thus the four basic scalar operations of addition, subtraction, multiplication, and division must be replaced by the four corresponding matrix operations; namely, matrix addition, matrix subtraction, matrix multiplication, and matrix inversion.

Multichannel filtering systems are generally designed to exploit the trace-to-trace redundancies that occur on the seismogram. In other words, seismic events ordinarily are identified if their presence is observed on a number of neighboring traces; hence, it makes a good deal of sense to develop filters which use information present on more than one input trace.

A large part of the treatment of this paper is concerned with the development of the basic mathematical tools, but much space has also been given to a discussion of the multichannel Wiener filter. Other linear multichannel processing systems have been proposed; the reader is referred to appropriate papers cited in the bibliography for further detail.

2.0 BASIC CONCEPTS

2.1 *The multichannel response function*

The theory of linear digital time-invariant filters relates an input x_t to a corresponding output y_t by means of the discrete convolution formula,

$$\begin{aligned} y_t &= f_0x_t + f_1x_{t-1} + \cdots + f_mx_{t-m} \\ &= \sum_{s=0}^{m} f_s x_{t-s}, \end{aligned} \tag{2.1}$$

where t and s are suitably chosen integer-valued time variables and where the filter f_t is described by its $(m+1)$-length response function,

$$f_t = (f_0, f_1, \cdots, f_m). \tag{2.2}$$

Let us assume that this series of filter weighting coefficients has been obtained by sampling the measured unit impulse response of some physical system. If the system contains one input channel and one output channel, the digital filter response function f_t is representable as the single-channel series of scalar-valued weighting coefficients of equation (2.2).

Consider now the case of a linear system that has two input and two output channels. If we excite this system with a unit impulse applied to input channel 1, we will in general be able to measure separate responses on both output channels 1 and 2. Similarly, if this system is excited with a unit impulse applied to input channel 2, separate responses will in general be measurable on both output channels. We are thus faced with four single-channel response functions, all of which may be different from each other. A simple example will help to clarify these concepts. Suppose that the four single-channel response functions have been determined and that they are each representable by two scalar-valued weighting coefficients. Then the overall response of the two-channel system can be displayed in the form of Table 2.1 below.

Table 2.1 Example of the response function of a two input and two output channel system

Output \ Input	Channel 1	Channel 2
Channel 1	(6, 1)	(0, 2)
Channel 2	(7, 1)	(1, 3)

Thus the response due to a unit impulse applied to input channel 1 is (6, 1) on output channel 1 and (7, 1) on output channel 2. Similarly, the response due to a unit impulse applied to input channel 2 is (0, 2) on channel 1 and (1, 3) on channel 2. Inspection of Table 2.1 suggests that this overall response could also be displayed in matrix form,

$$\begin{aligned} \mathbf{f}_t &= \begin{bmatrix} f_{11}(t) & f_{12}(t) \\ f_{21}(t) & f_{22}(t) \end{bmatrix} \\ &= \begin{bmatrix} (6, 1) & (0, 2) \\ (7, 1) & (1, 3) \end{bmatrix}, \end{aligned} \tag{2.3}$$

which is set in bold face to identify it as a matrix. However, each element of $\mathbf{f}_t$ consists of a series of sampled values rather than of a single sampled value, and hence care must be exercised in defining manipulations made with matrices of this

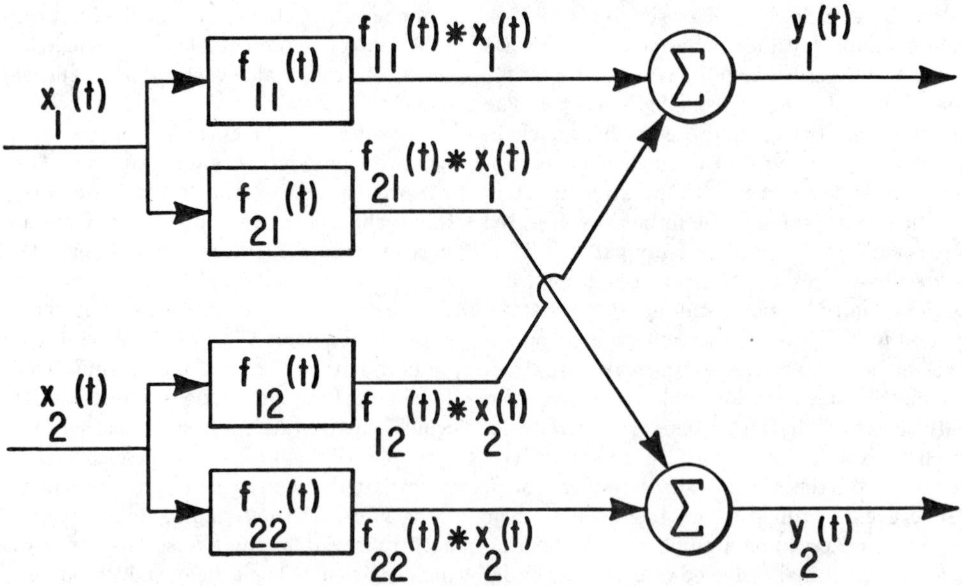

FIG. 2.1. The block diagram of a two input and two output linear multichannel system.

kind. The first subscript of each element identifies the output channel on which response is being measured, while the second subscript identifies the input channel that is being excited. The time variable t takes on the discrete values $t=0, 1, \cdots, m$. In the above case, $t=0$, 1 and all four elements are two-length single-channel series. For example, $f_{21}(t)=(7, 1)$ represents the response on output channel 2 due to a unit impulse excitation on input channel 1, and so on.

Let us next find the response of the system (2.3) to an arbitrary input. This problem is illustrated in Figure 2.1. The sampled data on input channel 1 is fed "in parallel" into the two single-channel filters $f_{11}(t)$ and $f_{21}(t)$. Similarly, the sampled data on input channel 2 is fed in parallel into the two single-channel filters $f_{12}(t)$ and $f_{22}(t)$. Figure 2.1 then indicates how these four intermediate outputs are combined to yield the two single-channel output series $y_1(t)$ and $y_2(t)$. We have

$$\begin{aligned} y_1(t) &= f_{11}(t) * x_1(t) + f_{12}(t) * x_2(t), \\ y_2(t) &= f_{21}(t) * x_1(t) + f_{22}(t) * x_2(t), \end{aligned} \tag{2.4}$$

where the symbol $*$ denotes convolution. Either output series is the result of summing filtered versions of both input series. In particular, the filters $f_{12}(t)$ and $f_{21}(t)$ allow for crossfeed between channels and enable the designer to combine input information on both channels in some desired manner. For example, if the two input channels are given by the two-length series,

$$x_1(t) = (2, -5),$$
$$x_2(t) = (4, 3),$$

and if the filter is given by equation (2.3), equation (2.4) yields

$$\begin{aligned} y_1(t) &= (6, 1) * (2, -5) + (0, 2) * (4, 3) \\ &= (12, -20, 1), \\ y_2(t) &= (7, 1) * (2, -5) + (1, 3) * (4, 3) \\ &= (18, -18, 4). \end{aligned}$$

It is not difficult to see that the notation of equation (2.4) will become cumbersome as the number of channels in the system increases. Relation (2.4) can be written

$$\begin{bmatrix} y_1(t) \\ y_2(t) \end{bmatrix} = \begin{bmatrix} f_{11}(t) & f_{12}(t) \\ f_{21}(t) & f_{22}(t) \end{bmatrix} * \begin{bmatrix} x_1(t) \\ x_2(t) \end{bmatrix}, \tag{2.5}$$

or simply

$$\mathbf{y}_t = \mathbf{f}_t * \mathbf{x}_t, \tag{2.6}$$

where $\mathbf{y}_t$ and $\mathbf{x}_t$ are column vectors representing the output and input channel data, respectively, and where $\mathbf{f}_t$ is the filter matrix of equation (2.3).

The elements of the above matrix equation are not scalars but single-channel series of sampled values; hence, corresponding elements of (2.5) are convolved rather than multiplied in order to recover (2.4). Each element of the column vector $\mathbf{x}_t$ represents the entire input data on a given channel. In order to make things simple, we suppose that the lengths of all input series are the same, say $(n+1)$, so that $\mathbf{x}_t$ is defined for $t=0, 1, \cdots, n$. If in actuality one input channel is described by fewer sampled values than another, the shorter series is augmented by a sufficient number of zeros such that both become of equal length $(n+1)$. Similarly, the elements of the filter matrix $\mathbf{f}_t$ are all made to be of length $(m+1)$; and as a result, the elements of the output vector $\mathbf{y}_t$ all become of length $(n+m+1)$. In this manner the definitions of series length carry over nicely from the single-channel case (Treitel and Robinson, 1964).

The matrix equation (2.5) can be interpreted in another way. Let us write $\mathbf{x}_t$ in the form

$$\mathbf{x}_t = \begin{bmatrix} x_1(t) \\ x_2(t) \end{bmatrix} = \left\{ \begin{bmatrix} x_1(0) \\ x_2(0) \end{bmatrix}, \begin{bmatrix} x_1(1) \\ x_2(1) \end{bmatrix}, \cdots, \begin{bmatrix} x_1(n) \\ x_2(n) \end{bmatrix} \right\}.$$

For every value of t, we obtain a two-rowed column vector of scalars, where each such vector constitutes a "snapshot" of the two input channels at a given time t. This method of multichannel data arrangement is known as "multiplexing." We say that $\mathbf{x}_t$ is a two-channel $(n+1)$-length input series, each of whose coefficients is a two-length column vector or, equivalently, a "two rows by one column" matrix. Similarly, we have

$$\mathbf{f}_t = \begin{bmatrix} f_{11}(t) & f_{12}(t) \\ f_{21}(t) & f_{22}(t) \end{bmatrix} = \left\{ \begin{bmatrix} f_{11}(0) & f_{12}(0) \\ f_{21}(0) & f_{22}(0) \end{bmatrix}, \begin{bmatrix} f_{11}(1) & f_{12}(1) \\ f_{21}(1) & f_{22}(1) \end{bmatrix}, \cdots, \begin{bmatrix} f_{11}(m) & f_{12}(m) \\ f_{21}(m) & f_{22}(m) \end{bmatrix} \right\}$$

and

$$\mathbf{y}_t = \begin{bmatrix} y_1(t) \\ y_2(t) \end{bmatrix} = \left\{ \begin{bmatrix} y_1(0) \\ y_2(0) \end{bmatrix}, \begin{bmatrix} y_1(1) \\ y_2(1) \end{bmatrix}, \cdots, \begin{bmatrix} y_1(m+n) \\ y_2(m+n) \end{bmatrix} \right\}.$$

In this way, the input, filter response, and output of our system are all expressible in multiplexed form; that is, they can be written in the form of a multichannel time series, *whose terms are matrix-valued rather than scalar-valued.* In fact, we may write

$$\begin{aligned} \mathbf{x}_t &= (\mathbf{x}_0, \mathbf{x}_1, \cdots, \mathbf{x}_j, \cdots, \mathbf{x}_n), \\ \mathbf{f}_t &= (\mathbf{f}_0, \mathbf{f}_1, \cdots, \mathbf{f}_j, \cdots, \mathbf{f}_m), \\ \mathbf{y}_t &= (\mathbf{y}_0, \mathbf{y}_1, \cdots, \mathbf{y}_j, \cdots, \mathbf{y}_{m+n}), \end{aligned} \tag{2.7}$$

where the matrix-valued coefficients are given by

$$\mathbf{x}_j = \begin{bmatrix} x_1(j) \\ x_2(j) \end{bmatrix}, \quad \mathbf{f}_j = \begin{bmatrix} f_{11}(j) & f_{12}(j) \\ f_{21}(j) & f_{22}(j) \end{bmatrix}, \quad \text{and} \quad \mathbf{y}_j = \begin{bmatrix} y_1(j) \\ y_2(j) \end{bmatrix}. \tag{2.8}$$

Substitution of (2.7) in (2.6) yields

$$(\mathbf{y}_0, \mathbf{y}_1, \cdots, \mathbf{y}_{m+n}) = (\mathbf{f}_0, \mathbf{f}_1, \cdots, \mathbf{f}_m) * (\mathbf{x}_0, \mathbf{x}_1, \cdots, \mathbf{x}_n).$$

Just as in the familiar single-channel case, this relation can also be put into the form,

$$\mathbf{y}_t = \mathbf{f}_0\mathbf{x}_t + \mathbf{f}_1\mathbf{x}_{t-1} + \cdots + \mathbf{f}_m\mathbf{x}_{t-m} \tag{2.9}$$

or

$$\mathbf{y}_t = \sum_{s=0}^{m} \mathbf{f}_s\mathbf{x}_{t-s}, \quad t=0, 1, \cdots, m+n. \tag{2.10}$$

We now observe that we have been able to express the multichannel convolution formula (2.10) almost as simply as the single-channel convolution formula (2.1). The important difference is, of course, that the coefficients of (2.10) are matrix-valued, while the coefficients of (2.1) are scalar-valued.

Our derivation thus far has been carried through for the case of two input and two output channels. However, the generalization of these results to systems with k input channels and l output channels can be accomplished by reasoning inductively from the above development. All we need do is to make appropriate changes in the dimensions of the matrices (2.8), and (2.10) will still hold. In the case of k input channels, the matrix-valued coefficient $\mathbf{x}_j$ is given by the $(k)\times(1)$ matrix, i.e. the column vector

$$\mathbf{x}_j = \begin{bmatrix} x_1(j) \\ x_2(j) \\ \vdots \\ x_k(j) \end{bmatrix}. \qquad (2.11)$$

If the system is to have l output channels, then $\mathbf{y}_j$ is given by the $(l)\times(1)$ matrix

$$\mathbf{y}_j = \begin{bmatrix} y_1(j) \\ y_2(j) \\ \vdots \\ y_l(j) \end{bmatrix}. \qquad (2.12)$$

We next note that the right member of equation (2.10) is a sum of matrix products of the form $\mathbf{f}_j\mathbf{x}_{t-j}$. Hence, it is necessary that the number of columns of the matrix $\mathbf{f}_j$ be equal to the number of rows of the vector $\mathbf{x}_j$, namely k. Since the vector $\mathbf{y}_j$ has l rows, the filter matrix coefficient $\mathbf{f}_j$ must have l rows also. Thus $\mathbf{f}_j$ is the $(l)\times(k)$ matrix

$$\mathbf{f}_j = \begin{bmatrix} f_{11}(j) & f_{12}(j) & \cdots & f_{1k}(j) \\ f_{21}(j) & & & \\ \vdots & & \ddots & \vdots \\ f_{l1}(j) & & \cdots & f_{lk}(j) \end{bmatrix} \qquad (2.13)$$

and the dimensions of the matrix equation (2.10) can be indicated explicitly by writing

$$\underset{(l)\times(1)}{\mathbf{y}_t} = \sum_{s=0}^{m} \underset{(l)\times(k)}{\mathbf{f}_s} \; \underset{(k)\times(1)}{\mathbf{x}_{t-s}} . \qquad (2.10)$$

In terms of the notation of equation (2.4), the k input and l output channel case becomes

$$y_1(t) = f_{11}(t) * x_1(t) + \cdots + f_{1k}(t) * x_k(t)$$
$$\vdots \qquad \cdots \qquad \cdots$$
$$y_l(t) = f_{l1}(t) * x_1(t) + \cdots + f_{lk}(t) * x_k(t)$$

or, using matrices, we have

$$\begin{bmatrix} y_1(t) \\ y_2(t) \\ \vdots \\ y_l(t) \end{bmatrix} = \begin{bmatrix} f_{11}(t) & f_{12}(t) & \cdots & f_{1k}(t) \\ f_{21}(t) & & & \\ \vdots & & \ddots & \vdots \\ f_{l1}(t) & f_{l2}(t) & \cdots & f_{lk}(t) \end{bmatrix} * \begin{bmatrix} x_1(t) \\ x_2(t) \\ \vdots \\ x_k(t) \end{bmatrix}, \qquad (2.14)$$

whose elements are all single-channel series. The above relation reduces to equation (2.5) when $k=l=2$. The multichannel convolution formula is thus either expressible in the multiplexed form of equation (2.10), whose matrix-valued coefficients are given by equations (2.11) through (2.13), or in the form of equation (2.14), where each element is itself a single-channel series.

2.2 *The multichannel z-transform*

The study of single-channel systems is much simplified through the introduction of z-transform theory (Robinson and Treitel, 1964). Here we shall extend these concepts to the multichannel situation. Consider the response function of equation (2.3). The elements of $\mathbf{f}_t$ are single-channel series, whose z-transforms can be written down by inspection. We obtain

$$\mathbf{F}(z) = \begin{bmatrix} F_{11}(z) & F_{12}(z) \\ F_{21}(z) & F_{22}(z) \end{bmatrix} = \begin{bmatrix} (6+z) & (0+2z) \\ (7+z) & (1+3z) \end{bmatrix}, \qquad (2.15)$$

where $F_{11}(z)$ is the z-transform of the two-length series $f_{11}(t)=(6, 1)$, where $F_{12}(z)$ is the z-transform of the two-length series $f_{12}(t)=(0, 2)$, and so on. The overall z-transform of the response function of this filter is thus expressible in terms of the $(2)\times(2)$ matrix $\mathbf{F}(z)$, whose elements are single-channel z-transforms and which here are ordinary polynomials in z. A matrix of this type is called a polynomial matrix or lambda (λ) matrix.

The above polynomial matrix may also be written in the form of a polynomial with matrix-valued coefficients,

$$\mathbf{F}(z) = \mathbf{f}_0 + \mathbf{f}_1 z = \begin{bmatrix} 6 & 0 \\ 7 & 1 \end{bmatrix} + \begin{bmatrix} 1 & 2 \\ 1 & 3 \end{bmatrix} z, \qquad (2.16)$$

where the coefficients $\mathbf{f}_0$ and $\mathbf{f}_1$ are $(2)\times(2)$ matrices with scalar elements. A polynomial of this type is called a matrix polynomial. Hence, the overall z-transform of our multichannel response function is expressible either as a polynomial matrix (equation 2.15) or as a matrix polynomial (equation 2.16). The polynomial matrix (2.15) constitutes a "channel by channel" representation of this z-transform, while the matrix polynomial (2.16) constitutes its multiplexed representation.

Let us next take the z-transforms of both sides of the input/output relations (2.4),

$$\begin{aligned} Y_1(z) &= F_{11}(z)X_1(z) + F_{12}(z)X_2(z), \\ Y_2(z) &= F_{21}(z)X_1(z) + F_{22}(z)X_2(z), \end{aligned} \tag{2.17}$$

where $X_1(z)$, $X_2(z)$, $Y_1(z)$, and $Y_2(z)$ are the single-channel z-transforms of the input and output channel series, respectively. We recall that convolution of time functions is now replaced by multiplication of corresponding z-transforms. Equation (2.17) can be written

$$\begin{bmatrix} Y_1(z) \\ Y_2(z) \end{bmatrix} = \begin{bmatrix} F_{11}(z) & F_{12}(z) \\ F_{21}(z) & F_{22}(z) \end{bmatrix} \begin{bmatrix} X_1(z) \\ X_2(z) \end{bmatrix} \tag{2.18}$$

or simply

$$\mathbf{Y}(z) = \mathbf{F}(z)\mathbf{X}(z), \tag{2.19}$$

where $\mathbf{X}(z)$ and $\mathbf{Y}(z)$ are column vectors. The reader should compare equations (2.18) and (2.19) with equations (2.5) and (2.6).

Returning to our previous numerical example [equation (2.4) et seq.], we write down the z-transform of the input series, which is

$$\mathbf{X}(z) = \begin{bmatrix} 2-5z \\ 4+3z \end{bmatrix} = \begin{bmatrix} 2 \\ 4 \end{bmatrix} + \begin{bmatrix} -5 \\ 3 \end{bmatrix} z.$$

The second member above is the polynomial matrix representation of $\mathbf{X}(z)$, while the third is the matrix polynomial representation of $\mathbf{X}(z)$. The output $\mathbf{Y}(z)$ can be given as a product of polynomial matrices,

$$\mathbf{Y}(z) = \begin{bmatrix} 6+z & 0+2z \\ 7+z & 1+3z \end{bmatrix} \begin{bmatrix} 2-5z \\ 4+3z \end{bmatrix}$$

$$= \begin{bmatrix} (6+z)(2-5z)+2z(4+3z) \\ (7+z)(2-5z)+(1+3z)(4+3z) \end{bmatrix}$$

$$= \begin{bmatrix} 12-20z+z^2 \\ 18-18z+4z^2 \end{bmatrix} = \begin{bmatrix} Y_1(z) \\ Y_2(z) \end{bmatrix}. \tag{2.20}$$

This output can also be written as a product of matrix polynomials,

$$\begin{aligned} \mathbf{Y}(z) &= \left\{ \begin{bmatrix} 6 & 0 \\ 7 & 1 \end{bmatrix} + \begin{bmatrix} 1 & 2 \\ 1 & 3 \end{bmatrix} z \right\} \left\{ \begin{bmatrix} 2 \\ 4 \end{bmatrix} + \begin{bmatrix} -5 \\ 3 \end{bmatrix} z \right\} \\ &= \begin{bmatrix} 6 & 0 \\ 7 & 1 \end{bmatrix}\begin{bmatrix} 2 \\ 4 \end{bmatrix} + \begin{bmatrix} 1 & 2 \\ 1 & 3 \end{bmatrix}\begin{bmatrix} 2 \\ 4 \end{bmatrix} z \\ &\quad + \begin{bmatrix} 6 & 0 \\ 7 & 1 \end{bmatrix}\begin{bmatrix} -5 \\ 3 \end{bmatrix} z \\ &\quad + \begin{bmatrix} 1 & 2 \\ 1 & 3 \end{bmatrix}\begin{bmatrix} -5 \\ 3 \end{bmatrix} z^2 \\ &= \begin{bmatrix} 12 \\ 18 \end{bmatrix} + \begin{bmatrix} -20 \\ -18 \end{bmatrix} z + \begin{bmatrix} 1 \\ 4 \end{bmatrix} z^2 \\ &= \mathbf{y}_0 + \mathbf{y}_1 z + \mathbf{y}_2 z^2. \end{aligned} \tag{2.21}$$

The equivalent results (2.20) and (2.21) should be compared with the output obtained for this example by convolution (equation [2.4] et seq.).

We observe that the multiplication of matrix polynomials proceeds like the multiplication of scalar polynomials, but with the important difference that the coefficients to be multiplied are matrix-valued rather than scalar-valued, so that the rules of matrix multiplication now hold. In particular, this means that the matrices in any product of the form **AB** must be conformable; that is, the number of columns of **A** must be equal to the number of rows of **B**.

The generalization of these concepts to systems with k input channels and l output channels is again accomplished inductively. Thus we may take the z-transform of both sides of equation (2.10),

$$\mathbf{Y}(z) = \mathbf{F}(z)\mathbf{X}(z), \tag{2.22}$$

where

$$\mathbf{X}(z) = \mathbf{x}_0 + \mathbf{x}_1 z + \cdots + \mathbf{x}_j z^j + \cdots$$

$$+ \mathbf{x}_n z^n = \sum_{j=0}^{n} \mathbf{x}_j z^j,$$

$$\mathbf{F}(z) = \mathbf{f}_0 + \mathbf{f}_1 z + \cdots + \mathbf{f}_j z^j + \cdots$$

$$+ \mathbf{f}_m z^m = \sum_{j=0}^{m} \mathbf{f}_j z^j, \qquad (2.23)$$

$$\mathbf{Y}(z) = \mathbf{y}_0 + \mathbf{y}_1 z + \cdots + \mathbf{y}_j z^j + \cdots$$

$$+ \mathbf{y}_{m+n} z^{m+n} = \sum_{j=0}^{m+n} \mathbf{y}_j z^j.$$

These z-transforms have been written as matrix polynomials, whose coefficients are given by the matrices $\mathbf{x}_j$, $\mathbf{y}_j$, and $\mathbf{f}_j$ of equations (2.11), (2.12), and (2.13), respectively.

On the other hand, the polynomial matrix representation of (2.22) is obtained by taking the z-transform of both sides of (2.14),

$$\begin{bmatrix} Y_1(z) \\ Y_2(z) \\ \vdots \\ Y_l(z) \end{bmatrix} = \begin{bmatrix} F_{11}(z) & F_{12}(z) & \cdots & F_{1k}(z) \\ F_{21}(z) & & & \vdots \\ \vdots & & \cdot & \vdots \\ F_{l1}(z) & F_{l2}(z) & \cdots & F_{lk}(z) \end{bmatrix} \begin{bmatrix} X_1(z) \\ X_2(z) \\ \vdots \\ X_k(z) \end{bmatrix}. \qquad (2.24)$$

We have indicated here only a few of the properties of polynomial matrices and matrix polynomials, and others will be presented as needed in the sequel. The reader who wishes to pursue these points further is referred to the excellent treatment by Gantmacher (1959, Vol. 1, Chapters 4 and 6).

2.3 *Multichannel correlation functions*

Let us now deal with the subject of multichannel correlation functions, which we shall introduce by means of an example. Suppose that we are given the two single-channel series $x_1(t)$ and $x_2(t)$. From this data we can calculate the single-channel correlation functions

$$\begin{aligned} \phi_{x_1x_1}(\tau) &= E\{x_1(t+\tau)x_1(t)\}, \\ \phi_{x_1x_2}(\tau) &= E\{x_1(t+\tau)x_2(t)\}, \\ \phi_{x_2x_1}(\tau) &= E\{x_2(t+\tau)x_1(t)\}, \\ \phi_{x_2x_2}(\tau) &= E\{x_2(t+\tau)x_2(t)\}, \end{aligned} \qquad (2.25)$$

where E is the expected value symbol, and where τ is the discrete lag variable. In the present two channel case, we are thus dealing with four single-channel correlation functions, where

$$\begin{aligned} \phi_{x_1x_1}(\tau) &= \text{autocorrelation of } x_1(t), \\ \phi_{x_1x_2}(\tau) &= \text{crosscorrelation of } x_1(t) \text{ with } x_2(t), \\ \phi_{x_2x_1}(\tau) &= \text{crosscorrelation of } x_2(t) \text{ with } x_1(t), \\ \phi_{x_2x_2}(\tau) &= \text{autocorrelation of } x_2(t). \end{aligned}$$

These correlation functions can be written as the elements of the $(2)\times(2)$ matrix

$$\begin{bmatrix} E\{x_1(t+\tau)x_1(t)\} & E\{x_1(t+\tau)x_2(t)\} \\ E\{x_2(t+\tau)x_1(t)\} & E\{x_2(t+\tau)x_2(t)\} \end{bmatrix}.$$

But if $\mathbf{x}(t)$ is the column vector

$$\mathbf{x}(t) = \begin{bmatrix} x_1(t) \\ x_2(t) \end{bmatrix},$$

its transpose $\mathbf{x}^T(t)$ is the row vector[1]

$$\mathbf{x}^T(t) = [x_1(t) \quad x_2(t)];$$

and we may write

$$\begin{aligned} \begin{bmatrix} E\{x_1(t+\tau)x_1(t)\} & E\{x_1(t+\tau)x_2(t)\} \\ E\{x_2(t+\tau)x_1(t)\} & E\{x_2(t+\tau)x_2(t)\} \end{bmatrix} &= E\left\{\begin{bmatrix} x_1(t+\tau) \\ x_2(t+\tau) \end{bmatrix} [x_1(t) \quad x_2(t)]\right\} \\ &= \mathbf{E}\{\mathbf{x}(t+\tau)\mathbf{x}^T(t)\}. \end{aligned}$$

Making use of the notation introduced in (2.25), we obtain

$$E\{\mathbf{x}(t+\tau)\mathbf{x}^T(t)\} = \begin{bmatrix} \phi_{x_1x_1}(\tau) & \phi_{x_1x_2}(\tau) \\ \phi_{x_2x_1}(\tau) & \phi_{x_2x_2}(\tau) \end{bmatrix} = \boldsymbol{\phi}_{xx}(\tau).$$

The $(2)\times(2)$ matrix $\boldsymbol{\phi}_{xx}(\tau)$ is called the autocorrelation matrix of the process $\mathbf{x}(t)$. This function is scalar-valued in the single-channel case, but is matrix-valued in the multichannel case. We note further that the autocorrelation of the multichannel process $\mathbf{x}(t)$ is not only a function of the autocorrelations of its single-channel components

[1] The transpose $\mathbf{A}^T$ of a matrix $\mathbf{A}$ is the matrix whose row(s) are identical to the column(s) of $\mathbf{A}$. If $\mathbf{A}$ is $(l)\times(k)$, then $\mathbf{A}^T$ is $(k)\times(l)$.

$x_1(t)$ and $x_2(t)$ but is also a function of the crosscorrelations between these two components.

Next let $\mathbf{y}(t)$ be the column vector,

$$\mathbf{y}(t) = \begin{bmatrix} y_1(t) \\ y_2(t) \end{bmatrix}.$$

Then the crosscorrelation between $\mathbf{x}(t)$ and $\mathbf{y}(t)$ is

$$E\{\mathbf{x}(t+\tau)\mathbf{y}^T(t)\} = \begin{bmatrix} \phi_{x_1y_1}(\tau) & \phi_{x_1y_2}(\tau) \\ \phi_{x_2y_1}(\tau) & \phi_{x_2y_2}(\tau) \end{bmatrix} = \boldsymbol{\phi}_{xy}(\tau).$$

The $(2)\times(2)$ matrix $\boldsymbol{\phi}_{xy}(\tau)$ is called the crosscorrelation of $\mathbf{x}(t)$ with $\mathbf{y}(t)$. Here $\phi_{x_1y_1}(\tau)$ is the single-channel crosscorrelation between $x_1(t)$ and $y_1(t)$, $\phi_{x_1y_2}(\tau)$ is the single-channel crosscorrelation between $x_1(t)$ and $y_2(t)$, and so on. Similarly, we have

$$E\{\mathbf{y}(t+\tau)\mathbf{x}^T(t)\} = \begin{bmatrix} \phi_{y_1x_1}(\tau) & \phi_{y_1x_2}(\tau) \\ \phi_{y_2x_1}(\tau) & \phi_{y_2x_2}(\tau) \end{bmatrix} = \boldsymbol{\phi}_{yx}(\tau),$$

where the $(2)\times(2)$ matrix $\boldsymbol{\phi}_{yx}(\tau)$ is called the crosscorrelation of $\mathbf{y}(t)$ with $\mathbf{x}(t)$.

The generalization of these concepts to a k-channel process $\mathbf{x}(t)$ and an l-channel process $\mathbf{y}(t)$ is straightforward. In fact, if

$$\mathbf{x}(t) = \begin{bmatrix} x_1(t) \\ x_2(t) \\ \cdot \\ \cdot \\ x_k(t) \end{bmatrix} \quad \text{and} \quad \mathbf{y}(t) = \begin{bmatrix} y_1(t) \\ y_2(t) \\ \cdot \\ \cdot \\ y_l(t) \end{bmatrix},$$

we obtain the formulae

$$E\{\mathbf{x}(t+\tau)\mathbf{y}^T(t)\} = \boldsymbol{\phi}_{xy}(\tau), \quad (2.26a)$$

$$E\{\mathbf{y}(t+\tau)\mathbf{x}^T(t)\} = \boldsymbol{\phi}_{yx}(\tau), \quad (2.26b)$$

$$E\{\mathbf{x}(t+\tau)\mathbf{x}^T(t)\} = \boldsymbol{\phi}_{xx}(\tau). \quad (2.26c)$$

Here $\boldsymbol{\phi}_{xy}(\tau)$ is the $(k)\times(l)$ rectangular matrix

$$\boldsymbol{\phi}_{xy}(\tau) = \begin{bmatrix} \phi_{x_1y_1}(\tau) & \phi_{x_1y_2}(\tau) & \cdots & \phi_{x_1y_l}(\tau) \\ \cdot & \cdot & \cdots & \cdot \\ \cdot & \cdot & & \cdot \\ \phi_{x_ky_1}(\tau) & \phi_{x_ky_2}(\tau) & \cdots & \phi_{x_ky_l}(\tau) \end{bmatrix}, \quad (2.27a)$$

$\boldsymbol{\phi}_{yx}(\tau)$ is the $(l)\times(k)$ rectangular matrix

$$\boldsymbol{\phi}_{yx}(\tau) = \begin{bmatrix} \phi_{y_1x_1}(\tau) & \phi_{y_1x_2}(\tau) & \cdots & \phi_{y_1x_k}(\tau) \\ \cdot & \cdot & \cdots & \cdot \\ \cdot & \cdot & & \cdot \\ \phi_{y_lx_1}(\tau) & \phi_{y_lx_2}(\tau) & & \phi_{y_lx_k}(\tau) \end{bmatrix}, \quad (2.27b)$$

and $\boldsymbol{\phi}_{xx}(\tau)$ is the $(k)\times(k)$ square matrix

$$\boldsymbol{\phi}_{xx}(\tau) = \begin{bmatrix} \phi_{x_1x_1}(\tau) & \phi_{x_1x_2}(\tau) & \cdots & \phi_{x_1x_k}(\tau) \\ \cdot & \cdot & \cdots & \cdot \\ \cdot & \cdot & & \cdot \\ \phi_{x_kx_1}(\tau) & \phi_{x_kx_2}(\tau) & \cdots & \phi_{x_kx_k}(\tau) \end{bmatrix}. \quad (2.27c)$$

We next derive some useful properties of the multichannel correlation functions given by (2.26). Transposing both sides of (2.26a), we obtain

$$E\{\mathbf{y}(t)\mathbf{x}^T(t+\tau)\} = \boldsymbol{\phi}^T_{xy}(\tau),$$

where we recall that for two conformable matrices $\mathbf{A}$ and $\mathbf{B}$, one has $(\mathbf{AB})^T = \mathbf{B}^T\mathbf{A}^T$ and that for any matrix $\mathbf{A}$, the relation $(\mathbf{A}^T)^T = \mathbf{A}$ always holds. Under the assumption that the multichannel processes are stationary, we can write

$$E\{\mathbf{y}(t-\tau)\mathbf{x}^T(t)\} = \boldsymbol{\phi}^T_{xy}(\tau).$$

Replacing τ by $-\tau$ on both sides of this equation, we have

$$E\{\mathbf{y}(t+\tau)\mathbf{x}^T(t)\} = \boldsymbol{\phi}^T_{xy}(-\tau).$$

If we combine this result with (2.26b), we obtain a relation between the two crosscorrelation functions $\boldsymbol{\phi}_{xy}(\tau)$ and $\boldsymbol{\phi}_{yx}(\tau)$, which is

$$\boldsymbol{\phi}_{yx}(\tau) = \boldsymbol{\phi}^T_{xy}(-\tau). \quad (2.28a)$$

Similarly, we have

$$\boldsymbol{\phi}_{xy}(\tau) = \boldsymbol{\phi}^T_{yx}(-\tau), \quad (2.28b)$$

and, in particular,

$$\boldsymbol{\phi}_{xx}(\tau) = \boldsymbol{\phi}^T_{xx}(-\tau). \quad (2.28c)$$

In the single-channel case, these expressions reduce to the familiar relations,

$$\phi_{yx}(\tau) = \phi_{xy}(-\tau),$$

$$\phi_{xy}(\tau) = \phi_{yx}(-\tau),$$

$$\phi_{xx}(\tau) = \phi_{xx}(-\tau).$$

It is also quite useful to consider multichannel correlation functions with the aid of the z-transform. Appendix A demonstrates that for the single-channel case we have

$$\Phi_{xy}(z) = X(z)Y(z^{-1}), \qquad (2.29)$$

where $\Phi_{xy}(z)$ is the z-transform of the crosscorrelation $\phi_{xy}(\tau) = E\{x(t+\tau)y(t)\}$. Here

$$X(z) = x_0 + x_1 z + \cdots + x_n z^n$$

is the z-transform of the $(n+1)$-length series $x(t)$, and

$$Y(z) = y_0 + y_1 z + \cdots + y_m z^m$$

is the z-transform of the $(m+1)$-length series $y(t)$. Hence, $Y(z^{-1})$ is the z-transform of the $(m+1)$-length series $y(-t)$,

$$Y(z^{-1}) = y_0 + y_1 z^{-1} + \cdots + y_m z^{-m},$$

where $y(-t)$ is the series $y(t)$ reflected about the origin $t=0$. We conclude that $\Phi_{xy}(z)$ must be of length $(m+n+1)$.

Let us generalize these concepts to the multichannel case. We take the z-transform of both sides of (2.27a), and notice that each element $\phi_{x_i y_j}(\tau)$ has a single-channel z-transform of the type (2.29). We obtain,

$$\mathbf{\Phi}_{xy}(z) = \begin{bmatrix} \Phi_{x_1y_1}(z) & \Phi_{x_1y_2}(z) & \cdots & \Phi_{x_1y_l}(z) \\ \vdots & \vdots & \cdots & \vdots \\ \Phi_{x_ky_1}(z) & \Phi_{x_ky_2}(z) & \cdots & \Phi_{x_ky_l}(z) \end{bmatrix},$$

where $\mathbf{\Phi}_{xy}(z)$ is the z-transform of $\phi_{xy}(\tau)$. Now (2.29) can be written in the form

$$\Phi_{x_iy_j}(z) = X_i(z)Y_j(z^{-1}),$$

where $X_i(z)$ and $Y_j(z)$ are the z-transforms of the ith channel of $\mathbf{x}(t)$ and the jth channel of $\mathbf{y}(t)$, respectively. Therefore,

$$\mathbf{\Phi}_{xy}(z) = \begin{bmatrix} X_1(z)Y_1(z^{-1}) & X_1(z)Y_2(z^{-1}) & \cdots & X_1(z)Y_l(z^{-1}) \\ \vdots & \vdots & \cdots & \vdots \\ X_k(z)Y_1(z^{-1}) & X_k(z)Y_2(z^{-1}) & \cdots & X_k(z)Y_l(z^{-1}) \end{bmatrix}$$

$$= \begin{bmatrix} X_1(z) \\ X_2(z) \\ \vdots \\ X_k(z) \end{bmatrix} [Y_1(z^{-1})\, Y_2(z^{-1}) \cdots Y_l(z^{-1})],$$

or simply,

$$\mathbf{\Phi}_{xy}(z) = \mathbf{X}(z)^T\mathbf{Y}(z^{-1}), \qquad (2.30a)$$

where we recall that $\mathbf{x}(t)$ and $\mathbf{y}(t)$ are k-channel and l-channel column vectors, respectively, and where, by definition, all x-channels are of length $(n+1)$ and all y-channels are of length $(m+1)$. The above relation reduces to (2.29) when $k=l=1$. In a similar manner, we obtain

$$\mathbf{\Phi}_{yx}(z) = \mathbf{Y}(z)\mathbf{X}^T(z^{-1}) \qquad (2.30b)$$

and

$$\mathbf{\Phi}_{xx}(z) = \mathbf{X}(z)\mathbf{X}^T(z^{-1}). \qquad (2.30c)$$

We shall now consider some numerical examples. Let $\mathbf{X}(z)$ and $\mathbf{Y}(z)$ be given by

$$\mathbf{X}(z) = \begin{bmatrix} (2-5z) \\ (4+3z) \end{bmatrix} \text{ and } \mathbf{Y}(z) = \begin{bmatrix} (1+z) \\ (3-z) \\ (2+\ ) \end{bmatrix}$$

so that $k=2$, $l=3$, $n=1$, and $m=1$. We have by equation (2.30a) that

$$\mathbf{\Phi}_{xy}(z) = \begin{bmatrix} (2-5z) \\ (4+3z) \end{bmatrix} [(1+z^{-1}) \quad (3-z^{-1}) \quad (2+z^{-1})]$$

$$= \begin{bmatrix} (2z^{-1}-3-5z) & (-2z^{-1}+11-15z) & (2z^{-1}-\ 1-10z) \\ (4z^{-1}+7+3z) & (-4z^{-1}+\ 9+\ 9z) & (4z^{-1}+11+\ 6z) \end{bmatrix}.$$

We observe that the elements of the above matrix are not polynomials, since they involve negative powers of z. This is a consequence of the fact that correlation functions are generally two-sided, that is, the lag variable τ takes on both negative as well as positive values. Following Robinson (1967), we call such elements "quasi-polynomials," and hence $\mathbf{\Phi}_{xy}(z)$ can be termed a "quasipolynomial matrix."

We may find $\mathbf{\Phi}_{xx}(z)$ with the aid of equation (2.30c),

$$\mathbf{\Phi}_{xx}(z) = \begin{bmatrix}(2-5z)\\(4+3z)\end{bmatrix}[(2-5z^{-1})\ \ (4+3z^{-1})] = \begin{bmatrix}(-10z^{-1}+29-10z) & (6z^{-1}-7-20z)\\(-20z^{-1}-7+6z) & (12z^{-1}+25+12z)\end{bmatrix}.$$

Obviously $\mathbf{\Phi}_{xx}(z)$ is also a quasipolynomial matrix. It is instructive to express $\mathbf{\Phi}_{xy}(z)$ and $\mathbf{\Phi}_{xx}(z)$ in the form of "matrix quasipolynomials," that is,

$$\mathbf{\Phi}_{xy}(z) = \begin{bmatrix}2 & -2 & 2\\4 & -4 & 4\end{bmatrix}z^{-1} + \begin{bmatrix}-3 & 11 & -1\\7 & 9 & 11\end{bmatrix} + \begin{bmatrix}-5 & -15 & -10\\3 & 9 & 6\end{bmatrix}z$$

and

$$\mathbf{\Phi}_{xx}(z) = \begin{bmatrix}-10 & 6\\-20 & 12\end{bmatrix}z^{-1} + \begin{bmatrix}29 & -7\\-7 & 25\end{bmatrix} + \begin{bmatrix}-10 & -20\\6 & 12\end{bmatrix}z.$$

The corresponding correlation functions can be written down by inspection of the above z-transforms. We obtain the matrix-valued sequences,

$$\phi_{xy}(\tau) = \left\{\begin{bmatrix}2 & -2 & 2\\4 & -4 & 4\end{bmatrix}, \begin{bmatrix}-3 & 11 & -1\\7 & 9 & 11\end{bmatrix}, \begin{bmatrix}-5 & -15 & -10\\3 & 9 & 6\end{bmatrix}\right\}$$

$$= \{\phi_{xy}(-1), \phi_{xy}(0), \phi_{xy}(+1)\},$$

where the coefficients $\phi_{xy}(\tau)$ are $(2)\times(3)$ matrices and

$$\phi_{xx}(\tau) = \left\{\begin{bmatrix}-10 & 6\\-20 & 12\end{bmatrix}, \begin{bmatrix}29 & -7\\-7 & 25\end{bmatrix}, \begin{bmatrix}-10 & -20\\6 & 12\end{bmatrix}\right\}$$

$$= \{\phi_{xx}(-1), \phi_{xx}(0), \phi_{xx}(+1)\},$$

where the coefficients $\phi_{xx}(\tau)$ are $(2)\times(2)$ matrices. We note in particular that the matrix-valued coefficients of $\phi_{xx}(\tau)$ satisfy the symmetry relations (2.28c). In the above example, these are

$$\phi_{xx}(0) = \overset{T}{\phi}_{xx}(0)$$

and

$$\phi_{xx}(1) = \overset{T}{\phi}_{xx}(-1).$$

These relations are easily verifiable from our example. More generally, a matrix-valued function which satisfies (2.28c) is called "reverse symmetric," and all real-valued multichannel autocorrelations are indeed reverse-symmetric functions. We observe also that the relation $\phi_{xx}(0)=\phi^T{}_{xx}(0)$ implies that the zeroth autocorrelation coefficient is a symmetric matrix. Finally, the autocorrelation coefficients $\phi_{xx}(\tau)$ are always square $(k)\times(k)$ matrices. On the other hand, we have seen that the crosscorrelation coefficients $\phi_{xy}(\tau)$ and $\phi_{yx}(\tau)$ are generally rectangular $(k)\times(l)$ and $(l)\times(k)$ matrices, respectively.

3.0 INVERSE MULTICHANNEL FILTERING

Consider the two input and two output channel system described by equations (2.18) or (2.19),

$$\begin{bmatrix}Y_1(z)\\Y_2(z)\end{bmatrix} = \begin{bmatrix}F_{11}(z) & F_{12}(z)\\F_{21}(z) & F_{22}(z)\end{bmatrix} \cdot \begin{bmatrix}X_1(z)\\X_2(z)\end{bmatrix}; \qquad (2.18)$$

that is,

$$\mathbf{Y}(z) = \mathbf{F}(z)\mathbf{X}(z), \qquad (2.19)$$

so that $k=l=2$. The question arises whether we can recover the input $\mathbf{X}(z)$, given the output

$\mathbf{Y}(z)$ and the filter $\mathbf{F}(z)$. If we deal with a single-channel system, (2.19) reduces to the scalar relation

$$Y(z) = F(z)X(z);$$

and hence $X(z)$ can be obtained by the operation

$$\frac{1}{F(z)} Y(z) = F^{-1}(z)\,Y(z) = X(z) \qquad (3.1)$$

where $F^{-1}(z) = 1/F(z)$ is the z-transform of the inverse filter (Treitel and Robinson, 1964).

On the other hand, the input/output relation (2.18) involves the matrix-valued filter $\mathbf{F}(z)$, which we have written as a polynomial matrix. If we premultiply both sides of the matrix equation (2.19) by the *inverse* polynomial matrix $\mathbf{F}^{-1}(z)$, we obtain

$$\begin{aligned} \mathbf{F}^{-1}(z)\mathbf{Y}(z) &= \mathbf{F}^{-1}(z)\mathbf{F}(z)\mathbf{X}(z) \\ &= \mathbf{I}\mathbf{X}(z) = \mathbf{X}(z), \end{aligned} \qquad (3.2)$$

where $\mathbf{F}^{-1}(z)$ is the z-transform of the inverse multichannel filter $\mathbf{F}(z)$ and where $\mathbf{I}$ is the $(k)\times(k)$ identity matrix. Our problem thus reduces to the calculation of the inverse of a polynomial matrix. Now the inverse of a square matrix $\mathbf{A}$ whose elements are scalars is given by the familiar relation,

$$\mathbf{A}^{-1} = \frac{\text{adj } \mathbf{A}}{\det \mathbf{A}},$$

where adj $\mathbf{A}$ and det $\mathbf{A}$ are respectively the adjoint and determinant of the matrix $\mathbf{A}$. The inverse of a square matrix $\mathbf{F}(z)$ whose elements are polynomials can be found in an entirely similar manner, except that the manipulation performed on the elements now involve operations with polynomials rather than with scalars. These matters are thoroughly described in the books by Frazer, Duncan, and Collar (1938, Chapter 3), and by Gantmacher (1959, Chapters 4 and 6). Accordingly, we may write

$$\mathbf{F}^{-1}(z) = \frac{\text{adj } \mathbf{F}(z)}{\det \mathbf{F}(z)}. \qquad (3.3)$$

This method is again best described by a small numerical example. Let $\mathbf{F}(z)$ be the filter given by equation (2.15),

$$\mathbf{F}(z) = \begin{bmatrix} 6+z & 2z \\ 7+z & 1+3z \end{bmatrix}.$$

The adjoint matrix adj $\mathbf{F}(z)$ is then

$$\text{adj } \mathbf{F}(z) = \begin{bmatrix} 1+3z & -2z \\ -7-z & 6+z \end{bmatrix},$$

while the determinant det $\mathbf{F}(z)$ is

$$\begin{aligned} \det \mathbf{F}(z) &= (6+z)(1+3z) - 2z(7+z) \\ &= (6+5z+z^2). \end{aligned}$$

Combination of these two results yields

$$\begin{aligned} \mathbf{F}^{-1}(z) &= \frac{1}{6+5z+z^2}\begin{bmatrix} 1+3z & -2z \\ -7-z & 6+z \end{bmatrix} \\ &= \frac{1}{\det \mathbf{F}(z)} \text{adj } \mathbf{F}(z). \end{aligned} \qquad (3.4)$$

The inverse filter $\mathbf{F}^{-1}(z)$ thus consists of two stages which are connected in cascade. The first stage is given by the filter

$$\frac{1}{\det \mathbf{F}(z)} = \frac{1}{6+5z+z^2}.$$

The determinant det $\mathbf{F}(z)$ of a polynomial matrix is always a scalar polynomial in powers of z, and thus the filter $1/\det \mathbf{F}(z)$ is the familiar *single-channel* inverse filter. On the other hand, the second stage consists of the polynomial matrix adj $\mathbf{F}(z)$, whose elements are obtained from the elements of $\mathbf{F}(z)$ by the operations of addition and multiplication but *not* division (Gantmacher, loc. cit.). Hence the elements of adj $\mathbf{F}(z)$ are always finite-length polynomials, so that no stability problems arise with the constituent single channel filters of this matrix. The stability of the inverse filter $\mathbf{F}^{-1}(z)$ is thus governed by the stability of its first stage, namely of the single-channel inverse filter $1/\det \mathbf{F}(z)$. It has been shown by Treitel and Robinson (1964) that the stability of this inverse filter is uniquely determined by its delay properties, which in turn are establishable from the position of the zeros of det $\mathbf{F}(z)$ in the complex z-plane. The following possibilities arise:

(1) The zeros of det $\mathbf{F}(z)$ all lie outside the unit circle ($|z| = 1$). The filter det $\mathbf{F}(z)$ is minimum-delay, and $1/\det \mathbf{F}(z)$ is ex-

pandable in terms of a stable memory function.

(2) The zeros of det $\mathbf{F}(z)$ all lie inside the unit circle. The filter det $\mathbf{F}(z)$ is maximum-delay and $1/\det \mathbf{F}(z)$ is expandable in terms of a stable anticipation function.

(3) The zeros of det $\mathbf{F}(z)$ lie both inside and outside the unit circle. The filter det $\mathbf{F}(z)$ is mixed-delay, and $1/\det \mathbf{F}(z)$ is expandable in terms of both a stable memory function and a stable anticipation function.

(4) One or more of the zeros of det $\mathbf{F}(z)$ lie on the unit circle. The zeros on the unit circle contribute to the equi-delay component of det $\mathbf{F}(z)$, which is not expandable in terms of a stable memory function nor of a stable anticipation function; hence such a filter has no stable exact inverse.

(5) The polynomial det $\mathbf{F}(z)$ either vanishes identically or is equal to a constant, say K. In the former case no inverse $1/\det \mathbf{F}(z)$ exists, while in the latter case $1/\det \mathbf{F}(z)$ is simply given by the scalar quantity $1/K$.

Returning to our numerical example, we find that det $\mathbf{F}(z)$ can be factored in the form,

$$\det \mathbf{F}(z) = (2+z)(3+z).$$

Both zeros of this scalar polynomial lie outside the unit circle. Therefore det $\mathbf{F}(z)$ is minimum-delay, and $1/\det \mathbf{F}(z)$ is expandable in terms of the stable memory function,

$$\frac{1}{6+5z+z^2} = \frac{1}{6}\left(1 - \frac{5}{6}z + \frac{19}{36}z^2 - \cdots\right).$$

The above power series expansion is infinite, and in practice must be truncated after the retention of a suitable number of terms (Treitel and Robinson, 1964). The inverse filter $\mathbf{F}^{-1}(z)$ is thus given by

$$\mathbf{F}^{-1}(z) = \frac{1}{6}\left(1 - \frac{5}{6}z + \frac{19}{36}z^2 - \cdots\right) \cdot \begin{bmatrix} 1+3z & -2z \\ -7-z & 6+z \end{bmatrix}. \tag{3.5}$$

In this manner we have been able to express $\mathbf{F}^{-1}(z)$ in the form of a two-stage cascaded convolutional filter. On the other hand, equation (3.4) expresses $\mathbf{F}^{-1}(z)$ in the form of a two-stage rational filter, where the first stage $1/\det \mathbf{F}(z)$ is recursive (Shanks, 1967), while the second stage adj $\mathbf{F}(z)$ is convolutional.

Let us attempt to recover the input $\mathbf{X}(z)$ from the output $\mathbf{Y}(z)$ that we obtained for the numerical example of equation (2.20). Use of equation (3.5) in equation (3.2) yields,

$$\begin{bmatrix} X_1(z) \\ X_2(z) \end{bmatrix} = \frac{1}{6}\left(1 - \frac{5}{6}z + \frac{19}{36}z^2 - \cdots\right) \cdot \begin{bmatrix} 1+3z & -2z \\ -7-z & 6+z \end{bmatrix} \begin{bmatrix} 12-20z+z^2 \\ 18-18z+4z^2 \end{bmatrix} = \begin{bmatrix} 2-5z+0z^2+E_1(z^3, z^4, \cdots) \\ 4+3z+0z^2+E_2(z^3, z^4, \cdots) \end{bmatrix},$$

where $E_i(z^3, z^4, \cdots), i=1, 2$ represents the truncation error which occurs because the single-channel filter $1/\det \mathbf{F}(z)$ has only been expanded into a 3-length stable memory function. These truncation errors will become smaller only as the number of terms in this expansion becomes larger. For this reason the form (3.5) is not too useful from a computational viewpoint. Instead, we may treat $1/\det \mathbf{F}(z)$ as a recursive filter, and then obtain

$$\begin{bmatrix} X_1(z) \\ X_2(x) \end{bmatrix} = \frac{1}{(6+5z+z^2)} \cdot \begin{bmatrix} 1+3z & -2z \\ -7-z & 6+z \end{bmatrix} \begin{bmatrix} 12-20z+z^2 \\ 18-18z+4z^2 \end{bmatrix} = \begin{bmatrix} 2-5z \\ 4+3z \end{bmatrix},$$

so that the input $\mathbf{X}(z)$ has been recovered with no error.

The filter det $\mathbf{F}(z)$ of the above example is minimum-delay. If det $\mathbf{F}(z)$ is maximum delay or mixed delay, the filter $1/\det \mathbf{F}(z)$ is implementable in its recursive form through use of the forward/reverse filtering techniques discussed by Shanks (loc. cit.).

The reader will have noticed that our discussion of inverse multichannel filters is restricted to the case for which the number of input channels equals the number of output channels (i.e., $k=l$). This means that the filter $\mathbf{F}(z)$ is a square polynomial matrix, whose exact inverse can be computed as just described. The case $k \neq l$ leads to rectangular polynomial matrices, which do not have inverses in the ordinary sense; they will not be considered here. Moreover, we have tacitly assumed that $\mathbf{F}(z)$ is a polynomial matrix; that is, its elements are polynomials in z. In some instances these elements turn out to be ratios of polynomials in z, so that $\mathbf{F}(z)$ becomes a so-called "rational" matrix. The stability of $\mathbf{F}^{-1}(z)$ is then no longer determined by the delay properties of det $\mathbf{F}(z)$ alone, since the elements of adj $\mathbf{F}(z)$ will generally be rational functions of z. However, a wide variety of multichannel computing algorithms does lead to filters for which $\mathbf{F}(z)$ is a polynomial matrix. This is true in particular of the multichannel time domain technique for the calculation of Wiener filters, which we describe in the sequel.

We remark that the inverse filter $\mathbf{F}^{-1}(z)$ can also be represented as a partial fraction expansion. Such a representation brings forth the general mathematical structure of the multichannel inverse. A paper by Robinson (1966) gives a summary of this approach.

The calculation of adj $\mathbf{F}(z)$ and det $\mathbf{F}(z)$ is not generally a simple matter when k is larger than 2. One technique that yields both these functions is due to Faddeev and is described by Gantmacher (1959, Chapter 4). In practice, we will often prefer to compute least squares rather than exact inverses, a matter which we take up in detail in Section 4.0.

4.0 THE MULTICHANNEL WIENER FILTER

4.1 *Introductory remarks*

Multichannel Wiener theory constitutes a natural extension of the more familar single-channel version of this subject. The single-channel Wiener filter is the keystone of many current seismic digital processing systems, but there are reasons to believe that the multichannel Wiener filter will in some instances become an even more powerful tool in our quest to achieve increased seismic signal enhancement and resolution. This is primarily due to the fact that multichannel processors make use of information present in a multiplicity of input channels. The seismic interpreter attempts to identify reflections by visual trace-to-trace correlation of coherent energy; rarely, if ever, does he make such an identification based only on what he sees on a single trace. It is therefore appropriate to consider the realization of filters which make use of the inherently redundant information that is present on a set of two or more neighboring seismic traces. In fact, this has been done for many years in the case of the common-depth-point method, and we show later that the trace summing technique can be viewed as a special case of a more general multichannel filtering system.

Since our present treatment is tutorial, we shall not deal with the various successful seismic applications that have already been made with the multichannel Wiener filter (see e.g. Burg, 1964; Schneider et al, 1964; Schneider et al, 1965; Laster and Linville, 1966; Davies and Mercado, 1968; Galbraith and Wiggins, 1968). Instead, we shall attempt to derive this filter from basic principles in the time domain and to illustrate its use by means of some simple numerical examples.

4.2 *The filter design model*

Our design model contains the generalizations of the entities that are already familiar to us from single-channel theory (Robinson and Treitel, 1967). These are:

(1) k channels of sampled input data $x_i(t)$,

$$i = 1, 2, \cdots, k,$$

(2) l channels of desired output data $d_i(t)$,

$$i = 1, 2, \cdots, l,$$

(3) l channels of actual output data $y_i(t)$,

$$i = 1, 2, \cdots, l.$$

We are thus at liberty to specify the information on l desired output channels and, hence, obtain l actual output channels once the filter has operated on the k channels of input data. The design criterion is again based on the least squares principle; namely, we wish to minimize the energy existing in the difference between the desired and the actual outputs. Thus we let $e_i(t)$ be the difference between the ith desired output and the ith actual output,

$$e_i(t) = d_i(t) - y_i(t), \quad i = 1, 2, \cdots, l. \qquad (4.1)$$

Following the length conventions introduced in Section 2.1, we let all input channels $x_i(t)$ be of length $(n+1)$; and we let all filter elements $f_{ij}(t)$ be of length $(m+1)$. Then, all actual output chan-

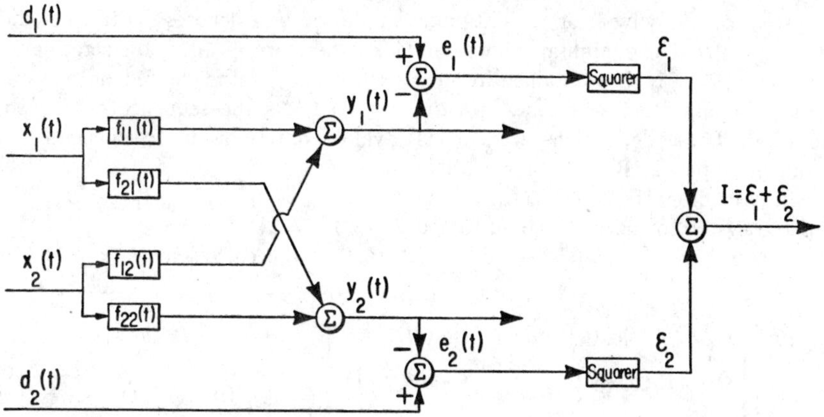

FIG. 4.1. The block diagram for a two input and two output channel Wiener filter.

nels $y_i(t)$ will be of length $(m+n+1)$, and hence we make all desired output channels $d_i(t)$ of this length as well. The differences or residuals, $e_i(t)$, are called error series. There are l such error series, each of which is of length $(m+n+1)$. The energy $\epsilon_i(i=1, 2, \cdots, l)$ in each error series is given by the sum of the squares of its sampled values,

$$\epsilon_i = e_i^2(0) + e_i^2(1) + \cdots + e_i^2(m+n) = \sum_{t=0}^{m+n} e_i^2(t). \tag{4.2}$$

The total error energy I is the sum of the l individual error energies ϵ_i,

$$I = \sum_{i=1}^{l} \epsilon_i. \tag{4.3}$$

As will be shown below, the smallest possible value of I is obtained when the elements $f_{ij}(t)$, $t=0, 1, \cdots, m$ are the constituents of the multichannel Wiener filter. We note in particular that the individual error energies ϵ_i and the total error energy I are both scalar quantities.

A block diagram for a two input channel $(k=2)$ and two output channel $(l=2)$ Wiener filter is shown in Figure 4.1. This diagram should be compared to the system depicted in Figure 2.1, which illustrates the general multichannel filter for the case $k=l=2$.

If we substitute (4.1) and (4.2) into (4.3), we obtain,

$$I = \sum_{i=1}^{l} \left\{ \sum_{t=0}^{m+n} [d_i(t) - y_i(t)]^2 \right\}. \tag{4.4}$$

But, by equation (2.14), we note that every actual output series $y_i(t)$ is the result of summing the outputs from k single-channel convolutions,

$$y_i(t) = f_{i1}(t) * x_1(t) + f_{i2}(t) * x_2(t) + \cdots + f_{ik}(t) * x_k(t) = \sum_{j=1}^{k} f_{ij}(t) * x_j(t). \tag{4.5}$$

Hence, I can be written in the form

$$I = \sum_{i=1}^{l} \left\{ \sum_{t=0}^{m+n} \left[d_i(t) - \sum_{j=1}^{k} f_{ij}(t) * x_j(t) \right]^2 \right\}. \tag{4.6}$$

In general there is a unique set of filter coefficients $f_{ij}(t)$ which minimizes the total error energy I. We achieve this minimum by solving the system of linear simultaneous equations that is obtained when we set the partial derivatives of I with respect to all filter coefficients $f_{ij}(t)$ equal to zero. In other words, we compute

$$\frac{\partial}{\partial f_{ij}(t)} I \quad \text{for } i = 1, 2, \cdots, l, \quad j = 1, 2, \cdots, k, \quad t = 0, 1, \cdots, m,$$

and equate each of these expressions to zero. The

resulting system can be solved for the optimum filter coefficients $f_{ij}(t)$. The minimum value of the total error energy, $I_{\min}$, is then obtained by substituting the optimum filter coefficients into the expression (4.6). These derivations are carried out in detail in Appendix B. It is shown there that the optimum filter $(\mathbf{f}_0, \mathbf{f}_1, \cdots, \mathbf{f}_m)$ is the solution of the system of normal matrix-valued equations

$$[\mathbf{f}_0 \quad \mathbf{f}_1 \cdots \mathbf{f}_m]\begin{bmatrix} \phi_{xx}(0) & \phi_{xx}(1) & \cdots & \phi_{xx}(m) \\ \phi_{xx}^T(1) & \phi_{xx}(0) & \cdots & \phi_{xx}(m-1) \\ \vdots & \vdots & & \vdots \\ \phi_{xx}^T(m) & \phi_{xx}^T(m-1) & \cdots & \phi_{xx}(0) \end{bmatrix} = [\phi_{dx}(0) \quad \phi_{dx}(1) \cdots \phi_{dx}(m)]. \quad (4.7)$$

The filter coefficients $\mathbf{f}_t$, the input autocorrelation coefficients $\phi_{xx}(t)$, and the desired output-with-input crosscorrelation coefficients $\phi_{dx}(t)$ are $(l)\times(k)$, $(k)\times(k)$, and $(l)\times(k)$ matrices, respectively.

The system (4.7) can be written in the convenient abbreviated form

$$\mathbf{FR} = \mathbf{G}. \quad (4.8)$$

Here $\mathbf{F}$ is the $(m+1)$-length row vector of matrix-valued filter coefficients $\mathbf{f}_t$; $\mathbf{R}$ is the $(m+1)\times(m+1)$ square matrix of matrix-valued autocorrelation coefficients $\phi_{xx}(t)$; and $\mathbf{G}$ is the $(m+1)$-length row vector of matrix-valued crosscorrelation coefficients $\phi_{dx}(t)$. The solution of (4.8) is

$$\mathbf{F} = \mathbf{GR}^{-1}, \quad (4.9)$$

where $\mathbf{R}^{-1}$ is the inverse of $\mathbf{R}$. Since the coefficients $\phi_{xx}(t)$ are $(k)\times(k)$ matrices with scalar entries, $\mathbf{R}$ is in fact a $k(m+1)\times k(m+1)$ square matrix in the scalar-valued autocorrelation coefficients,

$$\phi_{x_i x_j}(t), \quad i, j = 1, 2, \cdots, k,$$
$$t = 0, 1, \cdots, m.$$

The normalized minimum total error energy, or normalized mean square error E, is given by equation (B-20) of Appendix B,

$$\mathrm{E} = 1 - \frac{tr \sum_{t=0}^{m} \phi_{dx}(t)\mathbf{f}_t^T}{tr\phi_{dd}(0)}, \quad (4.10)$$

where $\mathrm{E} = I_{\min}/tr\phi_{dd}(0)$, and where $0 \leq \mathrm{E} \leq 1$. The notation $tr\,\mathbf{A}$ denotes the trace of the matrix $\mathbf{A}$.[2] Furthermore, $\phi_{dd}(0)$ is the zeroth autocorrelation coefficient of the desired output channels $\mathbf{d}(t)$, while $\mathbf{f}_t^T$ is the sequence of transposed matrix-valued optimum filter coefficients

$$\mathbf{f}_t^T = (\mathbf{f}_0^T, \mathbf{f}_1^T, \cdots, \mathbf{f}_m^T),$$

so that the coefficients $\mathbf{f}_t^T$ are $(k)\times(l)$ matrices.

4.3 *Some numerical examples*

We shall now present three simple numerical examples which will help to illustrate the theory developed in the preceding sections.

Example 1

Consider first the system,

$$\mathbf{x}(t) = \begin{bmatrix} x_1(t) \\ x_2(t) \end{bmatrix} = \begin{bmatrix} (-1, 2, 5, 2) \\ (3, 1, 1, 2) \end{bmatrix},$$
$$\mathbf{d}(t) = [d_1(t)] = [(1, 0, 0, 0, 0)],$$

so that $k=2$ and $l=1$. Let us compute the Wiener filters of lengths 1 and 2 (i.e., $m=0$ and $m=1$) which shape the two-channel input $\mathbf{x}(t)$ into the single-channel desired output $\mathbf{d}(t)$. Using the results of Section 2.3, we obtain

$$\phi_{xx}(0) = \begin{bmatrix} \phi_{x_1x_1}(0) & \phi_{x_1x_2}(0) \\ \phi_{x_2x_1}(0) & \phi_{x_2x_2}(0) \end{bmatrix} = \begin{bmatrix} 34 & 8 \\ 8 & 15 \end{bmatrix},$$

$$\phi_{xx}(1) = \begin{bmatrix} \phi_{x_1x_1}(1) & \phi_{x_1x_2}(1) \\ \phi_{x_2x_1}(1) & \phi_{x_2x_2}(1) \end{bmatrix} = \begin{bmatrix} 18 & 13 \\ 11 & 6 \end{bmatrix},$$

$$\phi_{dx}(0) = [\phi_{d_1x_1}(0) \quad \phi_{d_1x_2}(0)] = [-1 \quad 3],$$
$$\phi_{dx}(1) = [\phi_{d_1x_1}(1) \quad \phi_{d_1x_2}(1)] = [0 \quad 0].$$

For $m=0$, the system (4.7) becomes

$$[\mathbf{f}_0][\phi_{xx}(0)] = [\phi_{dx}(0)],$$

where $\mathbf{f}_0$ is the $(1)\times(2)$ matrix

[2] The trace of a square matrix is equal to the sum of the elements of its main diagonal.

$$\mathbf{f}_0 = [f_{11}(0) \quad f_{12}(0)].$$

Hence, we have

$$[f_{11}(0) \quad f_{12}(0)]\begin{bmatrix} 34 & 8 \\ 8 & 15 \end{bmatrix} = [-1 \quad 3],$$

whose solution is

$$\begin{aligned} \mathbf{f}_0 &= [f_{11}(0) \quad f_{12}(0)] \\ &= [-1 \quad 3]\begin{bmatrix} 34 & 8 \\ 8 & 15 \end{bmatrix}^{-1} \\ &= [-1 \quad 3]\begin{bmatrix} 0.03363 & -0.01794 \\ -0.01794 & 0.07623 \end{bmatrix} \\ &= [-0.08744 \quad 0.24664]. \end{aligned}$$

The filter $\mathbf{f}_0$ thus consists of two single-channel one-length filters; namely,

$$\begin{aligned} f_{11}(t) &= f_{11}(0) = -0.08744, \\ f_{12}(t) &= f_{12}(0) = 0.24664. \end{aligned}$$

The actual output, $\mathbf{y}(t)=y_1(t)$ is, by equation (4.5),

$$\begin{aligned} y_1(t) &= \{f_{11}(t) * x_1(t)\} + \{f_{12}(t) * x_2(t)\} \\ &= \{(-0.08744) * (-1, 2, 5, 2)\} \\ &\quad + \{(0.24664) * (3, 1, 1, 2)\} \\ &= (0.82736, \quad 0.07176, \\ &\quad -0.19056, \quad 0.31840). \end{aligned}$$

The normalized mean square error E is, by equation (4.10) with $m=0$,

$$\mathrm{E} = 1 - \frac{tr\boldsymbol{\phi}_{dx}(0)\mathbf{f}_0^T}{tr\boldsymbol{\phi}_{dd}(0)}\cdot$$

But

$$\begin{aligned} tr\boldsymbol{\phi}_{dx}(0)\mathbf{f}_0^T &= tr\left\{[-1 \quad 3]\begin{bmatrix} -0.08744 \\ 0.24664 \end{bmatrix}\right\} \\ &= tr[0.82736] = 0.82736, \end{aligned}$$

since in this case ($l=1$) the matrix product $\boldsymbol{\phi}_{dx}(0)\mathbf{f}_0^T$ is merely a $(1)\times(1)$ matrix. Similarly,

$$tr\boldsymbol{\phi}_{dd}(0) = tr[\phi_{d_1d_1}(0)] = tr[1] = 1,$$

and therefore,

$$\mathrm{E} = 1 - 0.82736 = 0.17264.$$

For $m=1$, the system (4.7) becomes

$$[\mathbf{f}_0 \quad \mathbf{f}_1]\begin{bmatrix} \boldsymbol{\phi}_{xx}(0) & \boldsymbol{\phi}_{xx}(1) \\ \boldsymbol{\phi}_{xx}^T(1) & \boldsymbol{\phi}_{xx}(0) \end{bmatrix} = [\boldsymbol{\phi}_{dx}(0) \quad \boldsymbol{\phi}_{dx}(1)].$$

Proceeding as before, we write

$$[f_{11}(0) \quad f_{12}(0) \quad f_{11}(1) \quad f_{12}(1)]$$

$$= [-1 \quad 3 \quad 0 \quad 0]\left[\begin{array}{cc|cc} 34 & 8 & 18 & 13 \\ 8 & 15 & 11 & 6 \\ \hline 18 & 11 & 34 & 8 \\ 13 & 6 & 8 & 15 \end{array}\right]^{-1},$$

which yields

$$\begin{aligned} [\mathbf{f}_0\mathbf{f}_1] &= [f_{11}(0) \quad f_{12}(0) \quad f_{11}(1) \quad f_{12}(1)] \\ &= [-0.05215 \quad 0.28375 \\ &\quad -0.05503 \quad -0.03896]. \end{aligned}$$

The filter $\mathbf{f}_t=(\mathbf{f}_0, \mathbf{f}_1)$ now consists of two single-channel two-length filters; namely,

$$\begin{aligned} f_{11}(t) &= \{f_{11}(0), \quad f_{11}(1)\} \\ &= (-0.05215, \quad -0.05503), \\ f_{12}(t) &= \{f_{12}(0), \quad f_{12}(1)\} \\ &= (0.28375, \quad -0.03896). \end{aligned}$$

The actual output $\mathbf{y}(t)=y_1(t)$ is in this case

$$\begin{aligned} y_1(t) &= \{f_{11}(t) * x_1(t)\} + \{f_{12}(t) * x_2(t)\} \\ &= (0.90339, \quad 0.11761, \\ &\quad -0.12601, \quad 0.14911, \quad -0.18796), \end{aligned}$$

while the normalized mean square error is

$$\begin{aligned} \mathrm{E} &= 1 - \frac{tr\{\boldsymbol{\phi}_{dx}(0)\mathbf{f}_0^T + \boldsymbol{\phi}_{dx}(1)\mathbf{f}_1^T\}}{tr\boldsymbol{\phi}_{dd}(0)} \\ &= 1 - 0.90339 = 0.09661. \end{aligned}$$

As expected, the error is smaller for the longer filter; however, even the one-length filter yields a good approximation of the desired output. These results are displayed graphically in Figure 4.2.

We observe that the solution for the two-length filter ($m=1$) already requires the inversion of a fourth order square matrix. In practice, direct matrix inversion is seldom used; instead, we apply the multichannel extension of the Wiener-Levinson algorithm (see Appendix B).

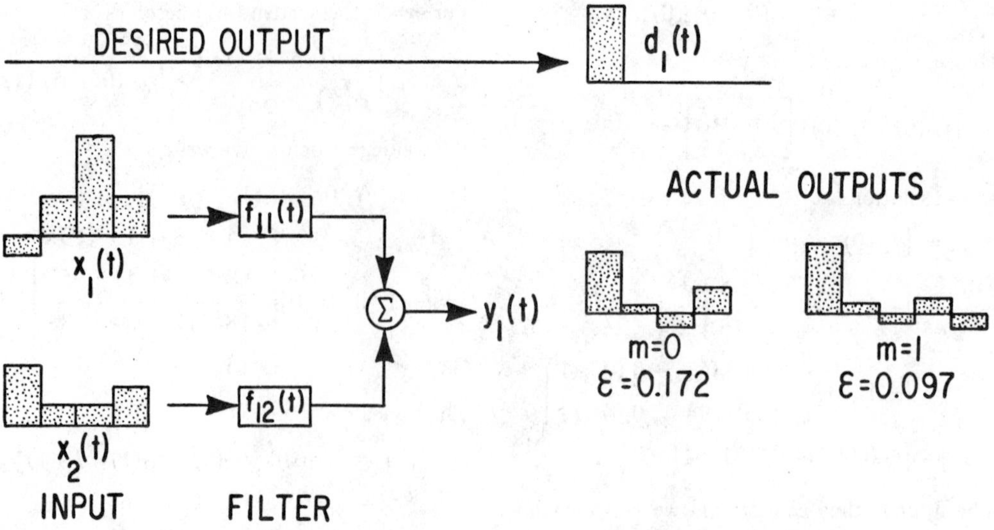

FIG. 4.2. Graphical display of Example 1. The two actual outputs are for the one-length ($m=0$) and for the two-length ($m=1$) Wiener filters.

Example 2

Let us next consider the simple summation model depicted in Figure 4.3. We are given two nine-length input channels, $x_1(t)$ and $x_2(t)$,

$$x_1(t) = (-1, 2, 3, 1, 4, 1, -2, -3, 2),$$
$$x_2(t) = (-5, -1, 6, 1, 4, 1, -5, -4, -1).$$

We assume that the "signal" is given by the sequence (0, 0, 0, 1, 4, 1, 0, 0, 0), which is common to both input channels. Our desire is to combine $x_1(t)$ and $x_2(t)$ in such a manner that a single output channel $y_1(t)$ exhibits the signal with minimum distortion. The first thing that comes to mind is to evaluate the average between $x_1(t)$ and $x_2(t)$,

$$y_1(t) = \tfrac{1}{2}[x_1(t) + x_2(t)].$$

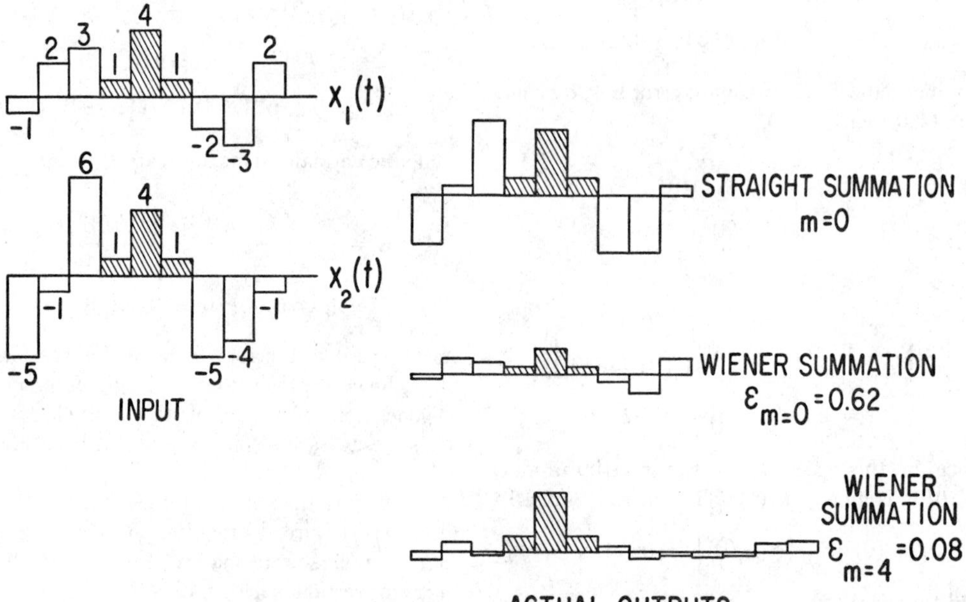

FIG. 4.3. Graphical display of Example 2: a simple summation experiment.

This expression can be written in the form,

$$y_1(t) = \{(0.5) * x_1(t)\} + \{(0.5) * x_2(t)\}$$
$$= \{f_{11}(t) * x_1(t)\} + \{f_{12}(t) * x_2(t)\}.$$

We see that straight summation is representable as a multichannel filtering operation, where the filters $f_{11}(t)$ and $f_{12}(t)$ are one-length ($m=0$) operators, whose single weighting coefficients are equal to each other,

$$f_{11}(0) = f_{12}(0) = 0.5.$$

The result is shown as the first actual output display of Figure 4.3, where it is labelled "Straight Summation." The hachured bars in the input diagrams indicate the common signal, while the hachuring in the output displays pinpoints the position of the signal estimates. It is evident that direct summation has not significantly helped to bring out the signal.

We may attack this problem with the help of the multichannel Wiener filter. For our particular case, $k=2$ and $l=1$, while the desired output $d_1(t)$ is

$$d_1(t) = (0, 0, 0, 1, 4, 1, 0, 0, 0).$$

Proceeding as in the previous example, we compute, for $m=0$,

$$\boldsymbol{\phi}_{xx}(0) = \begin{bmatrix} 49 & 59 \\ 59 & 122 \end{bmatrix} \text{ and } \boldsymbol{\phi}_{dx}(0) = [18 \quad 18],$$

so that

$$\mathbf{f}_0 = [18 \quad 18] \begin{bmatrix} 49 & 59 \\ 59 & 122 \end{bmatrix}^{-1}$$
$$= [0.45414 \quad -0.07209]$$
$$= [f_{11}(0) \quad f_{12}(0)].$$

The actual output for this one-length filter is

$$y_1(t) = \{(0.45414) * x_1(t)\} + \{(-0.07209) * x_2(t)\},$$

which is shown in Figure 4.3 as the diagram labelled "Wiener Summation, $m=0$." The two filter coefficients $f_{11}(0)$ and $f_{12}(0)$ are no longer equal; and, therefore, this case can be viewed as a "weighted summation" in the sense that the two coefficients $f_{11}(0)$ and $f_{12}(0)$ are computed by means of the least squares principle. Although the result is somewhat better than before, the desired signal is still masked by background "noise," and has in fact been distorted by the filter. The value of the normalized mean square error ϵ is, by equation (4.10),

$$E_{m=0} = 0.61794.$$

Just as is true for the single-channel case (Treitel and Robinson, 1966), the value of E decreases monotonically as m increases. For example, when $m=4$, we obtain a two-channel five-length filter $\mathbf{f}_t$, which can be written in the form of two single-channel five-length filters,

$$f_{11}(t) = \{f_{11}(0), f_{11}(1), \cdots, f_{11}(4)\},$$
$$f_{12}(t) = \{f_{12}(0), f_{12}(1), \cdots, f_{12}(4)\},$$

and whose weighting coefficients can be evaluated conveniently by means of the multichannel Wiener-Levinson algorithm (see Appendix B). The actual output is given by

$$y_1(t) = \{f_{11}(0), \cdots, f_{11}(4)\} * x_1(t) + \{f_{12}(0), \cdots, f_{12}(4)\} * x_2(t)$$

and is displayed in Figure 4.3 as the diagram labelled "Wiener Summation, $m=4$." The result is now quite satisfactory in that our objective of enhancing the signal at the expense of the background noise has been achieved. The value of E is here

$$E_{m=4} = 0.07915.$$

The error becomes progressively smaller as m continues to grow, and in fact for $m=7$, it is zero. This exact performance of the filter is a peculiarity of our multichannel model and has no counterpart in the single-channel case. In general, the error reduces to zero whenever

$$m = \frac{n+1-k}{k-1}, \qquad k \neq 1.$$

For $k=2$, we obtain $m=n-1$, so that perfect filter performance for the two input channel case is generally guaranteed whenever the filter length is one less than the length of the input time series. The above relation has been derived in a separate note (Treitel, 1970). In practice, this ideal performance is not often achievable because usually $n \gg m$; and hence unrealistically long filters would have to be computed in order for E to become zero.

Example 3

Consider the system

$$\mathbf{Y}(z) = \mathbf{F}(z)\mathbf{X}(z),$$

where both the input $\mathbf{X}(z)$ and the output $\mathbf{Y}(z)$ consist of k channels, so that $l=k$. The filter $\mathbf{F}(z)$ is a square kth order polynomial matrix. Suppose that we wish to recover the input $\mathbf{X}(z)$, given the output $\mathbf{Y}(z)$. It is then necessary to find a filter $\mathbf{H}(z)$ which operates on the output $\mathbf{Y}(z)$ such that the relation

$$\mathbf{H}(z)\mathbf{Y}(z) = \mathbf{X}(z)$$

be satisfied as closely as possible. Since $\mathbf{Y}(z) = \mathbf{F}(z)\mathbf{X}(z)$, we require that

$$\mathbf{H}(z)\mathbf{Y}(z) = \mathbf{H}(z)\mathbf{F}(z)\mathbf{X}(z) = \mathbf{X}(z).$$

Thus, we are led to the condition

$$\mathbf{H}(z)\mathbf{F}(z) = \mathbf{I}, \tag{4.11}$$

which must be satisfied as closely as possible by the sought inverse filter $\mathbf{H}(z)$. Here $\mathbf{I}$ is again the kth order identity matrix.

In Section 3.0 it was demonstrated that an *exact* inverse filter $\mathbf{F}^{-1}(z)$, which indeed satisfies

$$\mathbf{F}^{-1}(z)\mathbf{F}(z) = \mathbf{I},$$

can be found. Here we shall show that a least squares solution of this problem is also possible. The design procedure for the inverse Wiener filter requires the specification of an input and of a desired output. In the present context, the input is given by the response function of the filter $\mathbf{f}_t$ [whose z-transform is $\mathbf{F}(z)$],

$$\mathbf{f}_t = \begin{bmatrix} f_{11}(t) & \cdots & f_{1k}(t) \\ \vdots & & \vdots \\ f_{k1}(t) & \cdots & f_{kk}(t) \end{bmatrix}, \quad t = 0, 1, \cdots, n.$$

The reader should not confuse this "input" with the input data channels $\mathbf{X}(z)$ given above. In other words, the response function $\mathbf{F}(z)$ constitutes the input to the algorithm that will yield the required inverse Wiener filter $\mathbf{H}(z)$, while $\mathbf{X}(z)$ represents the multichannel input which it is desired to recover from $\mathbf{Y}(z)$ once $\mathbf{H}(z)$ has been calculated.

We remark also that the input $\mathbf{f}_t$ appears in the form of a $(k)\times(k)$ matrix, rather than as the $(k)\times(1)$ column vector used for the derivations of Appendix B. Nevertheless we shall show by means of a small example that the same formalism applies in this situation as well.

Since the least squares inverse filter $\mathbf{h}_t$ [whose z-transform is $\mathbf{H}(z)$] and the original filter $\mathbf{f}_t$ are kth order square matrices, it follows that the desired output $\mathbf{d}_t$ must also be a kth order square matrix. In particular, we have, in view of the requirement (4.11), that

$$\mathbf{d}_t = \begin{bmatrix} d_{11}(t) & \cdots & d_{1k}(t) \\ \vdots & & \vdots \\ d_{k1}(t) & \cdots & d_{kk}(t) \end{bmatrix} = \begin{cases} \mathbf{I}, & t = 0 \\ \mathbf{0}, & t > 0 \end{cases},$$

where $\mathbf{I}$ and $\mathbf{0}$ are the kth order identity and null[3] matrices, respectively. This formulation again differs from our earlier model, in which $\mathbf{d}_t$ is merely a $(k)\times(1)$ column vector.

Let us now illustrate these concepts by returning to the simple example already introduced in Section 3.0,

$$\mathbf{F}(z) = \begin{bmatrix} 6+z & 0+2z \\ 7+z & 1+3z \end{bmatrix},$$

so that $k=l=2$ and $n=1$. We wish to compute the inverse Wiener filter $\mathbf{H}(z)$ such that equation (4.11) be satisfied as closely as possible. The input is thus $\mathbf{F}(z)$, while the desired output is $\mathbf{D}(z)=\mathbf{I}$. Let $\mathbf{\Phi}_{ff}(z)$ and $\mathbf{\Phi}_{df}(z)$ be the z-transforms of the input autocorrelation and the input with desired output crosscorrelation, respectively. Then we have by equation (2.30c),

$$\begin{aligned}\mathbf{\Phi}_{ff}(z) &= \mathbf{F}(z)\mathbf{F}^T(z^{-1}) \\ &= \begin{bmatrix} 6+z & 2z \\ 7+z & 1+3z \end{bmatrix}\begin{bmatrix} 6+z^{-1} & 7+z^{-1} \\ 2z^{-1} & 1+3z^{-1} \end{bmatrix} \\ &= \begin{bmatrix} 6 & 6 \\ 9 & 10 \end{bmatrix} z^{-1} + \begin{bmatrix} 41 & 49 \\ 49 & 60 \end{bmatrix} + \begin{bmatrix} 6 & 9 \\ 6 & 10 \end{bmatrix} z,\end{aligned}$$

and by equation (2.30b),

$$\begin{aligned}\mathbf{\Phi}_{df}(z) &= \mathbf{D}(z)\mathbf{F}^T(z^{-1}) = \mathbf{I}\mathbf{F}^T(z^{-1}) \\ &= \begin{bmatrix} 6 & 7 \\ 0 & 1 \end{bmatrix} + \begin{bmatrix} 1 & 1 \\ 2 & 3 \end{bmatrix} z^{-1}.\end{aligned}$$

Therefore,

$$\boldsymbol{\phi}_{ff}(0) = \begin{bmatrix} 41 & 49 \\ 49 & 60 \end{bmatrix}; \quad \boldsymbol{\phi}_{ff}(1) = \begin{bmatrix} 6 & 9 \\ 6 & 10 \end{bmatrix};$$

and

$$\boldsymbol{\phi}_{df}(0) = \begin{bmatrix} 6 & 7 \\ 0 & 1 \end{bmatrix}.$$

[3] The null matrix is a matrix all whose elements are zero.

From the above, it is also evident that

$$\phi_{ff}(t) = \boldsymbol{\theta} \quad \text{for } |t| \neq 0, 1,$$
$$\phi_{df}(t) = \boldsymbol{\theta} \quad \text{for } t \neq 0.$$

For $m=0$, the normal equations (4.7) become

$$\mathbf{h}_0\phi_{ff}(0) = \phi_{df}(0),$$

and their solution is

$$\mathbf{h}_0 = \begin{bmatrix} h_{11}(0) & h_{12}(0) \\ h_{21}(0) & h_{22}(0) \end{bmatrix}$$
$$= \begin{bmatrix} 6 & 7 \\ 0 & 1 \end{bmatrix} \begin{bmatrix} 41 & 49 \\ 49 & 60 \end{bmatrix}^{-1}$$
$$= \begin{bmatrix} 0.28814 & -0.11864 \\ -0.83051 & 0.69491 \end{bmatrix}.$$

For $m=1$, the normal equations (4.7) are

$$[\mathbf{h}_0\mathbf{h}_1]\begin{bmatrix} \phi_{ff}(0) & \phi_{ff}(1) \\ \phi_{ff}^T(1) & \phi_{ff}(0) \end{bmatrix} = [\phi_{df}(0) \quad \phi_{df}(1)].$$

The correlation coefficients are second-order square matrices, and hence the above relation represents four linear simultaneous equations, whose solution is

$$[\mathbf{h}_0, \mathbf{h}_1] = \left\{\begin{bmatrix} 0.23127 & -0.06156 \\ -1.06399 & 0.90320 \end{bmatrix}, \begin{bmatrix} 0.18185 & -0.17294 \\ 0.52936 & -0.42325 \end{bmatrix}\right\}.$$

The normalized mean square error (4.10) must here be written in the form,

$$E = 1 - \frac{tr \sum_{t=0}^{m} \phi_{df}(t)\mathbf{h}_t^T}{tr\ \phi_{dd}(0)}.$$

But for the least squares inverse filter, $\mathbf{d}_t = \mathbf{I}$ and

$$\phi_{df}(t) = \begin{cases} \mathbf{f}_0^T, & t = 0, \\ \boldsymbol{\theta}, & t > 0, \end{cases}$$
$$tr\ \phi_{dd}(0) = tr\ \mathbf{I} = l,$$

and we recall that $\phi_{dd}(0)$ is a $(l)\times(l)$ matrix. Thus

$$E = 1 - \frac{1}{l}\, tr\ (\mathbf{f}_0^T\mathbf{h}_0^T)$$
$$= 1 - \frac{1}{l}\, tr\ (\mathbf{h}_0\mathbf{f}_0)^T.$$

But $\mathbf{h}_0\mathbf{f}_0 = \mathbf{y}_0$, where $\mathbf{y}_0$ is the zeroth matrix-valued actual output coefficient; therefore,

$$E = 1 - \frac{1}{l}\, tr\ \mathbf{y}_0, \tag{4.12}$$

where we recognize that the trace of a square matrix is equal to the trace of its transpose. Equation (4.12) is a particularly simple expression for the case of the zero-lag least squares inverse filter.

By direct computation, we find for $m=0$ that

$$(\mathbf{y}_0, \mathbf{y}_1) = (\mathbf{h}_0\mathbf{f}_0, \mathbf{h}_0\mathbf{f}_1)$$
$$= \left\{\begin{bmatrix} 0.89830 & -0.11864 \\ -0.11864 & 0.69491 \end{bmatrix}, \begin{bmatrix} 0.16949 & 0.22034 \\ -0.13559 & 0.42373 \end{bmatrix}\right\},$$

while the normalized mean square error is, by equation (4.11),

$$E = 1 - \tfrac{1}{2}(0.89830 + 0.69491) = 0.203.$$

Similarly, we obtain for $m=1$

$$(\mathbf{y}_0, \mathbf{y}_1, \mathbf{y}_2) = (\mathbf{h}_0\mathbf{f}_0, \mathbf{h}_0\mathbf{f}_1 + \mathbf{h}_1\mathbf{f}_0, \mathbf{h}_0\mathbf{f}_1)$$
$$= \left\{\begin{bmatrix} 0.95666 & -0.06156 \\ 0.06156 & 0.90320 \end{bmatrix}, \begin{bmatrix} 0.05022 & 0.10490 \\ 0.05265 & 0.15836 \end{bmatrix}, \begin{bmatrix} 0.00891 & -0.15512 \\ 0.10612 & -0.21102 \end{bmatrix}\right\}.$$

The normalized mean square error is

$$E = 1 - \tfrac{1}{2}(0.95666 + 0.90320) = 0.070.$$

As expected, the value of E decreases as the filter length increases. For ideal performance, the convolution of the inverse filter $\mathbf{h}_t$ with the given filter $\mathbf{f}_t$ should yield the identity matrix at zero delay; that is, as $m \to \infty$, we would obtain better and better approximations to the desired output

$$\mathbf{d}_t = \left\{\begin{bmatrix} 1 & 0 \\ 0 & 1 \end{bmatrix}, \begin{bmatrix} 0 & 0 \\ 0 & 0 \end{bmatrix}, \ldots\right\}.$$

We observe that even for $m=1$ the two-length least squares inverse filter yields a good approximation to this desired output.

As m grows larger, the value of E decreases

monotonically; and exact filter performance is achieved as $m \to \infty$. In that case, the least squares inverse filter $\mathbf{H}(z)$ and the exact inverse filter $\mathbf{F}^{-1}(z)$ become identical. The calculation of $\mathbf{F}^{-1}(z)$ has already been discussed in Section 3.0; we showed there that for our present example,

$$\mathbf{F}^{-1}(z) = \frac{1}{6+5z+z^2}\begin{bmatrix} 1+3z & -2z \\ -7-z & 6+z \end{bmatrix}.$$

If we divide each element of the above matrix by the polynomial $(6+5z+z^2)$, we obtain

$$\mathbf{F}^{-1}(z) = \begin{bmatrix} 0.16667 + 0.36111z + \cdots & 0 - 0.33333z + \cdots \\ -1.16667 + 0.80555z + \cdots & 1 - 0.66667z + \cdots \end{bmatrix},$$

where the dots signify that the expansions are infinite. On the other hand, the least squares inverse filter for $m=4$ is

$$\mathbf{H}(z) = \begin{bmatrix} 0.16876 + 0.35475z + \cdots & -0.00206 - 0.32740z + \cdots \\ -1.16437 + 0.79856z + \cdots & 0.99773 - 0.66015z + \cdots \end{bmatrix},$$

where the dots signify that only the first two terms of the component five-length single-channel filters have been retained. Comparison of corresponding coefficients in the expressions for $\mathbf{F}^{-1}(z)$ and $\mathbf{H}(z)$ shows close agreement; as $m \to \infty$, this agreement would be perfect.[4]

Finally, we remark that the multichannel zero-lag least squares inverses are minimum delay, a result which has been proved by Robinson (1962, chapter XI). Let us see whether the two-length least squares inverse filter computed previously satisfies this condition. The z-transform of the filter is

$$\mathbf{H}(z) = \begin{bmatrix} 0.23127 + 0.18185z & -0.06156 - 0.17294z \\ -1.06399 + 0.52936z & 0.90320 - 0.42325z \end{bmatrix}.$$

By the method introduced in Section 3.0, we find that

[4] The input for the present example is a square matrix, rather than a column vector; hence, exact filter performance is not possible for finite values of m, as was the case for Examples 1 and 2 above.

$$\text{Det}\, \mathbf{H}(z) = 0.14336 - 0.08504z + 0.01458z^2,$$

whose roots are

$$(z_1, z_2) = 2.916 \pm 1.153i.$$

Since both these roots lie outside the unit circle $|z|=1$ and since $\mathbf{H}(z)$ is a polynomial matrix, we conclude that $\mathbf{H}(z)$ is indeed minimum delay.

5.0 CONCLUDING REMARKS

The present treatment, in spite of its length, deals with only a few of the highlights of multichannel filter theory. A book by Robinson (1967) contains detailed discussions of more advanced topics in this field, such as multichannel spectral analysis and multichannel spectral factorization techniques. The Wiener filter is only one among several linear least squares multichannel processors which can be used for seismic data enhancement. Thus, the *minimum variance, unbiased* (or MVU) filter constitutes a related but alternate approach. It has been successfully applied to the reduction of data acquired through the LASA (Large Aperture Seismic Array) in Montana (Green et al, 1966; Capon et al, 1967).

The aim of this work has been primarily tutorial. For this reason, no multichannel analyses of real seismic data have been presented; the interested reader can find excellent discussions of such examples in the literature cited below. Not all of these applications have been successful, and in some instances improvement over single-channel processing has been marginal. No one processing philosophy offers a panacea for all problems at hand, however, and much further work needs to be done in order to determine how multichannel theory is best applied to seismic data analysis.

APPENDIX A

DERIVATION OF THE z-TRANSFORM OF THE SINGLE-CHANNEL CROSSCORRELATION FUNCTION

The single-channel crosscorrelation between the $(n+1)$-length series $x(t)$ and the $(m+1)$-length series $y(t)$ may be defined in the form,

$$\phi_{xy}(\tau) = \sum_{t=0}^{m} x(t+\tau)y(t),$$

$$\tau = -m, \cdots, 0, \cdots, +n.$$

Without loss of generality, we assume that $m \le n$. The cross-spectrum between $x(t)$ and $y(t)$, is

$$\Phi_{xy}(\omega) = \sum_{\tau=-m}^{+n} \phi_{xy}(\tau)e^{-i\omega\tau}$$

$$= \sum_{\tau=-m}^{+n} \sum_{t=0}^{+m} x(t+\tau)y(t)e^{-i\omega\tau},$$

where ω=angular frequency. If we make the change of variables $t+\tau=s$, the above becomes

$$\Phi_{xy}(\omega) = \sum_{s=t-m}^{t+n} \sum_{t=0}^{+m} x(s)y(t)e^{-i\omega(s-t)}.$$

But since the series $x(t)$ is defined only for $0 \le t \le n$, values of $x(t)$ outside this range are zero, and thus the limits on s are $0 \le s \le n$ as well. Accordingly, we may write,

$$\Phi_{xy}(\omega) = \sum_{s=0}^{n} x(s)e^{-i\omega s} \sum_{t=0}^{m} y(t)e^{+i\omega t}.$$

The z-transform of $\phi_{xy}(\tau)$ is obtained by substituting $z=e^{-i\omega}$ into the above relation. This substitution yields

$$\Phi_{xy}(z) = \sum_{s=0}^{n} x_s z^s \sum_{t=0}^{m} y_t z^{-t}$$

or simply,

$$\Phi_{xy}(z) = X(z)Y(z^{-1}),$$

which is equation (2.29) of the main text.

APPENDIX B

DERIVATION OF THE MULTICHANNEL WIENER FILTER

We seek the set of filter coefficients

$$f_{ij}(t), \qquad i = 1, 2, \cdots, l, \quad j = 1, 2, \cdots, k, \quad t = 0, 1, \cdots, m,$$

which minimizes the total error energy I. This quantity has been shown to be representable as the sum of l individual error energies ϵ_i,

$$I = \epsilon_1 + \epsilon_2 + \cdots + \epsilon_l,$$

[see equation (4.3)]. Differentiation of both sides with respect to the coefficients $f_{ij}(t)$ yields

$$\frac{\partial}{\partial f_{ij}(t)} I = \frac{\partial}{\partial f_{ij}(t)} \epsilon_1 + \frac{\partial}{\partial f_{ij}(t)} \epsilon_2 + \cdots + \frac{\partial}{\partial f_{ij}(t)} \epsilon_l. \qquad \text{(B-1)}$$

Use of equations (4.3) and (4.4) allows us to write

$$\epsilon_i = \sum_{t=0}^{m+n} e_i^2(t) = \sum_{t=0}^{m+n} \left[d_i(t) - y_i(t)\right]^2, \qquad \text{(B-2)}$$

where all desired output series $d_i(t)$ and all actual output series $y_i(t)$ are defined for $t=0, 1, \cdots, m+n$. From (4.5), we have

$$y_i(t) = \sum_{j=1}^{k} f_{ij}(t) * x_j(t),$$

or

$$y_i(t) = \sum_{j=1}^{k} \sum_{s=0}^{m} f_{ij}(s)x_j(t-s), \qquad \text{(B-3)}$$

where all input series $x_j(t)$ are defined for $t=0, 1, \cdots, n$. Let us successively set $i=1, 2, \cdots, l$ in equation (B-1). We obtain the system,

$$\frac{\partial}{\partial f_{1j}(t)} I = \frac{\partial}{\partial f_{1j}(t)} \epsilon_1 + \frac{\partial}{\partial f_{1j}(t)} \epsilon_2 + \cdots + \frac{\partial}{\partial f_{1j}(t)} \epsilon_l,$$

$$\frac{\partial}{\partial f_{2j}(t)} I = \frac{\partial}{\partial f_{2j}(t)} \epsilon_1 + \frac{\partial}{\partial f_{2j}(t)} \epsilon_2 + \cdots + \frac{\partial}{\partial f_{2j}(t)} \epsilon_l; \qquad \text{(B-4)}$$

$$\cdots\cdots\cdots$$

$$\frac{\partial}{\partial f_{lj}(t)} I = \frac{\partial}{\partial f_{lj}(t)} \epsilon_1 + \frac{\partial}{\partial f_{lj}(t)} \epsilon_2 + \cdots + \frac{\partial}{\partial f_{lj}(t)} \epsilon_l.$$

From equations (B-2) and (B-3), we establish that $y_i(t)$ and hence ϵ_i depend exclusively on the ith filter components f_{ij}, so that

$$\frac{\partial}{\partial f_{ij}(t)} \epsilon_\nu$$

vanishes for all ν except $\nu=i$. We follow the usual minimization procedure and set

$$\frac{\partial}{\partial f_{ij}(t)} I = 0,$$

so that the system (B-4) becomes

$$\frac{\partial}{\partial f_{1j}(t)} I = \frac{\partial}{\partial f_{1j}(t)} \epsilon_1 = 0,$$
$$\frac{\partial}{\partial f_{2j}(t)} I = \frac{\partial}{\partial f_{2j}(t)} \epsilon_2 = 0, \qquad \text{(B-5)}$$
$$\cdots\cdots\cdots\cdots$$
$$\frac{\partial}{\partial f_{lj}(t)} I = \frac{\partial}{\partial f_{lj}(t)} \epsilon_l = 0.$$

We now see that the minimization of the total error energy I is equivalent to the minimization of each of the individual error energies ϵ_i. It will thus suffice to derive the normal equations resulting from the minimization of a particular ϵ_i, since the overall system of normal equations can then be obtained by induction.

Let $i=1$, say. We have in this case,

$$\frac{\partial}{\partial f_{1j}(t)} \epsilon_1 = 0 \quad \text{for } j = 1, 2, \cdots, k, \qquad t = 0, 1, \cdots, m. \qquad \text{(B-6)}$$

In order to simplify matters further, let $j=1, 2$ and $t=0, 1$. By equation (B-3), we write the single output series $y_1(t)$ as the sum

$$\begin{aligned} y_1(t) &= \sum_{j=1}^{2} \sum_{s=0}^{1} f_{1j}(s) x_j(t-s) \\ &= f_{11}(0)x_1(t) + f_{12}(0)x_2(t) \\ &\quad + f_{11}(1)x_1(t-1) + f_{12}(1)x_2(t-1), \end{aligned} \qquad \text{(B-7)}$$

where s is a dummy variable for the discrete time. But from equation (B-2), we have

$$\begin{aligned} \epsilon_1 &= \sum_t [d_1(t) - y_1(t)]^2 \\ &= \sum_t d_1^2(t) - 2\sum_t d_1(t)y_1(t) + \sum_t y_1^2(t). \end{aligned}$$

Substitution of this expression into (B-6) yields, after differentiation with respect to $f_{1j}(s)$,

$$\begin{aligned} \frac{\partial}{\partial f_{1j}(s)} \epsilon_1 &= -2\sum_t d_1(t) \frac{\partial y_1(t)}{\partial f_{1j}(s)} \\ &\quad + 2\sum_t y_1(t) \frac{\partial y_1(t)}{\partial f_{1j}(s)} = 0, \end{aligned} \qquad \text{(B-8)}$$

where $s=0, 1$ and $j=1, 2$. Let us arrange our computations in such a way that we evaluate $\partial y_1(t)/\partial f_{1j}(s)$ in the order

$$\begin{aligned} s &= 0 \quad j = 1, \\ s &= 0 \quad j = 2, \\ s &= 1 \quad j = 1, \\ s &= 1 \quad j = 2. \end{aligned}$$

From (B-7) we then obtain

$$s = 0 \quad j = 1 \quad \frac{\partial y_1(t)}{\partial f_{11}(0)} = x_1(t),$$
$$s = 0 \quad j = 2 \quad \frac{\partial y_1(t)}{\partial f_{12}(0)} = x_2(t),$$
$$s = 1 \quad j = 1 \quad \frac{\partial y_1(t)}{\partial f_{11}(1)} = x_1(t-1),$$
$$s = 1 \quad j = 2 \quad \frac{\partial y_2(t)}{\partial f_{12}(1)} = x_2(t-1).$$

Substitution of the first of the above relations into equation (B-8) yields

$$\begin{aligned} \frac{\partial}{\partial f_{11}(0)} \epsilon_1 &= -2\sum_t d_1(t)x_1(t) \\ &\quad + 2\sum_t y_1(t)x_1(t) = 0. \end{aligned} \qquad \text{(B-9)}$$

But

$$\sum_t d_1(t)x_1(t) = \phi_{d_1x_1}(0), \qquad \text{(B-10)}$$

where $\phi_{d_1x_1}(\tau)$ is the crosscorrelation between the single-channel series $d_1(t)$ and $x_1(t)$. From (B-7) we have

$$\begin{aligned} \sum_t y_1(t)x_1(t) &= f_{11}(0) \sum_t x_1(t)x_1(t) \\ &\quad + f_{12}(0) \sum_t x_2(t)x_1(t) \\ &\quad + f_{11}(1) \sum_t x_1(t-1)x_1(t) \\ &\quad + f_{12}(1) \sum_t x_2(t-1)x_1(t), \end{aligned}$$

or

$$\sum_t y_1(t)x_1(t) = f_{11}(0)\phi_{x_1x_1}(0) + f_{12}(0)\phi_{x_2x_1}(0) + f_{11}(1)\phi_{x_1x_1}(-1) + f_{12}(1)\phi_{x_2x_1}(-1). \qquad \text{(B-11)}$$

Here we have made use of the general relation

$$\phi_{x_px_q}(\tau) = \sum_t x_p(t+\tau)x_q(t),$$

so that $\phi_{x_px_q}(\tau)$ is the crosscorrelation between the single-channel series $x_p(t)$ and $x_q(t)$ (see Section 2.3). Relations (B-10) and (B-11) may now be combined with equation (B-9). In this manner we obtain the first of our four required normal equations, namely the one for $s=0$ and $j=1$,

$$f_{11}(0)\phi_{x_1x_1}(0) + f_{12}(0)\phi_{x_2x_1}(0) + f_{11}(1)\phi_{x_1x_1}(-1) + f_{12}(1)\phi_{x_2x_1}(-1) = \phi_{d_1x_1}(0).$$

The remaining three normal equations are derivable in the same way, so that the reader will be spared further drudgery. The resulting system of four linear simultaneous equations in the unknown filter coefficients $f_{11}(0)$, $f_{12}(0)$, $f_{11}(1)$, and $f_{12}(1)$ can be expressed in matrix form,

$$\left[\begin{array}{cc|cc} f_{11}(0) & f_{12}(0) & f_{11}(1) & f_{12}(1) \end{array}\right] \left[\begin{array}{cc|cc} \phi_{x_1x_1}(0) & \phi_{x_1x_2}(0) & \phi_{x_1x_1}(1) & \phi_{x_1x_2}(1) \\ \phi_{x_2x_1}(0) & \phi_{x_2x_2}(0) & \phi_{x_2x_1}(1) & \phi_{x_2x_2}(1) \\ \hline \phi_{x_1x_1}(-1) & \phi_{x_1x_2}(-1) & \phi_{x_1x_1}(0) & \phi_{x_1x_2}(0) \\ \phi_{x_2x_1}(-1) & \phi_{x_2x_2}(-1) & \phi_{x_2x_1}(0) & \phi_{x_2x_2}(0) \end{array}\right] = \left[\begin{array}{cc|cc} \phi_{d_1x_1}(0) & \phi_{d_1x_2}(0) & \phi_{d_1x_1}(1) & \phi_{d_1x_2}(1) \end{array}\right]. \qquad \text{(B-12)}$$

Next we notice that this matrix equation can be partitioned into the submatrices as indicated by the dashed lines. If we compare these submatrices with equations (2.13), (2.27b), and (2.27c) of the main text, we see that the above matrix relation can be written

$$\left[\begin{array}{c|c} \mathbf{f}_0 & \mathbf{f}_1 \end{array}\right] \left[\begin{array}{c|c} \boldsymbol{\phi}_{xx}(0) & \boldsymbol{\phi}_{xx}(1) \\ \hline \boldsymbol{\phi}_{xx}(-1) & \boldsymbol{\phi}_{xx}(0) \end{array}\right] = \left[\begin{array}{c|c} \boldsymbol{\phi}_{dx}(0) & \boldsymbol{\phi}_{dx}(1) \end{array}\right]. \qquad \text{(B-13)}$$

The coefficients of this matrix equation are thus matrices rather than scalars. In particular, the filter weighting coefficients $\mathbf{f}_s(s=0, 1)$ are $(1)\times(2)$ matrices; the autocorrelation coefficients $\boldsymbol{\phi}_{xx}(s)$ are $(2)\times(2)$ matrices; and the crosscorrelation coefficients $\boldsymbol{\phi}_{dx}(s)$ are $(1)\times(2)$ matrices.

This completes our minimization of the individual error energy ϵ_1, and we must now do likewise for the remaining error energies $\epsilon_2, \cdots, \epsilon_l$. But it is not difficult to see that a system analogous to that of equation (B-12) will be obtained for every such minimization, except for the fact that we must deal with a new set of filter coefficients and a new set of crosscorrelation coefficients. The autocorrelation coefficients remain the same because the input series $x_i(t)$, $(i=1, \cdots, l)$, remain unchanged for all cases. For example, the minimization of ϵ_2 leads to the system,

$$[f_{21}(0) \quad f_{22}(0) \quad f_{21}(1) \quad f_{22}(1)] \cdot \left[\begin{array}{c|c} \quad & \quad \\ \hline \quad & \quad \end{array}\right] = [\phi_{d_2x_1}(0) \quad \phi_{d_2x_2}(0) \quad \phi_{d_2x_1}(1) \quad \phi_{d_2x_2}(1)]. \qquad \text{(B-14)}$$

The system of four simultaneous linear equations (B-12) can be combined with the system of four linear simultaneous equations (B-14) in order to yield an equivalent system in the eight unknown filter coefficients

$$f_{ij}(s), \qquad i = 1, 2, \quad j = 1, 2, \quad s = 0, 1,$$

which is

$$\begin{bmatrix} f_{11}(0) & f_{12}(0) & | & f_{11}(1) & f_{12}(1) \\ f_{21}(0) & f_{22}(0) & | & f_{21}(1) & f_{22}(1) \end{bmatrix} \left[\begin{array}{c|c} & \\ \hline & \end{array}\right] = \begin{bmatrix} \phi_{d_1x_1}(0) & \phi_{d_1x_2}(1) & | & \phi_{d_1x_1}(1) & \phi_{d_1x_2}(1) \\ \phi_{d_2x_1}(1) & \phi_{d_2x_2}(0) & | & \phi_{d_2x_1}(1) & \phi_{d_2x_2}(1) \end{bmatrix}. \tag{B-15}$$

However, the representation (B-13) again holds here, except that the coefficients $\mathbf{f}_s$ and $\boldsymbol{\phi}_{dx}$ are now $(2)\times(2)$ matrices.

The general case of k input channels, l output channels, and $(m+1)$-length filters follows by induction from (B-13). We thus obtain the *system of multichannel normal equations*, which we express in the form,

$$[\mathbf{f}_0 \quad \mathbf{f}_1 \cdots \mathbf{f}_m] \begin{bmatrix} \boldsymbol{\phi}_{xx}(0) & \boldsymbol{\phi}_{xx}(1) & \cdots & \boldsymbol{\phi}_{xx}(m) \\ \boldsymbol{\phi}_{xx}(-1) & \boldsymbol{\phi}_{xx}(0) & \cdots & \boldsymbol{\phi}_{xx}(m-1) \\ \vdots & & & \vdots \\ \boldsymbol{\phi}_{xx}(-m) & \boldsymbol{\phi}_{xx}(-m+1) & \cdots & \boldsymbol{\phi}_{xx}(0) \end{bmatrix} = [\boldsymbol{\phi}_{dx}(0) \quad \boldsymbol{\phi}_{dx}(1) \cdots \boldsymbol{\phi}_{dx}(m)], \tag{B-16}$$

where the filter coefficients $\mathbf{f}_s$ are $(l)\times(k)$ matrices, where the autocorrelation coefficients $\boldsymbol{\phi}_{xx}(s)$ are $(k)\times(k)$ matrices, and where the crosscorrelation coefficients $\boldsymbol{\phi}_{dx}(s)$ are $(l)\times(k)$ matrices.

The above $(m+1)\times(m+1)$ matrix of *matrix-valued* autocorrelation coefficients is called the multichannel autocorrelation matrix. It is structurally similar to the single-channel autocorrelation matrix, but with the important difference that its elements are $(k)\times(k)$ matrices rather than scalars. The relation (B-16) represents a total of $lk\ (m+1)$ linear simultaneous scalar equations in the scalar-valued filter coefficients $f_{ij}(s)$, $i=1, 2, \cdots, l$; $j=1, 2, \cdots, k$; $s=0, 1, \cdots, m$. Alternatively, it represents a system of $(m+1)$ linear simultaneous matrix equations in the matrix-valued filter coefficients $\mathbf{f}_s$, $(s=0, 1, \cdots, m)$.

The representation of the system (B-16) is not unique in the sense that its constituent scalar equations could have been arranged in a different order. However, we notice that all submatrices on any given diagonal of the multichannel autocorrelation matrix are identical. This is the so-called "block Toeplitz" property of the multichannel autocorrelation matrix. It was first recognized by E. A. Robinson that such an arrangement leads directly to a multichannel generalization of the familiar single-channel Wiener-Levinson algorithm (Levinson, 1947). We are thus in a position to solve the system (B-16) recursively for the matrix-valued filter coefficients $\mathbf{f}_s$. The details of this scheme have been described by Wiggins and Robinson (1965). The main advantages of the recursive technique are in both computing speed and in computer storage requirements. Thus speed is proportional to m^3 if the system (B-16) is solved by simultaneous equations methods, but is only proportional to m^2 if Robinson's generalized Wiener-Levinson algorithm is used. It should be pointed out that the treatment by Wiggins and Robinson (loc. cit.) contains a number of errors, but these have been corrected in Robinson (1967, Chapter 6).

By virtue of equation (2.28c) we can rewrite the system (B-16) in the alternate form,

$$
[\mathbf{f}_0 \quad \mathbf{f}_1 \cdots \mathbf{f}_m]
\begin{bmatrix}
\phi_{xx}(0) & \phi_{xx}(1) & \cdots & \phi_{xx}(m) \\
\overset{T}{\phi}_{xx}(1) & \phi_{xx}(0) & \cdots & \phi_{xx}(m-1) \\
\vdots & \vdots & & \vdots \\
\overset{T}{\phi}_{xx}(m) & \overset{T}{\phi}_{xx}(m-1) & \cdots & \phi_{xx}(0)
\end{bmatrix}
= [\phi_{dx}(0) \quad \phi_{dx}(1) \cdots \phi_{dx}(m)], \tag{B-17}
$$

so that $\phi_{xx}(\tau)$ need only be computed for $\tau \geq 0$. The submatrices below the main diagonal are then obtained by transposition of corresponding submatrices above the main diagonal.

The solution of the system (B-17) yields the set of optimum filter coefficients $\mathbf{f}_s (s=0, 1, \cdots, m)$. These coefficients are optimum in the sense that they minimize the total error energy, $I_{\min}$. As was shown earlier [equation (B-5) et seq.], the minimization of I implies the minimization of the individual error energies ϵ_i, and thus

$$
I_{\min} = \epsilon_{1,\min} + \epsilon_{2,\min} + \cdots + \epsilon_{l,\min}, \tag{B-18}
$$

where $\epsilon_{i,\min}$ is the minimum value of ϵ_i. An expression for $I_{\min}$ can be obtained by substituting solutions of the form (B-12) into equation (B-2) and summing over all output channels, $i=1, \cdots, l$. Robinson (1967, Chapter 6) gives a detailed derivation; we merely state the final result, which is

$$
I_{\min} = tr\, \phi_{dd}(0) - tr \sum_{s=0}^{m} \phi_{dx}(s)\mathbf{f}_s^T, \tag{B-19}
$$

where the notation $tr\, \mathbf{A}$ denotes the trace of the matrix $\mathbf{A}$.[5] $\phi_{dd}(0)$ is the zeroth autocorrelation coefficient of the desired output series $\mathbf{d}(t)$; that is,

$$
\phi_{dd}(0) = E\{\mathbf{d}(t)\mathbf{d}^T(t)\}
= \begin{bmatrix}
\phi_{d_1d_1}(0) & \cdots & \phi_{d_1d_l}(0) \\
\vdots & & \vdots \\
\phi_{d_ld_1}(0) & \cdots & \phi_{d_ld_l}(0)
\end{bmatrix},
$$

which is a square $(l)\times(l)$ matrix [cp. equation (2.27c)]. Thus,

$$
tr\, \phi_{dd}(0) = \sum_{i=1}^{l} \phi_{d_id_i}(0) = \sum_{i=1}^{l} \sum_{t=0}^{m+n} d_i^2(t).
$$

The crosscorrelation coefficients $\phi_{dx}(s)$ are $(l)\times(k)$ matrices, while the transposed filter coefficients $\mathbf{f}_s^T$ are $(k)\times(l)$ matrices. Therefore, the second term of (B-19) is the trace of a sum of $(m+1)$ square $(l)\times(l)$ matrices. It is convenient to normalize the expression for $I_{\min}$ in such a way that this quantity will always lie between zero and unity. We do this by dividing both sides of equation (B-19) through by $tr\, \phi_{dd}(0)$,

$$
\mathrm{E} = 1 - \frac{tr \sum_{s=0}^{m} \phi_{dx}(s)\mathbf{f}_s^T}{tr\, \phi_{dd}(0)}, \tag{B-20}
$$

where $\mathrm{E} = I_{\min}/tr\phi_{dd}(0)$. The quantity E is called the normalized minimum total error energy, or simply the normalized mean square error. Since E is a measure of energy, it cannot be negative. Neither can E be greater than 1, because the value $\mathrm{E}=1$ results when all filter coefficients $\mathbf{f}_s$ are null matrices. The similarities with the single-channel case are evident; the reader is referred to Treitel and Robinson (1966) for further elaboration.

[5] The trace of a square matrix is equal to the sum of the elements of its main diagonal.

REFERENCES

Burg, J. P., 1964, Three-dimensional filtering with an array of seismometers: Geophysics, v. 29, p. 693–713.

Capon, J., Greenfield, R. J., and Kolker, R. J., 1967, Multidimensional maximum-likelihood processing of a large aperture seismic array: Proc. of the IEEE, v. 55, p. 192–211.

Davies, E. B., and Mercado, E. J., 1968, Multichannel deconvolution filtering of field recorded seismic data: Geophysics, v. 33, p. 711–722.

Frazer, R. A., Duncan, W. J., and Collar, A. R., 1938, Elementary matrices: New York, Cambridge University Press.

Galbraith, J. N., and Wiggins, R. A., 1968, Characteristics of optimum multichannel stacking filters: Geophysics, v. 33, p. 36–48.

Gantmacher, F. G., 1959, The theory of matrices: New York, Chelsea Publishing Co., two vols., (translated from the Russian by K. A. Hirsch).

Green, P. E., Jr., Kelly, E. J., Jr., and Levin, M. J., 1966, A comparison of seismic array processing methods: Geophys. J. R. Astr. Soc., v. 11, p. 67–84.

Laster, S. J., and Linville, A. F., 1966, Application of multichannel filtering to the separation of dispersive modes of propagation: J. Geoph. Res., v. 71, p. 1669–1701.

Levinson, N., 1947, The Wiener RMS (root mean

square) error criterion in filter design and prediction: J. of Math. and Phys., v. 25, p. 261–278.
Robinson, E. A., 1962, Random wavelets and cybernetic systems: London, Charles Griffin and Co., Ltd.
——— 1966, Multichannel z-transforms and minimum delay: Geophysics, v. 31, p. 482–500.
——— 1967, Multichannel time series analysis with digital computer programs: San Francisco, Holden-Day.
Robinson, E. A., and Treitel, S., 1964, Principles of digital filtering: Geophysics, v. 29, p. 395–404.
——— 1967, Principles of digital Wiener filtering: Geophys. Prosp., v. 15, p. 311–333.
Shanks, J. L., 1967, Recursion filters for digital processing: Geophysics, v. 32, p. 33–51.
Schneider, W. A., Larner, K. L., Burg, J. P., and Backus, M. M., 1964, A new data processing technique for the elimination of ghost arrivals on reflection seismograms: Geophysics, v. 29, p. 783–805.
Schneider, W. A., Prince, E. R., and Giles, B. F., 1965, A new data processing technique for multiple attenuation exploiting differential normal moveout: Geophysics, v. 30, p. 348–362.
Treitel, S., 1970, An ideal performance property of the multichannel Wiener filter: Submitted for publication.
Treitel, S., and Robinson, E. A., 1964, The stability of digital filters: IEEE Trans. on Geosci. Electronics, v. GE-2, p. 6–18.
——— 1966, The design of high resolution digital filters: IEEE Trans. on Geosci. Electronics, v. GE-4, p. 25–38.
Wiggins, R. A., and Robinson, E. A., 1965, Recursive solution to the multichannel filtering problem: J. Geoph. Res., v. 70, p. 1885–1891.

VI. TIME-VARYING DECONVOLUTION

G. K. C. Clarke considers the problem of deconvolution when the signal waveform changes shape due to attenuation. To simplify the analysis, he considers the attenuation to be constant over the entire geologic section so that the change of a signal waveform is independent of its raypath and depends only on its age. A successful deconvolution is demonstrated for the case where both the initial waveform and the attenuation law are minimum phase. Further, if the trace is not strongly nonstationary for times of the order of the filter length, equivalent results can be obtained with far less computation.

R. J. Wang points out that most time-varying filter techniques involve the empirical division of a seismic trace into a number of gates and the design of a time-invariant filter for each gate. His paper describes a technique for the determination of optimum gate lengths. In a rather complicated analysis, the trace is divided into a number of short sections; a time-invariant autocorrelation is computed for each section; and these autocorrelations are then expanded into cosine series. For three adjacent sections, the coefficients of each cosine term can be expressed as a second-order function of time. Finally, the coefficients of the quadratic terms in these equations are used in an estimate of the optimum gate length.

D. D. Thompson and G. R. Cooper discuss the estimation of a time-varying autocorrelation by computing lagged product averages over a "sliding window." As they point out, one is faced with a tradeoff between a bias error resulting from excessive window size and a variance error resulting from insufficient window size. A rapidly convergent iterative procedure is presented which minimizes the bias error while preserving the small variance afforded by a large window.

In a very interesting paper, L. J. Griffiths et al show the results from using adaptive deconvolution on seismic data. In this procedure, a new operator is designed and applied at each point of the trace. Choice of the parameter which controls the rate of adaptation is critical. If the rate is too large, the output noise power is large; if it is too small, the operator may be unable to track the changes in reverberation time. The authors show that (fortunately) the adaptive time constant can be changed over a range of nearly 3 to 1 without significantly affecting the quality of the deconvolved trace. A three-step procedure is recommended: First, process in reverse time order, saving the final operator; then, starting with that operator, process the original data in forward time; and finally, average the reverse and forward-time deconvolved traces. Very impressive results are presented using both synthetic data and data from the notoriously difficult Celtic Sea area.

Reprinted from Geophysics v. 33, no. 6, p. 936-944

TIME-VARYING DECONVOLUTION FILTERS†

G. K. C. CLARKE*

Deconvolution or spiking filters are frequently employed to sharpen the character of seismograms and hence improve resolution of the earth's layering. In media having significant attenuation or scattering, the shape of an input waveform will change as it propagates. On a reflection seismogram this results in a change with time in the character and frequency content of the seismic signal. Since a statistical description of this situation would be time-dependent, the seismogram should be treated as a nonstationary random process. A spiking filter to deconvolve such a seismogram must be time-variable. Optimum filters for nonstationary inputs can be designed using the least-mean-square error criterion. For a zero-delay filter to give satisfactory results, the signal waveform must at all times be minimum phase. This will be true if the input waveform and the law governing the change of waveshape with time are both minimum phase. If these conditions are not met, a filter which has a time delay should be used. For a linear attenuation mechanism with frequency-independent Q, there is some justification for expecting a minimum-phase law. In the absence of a priori knowledge of the input waveform and the attenuation structure of the layered earth, it is necessary to estimate the time-dependent autocorrelation function of the recorded seismic trace.

INTRODUCTION

A reflection seismogram is often imagined to be formed by the convolution of the response function for a layered earth with some constant waveform. The goal of deconvolution filtering is to return from the recorded seismogram to the ground response function. If, however, the signal waveform changes shape due to attenuation or scattering of signal energy, a suitable deconvolution filter is necessarily time-variable, and the problem of filter design becomes more complicated. The usual (time-invariant) deconvolution problem is frequently regarded as a special case of optimum waveshaping, where the filter input is a known waveform with noise and the desired output is a spike-like function. Upon application of the Wiener filter theory, the time-invariant property of the filter follows directly from the assumed stationarity of the input process. If the input signal waveform is thought to change shape as it propagates, the filter input must be treated as a nonstationary random process, that is, a random process which requires a time-dependent statistical description. The simplifications which follow from stationarity are no longer available, and the optimum filter must be time-variable.

We shall be concerned with linear time-variable filters; for these the output is obtained by convolution of the input $i(t)$ with the filter response function $h(\tau, t)$. If $h(t-t', t)$ is defined as the response measured at time t due to an impulse originating at time t' then the output is

$$y(t) = \int_{-\infty}^{\infty} h(\tau, t)i(t - \tau)d\tau. \qquad (1)$$

The variable $\tau = t - t'$ measures the interval between initiation and observation of the disturbance; it is therefore the "age" of the output. If $i(t)$ vanishes for $t<0$, the upper integration limit can be replaced by t, and if the system is causal then $h(\tau, t)=0$ for $\tau<0$ so that the lower limit can be replaced by zero.

For the case of time-invariant deconvolution, we suppose that a geologic section which consists of laterally-homogeneous plane layers is excited at the free surface by an impulsive source which emits plane compressional waves vertically into the section. At every interface, upward- and downward-traveling rays result, and the amplitudes of reflected and transmitted pulses are determined by the reflection and transmission coefficients for the particular interface. Because the raypaths are normal to the layering, the complication of conversion between compressional and shear waves does not arise. A seismometer at the surface detects arrivals at times governed by the two-way transit times for the various

† Manuscript received by the Editor January 25, 1968; revised manuscript received May 17, 1968.

* Department of Geophysics, University of British Columbia, Vancouver, Canada.

layers and with amplitudes related to the reflection coefficients between layers. The train of impulses observed at the surface is called the ground response function $g(t)$. If the input signal waveshape at the surface is $s(t)$, then the observed surface disturbance is $r(t)=s(t)*g(t)$. This is the expression for an idealized reflection seismogram in the absence of noise. The aim of deconvolution filtering is to obtain the ground response function $g(t)$ from the recorded seismogram $r(t)$, and therefore to eliminate the effect of the signal waveshape $s(t)$. This is a reasonable objective since $s(t)$ yields no information concerning earth structure, and merely obscures the fine details of the ground response function.

The situation is greatly complicated when the layers have attenuation. If we again assume a discretely-layered, laterally-homogeneous half-space, each layer alters the pulse shape as well as contributes a time delay. The character of the observed impulse-like disturbance will depend on the path taken through the layered attenuating medium. To simplify the analysis we consider the attenuation to be constant over the entire section. In this case the change of signal waveform due to attenuation by the layers is independent of its path and depends only on its age, that is, how long it has been propagating through the attenuating medium. The ground response function can be obtained by passing the attenuation-free ground response $g(t)$ through a time-varying filter which can be determined from the attenuation law.

One way of expressing an attenuation law is by the change of shape of a displacement impulse as it propagates through an attenuating medium. For plane waves traveling in the positive z direction the attenuation law due to Knopoff (1956; 1959) could be written

$$\hat{u}(z,t)=\frac{\beta c z}{\pi}\,\frac{(1+1/Q^2)^{1/2}}{\tau^2c^2(1+1/Q^2)+z^2\beta^2},\qquad(2)$$

where Q is the specific attenuation factor, $c=[(\lambda+2\mu)/\rho]^{\frac{1}{2}}$ is the wave velocity in the absence of losses,

$$\beta=[\tfrac{1}{2}(-1+\sqrt{1+1/Q^2})]^{1/2},$$
$$\gamma=[\tfrac{1}{2}(1+\sqrt{1+1/Q^2})]^{1/2},$$

and

$$\tau=t-\gamma z/(c\sqrt{1+1/Q^2}).$$

Upon substitution of $z=c(t-\tau)(1+1/Q^2)^{\frac{1}{2}}/\gamma$, the expression for $\hat{u}(z,t)$ could be rewritten as some function $u(\tau,t)$ which would then be analogous to the impulse response of a time-varying linear system. The change of shape resulting from the attenuation of some known input waveform is also of interest. To reduce confusion the following notation will be adopted: $s(t,t')$ is the signal of "age" t' which is observed at time t; $s(t,0)$ is therefore the initial waveform. By holding t' fixed and letting t vary, the waveshape at time t' is obtained. This convention is reminiscent of the source and observer coordinates for Green's functions, where t' is the source time and t is the observer time. The amplitude at time t for a waveform of age t' can be expressed in terms of the initial or input waveform $s(t,0)$ and $u(\tau,t)$ because

$$s(t,t')=\int_{-\infty}^{\infty}s(t-t'-\tau,0)u(\tau,t')d\tau.\qquad(3)$$

If the signal waveshape does not change with time, then $s(t,t')$ depends on the difference $t-t'$ between initiation and observation of the signal. Finally, if the waveform is causal, then $s(t,t')=0$ when $t<t'$ since a signal of age t' cannot be observed before an elapsed time t'.

In the case of a constant attenuation, an idealized seismogram can be found either by forming an attenuated ground response function and convolving this with the input waveform, or by convolving the unattenuated ground response function with a time-dependent waveform $s(t,t')$. In the first method the attenuated ground response $k(t)$ can be found from the unattenuated function $g(t)$ by the convolution

$$k(t)=\int_{-\infty}^{\infty}u(\tau,t)g(\tau)d\tau.\qquad(4)$$

For the input signal waveform $s(t)$, the resulting seismogram is

$$r(t)=\int_{-\infty}^{\infty}s(t-\tau)k(\tau)d\tau.\qquad(5)$$

If we take the second point of view, the idealized seismogram results from the convolution

$$r(t)=\int_{-\infty}^{\infty}s(t-t',t')g(t')dt'.\qquad(6)$$

To obtain the unattenuated ground response function from the attenuated seismogram, a time-varying filter is required.

OPTIMUM WAVESHAPING

Optimum filters for nonstationary inputs can be found using an extension of the Wiener filter theory (Booton, 1952). As in the stationary case, one supposes that some input process $i(t)$ is passed through a linear filter to form an output $y(t)$. The optimum filter is chosen so that its actual output $y(t)$ approximates some desired output $d(t)$ in a manner which minimizes the mean-square error between $d(t)$ and $y(t)$. We write this as the expected or average value of the squared error $E\{[d(t)-y(t)]^2\}$. Because the input, output, and desired output processes can be nonstationary, this average is an ensemble average rather than a time average. The optimum linear filter satisfies the integral equation

$$\Phi_{di}(t, t-\tau) = \int_{-\infty}^{\infty} h(\tau', t)\Phi_{ii}(t-\tau, t-\tau')d\tau', \quad (7)$$

where $\Phi_{di}(t, t-\tau)=E[d(t)i(t-\tau)]$ is the crosscorrelation function between the desired filter output and the filter input, and $\Phi_{ii}(t, t-\tau)=E[i(t)i(t-\tau)]$ is the autocorrelation function of the filter input. Because $i(t)$ and $d(t)$ are nonstationary, the auto- and crosscorrelation functions depend on the observation time t as well as the lag τ. The derivation of equation (7) is presented in the Appendix.

When the form of $h(\tau, t)$ is restricted to that of a time-variable delay line (Cruz, 1959) the filter response may be written as

$$h(\tau, t) = \sum_n a_n(t)\delta(\tau - n), \quad (8)$$

and if t takes only discrete values k=1, 2, 3, . . . the integral equation (7) becomes the set of linear equations

$$\Phi_{di}(k, k-m) = \sum_n a_n(k)\Phi_{ii}(k-m, k-n). \quad (9)$$

The coefficients $a_n(k)$ of the filter are found by solving the set of equations for each value of k. When the correlation functions do not depend on k, the matrix $\Phi_{ii}(k-m, k-n)$ becomes equidiagonal and the familiar equations for the stationary case result. If $\Phi_{ii}(k-m, k-n)$ and $\Phi_{di}(k, k-m)$ do not change rapidly with k, the equations need not be solved for every value of k. As nonstationarity becomes less pronounced, the times between optimizations can be increased, and the input can be regarded as stationary over long intervals. In effect, the input is divided into stationary segments and a time-invariant filter is applied to each segment.

In the waveshaping approach to deconvolution, the filter input is imagined to consist of a signal process formed by convolving the input signal waveform with the ground response function (with or without attenuation) and some additive noise process. A filter is sought which will change the shape of the signal waveform $s(t)$ in the presence of noise into some desired spike-like output waveform $d(t)$. Here the desired output function does not change with time, but in the general waveshaping case both the input and desired-output waveshapes are time dependent. Before turning to the particular case of deconvolution, or spiking filters, we will consider the general case. Our aim is to design a filter $h(\tau, t)$ which will change the input waveshape $s(t, t')$ in the presence of noise $n(t)$ into some desired output waveshape $d(t, t')$. The optimum filter satisfies equation (7), so the only difficulty is to determine the functions $\Phi_{di}(t, t-\tau)$ and $\Phi_{ii}(t, t-\tau)$. For additive signal and noise, $i(t)=s(t)+n(t)$ and

$$\Phi_{ii}(t, t-\tau) = \Phi_{ss}(t, t-\tau)+\Phi_{sn}(t, t-\tau) + \Phi_{ns}(t, t-\tau)+\Phi_{nn}(t, t-\tau). \quad (10)$$

If the signal is uncorrelated with the noise, the crosscorrelations vanish; and if we assume white noise, the functions become

$$\Phi_{ii}(t, t-\tau) = \Phi_{ss}(t, t-\tau) + N\delta(\tau)$$

and

$$\Phi_{di}(t, t-\tau) = \Phi_{ds}(t, t-\tau).$$

If $s(t,t')$ and $d(t, t')$ are known, $\Phi_{ss}(t, t-\tau)$ and $\Phi_{ds}(t, t-\tau)$ can be found without difficulty. The autocorrelation $\Phi_{ss}(t, t-\tau)$ is simply the average of the possible values of the product $s(t, t')$ $s(t-\tau, t'-\tau)$ that can be observed at time t. To clarify this point, consider a discrete case for which the signal waveform has a duration of only two sample units (Figure 1). At time $t=k$ the only possible contributions are from $s(k, k-1)$ the wavelet of age $k-1$ observed at time k, and from $s(k, k)$ the wavelet of age k observed at time k, so that

$$\Phi_{ss}(k, k) = s(k, k-1)s(k, k-1) + s(k, k)s(k, k);$$

similarly,

$$\Phi_{ss}(k, k-1) = s(k, k)s(k, k-1) + s(k-1, k-2)s(k-1, k-1)$$

and $\Phi_{ss}(k-2, k) = 0$. The crosscorrelation $\Phi_{di}(k-m, k)$ can be determined in the same way.

The desired output for a deconvolution filter is a spike-like function. We shall consider discrete functions of time and take a unit spike as the desired filter output. An ideal filter which spikes after a p-unit time delay for a signal of age k' would have

$$d(k, k') = 1 \qquad k = k' + p$$
$$d(k, k') = 0 \qquad k \neq k' + p$$

as its output. The crosscorrelation between desired output and the input signal is then

$$\Phi_{ds}(k, k-m) = E[d(k, k')s(k-m, k'-m)] = s(k-m, k-p).$$

The input signal waveform is assumed to be causal

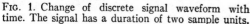

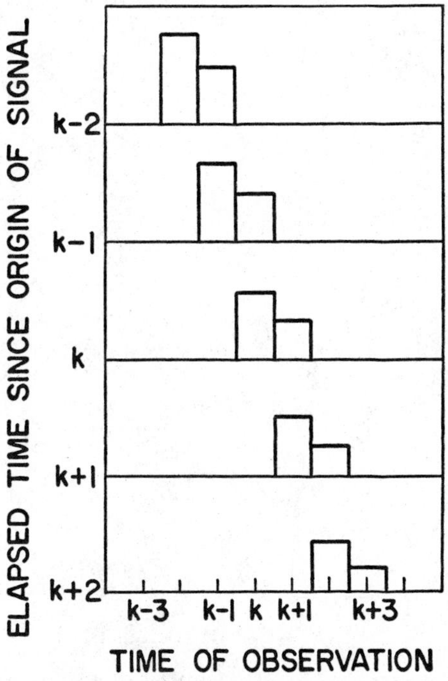

FIG. 1. Change of discrete signal waveform with time. The signal has a duration of two sample units.

so that a signal of age $k-p$ cannot be observed before the time $k-p$. Therefore $s(k-m, k-p) = 0$ for $m > p$ and the crosscorrelation will vanish for values of lag m exceeding p. The case of a zero-delay spiking filter is of special interest when the input signal waveform is minimum-phase, since it is then the optimum location for the output spike, and there is only one nonvanishing term of the crosscorrelation function for positive values of m. Thus

$$\Phi_{ds}(k, k-m) = s(k, k) \qquad m = 0$$
$$= 0 \qquad m > 0.$$

For nonminimum-phase signals a delay in the output spike is necessary for satisfactory filter performance. Treitel and Robinson (1966) have considered the question of optimizing the filter delay as well as the filter coefficients for the time-invariant case. For a time-dependent signal waveshape this would require continual optimizations of the filter lag. We shall therefore assume that the signal waveform is minimum phase at all times. For this to be true, the input signal waveform and the law governing its change of shape must be minimum phase.

In the presence of attenuation with accompanying dispersion, the signal waveform will change shape as it propagates, and an initially minimum-phase waveform will become nonminimum phase unless the attenuation law itself is minimum phase (Sherwood and Trorey, 1965). It is therefore of interest to consider the likelihood of a minimum-phase law. If a linear attenuation law is accepted, the change of shape of a plane wave input at the surface of a homogeneous half-space can be written

$$u(z, t) = \int_{-\infty}^{\infty} d\omega U(0, \omega) \cdot \exp(i[K(\omega)z - \omega t]), \qquad (11)$$

where

$$U(0, \omega) = \frac{1}{2\pi}\int_{-\infty}^{\infty} dt\, u(0, t) \exp(i\omega t)$$

and $K(\omega) = \omega n(\omega)/c$ is a complex function which describes the attenuation and dispersion properties of the medium. For $u(z, t)$ to be a minimum-phase function for all z, the function $U(z, \omega)$ can have no zeros in the upper half of the complex ω plane. This will be the case if $U(0, \omega)$ and

$\exp[iK(\omega)z]$ have no zeros in the upper half plane; zeros occur when $\omega Im[n(\omega)]/c \to \infty$, i.e. there is infinite attenuation for some complex frequency. Since such zeros are not necessarily located on the real frequency axis, they do not correspond to actual transmission zeros. However, along the real frequency axis these displaced zeros would probably cause relative maxima for $\omega Im[n(\omega)]$. For this reason a nonminimum-phase law is unlikely to be consistent with experimental evidence supporting the linear increase of attenuation with frequency, and hence the frequency-independence of Q for earth materials. The causal attenuation laws proposed by Futterman (1962) to fit experimental results are all minimum phase.

TEST OF SPIKING FILTERS

As a test of the optimum spiking filter, synthetic seismograms with attenuation were constructed, and the deconvolution of these seismograms was attempted. From a hypothetical well velocity log, the unattenuated ground response function including multiple reflections was computed using the method of Wuenschel (1960). The synthetic seismogram was obtained by convolving the ground response with the time-varying signal waveform; this is equivalent to computing the ground response function with attenuation and convolving with the input signal. The instantaneous waveform was found by assuming an initially minimum-phase waveform and computing $s(t, t')$ from (3) according to the assumed attenuation law. From this the functions $\Phi_{ss}(t,t-\tau)$ and $\Phi_{ds}(t, t-\tau)$ were found by assuming that the ground response function was stationary and uncorrelated over the region of interest so that all nonstationarity resulted from the presence of attenuation. A more complicated procedure for constructing synthetic seismograms is required if the attenuation properties vary throughout the medium because the resulting change in shape will depend on the raypath (Trorey, 1962).

We shall first consider the case where both the initial waveform and the attenuation law are minimum phase so that a zero-delay spiking filter is appropriate. Because the minimum-phase laws which have so far appeared in the literature would require lengthy procedures to evaluate $s(t, t')$, it was decided to bypass these and assume that the change of shape of an impulse with time for some attenuation law is

$$u(\tau, t) = \begin{cases} A(t)\exp[-b(t)\tau] & \tau \geq 0 \\ 0 & \tau < 0, \end{cases} \tag{12}$$

where $A(t)$ and $b(t)$ are positive functions which decrease with time in some simple way. This is certainly a minimum-phase function with respect to τ for all positive values of t but does not, in fact, correspond to a physically possible expression for a constant-Q attenuation law (unless perhaps $A(t)$ and $b(t)$ are chosen with great care). Despite this qualification, the form (12) is suitable for testing the time-varying deconvolution filters because the question of whether a physically possible attenuation law is being used has no bearing on the filter performance. Figure 2 shows the change of shape with time for a displacement spike and the assumed minimum-phase attenuation law. The maximum broadening of the spike was limited to five sample points by multiplying

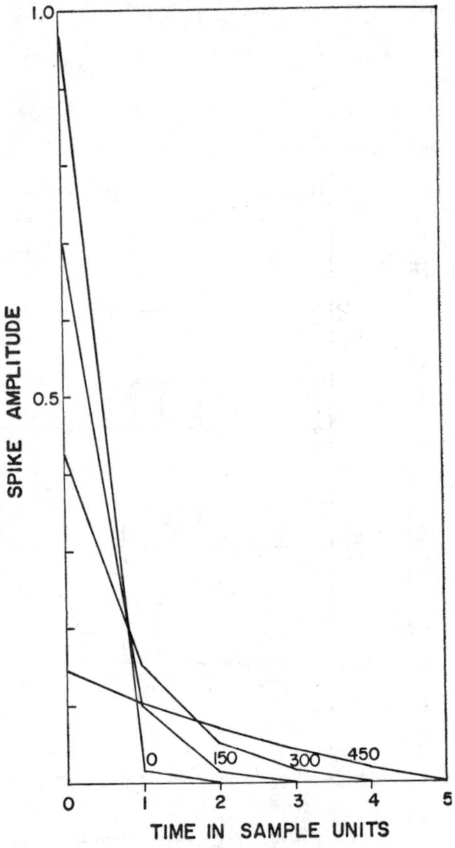

FIG. 2. Change of shape of displacement impulse with time for minimum-phase attenuation law. Indicated times are in sample units.

expression (12) by a triangular weighting function.

Expression (9) was solved for a 31-point, zero-delay, time-variable, spiking filter optimized at every third input point; the resulting deconvolved seismogram together with the attenuation-free ground response function are shown in Figure 3. The close correspondence of the ground response function and the spiking filter output indicates that a successful deconvolution has resulted. Because the time-dependent correlation functions $\Phi_{ds}(t, t-\tau)$ and $\Phi_{ss}(t, t-\tau)$ are not rapidly changing with time t, it is not necessary to optimize the filter for every input point.

One disadvantage of the method is that the matrix $\Phi_{ii}(k-m, k-n)$, which must be inverted to find the filter coefficients, is not equidiagonal (Toeplitz) if the input is a nonstationary process. For processes which are not highly nonstationary, the matrix can be approximated by an equidiagonal matrix so that a fast matrix inversion technique due to Levinson (1947) can be employed. There is an added advantage in making this approximation, since for an $N\times N$ symmetric matrix there are $N!$ independent elements, and, for an $N\times N$ equidiagonal matrix, only N independent elements must be found. Moreover when real data are used, the problem of ensuring that an estimate of $\Phi_{ii}(t, t-\tau)$ does in fact correspond to a physically possible nonstationary process is acute. The simplest way to reduce this difficulty is to assume that $\Phi_{ii}(t, t-\tau)$ is symmetric in τ, and then employ the standard weighting procedures used in forming autocorrelation estimates for stationary functions. The assumed symmetry in τ is actually impossible for a nonstationary process and corresponds to an assumption of instantaneous stationarity similar to the idea of thermodynamic reversibility. When this assumption was made, a deconvolution of the same synthetic seismogram produced results which were virtually indistinguishable from those obtained using the asymmetric $\Phi_{ii}(t, t-\tau)$. This indicates that the particular seismic trace is not strongly nonstationary for times of the order of the filter length. Even when this is untrue, the tremendous reduction in computation which results from the symmetry assumption may outweigh concern over a degraded filter output.

As a second illustration, a synthetic seismogram was computed for the nonminimum-phase (and acausal) attenuation law of expression (2). The

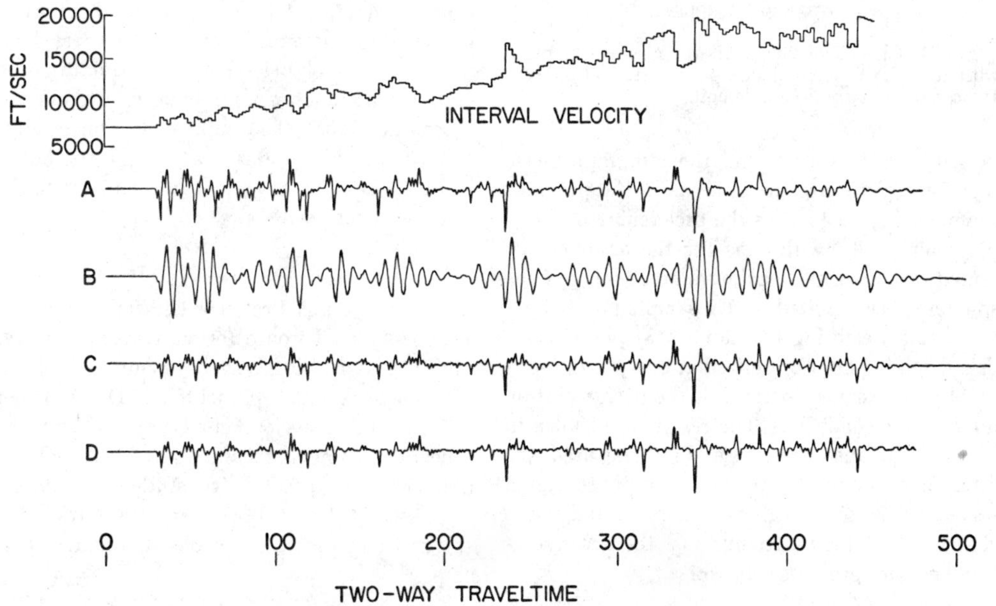

FIG. 3. Deconvolution of synthetic seismogram for a hypothetical geologic section having a minimum-phase attenuation law. The two-way traveltime is measured in sample units. (A) Ground response function with multiple reflections and minimum-phase attenuation law. (B) Synthetic seismogram with minimum-phase signal waveform and minimum-phase attenuation law. (C) Deconvolution of the synthetic seismogram using a 31-point, zero-delay, spiking filter optimized at every third sample point. (D) Ground response function in the absence of attenuation.

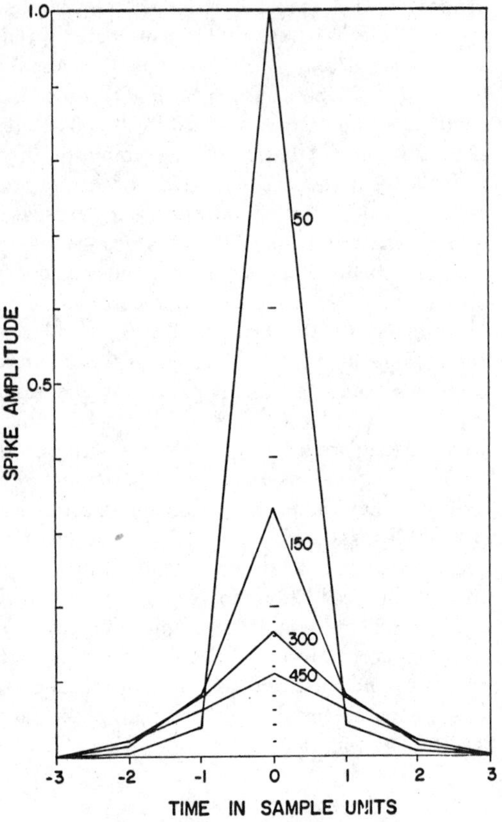

FIG. 4. Change of shape of displacement impulse with time for a nonminimum-phase attenuation law. Indicated times are in sample units.

velocity-depth relation and the minimum-phase input waveform are the same as in the previous example. Figure 4 shows the broadening of a displacement spike with time for the assumed attenuation. Again the maximum broadening of the spike has been limited to five sample points by a triangular weighting function. A 31-point, zero-delay, spiking filter optimized at every fifth input point was designed to attempt the deconvolution under these conditions. The results are shown in Figure 5. Because a zero-delay filter was used, the filter is optimum for a minimum-phase signal waveform; as the signal loses its minimum-phase character due to the attenuation, the deconvolution becomes progressively worse.

CONCLUDING REMARKS

The application of the methods described above to actual field seismograms is made difficult by the absence of a priori knowledge of the attenuation properties of the earth below the shotpoint. Other time-varying effects not due to attenuation are likely to be present. It is, therefore, essential to estimate the autocorrelation function $\Phi_{ii}(t,t-\tau)$ directly from the seismic trace. Because the input is considered to be nonstationary and, therefore, nonergodic, time averages may not provide good estimates of $\Phi_{ii}(t, t-\tau)$, and ensemble averages are suggested. In practice this is not possible since only one realization of the process is at hand, unless one wishes to consider other seismic traces as other realizations of the same process. It is incorrect, however, to interpret nonstationarity as absolutely forbidding the use of time averages to obtain estimates of the correlation functions. For a nonstationary process, some sort of estimate of $\Phi_{ii}(t, t-\tau)$ can be obtained by averaging $i(t)i(t-\tau)$ over a region centered at the time t. If the smoothing time T is too long, the nonstationarity will be smoothed out, and if T is too short, the estimate of $\Phi_{ii}(t, t-\tau)$ will be erratic. In this spirit Berndt and Cooper (1965; 1966) investigated methods of estimating correlation functions from a single realization of a nonstationary process by using time averages. In the first reference they derive an optimum smoothing length T and in the second an optimum weighting function $h(\tau, t, T)$ which must be used in performing the time average. An apparently more flexible procedure, involving optimum estimates of the correlation functions, has been proposed by Wierwille (1965). The implementation of any of these methods is difficult and may not be warranted by the improvement in the estimated correlation functions.

ACKNOWLEDGMENTS

I wish to thank Professor G. D. Garland of the University of Toronto for supervising this work and Professor J. C. Savage for a number of helpful discussions. I am also indebted to Dr. A. J. Seriff and the Shell Development Company for making reflection seismograms and related material available for some preliminary studies. Throughout this research I was supported by a National Research Council of Canada post-graduate scholarship.

REFERENCES

Berndt, H., and Cooper, G. R., 1965, An optimum observation time for estimates of time-varying correlation functions: IEEE Trans. on Information Theory, v. IT-11, p. 307–310.

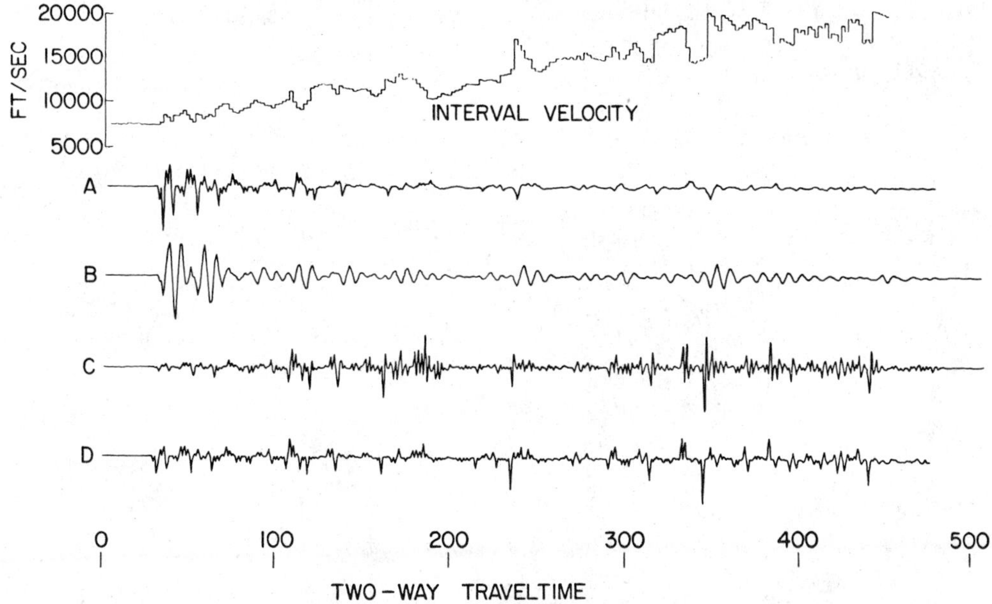

FIG. 5. Attempted déconvolution of synthetic seismogram for a hypothetical geologic section having a nonminimum-phase attenuation law. The two-way traveltime is measured in sample units. (A) Ground response function with multiple reflections and nonminimum-phase attenuation law. (B) Synthetic seismogram with minimum-phase signal waveform and nonminimum-phase attenuation law. (C) Output of 31-point, zero-delay, spiking filter optimized at every fifth sample point. In the interval $T=0$ to $T=200$, this trace has been amplified by a factor of 5 so that the details of the early part of the trace can be seen. (D) Ground response function in absence of attenuation.

—— 1966, Estimates of correlation functions of nonstationary random processes: IEEE Trans. on Information Theory, v. IT-11, p. 70–72.

Booton, R. C., 1952, An optimization theory for time-varying linear systems with nonstationary statistical inputs: Proc. IRE, v. 40, p. 977–981.

Cruz, J. B., Jr., 1959, A generalization of the impulse train approximation for time-varying linear system synthesis in the time domain: IRE Trans. on Circuit Theory, v. CT-6, p. 393–394.

Futterman, W. I., 1962, Dispersive body waves: J. Geophys. Res., v. 67, p. 5279–5291.

Knopoff, L., 1956, The seismic pulse in materials possessing solid friction, I: plane waves: Bull. Seismol. Soc. Am., v. 46, p. 175–183.

——— 1959, The seismic pulse in materials possessing solid friction, II: Lamb's problem: Bull. Seismol. Soc. Am., v. 49, p. 403–413.

Levinson, N., 1947, The Wiener RMS (root mean square) error criterion in filter design and prediction: J. Math. and Phys., v. 25, p. 261–278.

Sherwood, J. W. C., and Trorey, A. W., 1965, Minimum-phase and related properties of horizontally stratified absorptive earth to plane acoustic waves: Geophysics, v. 30, p. 191–197.

Simpson, S. M., Jr., Robinson, E. A., Wiggins, R. A., and Wunsch, C. I., 1963, Studies in optimum filtering of single and multiple stochastic processes: Sci. Rept. 7 of Contract AF 19(604)7378 ARPA Project VELA UNIFORM.

Treitel, S., and Robinson, E. A., 1966, The design of high-resolution digital filters: IEEE Trans. on Geoscience Electronics, v. GE-4, p. 25–38.

Trorey, A. W., 1962, Theoretical seismograms with frequency and depth dependent absorption: Geophysics, v. 27, p. 766–785

Wiener, N., 1949, Extrapolation, interpolation, and smoothing of stationary time series: Cambridge, M.I.T. Press.

Wierwille, W. W., 1965, A theory and method for correlation analysis of nonstationary signals: IEEE Trans. on Electronic Computers, v. EC-14, p. 909–919.

Wuenschel, P. C., 1960, Seismogram synthesis including multiples and transmission coefficients: Geophysics, v. 25, p. 106–129.

APPENDIX

The Wiener filter theory for inputs with time-invariant correlation functions has been generalized for nonstationary inputs by Booton (1952) and others. The derivation parallels Wiener's (1949) development exactly with only one difference—the optimum filter is time-varying and is obtained by solving a more complicated form of the Wiener-Hopf integral equation. There is an added problem in finding the correlation functions involved. Only the derivation for the single-channel case is included here although the generalization to the multichannel case is trivial.

Suppose the input $i(t)$ to a time-varying linear system consists of signal $s(t)$ and additive noise

$n(t)$, i.e., $i(t)=s(t)+n(t)$. If the impulse response of the system is $h(t-\tau, t)$, the system output is obtained by the convolution

$$y(t) = \int_{-\infty}^{\infty} h(\tau, t)i(t - \tau)d\tau, \qquad \text{(A-1)}$$

where the integration limits have been defined so as to permit acausal filters. If the desired filter output is $d(t)$, the error signal is

$$e(t) = d(t) - \int_{-\infty}^{\infty} h(\tau, t)i(t - \tau)d\tau, \qquad \text{(A-2)}$$

and the expected error power (mean-squared error) is

$$E[e^2(t)] = E\left\{\left[d(t) - \int_{-\infty}^{\infty} h(\tau, t)i(t - \tau)d\tau\right]^2\right\}. \qquad \text{(A-3)}$$

By assuming interchangeability of integration and averaging, and defining

$$\Phi_{di}(t, t - \tau) = E[d(t)i(t - \tau)],$$

$$\Phi_{dd}(t, t - \tau) = E[d(t)d(t - \tau)],$$

and

$$\Phi_{ii}(t, t - \tau) = E[i(t)i(t - \tau)],$$

we note that expression (A-3) becomes

$$E[e^2(t)] = \Phi_{dd}(t, t) - 2\int_{-\infty}^{\infty} h(\tau, t)\Phi_{di}(t, t-\tau)d\tau + \int_{-\infty}^{\infty}\int_{-\infty}^{\infty} h(\tau, t)h(\tau', t) \Phi_{ii}(t - \tau, t - \tau')d\tau d\tau'. \qquad \text{(A-4)}$$

The optimum filter $h(\tau, t)$ will minimize $E[e^2(t)]$ in (A-4).

If $h(\tau, t)$ is the response function for the optimum filter, the mean-square error will increase for any perturbation $\delta h(\tau, t)$ from the optimum. For the perturbed system

$$E\{[e(t)+\delta e(t)]^2\} = \Phi_{dd}(t, t) - 2\int_{-\infty}^{\infty} [h(\tau, t)+\delta h(\tau, t)]\Phi_{di}(t, t-\tau)d\tau + \int_{-\infty}^{+\infty}\int_{-\infty}^{+\infty} [h(\tau, t)+\delta h(\tau, t)][h(\tau', t) + \delta h(\tau', t)]\Phi_{ii}(t-\tau, t-\tau')d\tau d\tau'. \qquad \text{(A-5)}$$

When $E[e^2(t)]$ is a minimum, the difference Δ in mean-square error for equations (A-4) and (A-5) is always positive, being equal to

$$\Delta = -2\int_{-\infty}^{\infty} \delta h(\tau, t)\Phi_{di}(t, t-\tau)d\tau + 2\int_{-\infty}^{\infty}\int_{-\infty}^{\infty} h(\tau', t)\delta h(\tau, t) \cdot \Phi_{ii}(t-\tau, t-\tau')d\tau d\tau' + \int_{-\infty}^{\infty}\int_{-\infty}^{\infty} \delta h(\tau, t)\delta h(\tau', t) \cdot \Phi_{ii}(t-\tau, t-\tau')d\tau d\tau'. \qquad \text{(A-6)}$$

Since the last term in equation (A-6) can be written as a perfect square, it is always positive. Thus Δ will be positive if

$$\int_{-\infty}^{\infty} \delta h(\tau, t)\left[\Phi_{di}(t, t-\tau) - \int_{-\infty}^{\infty} h(\tau', t) \cdot \Phi_{ii}(t-\tau, t-\tau')d\tau'\right]d\tau = 0, \qquad \text{(A-7)}$$

that is, if the optimum filter response satisfies the integral equation

$$\Phi_{di}(t, t - \tau) = \int_{-\infty}^{\infty} h(\tau', t)\Phi_{ii}(t - \tau, t - \tau')d\tau'. \qquad \text{(A-8)}$$

Equation (A-8) is the nonstationary form of the Wiener-Hopf equation and involves time-dependent correlation functions and a time-varying linear filter. An equivalent matrix derivation of optimum time-variable filters for a more general class of inputs containing deterministic as well as random components has been published by Simpson et al, (1963).

Reprinted from Geophysics v. 34, no. 5, p. 683-695

GEOPHYSICS

THE DETERMINATION OF OPTIMUM GATE LENGTHS FOR TIME-VARYING WIENER FILTERING†

R. J. WANG*

The response function of a time-varying filter changes with the output signal, or observation time. Most existing time-varying filter techniques involve the empirical division of a seismic trace into a number of gates (or time windows) of given length, and a time-invariant filter is determined for each such gate. Few treatments have dealt with analytical methods to establish the gate lengths according to some optimum criterion.

This paper describes a technique for the determination of optimum gate lengths. It is based on the work of Berndt and Cooper, which is here applied to the calculation of time-varying Wiener filters. The Berndt and Cooper technique produces an upper bound for the mean-square error between the true and a given approximated time-varying correlation function. The minimization of this upper bound leads to a relation which enables one to establish gate lengths directly from the input trace. Thereafter, ordinary time-invariant Wiener filters can be computed for each gate. The overall filtered trace is obtained in the form of a suitably combined version of the individually filtered gates.

Experimentally it is shown that, with the Berndt and Cooper technique to determine optimum gate lengths, time-varying Wiener filters can be better than a time-invariant filter.

INTRODUCTION

In a nonstationary process, the usual entities that serve to characterize the input, such as mean, variance, correlation functions, etc., vary with time. If the process is nonstationary and if one wishes to design a filter which yields minimum mean-square error between some desired output and an actual output, Wiener's (1949) solution based on a stationary process is not applicable. Instead, one must consider an integral equation such as Booton's (1952) for a nonstationary process. Unfortunately, the general solution to Booton's integral equation is not yet known. A method of solving this for some special cases has been proposed by Shinbrot (1957, 1958). While Shinbrot's solution has the advantage that it is in closed form, the assumptions needed are rather restrictive.

For a stationary input, Wiener (1949) has shown that a necessary and sufficient condition for the mean-square error between some desired output $z(t)$ and an actual output $y(t)$ to be a minimum, the integral equation (called the Wiener-Hopf integral equation of the first kind)

$$\phi_{zx}(\tau) = \int_0^\infty g(\sigma)\phi_{xx}(\tau - \sigma)d\sigma \qquad (1)$$

must be satisfied, where $g(\sigma)$ is the impulse response function of the filter to be determined,

† Presented at the 38th Annual International SEG Meeting in Denver, Colorado, October 1, 1968. Manuscript received by the Editor October 16, 1968; revised manuscript received February 11, 1969.

* Research Center, Pan American Petroleum Corporation, Tulsa, Oklahoma 74102.

$\phi_{xx}(\tau)$ is the autocorrelation function of the input $x(t)$, and $\phi_{zx}(\tau)$ is the crosscorrelation function between the desired output $z(t)$ and the input $x(t)$. If the response time of the filter σ and the lag time of the correlation function τ take on only discrete values, equation (1) may be written as a set of linear algebraic equations called the normal equations (Robinson and Treitel, 1967). An efficient method of solving these equations was first given by Levinson (1947). The digital solution for $g(\sigma)$ with the aid of z-transform theory was illustrated by Robinson and Treitel (1964), and the recursive scheme to speed up digital computations was presented by Shanks (1967).

Once $g(\sigma)$ is determined, the output $y(t)$ is found from the well-known convolution equation

$$y(t) = \int_0^t g(t-\sigma)x(\sigma)d\sigma. \tag{2}$$

A simpler form of equation (2) is

$$y(t) = g(t) * x(t), \tag{3}$$

where * denotes convolution. In equations (2) and (3) it has been assumed that $x(t)$, $y(t)$, and $g(t)$ are all zero for $t<0$, i.e., they are causal.

Let us now turn our attention to a nonstationary input. In this case, equation (1) is no longer valid since both the autocorrelation and the crosscorrelation functions vary with output or observation time. We introduce the new notations $\phi_{xx}(t, \gamma)$ and $\phi_{zx}(t, \gamma)$ instead of $\phi_{xx}(\tau)$ and $\phi_{zx}(\tau)$, respectively. The time γ equals $\tau+t$, where τ, as before, represents the lag time, and t denotes the output, or observation time. In other words, for a nonstationary process, the correlation functions depend on both t and γ, whereas for a stationary process, the correlation functions depend only on the difference $\gamma-t=\tau$. For a nonstationary input, Booton (1952) has shown that a necessary and sufficient condition for the mean-square error between some desired output and an actual output to be a minimum is

$$\phi_{zx}(t, \gamma) = \int_0^\infty g(t, \sigma)\phi_{xx}(\sigma, \gamma)d\sigma, \tag{4}$$

where $g(t,\sigma)$ is the time-varying impulse response function of the filter to be determined, $\phi_{xx}(t,\gamma)$ is the time-varying autocorrelation function of the input $x(t)$, and $\phi_{zx}(t,\gamma)$ is the time-varying crosscorrelation function of the desired output $z(t)$ and the input $x(t)$. Given $g(t,\sigma)$, the actual output $y(t)$ is

$$y(t) = \int_0^\infty g(t, \sigma)x(\sigma)d\sigma. \tag{5}$$

Equation (4) is called Booton's integral equation, or the modified Wiener-Hopf integral equation of the second kind. (See Appendix I for its derivation.)

THE ESTIMATION OF TIME-VARYING CORRELATION FUNCTIONS AND THE IMPLEMENTATION OF TIME-VARYING WIENER FILTERING

Since the general solution of equation (4) is not known, one can only approximate $g(t,\sigma)$ in some way. The first obstacle in an attempt to approximate $g(t, \sigma)$ according to equation (4) is encountered in the determination of time-varying correlation functions $\phi_{xx}(t, \gamma)$ and $\phi_{zx}(t, \gamma)$. This problem does not arise in the stationary, ergodic[1] case, where the corresponding correlation functions $\phi_{xx}(\tau)$ and $\phi_{zx}(\tau)$ are time-invariant.

The time-varying correlation functions $\phi_{xx}(t,\gamma)$ and $\phi_{zx}(t, \gamma)$ are defined, respectively, as

$$\phi_{xx}(t, \gamma) = E[x(t)x(\gamma)]$$

and

$$\phi_{zx}(t, \gamma) = E[z(t)x(\gamma)],$$

where $E[\ ]$ denotes the expected value, or ensemble average, of the quantity within brackets. Before proceeding further, we note that we are here dealing only with a single channel, i.e., there is only one input and one output. Under these conditions it is impossible to determine $\phi_{xx}(t, \gamma)$ and $\phi_{zx}(t, \gamma)$ rigorously by computing the ensemble average of the quantities $[x(t)\ x(\gamma)]$ and $[z(t)\ x(\gamma)]$, respectively. Rather, it is necessary to assume that the process is "piecewise stationary." In other words, we propose to divide both the input $x(t)$ and the desired output $z(t)$ into sections, where each section may be considered to be one realization of some stationary and ergodic pro-

[1] A stationary random process is said to possess the ergodic property if one can equate ensemble averages and averages with respect to time performed on a single "representative" function of the ensemble. For a definition of ensemble averages, see Lee (1960), Chapter 7. Ergodicity will be assumed for the stationary process in the sequel.

cess. Thus, $x(t)$ may be expressed as a sum of stationary inputs $x_1(t), x_2(t), \cdots, x_N(t)$, namely,

$$x(t) = \sum_{k=1}^{N} x_k(t) \tag{6}$$

where

$$x_k(t) = x(t)\{u[t-(k-1)T] - u(t-kT)\}.$$

In the above expression, $u(t)$ denotes a unit step function, defined by

$$u(t) = \begin{cases} 1 & \text{if } t \geq 0 \\ 0 & \text{if } t < 0, \end{cases}$$

and T represents an interval over which $x(t)$ is assumed stationary. This operation is illustrated in Figure 1. The desired output $z(t)$ is similarly divided into N such sections of duration T. Once $x_k(t)$ and $z_k(t)$, $k=1, 2, \cdots, N$, are determined, the time-invariant autocorrelation and crosscorrelation functions are found from the relations

$$\phi_{xx}^{[k]}(\tau) = \frac{1}{T}\int_{(k-1)T}^{kT} x_k(t)x_k(t+\tau)dt$$

and

$$\phi_{zx}^{[k]}(\tau) = \frac{1}{T}\int_{(k-1)T}^{kT} z_k(t)x_k(t+\tau)dt, \quad k = 1, 2, \cdots, N. \tag{7}$$

This is illustrated in Figure 2.

We next propose to split equation (4) into the components

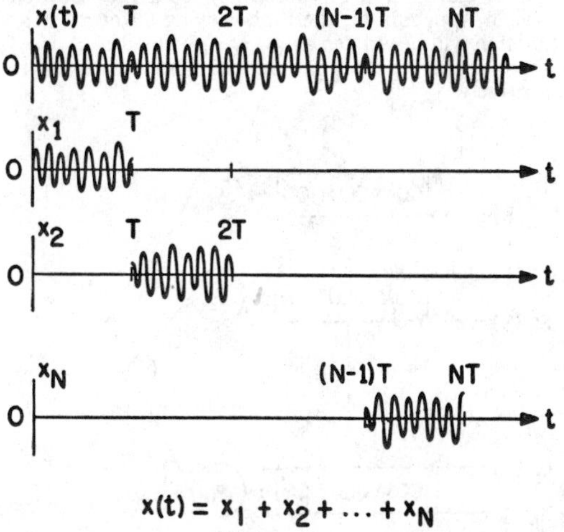

FIG. 1. Subdivision of a seismic trace.

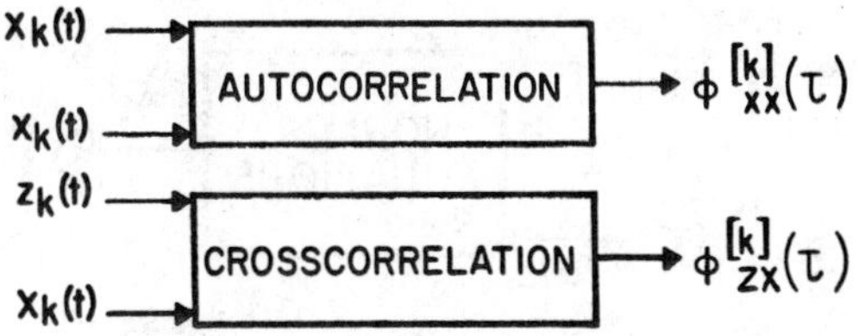

FIG. 2. Time-invariant correlation.

$$\phi_{zx}^{[k]}(\tau) = \int_0^{\infty} g_k(\sigma)\phi_{xx}^{[k]}(\tau-\sigma)d\sigma, \quad k = 1, 2, \cdots, N, \tag{8}$$

where the $g_k(\sigma)$ are the time-invariant components of $g(t,\sigma)$, i.e., we express $g(t,\sigma)$ in the form

$$g(t,\sigma) = \sum_{k=1}^{N} g_k(\sigma)\{u[t-(k-1)T] - u(t-kT)\}. \tag{9}$$

As can be seen, the form of equation (8) is exactly the same as that of equation (1).

Given $g_k(\sigma)$, $k=1, 2, \cdots, N$, the actual output $y(t)$ is computed by means of the formula

$$y(t) = \sum_{k=1}^{N} y_k(t) \tag{10}$$

where

$$y_k(t) = \int_0^t g_k(t-\sigma)x_k(\sigma)d\sigma. \tag{11}$$

Equation (10) implies that $y_k(t)$ may also be expressed in terms of $y(t)$ in the form

$$y_k(t) = y(t)\{u[t-(k-1)T] - u(t-kT)\}.$$

Let us show that equation (4) reduces to equation (8) for the stationary case. In this event we have

$$\left.\begin{aligned} \phi_{zx}(t,\gamma) &= \phi_{zx}(\tau) \\ g(t,\sigma) &= g(\sigma) \\ \text{and} \quad \phi_{xx}(\sigma,\gamma) &= \phi_{xx}(\tau-\sigma) \end{aligned}\right\}. \tag{12}$$

Because of the assumption that $x(t)$ is stationary over the interval $(k-1)T \leq t < kT$, $k=1, 2, \cdots, N$, we substitute equation (12) into equation (4) and obtain

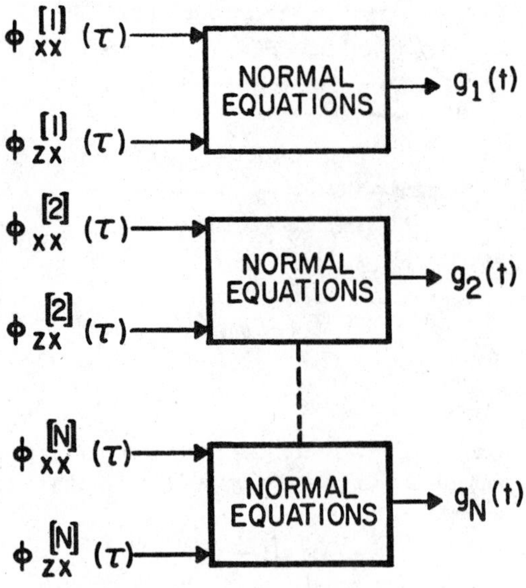

FIG. 3. Time-varying impulse response function.

$$\phi_{zx}^{[k]}(\tau) = \int_0^\infty g_k(\sigma)\phi_{xx}^{[k]}(\tau - \sigma)d\sigma,$$

$$k = 1, 2, \cdots, N. \quad (8)$$

For the discrete case, the functions $g_k(\sigma)$, $k=1, 2, \cdots, N$, are determined through solution of the N sets of normal equations as illustrated in Figure 3. The combined operation of equations (10) and (11) is illustrated in Figure 4.

Thus far, the concept involved is quite simple and straightforward. However, one question remains outstanding, namely: how closely do the time-invariant correlation functions $\phi_{xx}^{[k]}(\tau)$ and $\phi_{zx}^{[k]}(\tau)$, $k=1, 2, \cdots, N$, approximate the time-varying correlation functions $\phi_{xx}(t,\gamma)$ and $\phi_{zx}(t,\gamma)$? Evidently, there is some optimum value of T for which these time-invariant correlation functions best approximate the time-varying correlation functions. Now, for an ergodic process the variance of such an estimate vanishes as the observation interval T approaches infinity. Thus, it is desirable to make T as large as possible. On the other hand, the expected error between an estimated correlation function and a true correlation function decreases as T decreases [see equation (A-18), Appendix II]. What then is this optimum value of T? An equation derived by Berndt and Cooper (1965) seems to provide a promising criterion for the optimum selection of T. The following section describes certain aspects of Berndt and Cooper's theory and its application to seismic trace analysis.

THE DETERMINATION OF OPTIMUM GATE LENGTHS

Let us assume that a time-varying autocorrelation function may be expressed in the form

$$\phi_{xx}(t, \tau) = \sum_{i=1}^{n} a_i(t)b_i(\tau),^2 \quad (13)$$

and that the function $a_i(t)$ may be expanded in a Taylor series about a point $t=t_0$, up to the mth terms; that is,

$$a_i(t) = \sum_{l=0}^{m} c_{il}(t - t_0)^l,$$

$$i = 1, 2, \cdots, n, \quad (14)$$

where the Taylor coefficients are given by

$$c_{il} = \frac{1}{l!} a_i^{(l)}(t_0). \quad (15)$$

The superscript (l) denotes the lth derivative with respect to t. Furthermore, let ${}^0\phi_{xx}(t_0,\tau,T)$ denote the estimated autocorrelation function at $t=t_0$. For a Gaussian process, Berndt and Cooper (1965) have shown that the upper bound of the mean-square error defined by

$$S^2 = E\{[{}^0\phi_{xx}(t_0, \tau, T) - \phi_{xx}(t_0, \tau)]^2\} \quad (16)$$

is minimized when the relation

[2] For clarity and in conformity with the notations used in Appendix II, the time-varying autocorrelation function will be written as $\phi_{xx}(t, \tau)$ instead of $\phi_{xx}(t, \gamma)$, where τ and t denote lag and observation times, respectively.

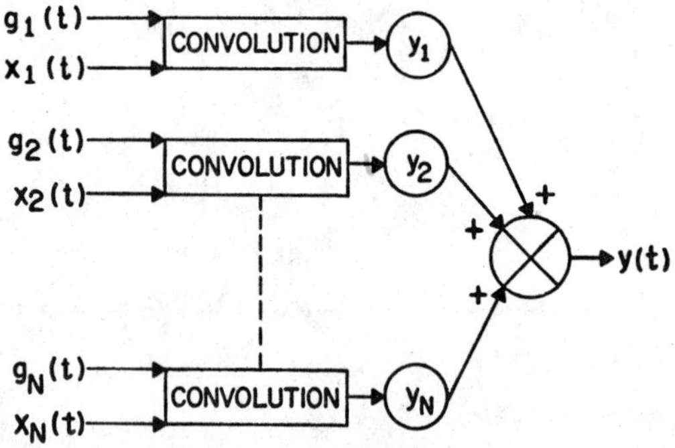

FIG. 4. Time-varying convolution.

$$\sum_{i=1}^{n}\sum_{j=1}^{n}\sum_{p=1}^{n}\sum_{q=1}^{n} c_{i2p}c_{j2q} \cdot \frac{(p+q)b_i(0)b_j(0)T^{2(p+q)+1}}{(2p+1)(2q+1)2^{2(p+q)}} = \int_{-\infty}^{\infty}[\phi_{xx}(t_0,\tau)]^2 d\tau \tag{17}$$

is satisfied, where the estimated autocorrelation function at t_0 is given by

$$^0\phi_{xx}(t_0,\tau,T) = \frac{1}{T}\int_{-T/2}^{T/2} x\left(t+t_0+\frac{\tau}{2}\right) \cdot x\left(t+t_0-\frac{\tau}{2}\right)dt. \tag{18}$$

Relation (17) is derived in Appendix II. For the discrete case, equation (18) becomes

$$^0\phi_{xx}(t_0,\tau,T) = \frac{1}{T}\sum_{t=-T/2}^{T/2} x(t+t_0+\tau/2) \cdot x(t+t_0-\tau/2). \tag{19}$$

In equation (17), n is the number of terms used in the expansion of $\phi_{xx}(t,\tau)$ as given in equation (13), and,

$$\mu = \begin{cases} \dfrac{m}{2} & \text{if } m = \text{even} \\ \dfrac{m-1}{2} & \text{if } m = \text{odd}. \end{cases}$$

In particular, if the Taylor expansion for $a_i(t)$ can be terminated after the second term, m is 2 and $\mu=1$. Then equation (17) reduces to

$$\sum_{i=1}^{n}\sum_{j=1}^{n} c_{i2}c_{j2}\frac{b_i(0)b_j(0)T^5}{72} = \int_{-\infty}^{\infty}[\phi_{xx}(t_0,\tau)]^2 d\tau. \tag{20}$$

Thus the optimum gate length for this case is

$$T = \left\{\frac{72\int_{-\infty}^{\infty}[\phi_{xx}(t_0,\tau)]^2 d\tau}{\sum_{i=1}^{n}\sum_{j=1}^{n} c_{i2}c_{j2}b_i(0)b_j(0)}\right\}^{1/5}. \tag{21}$$

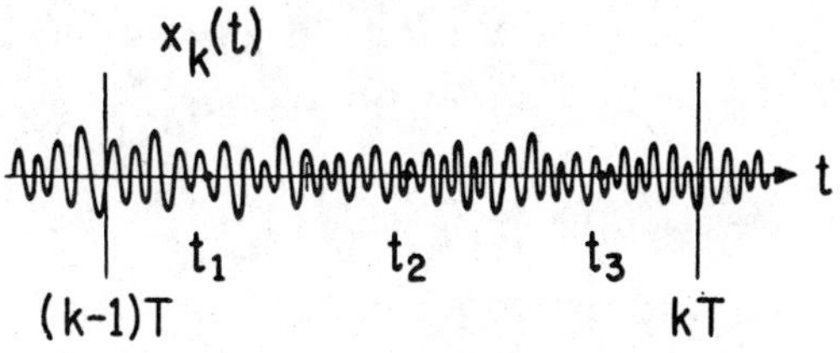

FIG. 5. Portion of a nonstationary time series.

Consider a portion of a nonstationary time series of length T, as shown in Figure 5. Divide T into three equal sections and label the center points of these three sections as t_1, t_2, and t_3. Furthermore, let[3]

$$\left.\begin{aligned} a_i(t) &= d_i + e_i(kT-t) + f_i(kT-t)^2 \\ &\text{and} \\ b_i(\tau) &= \cos\,(i-1)\frac{2\pi\tau}{T} \end{aligned}\right\}. \tag{22}$$

Then the time-varying autocorrelation function (13) may be approximated by three time-invariant autocorrelation functions

$$\phi_{xx}^{[j]}(\tau) = \sum_{i=1}^{n} a_i(t_j)b_i(\tau), \quad j = 1, 2, 3, \tag{23}$$

where

$$a_i(t_j) = d_i + e_i(kT-t_j) + f_i(kT-t_j)^2. \tag{24}$$

The letter k in expressions (22) and (24) is an index which identifies the kth section of the trace, where each section is of equal length T. The coefficients d_i, e_i, and f_i can be determined by solving the three simultaneous equations (24), which result when one successively sets $j=1$, 2, and 3. (See Appendix III for a numerical example, and the reason why the gate is further divided into three short sections.)

Now from equation (15) by differentiating

[3] The form

$$a_i(t) = d_i + e_i t + f_i t^2 \tag{24}$$

could also have been chosen for the function $a_i(t)$. For a given τ, the autocorrelation functions of seismic traces tend in general to decrease as t increases. The form of $a_i(t)$ in equations (22) will satisfy this condition if d_i, e_i, and f_i are all positive. Similarly, other forms for the function $b_i(\tau)$ could also have been chosen. With $b_i(\tau)=\cos\,(i\text{-}1)2\pi\tau/T$, the functions $a_i(t)$, $i=1, 2, \cdots, n$, simply become the coefficients of the cosine transform of the autocorrelation function (23).

$a_i(t)$ with respect to t and setting $t=t_j$ we obtain

$$c_{i0} = a_i(t_j) = d_i + e_i(kT - t_j) + f_i(kT - t_j)^2,$$

$$c_{i1} = a_i^{(1)}(t_j) = - e_i - 2f_i(kT - t_j),$$

$$c_{i2} = \tfrac{1}{2}a_i^{(2)}(t_j) = f_i,$$

$$c_{i3} = c_{i4} = \cdots = c_{im} = 0.$$

Since the Taylor expansion of $a_i(t)$ terminates after the second term ($m=2$), equation (21) is applicable for the determination of optimum observation time T_0. Also from equation (22), $b_i(0)=1$. Hence, equation (21) may be rewritten in the form,

$$T = \left\{ \frac{144 \int_0^{\infty} [\phi_{xx}(t_2, \tau)]^2 d\tau}{\sum_{i=1}^{n} \sum_{j=1}^{n} f_i f_j} \right\}^{1/5}, \qquad (25)$$

where the symmetry of $\phi_{xx}(t,\tau)$ about $\tau=0$ has been utilized to obtain the above expression. As previously stated, it is impossible to determine $\phi_{xx}(t,\tau)$ rigorously by computing the ensemble average of the quantity $[x(t)x(\gamma)]$ in the event that only one member of the ensemble is available. We must therefore find $\phi_{xx}(t_2,\tau)$ by taking a time average of the quantity $[x(t_2)x(\gamma)]$, i.e.,

$$\phi_{xx}(t_2, \tau) = \frac{1}{T} \int_{-T/2}^{T/2} x\left(t + t_2 + \frac{\tau}{2}\right) \cdot x\left(t + t_2 - \frac{\tau}{2}\right) dt. \qquad (26)$$

It is now clear that the right-hand side of expression (25) is a function of the selected gate length T. Thus we must resort to a trial-and-error method in determining the optimum gate length T_0, which satisfies the condition (25). An arbitrary but reasonable value of T is first selected and the right-hand side of equation (25) is then evaluated. If the value of the right-hand side of equation (25) thus established deviates by more than a predetermined amount from the value that was first assumed, another value of T is selected and the right-hand side is again computed. This procedure is repeated until the condition (25) is almost satisfied, i.e., until $T \cong T_0$.

A NUMERICAL STUDY OF BERNDT AND COOPER'S EQUATION APPLIED TO TIME-VARYING WIENER FILTERING

A synthetic trace was formed with the aid of the equation

$$x(t) = s(t) + n(t), \qquad (27)$$

where $x(t)$ is a nonstationary time series, $s(t)$ represents a signal trace consisting of time-varying wavelets, and $n(t)$ is white noise. The functions $x(t)$, $s(t)$, and $n(t)$ (see Figure 6) are each described by 420 equally spaced sample values. $s(t)$ is composed of six signal wavelets whose onsets are separated by 70 sampling increments.

In order to evaluate the performance of Berndt and Cooper's theory, we chose the mean-square error as a measure of error. For example, assume that the 420-length trace is divided into six segments of 70 sample values each. Then we will generally have six different errors from equation (25), each corresponding to one of the six segments. Let $\phi_{xx}(t_k,\tau)$ and f_i^k denote the autocorrelation function and the Taylor coefficients corresponding to the kth segment respectively. Then the normalized error between the right- and left-hand sides of equation (25) is defined to be

$$E_k = 1 - \frac{T_k}{T},$$

where T_k is computed from

$$T_k = \left\{ \frac{144 \sum_{\tau=0}^{T-1} [\phi_{xx}(t_k, \tau)]^2}{\sum_{i=1}^{n} \sum_{j=1}^{n} f_i^k f_j^k} \right\}^{1/5} \qquad (28)$$

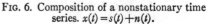

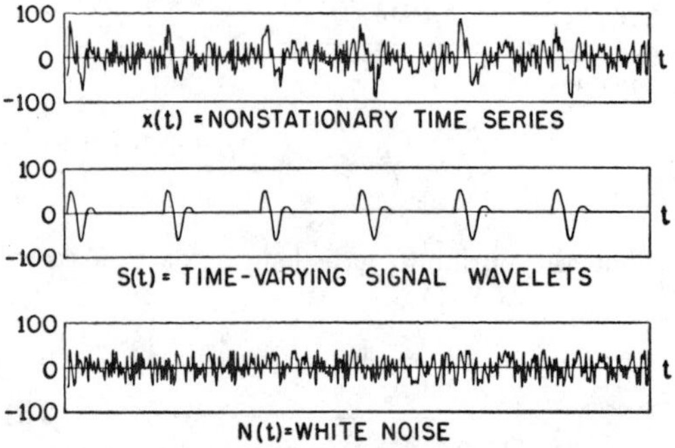

FIG. 6. Composition of a nonstationary time series. $x(t)=s(t)+n(t)$.

and where T is the selected gate length (see previous section). Since there are six segments ($T=70$), the mean-square error is defined to be

$$Q_T = \frac{1}{6} \sum_{k=1}^{6} E_k^2. \qquad (29)$$

A plot of Q_T versus T for the synthetic trace is shown in Figure 7. The values of T ranged from 30 to 100 sample points. The minima for Q_T occur at $T=35$ and in the neighborhood of $T=70$. Time-varying Wiener filtering[4] has also been applied to the trace for various values of T. The mean-square errors between the desired and the actual outputs versus T for Wiener filter lengths of 40 and 100 sample points are shown in Figure 8. The desired output is the signal trace, or $s(t)$, in this case. T ranges from 30 to 100 sample points for Figure 8, and one unit of the abscissa corresponds to five points in both Figures 7 and 8. It should be pointed out that the maximum number of lags used in the autocorrelation and cross-correlation functions has been made identical to the corresponding gate lengths. It has been observed empirically that the mean-square error tends to decrease as T decreases even though the filter length is unchanged.

[4] By time-varying Wiener filtering we mean that different Wiener filters are applied to different segments and the resultant segmented outputs are combined to form the actual output.

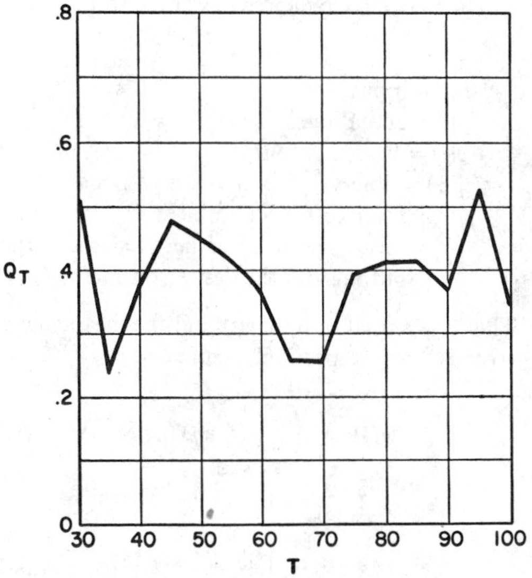

FIG. 7. Q_T (mean-square error from Berndt and Cooper's equation) versus T (gate length).

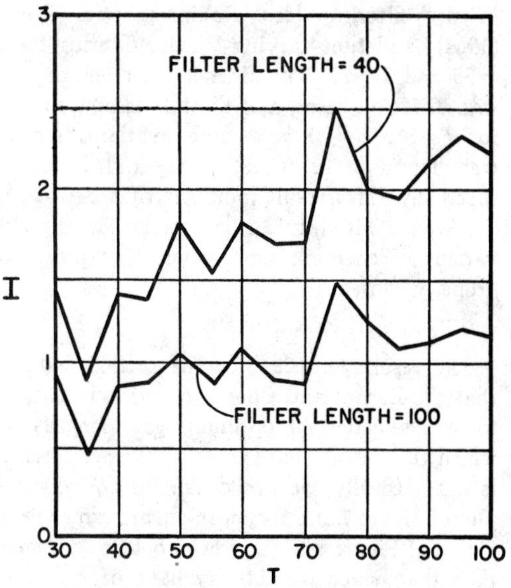

FIG. 8. I (mean-square error between the actual and desired outputs) versus T (gate length).

In Figure 8, even though the mean-square errors at $T=70$ are considerably greater than those at $T=35$, they both fall in troughs of the error curves. Since $T=35$ is one-half of $T=70$ and also turns out to be the average length of the signal wavelets, small mean-square errors should be expected at $T=35$. A comparison of mean-square errors at $T=70$ on Figure 8 showed that the mean-square error decreased by 49.3 percent when the filter length was changed from 40 to 100 sample points, while those at $T=65$ and 75 decreased by 47.3 percent and 43.3 percent, respectively.

Figure 9 shows the results of time-invariant

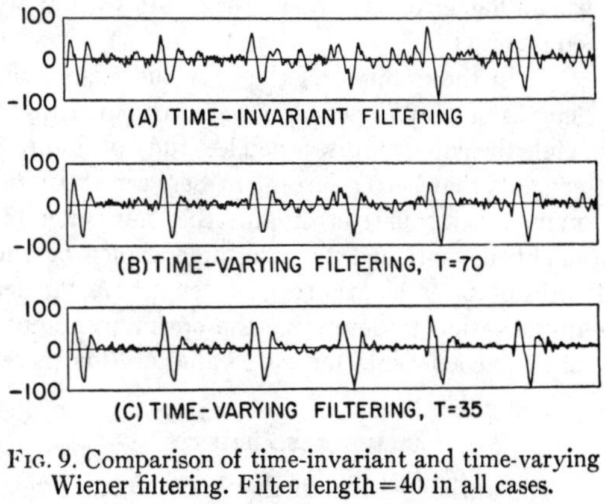

FIG. 9. Comparison of time-invariant and time-varying Wiener filtering. Filter length = 40 in all cases.

Wiener filtering [see Robinson and Treitel (1964)] and time-varying Wiener filtering for $T=35$ and $T=70$. The Wiener filter length used was 40 for all cases, and the mean-square errors for $T=35$, $T=70$, and time-invariant filtering were found to be 0.888, 1.713, and 2.808, respectively. This result indicates that time-varying Wiener filtering can be better than time-invariant Wiener filtering, when the gates are properly chosen.

CONCLUSIONS

The results of the previous section indicate that the Berndt and Cooper criterion is suitable for the selection of optimum gate lengths for which the mean-square error in Wiener filtering is small. Ideally, we would like the Q_T versus T plot of Figure 7 and the mean-square error versus T plot of Figure 8 to show even better agreement than that which we actually have been able to obtain. Among the reasons that we can give to explain the observed lack of agreement are the following:

a) The equation of Berndt and Cooper is derived for a Gaussian process (see Appendix II). The synthetic trace to which this equation was applied is of course not strictly Gaussian.

b) The performance criterion for the optimum observation time used in the derivation of equation (17) is that the upper bound of the mean-square error for the estimated autocorrelation function be minimum (see Appendix II). There is no assurance that this mean-square error itself is minimum at any time.

c) The function $a_i(t)$ was assumed to be of the form given by equation (22), so that equation (21) rather than equation (17) could be used. The approximation for $\phi_{xx}(t,\tau)$ can be improved if we retain higher order terms in $a_i(t)$ and then apply equation (17).

d) In the example of the previous section the lengths of the Wiener filter were kept unchanged while the gate lengths varied. A study of Figure 8 suggests that better agreement between the minimum mean-square error and the minimum Q_T might be obtained by a suitable choice of the ratio of filter length to gate length. A further investigation to determine the optimum number of filter coefficients for each value of T is therefore in order.

ACKNOWLEDGEMENTS

The author wishes to thank Dr. Sven Trietel of Pan American Petroleum Corporation for many helpful suggestions and discussions on the subject and the text of this paper. The author also wishes to express his thanks to Pan American Petroleum Corporation for permission to publish this paper.

REFERENCES

Berndt, H. and Cooper, G. R., 1965, An optimum observation time for estimates of time-varying correlation functions: IEEE Trans, on Inform. Theory, v. IT-11, p. 307–310.

Booton, R. C., 1952, An optimization theory for time-varying linear systems with nonstationary inputs: Proc. IRE, v. 40, p. 977–981.

Laning, J. H., and Battin, R. H., 1956, Random processes in automatic control: New York, McGraw-Hill.

Lee, Y. W., 1960, Statistical theory of communication: New York, John Wiley and Sons, Inc.

Levinson, N., 1947, The Wiener RMS (Root Mean Square) error criterion in filter design and prediction: J. Math. and Phys., v. 25, p. 261–278.

Robinson, E. A., and Treitel, S., 1964, Principles of digital filtering: Geophysics, v. 29, p. 395–404.

——— 1967, Principles of digital Wiener filtering: Geophys. Prosp., v. 15, p. 311–333.

Shanks, J. L., 1967, Recursion filters for digital processing: Geophysics, v. 32, p. 33–51.

Shinbrot, M., 1957, On the integral equation occurring in optimization theory with nonstationary inputs: J. Math. and Phys., v. 26, p. 121–129.

——— 1958, Optimization of time-varying linear systems with nonstationary inputs: Trans. ASME, v. 80, p. 457–462.

Wiener, N., 1949, Extrapolation, interpolation, and smoothing of stationary time series: New York, John Wiley and Sons, Inc.

APPENDIX I

DERIVATION OF BOOTON'S EQUATION

Let

$x(t)$ = input,

$y(t)$ = actual output,

$z(t)$ = desired output,

$g(t, \sigma)$ = time-varying impulse response of the filter to be determined,

$I(t)$ = mean-square error between the actual output and the desired output.

Then the actual output $y(t)$ and the mean-square error $I(t)$ are respectively expressed as

$$y(t) = \int_0^\infty g(t, \sigma)x(\sigma)d\sigma \qquad \text{A-1}$$

and

$$I(t) = E\{[z(t) - y(t)]^2\}, \qquad \text{A-2}$$

where the notation E{ } denotes the expected value of the function within the braces.

Substituting A-1 into A-2, we have

$$I(t) = E\left\{\left[z(t) - \int_0^\infty g(t,\sigma)x(\sigma)d\sigma\right]^2\right\}$$

$$= \phi_{zz}(t,t) - 2\int_0^\infty g(t,\sigma)\phi_{zx}(t,\sigma)d\sigma \quad \text{A-3}$$

$$+ \int_0^\infty g(t,\sigma)\int_0^\infty g(t,\gamma)\phi_{xx}(\gamma,\sigma)d\gamma d\sigma,$$

where

$$\phi_{zz}(t,t) = E\{[z(t)]^2\}$$
$$\phi_{xx}(\gamma,\sigma) = E\{x(\gamma)x(\sigma)\}$$
$$\phi_{zx}(t,\sigma) = E\{z(t)x(\sigma)\}.$$

Assume now that there exists an impulse response function of the system $h(t,\sigma)$ different from that of the optimum system $g(t,\sigma)$. Then $h(t,\sigma)$ may be written in the form

$$h(t,\sigma) = g(t,\sigma) + \epsilon f(t,\sigma), \quad \text{A-4}$$

where $f(t,\sigma)$ is any impulse response function, and ϵ any arbitrary number. If we denote the mean-square error due to the impulse response $h(t,\sigma)$ by $N(t)$, and replace $g(t,\sigma)$ by $h(t,\sigma)$ in A-3, we have

$$N(t) = \phi_{zz}(t,t) - 2\int_0^\infty h(t,\sigma)\phi_{zx}(t,\sigma)d\gamma$$

$$+ \int_0^\infty h(t,\sigma)$$

$$\cdot \int_0^\infty h(t,\gamma)\phi_{xx}(\gamma,\sigma)d\gamma d\sigma. \quad \text{A-5}$$

Substituting A-4 into A-5 and simplifying yield

$$N(t) = M(t) - 2\epsilon\int_0^\infty f(t,\sigma)\phi_{zx}(t,\sigma)d\sigma$$

$$+ \epsilon\int_0^\infty f(t,\sigma)\int_0^\infty g(t,\gamma)\phi_{xx}(\gamma,\sigma)d\gamma d\sigma$$

$$+ \epsilon\int_0^\infty g(t,\sigma)\int_0^\infty f(t,\gamma)\phi_{xx}(\gamma,\sigma)d\gamma d\sigma$$

$$+ \epsilon^2\int_0^\infty f(t,\sigma)$$

$$\cdot \int_0^\infty f(t,\gamma)\phi_{xx}(\gamma,\sigma)d\gamma d\sigma, \quad \text{A-6}$$

where $M(t)$ is the minimum mean-square error, given by A-3.

If we recognize that

$$\int_0^\infty f(t,\sigma)\int_0^\infty g(t,\gamma)\phi_{xx}(\gamma,\sigma)d\gamma d\sigma$$

$$= \int_0^\infty g(t,\sigma)\int_0^\infty f(t,\gamma)\phi_{xx}(\gamma,\sigma)d\gamma d\sigma,$$

A-6 can be reduced to

$$N(t) = M(t) - 2\epsilon I_1(t) + \epsilon^2 I_2(t), \quad \text{A-7}$$

where

$$I_1(t) = \int_0^\infty f(t,\sigma)$$

$$\cdot \left\{\phi_{zx}(t,\sigma) - \int_0^\infty g(t,\gamma)\phi_{xx}(\gamma,\sigma)d\gamma\right\} d\sigma$$

$$I_2(t) = E\left\{\left[\int_0^\infty f(t,\sigma)x(\sigma)d\sigma\right]^2\right\}.$$

Let us suppose that

$$I_1(t) \neq 0.$$

Then for some value of ϵ, and some $f(t,\sigma)$, it is possible to have

$$\left[\epsilon I_1(t) - \frac{\epsilon^2}{2} I_2(t)\right] > 0.$$

But this implies

$$M(t) > N(t).$$

By hypothesis, however,

$$N(t) > M(t) = \text{minimum mean-square error};$$

that is, for any nonoptimum impulse response function $h(t,\sigma)$, the actual mean-square error $N(t)$ must be greater than the minimum mean-square error $M(t)$. Therefore, the inequality

$$\left[\epsilon I_1(t) - \frac{\epsilon^2}{2} I_2(t)\right] > 0$$

must not hold. This can only mean that

$$I_1(t) = 0,$$

if $g(t,\sigma)$ is to be the optimum impulse response function. Thus, for an optimum system,

$$I_1(t) = \int_0^\infty f(t,\sigma)\left\{\phi_{zx}(t,\sigma) - \int_0^\infty g(t,\gamma)\phi_{xx}(\gamma,\sigma)d\gamma\right\}d\sigma = 0, \quad \text{A-8}$$

for any impulse response function $f(t, \sigma)$. But equation A-8 is satisfied if and only if

$$\phi_{zx}(t,\sigma) = \int_0^\infty g(t,\gamma)\phi_{xx}(\gamma,\sigma)d\gamma. \quad \text{A-9}$$

Equation A-9 is called Booton's integral equation or the modified Wiener-Hopf linear integral equation of the second kind; this is also equation (4) of the main text. Equation A-9 is the necessary and sufficient condition for the system to be optimum in the sense that the mean-square error

$$I(t) = E\{[z(t) - y(t)]^2\}$$

be minimum. By use of equations A-3 and A-9, the minimum mean-square error finally reduces to

$$M(t) = \phi_{zz}(t,t) - \int_0^\infty g(t,\sigma)\phi_{zx}(t,\sigma)d\sigma. \quad \text{A-10}$$

APPENDIX II

AN OPTIMUM OBSERVATION TIME FOR ESTIMATES OF TIME-VARYING CORRELATION FUNCTIONS

Let us assume that a time-varying autocorrelation function can be expressed as

$$\phi(t,\tau) = \sum_{i=1}^{n} a_i(t)b_i(\tau). \quad \text{A-11}$$

Let us further assume that $a_i(t)$, $i=1, 2, \cdots, n$ can be expanded in a Taylor series about an estimation point t_0, up to an mth term, or

$$a_i(t) = \sum_{l=0}^{m} c_{il}(t-t_0)^l, \quad i = 1, 2, \cdots, m, \quad \text{A-12}$$

where m is sufficiently large so that the remainder can be neglected. The Taylor coefficients are given by

$$c_{il} = \frac{1}{l!} a_i^{(l)}(t_0), \quad \text{A-13}$$

where the superscript (l) denotes the lth derivative of $a_i(t)$ with respect to t.

Substituting A-12 into A-11, we obtain

$$\phi(t,\tau) = \sum_{i=1}^{n}\sum_{l=0}^{m} c_{il}(t-t_0)^l b_i(\tau). \quad \text{A-14}$$

As an initial approximation, let us choose

$${}^0\phi(t_0,\tau,T) = \frac{1}{T}\int_{-T/2}^{T/2} x\left(t+t_0+\frac{\tau}{2}\right) \cdot x\left(t+t_0-\frac{\tau}{2}\right)dt. \quad \text{A-15}$$

The function ${}^0\phi(t_0, \tau)$ yields the desired estimate of $\phi(t_0, \tau)$ when T equals the optimum observation time T_0. Now the expected value of this function is

$$E\{{}^0\phi(t_0,\tau,T)\} = \frac{1}{T}\int_{-T/2}^{T/2}\phi(t+t_0,\tau)dt, \quad \text{A-16}$$

where

$$\phi(t+t_0,\tau) = E\left\{x\left(t+t_0+\frac{\tau}{2}\right)x\left(t+t_0-\frac{\tau}{2}\right)\right\}.$$

Substitution of A-14 into A-16 gives

$$E\{{}^0\phi(t_0,\tau,T)\} = \frac{1}{T}\int_{-T/2}^{T/2}\sum_{i=1}^{n}\sum_{l=0}^{m} c_{il}t^l b_i(\tau)dt$$

which can be written as

$$E\{{}^0\phi(t_0,\tau,T)\} = \sum_{i=1}^{n}\sum_{l=0}^{m} c_{il}b_i(\tau)\left[\frac{1}{T}\int_{-T/2}^{T/2} t^l dt\right]. \quad \text{A-17}$$

But,

$$\frac{1}{T}\int_{-T/2}^{T/2} t^l dt = \begin{cases} \dfrac{1}{l+1}\left(\dfrac{T}{2}\right)^{l+1} & \text{if } l \text{ is even} \\ 0 & \text{if } l \text{ is odd.} \end{cases}$$

Hence A-17 becomes

$$E\{{}^0\phi(t_0, \tau, T)\}$$
$$= \sum_{i=1}^{n} \sum_{l=0}^{\mu} c_{i2l} \frac{T^{2l}}{2^{2l}(2l+1)} b_i(\tau)$$
$$= \phi(t_0, \tau) + \sum_{i=1}^{n} \sum_{l=1}^{\mu} c_{i2l} \frac{T^{2l} b_i(\tau)}{2^{2l}(2l+1)}, \quad \text{A-18}$$

where

$$\mu = \begin{cases} m/2 & \text{if } m \text{ is even} \\ (m-1)/2 & \text{if } m \text{ is odd.} \end{cases}$$

The double summation in the last term of A-18 is the expected error in the approximation. As can be seen, it is a function of T^2. Thus, the observation interval must be small for the expected error to be small.

The variance σ^2 of ${}^0\phi(t_0, \tau, T)$ is given by

$$\sigma^2 = E\{[{}^0\phi(t_0, \tau, T) - {}^0\bar{\phi}(t_0, \tau, T)]^2\}$$

or

$$\sigma^2 = E\{{}^0\phi^2(t_0, \tau, T)\} - \{E[{}^0\phi(t_0, \tau, T)]\}^2, \quad \text{A-19}$$

where

$${}^0\bar{\phi}(t_0, \tau, T) = E\{{}^0\phi(t_0, \tau, T)\},$$

and

$${}^0\phi(t_0, \tau, T) = \frac{1}{T}\int_{-T/2}^{T/2} x\left(t + t_0 + \frac{\tau}{2}\right) \cdot x\left(t + t_0 - \frac{\tau}{2}\right) dt.$$

Substituting the above expressions for ${}^0\phi(t_0, \tau, T)$ and ${}^0\bar{\phi}(t_0, \tau, T)$ in equation A-19 yields

$$\sigma^2 = \frac{1}{T^2}\int_{-T/2}^{T/2}\int_{-T/2}^{T/2} \psi^2(t_1, t_2, t_0, \tau)\, dt_1 dt_2 - \frac{1}{T^2}\left[\int_{-T/2}^{T/2} \phi(t + t_0, \tau)\, dt\right]^2, \quad \text{A-20}$$

where $\psi^2(t_1, t_2, t_0, \tau)$ is the fourth product moment of $x(t)$, which is given by

$$\psi^2(t_1, t_2, t_0, \tau) = E\left\{x\left(t_1 + t_0 + \frac{\tau}{2}\right) x\left(t_1 + t_0 - \frac{\tau}{2}\right) \cdot x\left(t_2 + t_0 + \frac{\tau}{2}\right) x\left(t_2 + t_0 - \frac{\tau}{2}\right)\right\}. \quad \text{A-21}$$

In equations A-20 and A-21, t_1 and t_2 are the dummy variables for the integration. t_0 and τ are, respectively, the observed time and the lag time of the autocorrelation function. If $x(t)$ is a Gaussian process, $\psi^2(t_1, t_2, t_0, \tau)$ can be written as [see Laning and Battin (1956), p. 160–163.]

$$\psi^2(t_1, t_2, t_0, \tau) = \phi(t_1 + t_0, \tau)\phi(t_2 + t_0, \tau) + \phi(t_1 + t_0, t_1 - t_2) \cdot \phi(t_1 + t_0 - \tau, t_1 - t_2) + \phi(t_1 + t_0, t_1 - t_2 + \tau) \cdot \phi(t_1 + t_0 - \tau, t_1 - t_2 - \tau). \quad \text{A-22}$$

Equation A-20 may now be written in the form

$$\sigma^2 = \frac{1}{T^2}\int_{-T/2}^{T/2}\int_{-T/2}^{T/2} \{\phi(t_1 + t_0, t_1 - t_2) \cdot \phi(t_1 + t_0 - \tau, t_1 - t_2) + \phi(t_1 + t_0, t_1 - t_2 + \tau) \cdot \phi(t_1 + t_0 - \tau, t_1 - t_2 - \tau)\}\, dt_1 dt_2. \quad \text{A-23}$$

Let us assume that the observation time can be made small enough so that the time average of the autocorrelation function over T does not differ appreciably from the true value at t_0. Then A-23 can be approximated by

$$\sigma^2 \cong \frac{1}{T^2}\int_{-T/2}^{T/2}\int_{-T/2}^{T/2} \cdot \{\phi^2(t_0, t_1 - t_2) + \phi(t_0, t_1 - t_2 + \tau) \cdot \phi(t_0, t_1 - t_2 - \tau)\}\, dt_1 dt_2. \quad \text{A-24}$$

Equation A-24 is derived from equation A-23 by setting:

$$\phi(t_0, t_1 - t_2) \cong \phi(t_1 + t_0, t_1 - t_2),$$
$$\phi(t_0, t_1 - t_2) \cong \phi(t_1 + t_0 - \tau, t_1 - t_2),$$
$$\phi(t_0, t_1 - t_2 + \tau) \cong \phi(t_1 + t_0, t_1 - t_2 + \tau),$$
$$\phi(t_0, t_1 - t_2 - \tau) \cong \phi(t_1 + t_0 - \tau, t_1 - t_2 - \tau).$$

An upper bound of σ^2 may be established if τ is small compared to T, but T is large compared to the significant duration of the autocorrelation function. Replacing (t_1-t_2) by λ in equation A-24, we have [see Berndt and Cooper (1965)],

$$\sigma^2 \cong \frac{1}{T}\int_{-T}^{T}\left(1-\frac{|\lambda|}{T}\right) \cdot \{\phi^2(t_0,\lambda)+\phi(t_0,\lambda+\tau)\phi(t_0,\lambda-\tau)\}d\lambda. \qquad \text{A-25}$$

By using the Schwarz inequality and assuming that the "energy" of the autocorrelation function is bounded, the upper bound of σ^2 can be obtained from equation A-25 in the form

$$\sigma^2 \leq \frac{2}{T}\int_{\infty}^{\infty}\phi^2(t_0,\lambda)d\lambda. \qquad \text{A-26}$$

Let S^2 be the mean-square error of the approximation, which is

$$S^2 = E\{[{}^0\phi(t_0,\tau,T)-\phi(t_0,\tau)]^2\},$$

or

$$S^2 = \sigma^2 + \{{}^0\bar{\phi}(t_0,\tau,T)-\phi(t_0,\tau)\}^2. \qquad \text{A-27}$$

However, from equation A-18 we have

$${}^0\bar{\phi}(t_0,\tau,T)-\phi(t_0,\tau) = \sum_{i=1}^{n}\sum_{l=1}^{\mu} c_{i2l}\frac{T^{2l}}{2^{2l}(2l+1)}b_i(\tau),$$

and thus the use of inequality A-26 yields

$$S^2 \leq \frac{2}{T}\int_{-\infty}^{\infty}\phi^2(t_0,\lambda)d\lambda + \sum_{i=1}^{n}\sum_{j=1}^{n}\sum_{p=1}^{\mu}\sum_{q=1}^{\mu} c_{i2p}c_{j2q} \cdot \frac{T^{2(p+q)}b_i(0)b_j(0)}{(2p+1)(2q+1)2^{2(p+q)}}. \qquad \text{A-28}$$

Let R denote the upper bound of S^2. Then the optimum observation time is attained when the upper bound R is a minimum. This requires that the conditions

$$\left.\begin{aligned} \left.\frac{\partial R}{\partial T}\right|_{T=T_0} &= 0 \\ \left.\frac{\partial^2 R}{\partial T^2}\right|_{T=T_0} &> 0 \end{aligned}\right\} \qquad \text{A-29}$$

be satisfied.

Differentiating the right-hand side of inequality A-28, R with respect to T yields

$$\sum_{i=1}^{n}\sum_{j=1}^{n}\sum_{p=1}^{\mu}\sum_{q=1}^{\mu} c_{i2p}c_{j2q} \cdot \frac{(p+q)b_i(0)b_j(0)T^{2(p+q)+1}}{(2p+1)(2q+1)2^{2(p+q)}} = \int_{-\infty}^{\infty}\phi^2(t_0,\lambda)d\lambda. \qquad \text{A-30}$$

In particular, if the Taylor expansion of $a_i(t)$ can be terminated after the second term, equation A-30 reduces to

$$\sum_{i=1}^{n}\sum_{j=1}^{n} c_{i2}c_{j2}\frac{b_i(0)b_j(0)T^5}{72} = \int_{-\infty}^{\infty}\phi^2(t_0,\lambda)d\lambda \qquad \text{A-31}$$

or

$$T = \left\{\frac{72\int_{-\infty}^{\infty}\phi^2(t_0,\lambda)d\lambda}{\sum_{i=1}^{n}\sum_{j=1}^{n} c_{i2}c_{j2}b_i(0)b_j(0)}\right\}^{1/5} \qquad \text{A-32}$$

which is equation (21) of the main text.

APPENDIX III

A NUMERICAL EXAMPLE FOR DETERMINING THE COEFFICIENTS OF $a_i(t)$.

Let us assume that we have determined the three time-invariant autocorrelation functions $\phi_{xx}^{[1]}(\tau)$, $\phi_{xx}^{[2]}(\tau)$, and $\phi_{xx}^{[3]}(\tau)$, corresponding to t_1, t_2, and t_3, respectively (see Figure 5). We can then expand these autocorrelation functions in finite cosine series, that is,

$$\phi_{xx}^{[j]}(\tau) = \sum_{i=1}^{n} a_i(t_j)\cos(i-1)\frac{2\pi\tau}{T}, \qquad j = 1, 2, 3, \qquad \text{A-33}$$

where n is an appropriate number of terms to be used in the expansion and the $a_i(t_j)$ are the ith coefficients of the cosine transforms of $\phi_{xx}^{[j]}(\tau)$. Furthermore, if we assume that $a_i(t)$ in equation (7) of the main text varies as a second-order equation in t, and let

$$a_i(t) = d_i + e_i(kT - t) + f_i(kT - t)^2, \qquad i = 1, 2, \cdots, n, \qquad \text{A-34}$$

d_i, e_i, and f_i can easily be found by solving three simultaneous algebraic equations, namely,

$$\begin{aligned} a_i(t_1) &= d_i + e_i(kT - t_1) + f_i(kT - t_1)^2, \\ a_i(t_2) &= d_i + e_i(kT - t_2) + f_i(kT - t_2)^2, \\ a_i(t_3) &= d_i + e_i(kT - t_3) + f_i(kT - t_3)^2, \end{aligned} \qquad \text{A-35}$$

where k is an index identifying the kth section of the trace and T is the gate length.

For example, let $k=1$, $T=30$, $t_1=5$, $t_2=15$, $t_3=25$, and $i=1$. Furthermore, assume that $a_1(t_1)=5$, $a_1(t_2)=2$, and $a_1(t_3)=1$. Then, from equations A-35, we have

$$\begin{aligned} 5 &= d_1 + e_1(25) + f_1(25)^2, \\ 2 &= d_1 + e_1(15) + f_1(15)^2, \\ 1 &= d_1 + e_1(5) + f_1(5)^2. \end{aligned}$$

Solving the above equations for d_1, e_1, and f_1 yields $d_1=1.25$, $e_1=-0.10$, $f_1=0.01$, and $a_1(t)$ can now be expressed as

$$a_i(t) = 1.25 - 0.10(30 - t) + 0.01(30 - t)^2.$$

$a_1(t)$ tells us how the first coefficient in the cosine expansion of the time-varying autocorrelation function varies with observation time within the interval $(k-1)T \leq t < kT$.

From equation (25) of the main text, however, one sees that only the coefficients f_i are needed for determination of the optimum gate lengths. Cramer's rule, f_i can be found from A-35 as

$$f_i = \frac{1}{\Delta} \begin{vmatrix} 1 & (kT - t_1) & a_i(t_1) \\ 1 & (kT - t_2) & a_i(t_2) \\ 1 & (kT - t_3) & a_i(t_3) \end{vmatrix},$$

where

$$\Delta = \begin{vmatrix} 1 & (kT - t_1) & (kT - t_1)^2 \\ 1 & (kT - t_2) & (kT - t_2)^2 \\ 1 & (kT - t_3) & (kT - t_3)^2 \end{vmatrix}.$$

In the above expressions the vertical bars denote the determinant. By now it should be evident that given the form of $a_i(t)$ as A-34, three equations are needed to compute f_i. This, of course, is the reason why the gate is further divided into three short sections and the autocorrelation functions are then found at t_1, t_2, and t_3.

Reprinted from Geophysics v. 37, no. 6, p. 947-952

GEOPHYSICS

ESTIMATION OF A TIME-VARYING SEISMIC AUTOCORRELATION FUNCTION†

DAVID D. THOMPSON * AND GEORGE R. COOPER ‡

The autocorrelation function of a seismic trace provides information about the generating wavelet needed for optimal processing. The classical time-averaging technique used to estimate the autocorrelation, however, fails for the commonly encountered time-varying autocorrelation resulting from progressive wavelet distortion. To estimate a time-varying autocorrelation, an iterative computational procedure is proposed from which a sequence of progressively improved estimates of the autocorrelation is obtained. In the time-invariant case, the algorithm reduces to the usual time-average procedure applied to several different "windows" along the trace, and convergence occurs on the first step. For a "slowly" varying autocorrelation, however, the procedure tends to iteratively correct the bias error resulting from time averaging. It is proven that if the time variation for each "lag" value can be modeled by a polynomial of degree N, the sequence of estimates converges (in expected value) to the solution in no more than the integer part of $(N+2)/2$ steps. Examples are included to illustrate the procedure.

INTRODUCTION

The autocorrelation function plays an important role in processing seismic data by providing spectral information about the generating wavelet. Unfortunately, the simple time averaging of lagged products often employed to compute the autocorrelation does not provide the most accurate estimate possible in the seismic case. This follows because the wavelet, and, thus, also the autocorrelation are time varying as a result of frequency-dependent attenuation.

A commonly employed approach to this problem is to assume that the process is piecewise time invariant and simply compute lagged product averages over a "sliding window," thereby obtaining a time-varying estimate. The difficulty with such an approach, of course, is that the approximation is only valid for small windows, while the statistical variance of the averages is small only for large window sizes.

As a result, one is faced with a tradeoff between a bias error resulting from excessive window size, and a variance error resulting from an insufficient window size. Berndt and Cooper (1965) have de-

† Paper presented at the 41st Annual International SEG Meeting, November 11, 1971, Houston, Texas. Manuscript received by the Editor February 24, 1972; revised manuscript received July 26, 1972.

* Atlantic Richfield Company, Dallas, Texas 75221.

‡ Purdue University, Lafayette, Indiana 74907.

veloped a means for determining the optimal window width which minimizes the mean-square error of such estimates, and Wang (1969) has applied this result to seismic data. However, the question has not been raised in these works as to the optimality of the sliding-window approach itself in the time-varying case.

Here, a simple iterative correction procedure is presented for the sliding-window method which permits one to remove much of the bias error while still maintaining the small variance error resulting from the use of a large averaging window.

THE ALGORITHM

To describe the procedure, consider $x(t)$ to be a sample function from a random process X. The autocorrelation of X for time t and lag τ is defined as

$$R(t, \tau) = E[x(t + \tau/2)\, x(t - \tau/2)], \quad (1)$$

where $E[\cdot]$ is the expected value or ensemble average operator. In this paper, the following conditions on X will be assumed:

1) The process X has a fading autocorrelation defined as $R(t, \tau) \to 0$ as $\tau \to \infty$ for each t.
2) The interchange of time and ensemble averages is valid for the autocorrelation of X.

Let us first consider the sliding-window estimate $R_{sw}(t, \tau, T)$.

$$R(t, \tau) \approx R_{sw}(t, \tau, T) = (1/T) \int_{t-T/2}^{t+T/2} x(\xi + \tau/2) x(\xi - \tau/2)\, d\xi. \quad (2)$$

The mean and variance of R_{sw} are, respectively,

$$m_{sw}(t, \tau, T) = E[R_{sw}(t, \tau, T)] = (1/T) \int_{t-T/2}^{t+T/2} R(\xi, \tau)\, d\xi \quad (3)$$

and

$$\sigma^2_{sw}(t, \tau, T) = E\{[R_{sw}(t, \tau, T) - m_{sw}(t, \tau, T)\}^2] = (1/T^2) \int_{\xi_1=t-T/2}^{t+T/2} \int_{\xi_2=t-T/2}^{t+T/2} [M(\xi_1, \xi_2, \tau) - R(\xi_1, \tau) R(\xi_2, \tau)]\, d\xi_2\, d\xi_1, \quad (4)$$

where $M(\xi_1, \xi_2, \tau)$ is the fourth moment defined by

$$M(\xi_1, \xi_2, \tau) = E[x(\xi_1 + \tau/2)\, x(\xi_1 - \tau/2)\, x(\xi_2 + \tau/2)\, x(\xi_2 - \tau/2)]. \quad (5)$$

If X is Gaussian and zero mean, this fourth moment is (Laning and Batin, 1956, p. 160–163)

$$M(\xi \;\; \xi_2, \tau) = R(\xi_1, \tau) R(\xi_2, \tau) + R\left(\frac{\xi_1 + \xi_2 + \tau}{2}, \xi_2 - \xi_1\right) \cdot R\left(\frac{\xi_1 + \xi_2 - \tau}{2}, \xi_2 - \xi_1\right) + R\left(\frac{\xi_1 + \xi_2}{2}, \xi_2 - \xi_1 - \tau\right) \cdot R\left(\frac{\xi_1 + \xi_2}{2}, \xi_2 - \xi_1 + \tau\right). \quad (6)$$

Let us define the bias error

$$b_{sw}(t, \tau, T) = m_{sw}(t, \tau, T) - R(t, \tau). \quad (7)$$

In the stationary case, of course, the bias error is zero from equation (3), since time averaging the invariant autocorrelation has no effect. On the other hand, in the nonstationary case the magnitude of this time average will often (though not always) depart further from $R(t, \tau, T)$ as the window width is increased to take in values further from t.

Furthermore, at least for the nonstationary Gaussian case, equations (4) and (6) and the fading correlation assumption imply that the estimate variance σ^2_{sw} approaches zero as T approaches infinity. Thus, by choosing a large window T, the expected deviation of the estimate $R_{sw}(t, \tau, T)$ from $R(t, \tau)$ can be made small.

For the time-varying case, however, it is clear from equations (3) and (7) that the approximation $R_{sw}(t, \tau, T)$ assumes, in general, an increasing bias error with increasing T. However, the estimate variance σ^2_{sw} continues to be made small by increasing T, provided the correlation fade is rapid relative to the rate of time variation. (It is not necessarily true in all cases that the variance approaches zero though, as T approaches infinity.)

The obvious answer to this dilemma is to attempt to correct the sliding-window estimate for its bias error so that advantage can be taken of the small variance error afforded by a large window T.

For this purpose, consider the sequence of estimates $R_i(t, \tau, T)$ given by the recursive relations:

$$R_0(t, \tau, T) = 0, \quad \text{and}$$

$$R_{i+1}(t, \tau, T) = R_{sw}(t, \tau, T) + R_i(t, \tau, T) - (1/T)\int_{t-T/2}^{t+T/2} R_i(\xi, \tau, T)d\xi, \tag{8}$$

for $i = 0, 1, 2, \ldots,$

where $R_{sw}(t, \tau, T)$ is given by equation (2).

The intuitive motivation behind equation (8) is easily seen. Suppose $R_i(t, \tau, T)$ is a reasonably close estimate for $R(t, \tau)$. Then the last two terms of the right-hand side of equation (8) tend to correct the first term $R_{sw}(t, \tau, T)$ by an amount approximately equal to the difference between the value of the autocorrelation at t and its integral over the window $(t-T/2,\ t+T/2)$. In other words, the procedure should approximately correct the bias error with the next iteration.

The first question is whether such a process leads to repeated improvements in the estimate with each iteration. In this regard we state the following results.

Theorem

Suppose that the autocorrelation $R(t, \tau)$ can be expressed by the Nth-degree power series in t, for every τ, as

$$R(t, \tau) = \sum_{j=0}^{N} \bar{a}_j(\tau)t^j.$$

Further suppose that the ith estimate $R_i(t, \tau, T)$ for the autocorrelation can be expressed similarly, and that for some K, $0 \le K \le N+1$, the K highest-order coefficients agree with those of $R(t, \tau)$ (a vacuous assumption if $K=0$). In other words,

$$R_i(t, \tau, T) = \sum_{j=0}^{N} a_{ij}(\tau)t^j,$$

where (if $K \neq 0$),

$$\bar{a}_{ij}(\tau) = a_j(\tau) \quad \text{for } (N - K + 1) \le j \le N.$$

Then, for the next estimate $R_{i+1}(t, \tau, T)$ given by equation (8):[1]

[1] The notation $\bar{E}$ will be used here to designate the expected value under the condition that the estimate $R_i(t, \tau, T)$ is given.

$$\bar{E}[R_{i+1}(t, \tau, T)] = \sum_{j=0}^{N} \bar{a}_{(i+1),j}(\tau)t^j,$$

where

$$\bar{a}_{(i+1),j}(\tau) = a_j(\tau)$$

$$\text{for } \max[0, (N - K - 1)] \le j \le N.$$

Proof: See appendix for detailed proof by direct computation.

Corollary

If for each τ, $R(t, \tau)$ can be expressed as an Nth-degree power series in t, and if each estimate attains its expected value, the iterative sequence of estimates given by equation (8) converges to the autocorrelation $R(t, \tau)$ in, at most, the integer part of $[(N+2)/2]$ steps.

Proof:

By repeated application of the theorem we note that each iteration (if it attains its expected value as is assumed here) corrects two previously incorrect coefficients (except, of course, when fewer than two remain) while maintaining the previously correct coefficients. Thus, the conclusion follows directly.

Table 1 shows the "ideal" number of steps required for convergence for various values of N as given by the corollary.

Table 1. Ideal convergence rate

Degree of the polynomial in t needed to represent $R(t, \tau)$	Number of steps needed for convergence if each iterate achieves its expected value
0	1
1	1
2	2
3	2
4	3
5	3
6	4
7	4
.	.
.	.
.	.

To determine how closely we can expect to achieve these ideal results, we need only examine the estimate variance which determines the probable deviation of an estimate from its expected value. A straightforward computation reveals that given the estimate $R_i(t, \tau, T)$, the conditional variance of the next estimate $R_{i+1}(t, \tau, T)$ is

$$\begin{aligned}
&\bar{\sigma}^2_{i+1}(t, \tau, T) \\
&= \bar{E}[(R_{i+1}(t, \tau, T) - \bar{E}[R_{i+1}(t, \tau, T)])^2] \quad (9) \\
&= \sigma^2_{sw}(t, \tau, T)
\end{aligned}$$

as given by equation (4).

As discussed previously for the slowly time-varying case, $\sigma^2_{sw}(t, \tau, T)$ generally can be made small by choosing a large window T. Thus, the conditions of the corollary can be approximated closely with a sufficiently large T.

However, since in fact each estimate does not attain its expected value exactly, it has been found helpful to slightly modify the procedure of equation (8) to prevent amplification of the rapid perturbations associated with this error, while still properly correcting the slow time-varying bias error. To accomplish this, a simple modification of the correction terms (the last two terms) of equation (8) has been adopted which insures that the bias correction is slowly varying. In particular one simply replaces in these terms the previous estimate $R_i(t, \tau, T)$ by its least-squares p-degree polynomial approximation

$$R_i(\bar{t}, \tau, T) \approx \sum_{j=0}^{p} \alpha_{ij}(t, \tau, T) \left[\frac{\bar{t} - t}{T/2}\right]^j \quad (10)$$

determined on the interval

$$t - T/2 \leq \bar{t} \leq t + T/2. \quad (11)$$

If this is done, equation (8) becomes

$$\begin{aligned}
&R_0(t, \tau, T) = 0, \quad \text{and} \\
&R_{i+1}(t, \tau, T) = R_{sw}(t, \tau, T) \\
&\qquad - \sum_{\substack{j=2 \\ j \text{ even only}}}^{p} \left(\frac{1}{j+1}\right) \alpha_{ij}(t, \tau, T), \quad (12) \\
&\qquad i = 0, 1, 2, \cdots,
\end{aligned}$$

where $R_{sw}(t, \tau, T)$ is the sliding-window estimate given by equation (2) and $\alpha_{ij}(t, \tau, T)$, the least-squares coefficient appearing in equation (10). As a matter of practical consideration, a low value for p such as ($p=2$) is found to be quite adequate since a higher-order behavior can be handled readily by the fact that the coefficients are computed for each value of t (and τ) with the resulting coefficients used only to correct the midpoint on the interval of computation.

Equation (12) provides an additional advantage over the iterations of equation (8), in that it allows a straightforward and efficient extrapolation at the ends of the data. By simply estimating the coefficients $\alpha_{ij}(t, \tau, T)$ over that portion of the window on which values of $R_i(t, \tau)$ are available (always at least a window of $T/2$), we can prevent the contraction of the solution interval at each iteration which otherwise would result from equation (8). Admittedly the accuracy may be somewhat less on the ends as a result of this extrapolation, but at least a reasonable estimate is obtained there.

EXAMPLES

The procedure described by equation (12) was discretized and applied to a synthetic example of known solution. In particular, using a random number generator, a set of 5100 uncorrelated Gaussian samples was generated, having zero mean and a time-dependent variance (zero-lag autocorrelation) given for the ith sample by

$$\sigma_i^2 = 3/4 + (1/4) \cos\left(\frac{3 \cdot 2\pi i}{5100}\right)$$
$$i = 1, \cdots, 5100.$$

The window consisting of 1000 samples was "slid" in discrete steps of 100 samples from one end of the data to the other. For this test a quadratic fit ($p=2$) was used. Figure 1 gives a comparison

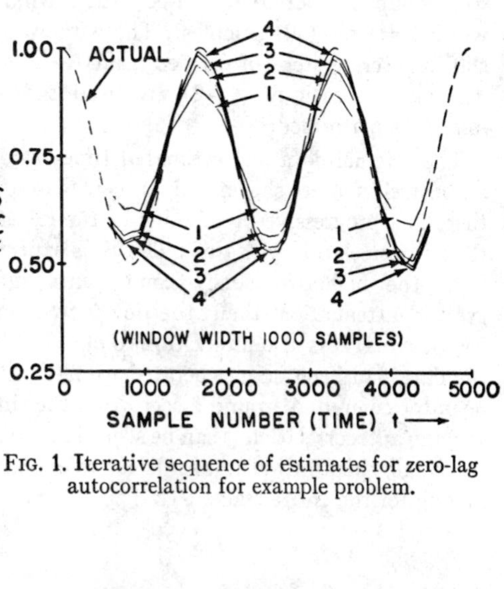

FIG. 1. Iterative sequence of estimates for zero-lag autocorrelation for example problem.

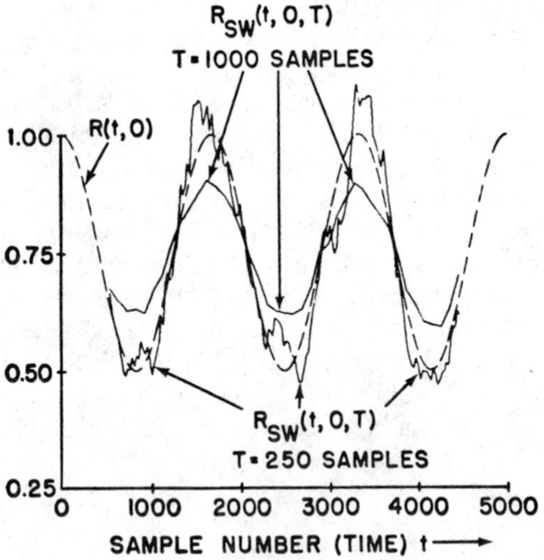

FIG. 2. Trade-off between bias error and variance error for example.

of the various iterates and the "true" solution. Note the rapid improvements on the first few iterations and the stability of the iterates as the solution is approached.

By contrast, Figure 2 shows the trade-off between bias error and variance error resulting from the application of the uncorrected sliding-window estimate with different window widths to this example.

As a final example, the algorithm was applied to a CDP stacked trace on a seismic line from the north slope of Alaska. The data consisted of 2-msec samples of which values from about 0.4 sec to 3.6 sec were used in the computation. Lags were computed out to 150 msec with a window width $T=1$ sec (501 samples). The window was slid in discrete steps of 0.1 sec from 1.−3. sec. Again a quadratic fit ($p=2$) was used and five iterations performed.

The original data and estimated time-varying autocorrelation are shown in Figure 3. Note that there is a progressive broadening of the lobes in the autocorrelation with time. This is as expected since the higher-frequency components suffer greater attenuation than the lower-frequency components. It is also interesting to observe that the data had been dereverberated using a single operator computed around 2 sec. From the time-varying autocorrelation it can be seen clearly that this single operator was not sufficient to eliminate ringing for the entire trace.

CONCLUSION

The algorithm presented here offers a simple modification to the sliding-window approach for estimation of a nonstationary autocorrelation. With this procedure it is possible in the slowly time-varying case to minimize the bias error while still maintaining the small estimate variance afforded by a large window width. Although the procedure is iterative, it has been shown both in theory and by example to be rapidly convergent. In addition, it should be noted that the same technique is equally applicable to the estimation of other nonstationary statistics as well as to the autocorrelation.

APPENDIX

To verify the theorem in the text, we suppose that for some integer $N \geq 0$,

$$R(t, \tau) = \sum_{j=0}^{N} a_j(\tau) t^j. \tag{A1}$$

Furthermore, we suppose that for some K, $0 \leq K \leq N+1$, the ith estimate of the autocorrelation can be expressed as[2]

$$R_i(t, \tau, T) = \sum_{j=0}^{N-K} \tilde{a}_{ij}(\tau) t^j + \sum_{j=N-K+1}^{N} a_j(\tau) t^j. \tag{A2}$$

[2] Throughout the proof it will be understood that if the upper limit is smaller than the lower limit on the index of a summation, the sum will be zero. This allows the proof to hold for the special case where no coefficients of R_i are correct ($K=0$) and the case where all coefficients of R_i are correct ($K=N+1$).

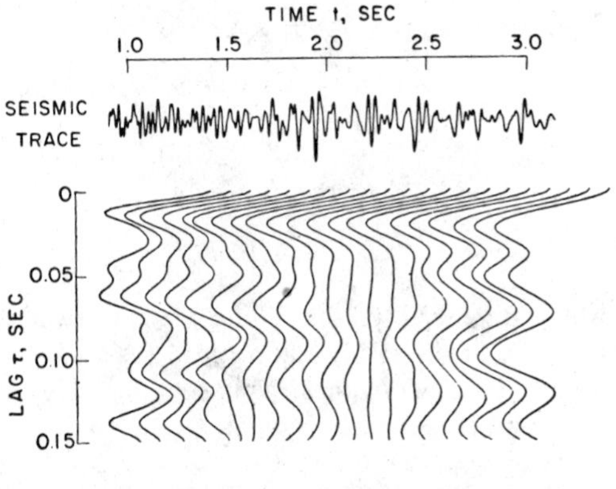

FIG. 3. Application of the algorithm to field data.

Then if $R_{i+1}(t, \tau, T)$ is the next estimate as given by equation (8), its expected value [conditioned on the given estimate $R_i(t, \tau, T)$] is

$$\bar{E}[R_{i+1}(t, \tau, T)] = R_i(t, \tau, T) + (1/T)\int_{t-T/2}^{t+T/2} R(\xi, \tau) - R_i(\xi, \tau, T)d\xi. \tag{A3}$$

Substituting equations (A1) and (A2) into (A3), we have

$$\bar{E}[R_{i+1}(t, \tau, T)] = h(t, \tau, T) + \sum_{j=0}^{N-K} \tilde{a}_{ij}(\tau)t^j + \sum_{j=N-K+1}^{N} a_j(\tau)t^j, \tag{A4}$$

where

$$h(t, \tau, T) = (1/T)\int_{t-T/2}^{t+T/2} \sum_{j=0}^{N-K} (a_j(\tau) - \tilde{a}_{ij}(\tau))\xi^j d\xi. \tag{A5}$$

After integration and binomial expansion, we obtain

$$h(t, \tau, T) = (1/T)\sum_{j=1}^{N-K+1}\left(\frac{a_{j-1}(\tau) - \tilde{a}_{i,(j-1)}(\tau)}{j}\right) \cdot \sum_{n=0}^{j}\binom{j}{n}[(T/2)^{j-n} - (-T/2)^{j-n}]t^n, \tag{A6}$$

where

$$\binom{j}{n} = \frac{j!}{(j-n)!n!}.$$

Note that all terms in equation (A6) for which $(j-n)$ is even are zero since

$$(T/2)^{j-n} = (-T/2)^{j-n}.$$

Thus, after interchanging the order of summation, removing the zero terms, absorbing the $1/T$, and shifting the index j down by 1, we have

$$h(t, \tau, T) = \sum_{n=0}^{N-K}\sum_{j=n}^{N-K}\binom{j+1}{n}(T/2)^{j-n} \cdot \left(\frac{a_j(\tau) - \tilde{a}_{ij}(\tau)}{j+1}\right)t^n. \tag{A7}$$

$(j-n)$ even only

Note that for $n = N-K$ we have only the term:

$$\binom{N-K+1}{N-K}\left(\frac{a_{N-K}(\tau) - \tilde{a}_{i,(N-K)}(\tau)}{N-K+1}\right)t^{N-K} = [a_{N-K}(\tau) - \tilde{a}_{i,(N-K)}(\tau)]t^{N-K}. \tag{A8}$$

Similarly, for $n = N-K-1$ we have only

$$\binom{N-K}{N-K-1} \cdot \left(\frac{a_{N-K-1}(\tau) - \tilde{a}_{i,(N-K-1)}(\tau)}{N-K}\right)t^{N-K-1} = [a_{N-K-1}(\tau) - \tilde{a}_{i,(N-K-1)}(\tau)]t^{N-K-1}. \tag{A9}$$

Thus, substituting equations (A8) and (A9) into (A7), and (A7) into (A4), we finally obtain

$$\bar{E}[R_{i+1}(t, \tau, T)] = \sum_{n=0}^{N-K-2}\left[\tilde{a}_{in}(\tau) + \sum_{j=n}^{N-K}\binom{j+1}{n}(T/2)^{j-n} \cdot \left(\frac{a_j(\tau) - \tilde{a}_{kj}(\tau)}{j+1}\right)\right]t^n \tag{A10}$$

$(j-n)$ even only

$$+ \sum_{n=\max[0, N-K-1]}^{N} a_n(\tau)t^n$$

where the first sum is understood to be vacuous if the upper limit $(N-K-2) < 0$. Equation (A10), thus, verifies the statement of the theorem that the power series coefficients of the new estimate $\tilde{a}_{i(+1)\cdot j}(\tau)$ satisfy

$$\tilde{a}_{(i+1),j}(\tau) = a_j(\tau)$$

for max $[0, (N-K-1)] \leq N$ and so completes the proof.

REFERENCES

Berndt, H., and Cooper, G. R., 1965, An optimum observation time for estimates of time-varying correlation functions: IEEE Trans. on Inform. Theory, v. IT-11, p. 307–310.

Laning, J. H., and Batin, R. H., 1956, Random processes in automatic control: New York, McGraw-Hill Book Co., Inc.

Wang, R. J., 1969, The determination of optimum gate lengths for time-varying Wiener filtering: Geophysics, v. 34, p. 683–695.

Reprinted from Geophysics v. 42, no. 4, p. 742-759

ADAPTIVE DECONVOLUTION: A NEW TECHNIQUE FOR PROCESSING TIME-VARYING SEISMIC DATA

L. J. GRIFFITHS*, F. R. SMOLKA‡ AND L. D. TREMBLY‡

A time-varying deconvolution method has been developed which is based upon adaptive linear filtering techniques. This adaptive deconvolution is applicable for use in processing reflection seismic data which contain multiples with periods that vary with traveltime.

Filter coefficients are designed for each sample of the input trace using an adaptive algorithm. Convergence properties of the adaptive processor are discussed and compared to conventional deconvolution techniques. The adaptive deconvolution method is illustrated using both synthetic and field reflection seismic data. By proper selection of parameters and by procesing the data in both time-reverse and time-forward directions, adaptive deconvolution removes multiples with varying periods while leaving primary reflections relatively undistorted.

INTRODUCTION

This paper presents a new time-varying deconvolution method for use in processing reflection seismograms. The technique is based on the use of a continuously adaptive linear prediction operator in which the operator coefficients are updated using a simple adaptive algorithm. New coefficient values are computed for each data sample in the seismic record so as to minimize a mean-square error criterion. This procedure differs significantly from time-varying deconvolution methods described by Clarke (1968), Wang (1969), and others. Previous methods have employed the well-known three-stage processes of first, computing autocorrelation estimates from the data; second, solving a set of appropriate normal equations to determine the operator coefficient values; and third, applying the operator to the data to obtain the deconvolved output trace. In the adaptive deconvolution procedure proposed in this paper, new coefficient values are computed directly from the seismic data values as the operator is applied to the data. In effect, the operator is designed as the deconvolved output is produced.

Deconvolution operators which employ a minimum mean-square error criterion—i.e., Wiener filters—have been widely used in reflection seismogram processing. The method generally assumes that the statistical structure of the data is stationary over the design gate used to generate the deconvolution operator. This is true even for the piecewise stationary time-varying methods discussed by Clarke (1968) and Wang (1969). In some seismic processing applications, the piecewise stationary assumption may be overly restrictive. An example of such data can be found in the marine, shallow, hard water-bottom environment. It is readily shown (Middleton and Whittlesey, 1968) that the time between successive water bottom reflections, for source-receiver spacings other than zero, changes markedly with reflection number. Because these multiples are relatively close together, the resulting seismogram exhibits nonstationary statistical properties over a short time interval, and conventional Wiener filtering methods do not yield satisfactory results.

Theoretical extensions of Wiener filtering theory to include time-varying systems have been proposed by Boonton (1952) but the resulting integral equations are extremely difficult to establish without a detailed a priori knowledge of the

Presented at the 45th Annual International SEG Meeting, October 15, 1975 in Denver. Manuscript received by the Editor June 10, 1976.
* University of Colorado, Boulder, CO 80302.
‡ Marathon Oil Co., Littleton, CO 80120.

nature of the underlying time variations. Kalman filtering methods have also been applied to the time-varying deconvolution problem (Crump, 1974), but the state-variable model used in this formulation also requires accurate estimates of the underlying time-varying parameters. Reliable estimates of the parameters can only be obtained under conditions of extremely high signal-to-noise ratio (SNR). Another approach, involving the use of homomorphic techniques, was recently proposed by Buhl et al (1974). Although this method is shown to handle successfully the time-varying multiple problem, the authors have not considered the effects of additive noise on the process and the procedure is also restricted to high SNR environments.

The adaptive deconvolution method we describe in this paper uses a time-continuous adaptive procedure similar to that used extensively in the communications and antenna array processing field (Widrow et al, 1967; Lacoss, 1968; Griffiths, 1969; Frost, 1972; Riegler and Compton, 1973; Widrow et al, 1975; Gabriel, 1976; Griffiths, 1976). These applications have demonstrated a substantial improvement in processed SNR in the presence of time-varying noise and interference. The algorithms used to obtain this performance can be grouped into two broad classes: those which adapt to minimize the average squared error between a given desired response signal and the adaptively filtered signal (Widrow et al, 1967; Wang and Treitel, 1971; Griffiths, 1975) and those which adapt to minimize the adapted output power, subject to a linear constraint on the filter coefficients (Booker and Ong, 1971; Frost, 1972).

The algorithm proposed for time-varying deconvolution in this paper is a member of the first class described above. An adaptive filter L samples in length is used to linearly combine L successive samples of the recorded reflection seismogram. The coefficients used in this filter are then adapted to minimize the mean-square difference between the filter output and the data value in the seismic trace which occurs γ samples later in time. In effect, L data values are used to predict the trace γ time samples later in a manner directly analogous to the predictive deconvolution procedure described by Peacock and Treitel (1969). The algorithm is then used to update the prediction coefficients as the filter moves along the seismogram. Details of the algorithm and processing examples illustrating its use on both synthetic and field-recorded seismic data are presented in the sections following.

ADAPTIVE DECONVOLUTION

Our adaptive deconvolution procedure utilizes a continuously time-varying operator to process the seismic trace. Thus, a different operator is used at each data point in the trace as the processor moves along the reflection seismogram. A simple adaptive algorithm, based on the method of steepest descent, is used to update the operator coefficients as each data point is deconvolved. In this section, we first present some background material on predictive deconvolution of stationary traces and on the development of adaptive techniques in general. We then describe the specific application of adaptive processing to the deconvolution problem.

We denote the set of N data points in the seismogram to be processed by $x(t)$, $t = 0,1,\cdots N - 1$. As shown by Peacock and Treitel (1969), predictive deconvolution of this trace for the special case of statistically stationary $x(t)$ is achieved by first solving the normal equations for an L-point prediction operator denoted by the vector $\mathbf{F}_o$. The appropriate equations are given by

$$\begin{bmatrix} r_x(0) & r_x(1)\cdots & r_x(L-1) \\ r_x(1) & r_x(0) & \cdot \\ \cdot & & \cdot \\ \cdot & & \cdot \\ r_x(L-1) & \cdots & r_x(0) \end{bmatrix} \begin{bmatrix} f_0(0) \\ f_0(1) \\ \cdot \\ \cdot \\ f_0(L-1) \end{bmatrix} = \begin{bmatrix} r_x(\gamma) \\ r_x(\gamma+1) \\ \cdot \\ \cdot \\ r_x(\gamma+L-1) \end{bmatrix}, \tag{1}$$

where $r_x(l)$ is the trace autocorrelation at lag l, $f_0(l)$ is the lth prediction coefficient and $\gamma \geq 1$ is the prediction distance. This equation may be expressed more compactly in matrix notation as

$$\mathcal{R}_{xx}\mathbf{F}_0 = \mathbf{P}_x(\gamma), \tag{2}$$

where $\mathcal{R}_{xx}$ is the expected value of $\mathbf{X}(t)\mathbf{X}^T(t)$ and $\mathbf{P}_x(\gamma)$ is the expected value of $x(t+\gamma)\mathbf{X}(t)$.

In the usual application of predictive deconvolution methods to field recorded data, the autocorrelation values required for $\mathcal{R}_{xx}$ and $\mathbf{P}_x(\gamma)$ are computed using time averages and the resulting prediction coefficients are then computed from equation (2) using the Levinson method—see

Robinson (1967). A deconvolved trace $y(t)$ is computed from the input $x(t)$ using the resulting Value of $\mathbf{F}_0$. One method for visualizing this computation which is particularly convenient for describing adaptive deconvolution proceeds as follows:

(a) The operator $\mathbf{F}_0$ is first used to generate a predicted or estimated trace $\hat{x}(t)$ from the data

$$\hat{x}(t+\gamma) = \sum_{l=0}^{L-1} f_0(l)x(t-l). \tag{3}$$

In matrix notation,

$$\hat{x}(t+\gamma) = \mathbf{F}_0{}^T\mathbf{X}(t) = \mathbf{X}^T(t)\mathbf{F}_0, \tag{4}$$

where

$$\mathbf{X}^T(t) = [x(t), x(t-1), \cdots x(t-L+1)]. \tag{5}$$

(b) The deconvolved trace $y(t)$ is then the difference between the input and predicted traces

$$y(t) = x(t) - \hat{x}(t). \tag{6}$$

A heuristic development of an adaptive method for obtaining $y(t)$ may be given directly from equation (2). We begin by assuming that a gradient descent procedure is used to find $\mathbf{F}_0$ which solves equation (2), rather than the computationally much more efficient Levinson algorithm. Using gradient descent, we start with an arbitrary initial guess $\mathbf{F}(0)$ for the prediction coefficients and then update $\mathbf{F}(t)$ using the following algorithm:

$$\mathbf{F}(t+1) = \mathbf{F}(t) + \mu[\mathbf{P}_x(\gamma) - \mathcal{R}_{xx}\mathbf{F}(t)]. \tag{7}$$

In this case, we denote the lth coefficient in $\mathbf{F}(t)$ by $F(l; t)$.

It is readily shown [see Wilde (1964)] that the term in square brackets is the negative of the gradient of the mean-square error between $x(t+\gamma)$ and $\hat{x}(t+\gamma)$ when $\mathbf{F}(t)$ is used to produce the estimate $\hat{x}(t+\gamma)$. Further, if $0 < \mu < 2/\lambda_{max}$, where λ_{max} is the largest eigenvalue of $\mathcal{R}_{xx}$, it can be shown that $\mathbf{F}(t)$ assymptotically approaches $\mathbf{F}_0$, i.e.,

$$\lim_{t\to\infty} \mathbf{F}(t) = \mathbf{F}_0. \tag{8}$$

In 1960, Widrow and Hoff suggested that a simple, noisy gradient descent algorithm could be derived from equation (7) by replacing the autocorrelation terms $\mathbf{P}_x(\gamma)$ and $\mathcal{R}_{xx}$ by their appropriate instantaneous values. Thus,

$$\mathbf{P}_x(\gamma) \to x(t+\gamma)\mathbf{X}(t),$$
$$\mathcal{R}_{xx} \to \mathbf{X}(t)\mathbf{X}^T(t),$$

and the algorithm in equation (7) becomes,

$$\mathbf{F}(t+1) = \mathbf{F}(t) + \mu[x(t+\gamma) - \hat{x}(t+\gamma)]\,\mathbf{X}(t), \tag{9}$$

where $\hat{x}(t+\gamma) = \mathbf{X}^T(t)\mathbf{F}(t)$ is the current estimate of the trace value $x(t+\gamma)$ based on the coefficients $\mathbf{F}(t)$.

Equation (9) has been termed the LMS algorithm by Widrow (1966) and others. Its simplicity is immediately apparent. Whereas the original gradient method in (7) required the order of L^2 operations per adaptation, the LMS algorithm in (9) requires only L operations because $x(t+\gamma)$ and $\hat{x}(t+\gamma)$ are scalar. An additional L operations are also required to compute the prediction $\hat{x}(t+\gamma)$, but these are also required in the normal equation method of predictive filtering described above.

Although the LMS algorithm requires less computation than the direct gradient approach, inspection of (9) reveals that the sequence of prediction coefficients $\mathbf{F}(t)$ are stochastic and, in general, correlated in time due to the fact that $x(t)$ is a stochastic process. In fact, the LMS algorithm is closely related to the stochastic approximation methods described by Robbins and Monroe (1951), Blum (1954), and Dupac (1965). The primary difference is that in stochastic approximation methods, the adaptive step size μ is made inversely proportional to t or to a power of t. It turns out that this difference is of significant practical importance in terms of the convergence properties of the algorithm. When μ decreases appropriately with increasing t, it can be shown (see the stochastic approximation references above) that $\mathbf{F}(t)$ approaches $\mathbf{F}_0$ with probability one when $x(t)$ is a stationary time series. Equivalently, the deconvolved output $y(t)$ will converge exactly to that obtained via the normal equations approach.

For the case of fixed μ, convergence analysis is considerably more difficult. Since the introduction of the LMS algorithm, a variety of authors, including Senne (1968), Daniell (1970), Gersho (1968), Kim and Davisson (1975), and Widrow et al (1976) have studied this problem and the interested reader is referred to these papers for a detailed discussion. Briefly, at this point it has been shown that for stationary input data, if μ is constant and satisfies $\mu < 2/\lambda_{max}$, then the mean coefficient vector converges to the optimal, i.e.,

$$\lim_{t \to \infty} E[\mathbf{F}(t)] = \mathbf{F}_0. \qquad (10)$$

In addition, the variances of the individual coefficients in F(t) converge to bounded values which decrease with decreasing μ.

At this point, one may question the use of the LMS adaptive algorithm in equation (7) for use as a predictive deconvolution tool, in that the normal equation approach avoids all of the previously mentioned convergence difficulties. Indeed, for those cases in which the input trace $x(t)$ can be modeled as a stationary time series, the best overall approach is undoubtedly the use of the normal equations. However, for those applications which occur all too frequently in practice (i.e., for the case of continuously time-varying input statistics), the LMS algorithm offers a potential advantage. Consider a simple reverberation model in which the period between successive reverberations is changing slowly in some unknown manner throughout the trace. The correlation measurements required for the normal equation approach necessitate time averages over window lengths and may involve averaging over significant time changes in the data. The deconvolution operator so obtained will then not provide effective multiple removal. (A numerical example illustrating this effect is presented below.)

If an adaptive deconvolution operator can converge in a time which is short compared with the time scale of the reverberation changes, it can then track these changes and provide multiple removal in the time-varying environment. It is shown in this paper that adaptive deconvolution can realize this potential advantage in practical reflection seismogram processing.

The specific adaptive deconvolution procedure suggested here is based directly on the LMS algorithm defined in equation (9). Two methods are employed: forward-time adaptation in which the iteration proceeds from earlier to later times along the input trace; and reverse-time adaptation corresponding to iteration in the opposite direction. In both cases, the predicted data point $\hat{x}(t+\gamma)$ is at a later time in the trace than the data which are presently stored in the $\mathbf{F}(t)$ operator. The corresponding algorithms are given by

$$\mathbf{F}(t+1) = \mathbf{F}(t) + \mu[x(t+\gamma) - \hat{x}(t+\gamma)]\mathbf{X}(t), \qquad (11)$$

for forward-time adaptation, and

$$\mathbf{F}(t-1) = \mathbf{F}(t) + \mu[x(t+\gamma) - \hat{x}(t+\gamma)]\mathbf{X}(t). \qquad (12)$$

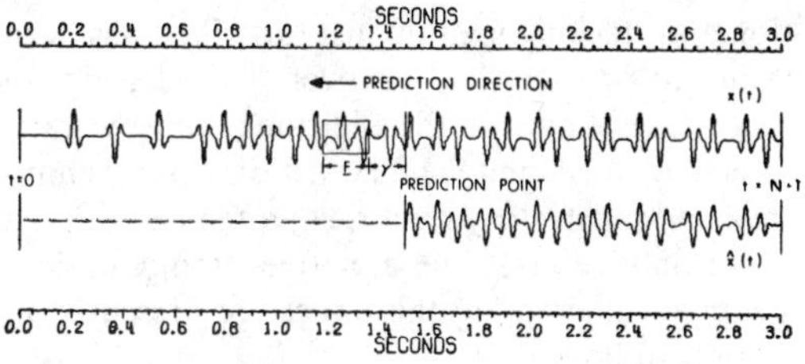

FIG. 1. Schematic representation of time-reverse adaptive deconvolution.

for reverse-time processing. In these expressions, the predicted trace value, $\hat{x}(t+\gamma) = \mathbf{F}^T(t)\mathbf{X}(t)$ is computed using the previous operator $\mathbf{F}(t)$ prior to adaptation. Note that the term in square brackets is the deconvolved trace which is generated as adaptation proceeds.

After updating the operator coefficients in this manner, the operator is moved either one sample further down the trace (time-forward processing) or one sample closer to the trace origin (time-reverse processing) and a new prediction value is generated using these coefficients. The prediction results in a new error which is then used to update the coefficients. Adaptation proceeds in this manner until the entire trace has been processed. In order to avoid having the operator $\mathbf{F}(t)$ overrun the ends of the trace, backward-time processing is carried out for $t = (N-1-\gamma), \cdots, L, L-1$. Selection of the proportionality constant μ for practical data is discussed below.

Evidence obtained using both experimental and synthetic data has shown that the best results are achieved when the trace is first processed in time-reverse order using an initial operator $\mathbf{F}^T(N-1-\gamma) = [0, 0, \cdots 0]$. Figure 1 illustrates time-reverse processing schematically. A deconvolved reverse-time output $y_r(t)$ is formed as the difference between the input trace $x(t)$ and the predicted trace $\hat{x}(t)$. This processing is then followed by a forward-time pass in which the initial operator $\mathbf{F}(L-1)$ is set equal to the final set of coefficients obtained from time-reverse adaptation. Note that the forward-time processing uses the original input trace $x(t)$ and not the deconvolved output produced by the reverse-time procedure. Denoting the deconvolved reverse and forward-time traces by $y_r(t)$ and $y_f(t)$, respectively, the final deconvolved trace $Y(t)$ is obtained as a simple stack.

$$y(t) = 1/2[y_f(t) + y_r(t)]. \qquad (13)$$

It should be noted that the double-deconvolution and stacking method just described has been found to provide the best overall processing. In many cases, however, the additional improvement offered by including the forward-time pass is minimal and satisfactory results can often be achieved using only reverse-time adaptive deconvolution. Examples illustrating this effect are presented in sections following.

One additional property of adaptive deconvolution merits discussion. The prediction coefficients are continually updated using the noisy gradient algorithm in (9) as the input trace is processed. As a result, each coefficient acts as a noisy modulator of the input trace and this coefficient noise appears directly at the filter output as an additional term in the prediction $\hat{x}(t + \gamma)$. Widrow (1966, 1976), Senne (1968), Daniell (1970), and others have studied this effect, which is termed misadjustment noise. It can be shown that the noise level is directly related to the value of the adaptive proportionality constant μ. A simple relationship has been given by Widrow et al (1976) for the case of adaptation on stationary data. If $\sigma_y^2(\mu)$ is used to denote the power in the deconvolved trace when adaptive processing is used, and $\sigma_y^2(\min)$ represents the output power achieved using a fixed, optimal, predictive operator, then

$$\frac{\sigma_y^2(\mu) - \sigma_y^2(\min)}{\sigma_y^2(\min)} = \mu L \sigma_x^2, \tag{14}$$

where $\sigma_x^2 = r_x(0)$ is the average power level of the input trace.

Clearly, the additional output noise power due to coefficient fluctuation can be made arbitrarily small by sufficiently small choice of μ. However, as shown in the following section, the rate of convergence of the adaptive processor is proportional to μ, and small values may lead to excessively long convergence times. The selection of appropriate values for μ for practical applications is the subject of the following section.

SELECTION OF ALGORITHM PARAMETERS

Inspection of the adaptive algorithm defined by equations (11) and (12) above shows that in addition to the initial coefficient vector, a total of three parameters are required to completely specify the procedure. These are: operator length L, prediction distance γ, and proportionality constant μ. The first two parameters are identical with those required for conventional predictive methods and the criteria used to select them are not changed in the adaptive procedure. Briefly, the filter length should be sufficiently long to encompass the basic wavelet, a value which is typically between 20 and 100 msec. If wavelet whitening is desired, a prediction distance of unity may be selected. For predictive deconvolution, γ should be chosen such that the distance from the midpoint of the operator to the prediction point is approximately equal to the multiple spacing.

The determination of an appropriate value for the proportionality constant μ depends upon the desired adaptive convergence properties. Previous theoretical work by Daniell (1970) and Kim and Davisson (1975) has described the convergence of this algorithm for the case of stationary input statistics. Griffiths (1975) has applied the algorithm with $\gamma = 1$ for purposes of short-term spectral estimation and has shown that the mean value of the operator converges to the solution given by the normal equations, provided that μ satisfies the following inequalities:

$$\mu = \frac{\alpha}{L \cdot \sigma_x^2}, \tag{15}$$

and

$$0 < \alpha < 2, \tag{16}$$

where σ_x^2 is the average power level of the input trace. Substituting equation (15) into the formula given in the previous section for misadjustment, i.e., equation (14), yields

$$\frac{\sigma_y^2(\mu) - \sigma_y^2(\min)}{\sigma_y^2(\min)} = \alpha \cdot \tag{17}$$

Thus, $\alpha = 1$ will produce a 100 percent increase in excess output noise power, and $\alpha = 0.1$ produces a 10 percent increase (for the case of stationary input data). In practical data containing amplitude fluctuations, however, the value of α must be set to ensure stability on the amplitude peaks, and the actual misadjustment levels observed in lower amplitude sections of the data will be lower by an amount determined by the peak-to-trough power ratios.

Values of α near 2 will provide rapid, noisy

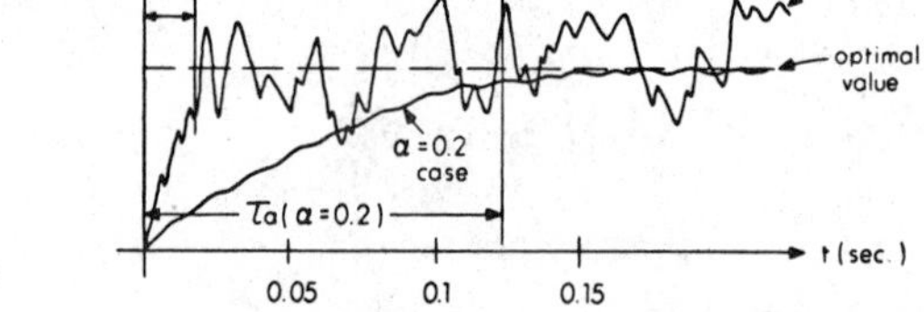

FIG. 2. Time behavior of adaptive operator coefficient for stationary trace statistics.

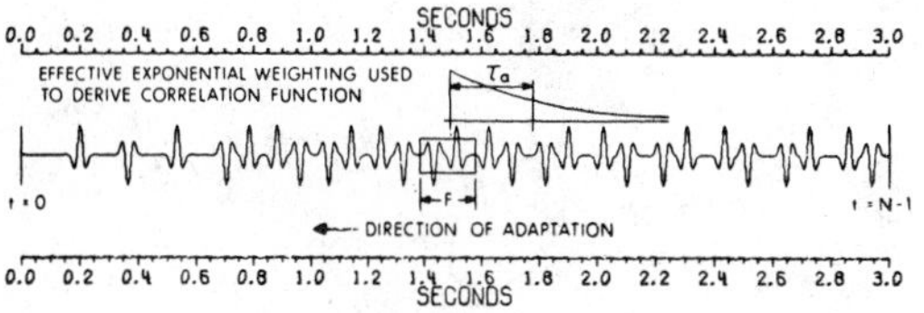

FIG. 3. Effective exponential weighting provided by adaptive deconvolution.

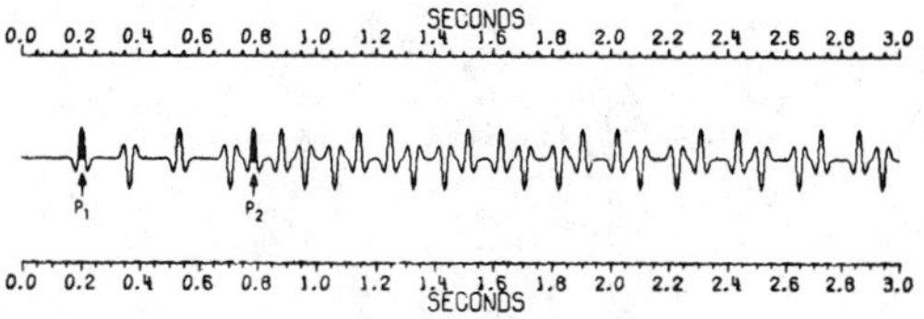

FIG. 4. Synthetic time-varying multiple trace with two primary events.

adaptation while those closer to zero produce slower, smoother results. This behavior is illustrated schematically in Figure 2 which shows the value of a hypothetical adaptive operator coefficient $f(i; t)$ as a function of adaptation time t (equivalently, time along the trace) for a stationary input example. Also shown for reference, is the optimal coefficient value, as would be computed from the normal equations for this case. Experimental results have indicated that α values in the range 0.05 to 0.2 are appropriate for stationary, constant power data and that little change in deconvolution output is observed for α in this range.

The time constant τ_a indicated in Figure 2 can be computed exactly if the input statistics are known exactly. The value obtained is related to both α and the number of coefficients L. An approximate value which has been verified experimentally by Griffiths (1975) is given by

$$\tau_a \simeq \frac{-\Delta T}{\ln(1-\mu\sigma_x^2)} \tag{18}$$

$$= \frac{-\Delta T}{\ln(1-\alpha/L)}, \tag{19}$$

where ΔT is the sampling interval.

In the case of stationary input statistics, adaptive deconvolution provides no advantages over conventional methods in which autocorrelation properties of the input waveform are computed and used to solve for the optimal operator using the normal equations. For inputs containing time-varying properties, however, adaptive methods offer a computationally simple procedure for tracking these variations. Although no theoretical results are available to date regarding the nature of convergence of the adaptive algorithm for the case of time-varying statistics, a great deal of experimental results have been accumulated in nonseismic applications which demonstrate the convergence properties in these cases. The results show that if the period of the time variations is the order of the convergence time shown in Figure 2, or longer, the algorithm can successfully track the variations. In effect, since the operator coefficients are continually updated, their value represents an average, with exponential weighting, over the statistics of previous trace samples. The time constant of this exponential is identical with the adaptation time constant τ_a defined by equation (19). Figure 3 shows this weighting diagrammatically.

Because the adaptive proportionality constant μ in equation (15) depends upon the average power level σ_x^2 of the input trace, care must be taken for those input waveforms which contain power level changes along the trace. One procedure for handling such waveforms is to first compute a power average using an averaging window which is equal to the length of the deconvolution operator. The averaging procedure is conducted for all such windows in the input trace. The value of σ_x^2 in equation (15) is then set equal to the maximum power average obtained in this manner. As a result, the algorithm will be guaranteed to provide stable processing, regardless of the amplitude variations present in the input trace. When α is set on data peaks in this manner, values in the range 0.3 to 0.8 have been shown to provide appropriate processing results for seismic data. This procedure was used for all processing results presented in the sections following.

SYNTHETIC DATA PROCESSING RESULTS

The adaptive deconvolution procedure described herein is designed for use in those seismic processing applications containing nonstationary statistical properties over relatively short-time intervals. A synthetic digital trace (4 msec sample period) representing an example of nonstationary behavior is presented in Figure 4. The trace contains two primary events, denoted by P_1 and P_2. The constant-amplitude multiples for each event have a spacing which increases by four msec at each order of the multiple. Thus, while the first

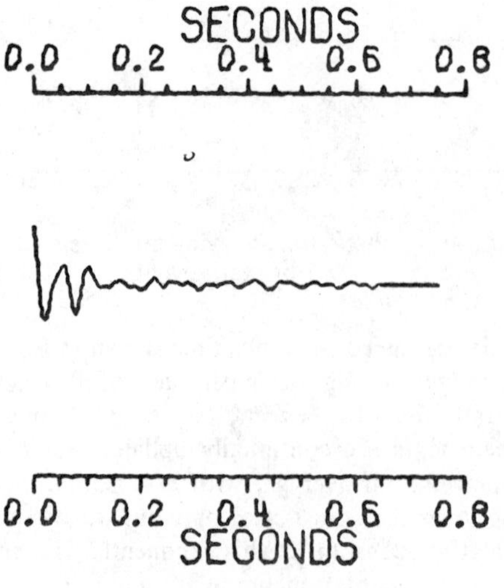

FIG. 5. Autocorrelation function calculated using 0 to 3.0 sec design window on synthetic time-varying trace.

multiple arrives about 160 msec after the primary, the multiple spacing near the end of the trace is 212 msec.

When conventional, fixed design gate deconvolution methods are applied to data of this type, the results are generally less than completely satisfactory. Because the correlation statistics of the trace vary with time along the trace, some form of time varying processor must be employed. For example, Figure 5 illustrates the autocorrelation function calculated using a 0 to 3.0 sec window on

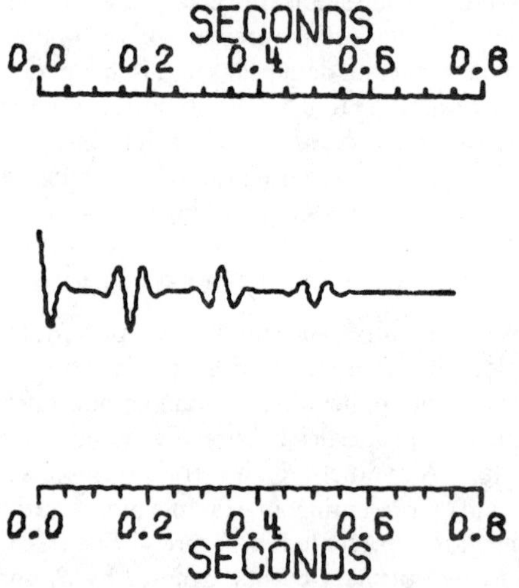

FIG. 6. Autocorrelation function calculated using 0 to 0.75 sec design window on synthetic time-varying trace.

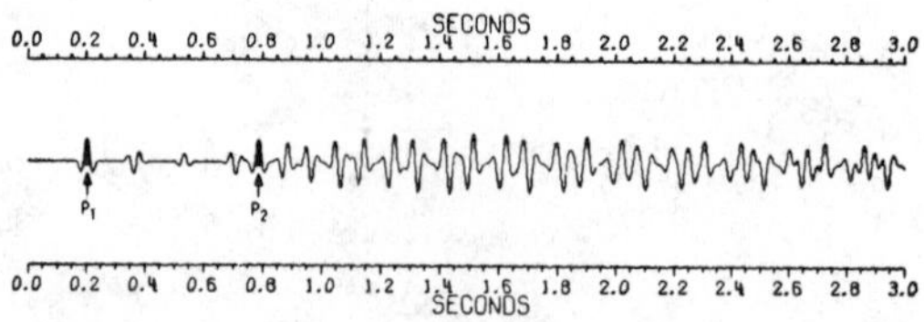

FIG. 7. Conventionally deconvolved trace obtained using 0 to 0.75 sec design window (80 msec operator with 120 msec prediction distance).

the trace in Figure 4. Little evidence of the multiples is observed due to their time-varying nature. Shorter design gates provide better evidence, however, as illustrated in Figure 6 using a 0 to 0.75 sec design gate. A predictive deconvolution operator of length 80 msec and prediction distance 120 msec was designed from this autocorrelation function using the normal equation solution method given by Peacock and Treitel (1969). The resulting operator was then applied to the entire trace and the deconvolved output is presented in Figure 7. As expected, the result indicates superior deconvolution results over the design gate relative to other regions of the trace. Better overall deconvolution may be achieved using the piecewise stationary deconvolution method described by Wang (1969). Figure 8 illustrates results obtained using a total of five overlapping design windows, each one sec in length, and an appropriately chosen prediction distance for each window.

Although the primary events are more readily distinguishable in Figure 8 than in Figure 7, a considerable amount of residual multiple energy remains in the deconvolved trace. Presumably, shorter design gates would provide improved performance. The computational load of this procedure, however, increases geometrically as shorter gates are selected and may well prevent a completely satisfactory deconvolution procedure.

The results obtained for this synthetic seismogram using adaptive deconvolution with reverse-time only, forward-time only, and reverse/

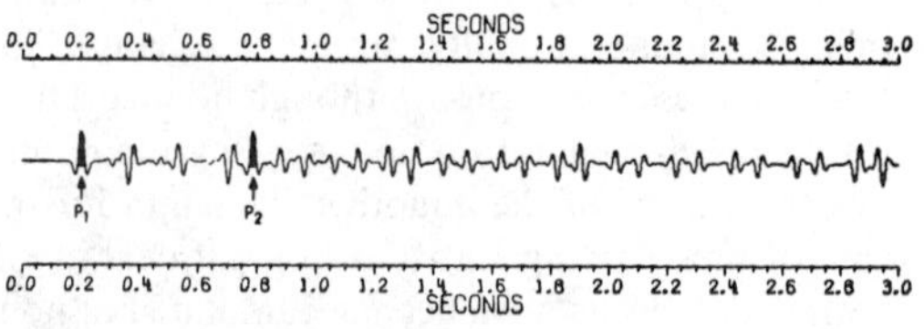

FIG. 8. Conventionally deconvolved trace using five design windows (80 msec operator with appropriate prediction distance for each window).

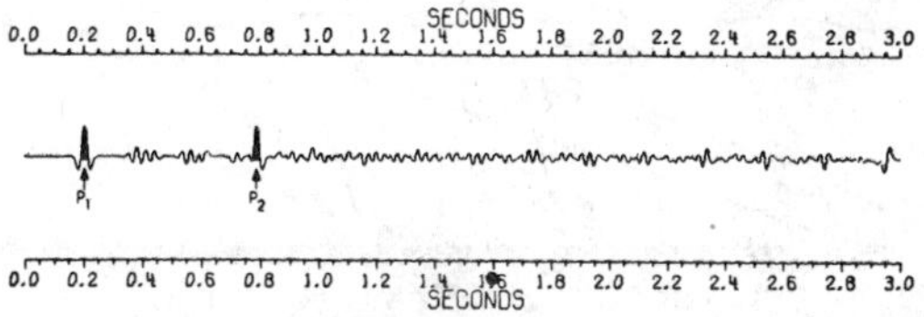

FIG. 9. Adaptively deconvolved trace obtained using only time-reverse processing.

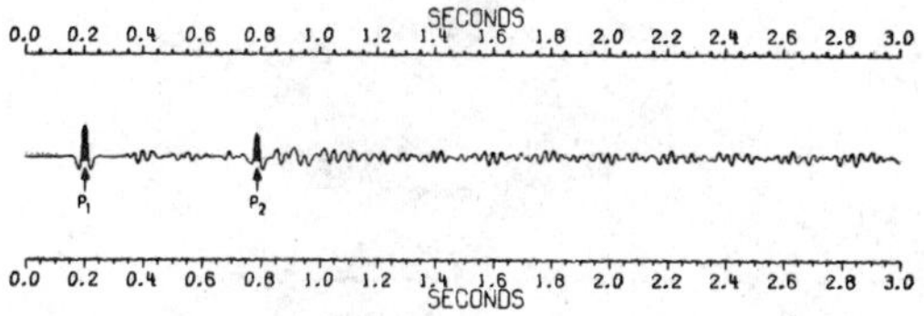

FIG. 10. Adaptively deconvolved trace obtained using time-forward processing.

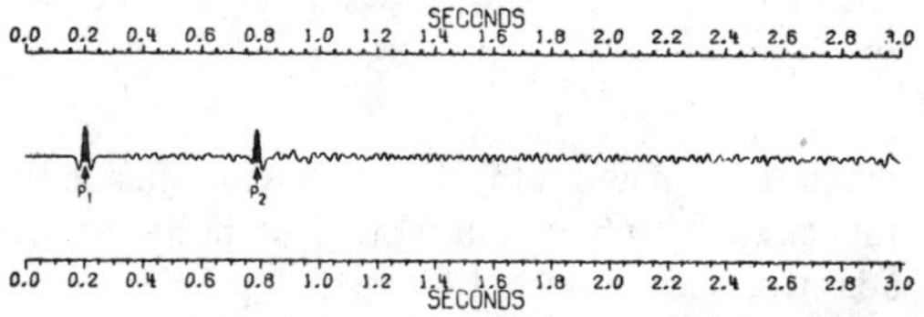

FIG. 11. Adaptively deconvolved trace obtained using time-reverse and time-forward, then sum processing.

forward stack are shown in Figures 9, 10, and 11, respectively. These results were achieved using an 80 msec length operator and constant 150 msec prediction distance throughout the trace. The value of α used to generate the deconvolved traces was 0.5. A comparison of Figures 9 and 11 indicates clearly the degree to which multiple rejection can be improved using the stacking procedure.

When adaptive deconvolution is applied to a trace containing noise only (i.e., unpredictable data), little modification of the input trace occurs—provided that the proportionality constant remains within the range previously given. This effect is illustrated here using the synthetic bandpass noise example shown as the original trace in Figure 12. This trace was generated by filtering white noise with a 5–60 Hz bandpass filter. The output traces observed after applying adaptive deconvolution to this example with different values of α are also shown in Figure 12. The operator parameters used for these calculations were similar to those used in the previous synthetic experiments, which were an operator length of 80 msec and a prediction distance of 120 msec.

Inspection of Figure 12 shows that little change in output is observed as α varies over a 50 to 1 range (0.01 to 0.50). However, at $\alpha = 0.50$ the effects of misadjustment noise are barely discernible which is to be expected since equation

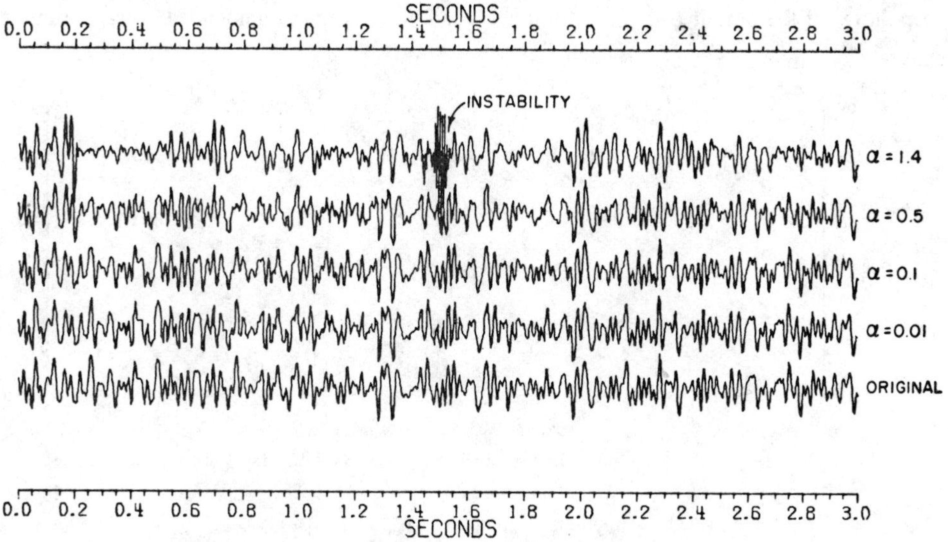

FIG. 12. The effects of adaptive deconvolution on filtered (6–60 Hz) white noise (i.e., unpredictable data) using varying values of α. The adaptive filter length was 80 msec and the prediction distance was 120 msec.

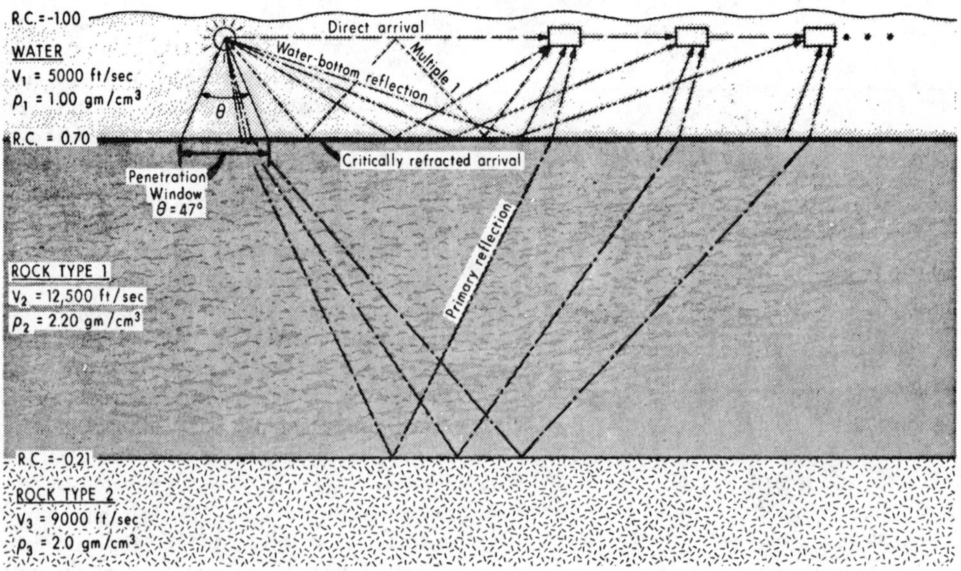

FIG. 13. A simplified diagram of a hard water-bottom model showing the raypaths for some of the arrivals.

(17) predicts a 50 percent increase in noise power at $\alpha = 0.5$. The output trace that results using a value of $\alpha = 1.4$ shows the filter response when driven to instability at 1.50 sec. Note that this value of α is well above the values recommended for practical application as discussed earlier. It is important to realize that the character of the unstable filter response is typically a rapid oscillating of the polarity of the output trace. We still have much to learn about the filter behavior outside the range on α which guarantees stability of the adaptive deconvolution filter.

In summary, synthetic experiments have demonstrated that adaptive deconvolution can provide effective multiple removal from a time-varying input trace, provided that these variations are slower than the time constant of adaptation. In addition, for recommended filter parameters it

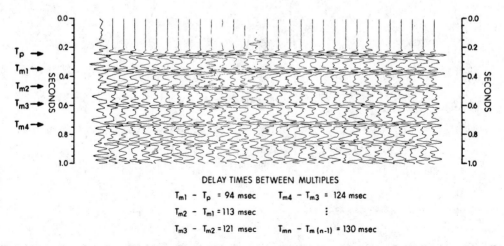

FIG. 14. A plot showing the near-offset traces for several adjacent shotpoints from line A with offset equal to 914 ft and water depth equal to 300 ft. T_p = traveltime of primary reflection and T_{M1} = traveltime of the first water-bottom multiple, etc.

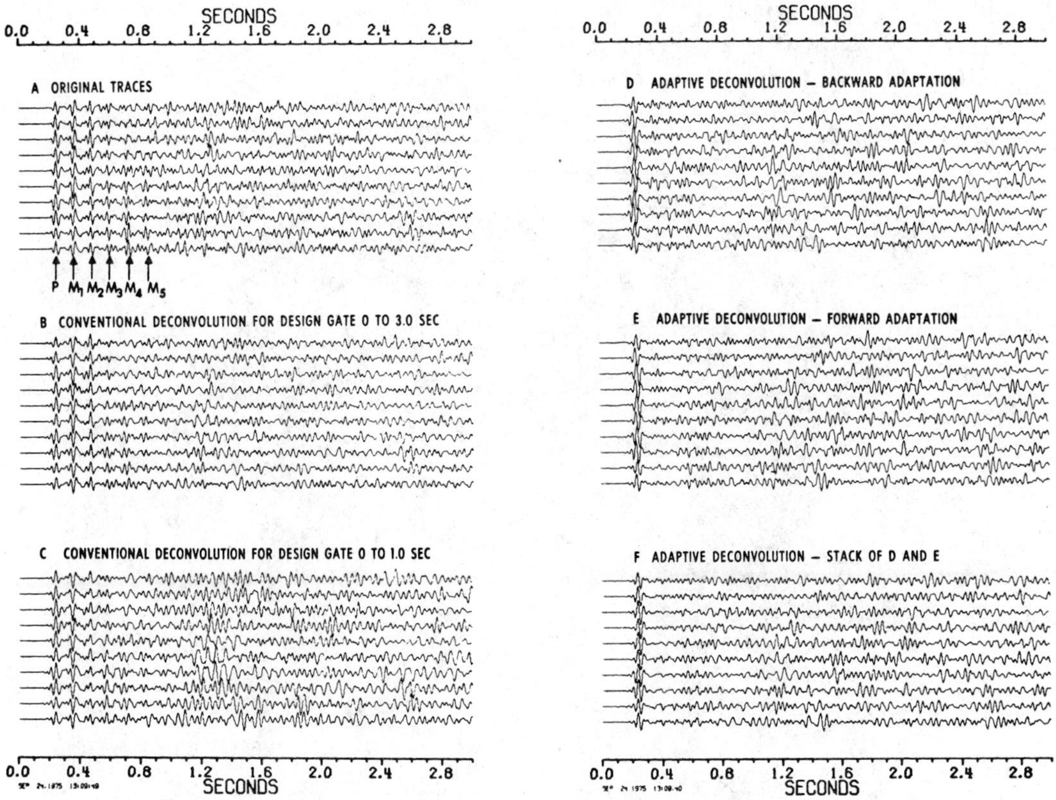

FIG. 15. Comparison of conventional and adaptive deconvolution results for near-offset traces from line B for ten adjacent shotpoints. (Offset = 914 ft, water depth = 300 ft, operator length = 80 msec, prediction distance = 120 msec, and α = 0.5.)

does not appreciably change noise-like (or primaries only) input traces which contain no multiple energy. We have also illustrated that adaptive deconvolution is a relatively robust procedure. That is, the output trace is not a critical function of the adaptive step size α, and increasing or decreasing its value by a factor of two will not significantly affect performance. Of course, many unanswered questions remain regarding the characteristics of adaptive processors in general. These include an appropriate performance model for those cases in which the input trace time variations are of the order of an adaptive time constant or less. Another relatively unknown area relates to the effects of gain variations on adaptive performance. While research on these problems is continuing, we feel that we have demonstrated that adaptive filtering is an effective deconvolution process. The next section completes the demonstration using field-recorded data.

FIELD-RECORDED DATA PROCESSING RESULTS

The purpose of this section is to illustrate the use of adaptive deconvolution methods on field-recorded marine seismic data which contain time-varying water-bottom reverberations. The seismic data shown are from the Celtic Sea offshore Ireland.

Reflection seismic data acquired in areas where the sediments at the water bottom possess high velocity and/or density are notoriously poor quality. Two of the main reasons for the poor quality of such reflection data are: (1) the small amount of energy transmitted through the water-sediment interface; and (2) the large amount of energy reflected between the water-sediment interface and the surface of the water which is nearly a perfect reflector. Under these conditions, the water layer is a very efficient wave guide where many types of

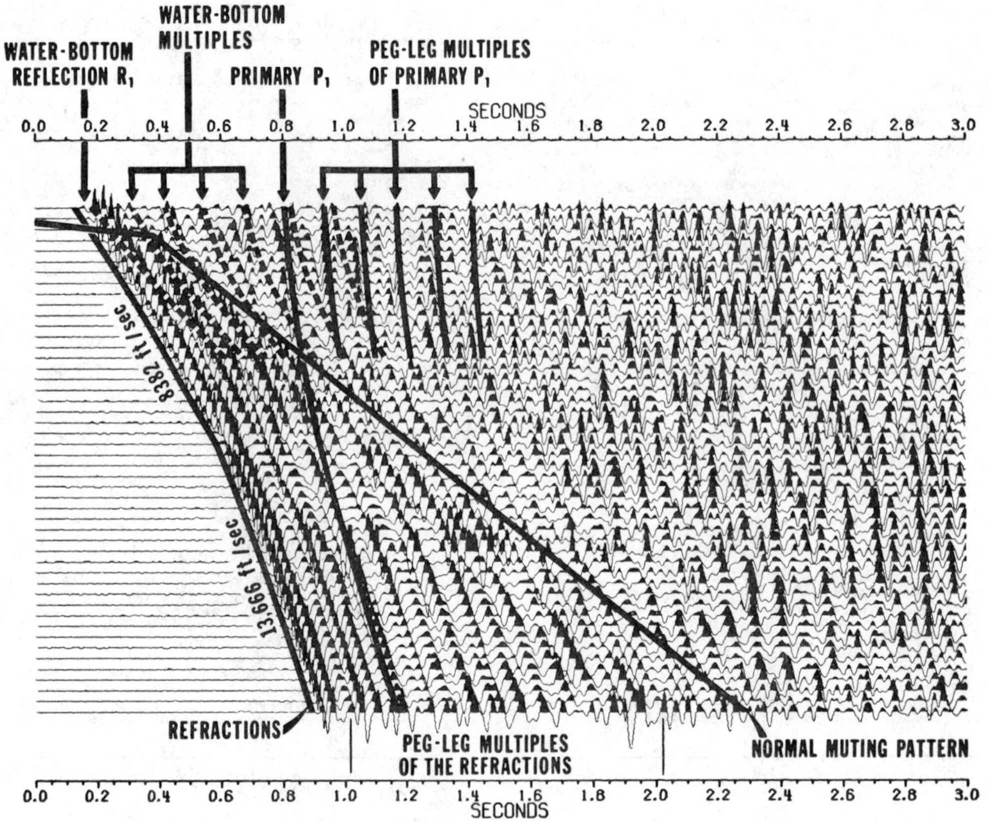

FIG. 16. An unprocessed 48-trace record from line C showing several of the primary reflection arrivals and refraction arrivals and their multiples. Near-offset distance = 669 ft, group interval = 164 ft, and water depth = 310 ft.

source-generated noises can propagate. The result of these conditions is usually low-amplitude primary reflections present in a background of very high amplitude and very coherent noise. Two of the most serious sources of noise are the refraction arrivals (and their multiples) and the water-bottom reverberations. These arrivals are shown on the simplified diagram in Figure 13. The reflection from the water bottom, the critical refraction along the water-sediment interface, and each of the primary reflections from interfaces in the subsurface materials are followed by multiples within the water layer. The simple reflection coefficients for compressional waves at each of the interfaces are shown in Figure 13. One can compute, using Snell's law, the angular extent of the penetration window, which for the simplified model is 47 degrees.

The water-bottom multiples referred to in this paper are those arrivals which follow the primary reflection from the water-sediment interface making additional two-way trips through the water layer (see for example, the multiple 1 in Figure 13). The character of the water-bottom multiples is illustrated in Figure 14 which shows the near-offset traces only from several adjacent records in line A. The offset of these traces was 914 ft, and the water depth was approximately 325 ft. Clearly, the water-bottom reverberations dominate these near-offset traces. The most troublesome characteristic of these multiples is that the period of multiples is not constant. That is, the delay time between each multiple is a function of the order of the multiple resulting in a time series of reverberations which is time variant. One can easily calculate what the variations in the periods

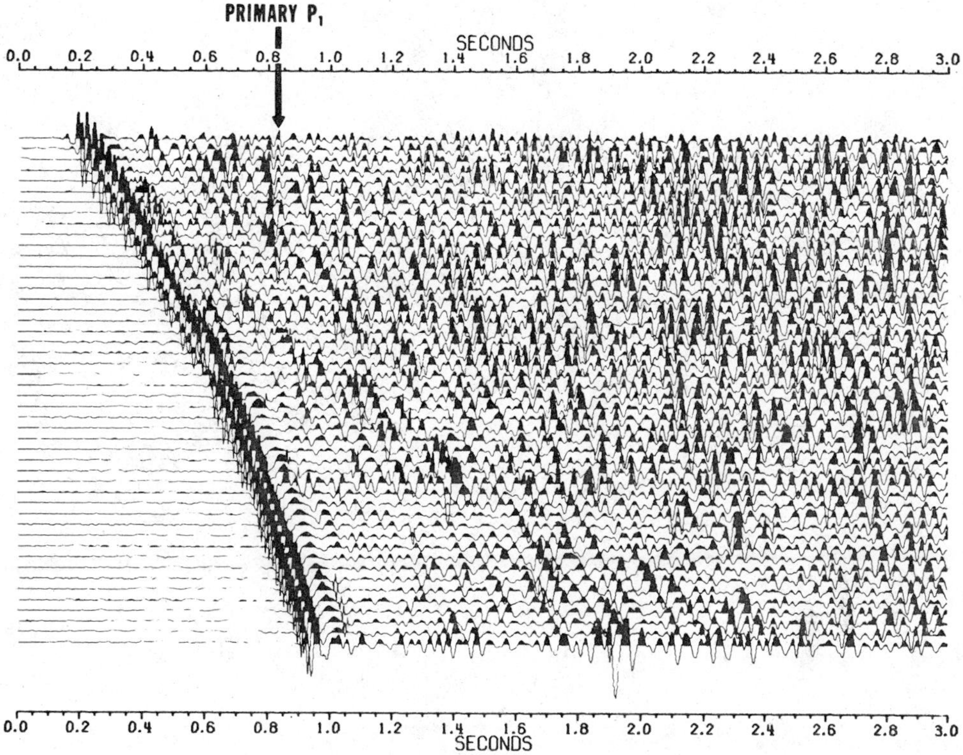

FIG. 17. An unmarked version of the unprocessed record shown in Figure 16.

of the multiples will be by assuming a cable geometry and a water depth and calculating the distances traveled by the various orders of multiples. The results of such calculations are shown in the lower portion of the figure. Note that the delay times increase with the order of the multiple and eventually become equal to the two-wav vertical traveltime in the water layer. While the differences in the delay times of the multiples are not large, they are great enough to violate the assumption of stationary statistics used in conventional deconvolution and thus cause conventional deconvolution to be ineffective.

A comparison of conventional and adaptive deconvolution results for field seismic data is shown in Figure 15 for the nearest offset traces from several adjacent shotpoints of field data on line B. As shown in Figure 15a, the delay time between the primary P and the first order multiple M_1 is shorter than the period between M_1 and M_2, etc. The results of conventional deconvolution with autocorrelation design windows 0.0 to 3.0 sec and 0.0 to 1.0 sec are shown in Figures 15b and 15c, respectively. The reverberations have not been satisfactorily removed by either of the conventional deconvolution filters. Experiments have also been conducted applying conventional deconvolution filters after normal-moveout corrections, but the results were also unsatisfactory.

The same near-offset traces are shown after the backward, forward, and summing implementations of adaptive deconvolution in Figures 15d, 15e, and 15f, respectively. Clearly, the water-bottom multiples have been satisfactorily removed. The parameter used in the processing shown in Figures 15d, 15e, and 15f were operator length = 40 msec, prediction distance = 84 msec, and α = 0.5. The data at traveltimes greater than 1.2 sec have not been altered by the filter because there were no periodic events present. That is, the adaptive deconvolution technique does not alter the data in regions not containing significant multiple structure.

Naturally, adaptive deconvolution must be applied to all traces of a common-depth-point (CDP) gather. An unprocessed 48-trace CDP gather is shown in Figure 16 with several of the events of interest identified. In this case, the sediments beneath the water layer are made up of two layers, one rock type having compressional veloci-

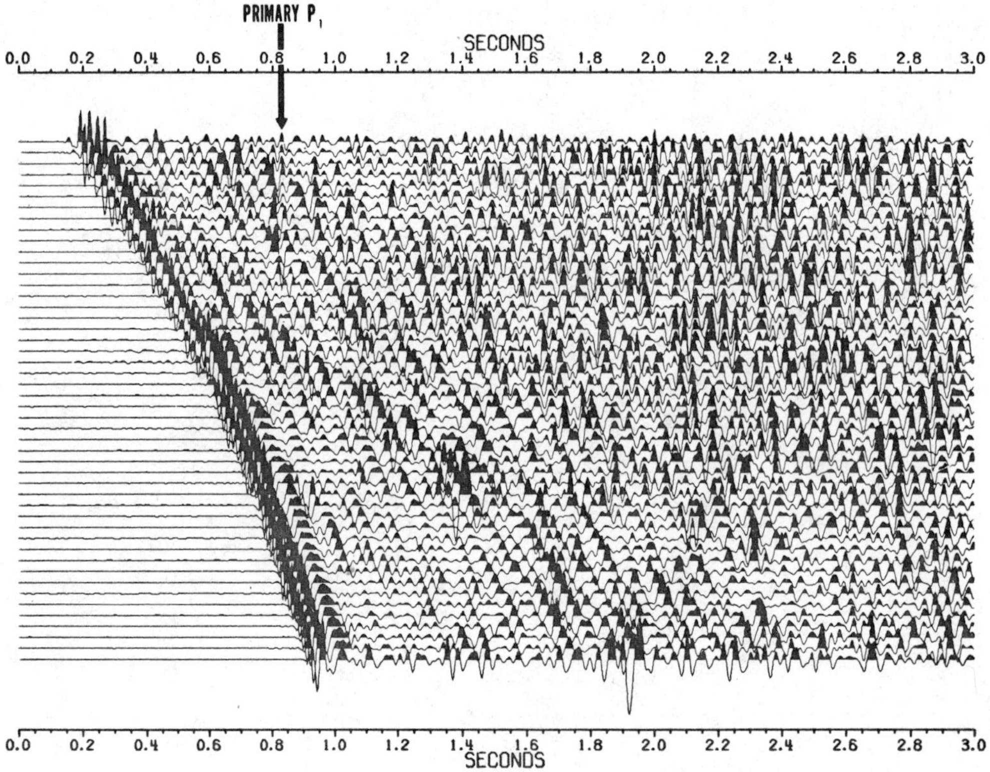

FIG. 18. The same 48-trace record from line C after adaptive deconvolution was applied. (40 msec operator length, 84 msec prediction distance, and $\alpha = 0.5$).

ties of 8382 ft/sec and the other having compressional wave velocity equal to 13,666 ft. The most obvious events on the record are the peg-leg multiples of the refraction arrivals (i.e., refraction arrivals which have made additional trips through the water layer). Note that the critical distance of the peg-leg multiples of the refraction appear to increase with the order of the multiples. This signature implies that each of the water-bottom multiples generate new refractions which are, in effect, closer to the detector array with each order of the multiple. The result of these peg-leg refraction multiples is that the data at offsets larger than the normal muting pattern shown on the record must be omitted from the stack. The result of this muting is to reduce the fold of stack at small traveltimes, which reduces the ability of the stacking process to remove the water-bottom multiples shown in Figure 16. Thus, it become imperative to remove the water-bottom multiples within the usable "data window" before stack. If these multiples are not removed by deconvolution, any primary reflection within the reduced-stack data window will be masked by the multiples.

If one calculates the normal-moveout hyperbola of a primary event near the base of the high-velocity layer (e.g., the event at 800 msec in Figure 16), the reflection approaches asymptotically the linear moveout pattern of the refraction multiples. Clearly it would be desirable to remove the refraction multiples while leaving the primary event, if possible, so as to increase the fold of effective CDP stack on the primary. Finally, the peg-leg multiples following the primary event at 800 msec are also shown on the 48-trace record. Peg-leg multiples are difficult to remove by conventional predictive deconvolution if there are only a few primary events in the data. This characteristic is due to the lack of evidence of these multiples in the autocorrelation function calculated from the data. Adaptive deconvolution, on the other hand, appears to recognize the presence of these peg-leg multiples and effectively removes them as shown in Figure 18 when compared to the unmarked

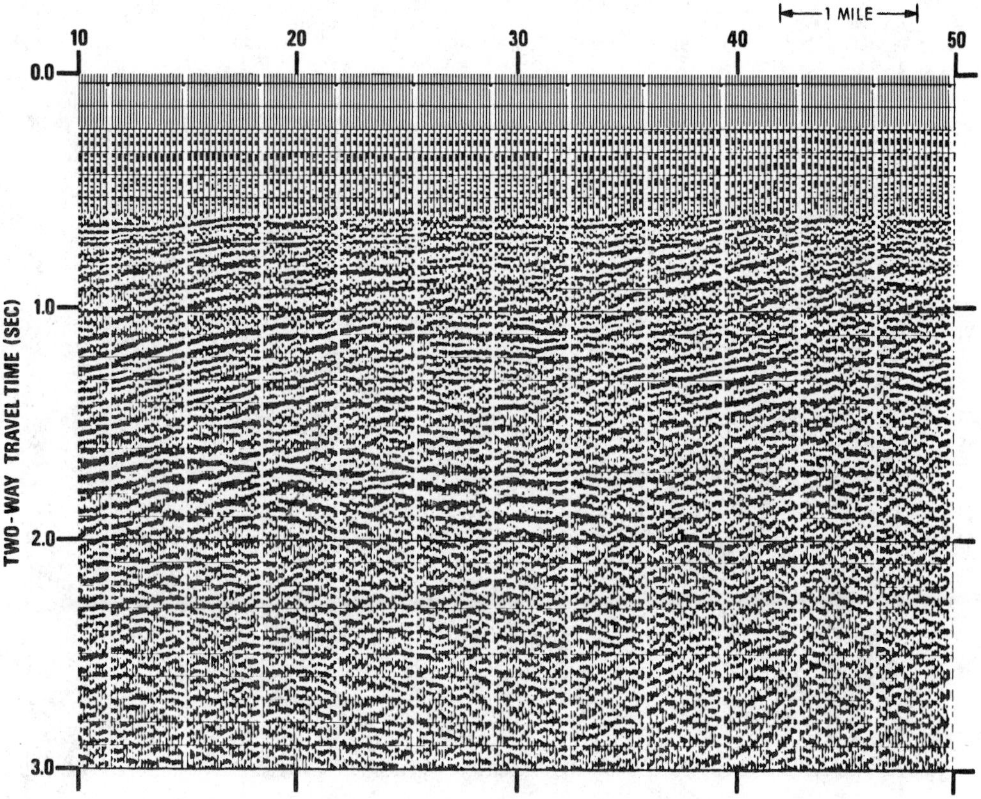

FIG. 19. 48-fold stacked section of line C with conventional deconvolution applied before stack.

original record shown in Figure 17. The reverse-then forward-time and stack implementation of adaptive deconvolution was used in processing the record shown in Figure 18. The values used for operator length, prediction distance, and α were 40 msec, 84 msec, and 0.5, respectively. It should be noted that, in some cases, the prediction distance may need to be made shorter for traces at increased offset distance (e.g., when arrivals with converging normal-moveout patterns, such as water-bottom multiples, are the dominant noise to be removed at large offsets). For the record shown in Figure 18, however, the refraction multiples are the most dominant arrivals to be removed and their linear moveout pattern requires a prediction distance which is constant with offset distance.

Clearly the adaptive deconvolution has removed most of the multiples of the refraction arrivals. However, it is questionable that any wide-angle primary reflections are left in the data. The conclusion drawn from these studies is that it is best to mute these far-offset data and not take the chance of stacking a refraction multiple and mistaking it for a primary reflection.

Comparison of the traces within the usable data window in Figures 16, 17, and 18 shows that adaptive deconvolution has effectively removed the water-bottom multiples and the peg-leg multiples from the primary event at 800 msec. It is also clear that the primary event *P* has been left undistorted in both amplitude and waveform.

The final stacked sections obtained using conventional deconvolution before stack are shown in Figures 19 and 21, and those obtained using adaptive deconvolution before stack are shown in Figures 20 and 22. While the plotting scales on the stacked sections may not be exactly the same, comparison of the amplitudes of primary events and first arrivals show they are fairly close. Thus, the appearance of lower overall amplitude on the sections is mostly due to the removal of predictable energy by the adaptive deconvolution process. The adaptive deconvolution parameters were the same as those used in Figure 18. We feel that

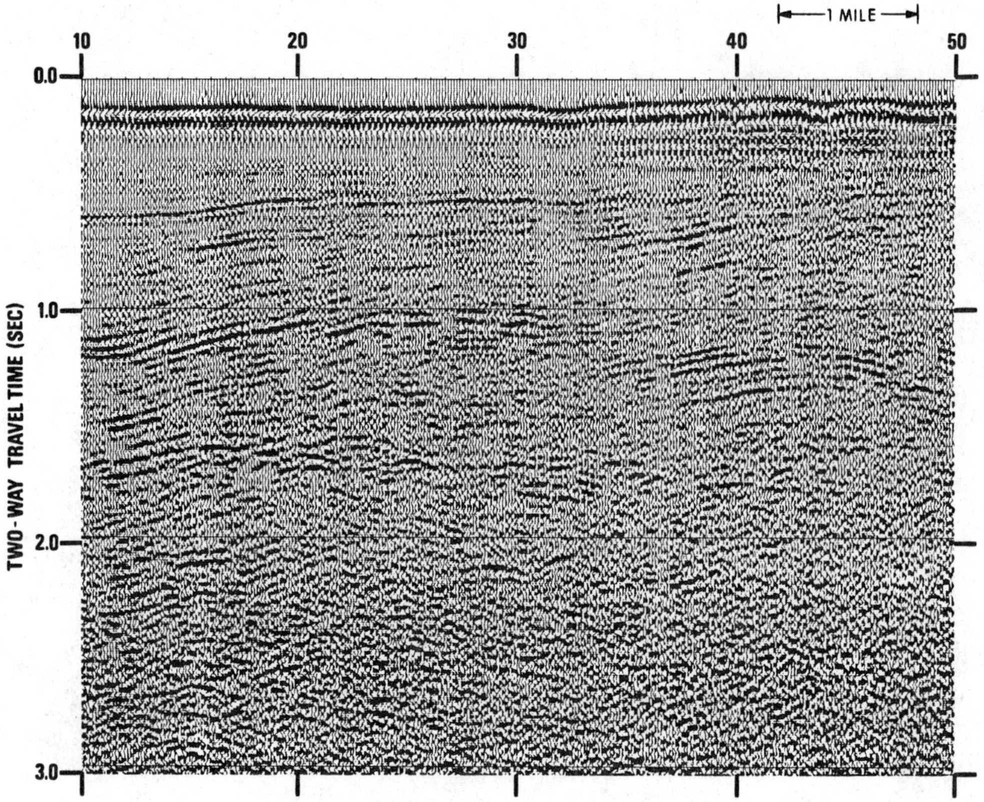

FIG. 20. 48-fold stacked section of line C with time-reverse, time-forward, then sum adaptive deconvolution applied before stack.

the use of adaptive deconvolution has made the sections more interpretable, especially at less than 1.0 sec traveltime where the base of the high-velocity surface layer is found. For example, the event at 650 msec at shotpoint 10 can be confidently mapped in Figure 20, at least to shotpoint 35, while it was not clearly mappable in Figure 19. The improvement in the shallow reflections on line D (compare Figures 21 and 22) is less apparent, but the water-bottom multiples have definitely been removed by the adaptive deconvolution. Comparisons on both lines C and D show that the peg-leg multiples associated with the deeper events have been more effectively removed by the adaptive deconvolution than by conventional deconvolution, thus making interpretation less ambiguous.

DISCUSSION AND CONCLUSIONS

This paper has presented a new time-varying deconvolution method for processing seismic reflection data. The method is based on the use of a simple adaptive algorithm which allows continuous updating of the deconvolution operator as the seismic trace is processed. Previously published work relating to the use of this algorithm, particularly in the communications field, has established that the procedure offers significant improvements in processed signal-to-noise ratio. In the present paper, the algorithm was applied to both synthetic and field-recorded reflection seismic data which contained a significant number of time-varying multiple reflections. The results obtained demonstrate that the multiple energy was effectively removed by the adaptive deconvolution procedure.

These experimental results have also demonstrated several advantages of the method. First, it was found that the effectiveness of the technique in removing undesired multiple energy is not a sensitive function of the parameters required to implement the procedure. Specifically, the scalar

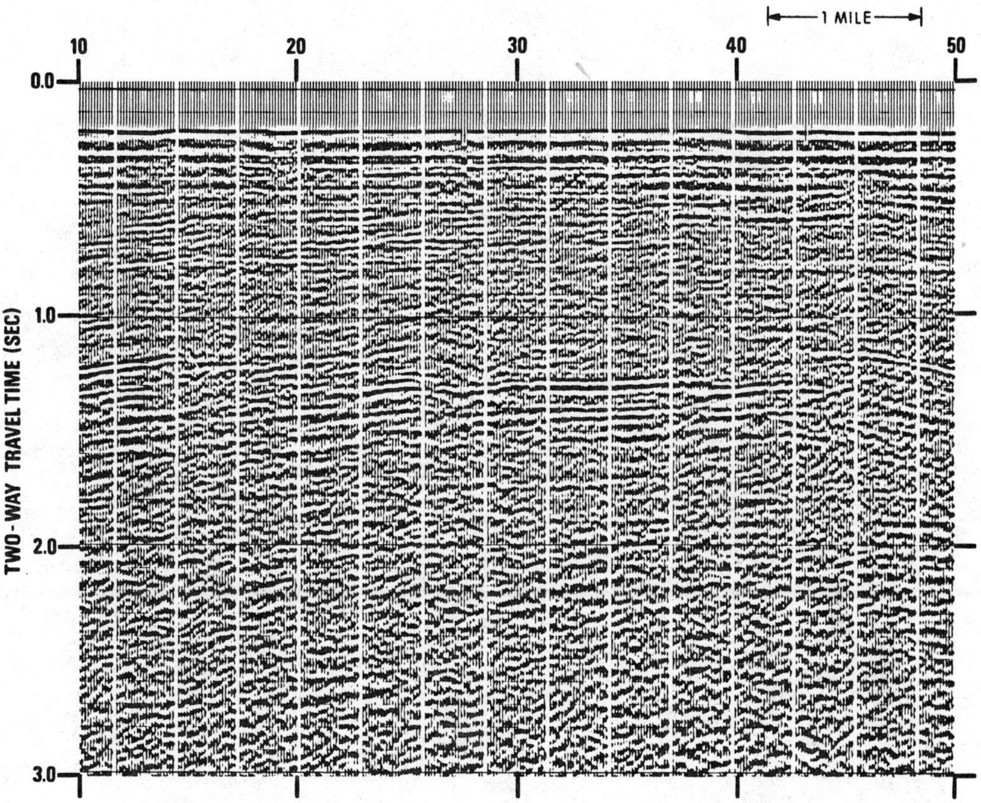

FIG. 21. 48-fold stacked section of line D with conventional deconvolution applied before stack.

constant α which controls the adaptive time constant can be varied between at least 0.3 and 0.8 without significantly affecting the quality of the deconvolved trace. A second advantage of adaptive deconvolution is that its use is not restricted to high signal-to-noise ratio input data. Of course, the ability of the procedure to remove time-varying multiples will be degraded as the quality of the input data is reduced. A third advantage noted is that the number of arithmetic operations required to implement adaptive deconvolution is actually smaller than the number required to implement conventional deconvolution methods. To illustrate, an input trace containing N data points can be deconvolved with an L point adaptive operator using $2NL$ multiplies and adds. Conventional deconvolution with five design gates would require the order of $2NL + 5L^2$ multiplies and adds.

Possibilities for further research on adaptive deconvolution include:

1) Implementation of the algorithm parameters (filter length, prediction distance, and adaptive time constant) in a time-varying manner.
2) A theoretical study of the convergence properties of the algorithm in nonstationary statistical environments.
3) The extension of adaptive deconvolution methods to multi-channel processing.

In summary, this paper has given a method for removing water-bottom multiples and refraction multiples. Clearly, the removal of these multiples is not the complete solution to obtaining good quality reflection seismic data in hard water-bottom areas. Much work remains to be done to develop the acquisition and processing techniques necessary to overcome all of the problems in such areas.

ACKNOWLEDGMENTS

We wish to thank Marathon Oil Co. for permission to publish this paper. In addition, we thank R. R. Burke, B. E. Merchant, and C. R.

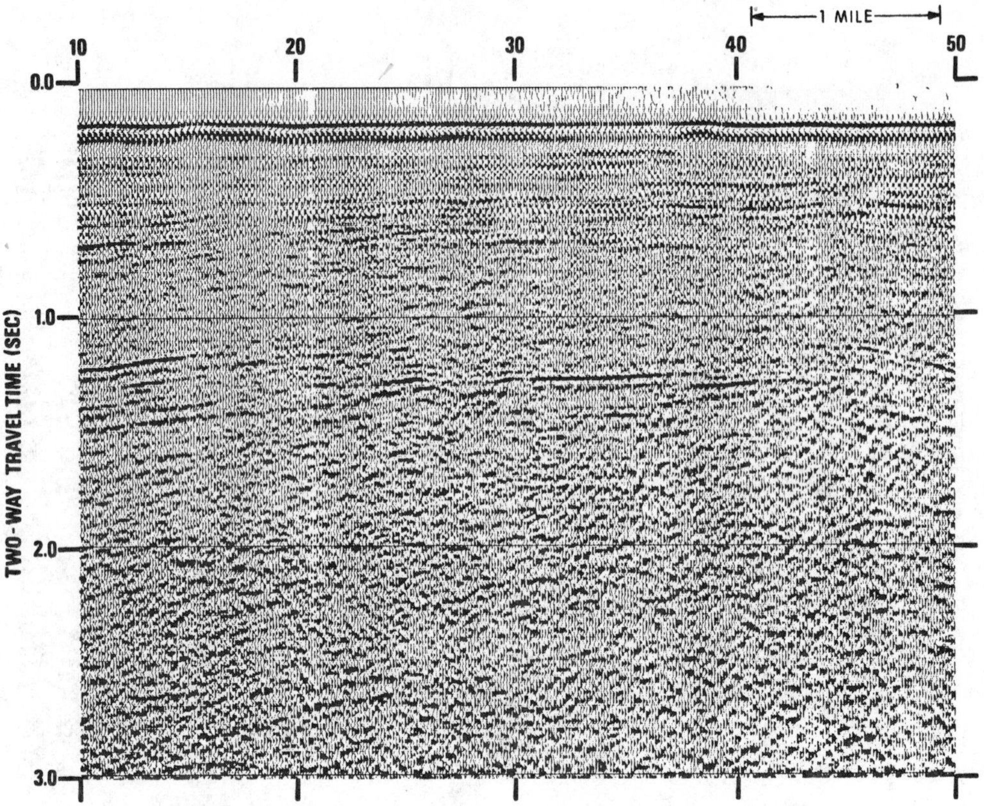

FIG. 22. 48-fold stacked section of line D with time-reverse, time-forward, then sum adaptive deconvolution applied before stack.

Harwood of Marathon Oil Co. Production International for supplying us with the seismic data.

REFERENCES

Blum, J. R., 1954, Multidimensional stochastic approximation methods: Annals Math. Stat., v. 25, p. 737–744.

Booker, A., and Ong, C., 1971, Multiple-constraint adaptive filtering: Geophysics, v. 36, p. 498–509.

Boonton, R. C., 1952, An optimization theory for time-varying linear systems with non-stationary statistical inputs: Proc. IRE, v. 40, p. 977–981.

Buhl, P., Stoffa, P. L., and Bryan, G. M., 1974, Application of homomorphic deconvolution to shallow-water marine seismology: Geophysics, v. 39, p. 401–426.

Clarke, G. K. C., 1968, Time-varying deconvolution filters: Geophysics, v. 33, p. 936–944.

Crump, N. D., 1974, A Kalman filter approach to the deconvolution of seismic signals: Geophysics, v. 39, p. 1–13.

Daniell, T. P., 1970, Adaptive estimation with mutually correlated training sequences: IEEE Trans. Systems Sci. and Cybern., p. 12–19.

Dupac, V., 1965, A dynamic stochastic approximation method: Annals Math. Stat., v. 36, p. 1695–1702.

Frost, O. L., III, 1972, An algorithm for linearly-constrained adaptive array processing: Proc. IEEE, v. 60, p. 926–935.

Gabriel, W. F., 1976, Adaptive arrays: An introduction: Proc. IEEE, v. 64, p. 239–271.

Gersho, A., 1968, Convergence properties of an adaptive filtering algorithm: Proc. 2nd Asilomar Conf. Circuits and Systems, p. 302–304.

Griffiths, L. J., 1969, A simple adaptive algorithm for real-time processing in antenna arrays: Proc. IEEE, v. 57, p. 1696–1704.

——— 1975, Rapid measurement of digital instantaneous frequency: IEEE Trans., v. ASSP-23, p. 207–222.

——— 1976, Time-domain adaptive beamforming of *H F* back-scatter radar signals: IEEE Trans., v. AP-24, no. 5.

Kim, J. K., and Davisson, L. D., 1975, Adaptive linear estimation for stationary *M*-dependent processes: IEEE Trans., v. IT-21, p. 23–31.

Lacoss, R. T., 1968, Adaptive combining of wideband array data for optimum reception: IEEE Trans. Geosci. Electron., v. 6, p. 78–86.

Middleton, D., and Whittlesey, J. R., 1968, Seismic models and deterministic operators for marine reverberation: Geophysics, v. 33, p. 557–583.

Peacock, K. L., and Treitel, S., 1969, Predictive deconvolution: Theory and practice: Geophysics, v. 34, p. 155–169.

Riegler, R. L., and Compton, R. T., Jr., 1973, An adaptive array for interference rejection: Proc. IEEE, v. 61, p. 748–758.

Robbins, H., and Monroe, S., 1951, A stochastic approximation method: Annals Math. Stat., v. 22, p. 400–407.

Robinson, E. A., 1967, Multichannel time series analysis with digital computer programs: San Francisco, Holden-Day, Inc.

Senne, K., 1968, Adaptive linear discrete-time estimation: Ph.D. dissertation, Stanford Univ.

Wang, R. J., 1969, The determination of optimum gate length for time-varying Wiener filtering: Geophysics, v. 34, p. 683–695.

Wang, R. J., and Treitel, S., 1971, Adaptive signal processing through stochastic approximation: Geophys. Prosp., v. 19, p. 718–727.

Widrow, B., 1966, Adaptive filters I: Fundamentals: Stanford Electronics Lab., rept. SEL-66-126.

Widrow, B., and Hoff, M. E., 1960, Adaptive switching circuits: IRE WESCON Conven. Rec., part 4, p. 96–104.

Widrow, B., Mantey, P. E., Griffiths, L. J., and Goode, B. B., 1967, Adaptive antenna systems: Proc. IEEE, v. 55, p. 2143–2159.

Widrow, B., et al, 1975, Adaptive noise cancelling: Principles and applications: Proc. IEEE, v. 63, p. 1692–1716.

Widrow, B., et al, 1976, Stationary and non-stationary learning characteristics of the LMS adaptive filter: Proc. IEEE, v. 64, p. 1521–1529.

VII. KALMAN FILTERING

In this section, the brave reader enters a new world: that of the Kalman filter. This approach to signal estimation is based on the state variable representation of a system. While the impulse response and frequency response of a system may be more familiar, it must be remembered that for physical systems these are derived from the differential equations which (with initial conditions) describe the system's behavior. J. W. Bayless and E. O. Brigham present an excellent tutorial explanation of state-space techniques (illustrated through simple filter circuits) and an introduction to the concepts of Kalman filter theory. In a final remark to their development of the Kalman filter (p. 419), the authors point out that for infinite observation time and stationary statistics, the optimum system of the Kalman filter must be identical to the Wiener filter. Consequently, the value of the Kalman method is likely to be in its capability for handling finite observation multichannel data from a time-varying system. Bayless and Brigham consider only continuous-time signals; seismic reflections are modeled as a series of random, sharply decaying exponential functions.

The paper by N. D. Crump, on the other hand, emphasizes the discrete Kalman filter. Briefly, in the discrete Kalman filter, an estimated state at each sample time is obtained as a weighted sum of two quantities: (1) an estimate predicted from the estimate at the previous sample time, and (2) the latest measurement. This paper, too, presents examples of Kalman filter application to the seismic deconvolution problem. It is assumed that the reflection coefficient at any sample time is the sum of (1) a random quantity and (2) a weighted sum of the reflection coefficients at previous sample times. Crump suggests that required wavelet models may be estimated by existing techniques such as combining computed autocorrelation functions with suitable phase assumptions; this technique is used in one example.

The development of the Kalman filter assumes that a message process is the output of a linear system with a white noise input and that this message is observed in the presence of additive noise. The principal difficulty in applying the method appears to be in obtaining the dynamics of the system and the statistics of the noise. In this connection, the reader might turn back to E. A. Robinson's thesis, section 5.4, "The filtering problem" (p. 91-95) and examine the description given of the message and the statistics of the random variables; the same difficulty appears in obtaining the information required for the design of a (stationary) Wiener filter. In practice, of course, deconvolution filters are not designed from ensemble averages. One designs a deterministic least-squares prediction-error filter for a specific trace. Application of Kalman filtering to seismic problems is still an active area of research; further work may lead to the use of equally deterministic filters based on the Kalman equations. The reader who wishes to pursue the topic of Kalman filtering will find further discussions and comparisons with Wiener filtering in the papers by Berkhout and Zaanen (1976) and by Mendel and Kormylo (1977).

REFERENCES

Berkhout, A. J., and Zaanen, P. R., 1976. A comparison between Wiener filtering, Kalman filtering, and deterministic least squares estimation: Geophysical Prospecting, v. 24, p. 141-197.

Mendel, J. M., and Kormylo, John, 1977, New fast optimal white-noise estimators for deconvolution: IEEE Transactions on Geoscience Electronics, v. GE-15, p. 32-41.

Reprinted from Geophysics v. 35, no. 1, p. 2-23

APPLICATION OF THE KALMAN FILTER TO CONTINUOUS SIGNAL RESTORATION†

J. W. BAYLESS* AND E. O. BRIGHAM‡

The Kalman filter is applied to the inverse filtering or deconvolution problem. The derivation given of the Kalman filter emphasizes the relationship between the Kalman and Wiener filter. This derivation is based on the representation of systems by state variables and the modeling of random processes as the output of linear systems excited by white noise. Illustrative results indicate the applicability of these techniques to a variety of geophysical data processing problems. The Kalman filter offers exploration geophysicists additional insight into processing-problem modeling and solution.

INTRODUCTION

The development of techniques for operation on a received signal to eliminate undesired filtering effects is called signal restoration or inverse filtering. Such techniques are of particular interest in geophysical research because signal enhancement is desired for smoothed measurements resulting from a sluggish measuring device. Considerable effort has been given (Brigham et al, 1968; Rice, 1962; Kunetz, 1962; George et al, 1962; Robinson, 1954) to design considerations of inverse filters based on the system model depicted in Figure 1.

In this model, the output $y(t)$ is related to the input $s(t)$ by the convolution integral

$$y(t) = \int_{-\infty}^{\infty} s(\tau)h(t-\tau)d\tau = s(t) * h(t), \quad (1)$$

where the impulse response $h(t)$ is considered for the present to be that of a linear time-invariant system. The function $h(t)$ is assumed known a priori and the function $y(t)$ or $y(t)+v(t)$ is measured; $v(t)$ is the measurement noise. It is desired to solve for $s(t)$.

Brigham et al (1968) solved equation (1) by the method of successive substitution; these results apply in the additive noise case if the variance of $v(t)$ is sufficiently small. Rice (1962) approached the solution of equation (1) by use of least-squares approximations; his results are applicable in the presence of low noise. Using least-squares techniques, Kunetz (1962) derived inverse filters by minimizing the sum of two terms; the first term effectively solving equation (1) and the second term accounting for the presence of noise. All of the above methods are time-domain solutions of the convolution integral (1).

Analysis of equation (1) is frequently made in the frequency or transform domain. Since convolution-multiplication form a Fourier transform pair, the transform of equation (1) is

$$Y(j\omega) = S(j\omega)H(j\omega), \quad (2)$$

where $Y(j\omega)$, $S(j\omega)$, and $H(j\omega)$ are the Fourier transforms of $y(t)$, $s(t)$, and $h(t)$, respectively. Note that equation (2) can be solved for $S(j\omega)$ and hence, $s(t)$ can be evaluated by using the inverse Fourier transformation. George et al (1962) solved equation (2) by making appropriate approximations in the frequency domain and applying the inverse Fourier transform.

Robinson (1954) computed the power spectrum

† Presented at the 38th Annual International SEG Meeting, Denver, Colorado, October 2, 1968. Manuscript received by the Editor July 10, 1969; revised manuscript received October 13, 1969.

* SMU Institute of Technology, Dallas, Texas 75222, and LTV Electrosystems, Inc. Garland, Texas 75040.

‡ LTV Electrosystems, Inc., Greenville, Texas 75401.

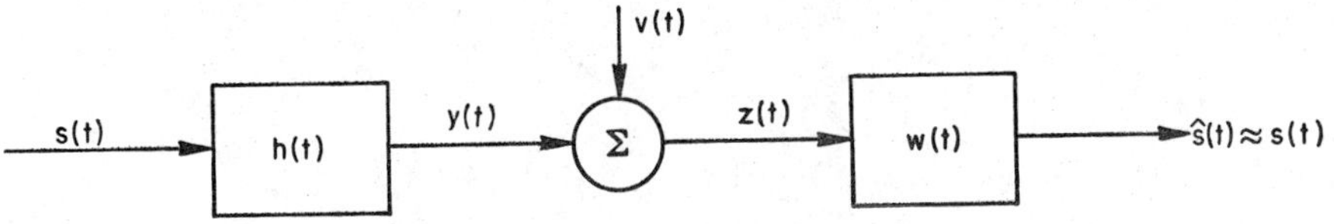

FIG. 1. System model for design of inverse filters.

of the measurement, factored this spectrum, and transformed it to obtain the impulse response $h(t)$. This approach assumes that the input signal $s(t)$ is averaged out in the computation of the power spectrum. The inverse filter was then determined from the relationship

$$\delta(t) = h(t) * w(t). \tag{3}$$

If the mean-square error between $\hat{s}(t)$ and $s(t)$ (see Figure 1) is minimized, one obtains the optimum Wiener (1949) inverse filter [see Turin, (1957) for derivation]

$$W(j\omega) = \frac{1}{H(j\omega)}\left[\frac{\Phi_s(\omega)}{\Phi_s(\omega) + \Phi_v(\omega)}\right],$$

where $\Phi_s(\omega)$ and $\Phi_v(\omega)$ are the power density spectra functions of the system input signal and additive noise, respectively.

Each of the above methods serves equally well for particular applications but none is applicable to more sophisticated inverse filter design problems. For example, if one follows the work of Robinson (1954) or attempts a solution of either equation (1) or (2), the solution is confined to time-invariant systems. Further, optimum Wiener filters are conventionally designed under the assumptions that the system is time invariant and that the statistics are stationary. Although the mathematical theory exists to extend Wiener theory to include time-varying systems with nonstationary statistical inputs (Booton, 1952), determination of the optimum system function requires solution of integral equations which often are not tractable and poorly suited to computer computation. In summary, a general theory for inverse filter design is not contained within the discussed approaches.

This problem of estimating a desired signal from noisy data observed over a time interval occurs in many disciplines such as communication theory (Price, 1956; Kelly et al, 1960), control theory (Magill, 1965; Wonham, 1965), as well as in many geophysical applications (Robinson, 1954; Treitel and Robinson, 1966). Much has been accomplished and much work continues (Kailath, 1968; Kailath and Frost, 1968; Hilborn and Lainiotis, 1969) in the theoretical development of optimum techniques for estimating signals embedded in a noisy environment. Much of this work, and in particular the significant contributions due to Kalman (1960) and Kalman and Bucy (1961), has been made possible by the relatively recent viewpoint of representing linear systems in terms of first order matrix differential equations (state variable representation). The results described in this paper depend on such a representation (Gupta, 1966) and could properly be described as a new look at the signal enhancement problem in light of the Kalman filtering theory. For exploration geophysical data processing problems, the Kalman filter can be viewed as a new technique which gives additional insight into problem modeling and solution. The formulation of the Kalman filter easily can be related to the wave equation, and for this reason it offers an immediate advantage in problem modeling. In addition, the theory presents a compact method of modeling multichannel data-acquisition systems and designing time-varying filters to be used in such systems.

The purpose of this discussion is to prevent tutorial development of the Kalman filter and its application to the signal enhancement problem. The Kalman filter derivation emphasizes system-state variable representation and the modeling of random processes as the output of linear systems excited by white (uncorrelated) noise. For these reasons, a review of state-variable techniques is presented, and random process models are then developed by use of state variables. From these concepts, a derivation of the Kalman filter is formulated, example problems are presented, and similarities and differences of the techniques to conventional Wiener theory are explored. The inverse Kalman filter is developed, example problems are considered, and illustrated results presented.

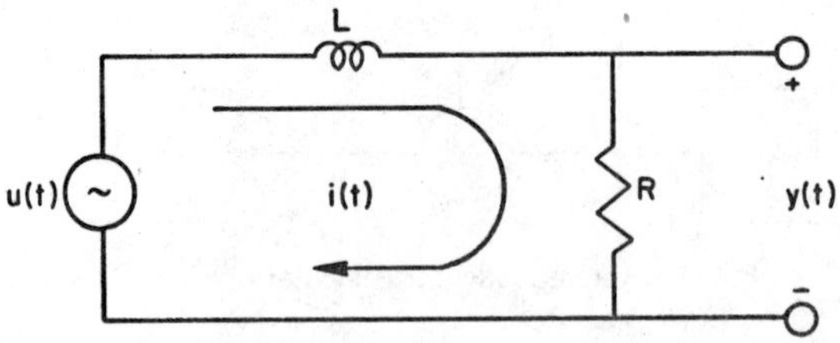

FIG. 2. Example of linear, time-invariant system.

STATE VARIABLES

State-variable (or state-space) techniques are based principally on the concept of stating a problem in the form of a particular differential equation and solving that equation. In particular, linear systems are specified by a set of first-order differential equations which, with initial conditions, uniquely define the system behavior. The state-variable approach to linear system analysis is formulated by the time-domain or analog-computer solution to this set of differential equations. Since the Kalman filter solution is derived in state notation, it is useful to relate the state-variable formulation to the more conventional methods of linear-system analysis. Differential equation and transfer function representations of simple filter circuits are utilized to motivate the discussion.

The most convenient way to exploit the relationship between the conventional transfer-function approach to system analysis and the state-variable formulation is through a consideration of the simple linear time-invariant system illustrated in Figure 2.

First, write the system differential equation and the equation defining the system output

$$u(t) = L\frac{di(t)}{dt} + Ri(t), \qquad y(t) = Ri(t), \tag{4}$$

respectively.

A solution for $y(t)$ for any input $u(t)$ is desired.

The classical method for analysis is to transform equation pair (4) with the Laplace transform $\mathcal{L}[\cdot]$ to the form

$$\mathcal{L}[u(t)] = \mathcal{L}\left[L\frac{di(t)}{dt}\right] + \mathcal{L}[Ri(t)], \qquad \mathcal{L}[y(t)] = \mathcal{L}[Ri(t)],$$

or (all initial conditions are assumed equal to zero)

$$U(s) = LsI(s) + RI(s), \qquad Y(s) = RI(s). \tag{5}$$

Equation pair (5) is solved for the transfer function $H(s)$ which relates the input and output transforms of the system

$$H(s) = \frac{Y(s)}{U(s)} = \frac{\frac{R}{L}}{s + \frac{R}{L}}. \tag{6}$$

As illustrated in Figure 3a, the output $y(t)$ can be evaluated by first forming the transform of the input, multiplying this transform by the transfer function to form the output transform, and then by applying the inverse Laplace transform $\mathcal{L}^{-1}[\cdot]$ to determine $y(t)$. The output $y(t)$ can also be determined by convolving the system impulse response (the inverse Laplace transform of the transfer function) with the system input (Figure 3b). These techniques for solving equation pair (4) are termed transform or frequency-domain analysis, and are conventionally employed on those problems where the coefficients of the system differential equation are constant.

The state-variable concept is simply a time-domain solution to equation pair (4) or equivalently, a time-domain realization of the transfer function $H(s)$. To develop this, let

$$x_1(t) = i(t),$$

and rewrite equation pair (4) as

$$\dot{x}_1(t) = -\frac{R}{L}x_1(t) + \frac{1}{L}u(t), \qquad y(t) = Rx_1(t). \tag{7}$$

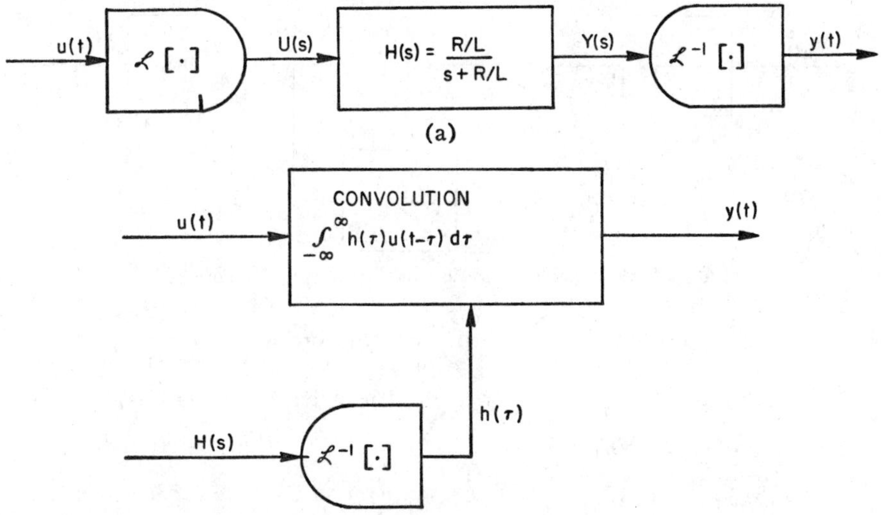

FIG. 3. a. Transfer function realization when $y(t)$ is evaluated by forming the transform of the input, multiplying this by the transfer function, and then applying the inverse Laplace transform. b. Transfer function realization when the system impulse response is convolved with the system input.

The variable $x_1(t)$ is termed the state variable of the example system; state variables are simply those variables which completely describe the system behavior at time t. For the example, if the current $i(t)$ is known at time t, then the system output is known at time t. Since t can be any value, a continuous solution for the state variable $x_1(t)$ is desired; such a solution is represented in Figure 4, and is simply the analog-computer realization of equation pair (7). Note that in the state-variable realization the system-differential equation can contain time-varying coefficients. For example, if the resistance in the system of Figure 2 is time-varying, equation pair (7) becomes

$$\dot{x}_1(t) = -\frac{R(t)}{L}x_1(t) + \frac{1}{L}u(t), \qquad y(t) = R(t)x_1(t). \tag{8}$$

The state-variable realization of equation (8) is illustrated in Figure 5. The problem of time-varying coefficients is more complex; however the state-variable realizations are of the same form (compare Figures 4 and 5). This constant form of realization is an important aspect of the state-variable formulation.

To investigate further the notation and concepts associated with state-variable techniques, consider the second-order system illustrated in Figure 6. The circuit resistance $R(t)$ is assumed to vary as a function of time. The circuit differential equation is

$$u(t) = \frac{1}{C}\int i(t)dt + L\frac{di(t)}{dt} + R(t)i(t). \tag{9}$$

Define the state variables

$$x_1(t) = \int i(t)dt = q, \quad \text{charge}, \tag{10}$$

$$x_2(t) = \frac{dx_1(t)}{dt} = \dot{x}_1(t) = \frac{dq}{dt} = i, \quad \text{current}, \tag{11}$$

and rewrite equation (9) as

$$u(t) = \frac{1}{C}x_1(t) + L\dot{x}_2(t) + R(t)x_2(t). \tag{12}$$

If equation (12) is rearranged and combined with equation (11), the differential equation of the example system is described in terms of state variables

$$\dot{x}_1(t) = x_2(t), \qquad \dot{x}_2(t) = -\frac{R(t)}{L}x_2(t) - \frac{1}{LC}x_1(t) + \frac{1}{L}u(t). \tag{13}$$

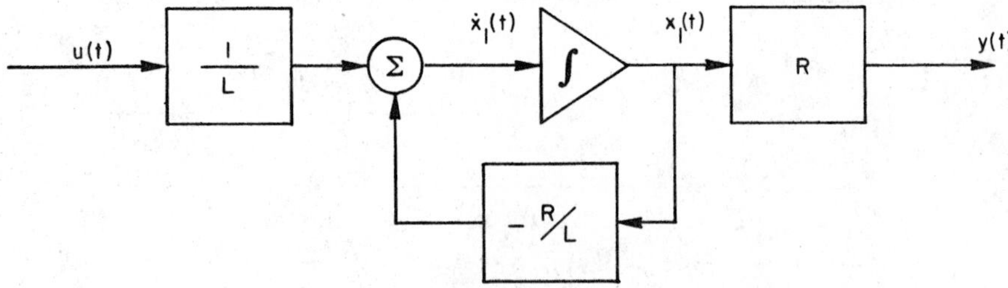

FIG. 4. State-variable realization for the example of Figure 2.

Normally, equation pair (13) is written in matrix form as

$$\begin{bmatrix}\dot{x}_1(t)\\ \dot{x}_2(t)\end{bmatrix} = \begin{bmatrix}0 & 1\\ -1/LC & -R(t)/L\end{bmatrix} \cdot \begin{bmatrix}x_1(t)\\ x_2(t)\end{bmatrix} + \begin{bmatrix}0\\ 1/L\end{bmatrix} u(t), \tag{14}$$

or in general.

$$\dot{\mathbf{x}}(t) = \mathbf{F}(t)\mathbf{x}(t) + \mathbf{G}(t)\mathbf{u}(t), \tag{15}$$

where all matrices can be determined by inspection of equation (14). (In the text the matrix notation is indicated by boldface; in the figures the matrix notation is indicated by a bar under the algebraic expression.) Matrices $\mathbf{F}(t)$ and $\mathbf{G}(t)$ have been written as functions of time for generality; for this example matrix $\mathbf{G}(t)$ is a constant. Matrix $\mathbf{F}(t)$ completely defines the characteristics or dynamics of the system and matrix $\mathbf{G}(t)$ describes how the forcing function or functions enter into the system.

The output $y(t)$ for this example

$$y(t) = R(t)i(t),$$

in terms of the state variable $x_2(t)$ is

$$y(t) = R(t)x_2(t),$$

which in general matrix form is

$$\mathbf{y}(t) = \mathbf{H}(t)\mathbf{x}(t). \tag{16}$$

Matrix $\mathbf{H}(t)$ describes how the output or outputs of the system are observed.

Equations (15) and (16) paired together

$$\begin{aligned}\dot{\mathbf{x}}(t) &= \mathbf{F}(t)\mathbf{x}(t) + \mathbf{G}(t)\mathbf{u}(t),\\ \mathbf{y}(t) &= \mathbf{H}(t)\mathbf{x}(t)\end{aligned} \tag{17}$$

define not only the behavior of the example second-order system but, in general, also describe the

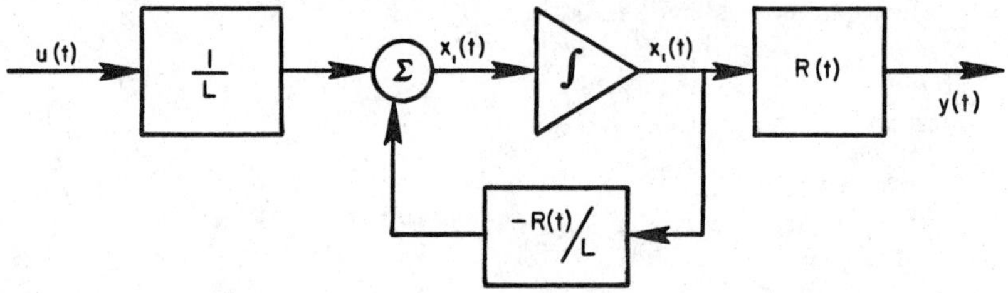

FIG. 5. State-variable realization when the resistance in the example system of Figure 2 is time-varying.

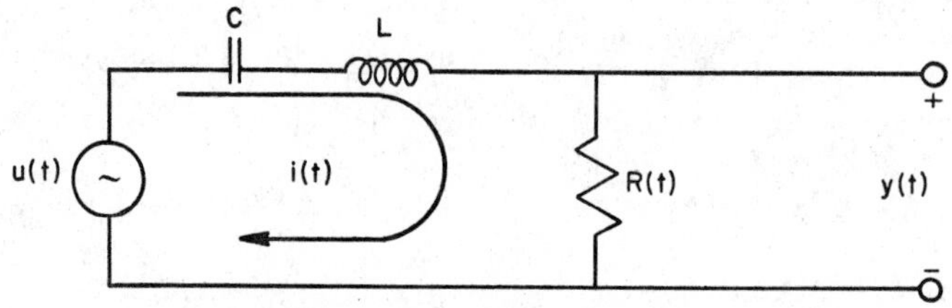

FIG. 6. Example of second-order system.

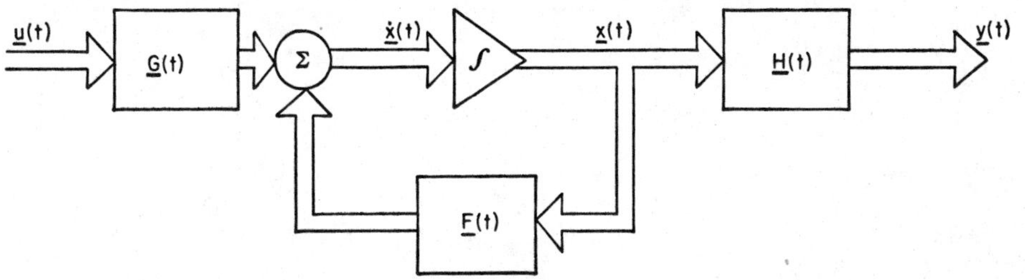

FIG. 7. State-variable matrix realization of the linear system model of Figure 6.

state-variable representation of any linear time-varying system. This generalized time-domain representation is analogous to the conventional transfer-function, $H(s)$, or impulse-response $h(t)$, representation of a linear system. Note, however, the ease with which time-varying and multi-dimensional systems are included in the state representation of equation (17). This state-variable realization of a linear system is illustrated in Figure 7. The double signal flow lines are used to indicate vector quantities. Matrix $\mathbf{x}(t)$ is termed the state vector and matrix $\mathbf{y}(t)$ the observation vector.

It has been shown that the state-variable formulation is merely a time-domain realization of the conventional transfer-function formulation. Further, it was shown that time-varying systems are included easily in a state-variable formulation. In general, a system which can be described by a set of linear first-order differential equations can be described by equation pair (17) and the state-variable realization illustrated in Figure 7; henceforth, linear systems will be described by this representation.

We now proceed to define a model for random processes in terms of state variables before proceeding to the Kalman filter development.

MODELING RANDOM PROCESSES

To derive optimum estimation procedures for random signals that are perturbed by measurement noise, it is necessary to specify in some manner the a priori knowledge of the signal and noise. The Kalman filter is based on a derivation which assumes that a random process can be modeled as the output of a linear system with a white-noise input. The intent here is to indicate the validity of this assumption.

Recall that if a signal $u(t)$ with power-spectral density function $\Phi_u(\omega)$ is the input to a linear system with a transfer function $H(j\omega)$, then the output signal $y(t)$ has a power-spectral density function (see Appendix A)

$$\Phi_y(\omega) = |H(j\omega)|^2 \Phi_u(\omega). \tag{18}$$

If $u(t)$ is a white-noise process, then $\Phi_u(\omega)=1$ and

$$\Phi_y(\omega) = |H(j\omega)|^2. \tag{19}$$

It can be shown (Davenport and Root, 1958) that if $\Phi_y(\omega)$ is a power spectrum of a nondeterministic random process, then $\Phi_y(\omega)$ can always be factored into

$$\Phi_y(\omega) = H(j\omega)H^*(j\omega), \tag{20}$$

where $H(j\omega)$ contains all poles of $\Phi_y(\omega)$ with positive imaginary parts and one-half the real zeros of $\Phi_y(\omega)$; the conjugate $H^*(i\omega)$ contains the remaining poles and zeros. The function $H(j\omega)$ is the transfer function of a realizable linear system. Because equations (19) and (20) are identical, a given power-spectral density function $\Phi_y(\omega)$ can be factored to determine $H(\omega)$. The process $y(t)$ is then visualized as having been generated by passing a random signal $u(t)$ with spectrum $\Phi_u(\omega)=1$ through a linear system with transfer function $H(j\omega)$. It is then possible to represent the mean and autocorrelation function of a random process as the output of a linear system excited by white noise (see Figure 8a). Recall that in the previous section it was shown that the state-variable formulation was just a time-domain realization of the transfer function $H(j\omega)$. It is therefore possible to model random processes (up to second-order statistics) as illustrated in Figure 8b. The input $\mathbf{u}(t)$ is a white-noise process and the output $\mathbf{y}(t)$ has the desired power spectrum. Note that nonstationary random processes can also be modeled since matrices $\mathbf{F}(t)$, $\mathbf{G}(t)$, and $\mathbf{H}(t)$ are in general time-varying quantities.

In summary, a state-variable model as illus-

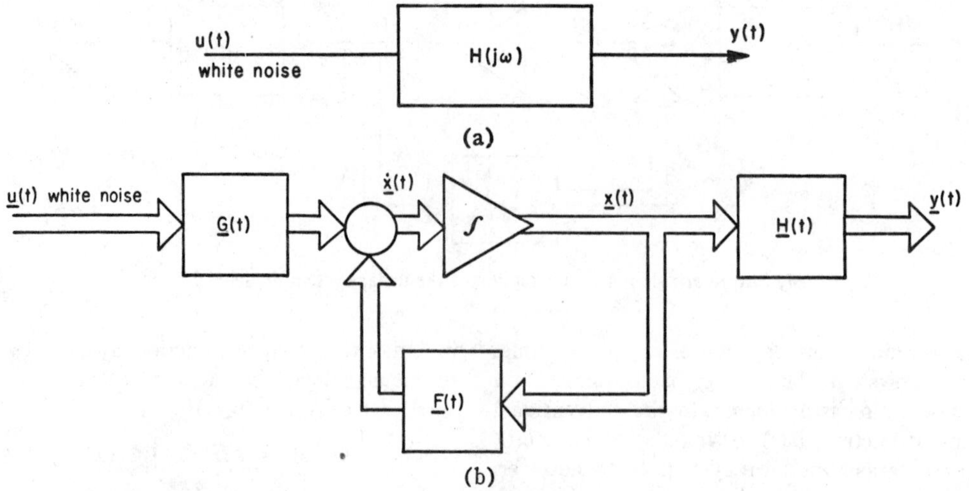

FIG. 8. a. Random-process model with white-noise exciting function. b. State-variable modeling of random process model of Figure 8a.

trated in Figure 8 can be constructed to model an arbitrary random process in terms of its mean and autocorrelation function. If the input white-noise process is Gaussian, the output will also be a Gaussian random process since the system is linear.

KALMAN FILTER DEVELOPMENT

The work of Kalman (1960, 1961) is often referred to as the single most significant contribution to filtering theory since the classical work of Wiener (1949). The intent here is to investigate the basic concepts of Kalman filter theory; the treatment is not to be construed as comprehensive but rather as an introduction to the design aspects of Kalman filters. In order to place Kalman's theory in a proper perspective to classical filtering and prediction theory, Wiener theory is briefly reviewed and the limitations of this theory are discussed. The Kalman filtering equations are then derived.

The classical Wiener problem is illustrated in Figure 9. It is desired to find the optimum linear time-invariant filter that will operate on the measureable signal $z(t)$ and minimize the error between the filter output $\hat{x}(t)$ and the desired signal $x(t)$. The equation

$$z(t) = x(t) + v(t) \tag{21}$$

models the problem. The random process $x(t)$ is termed the message process and $v(t)$ the measurement noise.

The solution to this problem is expressed in terms of the optimum filter impulse response $h_o(t)$ as

$$\hat{x}(t) = \int_{-\infty}^{\infty} h_o(\tau)z(t-\tau)d\tau. \tag{22}$$

The filter impulse response must satisfy the Wiener-Hopf integral equation,

$$\int_{-\infty}^{\infty} h_o(\tau)\phi_z(t-\tau)d\tau = \phi_{xz}(t), \quad t \geq 0, \tag{23}$$

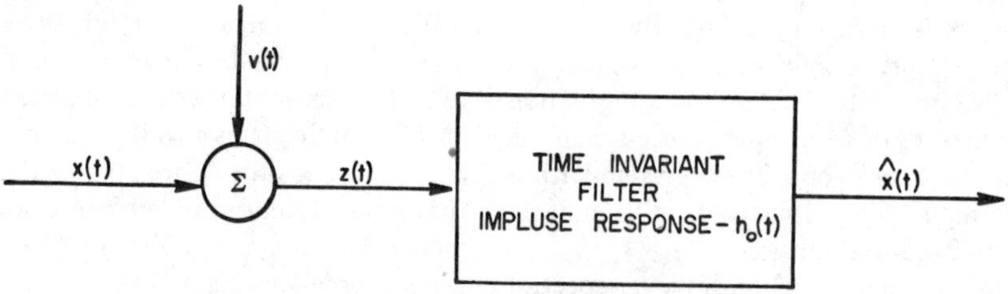

FIG. 9. Classical Weiner filter problem.

where ϕ_z and ϕ_{xz} are autocorrelation and cross-correlation functions, respectively. It is assumed that a statistical description of the stationary random signal $x(t)$ and the stationary random noise $v(t)$ are known a priori.

To specify the optimum Wiener filter it is then necessary to solve integral equation (23). This approach has a disadvantage in that the Wiener-Hopf integral equation is not easily solved; non-specialists find integral equations quite cumbersome. Numerical evaluation of the optimum filter is also complex, because the solution of an integral equation is not well suited to computer calculations. Although existing computational algorithms alleviate this problem to some degree, the situation becomes increasingly difficult as the complexity of the problem increases. Further, after a solution of the integral equation is obtained, it is not a simple task to synthesize the specified solution. It will be shown that these difficulties are eliminated in Kalman's theory.

The classical Wiener-filter theory is valid only for stationary one-dimensional processes. Davis (1963) has extended Wiener's results to include the case of n-dimensional signal plus noise structures for a certain class of problems and Booton (1952) has extended the original mathematical theory to include nonstationary random processes. Under the most general conditions, equation (23) then becomes

$$\int_{t_o}^{t} \mathbf{h}_o(t, \tau)\boldsymbol{\phi}_z(\tau, \sigma)d\tau = \boldsymbol{\phi}_{xz}(t, \sigma), \qquad t_o \leqq \sigma \leqq t, \tag{24a}$$

and the estimating equation is given by

$$\hat{\mathbf{x}}(t) = \int_{t_o}^{t} \mathbf{h}_o(t, \tau)\mathbf{z}(t, \tau)d\tau. \tag{24b}$$

Equation (24a) is extremely difficult to solve in this form. Note that this equation is valid for multidimensional, nonstationary random processes and that equation (24b) includes finite observation, time-varying estimates. Kalman (1961) converts the integral equation (24a) to a differential equation which is well suited to machine solution; the optimum filter is completely specified and synthesized with the solution of this differential equation.

The Kalman-filter equations are derived from many different viewpoints in the literature: Sage (1967) used least-squares techniques; Ho (1964) used the Bayesian approach to stochastic estimation; Athans and Tse (1969) utilized the maximum principle of Pontryagin to derive the optimal values of the filter coefficients; Papoulis (1965) used a simplified orthogonal projection theorem to determine the optimal estimate; and the Gauss-Markov theorem (Graybill, 1961) can be employed to form a minimum-variance linear estimator.

However, each of these viewpoints has the similar disadvantage that the true relationship of the derived filter and that obtained using classical Wiener-filter theory is lost in the development. For this reason, the derivation presented here follows that of Van Trees (1968) and is a simplified version of Kalman's (1961) development; the filtering equations are derived from the generalized Wiener-Hopf integral equation (24a).

For generality, equation (21) is rewritten to include vector-valued random processes

$$\mathbf{z}(t) = \mathbf{y}(t) + \mathbf{v}(t) = \mathbf{H}(t)\mathbf{x}(t) + \mathbf{v}(t). \tag{25}$$

Matrix $\mathbf{H}(t)$ is introduced to account for the possibility that various linear combinations of the components of the vector $\mathbf{x}(t)$ may be observed in the measurement $\mathbf{z}(t)$.

As discussed in the previous section, the random message process $\mathbf{y}(t)$ can be modeled by a linear system excited by white noise, that is

$$\dot{\mathbf{x}}(t) = \mathbf{F}(t)\mathbf{x}(t) + \mathbf{G}(t)\mathbf{u}(t), \tag{26a}$$

$$\mathbf{y}(t) = \mathbf{H}(t)\mathbf{x}(t), \tag{26b}$$

which give the system structure illustrated in Figure 8.

The random process $\mathbf{u}(t)$, which generates the message model, and the additive noise process $\mathbf{v}(t)$ are assumed to be zero-mean random processes with correlation functions (superscript T indicates matrix transpose).

$$\begin{aligned} \boldsymbol{\phi}_u(t, \tau) &= E[\mathbf{u}(t)\mathbf{u}^T(\tau)] = \mathbf{Q}(t)\delta(t-\tau), \\ \boldsymbol{\phi}_v(t, \tau) &= E[\mathbf{v}(t)\mathbf{v}^T(\tau)] = \mathbf{R}(t)\delta(t-\tau), \\ \boldsymbol{\phi}_{uv}(t, \tau) &= E[\mathbf{u}(t)\mathbf{v}^T(\tau)] = 0, \\ \boldsymbol{\phi}_{uz}(t, \tau) &= E[\mathbf{u}(t)\mathbf{z}^T(\tau)] = 0. \end{aligned} \tag{27}$$

Equations (25) and (26a) define the message model and additive-measurement noise. The assumption that the additive-noise process is white is not a restriction in the theory; the colored-noise case is easily included by simply writing $v(t)$ as the output of a linear system excited by a white-noise process (Van Trees, 1968). From equations (24), (25), (26), and (27), the Kalman filtering equations can be derived.

First, post multiply both sides of equation (26a) by $\mathbf{z}^T(\sigma)$ and take expectations

$$E\left[\frac{d\mathbf{x}(t)\mathbf{z}^T(\sigma)}{dt}\right] = \mathbf{F}(t)E[\mathbf{x}(t)\mathbf{z}^T(\sigma)] + \mathbf{G}(t)E[\mathbf{u}(t)\mathbf{z}^T(\sigma)], \tag{28}$$

and use equations (27) to obtain

$$\frac{\partial}{\partial t}[\boldsymbol{\phi}_{xz}(t, \sigma)] = \mathbf{F}(t)\boldsymbol{\phi}_{xz}(t, \sigma). \tag{29}$$

Second, post multiply equation (25) by $\mathbf{z}^T(\tau)$ and take expectations

$$E[\mathbf{z}(t)\mathbf{z}^T(\tau)] = \mathbf{H}(t)E[\mathbf{x}(t)\mathbf{z}^T(\tau)] + E[\mathbf{v}(t)\mathbf{z}^T(\tau)],$$

or

$$\boldsymbol{\phi}_z(t, \tau) = \mathbf{H}(t)\boldsymbol{\phi}_{xz}(t, \tau). \tag{30}$$

Third, differentiate the left-hand side of equation (24a)

$$\frac{\partial}{\partial t}\int_{t_o}^{t}\mathbf{h}_o(t, \tau)\boldsymbol{\phi}_z(\tau, \sigma)d\tau = \mathbf{h}_o(t, t)\boldsymbol{\phi}_z(t, \sigma) + \int_{t_o}^{t}\frac{\partial}{\partial t}\mathbf{h}_o(t, \tau)\boldsymbol{\phi}_z(\tau, \sigma)d\tau,$$

and substitute equation (30) to obtain

$$\frac{\partial}{\partial t}\int_{t_o}^{t}\mathbf{h}_o(t, \tau)\boldsymbol{\phi}_z(\tau, \sigma)d\tau = \mathbf{h}_o(t, t)\mathbf{H}(t)\boldsymbol{\phi}_{xz}(t, \tau) + \int_{t_o}^{t}\frac{\partial}{\partial t}\mathbf{h}_o(t, \tau)\boldsymbol{\phi}_z(\tau, \sigma)d\tau. \tag{31}$$

Hence, differentiation of equation (24a) and use of relationships (29) and (31) yield

$$\{\mathbf{F}(t) - \mathbf{h}_0(t, t)\mathbf{H}(t)\}\boldsymbol{\phi}_{xz}(t, \sigma) - \int_{t_o}^{t}\frac{\partial}{\partial t}\mathbf{h}_o(t, \tau)\boldsymbol{\phi}_z(\tau, \sigma)d\tau = 0. \tag{32}$$

Finally, for $\boldsymbol{\phi}_{xz}(t, \sigma)$ substitute into equation (32) the defining relation (24a) to get

$$\int_{t_o}^{t}\left\{\mathbf{F}(t)\mathbf{h}_o(t, \tau) - \mathbf{h}_o(t, t)\mathbf{H}(t)\mathbf{h}_o(t, \tau) - \frac{\partial}{\partial t}\mathbf{h}_o(t, \tau)\right\}\boldsymbol{\phi}_z(\tau, \sigma)d\tau = 0. \tag{33}$$

This integral equation is certainly satisfied if $\mathbf{h}_o(t, \tau)$ satisfies the differential equation

$$\frac{\partial}{\partial t}\mathbf{h}_o(t, \tau) = \mathbf{F}(t)\mathbf{h}_o(t, \tau) - \mathbf{h}_o(t, t)\mathbf{H}(t)\mathbf{h}_o(t, \tau). \tag{34}$$

Thus, it is necessary (Van Trees, 1968) and sufficient that the optimum impulse response $\mathbf{h}_o(t, \tau)$ satisfy equation (34). Integral equation (24a) has thus been reduced to differential equation (34). It remains to solve this differential equation and to determine the filtering equations.

To determine the optimum filtering equations, differentiate estimation equation (24b) with respect to time

$$\frac{d\hat{\mathbf{x}}(t)}{dt} = \mathbf{h}_o(t, t)\mathbf{z}(t) + \int_{t_o}^{t}\frac{\partial}{\partial t}\mathbf{h}_o(t, \tau)\mathbf{z}(\tau)d\tau. \tag{35}$$

Substitution of $\partial/\partial t\ \mathbf{h}_o(t, \tau)$ from equation (34) into equation (35) yields

$$\frac{d\hat{\mathbf{x}}(t)}{dt} = \mathbf{h}_o(t, t)\mathbf{z}(t) + \mathbf{F}(t)\int_{t_o}^{t}\mathbf{h}_o(t, \tau)\mathbf{z}(\tau)d\tau - \mathbf{h}_o(t, t)\mathbf{H}(t)\int_{t_o}^{t}\mathbf{h}_o(t, \tau)\mathbf{z}(\tau)d\tau. \tag{36}$$

If $\mathbf{K}(t) \triangleq \mathbf{h}_o(t, t)$ and equation (24b) is substituted into equation (36),

$$\frac{d\hat{\mathbf{x}}(t)}{dt} = \mathbf{K}(t)\mathbf{z}(t) + \mathbf{F}(t)\hat{\mathbf{x}}(t) - \mathbf{K}(t)\mathbf{H}(t)\hat{\mathbf{x}}(t), \tag{37}$$

which gives the optimal filter-system structure illustrated in Figure 10.

Note that the structure of the message-generating process is contained exactly in the filter

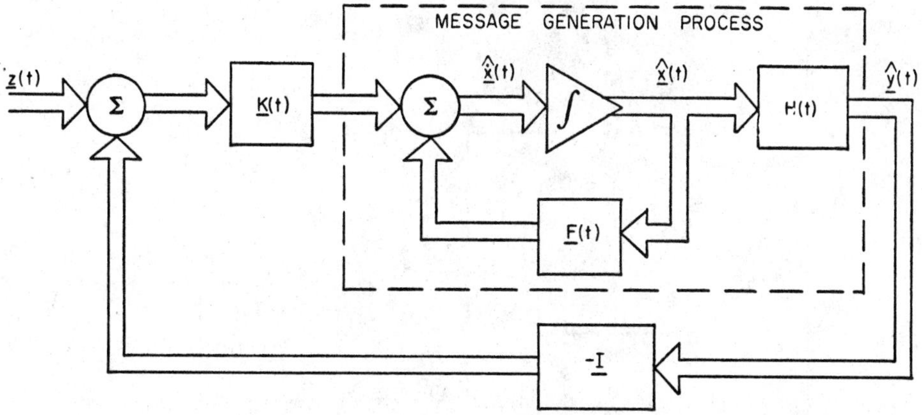

FIG. 10. Optimum system structure of Kalman filter for continuous estimation.

solution. Since this information is known a priori, the optimal filter is completely synthesized at this point with the exception of determining the system gain $\mathbf{K}(t)$. Equation (37) is then one of the celebrated Kalman filtering equations for continuous estimation. The second equation is the defining relationship for $\mathbf{K}(t)$.

The derivation of the defining relationships for $\mathbf{K}(t)$ lends little to the present development, but it is included in Appendix B for completeness. It is shown that the solution to equation (34) is given by the matrix equation

$$\mathbf{K}(t) = \mathbf{P}(t)\mathbf{H}^T(t)\mathbf{R}^{-1}(t), \qquad (38)$$

where $\mathbf{P}(t)$ is the error autocorrelation matrix

$$\begin{aligned} \mathbf{P}(t) &= E[\tilde{\mathbf{x}}(t)\tilde{\mathbf{x}}^T(t)], \\ \tilde{\mathbf{x}}(t) &= \hat{\mathbf{x}}(t) - \mathbf{x}(t), \end{aligned} \qquad (39)$$

which must satisfy the nonlinear matrix differential equation (matrix Riccati equation)

$$\begin{aligned} \frac{d\mathbf{P}(t)}{dt} &= \mathbf{F}(t)\mathbf{P}(t) + \mathbf{P}(t)\mathbf{F}^T(t) \\ &\quad - \mathbf{P}(t)\mathbf{H}^T(t)\mathbf{R}^{-1}(t)\mathbf{H}(t)\mathbf{P}(t) \qquad (40) \\ &\quad + \mathbf{G}(t)\mathbf{Q}(t)\mathbf{G}^T(t). \end{aligned}$$

Note that neither $\mathbf{K}(t)$ nor the differential equation involving $\mathbf{P}(t)$ contain the input $\mathbf{z}(t)$; that is, $\mathbf{K}(t)$ can be determined before data are taken and the filter completely synthesized. $\mathbf{K}(t)$ is commonly called the filter gain matrix. It should be noted that specification of the message-generating system in Figure 8 is essential to derivation of the optimum system in Figure 10; this explains the reason for the development of state-variable and random-process modeling concepts in the previous sections. Finally, if $t_o \to -\infty$ and stationary statistics are assumed, the optimum system in Figure 10 must be identical to the Wiener filter since the Wiener-Hopf integral equation was the initial equation of the derivation.

It is convenient to explain additional properties of the Kalman filter by means of an example. Figure 11 illustrates a piecewise, continuous time function which could represent a model of a well log measurement. This example will illustrate the problem of recovering this signal imbedded in noise under the assumption that a perfect measuring instrument is being used. An example to be discussed later will consider a sluggish or narrowband measuring instrument.

In Figure 11, the occurrence times t_i of the steps are assumed random and Poisson-distributed with parameter λ. The amplitude of the step a_i is also a random variable with mean zero and variance σ_a^2. The a_i are independent of the occurrence times of the steps. Under these conditions, the mean, autocorrelation, and power spectrum of the process are

$$\begin{aligned} m_x(t) &= E[x(t)] = 0, \\ \phi_x(t_1, t_2) &= E[x(t_1)x(t_2)] \\ &= \lambda\sigma_a^2 \min(t_1, t_2), \\ \min(t_1, t_2) &= \begin{cases} t_1 & 0 < t_1 \leqq t_2 \\ t_2 & 0 < t_2 \leqq t_1 \end{cases}, \qquad (41) \\ \Phi_x(\omega) &= \frac{\lambda\sigma_a^2}{\omega^2}. \end{aligned}$$

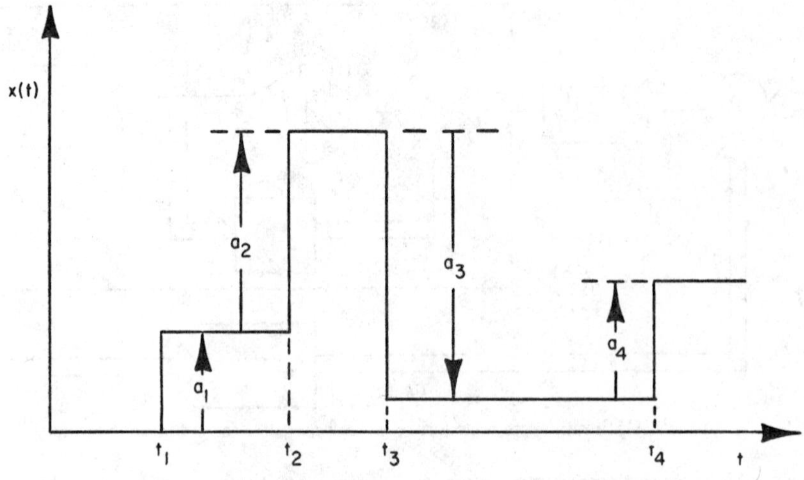

FIG. 11. Well log waveform as a sample function of a random process.

Assume that the signal $x(t)$ illustrated in Figure 11 is measured in the presence of additive white noise with mean zero and variance σ_v^2. To derive the Kalman filter, it is first necessary to model the message process $x(t)$ in state-variable form. As illustrated in Figure 12a, the transfer function $1/s$ is the appropriate system which when excited by white noise yields the correct output power spectrum. Figure 12b is the equivalent model in state-variable representation and the corresponding modeling equations are

$$\begin{aligned} \dot{x}(t) &= u(t) \\ y(t) &= x(t), \end{aligned} \tag{42}$$

where

$$\phi_x(t_1, t_2) = E[x(t_1)x(t_2)] = Q\delta(t_1 - t_2),$$
$$Q = \lambda\sigma_a^2.$$

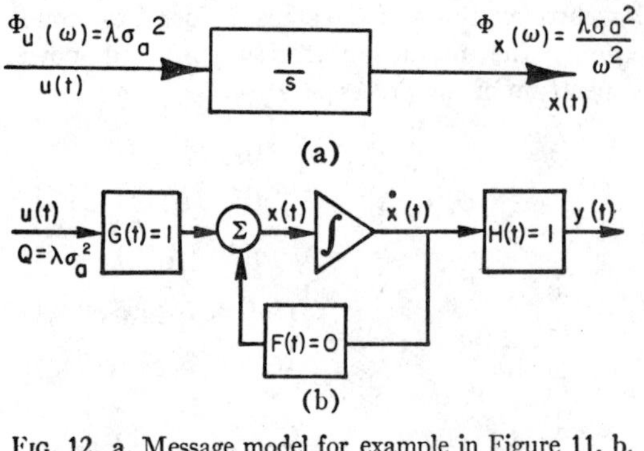

FIG. 12. a. Message model for example in Figure 11. b. Equivalent model in state-variable representation.

The Kalman filter (from Figure 10) is as illustrated in Figure 13.

Note that once the message process is modeled, matrices $\mathbf{F}(t)$, $\mathbf{G}(t)$, and $\mathbf{H}(t)$ are specified and the Kalman-filter solution is complete except for the specification of $\mathbf{K}(t)$.

The random process $x(t)$ is observed in the presence of the white-noise process $v(t)$ where

$$\phi_v(t_1, t_2) = E[v(t_1)v(t_2)] = R\delta(t_1 - t_2),$$
$$R = \sigma_v^2.$$

The gain $\mathbf{K}(t)$ can then be found from equation (38)

$$\mathbf{K}(t) = \mathbf{P}(t)\mathbf{H}^T(t)\mathbf{R}^{-1}(t) = \frac{p(t)}{\sigma_v^2}, \tag{43}$$

and from equation (40)

$$\begin{aligned} \frac{d\mathbf{P}(t)}{dt} &= \mathbf{F}(t)\mathbf{P}(t) + \mathbf{P}(t)\mathbf{F}^T(t) \\ &\quad - \mathbf{P}(t)\mathbf{H}^T(t)\mathbf{R}^{-1}(t)\mathbf{H}(t)\mathbf{P}(t) \\ &\quad + \mathbf{G}(t)\mathbf{Q}(t)\mathbf{G}^T(t) \end{aligned}$$

or

$$\dot{p}(t) = -\frac{p^2(t)}{\sigma_v^2} + \lambda\sigma_a^2. \tag{44}$$

The steady-state solution of equation (43) is obtained by setting $\dot{p}(t) = 0$ in equation (44) and solving the resulting algebraic equation

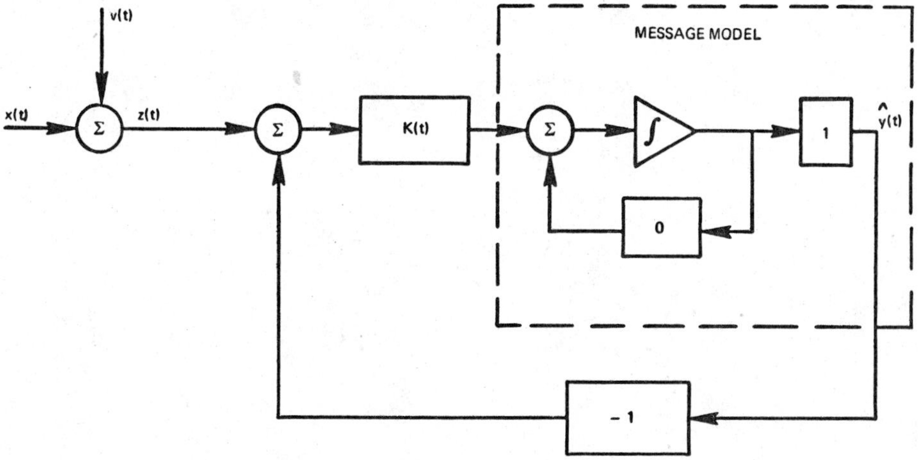

FIG. 13. Optimum Kalman filter for example of Figure 11.

$$K = \frac{p}{\sigma_v^2} = \left[\frac{\lambda\sigma_a^2}{\sigma_v^2}\right]^{1/2}. \tag{45}$$

The gain K is then a constant inversely proportional to the standard deviation of the measurement noise; intuitively, this type of result is anticipated.

The steady-state Kalman filter is as illustrated in Figure 13, where K is defined by equation (45). The filter is identical to that which can be obtained using conventional Wiener theory. To explore the advantage of the Kalman filter, it is necessary to determine the time-varying gain $K(t)$, i.e., to consider the transient solution rather than the steady-state solution.

To solve equation (44) for the transient case, it is necessary to determine the initial condition $p(0)$. Recall from equation (39) that $p(0)$ is the variance of the process at time t_o. The random process $x(t)$ may begin at time t_o with a known value, $p(0)$ is then zero, or may begin with a random value having a known variance. For purposes of discussion, assume $p(0)=0$. The solution for the gain $K(t)$ is then

$$K(t) = \frac{p(t)}{\sigma_v^2} = \left[\frac{\lambda\sigma_a^2}{\sigma_v^2}\right]^{1/2} \cdot \left[\frac{1-\exp(-2\gamma t)}{1+\exp(-2\gamma t)}\right], \tag{46}$$

where

$$\gamma = \frac{\lambda\sigma_a^2}{\sigma_v^2}.$$

Note that as $t\rightarrow\infty$, the solution approaches the steady-state solution (45),

$$K(t\rightarrow\infty) = \left[\frac{\lambda\sigma_a^2}{\sigma_v^2}\right]^{1/2}. \tag{47}$$

The optimum time-varying or transient filter is then as illustrated in Figure 13 with $K(t)$ defined by equation (46).

In summary, it has been shown that the Kalman-filter equations can be derived from the generalized Wiener-Hopf integral equation. The derived filter equations are based on a state-variable model of the message process. It has been shown that once the message process is modeled, the Kalman filter is completely synthesized with the exception of the filter gain $\mathbf{K}(t)$. The determination of $\mathbf{K}(t)$ requires the solution of the matrix Riccati equation (40). It is not to be inferred that variance equation (40) is easily solved for all filtering problems. However it is to be noted that this differential equation is well suited to either digital or analog computer solution. Using Kalman-filter theory, one has traded the difficulties associated with solving an integral equation for those associated with solving a differential equation. It can be shown (Kalman and Bucy, 1961) that if the message-generating process $\mathbf{u}(t)$ and the measurement noise $\mathbf{v}(t)$ are Gaussian, Figure 10 represents the best system that can be obtained not only for a mean-square optimality criterion but also for a number of other optimization criteria. If the random processes are not

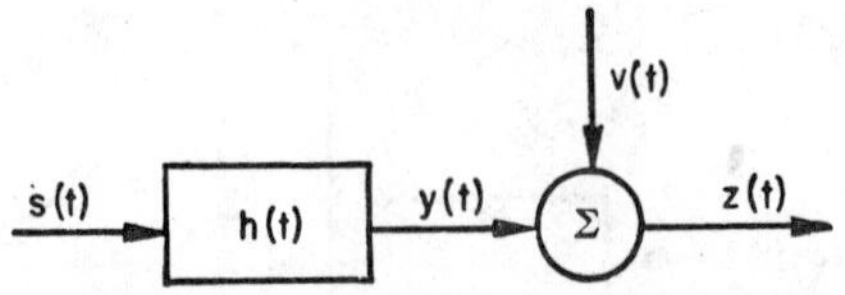

(a)

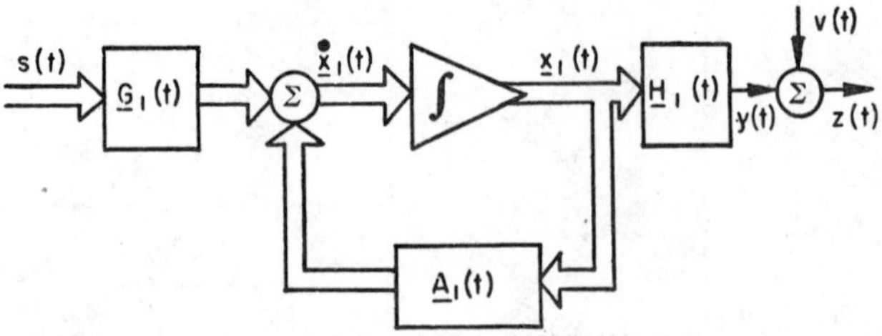

(b)

FIG. 14. a. Inverse filter system model which includes only noise at the output-measuring system. b. Inverse filter state-variable feedback model of distorting system.

Gaussian, then Figure 10 represents the optimum linear system.

The following derivation of the inverse or deconvolution Kalman filter is a straightforward manipulation of the derived Kalman-filter equations.

KALMAN INVERSE FILTER

The inverse filter problem is illustrated in Fig-14a. The input signal $s(t)$ is smeared by the measurement system and this measurement is observed in the presence of noise. Given the measurement $z(\tau)=y(\tau)+v(\tau)$, $t_o \leqq \tau \leqq t$, it is desired to find the best linear unbiased estimate of the input signal $s(t)$ assuming knowledge of the first- and second-order statistics of $s(t)$, $v(t)$, and the differential equation describing the distorting system. (The model indicated in Figure 14a includes only noise at the output of the measuring system. Noise on the input or noise at both input and output are easily included. This particular model is chosen to present as straightforward a presentation as possible.)

The solution readily follows from the previously derived Kalman filter. Recall from previous developments that linear systems can be modeled by the state-variable realization illustrated in Figure 7. Therefore, the state-variable feedback model of the distorting system is as illustrated in Figure 14b. This model is represented by the system equations

$$\dot{\mathbf{x}}_1(t) = \mathbf{A}_1(t)\mathbf{x}_1(t) + \mathbf{G}_1(t)s(t), \tag{48a}$$

$$z(t) = y(t) + v(t) = \mathbf{H}_1(t)\mathbf{x}_1(t) + v(t). \tag{48b}$$

Since the first- and second-order statistics of the random process $s(t)$ are assumed known, $s(t)$ can be modeled by passing white noise through a suitable linear system as shown in Figure 15 and as described by the equations

$$\dot{\mathbf{x}}_2(t) = \mathbf{A}_2(t)\mathbf{x}_2(t) + \mathbf{G}_2(t)\mathbf{u}(t), \tag{49a}$$

$$s(t) = \mathbf{H}_2(t)\mathbf{x}_2(t). \tag{49b}$$

Equation (49b) is substituted into equation (48a) and the resulting equation is combined with

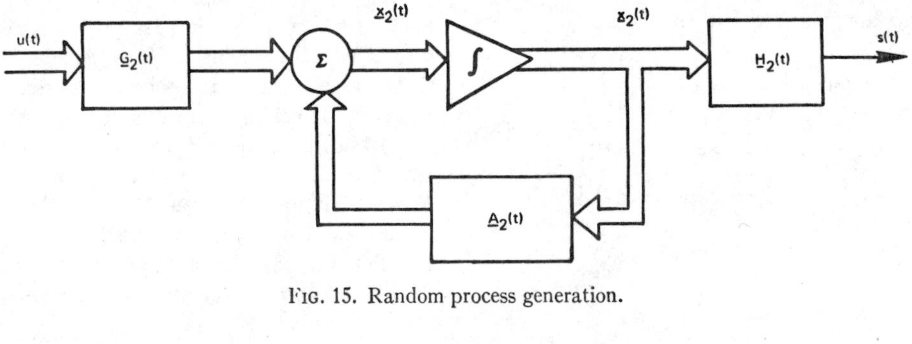

FIG. 15. Random process generation.

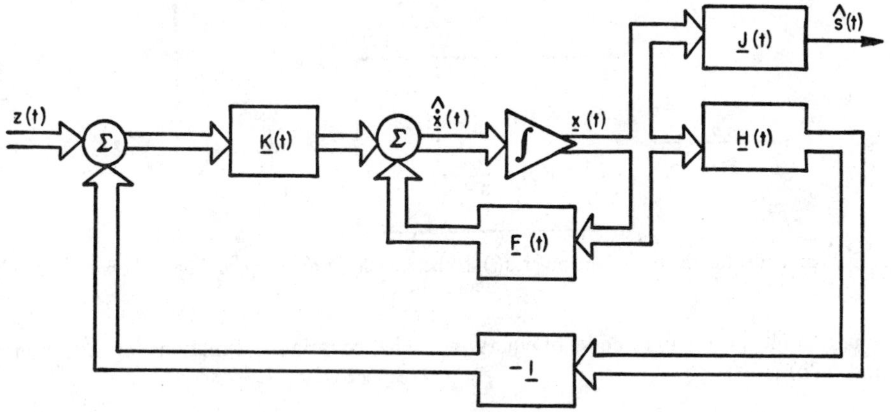

FIG. 16. Optimum continuous-time inverse Kalman filter.

equations (48b) and (49a) to give the standard-state formulation

$$\begin{aligned} \dot{\mathbf{x}}_1(t) &= \mathbf{A}_1(t)\mathbf{x}_1(t) + \mathbf{G}_1(t)\mathbf{H}_2(t)\mathbf{x}_2(t), \\ \dot{\mathbf{x}}_2(t) &= \mathbf{A}_2(t)\mathbf{x}_2(t) + \mathbf{G}_2(t)\mathbf{u}(t), \\ z(t) &= \mathbf{H}_1(t)\mathbf{x}_1(t) + v(t), \end{aligned} \tag{50}$$

or

$$\begin{aligned} \dot{\mathbf{x}}(t) &= \mathbf{F}(t)\mathbf{x}(t) + \mathbf{G}(t)\mathbf{u}(t), \\ z(t) &= \mathbf{H}(t)\mathbf{x}(t) + v(t), \end{aligned} \tag{51}$$

where

$$\mathbf{F}(t) = \begin{bmatrix} \mathbf{A}_1(t) & \mathbf{G}_1(t)\mathbf{H}_2(t) \\ 0 & \mathbf{A}_2(t) \end{bmatrix},$$

$$\mathbf{G}(t) = \begin{bmatrix} 0 \\ \mathbf{G}_2(t) \end{bmatrix}, \qquad \mathbf{x}(t) = \begin{bmatrix} \mathbf{x}_1(t) \\ \mathbf{x}_2(t) \end{bmatrix},$$

$$\mathbf{H}(t) = [\mathbf{H}_1(t) \quad 0].$$

Equation pair (51) is the exact form from which the Kalman-filter equation was derived previously and the optimum estimator of $\hat{\mathbf{x}}(t)$ is obtained as illustrated in Figure 10. [It should be emphasized that $\mathbf{F}(t)$ and $\mathbf{G}(t)$ are assumed known. In practice, these quantities must be determined. This is known as the system identification problem. Cuenod and Sage (1968) and Sage and Melsa (in press) are good references on this topic.]

Recall that

$$s(t) = \mathbf{H}_2(t)\mathbf{x}_2(t),$$

and since $\mathbf{H}_2(t)$ is a linear operator,

$$\hat{s}(t) = \mathbf{H}_2(t)\hat{\mathbf{x}}_2(t),$$

or

$$\hat{s}(t) = \mathbf{J}(t)\hat{\mathbf{x}}(t), \tag{52}$$

where

$$\mathbf{J}(t) = [0 \quad \mathbf{H}_2(t)].$$

Thus, the optimum continuous-time inverse filter is shown in Figure 16. Note in this formulation best estimators to both the smeared and non-smeared signal are obtained simultaneously. Analytical and simulation results are now presented to examine the Kalman inverse filter.

ILLUSTRATIVE EXAMPLES

Example 1

Consider the system model shown in Figure 17. The signal $s(t)$ is assumed to be the well log waveform which was modeled previously (Figure 11). The low-pass system indicated in Figure 17 is introduced to account for the blurring or smearing of the signal $s(t)$ as a result of measurement instrumentation; this model is used for purposes of discussion only and is not to be considered as the exact model of an actual system. The recorded or observed measurement $z(t)$ is assumed to be the sum of the low-pass system output $y(t)$ and the measurement noise $v(t)$. It is desired to determine in the optimum sense a best estimate of the signal $s(t)$ using the Kalman inverse filter previously developed.

The solution is obtained in a straightforward manner by: (a) determining by inspection of matrices $\mathbf{F}(t)$, $\mathbf{G}(t)$, and $\mathbf{H}(t)$ of equation (51); and (b) solving the matrix Riccati equation (40).

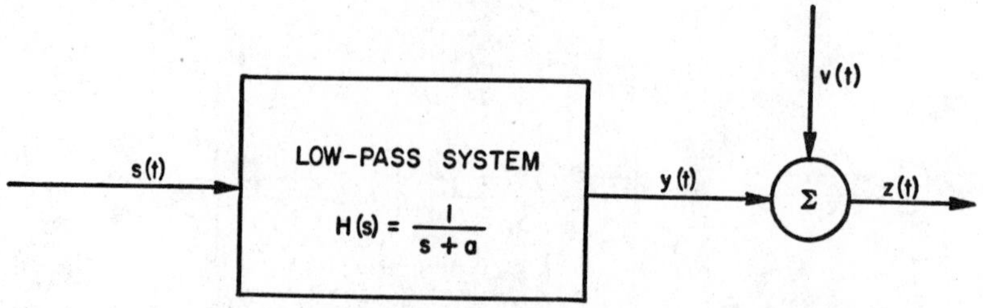

FIG. 17. System model for Example 1 in which $s(t)$ is the well log waveform of Figure 11 with system measurement noise $v(t)$.

From equation (42), the signal model of equation (49a) and (49b) is

$$\begin{aligned} \dot{x}_2(t) &= u(t) \\ s(t) &= x_2(t), \end{aligned} \tag{53}$$

where

$$\phi_u(t, \tau) = Q(t)\delta(t - \tau) = \lambda\sigma_a^2\delta(t - \tau).$$

The signal model is illustrated in Figure 18a. The low-pass system modeled by equations (48a) and (48b) is

$$\begin{aligned} \dot{x}_1(t) &= -ax_1(t) + s(t), \\ y(t) &= x_1(t), \end{aligned} \tag{54}$$

and is illustrated in Figure 18b. The output $y(t)$ of equation (54) is combined with additive white noise $v(t)$ (Figure 18c); $v(t)$ is assumed to be a zero mean random process with

$$\phi_v(t, \tau) = R\delta(t - \tau) = r\delta(t - \tau).$$

By inspection of Figure 18, the parameters of the standard-state formulation [equation (51)] are

$$\mathbf{F} = \begin{bmatrix} -a & 1 \\ 0 & 0 \end{bmatrix}, \quad \mathbf{G} = \begin{bmatrix} 0 \\ 1 \end{bmatrix}, \quad \mathbf{H} = [1 \quad 0]. \tag{55}$$

The covariance functions of $u(t)$ and $v(t)$ are respectively

$$\mathbf{Q} = [\lambda\sigma_a^2], \qquad \mathbf{R} = [r]. \tag{56}$$

Substitution of equations (55) and (56) into equation (40) yields

$$\begin{aligned} \dot{p}_{11} &= 2(p_{12} - ap_{11}) - p_{11}^2/r, \\ \dot{p}_{12} &= p_{22} - ap_{12} - p_{11}p_{12}/r, \\ \dot{p}_{22} &= \lambda\sigma_a^2 - p_{12}^2/r. \end{aligned} \tag{57}$$

The solution to the stationary problem (i.e., assume $z(t)$ has been observed since $t = -\infty$) is obtained by setting the right hand side of equation (57) to zero,

$$\begin{aligned} p_{11} &= -ar + \sqrt{a^2r^2 + 2r\sqrt{\lambda\sigma_a^2 r}}, \\ p_{12} &= \sqrt{\lambda\sigma_a^2 r}. \end{aligned} \tag{58}$$

Upon substituting these values into equation (38), one obtains the gain terms

$$\begin{aligned} k_{11} &= \frac{p_{11}}{r} = -a + \sqrt{a^2 + 2\sqrt{\frac{\lambda\sigma_a^2}{r}}}, \\ k_{21} &= \frac{p_{12}}{r} = \sqrt{\frac{\lambda\sigma_a^2}{r}}. \end{aligned} \tag{59}$$

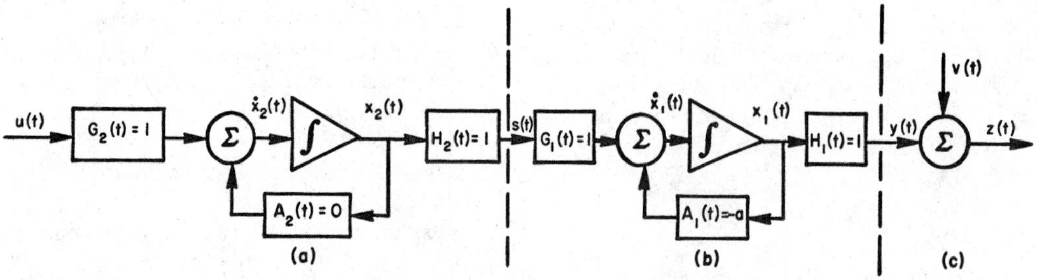

FIG. 18. The design model for Example 1. a) the signal model; b) low-pass system model; c) the output combined with additive white noise.

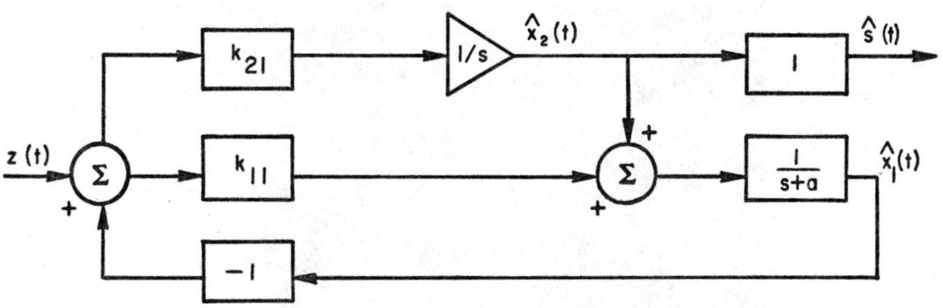

FIG. 19. Optimum Kalman inverse filter for Example 1.

Thus, the optimum Kalman inverse filter is as shown in Figure 19. It should be noted that the structure of this filter not only yields the best estimate of $\hat{s}(t)$, but also the best estimate of the low-pass system output $\hat{y}(t)$.

To indicate the quality of signal enhancement achievable by using the inverse Kalman filter, this example problem was simulated on an analog computer. The simulation parameters were: $\lambda=100/\pi$; $\sigma_a^2=1/2$; $a=220/4.7$ radians. Figures 20 and 21 illustrate the results of this simulation.

In Figure 20, the simulated well log signal (Figure 20a), the noise corrupted measurement (Figure 20b), and the restored signal (Figure 20c) are illustrated. Impressive results are expected since the measurement noise is low. Note in Figure 20c that even the smallest signal detail is recovered completely. Figure 21 illustrates the application of the Kalman inverse filter to the problem of medium level measurement noise. Signal detail is again almost perfectly restored. The increase in visible noise level in the restored signal (Figure 21c) is anticipated. In both Figures 20 and 21, the measurement noise level is not observable because of oscilloscope/camera resolution. Inverse filtering in the presence of high measurement noise is not in general practical.

The application of the Kalman inverse filter to the problem of medium level measurement noise as well as high level additive input noise is illustrated in Figure 22. The desired signal $s(t)$ (Figure 22a) and the additive input noise are smoothed by the low-pass filter. The filter output is observed in the presence of additive measurement noise (Figure 22c); the Kalman filter output is shown in Figure 22d. The inverse filter for this case differs from the cases above only in the gain terms k_{11} and k_{21}.

For purposes of discussion and illustration, a very simple model of a well logging system was assumed. However, the techniques illustrated can be applied to any chosen model.

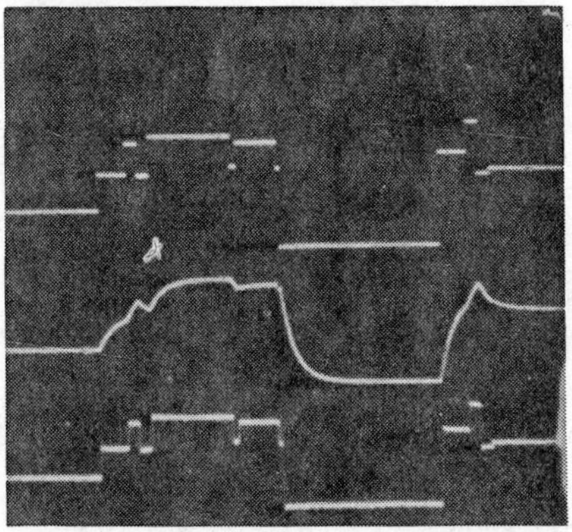

(a)
(b)
(c)

FIG. 20. Analog computer simulation of the application of the inverse Kalman filter to Example 1. a) The simulated well log signal; b) measured signal plus noise, $r=10^{-6}$; c) signal estimate.

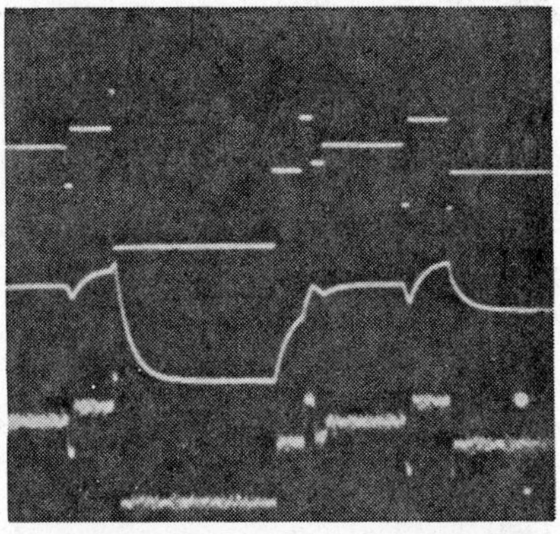

(a)
(b)
(c)

FIG. 21. Analog computer simulation of the application of Kalman inverse filter to signal restoration problem when medium level measurement noise is present. a) Desired signal; b) measured signal plus noise, $r=10^{-3}$; c) signal estimate.

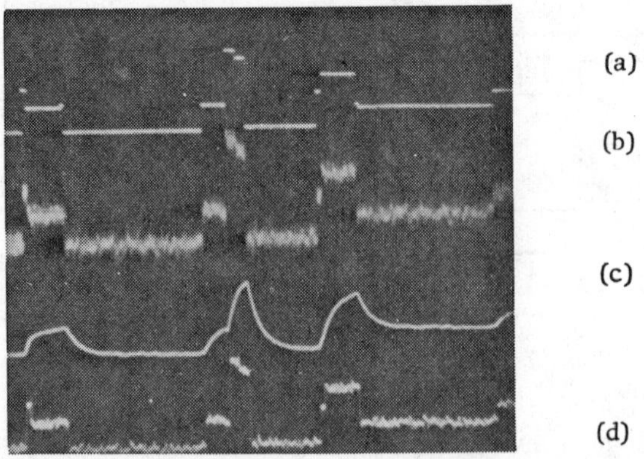

FIG. 22. Analog computer simulation of the application of Kalman inverse filter to the signal restoration problem when medium level measurement noise and high level additive input noise are present. a) Desired signal; b) desired signal plus noise, $r_{in}=10^{-1}$; c) measured signal plus noise; $r_{out}=10^{-6}$; d) signal estimate.

Example 2

To indicate the applicability of the inverse Kalman filter to the processing of seismic data, this example includes a basic model for estimating arrival times of impulses which produce a wavelet response. As in example 1, the model assumed is simplified for discussion purposes and is not to be considered as an exact model. The intent is to convey the application of the inverse Kalman filter to such problems.

Although both the wavelet amplitude and arrival time are random, for ease of presentation the wavelet amplitude will be assumed to be a known constant. This assumption can be removed at the expense of a more elaborate inverse filter development. The assumed wavelet model is the minimum-phase function

$$h_2(t) = e^{-at} \sin bt, \qquad t \geqq 0, \qquad (60)$$

where

$$a = 50, \qquad b = 2\pi \times 50.$$

The sharp impulsive reflections which give rise to the wavelet response are assumed to be of the form

$$h_1(t) = e^{-ct}, \qquad t \geqq 0, \qquad (61)$$

where

$$c = 1000.$$

The message or input signal model is then as illustrated in Figure 23. The input $u(t)$ is a white-noise process as required by the Kalman-filter theory; for this example $u(t)$ is a sequence of random Poisson-distributed impulses

$$u(t) = \sum_{i=-\infty}^{\infty} \delta(t-ti) - Q \qquad (62)$$

where

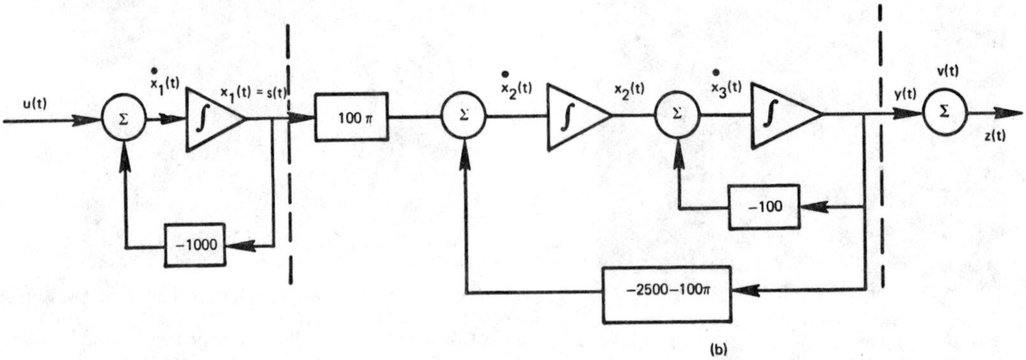

FIG. 23. Random wavelet model for Example 2. a) Signal model; b) wavelet model of equation (60).

$$\phi_{u}(t,\tau) = Q\delta(t-\tau).$$

The t_i, t_j are independent Poisson-distributed random variables with average occurence time of Q. The random process $u(t)$ is then a true model of a sequence of equal amplitude impulse function with a random time of occurrence. These are the arrival times which one desires to estimate; however, with the Kalman filter it is meaningless to estimate the generating process. For this reason $u(t)$ is passed through the impulsive reflection generator described by equation (61). (See Figure 23a.) The output $x_1(t) = s(t)$ is then a series of random, sharply decaying, exponential functions whose time of occurrence can be estimated using the inverse Kalman filter. This function $s(t)$ is the input to wavelet model equation (60) as illustrated in Figure 23b. The output $y(t)$ is a sequence of superimposed equal amplitude wavelets, the time of occurrence of which is random. It is assumed that $y(t)$ is observed in the presence of additive (Figure 21c) white noise with variance r. It is desired to process $z(t)$ and estimate the time of arrivals of the waveform $s(t)$ by removing the effect of noise and the wavelet model.

By inspection of Figure 23, equation (51) becomes

$$\frac{d}{dt}\begin{bmatrix} x_1(t) \\ x_2(t) \\ x_3(t) \end{bmatrix} = \begin{bmatrix} -1000 & 0 & 0 \\ 100\pi & -2500-100\pi & 0 \\ 0 & 1 & -100 \end{bmatrix} \cdot \begin{bmatrix} x_1(t) \\ x_2(t) \\ x_3(t) \end{bmatrix} + \begin{bmatrix} 1 \\ 0 \\ 0 \end{bmatrix} u(t), \tag{63}$$

$$z(t) = \begin{bmatrix} 0 & 0 & 1 \end{bmatrix} \begin{bmatrix} x_1(t) \\ x_2(t) \\ x_3(t) \end{bmatrix} + v(t).$$

Thus, the optimum Kalman filter is given by equation (37) and the general filter system structure illustrated in Figure 10. For this example the general structure of Figure 10 reduces to the optimum filter illustrated in Figure 24. The gain terms $k_{11}(t)$, $k_{21}(t)$, and $k_{31}(t)$ are determined by substituting the appropriate parameter into equation (38),

$$\begin{aligned}
\frac{dk_{11}(t)}{dt} &= -1100k_{11}(t) + k_{23}(t) - k_{11}(t)k_{31}(t), \\
\frac{dk_{21}(t)}{dt} &= k_{22}(t) + 100\pi k_{11}(t) + (-2600-100\pi)k_{21}(t) - k_{21}(t)k_{31}(t), \\
\frac{dk_{31}(t)}{dt} &= -200k_{31}(t) + 2k_{21}(t) - k_{31}^2(t), \\
\frac{dk_{22}(t)}{dt} &= -200\pi k_{23}(t), + 2(-2500-100\pi)k_{22}(t) - k_{21}^2(t), \\
\frac{dk_{33}(t)}{dt} &= -2000k_{33}(t) - k_{11}^2(t) + \frac{Q}{r}, \\
\frac{dk_{23}(t)}{dt} &= 100\pi k_{33}(t) + (-3500-100\pi)k_{23}(t) - k_{21}(t)k_{11}(t).
\end{aligned} \tag{64}$$

To solve for the steady-state gains one sets the right-hand side of equations (64) to zero; the resultant set of algebraic equations is extremely unwieldy. Hence, as is often necessary, differential equations (64) are solved on a digital computer and the steady-state solution determined.

An analog computer simulation of this example was performed and the results are indicated in Figures 25 and 26. The desired impulse, superimposed wavelets, observed waveforms, and estimated impulses are illustrated for the conditions of low and medium noise. As in Example 1, the measurement noise is not observable. A more realistic model can be obtained by assuming random amplitudes on the wavelets (Robinson, 1954). This assumption can be incorporated into this example by replacing equation (62) with

$$u(t) = \sum_{i=-\infty}^{\infty} a_i\delta(t-t_i) - E\left[\sum_{i=-\infty}^{\infty} a_i\delta(t-t_i)\right] \tag{65}$$

where the a_i and a_j are random variables. The correlation properties of this type waveform are

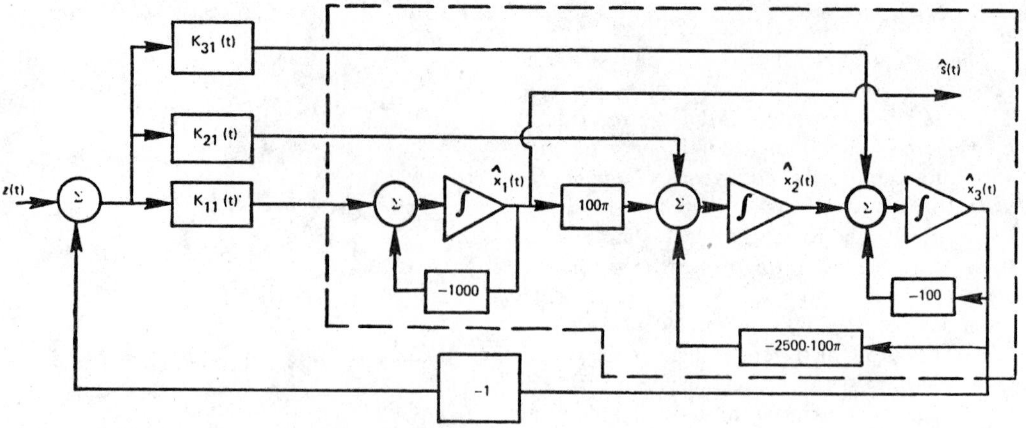

FIG. 24. Optimum Kalman filter for Example 2.

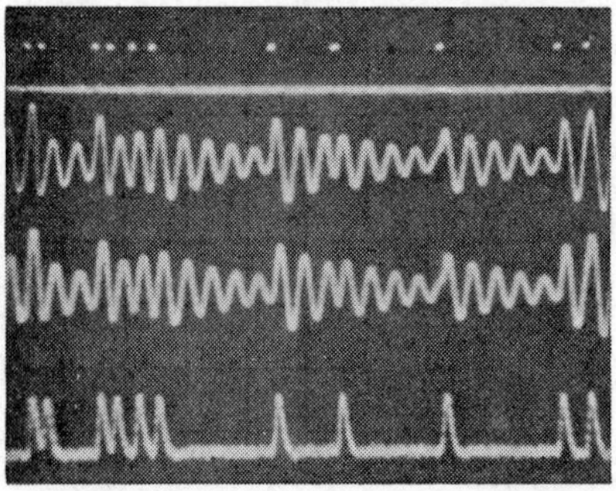

FIG. 25. Analog computer simulation of application of inverse Kalman filter to a seismic filter problem (Example 2). a) Desired impulses; b) superimposed wavelet; c) wavelets plus noise, $r=10^{-4}$; d) estimated impulse.

described by Beutler and Leneman (1967). The Kalman filter must then be derived for this new model.

CONCLUSIONS

A development of the application of the Kalman filter to the inverse filtering problem has been presented. This presentation included a tutorial development of system representation by state variables, and of the modeling of random processes as the output of linear systems excited by white noise. The Kalman filtering equations were derived with emphasis placed on the relationship between the Wiener and Kalman filters. Examples were presented to indicate the design steps necessary for structuring the Kalman filter. Simulation examples were presented to illustrate the applicability of the technique to the inverse filtering problem.

The examples presented were simplified for purposes of discussion. The well logging example considered the blurring or smearing of the mea-

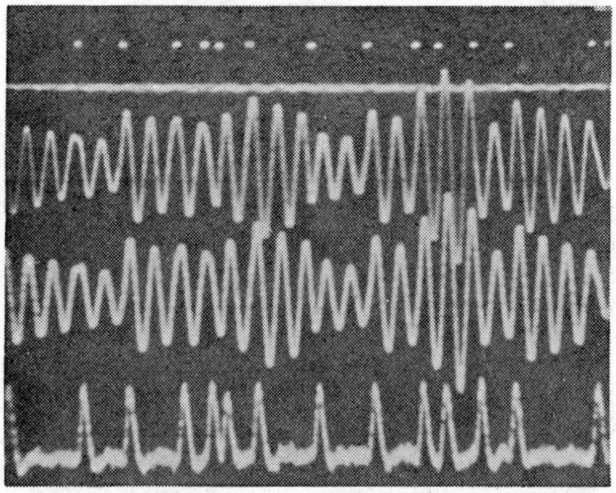

FIG. 26. Same as Figure 25, but with noise at $r=10^{-3}$. a) Desired impulse; b) superimposed wavelets; c) wavelets plus noise; d) estimated impulse.

surement system to be modeled by a physically realizable system; a nonphysically realizable system can be modeled by appropriately defining a causal system whose magnitude-squared frequency response is the noncausal system. Data processing for this case will require a magnetic tape recording since it is necessary to play the data backwards in time to realize the nonphysically realizable system. The seismic data processing example was simplified by assuming that all wavelets were of equal amplitude. The steps necessary for removing this assumption were indicated. The filter derived was a causal filter with real-time processing capability; this capability is valuable for "quick-look" analysis at the time of data acquisition. As noted in the simulation results, this filter produces a finite delay in the estimate of time-of-arrival. This delay can be eliminated by implementing the "smoothing" Kalman filter (Kailath and Frost, 1968). Generally, non-real-time processing is employed for "fine-grain" analysis; the digital Kalman filter appears to be quite applicable in this case.

This paper has considered only the analog processing problem. All of the main results can be carried over to the development of digital processing algorithms. A future paper will present these ideas in a tutorial development of the discrete Kalman filter and its application to the inverse filtering problem.

APPENDIX A

LINEAR SYSTEMS WITH RANDOM INPUTS

Suppose that the input to a linear system with impulse response $h(\tau)$ is a stationary random process $u(t)$. It is desired to determine the first and second order statistics of the output random process $x(t)$ in terms of the statistics of the input process $u(t)$. The output $x(t)$ is given by the convolution integral

$$x(t) = \int_{-\infty}^{\infty} h(\tau)u(t-\tau)d\tau. \qquad \text{(A-1)}$$

The expected value of $x(t)$ is found as

$$E[x(t)] = E\left[\int_{-\infty}^{\infty} h(\tau)u(t-\tau)d\tau\right]$$
$$= \int_{-\infty}^{\infty} h(\tau)E[u(t-\tau)]d\tau,$$

or since $u(t)$ is stationary

$$E[x(t)] = \int_{-\infty}^{\infty} h(\tau)E[u(t)]d\tau. \qquad \text{(A-2)}$$

The transfer function of the system is given by

$$H(j\omega) = \int_{-\infty}^{\infty} h(\tau)e^{-j\omega\tau}d\tau. \qquad \text{(A-3)}$$

Hence, Equation (A-2) can be rewritten as

$$E[x(t)] = H(o)E[u(t)]. \qquad \text{(A-4)}$$

Equation (A-4) states that the mean of the output random process is equal to the product of the dc gain of the system and the mean of the input random process.

The autocorrelation $\phi_x(t_1t_2)$ of the system output $x(t)$ is given by

$$\begin{aligned}\phi_x(t_1, t_2) &= E[x(t_1)x(t_2)] \\ &= E\left[\int_{-\infty}^{\infty} h(\alpha)u(t_1-\alpha)d\alpha \cdot \int_{-\infty}^{\infty} h(\beta)u(t_2-\beta)d\beta\right] \\ &= \int_{-\infty}^{\infty}\int_{-\infty}^{\infty} h(\alpha)h(\beta) \cdot \phi_u(t_1-\alpha, t_2-\beta)d\alpha d\beta.\end{aligned} \qquad \text{(A-5)}$$

Since the input process is stationary,

$$\phi_u(t_1-\alpha, t_2-\beta) = \phi_u(\tau+\beta-\alpha),$$

where $\tau = t_1 - t_2$. For this case, the output autocorrelation function becomes

$$\phi_x(\tau) = \int_{-\infty}^{\infty}\int_{-\infty}^{\infty} h(\alpha)h(\beta) \cdot \phi_u(\tau+\beta-\alpha)d\alpha d\beta. \qquad \text{(A-6)}$$

The power-spectral density $\Phi_x(\omega)$ of the output random process is defined as the Fourier transform of its autocorrelation function

$$\Phi_x(\omega) = \int_{-\infty}^{\infty} \phi_x(\tau)e^{-j\omega\tau}d\tau. \qquad \text{(A-7)}$$

Hence, from equation (A-6),

$$\Phi_x(\omega) = \int_{-\infty}^{\infty}\int_{-\infty}^{\infty}\int_{-\infty}^{\infty} [h(\alpha)h(\beta)d\alpha d\beta] \cdot \phi_u(\tau+\beta-\alpha)e^{-j\omega\tau}d\tau. \quad \text{(A-8)}$$

If $\gamma=\tau+\beta-\alpha$,

$$e^{-j\omega\tau} = e^{-j\omega\alpha}e^{+j\omega\beta}e^{-j\omega\gamma},$$

and equation (A-8) becomes

$$\Phi_x(\omega) = \int_{-\infty}^{\infty} h(\alpha)e^{-j\omega\alpha}d\alpha \int_{-\infty}^{\infty} h(\beta) \cdot e^{j\omega\beta}d\beta \int_{-\infty}^{\infty} \phi_u(\gamma)e^{-j\omega\gamma}d\gamma. \quad \text{(A-9)}$$

The integrals in equation (A-9) can be identified in terms of the system function and the input spectral density

$$\begin{aligned}\Phi_x(\omega) &= H(j\omega)H^*(j\omega)\Phi_u(\omega) \\ &= |H(j\omega)|^2\Phi_u(\omega).\end{aligned} \quad \text{(A-10)}$$

Thus, the power spectrum of a signal which is the output of a linear dynamic system with a random input is the power spectrum of the input multiplied by the magnitude squared of the system transfer function.

APPENDIX B

The optimum system gain matrix $\mathbf{K}(t)$ for the Kalman filter is derived in this appendix. From equation (24a) the Wiener-Hopf equation is [with $\phi_{zy}(t,\sigma)=\phi_{xy}(t,\sigma)$ and $\phi_z(\tau,\sigma)=\phi_{zy}(\tau,\sigma)+\phi_{zv}(\tau,\sigma)$]

$$\phi_{xy}(t,\sigma) - \int_{t_o}^{t} \mathbf{h}_o(t,\tau)\phi_{zy}(\tau,\sigma)d\tau = \mathbf{h}_o(t,\sigma)\mathbf{R}(\sigma). \quad \text{(B-1)}$$

For $\sigma=t$, equation (B-1) becomes

$$\phi_{xy}(t,t) - \int_{t_o}^{t} \mathbf{h}_o(t,\tau)\phi_{zy}(\tau,t)d\tau = \mathbf{K}(t)\mathbf{R}(t), \quad \text{(B-2)}$$

or

$$\phi_{\tilde{x}y}(t,t) = \mathbf{K}(t)\mathbf{R}(t), \quad \text{(B-3)}$$

where

$$\tilde{\mathbf{x}}(t) = \mathbf{x}(t) - \mathbf{x}(t).$$

Since

$$\mathbf{y}(t) = \mathbf{H}(t)\mathbf{x}(t)$$

and

$$\phi_{\tilde{x}x}(t,t) = \phi_{\tilde{x}1\tilde{x}}(t,t),$$

equation (B-3) reduces to

$$\phi_{\tilde{x}\tilde{x}}(t,t)\mathbf{H}^T(t) = \mathbf{K}(t)\mathbf{R}(t). \quad \text{(B-4)}$$

Let $\mathbf{P}(t) \triangleq \phi_{\tilde{x}\tilde{x}}(t,t)$. Then equation (B-4) becomes

$$\mathbf{K}(t) = \mathbf{P}(t)\mathbf{H}^T(t)\mathbf{R}^{-1}(t). \quad \text{(B-5)}$$

The matrix $\mathbf{P}(t)$ is the covariance matrix of the error term $\tilde{\mathbf{x}}(t)$. From the filter model (Figure 10), note that $\tilde{\mathbf{x}}(t)$ satisfies the equation

$$\frac{d\tilde{\mathbf{x}}(t)}{dt} = [\mathbf{F}(t) - \mathbf{K}(t)\mathbf{H}(t)]\tilde{\mathbf{x}}(t) + \mathbf{G}(t)\mathbf{u}(t) - \mathbf{K}(t)\mathbf{v}(t). \quad \text{(B-6)}$$

Consider the state transition matrix (Price, 1961) $\psi(t, t_o)$ which satisfies the equation

$$\frac{d}{dt}[\psi(t,t_o)] = [\mathbf{F}(t) - \mathbf{K}(t)\mathbf{H}(t)]\psi(t,t_o),$$

with

$$\psi(t,t_o) = \mathbf{I}.$$

Then

$$\tilde{\mathbf{x}}(t) = \psi(t,t_o)\tilde{\mathbf{x}}(t_o) + \int_{t_o}^{t} \psi(t,\tau)\{\mathbf{G}(\tau)\mathbf{u}(\tau) - \mathbf{K}(\tau)\mathbf{v}(\tau)\}d\tau. \quad \text{(B-7)}$$

Since $\mathbf{P}(t)=\phi_{\tilde{x}\tilde{x}}(t,t)$, multiplying equation (B-7) by its transpose and taking the expectation give

$$\mathbf{P}(t) = \psi(t,t_o)\mathbf{P}(t_o)\psi^T(t,t_o) + \int_{t_o}^{t} \psi(t,\tau) \cdot [\mathbf{G}(\tau)\mathbf{Q}(\tau)\mathbf{G}^T(\tau) + \mathbf{K}(\tau)\mathbf{R}(\tau)\mathbf{K}^T(\tau)] \cdot \psi^T(t,\tau)d\tau. \quad \text{(B-8)}$$

Differentiating this expression gives finally

$$\frac{d}{dt}\mathbf{P}(t) = \mathbf{F}(t)\mathbf{P}(t) + \mathbf{P}(t)\mathbf{F}^T(t) - \mathbf{P}(t)\mathbf{H}^T(t)\mathbf{R}^{-1}(t)\mathbf{H}(t)\mathbf{P}(t) + \mathbf{G}(t)\mathbf{Q}(t)\mathbf{G}^T(t), \quad \text{(B-9)}$$

with the initial condition

$$\mathbf{P}(t_o) = \phi_{\tilde{x}\tilde{x}}(t_o,t_o).$$

Equation (B-9) is a matrix Riccati equation for which considerable numerical solution techniques have been developed (Levin, 1959; Reid, 1959).

REFERENCES

Athans, M., and Tse, E., 1969, A direct derivation of the optimal linear filter using the maximum principle: IEEE Trans. on Automatic Control, v. AC-12, p. 690–698.

Beutler, F. J., and Leneman, O. A., 1967, The spectral analysis of impulse processes: International Symposium on Information Theory, San Remo, Italy, September 11–15.

Booton, R. C., 1952, An optimization theory for time-varying linear systems with nonstationary statistical inputs: Proc. IRE, v. 40, p. 977–981.

Brigham, E. O., Smith, H. W., Bostick, F. X., Jr., and Duesterhoeft, W. C., Jr., 1968, An iterative technique for determining inverse filters: IEEE Trans. on Geoscience Electronics, v. GE-6, p. 86–96.

Cuenod, M., and Sage, A. P., 1968, Comparison of some methods used for process identification: Automatica, v. 4, p. 235–269.

Davenport, W. B., Jr., and Root, W. L., 1958, Random signals and noise: New York, McGraw-Hill, 393 p.

Davis, M. C., 1963, Factoring the spectral matrix: IEEE Trans. on Automatic Control, v. AC-8, p. 196–225.

George, C. F., Smith, H. W., and Bostick, F. X., 1962, The application of inverse convolution techniques to improve signal response of recorded geophysical data: Proc. IRE, v. 50, p. 2313–2319.

Graybill, F. A., 1961, An introduction to linear statistical methods, v. 1: New York, McGraw-Hill, 463 p.

Gupta, S. C., 1966, Transforms and state variable methods in linear systems: New York, John Wiley and Sons, Ch. 8.

Hilborn, C. G., Jr., and Lainiotis, D. G., 1969, Optimal estimation in the presence of unknown parameters: IEEE Trans. on Systems Science and Cybernetics, v. SSC-5, p. 109–115.

Ho, Y. C., and Lee, R. C. K., 1964, A Bayesian approach to problems in stochastic estimation and control: IEEE Trans. on Automatic Control, v. AC-9, p. 333–334.

Kalman, R. E., 1960, A new approach to linear filtering and prediction problems, Trans. ASME, Series D, Journal of Basic Engineering, v. 82, p. 35–45.

——— and Bucy, R. S., 1961, New results in linear filtering and prediction theory: Trans. ASME, Series D, Journal of Basic Engineering, v. 83, p. 95–107.

Kailath, T., 1968, An innovations approach to the least-squares estimation-Part I: Linear filtering in additive white noise: IEEE Trans. on Automatic Control, v. AC-13, p. 646–655.

——— and Frost, P., 1968, An innovations approach to the least squares estimation-Part II: Linear smoothing in additive white noise: IEEE Trans. on Automatic Control, v. AC-13, p. 665–660.

Kelly, E. J., Reed, I. S., and Root, W. L., 1960, The detection of radar echoes in noise: J. SIAM, v. 8, p. 309–341.

Kunetz, G., 1962, Essai d' analyse de traces seismiques: Geophys. Prosp., v. 9, p. 317–341.

Levin, J. J., 1959, On the matrix Riccati equation: Proc. Amer. Math. Soc., v. 10, p. 519–524.

Magill, D. T., 1965, Optimal adaptive estimation of sampled stochastic signals: IEEE Trans. on Automatic Control, v. AC-10, p. 434–439.

Papoulis, A., 1965, Probability, random variables, and stochastic processes: New York, McGraw-Hill, 583 p.

Price, R., 1956, Optimum detection of random signals in noise with application to scatter multipath communication-I: Trans IRE-PGIT, v. IT-2, n. 4 p. 125–135.

Reid, W. T., 1959, Solutions of a Riccati matrix differential equation as functions of initial values: J. Math. Mech., v. 8, p. 897–905.

Rice, R. B., 1962, Inverse convolution filters: Geophysics, v. 27, p. 4–18.

Robinson, E. A., 1954, Predictive decomposition of seismic traces: Geophysics, v. 27, p. 767–778.

Sage, A. P., and Masters, G. W., 1967, Least-squares curve fitting and discrete optimum fitting: IEEE Trans. on Education, v. E-10, p. 29–36.

Sage, A. P., and Melsa, J. L., 1970, System identification: New York, Academic Press, in press.

Treitel, S., and Robinson, E. A., 1966, Seismic wave propagation in layered media in terms of communication theory: Geophysics, v. 31, p. 17–32.

Turin, George L., 1957, On the estimation in the presence of noise of the impulse response of a random linear filter: IRE Trans. on Information Theory, v. IT-3, p. 5–10.

Van Trees, H. L., 1968, Detection, estimation, and modulation theory, Part I: New York, John Wiley and Sons, Inc., 697 p.

Wiener, Norbert, 1949, Extrapolation, interpolation, and smoothing of stationary time series: New York, John Wiley and Sons, Inc., 163 p.

Wonham, W. M., 1965, Some applications of stochastic differential equations to optimal nonlinear filtering: SIAM J. Control, v. 2, p. 347–369.

Reprinted from Geophysics v. 39, no. 1, p. 1-13

GEOPHYSICS

A KALMAN FILTER APPROACH TO THE DECONVOLUTION OF SEISMIC SIGNALS†

NORMAN D. CRUMP*

It is common practice to model a reflection seismogram as a convolution of the reflectivity function of the earth and an energy waveform referred to as the seismic wavelet. The objective of the deconvolution technique described here is to extract the reflectivity function from the reflection seismogram.

The most common approach to deconvolution has been the design of inverse filters based on Wiener filter theory. Some of the disadvantages of the inverse filter approach may be overcome by using a state variable representation of the earth's reflectivity function and the seismic signal generating process. The problem is formulated in discrete state variable form to facilitate digital computer processing of digitized seismic signals. The discrete form of the Kalman filter is then used to generate an estimate of the reflectivity function. The principal advantages of this technique are its capability for handling continually time-varying models, its adaptability to a large class of models, its suitability for either single or multichannel processing, and its potentially high-resolution capabilities.

Examples based on both synthetic and field seismic data illustrate the feasibility of the method.

INTRODUCTION

It is common practice to model a reflection seismogram as a convolution of the reflectivity function of the earth and an energy waveform referred to as the seismic wavelet. The objective of deconvolution, as discussed in this paper, is to extract the reflectivity function from the reflection seismogram.

The most common approach to deconvolution has been the design of Wiener (1949) inverse filters which produce a desired output corresponding to each reflected wavelet. These inverse filters are generally based on the assumption of a white spectrum for the reflectivity function with a unit impulse function as the desired output and are derived from the estimated autocorrelation of the seismogram. The application of Wiener filter theory to deconvolution has been adequately described by Robinson (1954), Rice (1962), Robinson and Treitel (1967), Clarke (1968), Peacock and Treitel (1969), and many others.

The Wiener filter method is conventionally applied under the assumption of a time invariant system and stationary statistics. It is possible to extend the Wiener filter approach to include time-varying systems and nonstationary statistics (Booton, 1952), but solution of the resulting integral equations can be quite difficult to obtain or approximate.

Other methods for solving the deconvolution problem have also been proposed. One of these was the application by Ulrych (1971) of the technique of homomorphic filtering. This approach, based on a generalized superposition principal originally proposed by Oppenheim (1965a, 1965b), is a nonlinear processor which attempts to sep-

† Presented at the 42nd Annual International SEG Meeting, November 28, 1972, Anaheim, California. Manuscript received by the Editor December 12, 1972; revised manuscript received June 23, 1973.

* Atlantic Richfield Co., Dallas, Texas 75221.

arate the wavelet, and reflection coefficient function by suitable operations on a time domain function called the complex cepstrum. The homomorphic filtering deconvolution method is deterministic in nature, and does not require explicit assumptions regarding the phase of the wavelet or the character of the reflection coefficients. However, the method's success is dependent upon the degree of separability of the wavelet and reflection coefficient function in the complex cepstrum.

Another way of looking at the deconvolution problem was proposed by Bayless and Brigham (1970). They presented a tutorial explanation of state space modeling and Kalman (Kalman, 1960, and Kalman and Bucy, 1961) filtering and suggested seismic data processing applications including deconvolution. The treatment by Bayless and Brigham was limited to continuous signals only, and their Kalman filter solution to the deconvolution problem required an analytical description of the wavelet as well as a statistical model for the reflection coefficient function.

The discrete Kalman filter is the principal component of the deconvolution technique described below. The relatively new practice of modeling systems in state variable representation has resulted in widespread application of Kalman filter theory, since the Kalman filter is formulated to solve a problem described by first order matrix differential (or difference) equations. The expanding use of the Kalman filter in current technology has established the adaptability of state space techniques to a variety of problems. The primary advantages of state space modeling for geophysical applications are its suitability for handling a variety of physical and mathematical models, the ease with which it handles time-varying effects, and its compatibility with digital computer processing. In addition, state variable representation is capable of compactly modeling the multichannel data configuration so common in geophysical exploration methods.

This paper describes an application of state space modeling and Kalman filtering to the problem of deconvolution of seismic signals. Basic concepts of discrete state space representation and the discrete Kalman filter are first introduced. The seismic signal is then defined in state space form by the usual discrete convolution formula, and the discrete Kalman filter is applied by making use of a discrete seismic wavelet model and a stochastic model for the unknown reflection coefficient function. Examples based on both synthetic and field data are then presented to illustrate the results obtainable from the method.

BASIC CONCEPTS: STATE VARIABLES AND THE DISCRETE KALMAN FILTER

It is not the purpose of this paper to provide a detailed tutorial presentation of state space techniques and Kalman filtering. However, a few comments on the qualitative aspects of state space representation and references to the literature are given in order to provide a background for the development of the state variable signal model used later for application of the discrete Kalman filter equations. The Kalman filter equations and some comments on their nature are also given without derivation. This presentation emphasizes linear and sampled data systems and signals. The reader will have no difficulty in finding an abundance of articles and books on both state space techniques and Kalman filtering in the literature.

The state approach, or the use of state variables for system modeling, came into popularity over a decade ago. It offers a compact and mathematically tractable means for modeling linear or nonlinear multivariable time-varying systems describable by differential or difference equations. A mathematical model of a system can be considered to consist of a set of input variables, or forcing functions, a set of output variables, and a mathematical description of the effect of changes in the input variables on the output variables. A system can then be completely described by any set of variables which contain sufficient information about the system to determine both the output and future states of the system, assuming that future input variables and the equations describing the system behavior are known. A minimal set of such variables are called state variables and together constitute the state vector for the system. For a given system, the choice of state variables is not unique, but the number of state variables n, called the order of the system, is unique.

The dynamic equations describing a multivariable, linear, time-varying system may be written in the form:

$$\dot{\mathbf{x}}(t) = \mathbf{A}(t)\mathbf{x}(t) + \mathbf{B}(t)\mathbf{u}(t) \quad \text{and} \tag{1}$$

$$\mathbf{y}(t) = \mathbf{H}(t)\mathbf{x}(t) + \mathbf{D}(t)\mathbf{u}(t), \tag{2}$$

where $\mathbf{x}(t)$ is the state vector, $\mathbf{u}(t)$ is the input vector, $\mathbf{y}(t)$ is the output vector, and $\mathbf{A}(t)$, $\mathbf{B}(t)$, $\mathbf{H}(t)$, and $\mathbf{D}(t)$ are matrices having time-varying elements. The vector $\mathbf{x}(t)$ is of order n, while the vectors $\mathbf{u}(t)$ and $\mathbf{y}(t)$ may be of any order. The matrices $\mathbf{A}(t)$, $\mathbf{B}(t)$, $\mathbf{H}(t)$, and $\mathbf{D}(t)$ are $n \times n$, $n \times p$, $m \times n$, and $m \times p$ matrices, respectively, where p is the order of the input vector and m is the order of the output vector. Equation (1) is called the state equation of the system, and equation (2) is the output equation. The solution to equation (1) can be written in the form

$$\mathbf{x}(t) = \Phi(t, t_0)\mathbf{x}(t_0) + \int_{t_0}^{t} \Phi(t, \tau)\mathbf{B}(\tau)\mathbf{u}(\tau)d\tau, \quad (3)$$

where t_0 is some initial time reference and $\Phi(t, t_0)$ is called the state transition matrix for the system. Now assume that the input vector is the output of a zero-order sample-and-hold device, so that

$$\mathbf{u}(t) = \mathbf{u}[(k-1)T] \quad \text{for } (k-1)T \leq t \leq kT, \quad (4)$$

where T is the sampling period. Let $t_0 = (k-1)T$; the system solution at $t = kT$ results in the following difference equation (Kuo, 1970):

$$\mathbf{x}(kT) = \Phi[kT, (k-1)T]\mathbf{x}[(k-1)T] + \mathbf{G}[kT, (k-1)T]\mathbf{u}[(k-1)T], \quad (5)$$

where

$$\mathbf{G}[kT, (k-1)T] = \int_{(k-1)T}^{kT} \Phi[kT, \tau]\mathbf{B}(\tau)d\tau. \quad (6)$$

If the output $\mathbf{y}(t)$ is sampled in the same manner as the input then the output equation becomes

$$\mathbf{y}(kT) = \mathbf{H}(kT)\mathbf{x}(kT) + \mathbf{D}(kT)\mathbf{u}(kT). \quad (7)$$

Equations (5) and (7) constitute a discrete representation of the linear time-varying system defined by equations (1) and (2). It is this discrete representation of the dynamic equation of a system which will be treated in this paper.

A simple example will illustrate both the concept of state variable modeling and a procedure for obtaining a discrete time model for a continuous system having uniformly sampled output. In this example the resulting model is time-invariant, but time-varying system models are derived using generally the same techniques. More complete treatments of state variable techniques can be found in the literature [Kuo (1970), Zadeh and Desoer (1963), and DeRusso et al (1965)].

A mathematical model of the dynamic voltage response of a seismometer consists of the following differential equation (Silverman, 1939):

$$\ddot{e}(t) + C/M\dot{e}(t) + S/Me(t) = K\dot{v}(t), \quad (8)$$

where

- $e(t)$ is the seismometer output voltage,
- $v(t)$ is the amplitude of the impressed velocity,
- C is the coefficient of damping,
- M is the suspended mass,
- S is the spring constant, and
- K is a constant of the system.

With $\dot{v}(t)$ as the input, or forcing function, and $e(t)$ as the output of the system, equation (8) can be represented by the block diagram of Figure 1. With a block diagram of this form, it is common

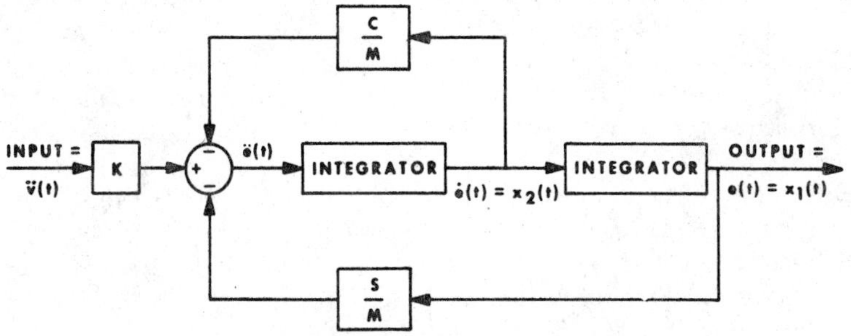

$$\ddot{e}(t) = -\frac{C}{M}\dot{e}(t) - \frac{S}{M}e(t) + K\dot{v}(t)$$

FIG. 1. Model for seismometer voltage response.

practice to choose the outputs of integrators as a suitable set of state variables. Therefore, we define state variables $x_1(t)$, $x_2(t)$ by

$$x_1(t) = e(t) \quad \text{and} \qquad x_2(t) = \dot{e}(t). \tag{9}$$

Then, from equations (8) and (9),

$$\dot{x}_1(t) = x_2(t) \quad \text{and} \qquad \dot{x}_2(t) = -S/Mx_1(t) - C/Mx_2(t) + K\ddot{v}(t). \tag{10}$$

Then the state variable equations for the continuous seismometer system become

$$\dot{\mathbf{x}}(t) = \mathbf{A}\mathbf{x}(t) + \mathbf{B}\ddot{v}(t) \quad \text{and} \qquad y(t) = \mathbf{H}\mathbf{x}(t), \tag{11}$$

where we define

$$\mathbf{A} = \begin{bmatrix} 0 & 1 \\ \dfrac{-S}{M} & -\dfrac{C}{M} \end{bmatrix}, \tag{12}$$

$$\Phi(T) = e^{AT} = \{\mathcal{L}^{-1}[(sI - \mathbf{A})^{-1}]\}\Big|_{t=T}$$

$$= \mathcal{L}^{-1}\left\{\begin{bmatrix} \left(\dfrac{s + \dfrac{C}{M}}{s^2 + s\dfrac{C}{M} + \dfrac{S}{M}}\right) & \left(\dfrac{1}{s^2 + s\dfrac{C}{M} + \dfrac{S}{M}}\right) \\ \left(\dfrac{-S/M}{s^2 + s\dfrac{C}{M} + \dfrac{S}{M}}\right) & \left(\dfrac{s}{s^2 + s\dfrac{C}{M} + \dfrac{S}{M}}\right) \end{bmatrix}\right\}\Bigg|_{t=T} \tag{20}$$

$$\mathbf{B} = \begin{bmatrix} 0 \\ K \end{bmatrix}, \tag{13}$$

$$\mathbf{H} = [1 \quad 0], \quad \text{and} \tag{14}$$

$$y(t) = e(t). \tag{15}$$

Note that, since S, M, K, and C are constants, the system is not time-varying and the matrices **A**, **B**, and **H** are written without time dependency. It remains to transform equations (11) into the general discrete form of equations (5) and (7). Assuming the output $y(t)$ is sampled at a constant rate with sample period T and $\ddot{v}(t)$ varies slowly relative to the sampling rate, it can be shown by using equation (3) that equations (11) can be written in the discrete form:

$$\mathbf{x}(kT) = \Phi(T)\mathbf{x}[(k-1)T] + G(T)\ddot{v}[(k-1)T] \quad \text{and} \qquad y(kT) = \mathbf{H}\mathbf{x}(kT), \tag{16}$$

where

$$\Phi(T) = e^{AT}, \tag{17}$$

$$G(T) = \int_0^T \Phi(T - \tau)\mathbf{B}d\tau, \tag{18}$$

and e^{AT} is defined by

$$e^{At} = \sum_{i=0}^{\infty} A^i \frac{t^i}{i!}. \tag{19}$$

The function e^{AT} may be evaluated in several ways, but probably the simplest method for this example is to evaluate $\Phi(T)$ by obtaining the inverse Laplace transform:

The seismometer system represented in the discrete state variable form of equation (16) can now be described by the block diagram of Figure 2. Double lines are used in the diagram to denote vector quantities. Equation (16) represents a discrete state variable model which is amenable to analysis by state space techniques.

The discrete state variable representation of a system is the proper form for application of the discrete Kalman filter. The estimation problem solved by the discrete Kalman filter is that of computing a "best" estimate, $\hat{\mathbf{x}}(kT)$, of the state vector $\mathbf{x}(kT)$, based on the measurements $\mathbf{y}(T)$, $\mathbf{y}(2T), \cdots, \mathbf{y}(jT)$. The estimation process is

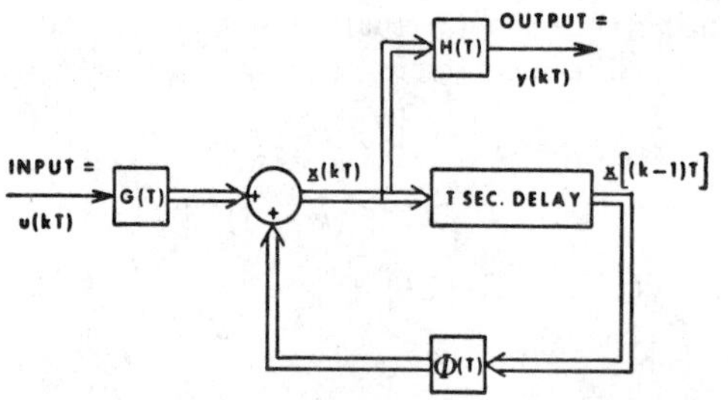

FIG. 2. Block diagram of seismometer voltage response in discrete state variable form.

called smoothing if $j>k$, filtering if $j=k$, and predicting if $j<k$. The case treated in this paper is that of filtering, but all three cases result in similar solutions. The criterion used to define what is meant by "best" is the minimization of a loss function which may belong to a large class of functions if the random quantities are Gaussian. Even without Gaussian statistics, the Kalman filter is the filter which minimizes the mean square error, $E\{[\hat{\mathbf{x}}(kT)-\mathbf{x}(kT)]^T[\hat{\mathbf{x}}(kT)-\mathbf{x}(kT)]\}$, under the constraint that the filter output be a linear combination of the measurements.

The Kalman filter equations have been derived in many ways using a variety of loss functions. Some of the possible approaches which have been published include least-squares estimation, variational techniques, invariant imbedding, and orthogonal projection. Probably one of the most appealing methods is that of minimizing the conditional expectation using the assumption of Gaussian statistics. This is the Bayesian estimation approach, and consists of defining $\hat{\mathbf{x}}(kT)$ as the value of the conditional expectation $E[\mathbf{x}(kT)|\mathbf{y}(T), \mathbf{y}(2T), \cdots, \mathbf{y}(kT)]$. The discrete Kalman filter equations apply to a system described in the discrete state variable representation

$$\begin{aligned}\mathbf{x}(k) &= \Phi(k, k-1)\mathbf{x}(k-1) \\ &\quad + G(k, k-1)\mathbf{u}(k-1) \qquad (21) \\ \mathbf{y}(k) &= \mathbf{H}(k)\mathbf{x}(k) + \mathbf{v}(k).\end{aligned}$$

Variables in these equations are the same as those given in equations (5) and (7), except for the addition of the output noise vector $\mathbf{v}(k)$ and the absence of the matrix $\mathbf{D}(k)$. For simplicity of notation, it is assumed that $T=1$ in the remainder of this paper.

It is assumed that $\Phi(k, k-1)$, $G(k, k-1)$, and $H(k)$ are known for all values of k to be encountered. The input vector $\mathbf{u}(k-1)$ is a random sequence with

$$E[\mathbf{u}(k)] = \mathbf{0} \qquad \text{for all } k$$

and

$$E[\mathbf{u}(k)\mathbf{u}^T(j)] = Q(k)\delta_{kj},$$

where δ_{kj} is the Kronecker delta and $\mathbf{Q}(k)$ is a known diagonal matrix. The vector $\mathbf{v}(k)$ is also a random sequence with

$$\mathbf{E}[\mathbf{v}(k)] = \mathbf{0} \qquad \text{for all } k$$

and

$$\mathbf{E}[\mathbf{v}(k)\mathbf{v}^T(j)] = \mathbf{V}(k)\delta_{kj},$$

where $\mathbf{V}(k)$ is a known diagonal matrix. The processes $\mathbf{u}(k)$ and $\mathbf{v}(k)$ are assumed to be uncorrelated, so that

$$\mathbf{E}[\mathbf{u}(k)\mathbf{v}^T(j)] = \mathbf{0} \qquad \text{for all } k, j.$$

An initial state $\mathbf{x}(0)$ is considered to be a vector random variable with

$$\begin{aligned}E[\mathbf{x}(0)] &= \mathbf{x}_0 \\ E[\mathbf{x}(0)\mathbf{x}^T(0)] &= \mathbf{M}_0,\end{aligned}$$

where $\mathbf{x}_0$ and M_0 are known. In addition, $\mathbf{x}_0$ is assumed to be uncorrelated with $\mathbf{u}(k)$ and $\mathbf{v}(k)$ for all k.

The application of the Kalman filter to the measurements $\mathbf{y}(k)$ defined in equations (21) consists of a sequence of operations performed for each sample point. These operations which comprise the Kalman filter are summarized as follows:

(1) Prior to the first sample ($k=1$), initialize the filter by defining

$$\begin{aligned}\mathbf{P}(0) &= \mathbf{M}_0 \\ \hat{\mathbf{x}}(0) &= \mathbf{x}_0\end{aligned} \qquad (22)$$

where $\mathbf{P}(k)$ is an $n\times n$ matrix discussed below.

(2) Compute the matrix $\mathbf{P}'(k)$ by

$$\begin{aligned}\mathbf{P}'(k) = {}&\Phi(k, k-1)\mathbf{P}(k-1)\Phi^T(k, k-1) \\ &+\mathbf{G}(k, k-1)\mathbf{Q}(k-1)\mathbf{G}^T(k, k-1).\end{aligned} \qquad (23)$$

(3) Compute the matrix $\mathbf{K}(k)$ by

$$\begin{aligned}\mathbf{K}(k) = \mathbf{P}'(k)\mathbf{H}^T(k)\{&\mathbf{H}(k)\mathbf{P}'(k)\mathbf{H}^T(k) \\ &+ \mathbf{V}(k)\}^{-1}.\end{aligned} \qquad (24)$$

(4) Compute the estimated state vector, $\hat{\mathbf{x}}(k)$, by

$$\hat{\mathbf{x}}(k) = \hat{\mathbf{x}}'(k) + \mathbf{K}(k)[\mathbf{y}(k) - \mathbf{y}'(k)], \quad (25)$$

where

$$\hat{\mathbf{x}}'(k) = \Phi(k, k-1)\hat{\mathbf{x}}(k-1) \quad (26)$$

and

$$\mathbf{y}'(k) = \mathbf{H}(k)\hat{\mathbf{x}}'(k). \quad (27)$$

(5) Compute the matrix $\mathbf{P}(k)$ by

$$\mathbf{P}(k) = \mathbf{P}'(k) - \mathbf{K}(k)\mathbf{H}(k)\mathbf{P}'(k). \quad (28)$$

(6) Take the next sample, incrementing k by one, and return to step (2).

The equations given above represent the discrete Kalman filter equations for the system described by equations (21). Kalman filter theory can also be applied to more general forms of state space models (such as inclusion of deterministic forcing functions) with more relaxed assumptions regarding the random processes. Such extensions are abundant in the literature. Very good treatments of the discrete Kalman filter are given by Sorenson (1966), Meditch (1969), and Kuo (1970).

In the Kalman filter equations shown above, some of the terms have a significance which should be noted. First, it can be shown that $\mathbf{P}(k)$ is the error covariance matrix of the estimate vector $\hat{\mathbf{x}}(k)$, so that

$$\mathbf{P}(k) = \mathbf{E}\{[\hat{\mathbf{x}}(k) - \mathbf{x}(k)][\hat{\mathbf{x}}(k) - \mathbf{x}(k)]^T\}. \quad (29)$$

Therefore, the Kalman filter equations are minimizing the trace of $\mathbf{P}(k)$. Also, the vector $\hat{\mathbf{x}}'(k)$ is the best "projected" estimate of $\mathbf{x}(k)$, that is, it is the best estimate for $\mathbf{x}(k)$ based on data through $\mathbf{y}(k-1)$. The matrix $\mathbf{P}'(k)$ is then the error covariance matrix of the projected estimate $\hat{\mathbf{x}}'(k)$. The matrix $\mathbf{K}(k)$ is generally called the "gain matrix," because it determines the extent to which the estimate $\hat{\mathbf{x}}(k)$ is influenced by the most recent data sample $\mathbf{y}(k)$. Finally, $\mathbf{y}'(k)$ is the best estimate of the measurement vector $\mathbf{y}(k)$ based on samples through $\mathbf{y}(k-1)$.

The nature of the discrete Kalman filter equations can best be illustrated by the block diagram shown in Figure 3. This diagram includes the signal and measurement model of equations (21) as well as the discrete Kalman filter. It can be seen that the Kalman filter diagram contains the same delay feedback path as the signal model. In addition, it is clear that the estimate $\hat{\mathbf{x}}(k)$ is obtained as a weighted sum of two quantities: (1) the best predicted estimate, $\hat{\mathbf{x}}'(k)$, and (2) the latest measurement sample, $\mathbf{y}(k)$. The relative weighting of these quantities is determined by the system dynamics model through the definition of the matrices Φ, $\mathbf{H}$, and $\mathbf{G}$, and by the statistical variances defined by $\mathbf{Q}$ and $\mathbf{V}$.

The recursive nature of the Kalman filter makes it readily adaptable to digital computer implementation. Most of the difficulty in applying the filter lies in obtaining the mathematical models describing the dynamics and statistics of the system. In the following section a model suitable for implementing a deconvolution filter by the discrete Kalman filter is proposed.

THE DECONVOLUTION PROBLEM AND A KALMAN FILTER SOLUTION

A common mathematical model of a seismic signal consists of the sum of a convolution and noise:

$$y(t) = \int w(t, \tau)r(\tau)d\tau + v(t), \quad (30)$$

where $y(t)$ is the seismic trace, $w(t, \tau)$ is the time-varying seismic wavelet, $r(t)$ is the reflectivity function of the earth, and $v(t)$ is noise.

Typically, the seismic signal is sampled at uniformly spaced discrete instants of time in order to use digital computer processing, so the discrete seismic signal model corresponding to equation (30) can be written

$$y(k) = \sum_{\gamma=N}^{k} [w(k, \gamma)r(\gamma)] + v(k)$$

$$k = 1, 2, 3, \cdots,$$

where k is the sample number and the index N is defined by $N=1$ for $k<L$ and $N=k-L+1$ for $k \geq L$, where L is the wavelet length in number of sample periods. The indices on the summation result from the assumptions that the wavelet and reflectivity function are zero for negative time (causality) and that the time-varying wavelet is of finite time duration. The term $w(k, \gamma)$ represents the $(k-\gamma+1)$th sample of the wavelet model corresponding to a time of k sample periods. A change in the variable of summation, defined by $i=k-\gamma+1$, transforms the above summation to the form

$$y(k)=\sum_{i=1}^{J}[w(k,k-i+1)r(k-i+1)]+v(k) \qquad k=1,2,3,\cdots, \tag{31}$$

where $J=k$ for $k<L$ and $J=L$ for $k\geq L$. In general, more than one seismic trace is available for processing, so the input to the deconvolution filter is assumed to be a vector $\mathbf{y}(k)$ consisting of components of the form

$$y_j(k)=\sum_{i=1}^{J}[H_{ji}(k)r(k-i+1)]+v_j(k), \qquad j=1,2,\cdots,M \quad k=1,2,\cdots,K. \tag{32}$$

This equation is based on the assumption that all traces are generated by the same reflectivity function, but by possibly different wavelets. Each of the M traces consists of K samples. The $M\times L$ matrix $\mathbf{H}(k)$ is a matrix of time-varying wavelet values, with the jth row containing the L samples of the wavelet which generates the jth trace. That is,

$$H_{ji}(k)=w_j(k,k-i+1).$$

Inspection of equations (21) and (32) reveals that (32) has the same form as the Kalman filter input measurement vector if the state vector $\mathbf{x}(k)$ is defined by

$$\mathbf{x}(k)\triangleq\begin{bmatrix} r(k)\\ r(k-1)\\ r(k-2)\\ \vdots\\ r(k-L+1)\end{bmatrix}. \tag{33}$$

If we define $\mathbf{y}(k)\triangleq[y_1(k)y_2(k)\cdots y_M(k)]^T$ and $\mathbf{v}(k)\triangleq[v_1(k)v_2(k)\cdots v_M(k)]^T$, equation (32) becomes

$$\mathbf{y}(k)=\mathbf{H}(k)\mathbf{x}(k)+\mathbf{v}(k), \tag{34}$$

which is exactly the form of the second of equations (21). Note that the state vector defined by equation (33) transforms the discrete convolution of equation (31) into a product form. In general, the discrete output vector representation of equation (21) does not imply a discrete convolution; it is the particular state vector definition given above which transforms the seismic signal model into a form suitable for direct application of the discrete Kalman filter.

The procedure described above defines a state vector as the set of reflection coefficients which influence each particular sample of the seismic signal vector. However, to apply the discrete Kalman filter one must have a model in the form of the first of equations (21) which describes the recursive process generating the state vector. The approach taken here is to assume a relatively general relationship for the reflection coefficients defining the state vector. This relationship is assumed to be of the form:

$$r(k)=\sum_{i=1}^{L}[b_i(k-1)r(k-1)] + u(k-1), \tag{35}$$

where $u(k-1)$ is a white random process. This equation simply means that the reflection coefficient at any sample time is assumed to be the sum of a random quantity and a weighted sum of the reflection coefficients at the L previous sample times. Values must be assigned to the $b_i(k-1)$ coefficients in order to implement the discrete Kalman filter. Procedures for defining the $b_i(k-1)$ coefficients will be discussed below and in the examples.

Inspection of equations (33) and (35) shows that the state variable model can now be written as

$$\mathbf{x}(k)=\Phi(k,k-1)\mathbf{x}(k-1) + \mathbf{g}u(k-1), \tag{36}$$

where

$$\Phi(k,k-1)=\begin{bmatrix} b_1(k-1) & b_2(k-1) & b_3(k-1) & \cdots & b_{L-1}(k-1) & b_L(k-1)\\ 1 & 0 & 0 & \cdots & 0 & 0\\ 0 & 1 & 0 & \cdots & 0 & 0\\ 0 & 0 & 1 & \cdots & 0 & 0\\ \vdots & & & & & \vdots\\ 0 & 0 & 0 & \cdots & 1 & 0\end{bmatrix} \tag{37}$$

and

$$\mathbf{g} = [1 \quad 0 \quad 0 \cdots 0]^T. \tag{38}$$

Equations (34) and (36) constitute the state variable model for the seismic deconvolution problem. In order to implement the discrete Kalman filter, it is necessary to first determine numerical quantities for the variables associated with the model. These variables are summarized as follows:

(1) $\mathbf{b}(k)$. This vector, possibly time-varying, is a measure of the manner in which the reflectivity function is expected to change with depth. The problem of optimally estimating $\mathbf{b}(k)$ is unsolved, but the examples presented later in this paper illustrate that useful results may be obtained by assuming a reasonable form for $\mathbf{b}(k)$. A method for estimating $\mathbf{b}(k)$ might be based on regression analysis of reflection coefficients obtained from well log data in an area of interest, but this has not been pursued by the author. Of course, assigning $\mathbf{b}(k) \equiv \mathbf{0}$ is equivalent to assuming that the reflection coefficients form a white random sequence.

(2) $\mathbf{Q}(k) = \mathbf{E}[u^2(k)]$. This unitless quantity is simply the variance of the random component associated with each reflection coefficient as shown in equation (35), and is therefore a measure of the expected fluctuations in reflection coefficient values. In practice, its value also influences the frequency response of the Kalman filter, with the high-frequency cutoff of the filter being roughly proportional to $\mathbf{Q}(k)$.

(3) $\mathbf{H}(k)$. This matrix of time-varying wavelet models may be estimated by existing techniques, such as combining estimated autocorrelation functions with suitable phase assumptions.

(4) $\mathbf{V}(k)$. The diagonal elements of this diagonal matrix represent the random measurement noise variance associated with each input seismic trace. In practice, the choice of $\mathbf{V}(k)$ also influences the frequency response of the Kalman filter, with the high-frequency cutoff of the filter being roughly inversely proportional to $V(k)$.

(5) $\hat{\mathbf{x}}(0)$, $\mathbf{P}(0)$. These parameters must be chosen to initialize the Kalman filter. In practice, choosing $\hat{\mathbf{x}}(0) = \mathbf{0}$ and $\mathbf{P}(0) = \gamma I$ has been found to be adequate, where I is the identity matrix and γ is a scaler quantity representing an expected upper bound on the variance of the estimated reflection coefficients.

After the model is completely defined by the variables summarized above, the seismic data are sequentially processed by the discrete Kalman filter equations summarized by equations (23) through (28) and illustrated in Figure 3.

For each sample point k, the discrete Kalman filter produces a vector estimate $\hat{\mathbf{x}}(k)$ consisting of estimates of L reflection coefficients. Inspection of equation (33) shows that this procedure results in a set of overlapping "moving windows" of estimated reflection coefficients, and the reflection coefficient for a particular sample point is actually estimated L times (except for those near the end of the traces). Experience has shown that

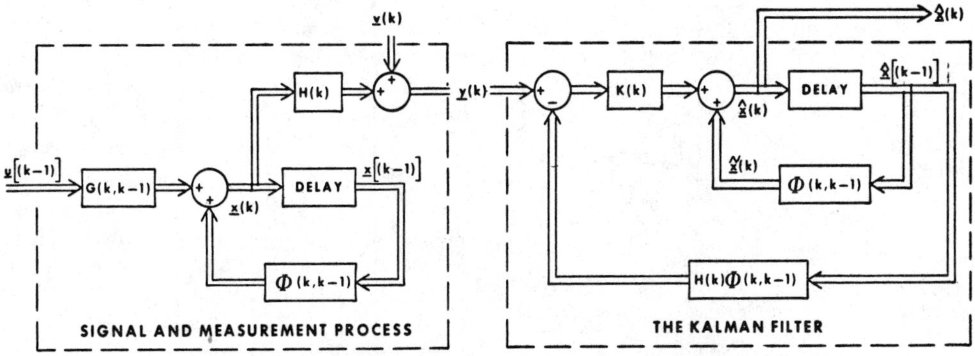

FIG. 3. Block diagram of signal model and discrete Kalman filter.

the choice of which estimate to use has very little effect on the shape of the resulting estimated reflectivity function.

The discrete Kalman filter solution described above represents a flexible and relatively general solution to the seismic signal deconvolution problem. The filter equations are in a form suitable for direct processing by digital computer. In addition, any of the parameters specified by the user may be time-varying to accommodate a changing model. Experience with the technique has shown that estimates of the parameters required for implementation of the filter can be obtained in relatively straightforward ways, and in fact the results are not extremely sensitive to the choices for values of some of the parameters. Examples described in the following section illustrate procedures for defining the necessary parameters. These examples include results from both synthetic and field data.

ILLUSTRATIVE EXAMPLES

Example 1

To illustrate the potential capability of the discrete Kalman filter deconvolution technique, it is useful to consider the results obtainable with an accurate signal model. For this purpose, a synthetic trace was generated by convolving a known wavelet and reflectivity function. Figure 4 shows the wavelet and reflectivity function which were combined by discrete convolution to produce a noise-free synthetic trace. The wavelet used was the result of applying a smoothing filter to a minimum phase wavelet obtained by computing the estimated autocorrelation of an actual seismic trace used in example 2 below. The wavelet consists of 40 sampled values with a 2 msec sample period. The reflectivity function was obtained from a digitized sonic well log. It consists of 463 sampled values, also with a 2 msec sample period. The trace amplitudes were normalized to 1.0.

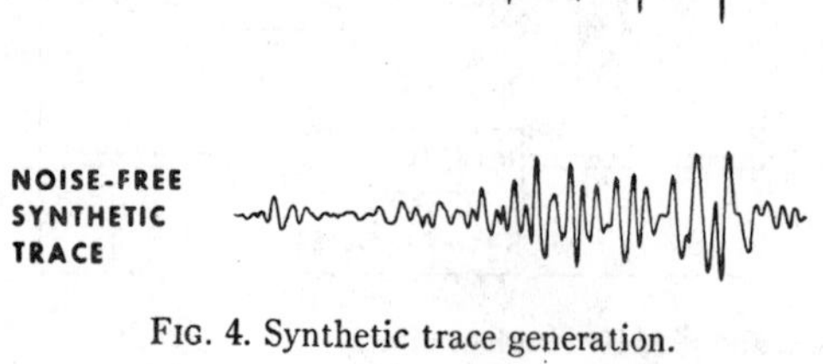

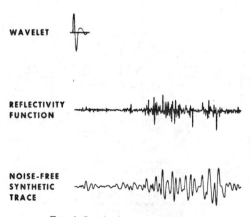

FIG. 4. Synthetic trace generation.

As described in the previous section, numerical values for certain parameters must be chosen in order to process the data. For this example, the values for these parameters were chosen as follows:

Order of the state vector $= n = 40$

Order of the measurement process $= M = 1$

$\mathbf{b}(k) \equiv 1.0/40.0[1 \quad 1 \cdots 1]$

$\mathbf{Q}(k) \equiv 0.002$

$\mathbf{H}(k) \equiv \{\text{row vector of actual wavelet values}\}$

$\mathbf{V}(k) \equiv 0.0$

$\hat{\mathbf{x}}(0) = \mathbf{0}$

$\mathbf{P}(0) = 1.0\mathbf{I}$

The value chosen for $P(0)$ and $\hat{\mathbf{x}}(0)$ were based on experience with the filter behavior. The value of $V(k)$ models the absence of measurement noise. $Q(k)$ was chosen to be the value of the sample variance of a set of reflection coefficients computed from the sonic well log. The choice for $\mathbf{b}(k)$ results from the assumption that each of the previous L reflection coefficients contributes equally to the value of a reflection coefficient at a particular sample time. While this model is merely assumed and has not been related to physical principles, it has been found to give useful results in the deconvolution of synthetic traces and field data from a variety of areas. Previous discussion in this paper has pointed out other possibilities for arriving at values for $\mathbf{b}(k)$, but in general the problem of estimating $\mathbf{b}(k)$ is one requiring significant additional research.

Figure 5(a) shows the discrete Kalman filter estimate of the reflectivity function used to generate the synthetic trace. The synthetic trace and the wavelet and reflectivity function used to generate it are repeated for comparison. The deconvolution appears to be quite accurate in this case, despite the fact that the reflectivity function model defined by choices for $\mathbf{b}(k)$ and $\mathbf{Q}(k)$ were rather arbitrary. Figure 5(b) shows the result obtained when $\mathbf{V}(k)$ was changed from 0.0 to 0.001 and all other parameters were left unchanged. The

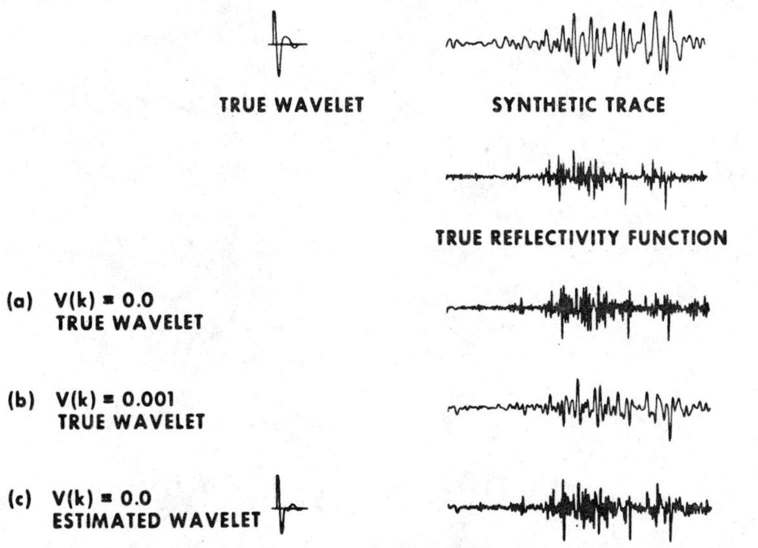

FIG. 5. Kalman filter estimates from noise-free synthetic trace.

decrease in high-frequency content in the estimated reflectivity function is obvious, although its character remains correct. Similar changes in frequency content occur with changes in $\mathbf{Q}(k)$, with the high-frequency content being reduced as $\mathbf{Q}(k)$ decreases. Thus, $\mathbf{V}(k)$ and $\mathbf{Q}(k)$ can be used to "tune" the filter frequency response.

To illustrate the more realistic case in which there is a lack of information regarding the generating wavelet, the noise-free synthetic trace was processed by a discrete Kalman filter in which the row vector $\mathbf{H}(k)$ contained a wavelet model computed from the noise-free synthetic trace. The wavelet model was defined as the minimum phase waveform having an autocorrelation function identical to that computed for the synthetic trace. The resulting wavelet model and estimated reflectivity function are shown in Figure 5(c). All parameters other than $\mathbf{H}(k)$ are identical to those for Figure 5(a). The estimated reflectivity function remains accurate despite some differences in the shapes of the true and estimated wavelets.

To illustrate results obtainable from noisy data, white random Gaussian noise was added to the synthetic trace of Figure 4. The noisy data were then processed by the discrete Kalman filter having the same parameter values as used for Figure 5(a), except for $\mathbf{V}(k)$ which was chosen to be $\mathbf{V}(k) \equiv 0.0001$. The results are shown in Figure 6 for various values of signal-to-noise ratio S/N. In this case, S/N is defined by

$$\mathrm{S/N} = \frac{\dfrac{1}{K}\sum_{k=1}^{K} s^2(k)}{\sigma_N^2},$$

where $s(k)$ is the noise-free synthetic trace, K is the number of trace samples, and σ_N^2 is the variance of the added noise. Normally one would define $\mathbf{V}(k) = \sigma_N^2$, but in these examples no attempt was made to adjust $\mathbf{V}(k)$ to account for the amount of noise present. In practice, one would choose $\mathbf{V}(k)$ to correctly model such noise energy present in the data.

Example 2

This example illustrates the application of the discrete Kalman filter to the deconvolution of seismic data generated by dynamite. The data used for this example came from the same geographical area as the sonic well log which was used for generating the synthetic data for example 1.

Figure 7 illustrates the deconvolution of a single trace from a CDP stacked section. This trace was the source data for computation of the wavelet used for the synthetic trace of example 1,

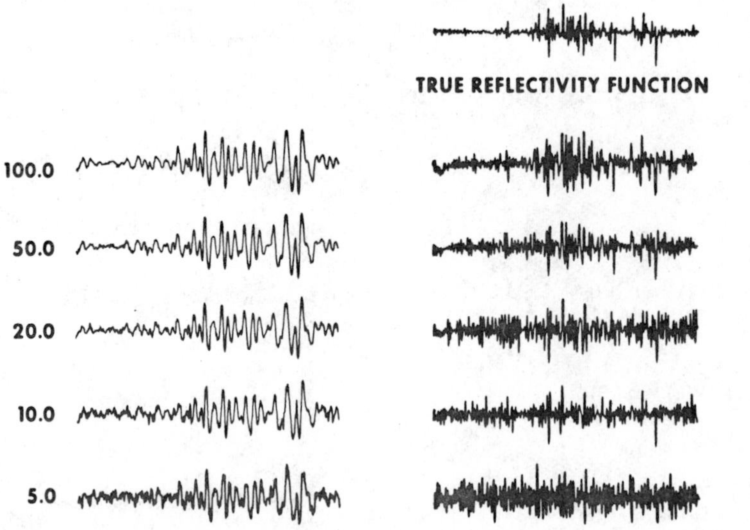

FIG. 6. Kalman filter estimates from noisy synthetic traces.

and the same wavelet was used for defining $\mathbf{H}(k)$ for the results of Figure 7. $\mathbf{V}(k)$ was chosen to be 0.0001 for all of the field data results described here. Other parameters were chosen to be identical to those used for example 1. Figure 7 reveals that the events on the seismic trace do appear to be more sharply defined on the deconvolved trace. However, a more interpretable result is obtained by processing a seismic section.

The stacked section shown in Figure 8a was deconvolved by the discrete Kalman filter with the result shown in Figure 8b. The parameters used to obtain the deconvolved section were the same as for Figure 7, with even the wavelet model

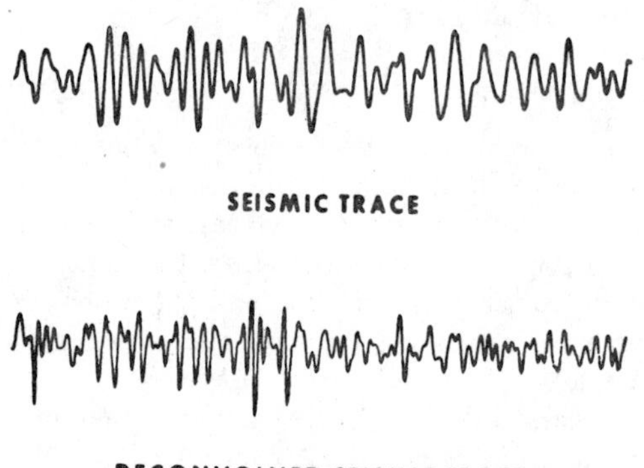

FIG. 7. Kalman filter deconvolution of seismic data.

remaining unchanged. The trace shown in Figure 7 is the fifth trace from the right end of the section shown in Figure 8a.

The deconvolved section in Figure 8b appears to contain fewer correlating events and higher frequency content than the original seismic section. These are the expected characteristics of a deconvolved section. An example of the effectiveness of the deconvolution is illustrated by the change in the left half of the section between .67 and .70 seconds. The seismic section shows several equally spaced events correlating in this time interval, indicating a possible ringing problem. The deconvolved section clearly defines the occurrence of one major reflection. In this example, as in the case of example 1, no attempt was made to optimize the values selected for parameters in the signal model, and all parameters were assumed to be constant rather than time-varying. These assumptions fail to take advantage of the flexibility inherent in the proposed technique, but the object here was to illustrate the behavior of a very simple configuration of the filter in terms of its response to typical seismic signals.

For comparison purposes, Figure 8c shows the result of deconvolving the same section using a conventional deconvolution procedure based on Wiener filter theory. Comparison of the two de-

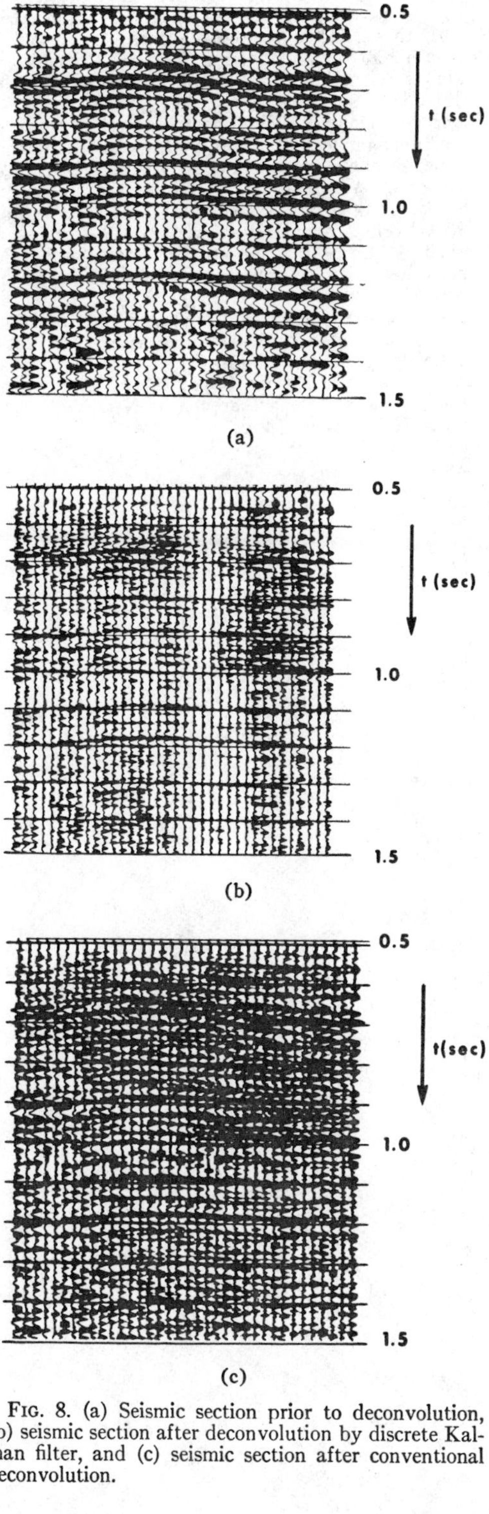

FIG. 8. (a) Seismic section prior to deconvolution, (b) seismic section after deconvolution by discrete Kalman filter, and (c) seismic section after conventional deconvolution.

convolved sections indicates that the discrete Kalman filter yields results which are similar to those obtained by the conventional method. It is therefore reasonable to conclude that superior results may be anticipated from using the discrete Kalman filter based on more accurate time-varying models.

CONCLUSIONS

A new approach to the problem of deconvolution of seismic signals has been proposed in this paper. The method involves application of the discrete Kalman filter equations to a state space model of the seismic signal generating process. Examples were presented to illustrate the procedure involved in implementing the filter as well as some of the characteristics and potential capability of the method.

In the experience of the writer, it appears that the deconvolution results obtained by using this method with model parameters based on the simplest possible assumptions are at least as good as those obtained by conventional methods. The real potential of the method lies in its capability for handling continually time-varying values for all parameters which are required. The accurate estimation of these time-varying parameters is the major problem in taking full advantage of the potential of the method. Further investigation of this problem is needed.

It is important also to note one feature of the method which was not illustrated here. That feature is its ability to process any number of traces simultaneously to arrive at a single estimated reflectivity function. This feature has not been thoroughly explored by the writer, but it is felt that the multichannel input configuration may be a very useful one for processing noisy traces obtained in an area without dip or lateral discontinuities.

REFERENCES

Bayless, J. W., and Brigham, E. D., 1970, Application of the Kalman filter to continuous signal restoration: Geophysics, v. 35, p. 2–23.

Booton, R. C., 1952, An optimization theory for time-varying linear systems with nonstationary statistical inputs: Proc. IRE, v. 40, p. 977–981.

Clarke, G. K. C., 1968, Time-varying deconvolution filters: Geophysics, v. 33, p. 936–944.

DeRusso, P. M., Roy, R. J., and Close, C. M., 1965, State variables for engineers: New York, John Wiley and Sons, Inc.

Kalman, R. E., 1960, A new approach to linear filtering

and prediction problems: Trans. ASME, Series D, J. of Basic Engr., v. 82, p. 35–45.
Kalman, R. E., and Bucy, R. S., 1961, New results in linear filtering and prediction theory: Trans. ASME, Series D, J. of Basic Engr., v. 83, p. 95–107.
Kuo, B. C., 1970, Discrete-data control systems: New Jersey, Prentice-Hall, Inc.
Meditch, J. S., 1969, Stochastic optimal linear estimation and control: New York, McGraw-Hill Book Co., Inc.
Oppenheim, A. V., 1965a, Superposition in a class of non-linear systems: Res. Lab. of Electr. MIT, Tech. Rep. 432.
——— 1965b, Optimum homomorphic filters: Res. Lab. of Electr. MIT, Quart. Prog. Rep. 77, p. 248–260.
Peacock, K. L., and Treitel, S., 1969, Predictive deconvolution: Theory and practice: Geophysics, v. 34, p. 155–169.
Rice, R. B., 1962, Inverse convolution filters: Geophysics, v. 27, p. 4–18.
Robinson, E. A., 1954, Predictive decomposition of time series with application to seismic exploration: PhD thesis, MIT, Cambridge, Mass.
Robinson, E. A., and Treitel, S., 1967, Principles of digital Wiener filtering: Geophys. Prosp., v. 15, p. 311–333.
Silverman, Daniel, 1939, The frequency response of electromagnetically damped dynamic and reluctance type seismometers: Geophysics, v. 4, p. 53–68.
Sorenson, N. W., 1966, Kalman filtering techniques, *in* Advances in control systems: C. T. Leondes, editor, New York, Academic Press, v. 3, p. 219–292.
Ulrych, T. J., 1971, Application of homomorphic deconvolution to seismology: Geophysics, v. 36, p. 650–660.
Wiener, Norbert, 1949, Extrapolation, interpolation, and smoothing of stationary time series: New York, John Wiley and Sons, Inc.
Zadeh, L. A., and Desoer, C. A., 1963, Linear system theory, the state space approach: New York, McGraw-Hill Book Co., Inc.

VIII. HOMOMORPHIC DECONVOLUTION

Previous sections have been devoted to the design and application of linear filters. Further, the model of seismic data has assumed minimum-phase wavelets for which minimum-phase inverses exist that can be used to recover the series of impulses. The paper by T. J. Ulrych discusses the application of homomorphic deconvolution to separating signals which have been combined through convolution. This nonlinear technique does not require the assumption of a minimum-phase wavelet. Instead, the model for seismic data is that of a source wavelet (possibly mixed phase) convolved with a minimum-phase impulse train. Since convolution in the time domain corresponds to multiplication in the frequency domain, the logarithm of the spectrum contains the convolutional components as a sum, as does the inverse transform of the log spectrum. Ulrych presents several examples of using the inverse transform (called the complex cepstrum) to separate the convolutional components. Ideally, the terms of the complex cepstrum near $n = 0$ (both positive and negative n) are contributed by the wavelet, while terms due to the impulse train occur for larger well-separated n values (positive n only). The difficulties of "phase unwrapping" are noted, as is the advantage of exponential weighting (namely, that by exponential weighting a mixed phase impulse sequence can be made into a minimum-phase sequence). Ulrych also points out the importance of time shifting so as to remove the linear phase component of the spectrum before computing the complex cepstrum. Finally, very encouraging results are presented for a noiseless synthetic simulating teleseismic events.

A number of applications of homomorphic deconvolution are presented in a two-part paper by P. L. Stoffa, Peter Buhl, and G. M. Bryan. The first part is devoted primarily to model studies. An instructive table of time functions, their inverses, and the complex cepstrum of each is included. As the table shows, the complex cepstrum of even a two-point time function is of infinite extent. The authors show how the resulting aliasing problem (which arises from the use of discrete Fourier transforms) can be relieved through careful exponential weighting. Also presented is an interesting approach to computing a continuous phase curve that uses the derivative of the phase with respect to frequency.

Part II emphasizes application to data from the Argentine continental shelf. It is shown that the primary contaminant in the data is the air gun bubble pulse (rather than water-column reverberation). By judicious exponential weighting, the authors develop a time series where they believe that the maximum-phase component is associated solely with the source. Based on an examination of the data, the authors also associate the first 44 minimum-phase terms of the complex cepstrum with the source. The resulting deconvolution (after pulse shaping and programmed gain control) is quite impressive, and compares very favorably with minimum-phase deconvolution. In homomorphic deconvolution, there is unfortunately a difficulty in assigning the proper absolute time origin to the output, since the time shift in "ramp removal" cannot be associated solely with the reflector series. Buhl, Stoffa, and Bryan solve this problem by aligning the processed traces on the large initial arrival.

Reprinted from Geophysics v. 36, no. 4, p. 650-660

APPLICATION OF HOMOMORPHIC DECONVOLUTION TO SEISMOLOGY†

T. J. ULRYCH*

Homomorphic systems (Oppenheim, 1965a and 1965b) are a class of nonlinear systems which satisfy a generalized principle of superposition. Such systems are particularly useful in separating signals which have been combined through convolution. This paper deals with the application of homomorphic deconvolution to the recovery of the seismic wavelet from a time series formed by the convolution of this wavelet with an impulse train. The unique point about this approach is that it does not require the usual assumptions of a minimum-phase wavelet and a random distribution of impulses.

INTRODUCTION

A seismic record is often represented as the convolution of a wavelet with the impulse response of the transmission path. The process of separating these two components of the convolution is termed deconvolution and finds considerable application in seismology. Deconvolution as commonly performed by means of inverse filtering or optimum zero-lag Wiener filtering (Rice, 1962; Robinson and Treitel, 1967) suffers from the limitation that either the shape of the seismic wavelet to be removed must be known or the assumption that the wavelet is minimum-phase (Robinson, 1966; Ulrych and Lasserre, 1966) must be made. Although predictive deconvolution with a prediction distance greater than unity (Peacock and Treitel, 1969) does not require this assumption, that process is effective in removing repetitive events and would not be applied to the deconvolution problem considered in this paper. Peacock and Treitel (1969) have pointed out that the unit prediction deconvolution filter is equivalent within a scale factor to the least squares, zero-lag inverse filter and does require the minimum-phase assumption.

It is the purpose of this paper to investigate the application to seismic signals of a nonlinear deconvolution technique originally proposed by Oppenheim (1965a, 1965b) as an application of the theory of generalized superposition. This method, known as homomorphic deconvolution and recently applied by Schafer (1969) and by Oppenheim et al (1968) to the problem of echo removal, offers the considerable advantage that no prior assumption about the nature of the seismic wavelet or the impulse response of the transmission path need be made. The seismic wavelet which is recovered by homomorphic deconvolution is of importance in studies of elastic wave attenuation and dispersion in the earth (Futterman, 1962; Strick, 1970).

THEORY

The theoretical basis of homomorphic deconvolution has been dealt with fully in the publications cited above; hence, only a summary of some relevant points will be presented here.

The analytical convenience of linear filtering is a result primarily of the principle of superposition which linear systems satisfy. Oppenheim (1965a) has suggested the generalization of this principle to a certain class of nonlinear systems for which linear filtering in the usual sense is not meaningful. We will restrict ourselves in this paper to the consideration of signals which have been combined by means of convolution and specifically to

† Manuscript received by the Editor September 23, 1970; revised manuscript received February 22, 1971.

* University of British Columbia, Vancouver, Canada.

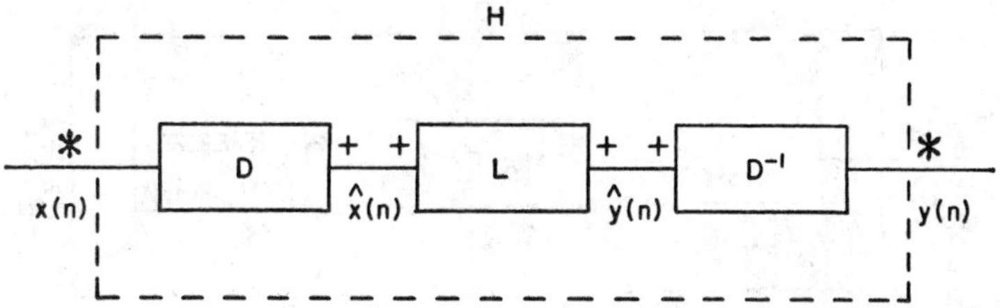

FIG. 1. Canonic representation for homomorphic deconvolution.

the case where one of the signals is an impulse train.

Consider the transformation defined by

$$y = T[x].$$

If T is a linear system, it satisfies the superposition relationship defined by

$$T[ax_1 + bx_2] = aT[x_1] + bT[x_2], \quad (1)$$

where a and b are constants. We can see from equation (1) why linear systems are particularly convenient for separating signals which have been additively combined.

If we wish to generalize the notion expressed by equation (1) to a signal resulting from the convolution of components $x = x_1 * x_2$, we look for a system with transformation H such that

$$H[{}^{(a)}x_1 * {}^{(b)}x_2] = {}^{(a)}H[x_1] * {}^{(b)}H[x_2], \quad (2)$$

where ${}^{(a)}$ denotes scalar multiplication. The formalism for this representation, which has been studied in detail by Oppenheim (1965a), lies in interpreting the system inputs and outputs as vector spaces where vector addition is defined as convolution. ${}^{(a)}$ denotes a rule for combining inputs with scalars in the convolutional space. For example, if a is an integer, ${}^{(a)}x$ signifies the convolution of x with itself a times. That a system H, known as a homomorphic system and defined by equation (2), does satisfy the postulates of vector addition has been formally shown by Oppenheim (1965a).

The advantage in the representation of a system by equation (2) is that systems of this class have been shown by Oppenheim (1965a) to have the canonic representation shown in Figure 1.

The system D, referred to as the characteristic system, is a homomorphic system with transformation from a convolutional space to an additive space. It is defined by the relationship

$$D[{}^{(a)}x_1 * {}^{(b)}x_2] = aD[x_1] + bD[x_2]. \quad (3)$$

The system L is a familiar linear system, and the system D^{-1}, the inverse of D, is a homomorphic system and performs the transformation from an additive space back to the output convolutional space.

The great flexibility of the arrangement shown in Figure 1 is that, once the characteristic system D has been determined, it remains fixed for all deconvolution problems; and the process reduces to one of linear filtering.

The characteristic system D

Since the realization of homomorphic deconvolution is performed on sampled data, it will be convenient in this discussion to deal with z transforms of input sequences $x(n)$ as defined in equation (4).

$$X(z) = \sum_{n=-\infty}^{\infty} x(n)z^{-n}. \quad (4)$$

The function of the characteristic system D is to transform signals from a convolutional space to an additive space. Since the z transform of two convolved signals is equal to the product of their z transforms, the required transformation may be accomplished as illustrated in Figure 2. The relevant steps are as follows:

The input sequence $x(n)$ is z transformed to obtain $X(z)$.

The logarithm of $X(z)$, $\hat{X}(z)$, separates the product into a sum.

The inverse z transform of $\hat{X}(z)$ gives $\hat{x}(n)$, the input to the linear system L in Figure 1.

$\hat{x}(n)$ has been termed the complex cepstrum. The word cepstrum originates from the work of Bogert et al (1962), who termed the power spectrum of the logarithm of the power spectrum the cepstrum by direct paraphrase of the word spec-

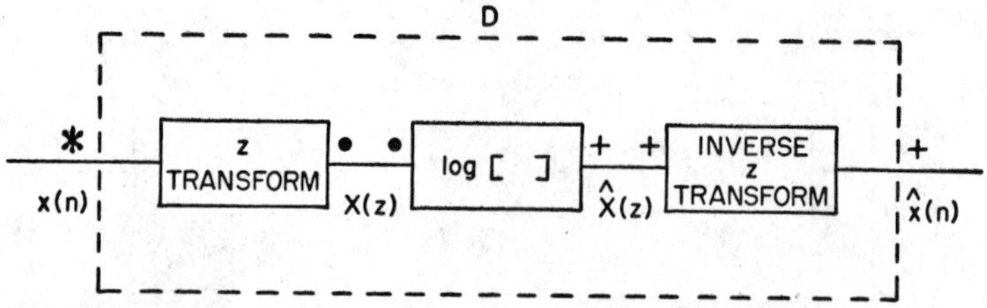

FIG. 2. Canonic representation of the characteristic system D.

trum. Complex has been added by Oppenheim et al (1968) to emphasize the point that the cepstrum has been computed utilizing both the amplitude and phase information contained in $x(n)$.

The complex cepstrum is defined by

$$\hat{x}(n) = \frac{1}{2\pi j}\int_C \log[X(z)]z^{n-1}dz, \qquad (5)$$

where C is a circular contour specified by

$$z = e^{\sigma+j\omega}, \quad -\pi < \omega < \pi.$$

The complex cepstrum $\hat{x}(n)$

There are some important considerations in regard to the actual computation of the complex cepstrum that result from the requirement the transformation D be unique. However, before discussing these, let us consider a simple example which will illustrate the concept of the complex cepstrum as applied to homomorphic deconvolution.

Consider a sampled signal $x(n)$, which is composed of a wavelet $s(n)$ and an echo n_o samples later.

$$x(n) = s(n) + as(n - n_o),$$

where a is a constant.

If $\delta(n)$ is the Dirac delta function,

$$x(n) = s(n) * [\delta(n) + a\delta(n - n_o)]. \qquad (6)$$

Evaluating the z transform of $x(n)$ on the unit circle gives

$$X(e^{j\omega}) = S(e^{j\omega})[1 + ae^{-j\omega n_o}],$$

and taking the logarithm of $X(e^{j\omega})$, we find

$$\hat{X}(e^{j\omega}) = \log S(e^{j\omega}) + \log[1 + ae^{-j\omega n_o}]. \qquad (7)$$

Therefore, since as seen from equation (7) the echo is represented in the log spectrum as an additive periodic component, the complex cepstrum $\hat{x}(n)$ will exhibit a peak at the echo delay.

Figure 3 shows the seismic wavelet which has been used in the examples which follow. It is approximately 35 samples long; the only point to note about this wavelet is that it is not a minimum-phase function. Figure 4(a) represents $x(n)$ of equation (6). The peak-to-peak separation of the wavelet and its echo is 12 samples; $a=0.9$. The complex cepstrum corresponding to this input sequence is shown in Figure 4(b).

This example illustrates several important points. In the first place, we can see that the contribution of the wavelet $s(n)$ in equation (6) to the complex cepstrum is concentrated near $n=0$, whereas the contribution of $[\delta(n)+a\delta(n-n_o)]$

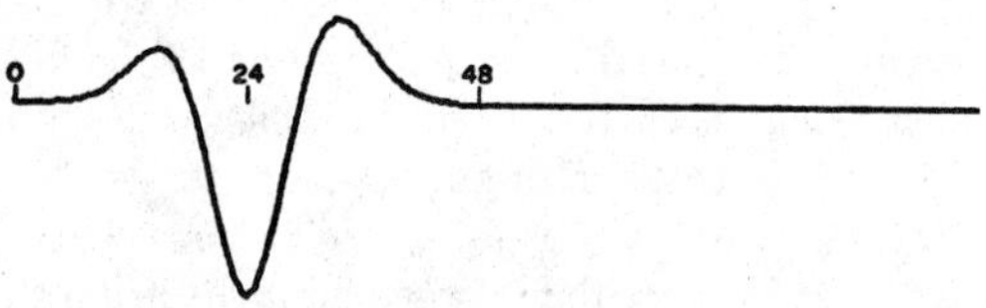

FIG. 3. A mixed-phase seismic wavelet.

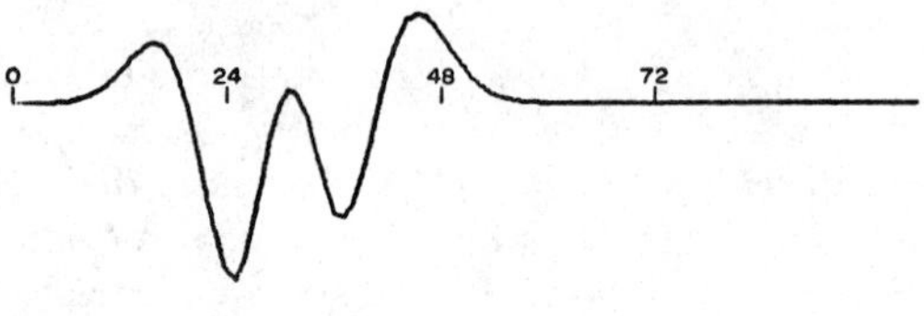

FIG. 4(a). A simple echo.

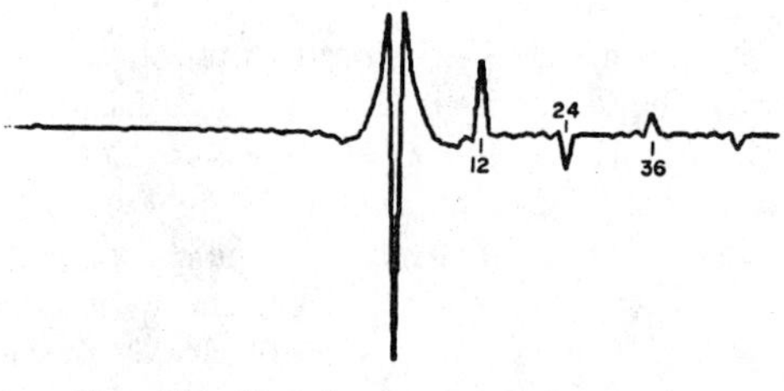

FIG. 4(b). Complex cepstrum of Fig. 4(a).

occurs for higher n values and is well separated from the wavelet component. We will see that, for seismic signals which are considered to be the result of the convolution of a wavelet with an impulse train, this observation is generally true. Secondly, it is clear from Figure 4(b) that, depending on the length of the delay, the echo may be simply removed by means of linear filtering.

Computational considerations.—There are four important considerations in the computation of the complex cepstrum:

I. The complex cepstrum involves the computation of the inverse z transform of a logarithmic function. We may write this function in terms of its magnitude and argument as

$$\log [X(z)] = \log | X(z) | + j \, arg [X(z)]. \quad (8)$$

In turn,

$$arg[X(z)] = ARG[X(z) \pm j2\pi k], \quad (9)$$

where

$$k = 0, 1, 2 \cdots \quad \text{and}$$
$$-\pi < ARG[X(z)] < \pi.$$

It can be seen from equations (8) and (9) that the complex logarithm is multivalued. Further, since $ARG[X(z)]$ is a discontinuous function, $\log[X(z)]$ will not, in general, be an analytic function. The homomorphic system D can, however, be unique only if in equation (5) $\log[X(z)]$ is analytic in an annular region containing the contour C. We can achieve this condition by computing $ARG[X(z)]$ and then unwrapping it to produce $arg[X(z)]$, which is continuous, providing that the phase curve has been sampled at sufficiently small intervals.

II. The requirement that the complex cepstrum $\hat{x}(n)$ be real for real input sequences $x(n)$ implies that (a) $arg[X(z)]$ is an odd function of ω and periodic in ω with a period of 2π and (b) $\log|X(z)|$ is an even function of ω and periodic in ω with a period of 2π.

III. The input sequences with which we will be concerned are always restricted by the condition

$$x(n) = 0, \qquad M > n > 0.$$

Such sequences are characterized by z transforms which have no singularities in the z plane and which are polynomials in z^{-1}. The z transform may be represented by equation (10) (Schafer, 1969).

$$X(z) = A_{z^{-M_0}} \prod_{k=1}^{m_i} (1 - a_k z^{-1}) \prod_{k=1}^{m_o} (1 - b_k z), \quad (10)$$
$$| a_k | < 1, \; | b_k | < 1,$$

where the a_k's are the m_i zeroes inside the unit circle, the b_k's are the m_o zeroes outside the unit circle, and z^{-m_o} represents a shift of the input sequence.

Let us consider the effect of the term z^{-m_o} on the computation of the complex cepstrum. Since the contour C in equation (5) is specified by $z = e^{\sigma + j\omega}$, equation (5) may be written as

$$\hat{x}(n) = \frac{1}{2\pi} \int_{-\pi}^{\pi} \log [X(e^{\sigma+j\omega})] e^{\sigma n} e^{j\omega n} d\omega. \quad (11)$$

Let the contribution of z^{-m_o} to $\hat{x}(n)$ be $\hat{\phi}(n)$. Then if for convenience we choose the contour C in equation (5) to be the unit circle,

$$\hat{\phi}(n) = -\frac{1}{2\pi} \int_{-\pi}^{\pi} m_0 \log [e^{j\omega}] e^{j\omega n} d\omega,$$

which integrates to

$$\hat{\phi}(n) = -\frac{m_o \cos \pi n}{n} \cdot \quad (12)$$

In a practical case, $X(z)$ may have many zeroes outside the unit circle. m_0 is thus large and the effect of $\hat{\phi}(n)$ is to swamp the interesting information contained in the complex cepstrum. It is, therefore, of importance to remove the linear phase component prior to the computation of $\hat{x}(n)$. This is easily achieved and, in fact, the computation of the unwrapped phase curve is combined with this operation. Since the removal of the linear phase component is equivalent to a shift of the output sequence, the final step in the deconvolution process is to reposition the output sequence.

IV. The input sequences of interest are always of finite length. $X(z)$ thus has a region of convergence which includes the unit circle. This allows $X(z)$ and the inverse transform to be evaluated for $z = e^{j\omega}$; therefore, implementation of the z transform and its inverse is by means of the discrete Fourier transform pair computed using

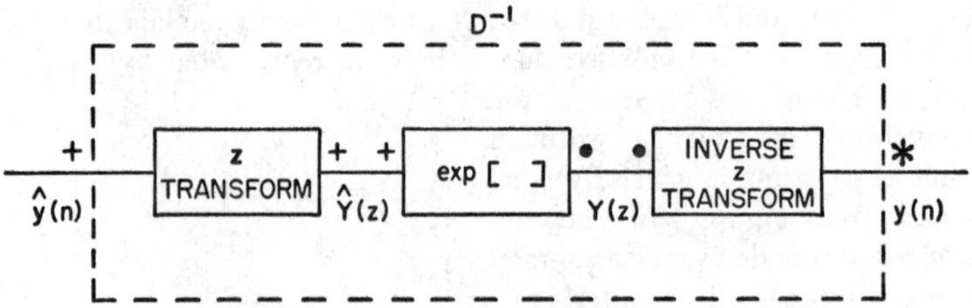

FIG. 5. Canonic representation of the inverse system D^{-1}.

the Cooley-Tukey fast Fourier transform algorithm (Cooley and Tukey, 1965).

The System D^{-1}

The transformation performed by the homomorphic system D^{-1} is from an additive to a convolutional space. It is the inverse system to D and its canonic representation is shown in Figure 5.

Linear filtering of the complex cepstrum

As we have seen in the example illustrated in Figure 4, the complex cepstrum contains the additive contributions of the wavelet and of the impulse response of the transmission channel. In this particular case, these contributions may be very easily separated by means of ideal low-pass and high-pass filters. The results of low-pass and high-pass filtering of the cepstrum of Figure 4(b) with a cutoff length equal to 12 samples, followed by processing with the system D^{-1}, are shown in Figure 6(a) and Figure 6(b). A comparison of the input and deconvolved wavelets shows that even in the case of this rather crude filtering, the shape of the wavelet is essentially preserved. A "comb" filter designed to have zero response at the echo peaks would, of course, provide a more exact component separation. For convenience, however, simple filtering has been used throughout the paper.

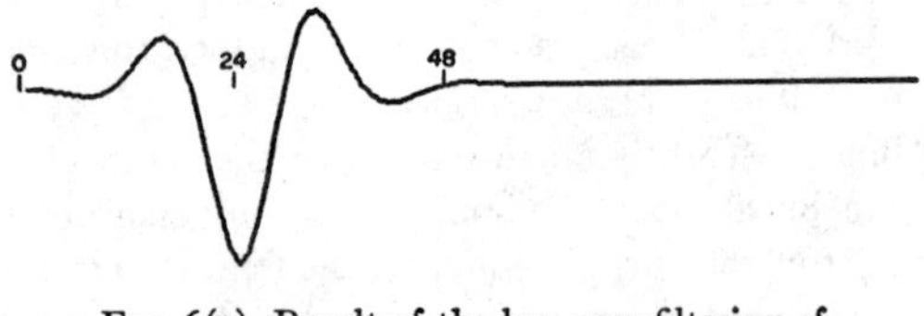

FIG. 6(a). Result of the low-pass filtering of the complex cepstrum of Fig. 4(b).

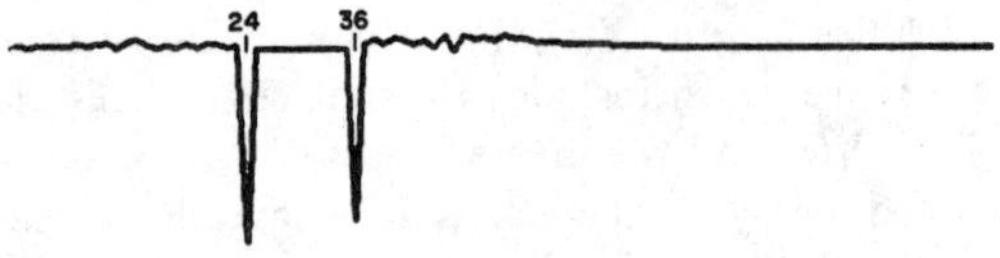

FIG. 6(b). Result of the high-pass filtering of the complex cepstrum of Fig. 4(b).

Examples of complex cepstra

The example of Figure 4 illustrates a very simple case and one to which linear filtering may be very easily applied. In seismology, we are generally concerned with an input sequence $x(n) = s(n)*i(n)$ where $i(n)$ may be a very complex impulse series. It turns out that minimum-phase impulse sequences play a very important part in homomorphic deconvolution. Let us therefore briefly consider the properties of the complex cepstra of such sequences. [Some analytical expressions for complex cepstra of impulse trains are given by Schafer (1969).]

Minimum-phase sequences.—Let $i(n)$ be a finite-length sequence such that $I(z)$, the z-transform of $i(n)$, has all its zeroes inside the unit circle. Then,

$$I(z) = A \prod_{k=1}^{m_i} (1 - a_k z^{-1}), \text{ where } |a_k| < 1,$$

and taking the contour C to be the unit circle, we have

$$\hat{\imath}(n) = \frac{1}{2\pi} \cdot \int_{-\pi}^{\pi} \log \left\{ A \prod_{k=1}^{m_i} (1 - a_k z^{-1}) \right\} e^{j\omega n} d\omega, \tag{13}$$

where $z = e^{j\omega}$.

Each of the terms inside the integral sign may be expanded in a Laurent series about $z=0$ to give

$$\log (1 - a_k z^{-1}) = - \sum_{n=1}^{\infty} \frac{a_k^n}{n} z^{-n} \tag{14}$$

for $|z| > |a_k|$.

We see from equations (13) and (14) that $\hat{\imath}(n)=0$ for $n<0$. This property of minimum-phase sequences is illustrated in Figure 7. Figure 7(a) shows the input trace, which is the convolution of the wavelet of Figure 3 with a minimum-phase impulse train composed of four unequally spaced impulses. The complex cepstrum, Figure 7(b), shows clearly the contributions of the two components which may be easily separated by linear filtering as shown in Figures 7(c) and 7(d). As a general comment, we can say that when the impulse train is minimum-phase, the two convolved components of the seismic trace are separated in the complex cepstrum by an amount equal to the separation of the first two impulses.

Mixed-phase sequences.—For finite-length mixed-phase sequences, $I(z)$ may be expressed by equation (10) and, since the Laurent series about $z=0$ for terms $\log(1-b_k z)$ is

$$\sum_{n=-\infty}^{-1} \frac{b_k^{-n}}{n} z^{-n} \quad \text{for } |z| < |b_k^{-1}|,$$

we can see that $\hat{\imath}(n)$ will have values in the range $-\infty<n<\infty$. The complex cepstra of sequences of unequally spaced impulses which are mixed phase are generally very complicated and the two components of the convolution are no longer separated. Figure 8(a) shows a mixed-phase input trace obtained by convolving the wavelet of Figure 3 with a mixed-phase series of five unequally spaced impulses. The separation of the first and second impulse is the same as in Figure 7. The complex cepstrum of this trace, shown in Figure 8(b), illustrates the complexity which may arise. Low-pass filtering to recover the seismic wavelet, with the same filter used to obtain the wavelet in Figure 7(c), produces a wavelet, Figure 8(c), which bears little resemblance to the

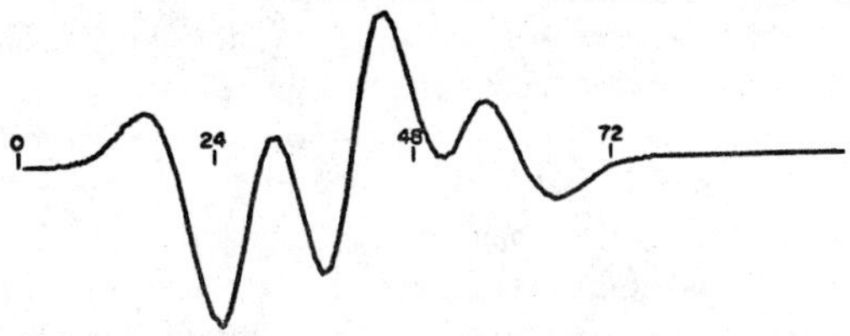

FIG. 7(a). A minimum-phase input sequence.

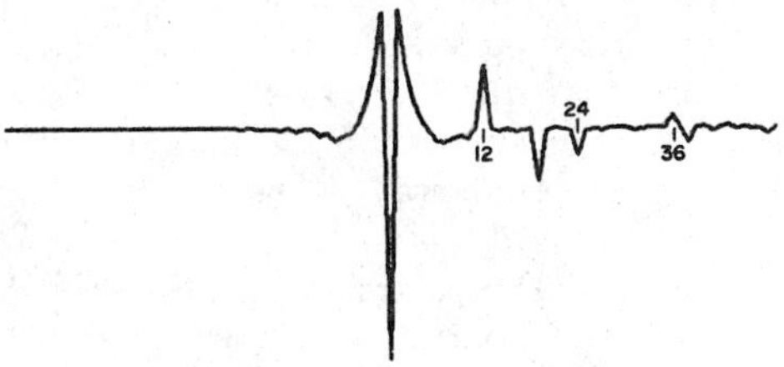

FIG. 7(b). Complex cepstrum of Figure 7(a).

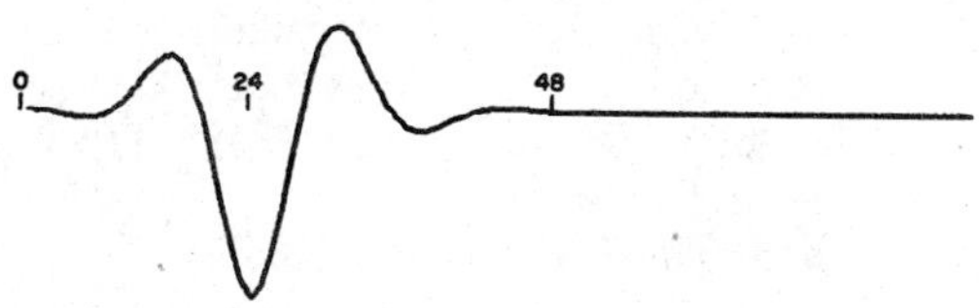

FIG. 7(c). Low-pass output.

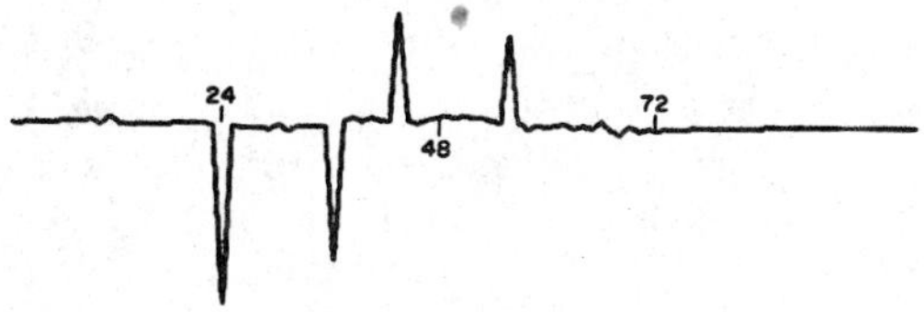

FIG. 7(d). High-pass output.

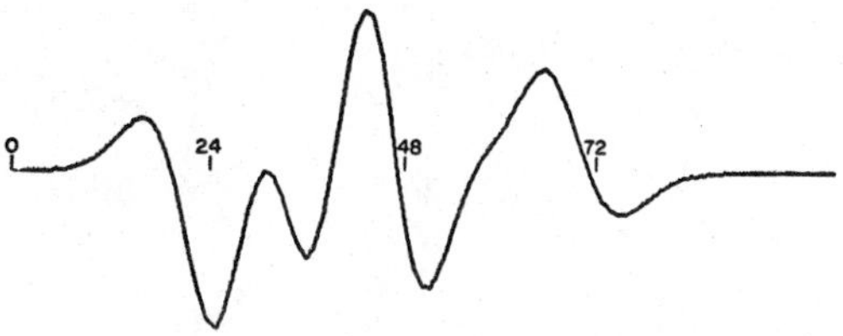

FIG. 8(a). A mixed-phase sequence.

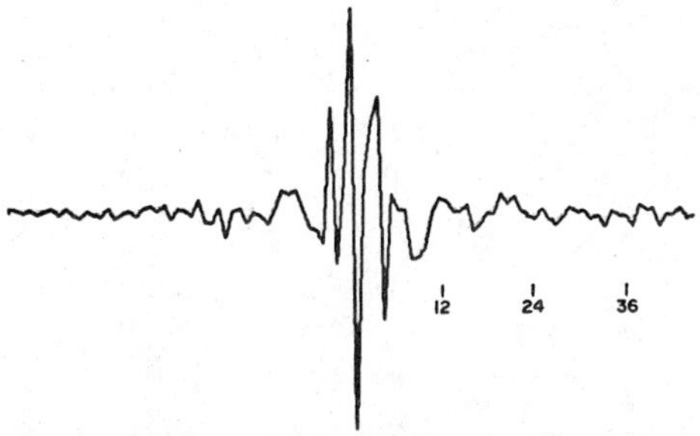

FIG. 8(b). Complex cepstrum of Figure 8(a).

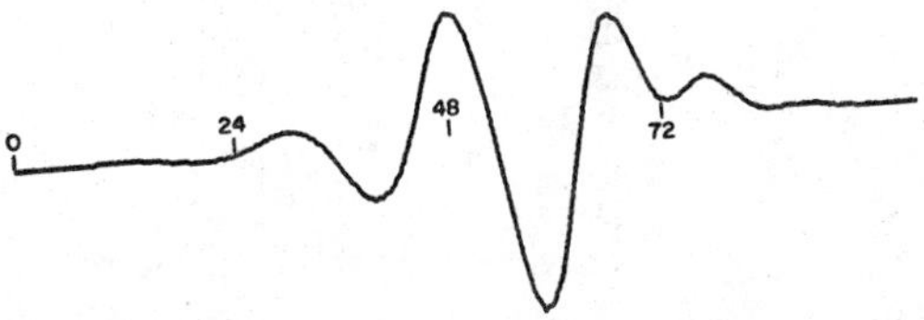

FIG. 8(c). Low-pass output.

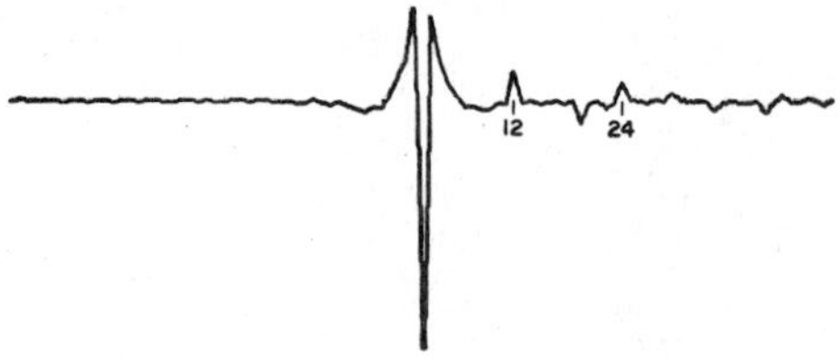

FIG. 9(a). Complex cepstrum of the sequence of Figure 8(a) exponentially weighted with $\alpha=0.965$.

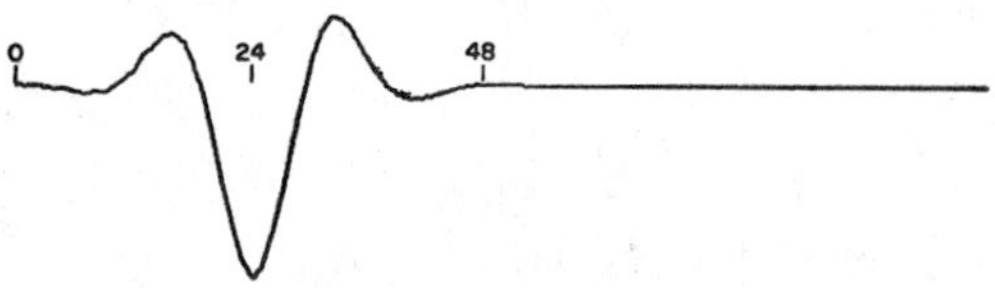

FIG. 9(b). Low-pass output.

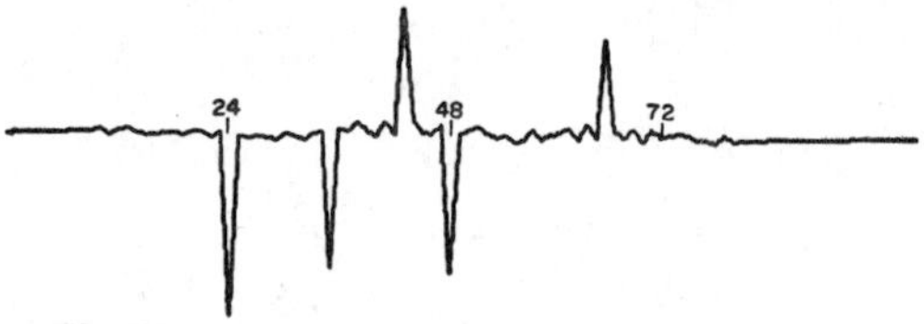

FIG. 9(c). High-pass output.

original. This is due to the fact that the region of the cepstrum near $n=0$ now contains the combined contributions of the wavelet and the impulse train.

Exponential weighting

Schafer (1969) has suggested an ingenious method by which, in order to exploit the special properties of minimum-phase sequences, a mixed-phase sequence may be converted to a minimum-phase sequence.

Suppose that the furthest zero of $I(z)$, the z transform of the mixed-phase impulse train $i(n)$, is at z_o, where $|z_o|>1$. We wish to transform $i(n)$ into a minimum-phase impulse train $j(n)$. Thus, the furthest zero of $J(z)$ must be at αz_o, where $|\alpha z_o|<1$, i.e.,

$$J(z) = I(\alpha^{-1}z);$$

and, therefore,

$$j(n) = \alpha^n i(n), \quad \alpha < 1.$$

In other words, a mixed-phase sequence may be made into a minimum-phase sequence by means of exponential weighting. To illustrate this point, the input sequence of Figure 8(a) was exponentially weighted with $\alpha=0.965$. The resulting complex cepstrum is shown in Figure 9(a) and the deconvolved wavelet and impulse train are shown in Figures 9(b) and 9(c).

Effect of noise on homomorphic deconvolution

The examples considered thus far are ideal in the sense that the input sequences are noise free. An actual seismic trace, $x(n)$, may be represented as

$$x(n) = s(n) * i(n) + m(n),$$

where the noise $m(n)$ may be decomposed into $m(n)=\eta(n)*s(n)+\mu(n)$. $\eta(n)$ is the part of the noise convolved by the wavelet $s(n)$ and $\mu(n)$ is a noise superimposed on the seismic trace. We will consider each component in turn.

Additive noise component $\mu(n)$.—The addition of $\mu(n)$ complicates the computation of a smooth-phase curve, and since the complex cepstrum depends on the contribution of the phase component of the input sequence, the simplicity of the complex cepstrum of a noise-free sequence, such as illustrated in Figure 7, is destroyed. However, providing that the impulse train has been made minimum-phase by exponential weighting, the portion of the complex cepstrum near $n=0$ (or the "short time" portion) may still be used to recover the seismic wavelet. Figure 10(a) shows the input sequence of Figure 7(a), together with white noise with a signal-to-noise amplitude ratio of 10 to 1. After exponential weighting with $\alpha=0.965$, the recovered seismic wavelet is shown in Figure 10(c). A considerably better result is achieved if the input sequence is filtered first of all. An important consideration in homomorphic deconvolution of filtered signals is the rate at which the continuous time signal is sampled (Schafer, 1969). This point is discussed below.

The high-frequency content of a filtered seismic signal is low and the Fourier transform of such signals may be considered to be zero for frequencies greater than f_c say. If the signal is sampled at the Nyquist rate $1/(2f_c)$, aliasing is avoided and $|X(e^{j\omega})|$ is finite at all frequencies. If the sampling interval is less than $1/(2f_c)$, aliasing is, of course, also avoided but $|X(e^{j\omega})|$ is zero and, hence, $\log|X(e^{j\omega})|$ is undefined over a finite interval. In general, if the sampling rate is greater than the Nyquist rate, there exists an interval in

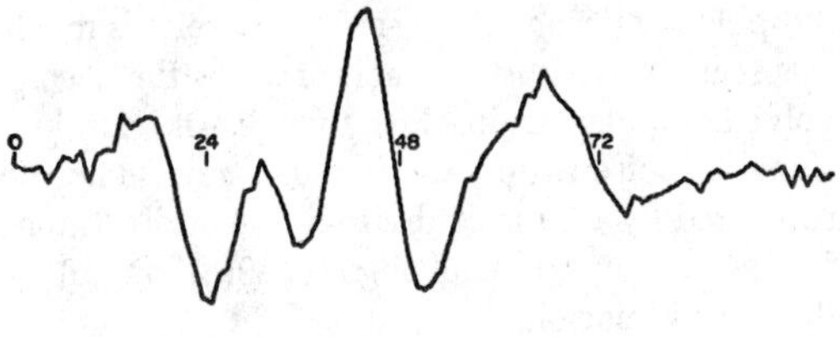

FIG. 10(a). Input sequence of Figure 8(a) with an added noise component.

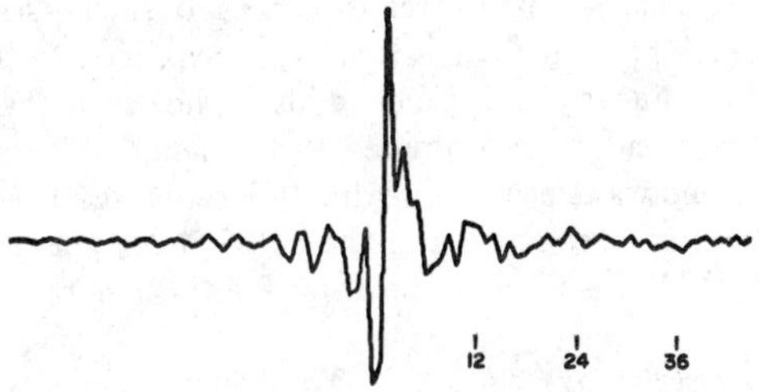

FIG. 10(b). Complex cepstrum of the sequence of Figure 10(a) exponentially weighted with $\alpha=0.965$.

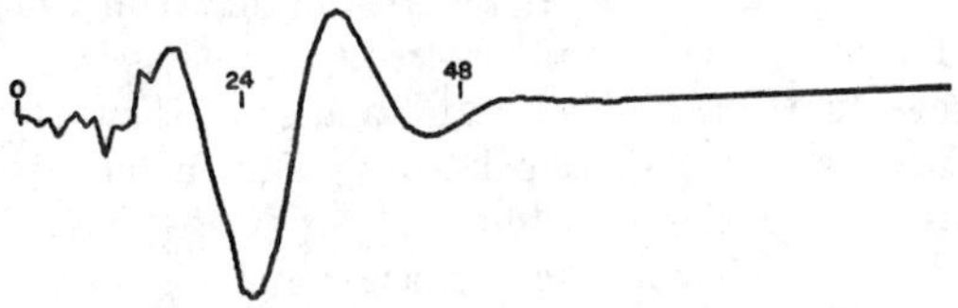

FIG. 10(c). Low-pass output.

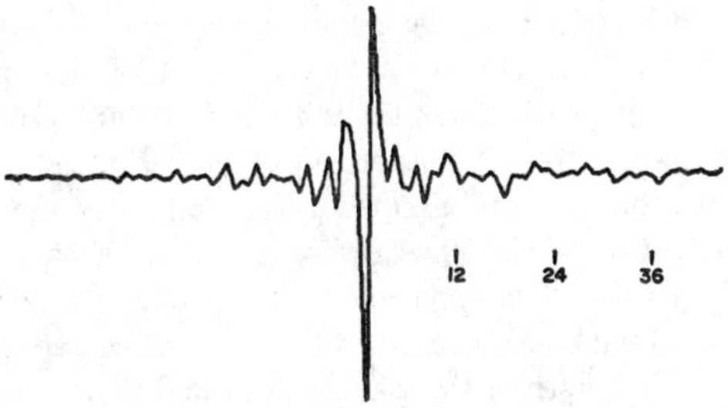

FIG. 10(d). Complex cepstrum of sequence of Figure 10(c) after optimum filtering and exponential weighting.

which the real and imaginary parts of $X(e^{j\omega})$ are small and considerable errors may arise in the computation of $\log|X(e^{j\omega})|$ and $ARG|X(e^{j\omega})|$. It is important, therefore, that, following prefiltering of the signal, the sampling rate be made equal to or slightly higher than the Nyquist rate.

The above discussion is illustrated in Figure 10. The input sequence of Figure 10(a) was filtered

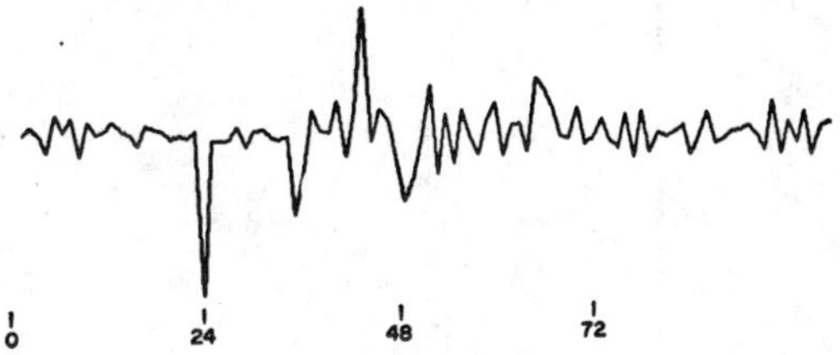

FIG. 10(e). High-pass output.

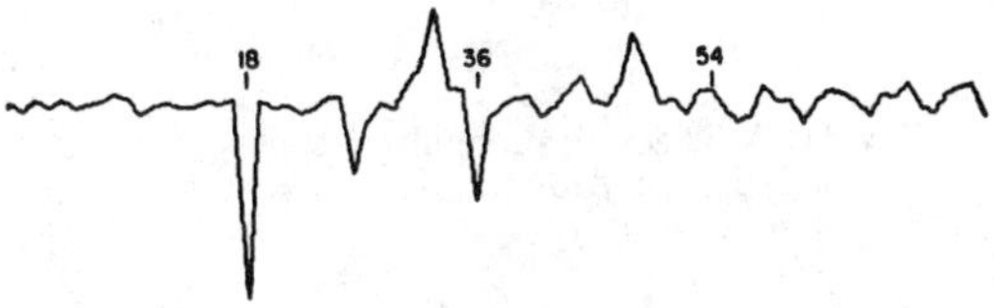

FIG. 10(f). Deconvolved impulse train showing the effect of increasing sample interval by 50 percent.

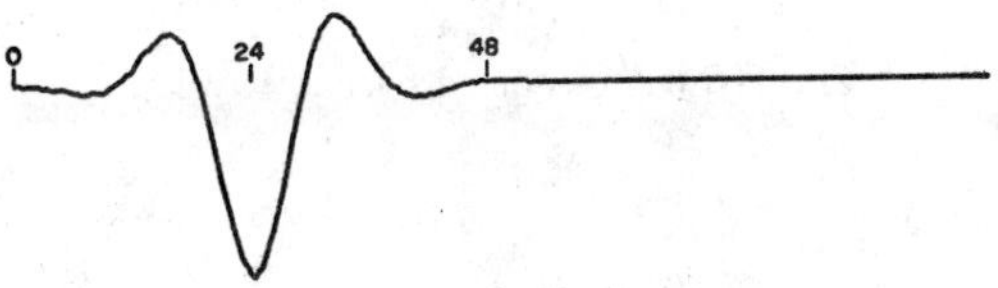

FIG. 10(g). Low-pass output of Figure 10(d).

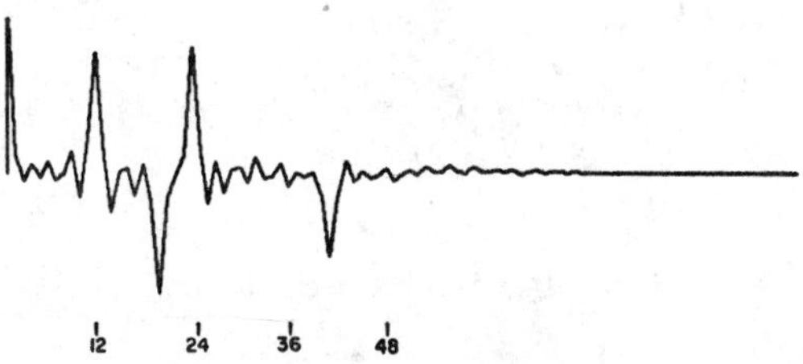

FIG. 10(h). Impulse train obtained by means of division in frequency domain.

using an optimum Wiener filter.[1] The complex cepstrum corresponding to this filtered sequence appears in Figure 10(d). Due to irregularities in the unwrapped phase curve, the deconvolved impulse train, shown in Figure 10(e), is very noisy. These irregularities are decreased by increasing the sampling interval, as discussed above;

[1] The Wiener filter, $H_{\text{opt}}(\omega)$, was designed on the basis of the power spectrum of the input signal.

$$H_{\text{opt}}(\omega) = \frac{P_s(\omega)}{P_s(\omega) + P_n(\omega)}$$

The noise power spectrum $P_n(\omega)$ was estimated by inspection of the input power spectrum to be 0.4 percent of the maximum input power. Since the noise is white, the noise power is a constant.

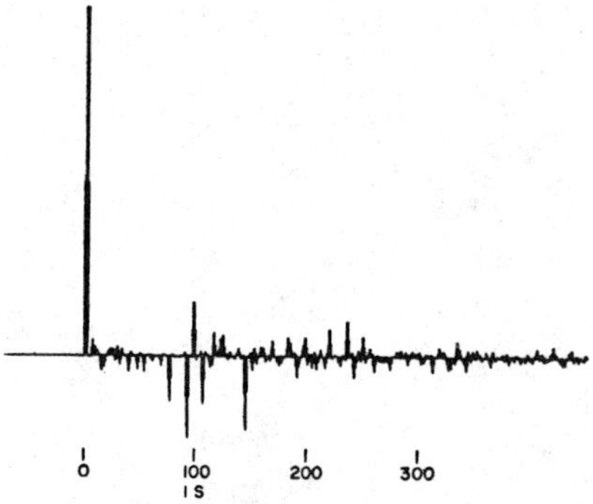

FIG. 11(a). Theoretical impulse response of crust near Leduc, Alberta (after O. Jensen).

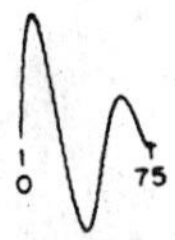

FIG. 11(b). An assumed seismic wavelet.

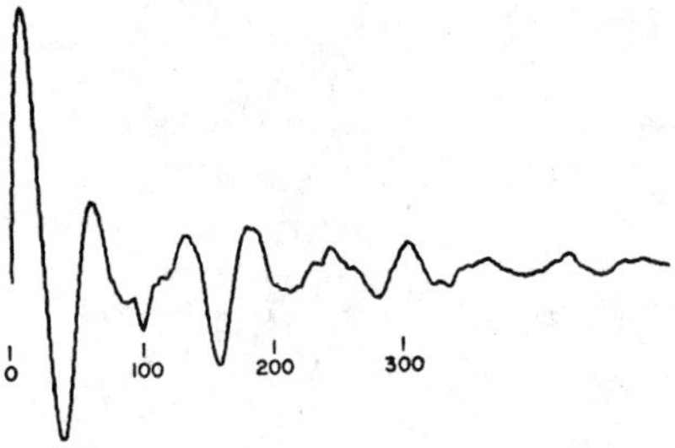

FIG. 11(c). Synthetic seismogram.

and the deconvolution is improved. Figure 10(f) shows the deconvolved impulse train after filtering but with a sampling interval 50 percent greater than that of the previous example. The improvement is marked.

The seismic wavelet which is recovered from the short-time portion of Figure 10(d) [Figure 10(g)] illustrates an important aspect of homomorphic filtering. It appears that noise in the phase curve influences the long-time portion of the complex cepstrum to a much larger degree than it does the short-time portion. Consequently, it is possible and may be preferable, once the wavelet has been recovered, to obtain the impulse train by means of division of Fourier transforms. The impulse train recovered in this manner using the wavelet of Figure 10(g) is shown in Figure 10(h).

We remark, in summary, that the homomorphically deconvolved seismic wavelet is much less sensitive to additive noise than is the deconvolved impulse train. For this reason, for low signal-to-noise ratios, the impulse train is better recovered by means of division of Fourier transforms. A comparison of Figures 10(e) and 10(h) illustrates this point.

The complex cepstra of sequences which have been filtered to remove an additive noise component may not enjoy the simplicity of the complex cepstra of noise-free sequences. In such cases, it may be difficult to estimate the required length of the linear filter. We have found, however, that good results are obtained if the length of the seismic wavelet is approximately estimated from the input sequence and the cutoff length of the ideal filter is taken to be a third of this length.

Convolutional noise $\eta(n)$.—To illustrate the effect of convolutional noise, homomorphic deconvolution has been applied to a synthetic seismogram prepared by O. Jensen and thought to represent the response of the earth's crust in the vicinity of Leduc, Alberta to low-frequency teleseismic events. Figures 11(a), 11(b), and 11(c) show the impulse response calculated by Jensen, an assumed wavelet, and the resulting synthetic seismogram. The impulse response contains all minor impulses which would result from multiple reflections. The length of the wavelet was estimated to be 75 samples from Figure 11(a) and, consequently, the cutoff length of the simple low-pass filter was chosen to be 25 samples. Exponential weighting with $\alpha=0.985$ was used. Figure 12(a) shows the complex cepstrum of the weighted input sequence and Figures 12(b) and 12(c) show the recovered impulse response and wavelet, respectively. It is apparent that the effect of convolutional noise is confined to the first 75 samples, i.e., the length of the wavelet, of the deconvolved impulse response. The striking similarity between the actual impulse response and the deconvolved response for $n>75$ is extremely encouraging, although it must be remembered that this example is free of additive noise.

DISCUSSION

A major problem in exploration and earthquake seismology is the identification of the seismic wavelet. A knowledge of the shape of this wavelet allows the determination of the attenuation and dispersion properties of the transmission path, a problem of considerable interest in seismology (Strick, 1970). Although it is probable that in exploration seismology the wavelet may be often assumed to be minimum-phase, this assumption may not be made in the case of earthquake seismology. Homomorphic deconvolution appears to be a very powerful method of recovery of the seismic wavelet and, hence, also of the impulse response. Most important, homomorphic deconvolution obviates the necessity of making the usual assumptions of a minimum-phase wavelet and a random impulse train. Failure of these assumptions, which are required by methods commonly used in the deconvolution of sequences of the type considered in this paper, may lead to gross errors.

We are at present (Ulrych et al, 1971) applying homomorphic deconvolution to a series of teleseismic events recorded at Leduc, Alberta with encouraging results. A preliminary example of this work is presented here.

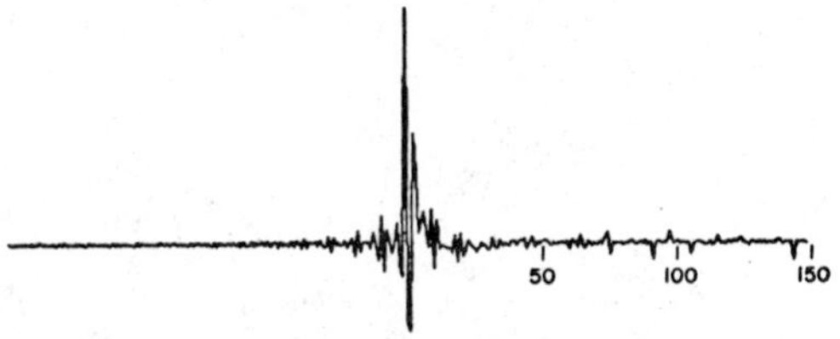

FIG. 12(a). Complex cepstrum of the trace of Figure 11(c) exponentially weighted with $\alpha=0.985$.

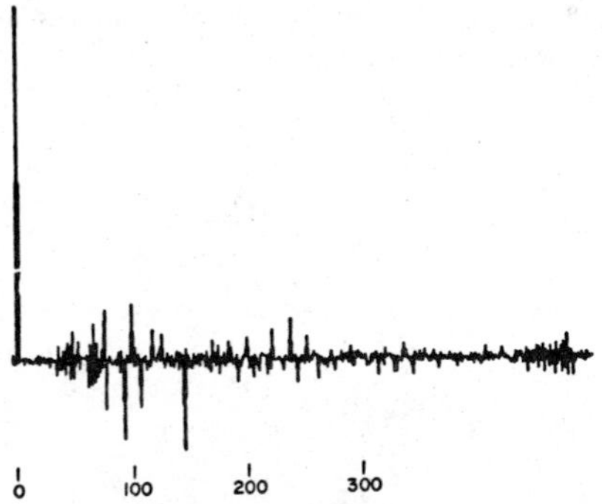

FIG. 12(b). High-pass output.

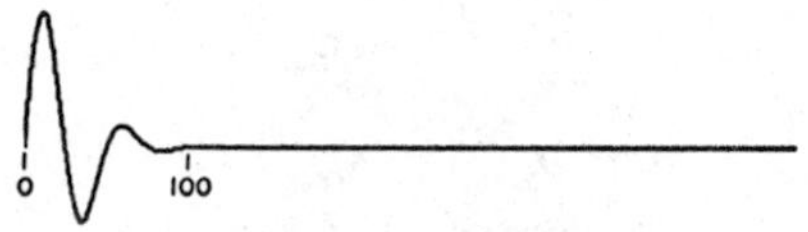

FIG. 12(c). Low-pass output.

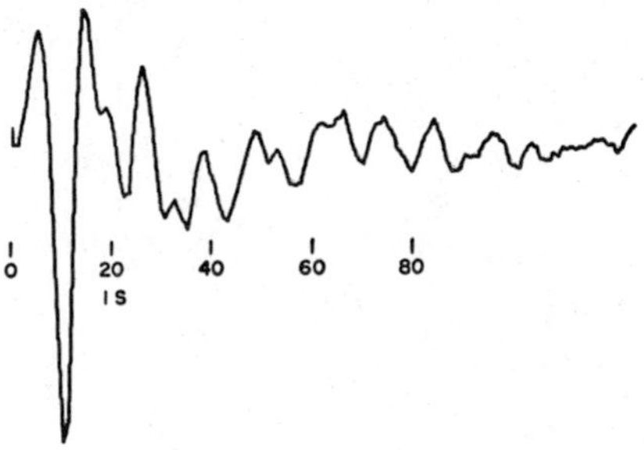

FIG. 13(a). Teleseismic event recorded in 1968 at Leduc, Alberta and originating in Venezuela.

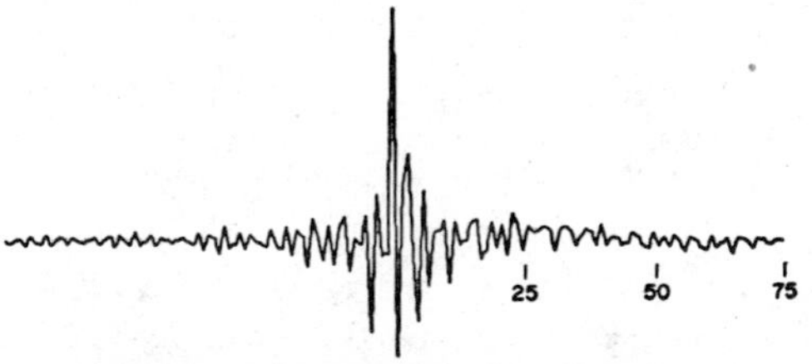

FIG. 13(b). Complex cepstrum of Figure 13(a) after exponential weighting with $\alpha=0.985$.

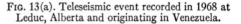

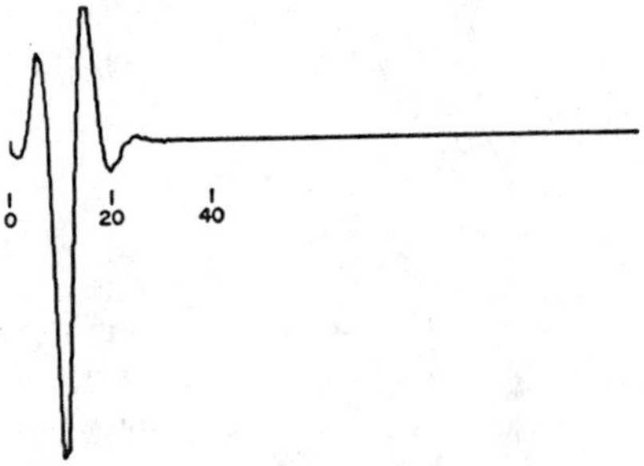

FIG. 13(c). Deconvolved seismic wavelet

Figure 13(a) shows an event which originated in Venezuela in 1968. The estimated length of the seismic wavelet is approximately 20 samples and the cutoff length for the simple low-pass filter which was used in the linear filtering was 8 samples. Following exponential weighting with $\alpha=0.985$, we recovered the seismic wavelet shown in Figure 13(c). An approximate check on the deconvolution is provided by convolving the wavelet in Figure 13(c) with the synthetic impulse response of Figure 11(a). The resultant "earthquake" shown in Figure 13(d) compares favorably with the observed event of Figure 13(a).

Many interesting questions remain to be explored. For example, perhaps some type of phase filtering may be employed to overcome the effect

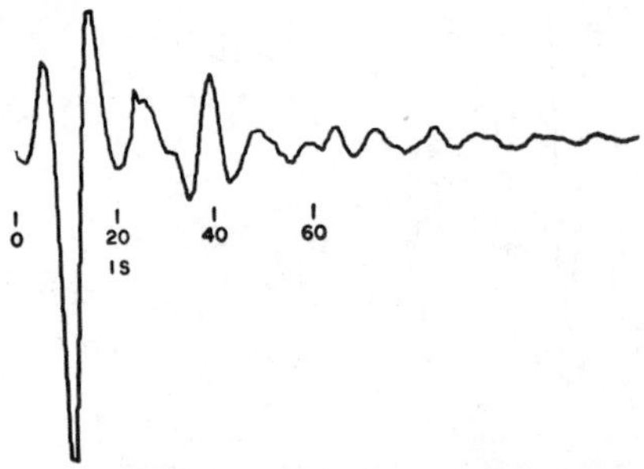

FIG. 13(d). Trace resulting from the convolution of Figure 13(c) with Figure 11(a).

of additive noise on the unwrapped phase curve. The choice of α depends at present on experience only, whereas a quantitative method of determining it is obviously preferable.

Schafer (1969) has suggested certain problems to which homomorphic deconvolution may be usefully applied. Our research has indicated the importance of this technique in processing seismic signals; it is hoped that this work will encourage further research into the application of homomorphic filtering to problems of importance in geophysics.

ACKNOWLEDGMENTS

I am extremely grateful to Oliver Jensen of the University of British Columbia for providing me with the synthetic seismogram used in this work, for stimulating thoughts, and for critically reading this manuscript. I am also obliged to Sven Treitel for an enlightening discussion of predictive deconvolution. This research was financed by the National Research Council grant-in-aid.

REFERENCES

Bogert, B. P., Healey, M. J., and Tukey, J. W., 1963, The quefrequency analysis of time series for echoes; cepstrum, pseudo-autocovariance, cross-cepstrum and saphe cracking: Proc. Symp. on Time Series Analysis, M. Rosenblatt, Ed., New York, Wiley, p. 209–243.

Cooley, J. W., and Tukey, J. W., 1965, An algorithm for the machine calculation of complex Fourier series: Math. of Comput., v. 19, p. 297–301.

Futterman, W. I., 1962, Dispersive body waves: J. Geophys. Res., v. 67, p. 5279–5291.

Oppenheim, A. V., 1965a, Superposition in a class of non-linear systems: Research Lab. of Electronics MIT, Tech. Rep. 432.

——— 1965b, Optimum homomorphic filters: Research Lab. of Electronics MIT, Quart. Progr. Rep. 77, p. 248–260.

Oppenheim, A. V., Schafer, R. W., and Stockham, T. G., 1968, Nonlinear filtering of multiplied and convolved signals: Proc. IEEE, v. 65, p. 1264–1291.

Peacock, K. L., and Treitel, S., 1969, Predictive deconvolution: Theory and practice: Geophysics, v. 34, p. 155–169.

Rice, R. B., 1962, Inverse convolution filters: Geophysics, v. 27, p. 4–18.

Robinson, E. A., 1966, Multichannel z transforms and minimum delay: Geophysics, v. 31, p. 473–500.

Robinson, E. A., and Treitel, S., 1967, Principles of digital Wiener filtering: Geophys. Prosp., v. 15, p. 311–333.

Schafer, R. W., 1969, Echo removal by discrete generalized linear filtering: Research Lab. of Electronics MIT, Tech. Rep. 466.

Strick, E., 1970, A predicted pedestal effect for pulse propagation in constant-Q solids: Geophysics, v. 35, p. 387–403.

Ulrych, T. J., and Lasserre, M., 1966, Minimum-phase: J. Can. Soc. Expl. Geophysicists, v. 2, p. 22–32.

Ulrych, T. J., Jensen, O., Ellis, R. M., and Summerfield, P. S., 1971, to be submitted to Bull. Seism. Soc. Am.

Reprinted from Geophysics v. 39, no. 4, p. 401-416

THE APPLICATION OF HOMOMORPHIC DECONVOLUTION TO SHALLOW-WATER MARINE SEISMOLOGY—PART I: MODELS

PAUL L. STOFFA,*§ PETER BUHL,*§ AND GEORGE M. BRYAN*

The complex cepstrum is investigated mathematically and through models for functions of interest in shallow-water marine seismology. Association of the slowly varying components of the phase spectrum with the source replaces the usual minimum-phase assumption. This is analogous to the usual treatment of the amplitude spectrum. Complex cepstrum expressions are developed for an arbitrary (but minimum-phase) reflector series, water-column multiple generator, and simplified bubble-pulse oscillation. While the complex cepstrum of all functions is of infinite extent, removing only the first n nonzero complex-cepstrum contributions of a decaying, impulsive, periodic time function (such as the water-column multiple generator) serves to eliminate the first n multiples entirely in the time domain and reduces the remaining multiples to at most $1/(n+1)$ of their original value.

A new method of computing the continuous, ramp-free phase spectrum required for complex-cepstrum analysis is developed on the basis of the derivative of the phase curve.

INTRODUCTION

In least-squares, time-domain inverse filtering (Robinson, 1957; Rice, 1962), removal of the seismic source begins with the autocorrelation function (ACF) of the seismic trace. The trace ACF is assumed equal to the ACF of the source. By taking only a small number of lags, relative to the trace length, we are effectively restricting our source to a short time duration. Since the ACF and the power spectrum are Fourier transform pairs, this is equivalent to saying that compared to the reflector series the power spectrum of the source varies slowly with respect to frequency. Thus time-domain deconvolution removes the slowly varying components from the power spectrum leaving the rapidly varying components, which presumably represent the reflector series. Thus we see that least-squares inverse filtering exploits a difference between the source and the trace, namely, the way their power spectra vary with frequency, and its success is in part determined by the size of this difference. It should be pointed out that a "random" reflector series has a white or flat amplitude spectrum only in a statistical sense. The spectrum of a particular random series, however, will not be white; it has both slowly and rapidly varying amplitude spectrum components.

In least-squares inverse filtering we must make an assumption about the phase spectrum of the source, since by taking the ACF we have destroyed all phase information. The usual realizations assume the source to be minimum phase. It might be better to treat the phase spectrum as we treat the power spectrum, i.e., associate the slowly varying components with the source and

Lamont-Doherty Geological Observatory contribution No. 2088.

Presented at the 43rd Annual International SEG Meeting, October 24, 1973, Mexico City. Manuscript received by the Editor July 23, 1973; revised manuscript received September 11, 1973.

* Lamont-Doherty Geological Observatory, Palisades, NY 10964.

§ Columbia University, New York, N.Y. 10027.

the rapidly varying components with the reflector series. It will be shown that this is indeed the case only if the reflector series is minimum-phase.

The above discussion leads us to the cepstrum (Bogert et al, 1963) and then the complex cepstrum which is one realization of a homomorphic system (Oppenheim, 1965; Schafer, 1969). This approach treats the complex natural logarithm of the amplitude and phase spectra, $\log [A(\omega)e^{i\phi(\omega)}]$, as a complex time series and takes the inverse Fourier transform of this series to produce the complex cepstrum. Then by removing the values of the complex cepstrum near the origin, we can eliminate the slowly varying components of both the phase and log-amplitude spectra.

In this paper, model studies are discussed which indicate that homomorphic deconvolution, as developed by Oppenheim (1965) and Schafer (1969), has application to the marine seismic deconvolution problem. [Ulrych (1971, 1972) has already had success with this method in extracting the source function from teleseismic events.] The marine seismic source and short-period, water-column multiples may be removed from marine seismic records by fairly simple operations in the complex cepstrum without the minimum-phase source assumption. Aliasing in the complex cepstrum, produced by the nonlinear complex logarithm operation, can be suppressed by judicious exponential weighting of the seismic trace.

Two methods of computing the continuous, ramp-free phase spectrum required for complex-cepstrum analysis are discussed. The first method is an iterative application of Schafer's (1969) algorithm, which "unwraps" the principal value of the phase spectrum. The second method is based on the derivative of the phase spectrum.

In Part II we will apply the methods of complex-cepstrum analysis outlined in this paper to real data.

THE COMPLEX CEPSTRUM—WEIGHTING, ALIASING, AND MINIMUM PHASE

Using z-transforms, Schafer (1969, p. 12–13) gives the following three-step definition of the complex cepstrum for discrete functions of unit sample interval:

$$X(z) = \sum_{t=-\infty}^{\infty} x(t)z^{-t}, \quad \text{where } z = e^{\sigma+i\omega}, \quad (1a)$$

$$\begin{aligned}\hat{X}(z) &= \log X(z)\\ &= \log |X(z)| + i \arg [X(z)],\end{aligned} \quad (1b)$$

$$\hat{x}(T) = \frac{1}{2\pi i}\oint_C \hat{X}(z)z^{T-1}dz, \quad \text{where } T = 0, \pm 1, \pm 2, \cdots. \quad (1c)$$

The three-step inverse definition for return to the time domain is:

$$\hat{X}(z) = \sum_{T=-\infty}^{\infty} \hat{x}(T)z^{-T}, \quad (2a)$$

$$X(z) = \exp [\hat{X}(z)], \quad (2b)$$

$$x(t) = \frac{1}{2\pi i}\oint_{C'} X(z)z^{T-1}dz. \quad (2c)$$

The complex natural logarithm defined in equation (1b) is a multivalued function since $\arg [X(z)]$ has a multiplicity of $2n\pi$ where $n=0, 1, 2 \cdots$. Since $\hat{X}(z)$ must be continuous, $\arg [X(z)]$ cannot be restricted to its principal value ($n=0$). In equation (1c) the values T define the quefrencies of Bogert et al (1963). Because of analogies developed later between the time domain and the complex cepstrum for impulsive time series, we prefer to associate the word period with the complex-cepstrum variable T. To emphasize this, we use the variable T in equation (1c). When performing the inverse transform [equations (2)], we choose the same contour of integration for equation (2c) as was used in equation (1c). Then the forward transform [equations (1)], followed by the reverse transform [equations (2)], will yield the original time series $x(t)$.

Weighting

In equation (1a) we have transformed a real-time function into its z-transform $X(z)$, which exists throughout the complex z-plane as does $\hat{X}(z)$ from equation (1b). Equation (1c) involves only those values of $\hat{X}(z)$ which lie on the closed-path integration contour C. Thus $\hat{x}(T)$ is a function of the particular contour. In computing the complex cepstrum, we use the discrete Fourier transform (DFT) for real frequencies in place of the z-transform, thus restricting the contour to the unit circle ($\sigma=0$). However, if we multiply our original time function by the weighting function a^t, where $0<a<1$, we have effectively moved our integration contour to a circle of radius e^{σ}, where

$$\sigma = -\log a. \qquad (3)$$

Since $a<1$, σ is positive, and $e^{-\sigma t}$ is an exponentially decaying function. In equation (1a) we have defined $z=e^{\sigma+i\omega}$ to include the weighting function as part of the definition so that the contour in equation (1c) will be determined by our selection of σ. This is equivalent to off-axis integration in the complex ω-plane. In returning to the time domain via equations (2), we should unweight the result by a^{-t} which guarantees that $C=C'$ in equations (1c) and (2c). In the DFT, since we are confined to evaluating $\hat{X}(z)$ around the unit circle, we can consider that the effect of weighting is to move the poles and zeroes of $\hat{X}(z)$ radially inward by the factor e^{σ}; and unweighting with the inverse function restores them to their original location.

The logarithm introduces an additional number of zeros and poles into our function. If z_0 is a zero of the original function $X(z)$, it becomes a pole of $\hat{X}(z)$. In addition, if for z_1, $X(z_1)=1$, then z_1 becomes a zero of $\hat{X}(z)$. Thus poles and zeros become poles, and "ones" become zeros.

If a function is minimum-phase, its Fourier transform will have all its poles and zeros in the left-half complex ω-plane (Ulrych and Lasserre, 1966). Under the definition of the z-transform used here, this corresponds to having all poles and zeros within the unit circle. Thus, if the zero of $X(z)$ farthest from the origin is at $z_0=e^{\tau_0+i\omega_0}$, then for $\sigma>\tau_0$, we will have moved all the zeros of $X(z)$ inside the unit circle, and the weighted function will be minimum-phase. Conversely, a maximum-phase function will have all its poles and zeros outside the unit circle. Schafer (1969) has shown that the complex cepstrum of a minimum-phase function is zero for $T<0$, and the complex cepstrum of a maximum-phase function is zero for $T>0$.

Aliasing

Aliasing is introduced into the complex cepstrum when the nonlinear logarithm operation of equation (1b) is followed by the discrete Fourier transform. Although $X(z)$ is adequately sampled, the nonlinear operations: logarithm, absolute value, and arctangent introduce harmonics into $\hat{X}(z)$. Thus $\hat{X}(z)$ is undersampled when $z=e^{in2\pi/N}$, $n=0, \pm 1, \pm 2, \cdots, \pm N/2$, as in the DFT. Since all harmonics out to infinite quefrencies or periods are present, the complex cepstrum will in general be nonzero out to infinity. The effect of using the DFT is to alias these periods into the principal period range: $-1/2\Delta f<T\leq 1/2\Delta f$. Schafer (1969) has shown that an a^T weighting is impressed on the complex cepstrum by the a^t time-domain weighting. Thus weighting can help suppress aliasing. This is because $e^{-\sigma t}$ smooths $\hat{X}(z)$, when it is evaluated on the unit circle, by moving its poles (if the function is already minimum-phase) further inward away from the unit circle. This reduces the amplitude of the rapid fluctuations in $\hat{X}(z)$. Thus the high quefrencies or periods of $\hat{x}(T)$ and their harmonics are reduced. The shorter periods are less suppressed, but their harmonics have fallen off considerably by the folding period, $1/2\Delta f$. Figures 1a and b show how weighting reduces aliasing for a simple three-impulse time series. Figure 1a[1] is the complex cepstrum of the weighted trace, $a=0.975$, which is minimum-phase. The aliasing is severe. Increasing the weighting to $a=0.960$ (Figure 1b), results in a complex cepstrum which is not aliased. It has been our experience in complex-cepstrum analysis that it is convenient to weight heavily initially (e.g., $a=0.94$), to guarantee an unaliased complex cepstrum, and then try less weighting until aliasing becomes a problem.

DEVELOPMENT OF THE COMPLEX CEPSTRUM

Before exploiting the properties of the complex cepstrum for deconvolution, we shall investigate the way in which the complex cepstrum develops for some signals of interest.

The reflector series

We model the reflector structure by a causal series of n impulses distributed arbitrarily in time at the unit sampling interval.

Let

$$r_o(t) = \sum_{i=1}^{n} \alpha_i \delta(t-t_i) \qquad t_i \geq 0. \qquad (4)$$

[1] The complex cepstrum is a real function. All complex-cepstrum plots are in terms of power, where we maintain the sign of the original real function. Thus a point at -60 db is a negative complex-cepstrum contribution whose power is 60 db. All contributions, positive and negative, whose power is less than zero db plot to zero db. The db axis is arbitrary and scaled with respect to the maximum absolute value power point of each complex cepstrum.

The first step in finding a complex cepstrum of this function is to weight it exponentially.

$$r(t) = r_0(t)a^t \tag{5}$$

$$= \sum_{i=1}^{n} \alpha_i a^{t_i}\delta(t - t_i), \tag{6}$$

where $0 < a \leq 1$.

The z-transform is

$$R(z) = \sum_{i=1}^{n} \alpha_i a^{t_i} z^{-t_i}. \tag{7}$$

We leave the weighting a^t as an integral part of the function and do not absorb it into z, so that $\sigma = 0$, and $z = e^{i\omega}$. Factoring $R(z)$ gives:

$$R(z) = \alpha_1 a^{t_1} z^{-t_1}\left[1 + \sum_{i=2}^{n} \beta_i a^{T_i} z^{-T_i}\right], \tag{8}$$

where

$$\beta_i = \alpha_i/\alpha_1$$

and

$$T_i = t_i - t_1; \qquad i = 2, 3, \cdots, n.$$

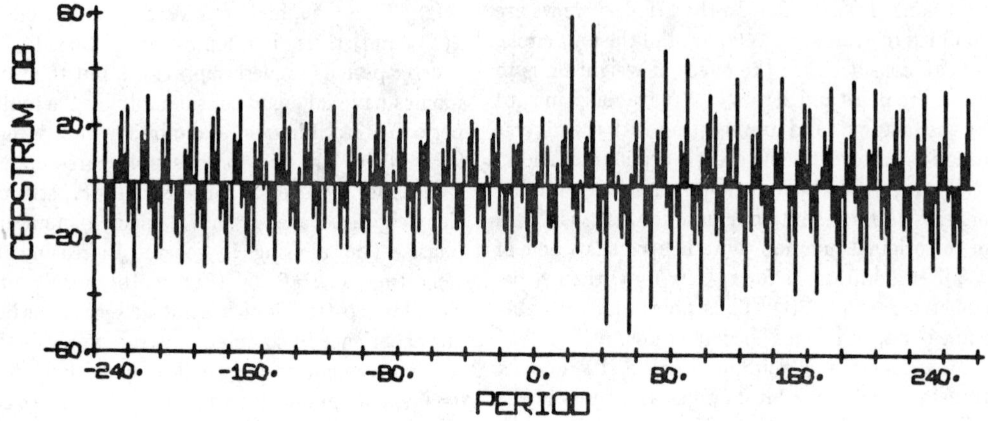

(a)

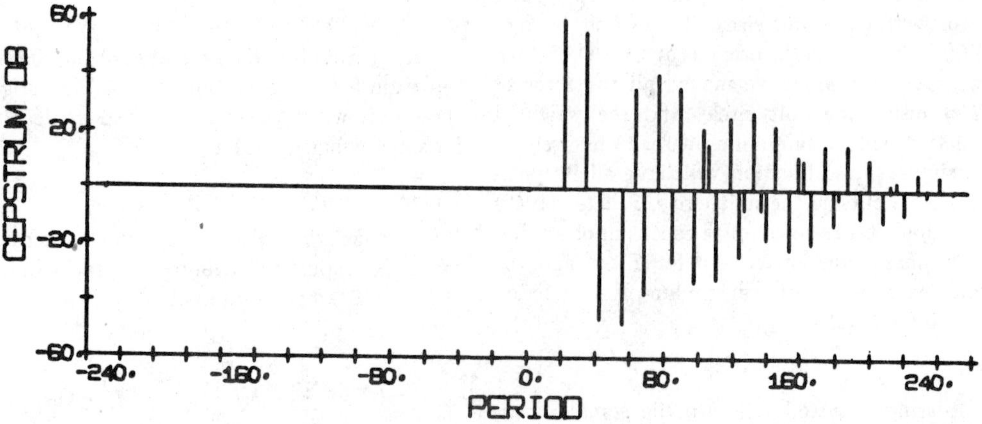

(b)

FIG. 1. Complex cepstra of the three impulse time series: $r_0(t) = \delta(t) + \delta(t-21) + \delta(t-34)$. (a) Weighting with a = 0.975 is sufficient to make the time series purely minimum phase (complex cepstrum fall off is toward increasing periods), but, the complex cepstrum is severely aliased. (b) Weighting with a = 0.96 reduces the complex cepstrum aliasing so that for all negative periods it is at least 60 db down from the largest cepstrum value. In this complex cepstrum it is quite clear that we get contributions at $T_2 = 21$, $T_3 = 34$, their multiples, and all combinations of their multiples. (All complex-cepstrum plots are scaled to the maximum complex-cepstrum contribution and plotted in db. However, we retain the sign of each complex-cepstrum contribution and plot it accordingly as a positive or negative contribution. Any complex-cepstrum contribution, positive or negative, which is 60 db or more down from the maximum plots to 0 db.)

Next, take the complex natural logarithm

$$\hat{R}(z) = \log(\alpha_1 a^{t_1} z^{-t_1}) + \log\left[1 + \sum_{i=2}^{n} \beta_i a^{T_i} z^{-T_i}\right]. \quad (9)$$

We remove a linear phase term (ramp), if present, since its transform has contributions for all complex-cepstrum periods and may swamp the desired information. The z^{-t_1} term is the linear phase shift. We can interpret its removal as shifting our weighted function t_1 locations toward the time origin, so that the first impulse is now at the time origin (i.e., $t_1=0$). With respect to the z-transform $R(z)$ [equation (8)], this is equivalent to removing the t_1 zeros at infinity in the z-plane. Thus, we have

$$\hat{R}(z) = \log(\alpha_1) + \log\left[1 + \sum_{i=2}^{n} \beta_i a^{T_i} z^{-T_i}\right]. \quad (10)$$

The final step of the transform is the contour integration of $\hat{R}(z)$ as indicated by the defining equation (1c). However, rather than perform the indicated integration, we follow the method outlined by Schafer (1969). If we can manipulate $\hat{R}(z)$ into a form such that it is recognizable as the z-transform of a known function, we then know that function. In this case, finding the Laurent-series expansion about $z=0$ for the second logarithm of equation (10) is the proper step. The Laurent-series expansion of $\log(1+x)$ about $x=0$ is:

$$\log(1+x) = \sum_{m=1}^{\infty} \frac{(-1)^{m+1}}{m} x^m, \quad \text{for } |x| < 1. \quad (11)$$

We let

$$x = \sum_{i=2}^{n} \beta_i a^{T_i} z^{-T_i}, \quad (12)$$

and require that

$$\left| \sum_{i=2}^{n} \beta_i a^{T_i} z^{-T_i} \right| < 1. \quad (13)$$

This requirement insures that the reflector series will be minimum-phase. While this appears overly restrictive, we are always able to satisfy equation (13) as we are free to choose an appropriate weighting. For real data, due to the geometric spreading loss and attenuation, little weighting should be necessary to make the reflector series (but not necessarily the trace) minimum-phase. We have then,

$$\hat{R}(z) = \log(\alpha_1) + \sum_{m=1}^{\infty} \frac{(-1)^{m+1}}{m} \cdot \left(\sum_{i=2}^{n} \beta_i a^{T_i} z^{-T_i} \right)^m. \quad (14)$$

The expression within parentheses is a polynomial of $n-1$ terms raised to the power m. We can expand this by use of the multinomial expansion (Morse and Feshbach, 1953, p. 412). The resulting $\hat{R}(z)$ is now recognizable as the z-transform of the following function:

$$\hat{r}(T) = \log(\alpha_1)\delta(T) + \sum_{m=1}^{\infty} \frac{(-1)^{m+1}}{m} \sum_{l_2, l_3, \cdots, l_n} \left[\frac{m!}{l_2! l_3! \cdots l_n!} \beta_2^{l_2} \beta_3^{l_3} \cdots \beta_n^{l_n} a^{\sum_{j=2}^{m} T_j l_j} \right] \cdot \delta\left(T - \sum_{j=2}^{m} T_j l_j\right), \quad (15a)$$

where

$$\sum_{j=2}^{n} l_j = m. \quad (15b)$$

The second sum in equation (15a) is over all possible combinations of the l_j which satisfy (15b). Thus for $m=1$, we will have a sum over $n-1$ terms in which each of the l_j is equal in turn to one, and the other l_j are zero. Little insight into the complex cepstrum is gained from equation (15a), except for some picture of the way the original impulse locations, individually and in combination, generate impulses in the complex cepstrum. We see from the δ-function in equation (15a) that we have complex-cepstrum contributions for all the original periods and all their multiples and also for all combinations of these multiples. The complex cepstrum is zero for negative T since the weighted reflector series is minimum-phase. Note that the complex cepstrum is zero between $T=0$ and $T=T_2$, a property we shall use later when deconvolving the source. The complex cepstrum is of infinite extent because of the logarithm operation. Therefore, calculation of the complex cepstrum using the DFT results in an aliased version of equation (15a). However, the weighting will reduce the contribution of periods greater than the folding period, $|T|>T_n$. Thus the weighting operation has served a dual purpose: it guarantees that our reflector series will be minimum-phase and it reduces the aliasing.

Consider a specific example, the three-impulse case. Let

$$r_0(t) = \sum_{i=1}^{3} \alpha_i \delta(t - t_i) \qquad t_i > 0. \qquad (16)$$

Proceeding as above and using the binomial expansion, we obtain the complex cepstrum of a ramp-free, weighted version of equation (16):

$$\begin{aligned} \hat{r}(T) &= \log(\alpha_1)\delta(T) \\ &+ \sum_{m=1}^{\infty} \frac{(-1)^{m+1}}{m} \sum_{l=0}^{m} \frac{m!}{l!(m-l)!} \\ &\cdot \left(\frac{\alpha_2}{\alpha_1}\right)^{m-l} \left(\frac{\alpha_3}{\alpha_1}\right)^{l} a^{T_2(m-l)+T_3 l} \\ &\cdot \delta(T - (m-l)T_2 - lT_3), \end{aligned} \qquad (17)$$

where

$$T_2 = t_2 - t_1$$
$$T_3 = t_3 - t_1.$$

Figure 1a illustrates this case for $T_2=21$, $T_3=34$, $\alpha_1=\alpha_2=\alpha_3$, and $a=0.975$. Notice that we have chosen "a" such that the weighted function is minimum-phase, but aliasing is still present. In Figure 1b the same impulse sequence is weighted with $a=0.96$, and there is no visible aliasing present. There are nonzero complex-cepstrum contributions at all multiples of the two original time-domain periods, T_2 and T_3, as well as all combinations of the *sums* of these two periods, but not the differences. It is important to notice that in all cases where we have weighted the original function sufficiently to make it minimum-phase there are no unaliased complex-cepstrum contributions for $0<T<T_2$. When performing deconvolution, which is subtraction in this domain, we can remove all complex-cepstrum contributions for $0<T<T_2$, without affecting the impulse train, provided we have weighted the trace sufficiently to make the impulse train minimum-phase.

Water column reverberation

Following the discussion by Pfleuger (1972), we represent the multiple generator in the water column by:

$$m(t) = \sum_{n=0}^{\infty} (-1)^n R^n \delta(t - nT_w), \qquad (18)$$

where R is the reflection coefficient, $0<R<1$, and T_w is the two-way traveltime through the water column. This is only the source multiple generator and would correspond to a 2-point filter (Backus, 1959). The function is minimum-phase since R is less than unity.

The z-transform of $m(t)$ is:

$$M(z) = \sum_{m=0}^{\infty} (-1)^m R^m z^{-mT_w}, \qquad (19)$$

which can be written as

$$M(z) = (1 + Rz^{-T_w})^{-1}. \qquad (20)$$

Taking the complex logarithm and using the Laurent-series expansion about $z=0$ gives

$$\hat{M}(z) = \sum_{m=1}^{\infty} (-1)^m \frac{R^m}{m} z^{-mT_w}, \qquad (21)$$

so that the complex cepstrum is

$$\hat{m}(T) = \sum_{m=1}^{\infty} (-1)^m \frac{R^m}{m} \delta(T - mT_w). \qquad (22)$$

Figure 2 illustrates this case for $R=0.8$ and $T_w=13\Delta t$. If Δt is 4 msec, this corresponds to a water depth of 39 m.

The bubble pulse

We shall represent the periodic part of the bubble-pulse oscillation by the following function:

$$b(t) = \sum_{n=0}^{\infty} R^n \delta(t - nT_b), \qquad (23)$$

where $0<R<1$; T_b is the period of the oscillations. Since R is less than unity, the oscillations are damped. The convolution of $b(t)$ [equation (23)], with a short-duration seismic source function would produce a reasonable approximation to the theoretical and experimental results of Ziolkowski (1970) and Schulze-Gattermann (1972). While equation (23) is a simplification even for the periodic part of the bubble-pulse function, it is sufficient for the purposes of our discussion.

Following exactly the method used above for the source water-column multiple generator, we obtain the complex cepstrum of $b(t)$:

$$\hat{b}(T) = \sum_{m=1}^{\infty} \frac{R^m}{m} \delta(T - mT_b). \qquad (24)$$

Table 1 summarizes some important transform pairs. The weighting has been chosen to insure that all functions are minimum-phase and unaliased. The last transform pair illustrates the case of the positive, impulsive, periodic, decaying time function described above.

DECONVOLUTION

Deconvolution in the time domain becomes subtraction in the complex cepstrum. An exact deconvolution of a convolutional component results in that component's complex-cepstrum contributions being set equal to zero. We have emphasized the correspondence between the time domain and the complex cepstrum for appropriately weighted traces. In the discussion which follows we assume that the trace has been weighted sufficiently so that the *reflector series* (but not necessarily the source) is minimum-phase and contains the shot at $t=0$. Since the minimum-phase reflector series has contributions in the

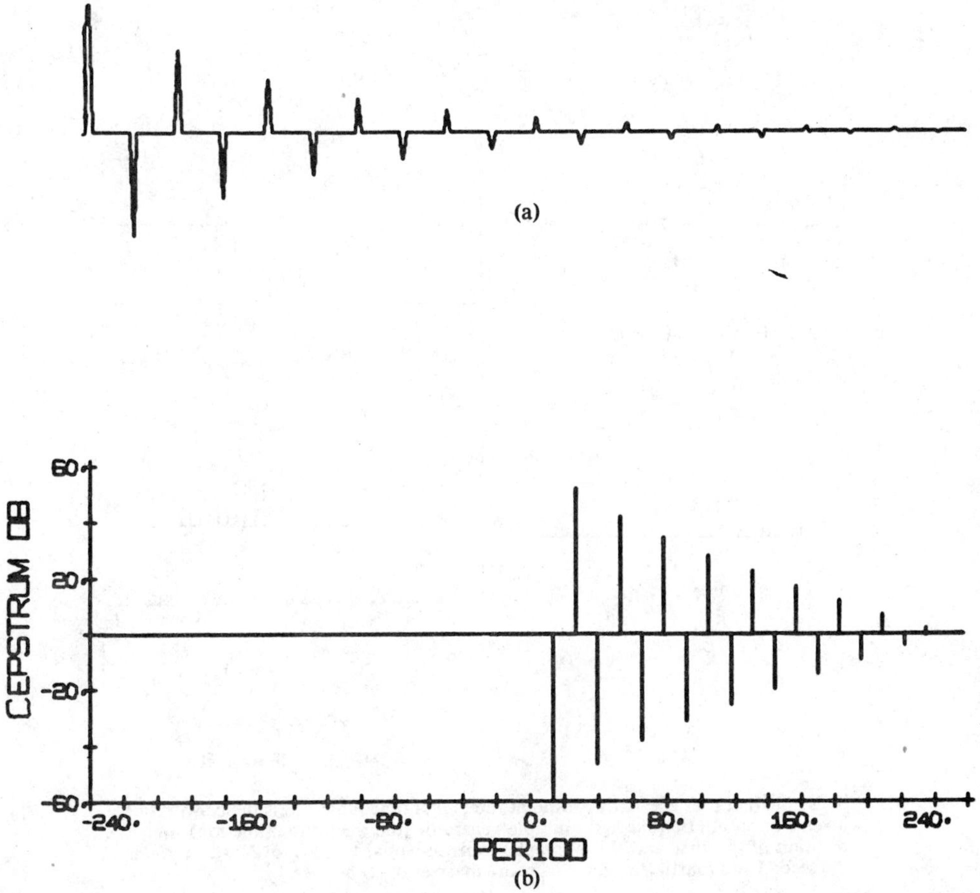

FIG. 2. (a) Multiple generator, where $T_w = 13$ and $R = 0.8$. (b) Corresponding complex cepstrum.

Table 1. Time and complex cepstrum transform pairs.[1]

$$a^t \left[\delta(t) + \delta(t-\tau) \right]$$
$$a = .94$$

$$\sum_{m=1}^{\infty} \frac{(-1)^{m+1}}{m} a^{m\tau} \delta(T - m\tau)$$

$$a^t \left[\delta(t) - \delta(t-\tau) \right]$$
$$a = .94$$

$$\sum_{m=1}^{\infty} \frac{a^{m\tau}}{m} \delta(T - m\tau)$$

$$\sum_{n=0}^{\infty} (-1)^n R^n \delta(t - n\tau)$$
$$R = .75$$

$$\sum_{m=1}^{\infty} (-1)^m \frac{R^m}{m} \delta(T - m\tau)$$

$$\sum_{n=0}^{\infty} R^n \delta(t - n\tau)$$
$$R = .75$$

$$\sum_{m=1}^{\infty} \frac{R^m}{m} \delta(T - m\tau)$$

$$f(t)$$
TIME

$$\hat{f}(T)$$
COMPLEX CEPSTRUM

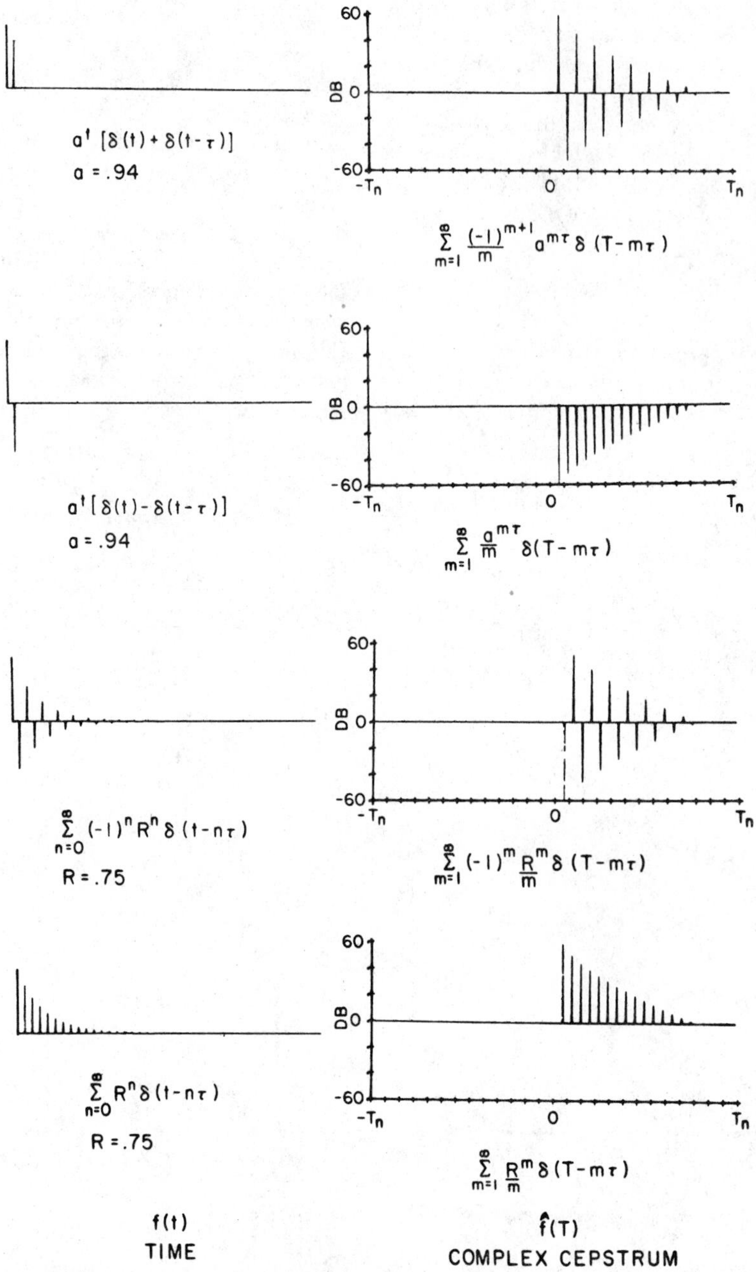

[1]It is apparent that a Backus two-point filter (first time function) is the convolutional inverse to an alternating sign multiple generator (the third time function) since the addition of the first and third complex cepstra would be zero for $R = a^\tau$. Similarly, the second and fourth transform pairs are inverses of each other.

complex cepstrum which are confined to periods greater than the two-way traveltime between the shot and the first reflector, one deconvolution we can perform is to set all these periods equal to zero. While this is a useful first attempt, we have shown that the bubble pulse and multiple generator of the water column are of infinite extent in the complex cepstrum. The higher-order complex-cepstrum terms of the source and multiple generator will be in the region of the reflectors.

Deconvolving a band-pass wavelet

First, let us consider a time function, Figure 3a, whose phase spectrum, power spectrum, and com-

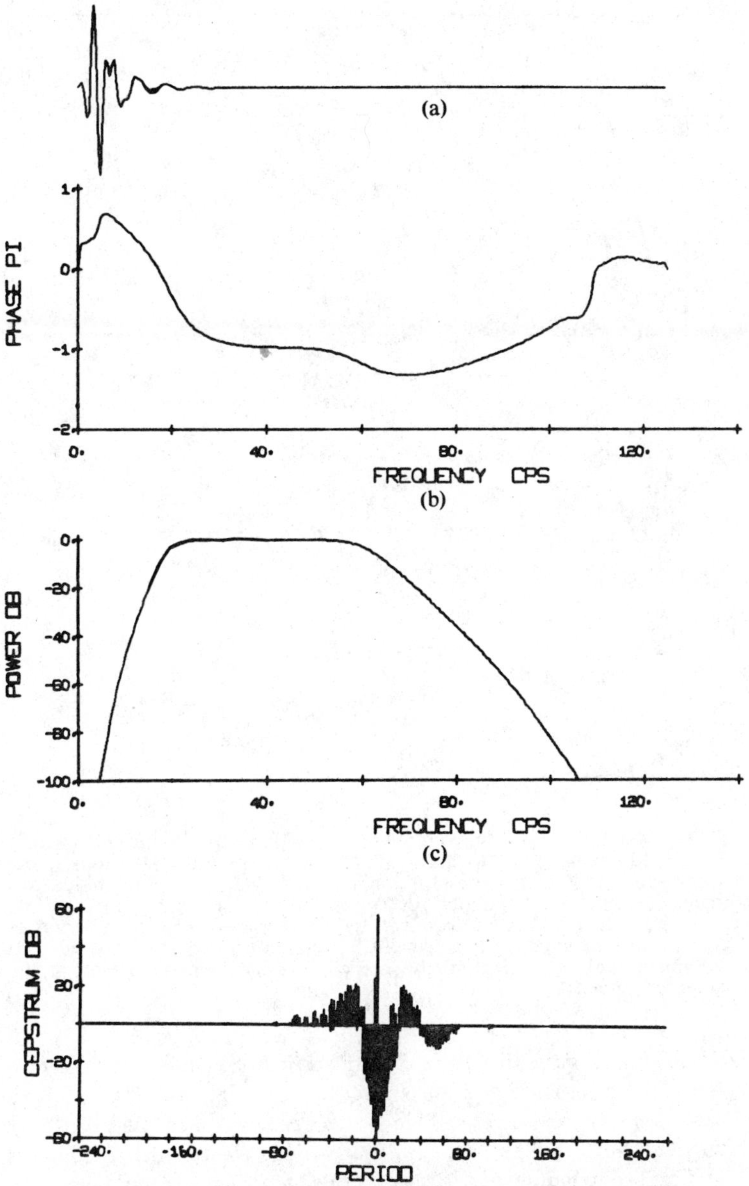

FIG. 3. (a) 20-60 hz, 8-pole band-pass wavelet. (b) Phase spectrum. (c) Power spectrum. (d) Corresponding complex cepstrum. (Note the nonimpulsive nature of the complex cepstrum in contrast to Figures 1 and 2.)

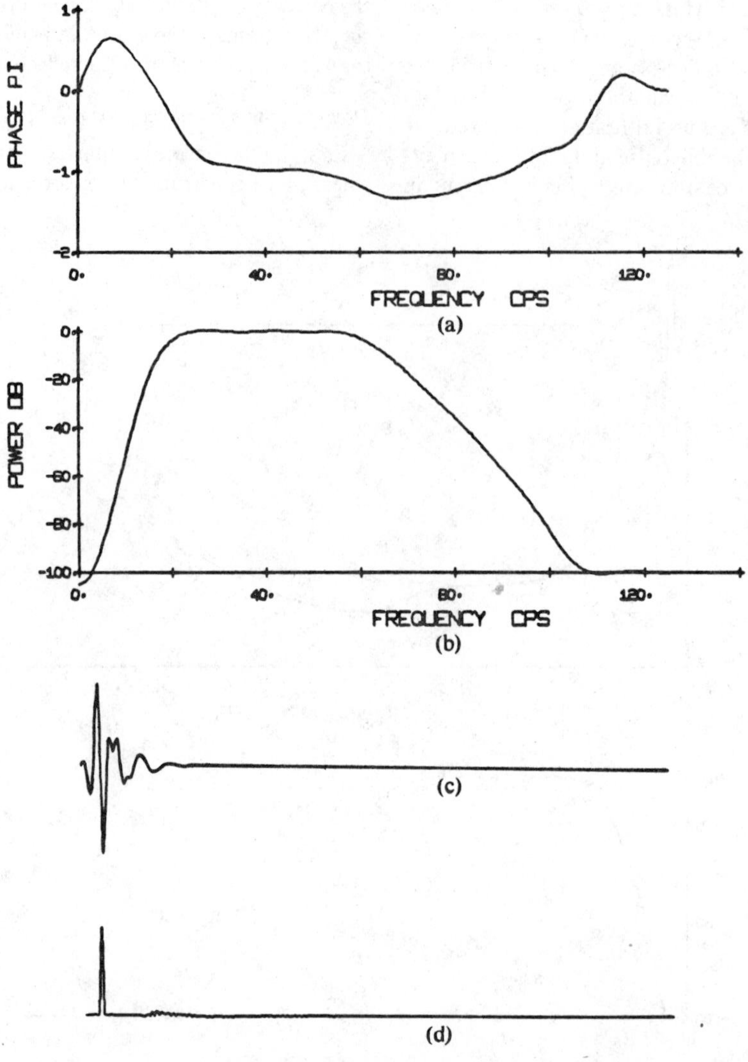

FIG. 4. (a) Phase spectrum obtained by defining the complex cepstrum in Figure 3d equal to zero for $|T| > 18$. (b) Power spectrum for same. (Note the similarity to Figures 3b and c indicating the dominance of the central complex-cepstrum contributions for this wavelet.) (c) Corresponding time domain wavelet. (d) Wave train given by zeroing the complex cepstrum of Figure 3d in the range $-18 \leq T \leq 18$. This wave train is the convolutional noise introduced by neglecting the complex cepstrum contribution of the wavelet for $|T| > 18$.

plex cepstrum are shown in Figures 3b, c and d. The function is the impulse response of a 20–60 hz, 8-pole band-pass filter representing a possible seismic source function. Note that the complex cepstrum has a significant contribution at all negative periods indicating a substantial maximum-phase component. Any attempt to recover a reflector series from a trace formed by convolving this function with the reflector series will fail if one uses a minimum-phase source assumption.

We remove all complex-cepstrum contributions at periods greater than 18. That is, all contributions at periods greater than $+18$ and less than -18 are set equal to zero. In Figure 4a and b we see the resulting phase and power spectra and in Figure 4c the time-domain function, which is al-

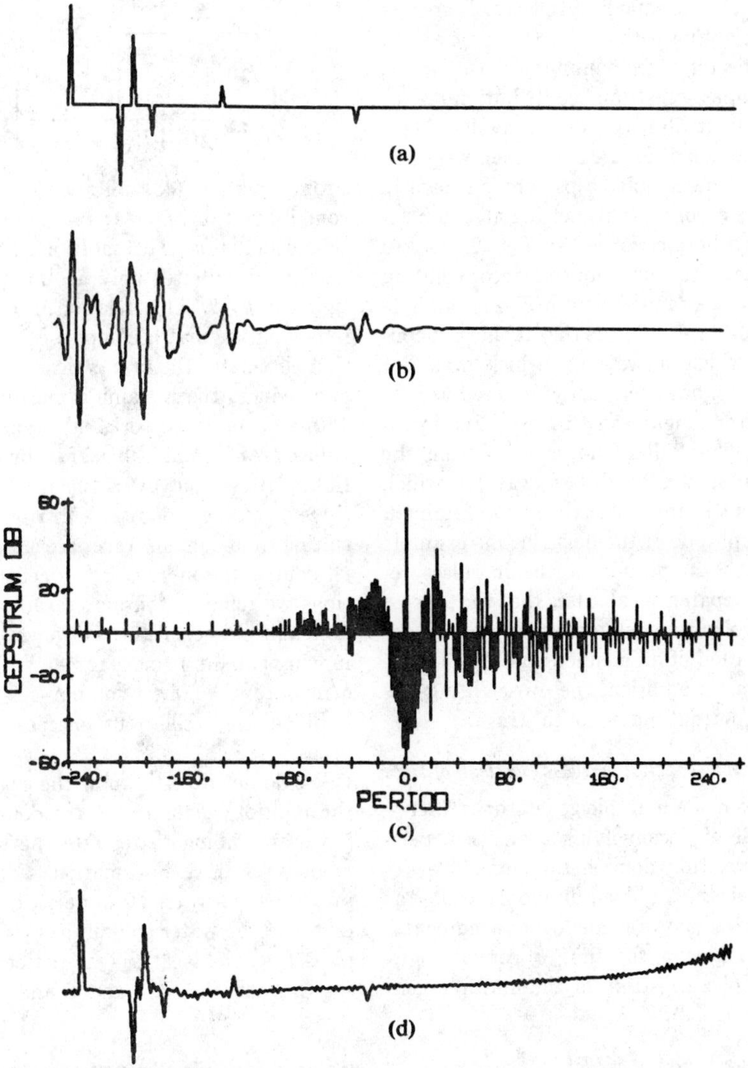

FIG. 5. (a) Mixed phase impulse train which is arbitrary except that separation of the first two impulses is 19. (b) Convolution of this impulse train with the wavelet of Figure 3a. (c) Complex cepstrum of the time series of Figure 5b, weighted with $a=0.98$. This weighting is sufficient to make the reflector series, but not the trace, minimum phase. (Note that a small amount of aliasing is present.) (d) Deconvolved time series obtained by defining all complex cepstrum contributions for $-18 \leq T \leq 18$ as zero. (Note the convolutional noise, introduced as in Figure 4d, gets amplified at the end of the time series due to the unweighting operation.)

most identical to the original in Figure 3a. Therefore, removing complex-cepstrum contributions at locations $-18 \leq T \leq 18$ should remove most of the effects of a convolution with this function as shown in Figure 4d, which is our deconvolved source. Of course, any reflector whose first complex-cepstrum contribution lies between 0 and 18 will be removed along with the source (and, incidentally, will introduce a small amount of convolutional noise). Hence, it may be necessary to sacrifice a small amount of reflector information in order to adequately deconvolve the source.

For purposes of illustration, we choose a mixed-phase reflector series, Figure 5a, where the separation between the shot and first reflector is greater than 18. (Weighting with $a=0.98$ is sufficient to make this reflector series minimum-phase, hence, its first complex-cepstrum contribution is at $T>18$.) In Figure 5b it is shown convolved with the source function of Figure 3 and then weighted with $a=0.98$. The complex cepstrum is shown in Figure 5c. We deconvolve by setting all complex-cepstrum contributions for $-18 \leq T \leq 18$ equal to zero. The trace resulting from this deconvolution is then unweighted and shown in Figure 5d. The importance of this example is that it shows a convolution in which a wavelet, which was not minimum-phase, has been easily deconvolved. It is apparent from Figure 5d that we have introduced noise. This "filter" noise arises from the contributions of the band-pass wavelet which are still present in the periods we have retained. While the amplitude of the noise in this example is less than the amplitude of the impulses recovered, it is apparent in all cases that the success of the deconvolution is related directly to the degree of separation of all components in the complex cepstrum. In addition, the unweighting has amplified the noise at the end of the trace.

Deconvolving an impulsive, periodic function of time

Consideration of a multiple generator will serve as an example of deconvolving a damped, periodic, impulsive, time-domain function. The deconvolution of the bubble-pulse oscillations as developed above follows an analogous argument.

Consider removing the first n contributions from the complex cepstrum of the multiple generator, equation (22).

$$\hat{p}_n(T)=\hat{m}(T)-\sum_{l=1}^{n}(-1)^l\frac{R^l}{l}\delta(T-lT_w). \quad (25)$$

Performing the subtraction and taking the z-transform gives

$$\hat{P}_n(z) = \sum_{l=n+1}^{\infty}\frac{\zeta^l}{l}, \quad (26)$$

$$\text{where } \zeta = -Rz^{-T_w}.$$

Exponentiating and expanding the exponential in a power series, we obtain

$$P_n(z)=\prod_{l=n+1}^{\infty}\sum_{j=0}^{\infty}\frac{1}{j!l^j}\zeta^{lj} \quad (27)$$

$$=1+\sum_{k=1}^{\infty}\frac{\zeta^{n+k}}{n+k}+\sum_{k=2}^{\infty} \cdot\left[\sum_{l=1}^{k}\frac{1}{2(n+l)(n+k-l)}\right]\zeta^{(2n+k)}+\cdots. \quad (28)$$

From equation (28) we see that except for the contribution of unity at $t=0$, the first contribution is at the $(n+1)$th multiple. The second summation contributes only at the $(2n+k)$th multiple for $k\geq 2$. This relation clearly shows that removing the first n complex-cepstrum contribution eliminates the first n time-domain multiples and reduces the remaining multiples to at most $1/(n+1)$ of their original amplitude. Actual values are given in Table 2 for the first eight multiples, after removal of the first one, two, and three complex-cepstrum contributions. The resulting time-domain traces are shown in Figure 6. Of course, the more complex-cepstrum contributions we remove, the more exactly we eliminate the multiple generator. However, by simply removing two, or at most three, we can sufficiently attenuate the water-column reverberations or bubble-pulse oscillations for most practical purposes.

We do not need to know the exact locations of the multiple generator in the complex cepstrum if we are willing to tolerate the loss of a small amount of trace information. For example, we know that in water 60 m deep the multiple generator of the water column has its first two complex-cepstrum locations at periods less than .2 sec (in fact, they are at .08 and .16 sec). Addi-

Table 2. Multiple elimination and reduction for the first 8 multiples in time obtained by removing the first $p_1(t)$, first two $p_2(t)$, and first three $p_3(t)$ nonzero complex cepstrum contribution of the multiple generator $m(t)$.

Multiple	m(t)	$p_1(t)$	$p_2(t)$	$p_3(t)$
0	1	1.0000	1.0000	1.0000
1	$-R$	0.0000	0.0000	0.0000
2	R^2	$0.5000\ R^2$	0.0000	0.0000
3	$-R^3$	$-0.3333\ R^3$	$-0.3333\ R^3$	0.0000
4	R^4	$0.3750\ R^4$	$0.2500\ R^4$	$0.2500\ R^4$
5	$-R^5$	$-0.3666\ R^5$	$-0.2000\ R^5$	$-0.2000\ R^5$
6	R^6	$0.3681\ R^6$	$0.2222\ R^6$	$0.1667\ R^6$
7	$-R^7$	$-0.3678\ R^7$	$-0.2262\ R^7$	$-0.1429\ R^7$
8	R^8	$0.3679\ R^8$	$0.2229\ R^8$	$0.1562\ R^8$

tionally, if we use an air gun whose spectrum peaks at 18 hz, the period of the air gun is .055 sec, so that we have three complex-cepstrum contributions of the bubble pulse at periods less than .2 sec. By removing all complex-cepstrum locations for $T<.2$ sec, we eliminate the first two water-column multiples and reduce the remainder to at most one-third of their original amplitude. Also, we eliminate the first three bubble-pulse oscillations and reduce the remaining oscillations to at most one-fourth of their original amplitudes. In effect, we have reduced the amplitude of the remaining terms to the point where they are negligible for most purposes.

Figure 7a shows the impulse train used previously convolved with the multiple generator and the band-pass wavelet. The trace is weighted with $a=0.98$, and Figure 7b is the resulting complex cepstrum. We deconvolve by removing all complex-cepstrum contributions for $-18 \leq T \leq 18$. This deconvolution includes only one contribution of the multiple generator. In Figure 7c we see the deconvolved trace and in Figure 7e the original impulses which we are attempting to recover. A deconvolution which includes the second complex-cepstrum contribution of the multiple generator greatly improves the result, as seen in Figure 7d. The power of this type of deconvolution is that one can improve the trace substantially by simply excluding one or two multiple-generator, complex-cepstrum contributions. In addition, any source convolution products which are maximum-phase can be easily deconvolved by setting the complex cepstrum equal to zero for negative periods. This will have no effect on the reflector series if the trace has been weighted sufficiently to make the reflector series minimum-phase and unaliased.

THE CONTINUOUS RAMP-FREE PHASE SPECTRUM

Given a function of time, its Fourier transform $F(\omega)$ consists of a real part $u(\omega)$ and an imaginary part $v(\omega)$. For properly sampled, time-limited data the complex Fourier coefficients will be samples of the continuous and unaliased $F(\omega)$.

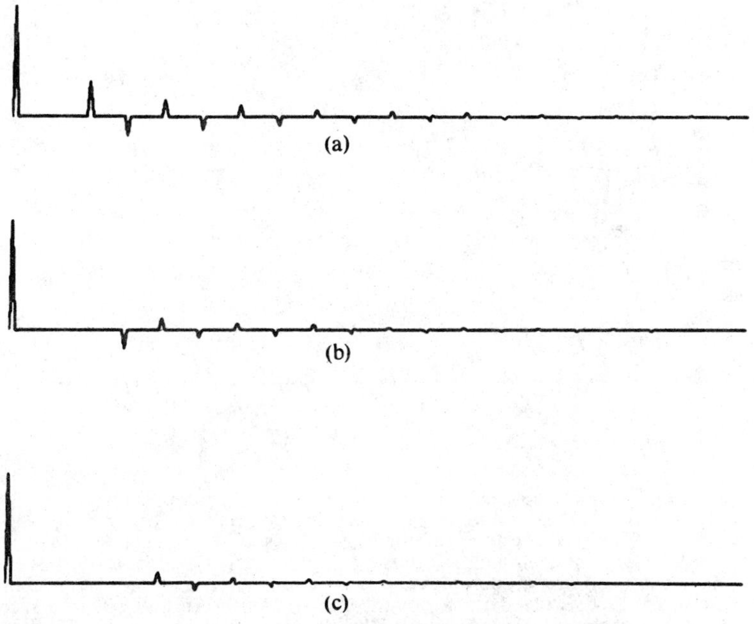

FIG. 6. The deconvolved time series obtained by removing (a) the first one, (b) the first two, and (c) the first three nonzero complex cepstrum contributions of the multiple generator of Figure 2. [Note that eliminating two (or three) complex cepstrum contributions not only eliminates the first two (or three) multiples in time but results in a substantial reduction of the remaining multiples.]

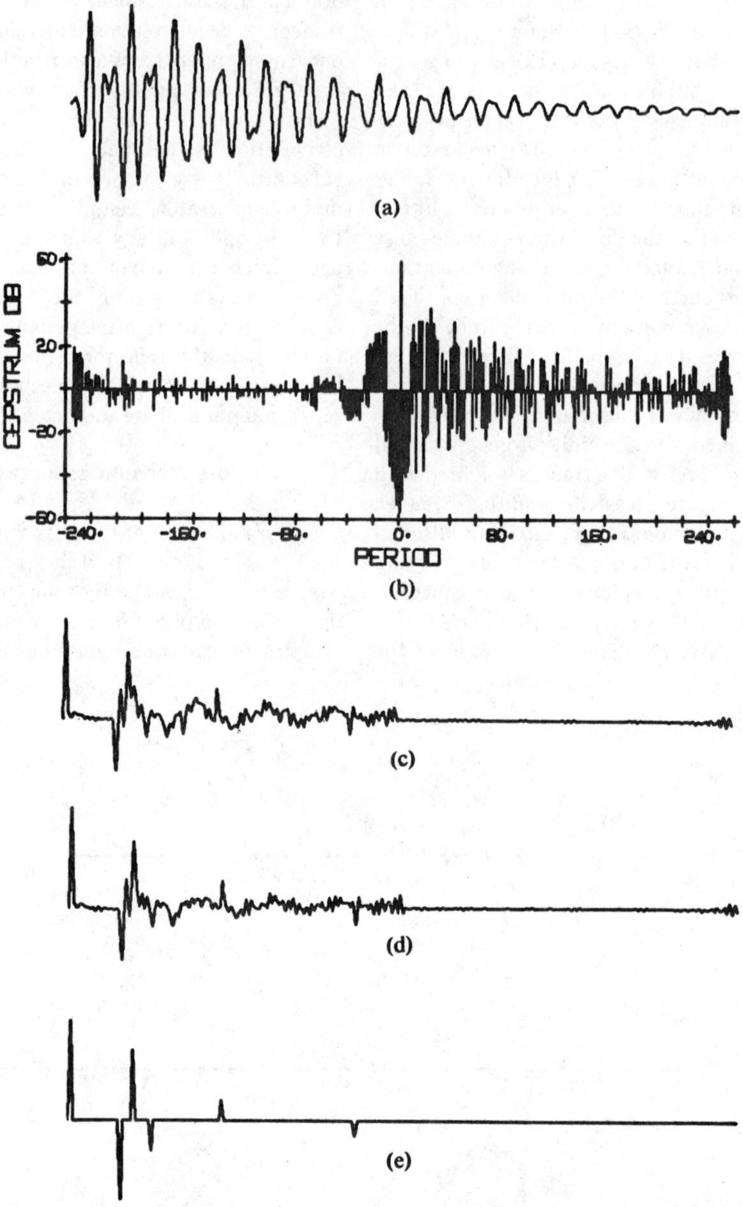

FIG. 7. (a) Time series formed by convolving the impulse train of Figure 5a with the source wavelet of Figure 3a and the multiple generator of Figure 2a. (b) Complex cepstrum of the time series of Figure 5a weighted with $a=0.98$. A small amount of aliasing is present. (c) Deconvolved trace obtained by defining all complex-cepstrum contributions for $-18 \leqslant T \leqslant 18$ as zero. (d) Deconvolved trace obtained as in Figure 7c but with the second complex-cepstrum contribution of the multiple generator at $T=26$ also removed. Notice the substantial improvement over Figure 7c. (Note in 7c and d we have only unweighted half the trace in order to eliminate the amplification of the convolutional noise which occurs at the end of the time series as in Figure 5d.) (e) The impulse train we are attempting to recover from the time series of Figure 7a.

The next step in performing the transformation into the complex cepstrum is to take the complex natural logarithm.

For example, $f(t)$ transforms to $F(\omega)$, where $F(\omega)=u(\omega)+iv(\omega)$. Writing this in terms of amplitude and phase, we obtain

$$\begin{aligned} F(\omega) &= [u^2(\omega)+v^2(\omega)]^{1/2} \\ &\quad \cdot \exp\left\{i \tan^{-1}\left[\frac{v(\omega)}{u(\omega)}\right]\right\} \\ &= A(\omega)\exp[i\phi(\omega)]. \end{aligned} \tag{29}$$

The complex natural logarithm of this expression is

$$\hat{F}(\omega) = \log[A(\omega)] + i\phi(\omega). \tag{30}$$

The greatest difficulty in computing the complex cepstrum is computing the proper phase curve. While $u(\omega)$ and $v(\omega)$ are continuous, $\tan^{-1}[v(\omega)/u(\omega)]$ is in general not continuous. Since the inverse tangent function is multivalued, we must choose the branches which make it continuous. If the principal value is determined, the resulting phase curve is restricted to $-\pi<\phi\leq\pi$. Schafer (1969) has suggested an algorithm for locating the resulting discontinuities or "jumps" in the phase curve and a procedure for "unwrapping" the phase curve. The success of this approach is limited by the fineness with which the original phase curve is sampled. Usually it is necessary to extend the original time function by appending zeros. This results in increased frequency-domain sampling since padding with zeros corresponds to smoothly interpolating between the original frequency-domain coefficients. However, in the presence of a large ramp (linear phase component), this unwrapping procedure still encounters difficulty. This can be alleviated by using an iterative unwrapping algorithm. We unwrap the principal value of the phase curve as suggested by Schafer (1969) and determine the apparent ramp. Next, remove the apparent ramp from the original principal value of the phase and unwrap the resultant. We continue in this manner until there is no apparent ramp.

We have also used an alternative approach to computing a continuous curve which considers the derivative of the phase with respect to frequency. We note that

$$\frac{d}{dx}[\tan^{-1}x] = \frac{1}{1+x^2}. \tag{31}$$

If $x=v(\omega)/u(\omega)$, and $\phi(\omega)=\tan^{-1}(x)$, then

$$\frac{d\phi}{d\omega} = \frac{1}{1+\dfrac{v^2(\omega)}{u^2(\omega)}}\frac{d}{d\omega}\left[\frac{v(\omega)}{u(\omega)}\right], \tag{32}$$

and

$$\begin{aligned} &\frac{d}{d\omega}\left[\frac{v(\omega)}{u(\omega)}\right] \\ &= \left[u(\omega)\frac{dv(\omega)}{d\omega} - v(\omega)\frac{du(\omega)}{d\omega}\right] \Big/ u^2(\omega), \end{aligned} \tag{33}$$

so that

$$\begin{aligned} \frac{d\phi}{d\omega} &= \frac{1}{u^2(\omega)+v^2(\omega)} \\ &\quad \cdot\left[u(\omega)\frac{dv(\omega)}{d\omega} - v(\omega)\frac{du(\omega)}{d\omega}\right]. \end{aligned} \tag{34}$$

Integration of equation (34) yields a continuous phase curve.

The Fourier coefficients are continuous, well behaved functions, which are properly sampled. The derivatives $dv/d\omega$ and $du/d\omega$ can be found to any accuracy desired by using difference methods or by finding the Fourier transform of $t\cdot f(t)$. The integration can also be performed to the desired accuracy. In addition to computing a continuous phase curve, this derivative approach offers a method of removing the linear phase shift. The mean value of $d\phi/d\omega$ is the linear phase shift. Determining this mean, removing it from $d\phi/d\omega$, and then integrating gives the continuous, ramp-free phase curve which we desire.

SUMMARY

Homomorphic deconvolution promises to be an important method of removing the unwanted convolutional components from the shallow-water, single-channel marine seismic trace. The complex cepstrum, the particular homomorphic system considered, offers several advantages over conventional deconvolution. Deconvolution in the complex cepstrum does not require the usual minimum-phase source assumption. Maximum or mixed-phase source components are deconvolved as easily as minimum-phase components. The relation of the complex cepstrum to the original time function is in terms of the periodicities of the time function. Since subbottom reflectors always arrive later in the trace than the bottom reflection, their complex-cepstrum contributions are at correspondingly greater complex cepstrum periods if the reflector series is minimum-phase. Most shallow-water reflector series are minimum-phase (Robinson, 1966). In cases where they are not, a small amount of weighting is usually sufficient to make them minimum-phase.

The bubble-pulse oscillation of an air gun is of short period, often less than the period of the water-column reverberation. All complex cepstra are of infinite extent. However, for an impulsive, periodic time function such as the bubble oscillation or the water-column reverberation, we have shown that merely removing the first few complex-cepstrum contributions reduces their time-domain contributions considerably. Deconvolution via the complex cepstrum allows one to choose readily the extent of the deconvolution and the expense, if any, of the desired reflector information.

Complex-cepstrum analysis requires the computation of a continuous, ramp-free phase spectrum. Algorithms which detect discontinuities in the principal value of the phase curve encounter difficulty in the presence of a large linear term (ramp). This difficulty is eliminated by the methods outlined in this paper.

ACKNOWLEDGMENTS

This work was supported by the National Science Foundation through the Office for the International Decade of Ocean Exploration under grant no. GX-34410, and through the Division of Environmental Sciences under grant GA-27281; and by the Office of Naval Research under grant no. N00014-67-A-0108-0004.

We are grateful to K. McCamy and P. Richards for critically reviewing the manuscript and offering many useful suggestions.

REFERENCES

Backus, Milo B., 1959, Water reverberations—their nature and elimination: Geophysics, v. 24, no. 2, p. 233–261.

Bogert, B. P., Healey, M. J., and Tukey, J. W., 1963, The quefrency analysis of time series for echoes; cepstrum, pseudo-autocovariance, cross-cepstrum, and saphe cracking: Proc. Symp. on Time Series Analysis, M. Rosenblatt, editor, New York, John Wiley and Sons, p. 209–243.

Morse, P. M., Feshbach, H., 1953, Methods of theoretical physics: New York, McGraw-Hill Book Co., Inc., 1977 p.

Oppenheim, A. V., 1965, Superposition in a class of non-linear systems: Tech. Rep. 432, MIT, Res. Lab. of Electr., 62 p.

Pfleuger, J., 1972, Spectra of water reverberations for primary and multiple reflections: Geophysics, v. 37, p. 788–796.

Rice, R. B., 1962, Inverse convolution filters: Geophysics, v. 27, no. 1, p. 4–18.

Robinson, E. A., 1957, Predictive decomposition of seismic traces: Geophysics, v. 22, no. 4, p. 767–778.

——— 1966, Multichannel z transforms and minimum delay: Geophysics, v. 31, p. 482–500.

Schafer, R. W., 1969, Echo removal by discrete generalized linear filtering: Tech. Rep. 466, MIT, Res. Lab. of Electr.

Schulze-Gatterman, R., 1972, Physical aspects of the "air pulser" as a seismic energy source: Geophys. Prosp., v. 20, p. 155–192.

Ulrych, T. J., 1971, Application of homomorphic deconvolution to seismology: Geophysics, v. 36, no. 4, p. 650–660.

——— 1972, Homomorphic deconvolution of some teleseismic events: Bull. SSA, v. 62, p. 1253–1265.

Ulrych, T. J., and Lasserre, M., 1966, Minimum phase: J. Canadian SEG, v. 2, no. 1, p. 22–32.

Ziolkowski, A., 1970, A method of calculating the output pressure waveform from an air gun: Geophys. J. Roy. Astr. Soc., v. 21, p. 137–161.

Reprinted from Geophysics v. 39, no. 4, p. 417-426

THE APPLICATION OF HOMOMORPHIC DECONVOLUTION TO SHALLOW-WATER MARINE SEISMOLOGY—PART II: REAL DATA

PETER BUHL,*§ PAUL L. STOFFA,*§ AND GEORGE M. BRYAN*

The application of the techniques of Part I to seismic reflection data acquired on the Argentine continental shelf yields results which appear superior to time-domain, minimum-phase inverse filtering via the auto-correlation function. This is in part because of a narrow-band, maximum-phase source component. Minimum-phase deconvolution disperses this component rather than compressing it. Very slight exponential weighting (a^t, $a=0.998$) appears to make the reflector series minimum phase. This weighting in conjunction with quadrupling the trace length by extending it with zeros virtually eliminates aliasing in the complex cepstrum. Simple zeroing of the complex-cepstrum terms works well as a deconvolution technique even though for exactness their harmonics at longer cepstrum periods should also be removed.

INTRODUCTION

In Part I (Stoffa et al, 1974) we have investigated the complex cepstrum theoretically and through the use of models for signals of interest in shallow-water marine seismology. We have emphasized the fact that any finite time series has a complex cepstrum which is of infinite extent. This presents two problems: First, since we use the discrete Fourier transform (DFT) to determine the complex cepstrum, the cepstrum is an aliased function. This problem can be alleviated by appending zeros to the original time series and by a judicious exponential weighting of the trace.

Second, any deconvolution via the complex cepstrum which neglects the fact that each convolutional component in time has an infinite complex cepstrum cannot result in an exact deconvolution. However, if the reflector series is minimum phase, or has been made so by an appropriate exponential weighting, we can remove the maximum-phase trace component exactly by setting to zero all complex-cepstrum contributions for $T<0$.

We deconvolve the minimum-phase source component approximately by setting to zero all complex-cepstrum contributions less than a carefully chosen short-pass period T_{sp}. In the case of an impulsive, damped, periodic time function (such as the minimum-phase component of the bubble pulse oscillation), we have shown that if T_{sp} includes this function's first n complex-cepstrum contributions, we will eliminate the first n multiples in time and reduce the remaining multiples to, at most, $1/(n+1)$ of their original value. This should result in a substantial improvement of the single-channel, shallow-water trace.

DATA ACQUISITION

The data used in this paper were collected on board the R/V *Robert D. Conrad* on the Argentine continental shelf near Bahía Blanca in April 1972. The seismic system included a single 30 cu in,

Lamont-Doherty Geological Observatory contribution No. 2089.

Presented at the 43rd Annual International SEG Meeting, October 24, 1973, Mexico City. Manuscript received by the Editor July 23, 1973; revised manuscript received September 11, 1973.

* Lamont-Doherty Geological Observatory, Palisades, NY 10964.

§ Columbia University, New York, N.Y. 10027.

free-firing air gun and a single-channel detector streamer with an active length of 80 m, located approximately 250 m behind the air gun and 20 m below the surface. The seismic signal was prefiltered with an anti-alias filter with 60 db/octave rolloff and corner frequency of 62.5 hz and a high pass filter set at 5 hz. Due to the rapid rolloff of the anti-alias filter, considerable phase shifting is introduced near the corner frequency and above.

The signal was sampled at 4 msec intervals with a 14 bit A/D converter and recorded on digital magnetic tape. Four seconds of data were digitized starting 0.1 sec after the shots, which occurred every 50 m (10 sec).

In this paper, we have used only the first 2 sec of each trace. With a constant water depth of 30 m there is considerable moveout in the early part of the trace due to the separation of the air gun and the streamer. We did not correct the data for this effect since the primary contaminant in the data is the air-gun bubble pulse and not water-column reverberation.

DECONVOLUTION EXAMPLE

A representative trace of 512 points is shown in Figure 1a. To reduce aliasing in the complex cepstrum we append zeros to the trace to yield 2048 data points and apply an a^t weighting function with $a=0.998$. We have also applied a cosine taper to the beginning of the trace to mute the direct arrival. This resultant trace, Figure 1b, yields the complex cepstrum shown in Figure 1c.

Although at first glance the complex cepstrum is not very elucidating, several observations can be made. First, due to the zero pad and weighting, the aliasing is negligible. On both the minimum-phase side ($T>0$) and the maximum-phase side ($T<0$) of the complex cepstrum, the complex-cepstrum terms are 60 db down well before we reach the folding periods at $|T|=1024$. Second, the center of the complex cepstrum, $|T|\gtrsim 20$, is predominantly odd, mirroring the dominance of the phase curve in this region, as a result of the effects of the anti-alias filter.

Input power and phase spectra

The intermediate step in the calculation of the complex cepstrum, the unwrapped, ramp-free phase and power spectra, is shown by the curves marked "a" in Figures 2a and 2b, respectively. In the phase spectrum the effect of the anti-alias filter is clearly seen out to approximately 90 hz. Above this point the power has been reduced by the anti-alias filter to the digitizing noise level yielding an unreliable phase curve. The power curve (Figure 2b, line a) is dominated by the anti-alias and low-cut filters and by the spectral peaks of the air gun at 18, 36, and 54 hz.

One piece of information that comes from the phase-curve unwrapping schemes described in Part I is the ramp in the original weighted data. The magnitude of this ramp for positive frequencies in units of π phase shifts equals the time duration minus one of the *maximum-phase* component (Robinson, 1967, p. 186–187). Typical ramp values for this data set are 45 to 55 π. This implies a time duration of $\sim$50 time units for the *maximum-phase* component of the trace (Figure 1b). This component includes the virtually zero-amplitude lead-in portion of the trace. Since the total length is 512, the time duration of the *minimum-phase* component is approximately 450 (ignoring the zero pad). From the above argument we conclude that only the first 50 terms on the *maximum-phase* side are fundamentals, even though the complex cepstrum does not finally fall to 0 db until $T=-560$. These additional contributions are harmonics of the first 50. The slightly less rapid fall-off of the *minimum-phase* side indicates fundamental information out to almost the length of the trace (512).

The minimum-phase reflector-series requirement

One of the fundamental requirements for deconvolution via the complex cepstrum is that the *reflector series*, but not necessarily the trace, be minimum phase or be made minimum phase by weighting. Robinson (1966) has indicated that it may be reasonable to assume that a reflector series is minimum phase. Since the reflection coefficients of the series are less than unity and the effect of transmission and reflections is to attenuate the seismic source with time, the derived reflector series will have its energy concentrated near the beginning of the series.

For the data considered, the unweighted reflector series may be slightly mixed phase. However, only a small amount of weighting, $a=0.998$, is considered necessary to make the reflector series minimum phase since we believe that with this weighting the maximum-phase component can be associated solely with the source.

To show this we have inverse transformed the

maximum-phase part of Figure 1c. The resulting time series is shown in Figure 2d. The time duration is approximately 0.19 sec as predicted by the 47 π linear phase term of this weighted trace. The dominant frequency is 54 hz, corresponding to the second harmonic of the 18 hz peak-frequency air gun. Thus the entire maximum-phase time series is probably due only to the source. This suspicion is given further strength when time-domain, minimum-phase deconvolution is applied to the original data (Figure 5). We will discuss the details of this process later.

However, in these traces we can see a large 54 hz component in the processed traces. This is not due to a failure to whiten the spectrum, but rather to the removal of an incorrect source phase curve in this frequency range where the source (in this band of frequencies) is dominantly maximum, not minimum, phase. Thus, we have dispersed rather than compressed this band of frequencies.

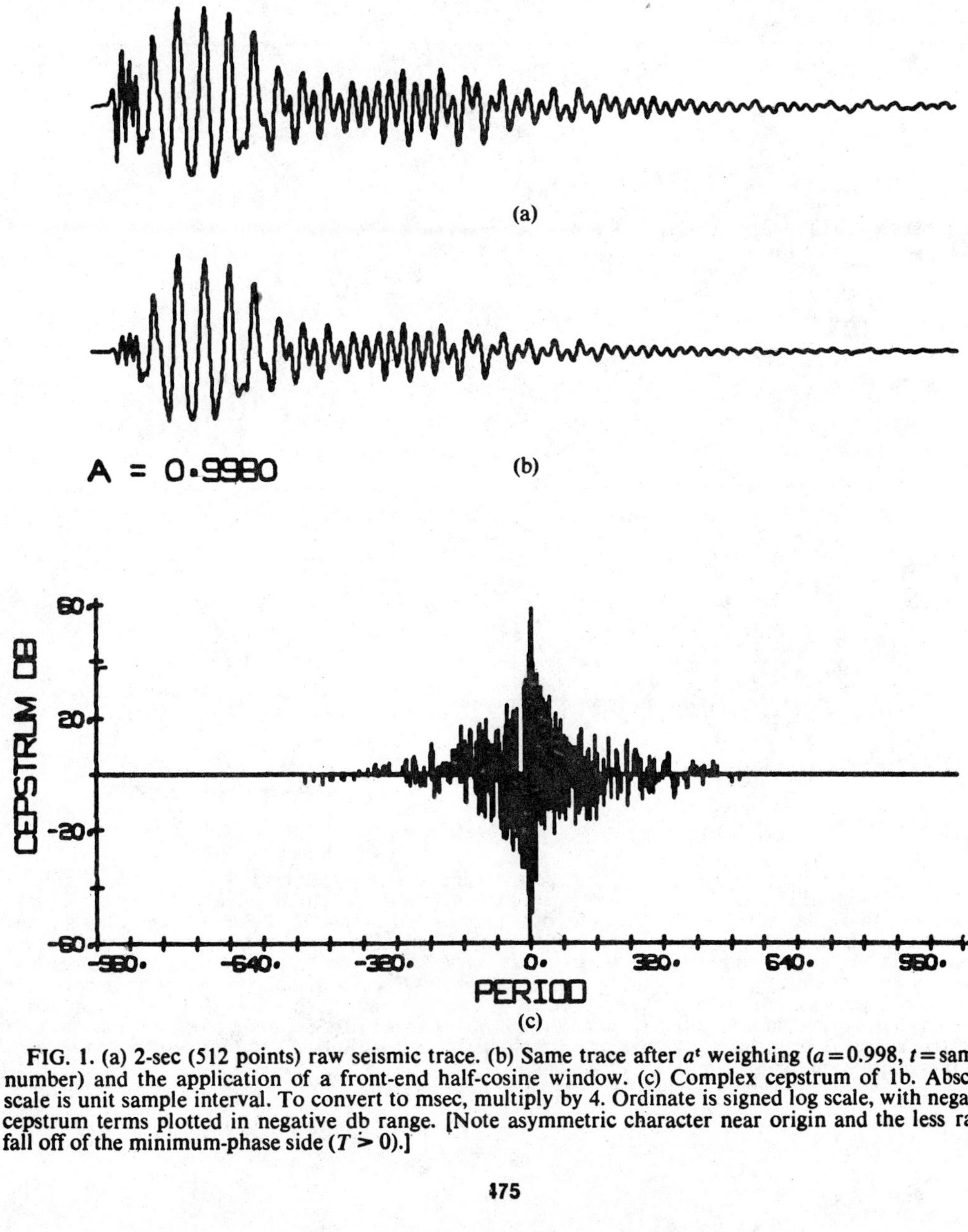

FIG. 1. (a) 2-sec (512 points) raw seismic trace. (b) Same trace after a^t weighting ($a = 0.998$, $t =$ sample number) and the application of a front-end half-cosine window. (c) Complex cepstrum of 1b. Abscissa scale is unit sample interval. To convert to msec, multiply by 4. Ordinate is signed log scale, with negative cepstrum terms plotted in negative db range. [Note asymmetric character near origin and the less rapid fall off of the minimum-phase side ($T > 0$).]

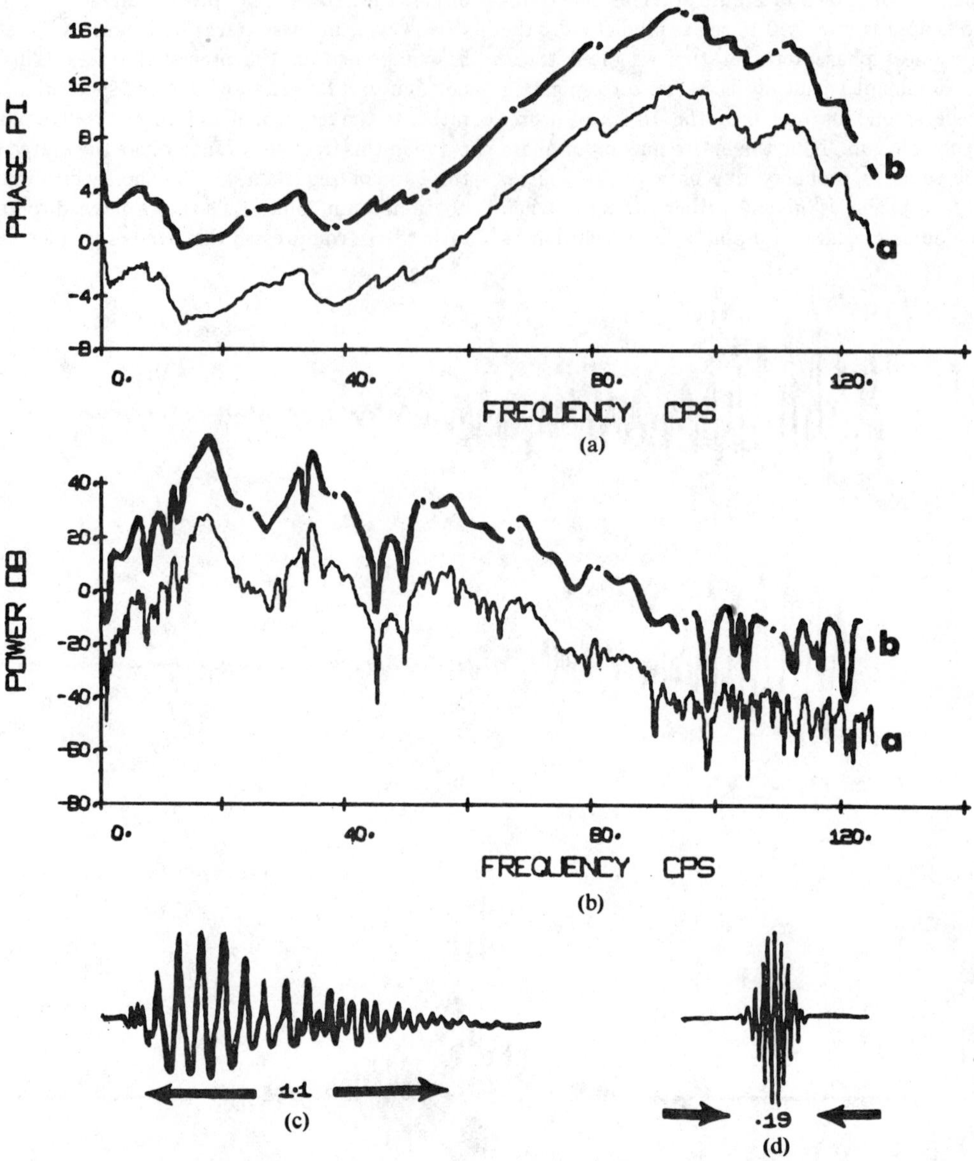

FIG. 2. (a) Unwrapped, ramp-free phase spectra. Line a, phase spectrum of input trace, Figure 1b. Line b, phase spectrum of effective source derived from complex cepstrum in Figure 1c. Note the effect of the anti-alias filter out to 90 hz. Beyond 90 hz phase is unreliable. (b) Power spectra. Line a, power spectrum of input trace, Figure 1b. Line b, power spectrum of effective source derived from complex cepstrum in Figure 1c. Note air-gun spectral peaks at 18, 36, and 54 hz and filter roll-offs. Lines b are offset by 6π in the phase plot and 30 db in the power plot. (c) Effective source derived by zeroing complex cepstrum, Figure 1c, for periods $44 \leq T \leq 1024$. (Note the ringing character due to air-gun bubble fundamental. Arrows indicate approximate duration in seconds.) (d) Maximum-phase component of the effective source derived by zeroing positive (minimum phase) periods of complex cepstrum, Figure 1c. Arrows indicate approximate duration in seconds.

The effective source

In our attempts to recover a satisfactory effective source from the complex cepstrum we have used two criteria: the source-function's power and phase spectra and the time series deconvolved with this source.

As discussed above, it is quite probable that the entire maximum-phase part of the complex cepstrum, $-T_n < T < 0$, is due to the source, and thus the weighted reflector series is minimum-phase. However, the minimum-phase side of the complex cepstrum presents a problem. We must decide which periods we are going to include as predominantly source contributions.

As shown in Part I, if we include two or three complex-cepstrum contributions due to an impulsive, decaying, periodic function (such as the water-column multiple generator or, perhaps, the minimum-phase component of the bubble pulse), we can achieve substantial elimination and suppression of this function in the time domain. (It should be pointed out that due to the uncorrected moveout, the water-column multiple generator is not periodic.)

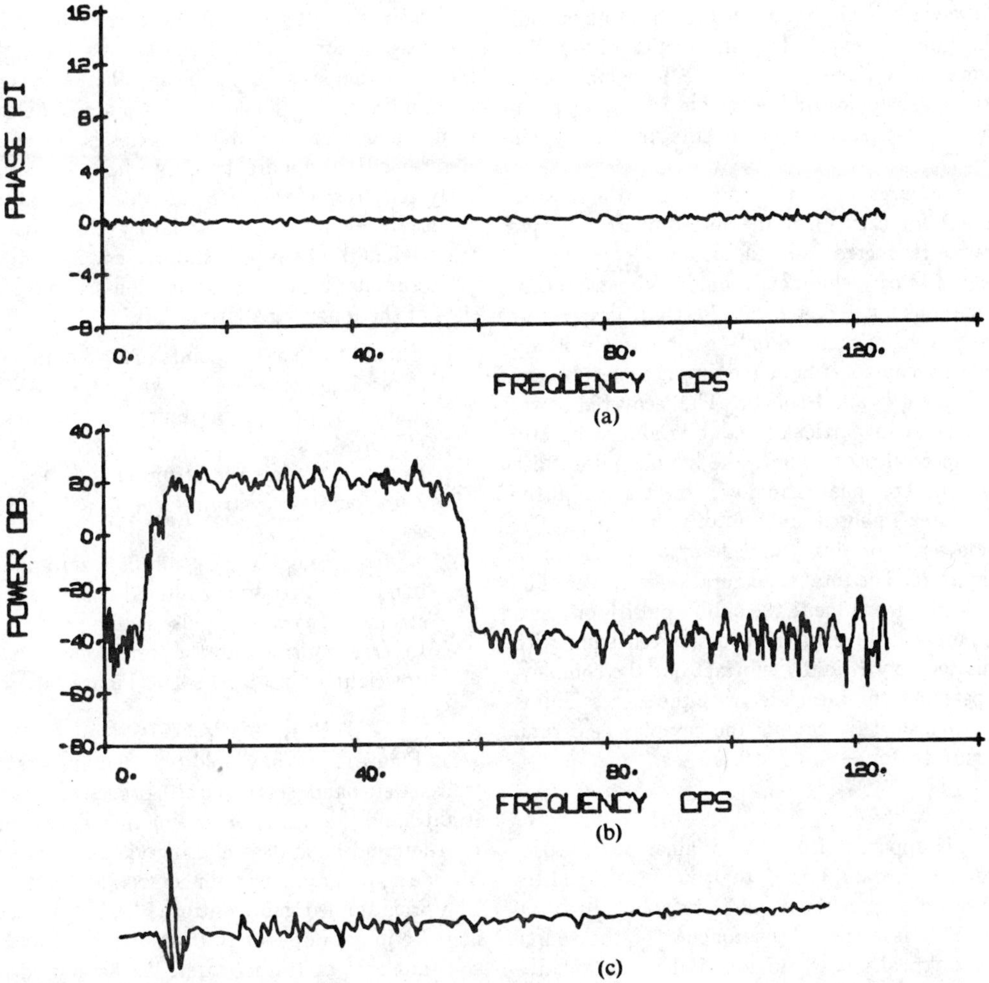

FIG. 3. (a) Output phase (b) Output power spectra. Both (a) and (b) are spectra after deconvolution of Figure 1b with effective source of Figure 2c and band-pass filtering via the complex cepstrum. (c) Output trace. (Note the first arrival which is the approximate output pulse shape.)

The bubble pulse oscillation for these data has a period of 0.055 sec. In the complex cepstrum there are local highs of about 10 db at $|T|=0.052$, 0.104, and 0.156 sec (for 4 msec data this corresponds to $n=13$, 26, and 39). Consequently, we have chosen $T_{sp}=0.176$ sec ($n=44$), which will include the first three complex-cepstrum contributions of the bubble pulse. This will guarantee a substantial reduction of the effects of this component in the time domain.

However, we may have included some complex cepstrum contributions of the reflector series. If the reflector series is minimum-phase, we lose all reflector information out to 0.176 sec two-way traveltime when we remove these contributions. However, we will not affect the reflectors outside this range except for the introduction of convolutional noise. This noise is due to complex-cepstrum harmonics in the region $T>T_{sp}$ of minimum-phase source components in the region $T\leq T_{sp}$.

In Figures 2a and b, lines b are the spectra (offset for clarity) of the mixed-phase effective source recovered from the complex cepstrum of Figure 1c by including all complex-cepstrum contributions for $-T_n<T<0.176$ sec. It is apparent that we have retained only the relatively slowly varying components of the original spectra seen in Figures 2a and b, lines a. The recovered power spectrum, in particular, exhibits almost exactly the spectral peaks due to the bubble pulse. Both the recovered phase and power spectra exhibit the character imposed by the anti-alias filter. The time series of this source function is shown in Figure 2c. The total time duration is 1.1 sec. The periodic nature due to the bubble pulse is obvious.

To deconvolve in the complex cepstrum with this source we simply subtract out the complex-cepstrum terms associated with the source. This is accomplished by zeroing the complex cepstrum, Figure 1c, for $-T_n<T<0.176$ sec.

Pulse shaping

It is apparent from the original power spectrum, Figure 2b, line a, that our best signal-to-noise ratio is in the band from about 10 to 60 hz. This is a direct consequence of the source spectral peaks at 18, 36, and 54 hz and the anti-alias filter. Therefore, we tailor the output spectrum of our deconvolved trace with a zero phase spectral window which is flat from 15 to 50 hz and has a Hanning taper such that at 6 hz and 59 hz the window is 60 db down from the passband and remains at this level. The even complex cepstrum corresponding to this spectral window is added to the filtered complex cepstrum of the trace. Filtering in this way results in a considerable saving of time since we are performing simple addition rather than complex multiplication or convolution.

Output trace and spectra

The processed output phase and power spectra are shown in Figures 3a and b. The power spectrum does not exhibit any spectral peaks due to the bubble pulse, and the phase curve is largely flat. (Figures 3a and b are plotted to the same scale as the input spectra in Figures 2a and b for purposes of comparison. Even at this scale, the phase curve does show significant fluctuations although they are confined to $-\pi\leq\phi\leq\pi$.) The corresponding deconvolved time series is shown in Figure 3c. It is apparent that we have lost some early reflector information; however, several arrivals are evident. The oscillatory character of the original trace has been suppressed.

To summarize, our deconvolution scheme consists of the following steps:

1. Extend trace of 512 points to 2048 points by appending zeros; weight with a^t, $a=0.998$; and window the beginning of the trace (Figure 1b).
2. Compute complex cepstrum (Figure 1c).
3. Zero complex cepstrum for $-T_n\leq T\leq 0.176$ sec.
4. Add the complex cepstrum of a zero-phase band-pass filter to the filtered complex cepstrum of the trace for pulse shaping.
5. Inverse transform to the time domain, and unweight with a^{-t}, $a=0$ 998 (Figure 3c).

SAMPLE RECORD SECTION

In Figure 4 we have plotted 10 unprocessed traces (left-hand side) and 10 processed traces (right-hand side). The processing includes deconvolution and band-pass filtering via the complex cepstrum, programmed gain control, and mixing. The programmed gain control (PGC) is a linear increase in gain of 1 to 10 from 0 to 1 sec and a constant gain of 10 thereafter. We have used an equal-weight running mix of five processed traces as indicated in the figure. Thus processed traces 1, 2, and 3 are completely independent of traces 8, 9, and 10. Comparing these independent traces

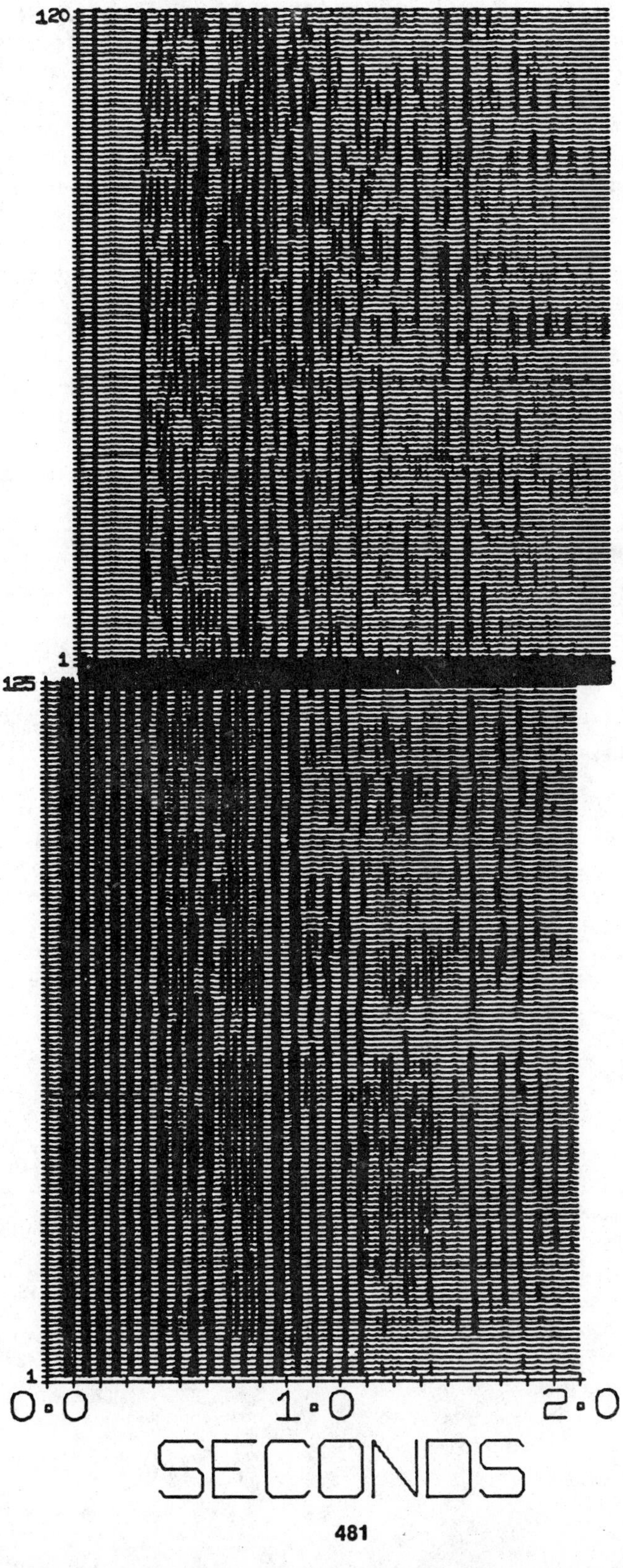

FIG. 6. Record section; original (left) and complex-cepstrum deconvolved (right). (Note removal of oscillations with retention of reflector information.) Both sections have a five-fold mix and PGC as in Figure 4. The initial spikes were aligned and, based on known reflectors, the processed section was shifted to correlate with the unprocessed.

PGC as described above and a running mix of five to both the original and processed traces. Thus, the only difference between the two profiles is the deconvolution via the complex cepstrum and subsequent pulse shaping. We have completely eliminated the contributions due to the bubble pulse, which dominate the original data. Several arrivals which were obscured by the bubble pulse are readily recognizable. We have confidence in this deconvolution technique (which means we have confidence in our determination of the continuous, ramp-free phase spectrum) since the results are repeatable. Moreover, as predicted, we have not eliminated any reflectors that are obvious in the original data except those at the beginning of the trace.

SUMMARY

In Part I of this paper we found that any maximum-phase source component can be readily deconvolved via the complex cepstrum if the reflector series is minimum phase. In addition, any impulsive, damped, periodic time function can be substantially reduced by simply removing the first two or three of its complex cepstrum contributions. In Part II we have applied the method of Part I to real, single-channel, marine seismic data.

For the data considered, we have shown that the reflector series can be made minimum phase by a slight amount of exponential weighting. This allows us to deconvolve the maximum-phase source component exactly. The minimum-phase source component was deconvolved by removing all complex cepstrum contributions from 0 to 0.176 sec. This guarantees the removal of the fundamental and first two harmonics of the bubble pulse from the complex cepstrum and virtually eliminates the bubble-pulse contribution in the time domain. In addition, we have applied the spectral window of a band-pass filter via the complex cepstrum which has resulted in a considerable saving of computation time.

We have not investigated the effect of additive noise on complex cepstrum analysis. However, the successful application to real data with a good signal-to-noise ratio indicates that additive noise is probably no more significant in deconvolution via the complex cepstrum than it is in other methods. Certainly, if there is a poor signal-to-noise ratio, difficulty will be encountered in generating the proper phase curve.

The fundamental noise problem associated with the recovery of the reflector series from the long-pass portion of the complex cepstrum is that of convolutional noise introduced by neglecting to remove the complex-cepstrum harmonics of the source from the reflector region of the complex cepstrum. A report on this problem is in preparation by two of the authors (Stoffa and Bryan). The elimination or minimization of this type of noise must be accomplished before one can properly ascertain the effect of additive noise on complex cepstrum analysis.

Some authors (Ulrych, 1971, 1972; Kemerait and Childers, 1971) have indicated that additive noise seems to affect the direct recovery of the long-pass region from the complex cepstrum. While this may indeed be the case, it cannot be proven until the harmonics of the short-pass region that extend into the long-pass region of the complex cepstrum have been taken into account. For our data we have found that deconvolution directly via the complex cepstrum produces results identical to those obtained by source determination in the complex cepstrum and subsequent frequency-domain deconvolution.

ACKNOWLEDGMENTS

This work was supported by the National Science Foundation through the Office for the International Decade of Ocean Exploration under grant no. GX-34410 and through the Division of Environmental Sciences under grant no. GA-27281; and by the Office of Naval Research under grant no. N00014-67-A-0801-0004.

We would like to thank Dennis Carmichael for assistance in implementing our sea-going digital seismic acquisition and processing system. We would also like to thank Keith McCamy and Paul Richards for critically reviewing the manuscript.

REFERENCES

Kemerait, R. C., and Childers, D. G., 1972, Signal extraction by cepstrum techniques: IEEE Trans., Information Theory, v. IT-18, no. 6.

Robinson, E. A., 1966, Multichannel *z* transforms and minimum delay: Geophysics, v. 31, p. 482–500.

——— 1967, Statistical communication and detection: London, Charles Griffin and Co., Ltd.

Stoffa, P. L., Buhl, P., and Bryan G. M. 1974 The application of homomorphic deconvolution to shallow-water marine seismology—Part 1: Models: Geophysics (this issue) p. 401–416.

Ulrych, T. J., 1971, Application of homomorphic deconvolution to seismology: Geophysics, v. 36, no. 4, p. 650–660.

——— 1972, Homomorphic deconvolution of some teleseismic events: Bull. SSA, v. 62, p. 1253–1265.

NOTES

NOTES

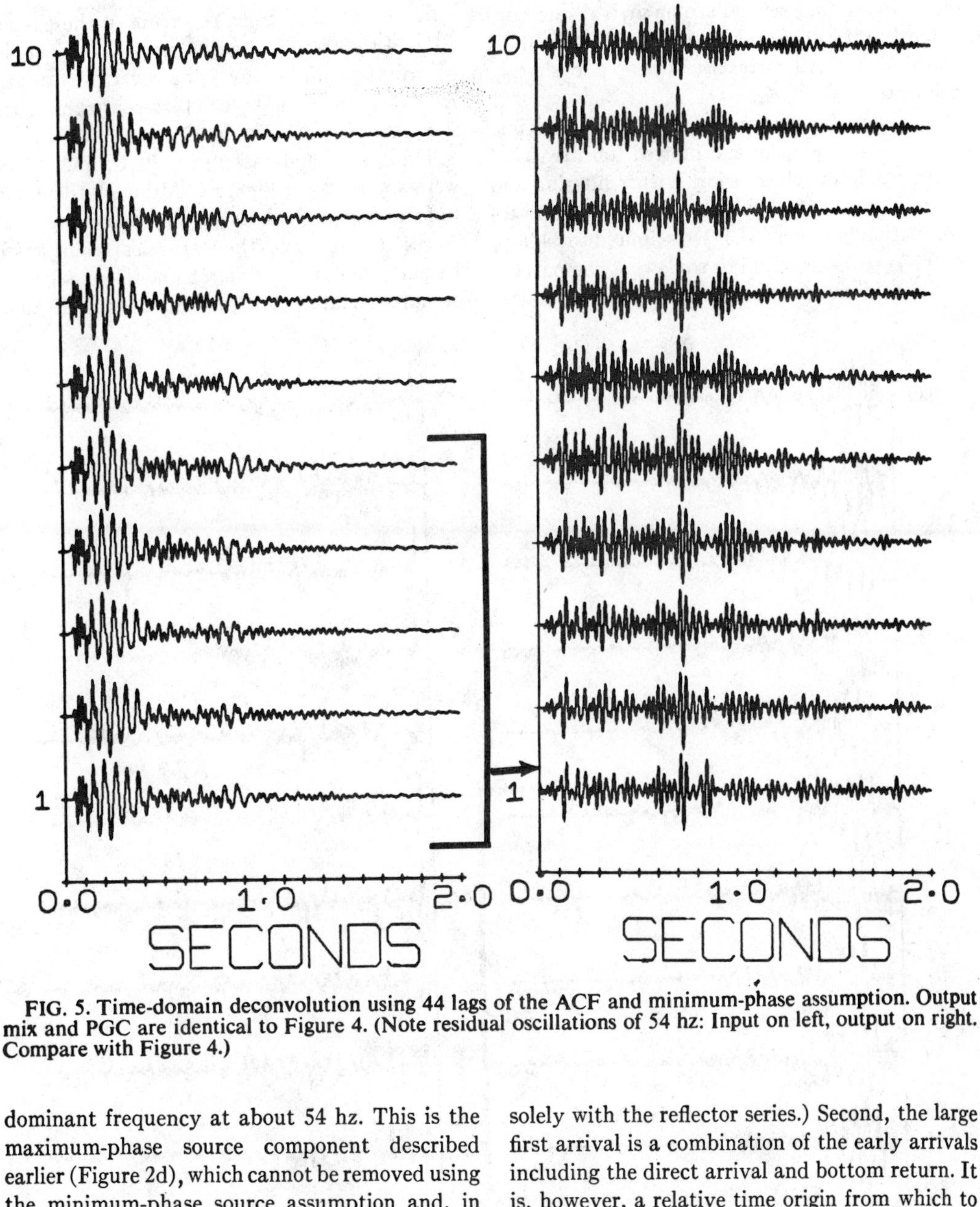

FIG. 5. Time-domain deconvolution using 44 lags of the ACF and minimum-phase assumption. Output mix and PGC are identical to Figure 4. (Note residual oscillations of 54 hz: Input on left, output on right. Compare with Figure 4.)

dominant frequency at about 54 hz. This is the maximum-phase source component described earlier (Figure 2d), which cannot be removed using the minimum-phase source assumption and, in fact, is dispersed by it.

Two difficulties are encountered with the output traces obtained by deconvolution via the complex cepstrum. First, we must assign the proper absolute time origin to the output since we have removed the ramp from the phase spectrum. (Since this ramp may have source and reflector series components, we can not associate it solely with the reflector series.) Second, the large first arrival is a combination of the early arrivals including the direct arrival and bottom return. It is, however, a relative time origin from which to measure subbottom reflection times; hence, in water of constant depth, it can be used to align the individual traces. In Figure 4 the processed traces were aligned on the large initial arrival which was arbitrarily placed at 0.048 sec.

In Figure 6 we present 125 original traces (left-hand side) and 120 processed traces (right-hand side) in profile fashion. We have applied the same only, we see that several strong arrivals are consistently present, for example at 0.25, 0.58, 0.65, and 0.85 sec. All arrivals are clear and sharply delineated.

In Figure 5 we have replotted the 10 original traces (left-hand side) and plotted (on the right) 10 traces deconvolved using Wiener filtering and the minimum-phase source assumption. The autocorrelation function (ACF) was limited to 44 lags, or 0.176 sec, so it would be somewhat comparable to our minimum-phase, complex-cepstrum source estimate. (The complex cepstrum of the circular ACF of the trace is the even part of the complex cepstrum multiplied by 2; however, a truncated ACF has no simple complex-cepstrum correspondence.)

The same band-pass filter, PGC, and mixing were applied as in Figure 4. Although the amplitude of the arrivals is larger in the processed traces of Figure 5, they are clearly dispersed. In particular, the arrival at about one second is of approximately 0.19 sec duration and has a

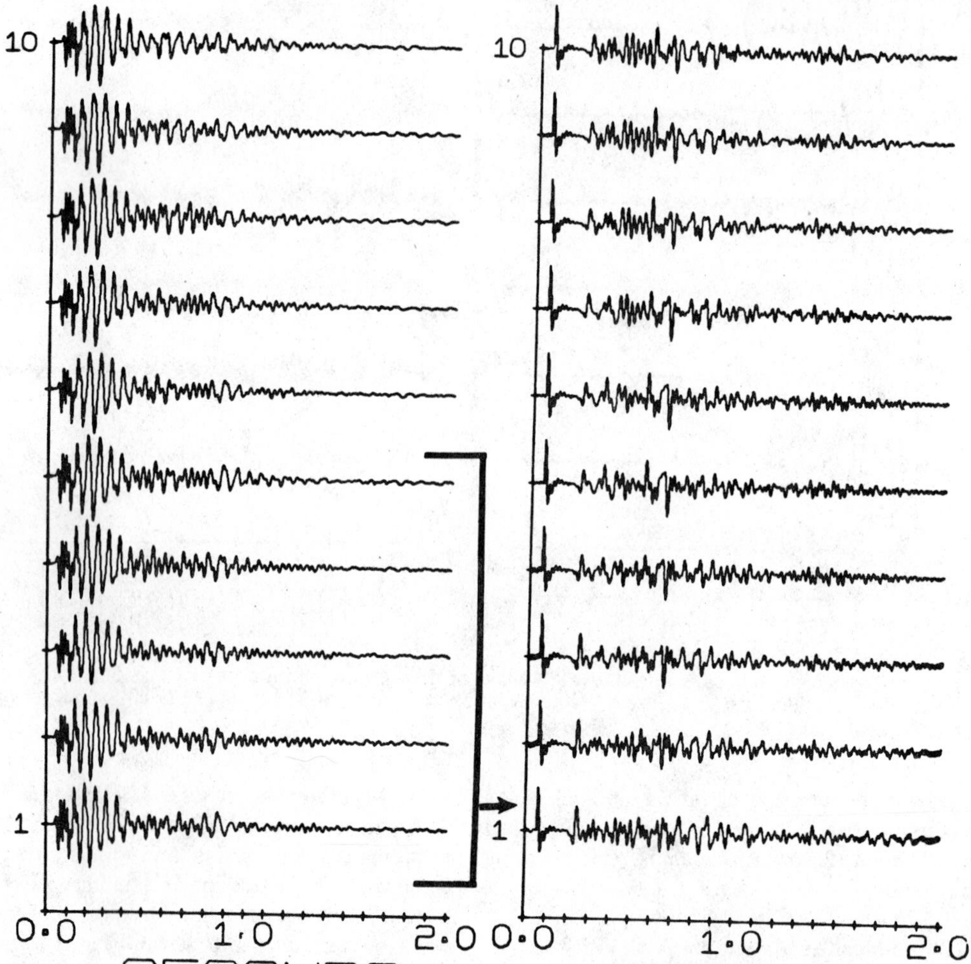

FIG. 4. Complex cepstrum deconvolution. Comparison of ten consecutive input traces (left) and output traces (right). Processing included complex-cepstrum deconvolution and band-pass filtering, and PGC. Traces were processed individually, and the results combined in a five-fold running mix as shown by the racket and arrow. Thus, output trace 1 is independent of output traces 6 – 10, etc. (Note the lack of dominant frequencies. The initial symmetrical pulse is distorted by PGC. The large initial spikes aligned and placed arbitrarily at 0.048 sec.)